IoT for Energy Management Systems and Smart Cities

IoT for Energy Management Systems and Smart Cities

Editors

Antonio Cano-Ortega
Francisco Sánchez-Sutil

MDPI

Basel • Beijing • Wuhan • Barcelona • Belgrade • Novi Sad • Cluj • Manchester

Editors
Antonio Cano-Ortega
University of Jaen
Jaen, Spain

Francisco Sánchez-Sutil
University of Jaen
Jaen, Spain

Editorial Office
MDPI
St. Alban-Anlage 66
4052 Basel, Switzerland

This is a reprint of articles from the Topic published online in the open access journals *Energies* (ISSN 1996-1073), *Sensors* (ISSN 1424-8220), *Electronics* (ISSN 2079-9292), *Smart Cities* (ISSN 2624-6511), and *IoT* (ISSN 2624-831X) (available at: https://www.mdpi.com/topics/iot_energy).

For citation purposes, cite each article independently as indicated on the article page online and as indicated below:

Lastname, A.A.; Lastname, B.B. Article Title. *Journal Name* **Year**, *Volume Number*, Page Range.

ISBN 978-3-0365-9152-0 (Hbk)
ISBN 978-3-0365-9153-7 (PDF)
doi.org/10.3390/books978-3-0365-9153-7

Contents

About the Editors

Antonio Cano-Ortega

Antonio Cano-Ortega received an M.S. in industrial engineering from the National Distance Education University (UNED), Spain, in 2000, as well as a Ph.D. from UNED in 2004. He has worked as an associate professor at the Department of Electrical Engineering, the University of Jaen, since 1996. He is the author of more than 40 papers published in journals included in the *Journal Citation Report* (JCR) and about 15 papers published in the proceedings of international conferences. He has been involved in research projects funded by Spanish Ministries.

Francisco Sánchez-Sutil

Francisco Sánchez-Sutil received an M.S. in industrial organization engineering from the University of Jaen, Spain, in 2007, as well as a Ph.D. from the University of Jaen in 2016. He has worked as an associate professor at the Department of Electrical Engineering, the University of Jaen, since 1999. He is the author of 27 papers published in journals included in the *Journal Citation Report* (JCR) and about 6 papers in the proceedings of international conferences. He has been involved in research projects funded by Spanish Ministries.

Preface

This book, titled IoT for Energy Management Systems and Smart Cities, aims to include the use of the latest technologies applied to the electricity sector, more specifically to smart cities/grids. IoT plays a fundamental role in the development of these networks.

The Internet of Things makes it possible for the information collected by the different sensors and actuators in real time to be available in the cloud so that it can be used instantly. This information will be of vital importance for the correct functioning of the algorithms and software that control smart cities/grids.

Communications play an important role in working with IoT devices. In addition to traditional networks, such as Wi-Fi, Bluetooth and Ethernet, new low-energy networks are indispensable for IoT devices. LPWANs such as LoRa, SigFox, Nb-IoT, Bluetooth LE, etc., make it possible to build devices that can operate in remote locations, thanks to their long range and low power consumption, without the need for a high power source.

With the information available in the cloud, in addition to the algorithms, users can undertake real-time monitoring of their installations in any location worldwide on their mobile devices or web browsers. It is also possible to act on installations in real time by sending commands to IoT devices.

Based on the previous paragraphs, it is possible to recognize the potential of this type of device for control and monitoring in industrial, domestic and energy generation environments. Distributed generation can be fed by all these technologies in such a way that they can be used for the progress and development of smart cities/grids.

This topic is divided into two main subjects: (i) the Internet of Things, cloud computing and wireless networks; (ii) smart cities and grids. The first of these deals with device communications, the networks they use and how they access information in the cloud. This allows data flows to be generated correctly. The second part of the topic is dedicated to the development of techniques, procedures and algorithms that can control smart cities/grids. This is one of the main applications of the devices created in this study.

Therefore, this book includes recent advances in the development and implementation of IoT devices applied to electrical engineering for use in smart cities/grids, so that readers can become familiar with new methodologies directly explained by experts in this scientific field.

Antonio Cano-Ortega and Francisco Sánchez-Sutil
Editors

sensors

MDPI

Article

EggBlock: Design and Implementation of Solar Energy Generation and Trading Platform in Edge-Based IoT Systems with Blockchain

Subin Kwak [1], Joohyung Lee [2,]*, Jangkyum Kim [3,]* and Hyeontaek Oh [4]

1 Samsung Electronics, Yeongtong-Gu, Suwon 10285, Korea; starfishda54@gmail.com
2 School of Computing, Gachon University, Seongnam 13120, Korea
3 School of Electrical Engineering, KAIST, Daejeon 34141, Korea
4 Institute for Information Technology Convergence, KAIST, Daejeon 34141, Korea; hyeontaek@kaist.ac.kr
* Correspondence: j17.lee@gachon.ac.kr (J.L.); wkdruadl@kaist.ac.kr (J.K.)

Abstract: In this paper, to balance power supplement from the solar energy's intermittent and unpredictable generation, we design a solar energy generation and trading platform (EggBlock) using Internet of Things (IoT) systems and blockchain technique. Without a centralized broker, the proposed EggBlock platform can promote energy trading between users equipped with solar panels, and balance demand and generation. By applying the second price sealed-bid auction, which is one of the suitable pricing mechanisms in the blockchain technique, it is possible to derive truthful bidding of market participants according to their utility function and induce the proceed transaction. Furthermore, for efficient generation of solar energy, EggBlock proposes a Q-learning-based dynamic panel control mechanism. Specifically, we set the instantaneous direction of the solar panel and the amount of power generation as the state and reward, respectively. The angle of the panel to be moved becomes an action at the next time step. Then, we continuously update the Q-table using transfer learning, which can cope with recent changes in the surrounding environment or weather. We implement the proposed EggBlock platform using Ethereum's smart contract for reliable transactions. At the end of the paper, measurement-based experiments show that the proposed EggBlock achieves reliable and transparent energy trading on the blockchain and converges to the optimal direction with short iterations. Finally, the results of the study show that an average energy generation gain of 35% is obtained.

Keywords: solar energy generation; energy trading; auction theory; testbed; measurement study; Internet of Things; blockchain; reinforcement learning

Citation: Kwak, S.; Lee, J.; Kim, J.; Oh, H. EggBlock: Design and Implementation of Solar Energy Generation and Trading Platform in Edge-Based IoT Systems with Blockchain. *Sensors* **2022**, *22*, 2410. https://doi.org/10.3390/s22062410

Academic Editors: Antonio Cano-Ortega, Francisco Sánchez-Sutil and Aurora Gil-de-Castro

Received: 9 February 2022
Accepted: 10 March 2022
Published: 21 March 2022

1. Introduction

The use of solar energy is considered as a promising renewable energy source. Solar energy has various advantages in terms of increasing energy efficiency and reducing greenhouse gas emissions [1]. In the conventional centralized power grid system, several problems can occur, such as energy loss owing to the transmission process and power instability due to the peak power demand [2]. Using solar energy as a distributed energy resource, it is possible to minimize the transmission loss and supply energy to the consumer more efficiently. However, the solar panel has a problem in that it is sensitive to changes in the surrounding environment, and it is difficult to arbitrarily control the amount of energy generation [3,4].

To address these challenges, various studies have been conducted in the literature, aiming for efficient energy trading mechanisms [5–7]. Specifically, using a centralized system manager, market-based energy trading models have been suggested for balancing demand and generation. However, such interventions of the centralized system cause additional participation fees and security issues in transaction records. Thus, to alleviate these problems, blockchain technology-based distributed energy trading models have been proposed as one of the promising technologies in the smart grid system [8–10]. Nevertheless, research on implementation-based energy trading in distributed solar energy generation is limited. Most of the studies were confined to numerical or simulation-based studies. As a result, no conventional studies have addressed practical designs, such as the implementation of solar energy generation and trading platforms that consider AI-based automatic energy generation and blockchain-based secure energy trading at the same time. Therefore, we deal with an operation method of solar energy as one of promising renewable energy source, considering the actual environment and the residual energy transaction platform using the actual dataset.

In this study, we design a solar energy generation and trading platform in Internet of Things (IoT) systems using blockchain (EggBlock), which promotes distributed energy trading without introducing a centralized broker and supports efficient energy generation. The contributions of this study are summarized as follows:

- To support reliable energy trading among users without the participation of a centralized broker, the EggBlock platform using Ethereum for blockchain is proposed in this paper. In order to determine a reasonable transaction of the generated renewable energy, we infer the determination of transaction price and the amount of energy in the market according to the second price sealed-bid auction mechanism. Furthermore, for efficient generation of solar energy, the EggBlock proposes the Q-learning based dynamic panel control mechanism. Specifically, we consider a model-free-based Q-learning algorithm, where the state and the action are the current position of the solar panel and the angle at which the solar panel will move to the next time step, respectively. Finally, the reward is designed as the amount of solar power that the solar panels can obtain under the given state.
- We implement the proposed EggBlock platform using Ethereum for blockchain, which enables Android smartphones to monitor contracts of energy trading in a real-time manner. Furthermore, we build an IoT hardware testbed equipped with a solar panel to generate and deliver energy for trading.
- The measurement-based experiments in the testbed show that the proposed EggBlock achieves reliable and transparent energy trading using blockchain, and it converges to the optimal direction with short-iterations. Finally, the results of the study show that an average energy generation gain of 35% is obtained.

The remainder of this study is organized as follows. We review existing research in Section 2. After the introduction of the overall system model in Section 3, the Q-learing algorithm is introduced in detail in Section 4. Section 5 contains a detailed description of the actual implementation, while Section 6 contains a detailed description of the experimental results. Finally, Section 7 concludes with a detailed description of our system's usage area and future plans.

2. Related Works

Currently, renewable energy accounts for a large portion of the total energy generation [11]. It is advantageous in terms of environmental sustainability, low maintenance cost, and ease of installation in urban areas [12]. However, the energy generation of renewable energy is unstable due to various uncertain factors, such as weather, cloud movement, or solar irradiation [13]. Therefore, various studies have been conducted to manage unstable energy generation with the installation of additional facilities, such as energy storage systems or auxiliary generators. However, the construction of such additional facilities requires considerable cost and time [14]. Thus, various energy trading approaches in smart grid systems have been proposed for reliable and efficient energy use by balancing demand and generation [5–7]. The work in [5] provided a game-theoretical analysis for a distributed energy trading mechanism in which multiple microgrids re-sell or store surplus energy to maximize their own profit. By representing the Nash equilibrium of all players, the authors analyzed individual players' strategies in each market environment. In [6], Zhang et al.

proposed a peer-to-peer (P2P) energy trading model to improve the local energy balance. Here, the authors considered a practical low-level voltage microgrid environment to analyze its effectiveness. In [7], Pei et al. introduced the concept of a two-stage market model to maintain the balance of energy supply and demand in the entire system and region. Specifically, using a Monte Carlo sampling method, the authors alleviated the effect of the uncertainties of renewable energy. Additionally, by showing the various simulation results, the authors verified that the proposed market framework applies to an actual system.

Nevertheless, in most of the previous energy trading studies, interventions of the system manager should manage such an energy trading market. However, such interventions also resulted in additional participation fees and security issues of transaction records. Therefore, blockchain technology-based distributed energy trading model without a centralized manager have been proposed as solutions to alleviate such issues in the smart grid system [8–10,15–17]. In [8], Mihaylov et al. proposed an energy trading method for energy prosumers and consumers using the proposed NRGcoin. The authors argued that the introduction of NRGcoin can solve the security issues associated with energy trading and the fiat money conversion problem. In [9], Mengelkamp et al. developed energy transaction models and demonstrated the simulation environment in the region using the private blockchain method. The authors show the market operation results by analyzing the relationship between the proportion of solar energy generation in the total energy supplement and market price from an economic perspective. On the other hand, with the implementation of an actual experimental system, Zhang et al. proposed a real-time energy transaction system by incorporating the concepts of prioritization and cryptocurrency [10]. By presenting the numerical results, the authors showed that the proposed system can be beneficial to market participants from both monetary and non-monetary perspectives. In addition, unlike most conventional studies that use cryptocurrency as an alternative currency, the proposed cryptocurrency in the paper has the abilities to convert fiat currency and change the physical power flow. Vehicle to grid (V2G) network is one of the environments in smartgrid system that is appropriate to use blockchain technology in, as depicted in [15]. Hassija et al. addressed the direct acyclic graph-based V2G network (DV2G), which is organized with lightweight blockchain protocol. Showing the mechanism and numerical results, authors proved that the proposed method is highly scalable and supports the micro-transaction that is required in V2G network. By introducing the blockchain technique, it is possible to operate the system according to the consensus of smart contract between the participants without the intervention of a third party. In [16], Xi Chen et al. proposed energy transaction among the renewable energy and electric vehicle to minimize the burden of the power system operator. Through the blockchain-based EV incentive system, authors addressed that it is possible to maximize the utilization of renewable energy and manage the power system more effectively. A similar approach to this paper is dealt in [17]. In the paper, Hassan et al. proposed auction and blockchain-based energy transaction scheme to maximize the energy producer's revenue in the system. Here, the authors insist that it is possible to provide moderate cost, secure, and private auction schemes for microgrids using the blockchain technology. Applying the advantages of blockchain to maximize the profit of various market participants and increase the stability of energy transaction is similar to this paper. However, our paper has the distinction of maximizing the amount of power generation by controlling the angle of the solar panel and considering the actual electricity tariff to facilitate application in the real environment.

3. Proposed System

As illustrated in Figure 1, the proposed EggBlock system model considers that there are multiple edge-based IoT devices as users who participate in energy trading using the generated energy. We set the environment in which the edge-based IoT devices are equipped with solar panels for energy generation and energy storage to store redundant energy for future use. Therefore, the proposed EggBlock system consists of two parts: (1) energy generation and (2) energy trading.

In the energy generation part, we deal with a method that can maximize energy generation by controlling the angle of the solar panel. Here, the intensity of light over various locations is measured to determine the direction of the solar panel where a BH1750 photo resistor controlled by Raspberry Pi is used. Such measured data are used for the Q-learning algorithm, which maximizes the expected reward (i.e., energy generation) by determining the optimal solar panel angle under the given state space.

In the energy trading part, we propose a blockchain-based decentralized energy trading model that is organized without a centralized broker. In this study, we design a transaction environment using an Ethereum-based decentralized application (Dapp) platform. Here, the use of Dapp is suitable for the proposed energy transaction model because it has various advantages (e.g., zero downtime, privacy, resistance to censorship, complete data integrity, and trustless computation/verifiable behavior) [18,19]. Based on these advantages, participants in the market cannot cheat or falsify transactions and have a reliable transaction service.

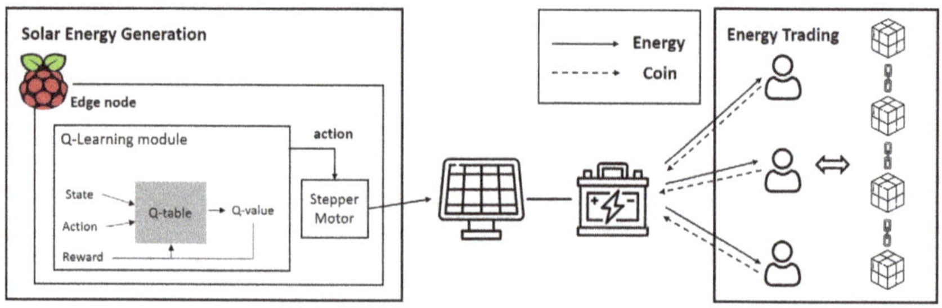

Figure 1. System model.

3.1. Energy Generation

To maximize energy generation by adjusting the position of the solar panels, we use a reinforcement learning (RL) framework. Here, we use two Raspberry Pi boards as edge nodes to control the solar panel: (1) collecting solar energy data and (2) using the collected data to train Q-Learning and adjusting the position of solar panels through stepper motors.

In the first edge node, solar energy data collection is performed using a photo register, which can be used for training the RL model. The collected data are sent on every time step to the second edge node.

To use the data collected from the first edge node, a Q-table is defined using a model-free RL framework at the second edge node [20]. For every time step, the second edge node selects an optimal action by referring to the Q-table and transmits the action value to a stepper motor to efficiently control the position of the solar panel. The Q-table is updated at every time step based on the received reward corresponding to the state so that it converges to the optimal action corresponding to the state. Here, when the stepper motor moves the solar panel based on the received action value, the reward occurs according to the action process and updates the Q-table. The detailed process of the RL framework will be covered in Section 4.

3.2. Energy Trading

The energy obtained from solar panel is sold to bidders at a price determined based on the auction mechanism. For reliable energy transactions between sellers and buyers, the energy trading platform is implemented through a blockchain using Ethereum. According to the Ethereum-based smart contract, transactions proceed among sellers and buyers with their purpose and status. Accordingly, if the transaction is conducted via the smart contract, the blockchain guarantees the authenticity and the reliability of the transaction execution without requiring a centralized broker [21].

3.2.1. Determination of Transaction Based on Sealed-Bid Auction Mechanism

The energy generated by the photovoltaic (PV) generator is sold at an appropriate price to buyers who need it. At this time, in order to determine the transaction price, a second price sealed-bid auction mechanism is proposed in this paper. There are two reasons for choosing the second price sealed-bid auction mechanism in this paper. First, it is possible to minimize the exposure of information about the buyer's bidding strategy (e.g., bidding price, current status, policy, etc.). Second, buyers adopt a truthful bidding strategy in which their expected price is equal to their bidding price. Since it can be seen that the advantages of such mechanism are efficient in the blockchain-based energy transaction model, we use the method and predict the bidding price of buyers.

In Figure 2, we represent the overall structure of auction-based market model. Since the proposed market is operated separately from the conventional energy market, it is necessary to consider the market data in actual system. In the figure, we could check that there are two different participants in the market. The energy producer is a seller who has renewable energy with a storage device to sell residual power. The *energy consumer* is a buyer who wants to purchase energy from the producer to minimize the overall cost. Here, the consumer might be a prosumer with solar energy or simple electricity consumer who wants to purchase energy in a cost-effective manner through EggBlock rather than purchasing from the conventional market. As depicted in the above statement, a transaction of the proposed scheme is the progressed auction mechanism. Therefore, the auction is began through the disclosure of the amount of energy sold by the seller.

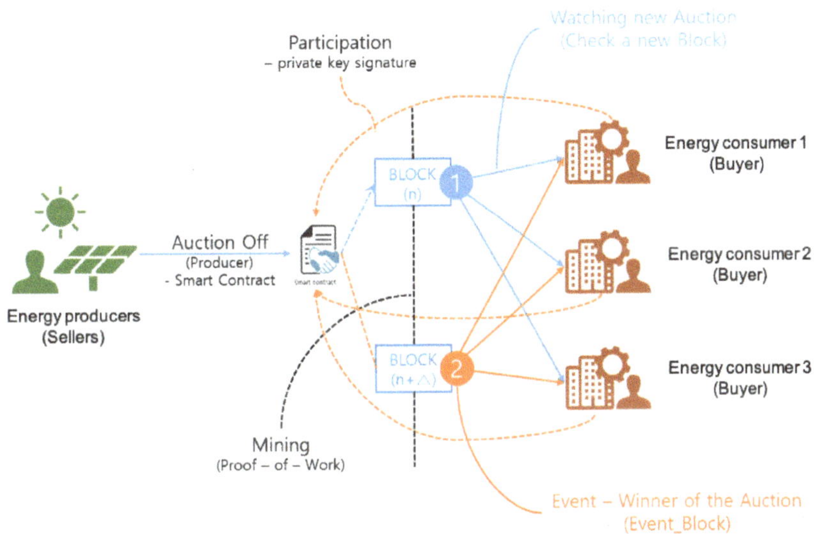

Figure 2. Auction-based energy transaction model.

In this paper, we use a second price sealed-bid auction mechanism that the seller sells the energy to the buyer with the highest bidding price for the second highest price. The characteristics of the auction called sealed-bid makes the buyer decide the bidding price by only considering it's own profit. In addition, considering the bidding mechanism, it can be seen that the buyer makes a truthful bidding in the proposed system [22,23]. Therefore, the details of the utility function to determine each participant's strategy are depicted as follow:

Seller: In the case of a seller, it is possible to determine the amount and price of energy to be sold considering the current profit and the future value achieved by storing the energy. Here, the utility function of the seller could be depicted as:

$$\max_{p} \quad \hat{U}(p) = C_{sell}\hat{E}p + C_{op}ln(1 + \hat{E}\eta(1-p)] \tag{1}$$

$$s.t. \quad 0 \le p \le 1. \tag{2}$$

In Equation (1), the first term $C_{sell}\hat{E}p$ refers to the profit that the seller achieves through selling energy. Here, C_{sell} refers to the unit price for selling energy, and $\hat{E}$ is the energy generated from the solar panel. In addition, p is the decision variable of the seller that means the portion of energy to sell in the market. Furthermore, the second represents the satisfaction the seller attained by storing the energy $\hat{E}(1-p)$ in the storage system [24]. Here, a weight factor C_{op} is added to transform the seller's satisfaction level into the monetary aspect. Since the formula is a concave model consisting of one decision variable, the optimal value can be calculated as follows:

Derivation: To obtain an optimal solution of the problem, we differentiate Equation (1) by p as:

$$\frac{\partial \hat{U}(p)}{\partial p} = \alpha\hat{E} - \frac{\hat{E}\eta}{1 + \hat{E}\eta(1-p)}, \tag{3}$$

$$\frac{\partial^2 \hat{U}(p)}{\partial p^2} = -\frac{\hat{E}\eta^2}{(1 + \hat{E}\eta(1-p))^2}. \tag{4}$$

To organize the formula more easily, we transform the $\frac{C_{sell}}{C_{op}}$ into the auxiliary variable α. Assuming that the value of C_{op} is larger than C_{sell}, we could set $0 \le \alpha \le 1$ in the equation. In addition, since the auxiliary variable η is positive, the second derivation of utility function in Equation (4) becomes a negative value. Since the utility function of the seller is organized with a single decision variable, it is possible to prove that the utility function of the seller is concave. Therefore, the optimal value p^* is calculated as 0 according to the first derivation in Equation (3). In this way, we could calculate the optimal value as follows:

$$p^* = \max(\min(1 - \frac{1}{\hat{E}}(\frac{(\eta - \alpha)}{\alpha\eta}), 1), 0). \tag{5}$$

Buyer: In the case of a buyer, the main purpose is minimizing the overall cost by submitting the proper bidding price to the auction market.

$$U_i(b_i) = (1 - e^{-\gamma_i b_i})((p_s - b_i)\min(D_i, E) - b_i \max(E - D_i, 0)) - e^{-\gamma_i b_i} p_s D_i. \tag{6}$$

In Equation (6), $(1 - e^{\gamma_i b_i})((p_s - b_i)\min(D_i, E) - b_i \max(E - D_i, 0))$ refers to the profit that a buyer i can achieve from the auction market. In the equation, $(1 - e^{\gamma_i b_i})$ refers to the probability that the buyer i could achieve the energy from the auction progress. Here, γ_i is an auxiliary variable that determines the probability of successful bidding according to the bidding price submitted by the buyer i. In addition, $(p_s - b_i)\min(D_i, E)$ is the profit that the buyer i achieved from the auction process. For the case that the buyer i successes the bidding process, the buyer could achieve the monetary profit $(p_s - b_i)\min(D_i, E)$ and get the loss $b_i \max(E - D_i, 0)$ when the purchasing energy E is higher than required energy D_i. Furthermore, the last term $-e^{\gamma_i b_i} p_s D_i$ is the cost that the buyer i has to pay when the buyer fails in the auction.

Case 1: For the case that a buyer i's required energy D_i is larger than purchasing energy E, we could set the utility function of buyer i as follows:

$$U_i(b_i) = (1 - e^{-\gamma_i b_i})((p_s - b_i)E) - e^{-\gamma_i b_i} p_s D_i. \tag{7}$$

According to the second derivative result of the buyer's utility function, it is possible to check that the utility function is organized with the concave function. To find the optimal b_i^* that maximizes the profit of buyer i's profit, we could differentiate the Equation (11) as:

$$\frac{\partial U(b_i)}{\partial b_i} = \gamma_i e^{-\gamma_i b_i}(p_s E - b_i E) - (1 - e^{-\gamma_i b_i})E + e^{-\gamma_i b_i}\gamma_i p_s D_i = 0. \tag{8}$$

Here, we could reorganize the formula as:

$$e^{\gamma_i b_i} = 1 + \gamma_i(p_s - b_i) + \frac{1}{E}(\gamma_i p_s D_i). \tag{9}$$

According to the Taylor series equation, we could change $e^{\gamma_i b_i} = 1 + \gamma_i b_i + \frac{\gamma_i^2 b_i^2}{2!} + \frac{\gamma_i^3 b_i^3}{3!} + \cdots$. Here, we assume that values after $\frac{\gamma_i^3 b_i^3}{3!}$ are too small to be 0. Therefore, we could find the optimal value through the following equation.

$$\begin{aligned} 1 + \gamma_i b_i + \frac{\gamma_i^2 b_i^2}{2} &= 1 + \gamma_i(p_s - b_i) + \frac{1}{E}(\gamma_i p_s D_i) \\ (b_i + \frac{2}{\gamma_i})^2 - \frac{4}{\gamma_i^2} - \frac{2p_s}{\gamma_i} - \frac{2}{E\gamma_i} p_s D_i &= 0 \\ \therefore b_i^* &= \sqrt{\frac{4}{\gamma_i^2} + \frac{2p_s}{\gamma_i} + \frac{2}{E\gamma_i} p_s D_i} - \frac{2}{\gamma_i} \end{aligned} \tag{10}$$

Case 2: For the case that purchasing energy E is larger than the buyer i's required energy D_i, we could set the utility function as follows:

$$U_i(b_i) = (1 - e^{-\gamma_i b_i})((p_s D_i - b_i E) - e^{-\gamma_i b_i} p_s D_i. \tag{11}$$

By differentiating the utility function, we could get the optimal bidding price as:

$$\gamma_i e^{-\gamma_i b_i}(p_s D_i - b_i E) - (1 - e^{-\gamma_i b_i})E + \gamma_i e^{-\gamma_i b_i} p_s D_i = 0 \tag{12}$$

$$\begin{aligned} \gamma_i e^{-\gamma_i b_i}(2p_s D_i - b_i E + \frac{E}{\gamma_i}) - E &= 0 \\ e^{\gamma_i b_i} &= \frac{2p_s D_i \gamma_i}{E} - \gamma_i b_i + 1 \\ 1 + \gamma_i b_i + \frac{\gamma_i^2 b_i^2}{2} &= \frac{2p_s D_i \gamma_i}{E} - \gamma_i b_i + 1 \\ b_i^2 + \frac{4b_i}{\gamma_i} - \frac{4p_s D_i}{E\gamma_i} &= 0 \\ (b_i + \frac{2}{\gamma_i})^2 - \frac{4}{\gamma_i^2} - \frac{4p_s D_i}{E\gamma_i} &= 0 \\ \therefore b_i^* &= \sqrt{\frac{4}{\gamma_i^2} + \frac{4p_s D_i}{E\gamma_i}} - \frac{2}{\gamma_i} \end{aligned} \tag{13}$$

In this way, it is possible to predict the strategies of seller and buyers in the market. Since we use the second price sealed-bid auction, each participant conducts a truthful bidding according to the optimal value of its utility function. After participants decide their strategy, the amount of transaction energy set by the seller is sold to the person who bids the highest price by paying the second highest bidding price. Here, information about the whole of the participants' strategy and transaction results are stored in the block.

3.2.2. Architecture of Smart Contract

Ethereum is a global, decentralized open-source blockchain featuring smart contract functionality, where ether (ETH) in Ethereum is the cryptocurrency generated by Ethereum miners as a reward for computations for adding blocks to the blockchain. The detailed Ethereum architecture is illustrated in Figure 3. The term user in Figure 3 represents both buyers and sellers who participate in a transaction. Correspondingly, when a user transacts via Ethereum, the user must first connect to a node and refer to the Ethereum client through a web browser powered by MetaMask [25].

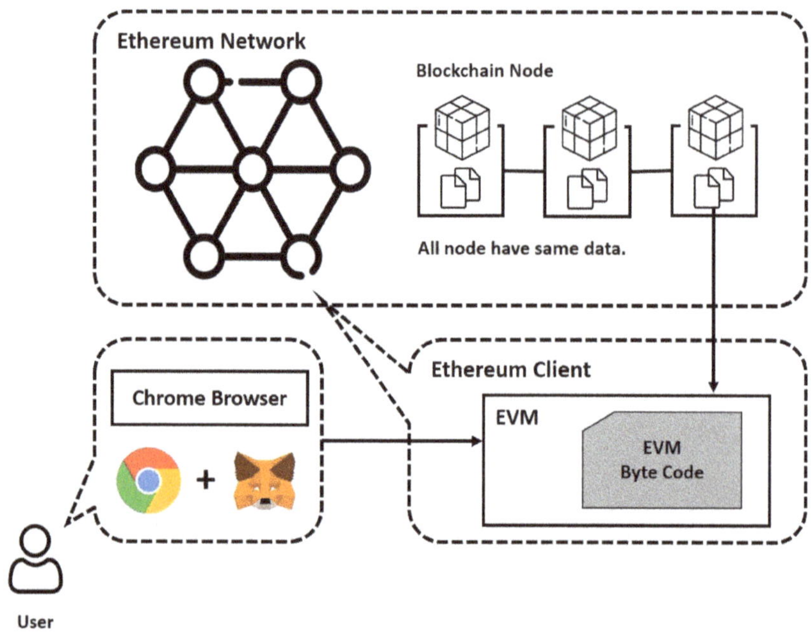

Figure 3. Ethereum architecture.

The Ethereum clients, who have a distributed database and are interconnected to the Ethereum Network, receive information when the block is created owing to a new transaction. In other words, Ethereum clients are nodes of a blockchain network, called blockchain nodes, which allow general users to connect to the blockchain. Users will be able to obtain blockchain information or use smart contracts through the Ethereum client. The smart contract is a script that implements the contents of a contract using the code and allows the contract to be automatically fulfilled when the conditions are met. The smart contract created by one user is stored in blocks through the Ethereum client. Therefore, all nodes in the blockchain network have the same smart contract code. The process of creating smart contracts includes the following:

- Smart contract coding: Code the contents of the contract you want to include in the smart contract.
- Connect with Ethereum client: The written code is placed on the Ethereum virtual machine (EVM) of the Ethereum client.
- Compile the implemented code: EVM byte code will provide the compilation result.
- Smart contract distribution: Add the compiled EVM byte code to the block as a transaction and register to the blockchain.

In this process, when the smart contract is registered to the blockchain, all Ethereum clients on the blockchain will have the byte code of the smart contract. Then, the Ethereum client will be able to run a registered smart contract on its EVM. Additionally, if a transaction occurs, the content of a smart contract will change. Thus, after the transaction, other nodes can access the smart contract so that they can change the content.

3.2.3. Transaction Process

Figure 4 shows the sequence diagram for the energy trading. In the energy trading platform, some participants have superfluous energy that they wish to sell to the platform as sellers, whereas others do not have sufficient energy to meet their demands and must buy the shortfall from the platform as buyers. Specifically, first, the buyer defines and sends Ether and code so that smart contracts can be created.

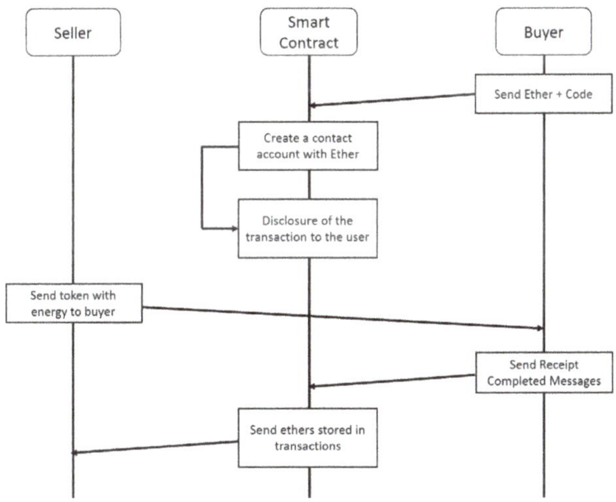

Figure 4. Sequence diagram for the energy trading.

A smart contract is run on top of the EVM. The terms of the transaction stored in the smart contract are disclosed to everyone who can participate in the transaction. For reliable transactions, sellers should prove the amount of energy they have. The amount of proven energy is returned to the token, which is held by the seller. At this time, the token is used only to prove the amount of energy without monetary value. If there is a seller who has sufficient energy to sell as much as the buyer needs, the seller sends a token with energy to the buyer. After the buyer receives the energy as required, it sends the completed message receipt to the smart contract account. Finally, the smart contract delivers the ETHs stored in the transactions to the seller. Simultaneously, the token sent to the purchaser is extinguished.

Through this process, two participants can make a safe transaction without third-party intervention, and this process is transparently recorded in the blockchain.

4. Model-Free Q-Learning

Model-free RL aims to achieve efficient control of the solar panel's movement so that the solar panel can collect the maximum energy when compensation is unknown. To design the model-free RL algorithm, a Markov decision process (MDP) framework is considered, which requires a description of states, actions, and rewards [20]. We consider the discrete state and action spaces where the state and action refer to the current location of the solar panel and the amount of angle to be moved. Here, the goal of the agent (i.e., solar panel) is to learn the policy π, which allows the agent to choose actions that can maximize long-term cumulative rewards (i.e., energy generation) where states are given.

4.1. Markov Decision Process

The environment for the MDP consists of an agent, a BH1750 photo register, and a solar energy battery capable of storing energy. In this study, we assume an environment in which time is divided into consecutive fixed-length intervals. In each time interval, the agent, which is a solar panel, determines the action. Here, we define a set of time steps that identifies the intervals as $T = \{0, 1, 2, ...\}$. Then, the agent determines the action at the next time step to maximize the expected cumulative rewards based on photo resistor observation and previous experience. Specifically, the angle movement of the solar panel is controlled through the stepper motor. Then, the chosen action is transmitted to the stepper motor to control the angle of the solar panel. A detailed description of the states, actions, and rewards for the proposed MDP framework is provided as follows.

State: State means the current location of the agent. The location of the agent in the t time step is received as a scalar value of s_t. The set of states is denoted by S and marked as $S = \{s_1, s_2, s_3, ..., s_t\}$. We denote the state of the system in time step t as s_t, which is given as follows:

$$s_t = [\textit{solar panel's angle in time step } t] \tag{14}$$

Action: Action means the angle at which the motor must move at s_t. The action value in the time step t can be expressed as a scalar value a_t. The set of actions is named A and marked as $A = \{a_1, a_2, a_3, ..., a_t\}$. Consider the state of the system in time step t denoted by a_t, which is given as follows:

$$a_t = [\textit{angle to move in time step } t] \tag{15}$$

Reward: Reward $R(s_t, a_t)$ obtained by taking action a_t at state s_t is the total amount of energy that can be obtained from state s_t. Our goal is to maximize the total energy generation, which can be expressed in a formula to maximize the expected rewards of the solar panel.

$$R(s_t, a_t) = [\textit{amount of energy can be obtained in time step } t]. \tag{16}$$

4.2. Q-Learning

The state and action pairs (s, a) applied in this study are mapped through policy π derived using Q-learning. $Q^\pi(s, a)$ is the cumulative reward for taking action a from a state s and follows policy π accordingly. Q^π is specified as follows:

$$Q^\pi(s, a) = \mathbb{E}\left[\sum_{t=0}^{\infty} \gamma_i^t R(s_t, \pi(s_t)) | s_0 = s, a_0 = a\right]. \tag{17}$$

where $\gamma_i \in (0, 1)$, which is a hyper-parameter of the Q-function, is the discount factor. Therefore, maximizing the cumulative reward is the same as finding a policy that maximizes the Q-function. The optimized Q-function is called Q^*, which satisfies the Bellman equation.

$$Q^*(s, a) = R(s, a) + \gamma_i \cdot \mathbb{E}_{s'}[\max_a Q^*(s', a)]. \tag{18}$$

The purpose of the Q-learning algorithm is to learn the optimal Q^* in the observation sequence $(s_t, a_t, R_{t+1}, s_{t+1})$. The optimal policy π^* can be computed using Q^* as follows:

$$\pi^*(s) = \arg\max_a Q^*(s, a). \tag{19}$$

The Q-learning algorithm was implemented as follows. At each time step t, the agent updates the Q-function Q_t as follows:

$$Q(s_t, a_t) = (1 - \alpha_t)Q(s_t, a_t) + \alpha_t(R_t + \gamma_i \max_a Q(s_{t+1}, a_t)), \tag{20}$$

where α_t denotes the learning rate. If state-action pairs are sampled infinitely and under suitable conditions on the learning rate, Q_t will converge to the optimal Q-function Q [26].

As described in Algorithm 1, we set the parameter K as the time range representing the daily energy collection time. Furthermore, ϵ is the hyper-parameter value of ϵ-greedy, which indicates the probability that a solar panel randomly selects an action. As described in the pseudocode, each episode consists of a one-day energy generation. t is the time interval for calculating the reward. Therefore, in each time step t, the solar panel obtains the current state and selects the action. There are two choices when selecting an action. Exploring the policy may speed up training, but it may cause problems of falling into a local optimum due to the inability to explore new routes. Therefore, we apply an e-greedy method that can select a random action with a certain probability. After selecting the action, the reward is calculated according to the state and action. Subsequently, the Q-table is updated according to Equation (20). Then, we go to the next time step and iterate this process.

Algorithm 1 Q-Learning based solar panel control method.

Setting K range
Initial γ_i, ϵ
Initial action-value function Q_0 and t
for each episode **do**
  while $t \leq K$ **do**
    Get current state s_t
    Select action

$$a_t = \begin{cases} \text{random } a, & \text{if probability } \epsilon, \\ \arg\max_a Q(s_t, a_t), & \text{else,} \end{cases}$$

    Execute action a_t
    Calculate reward $R(s_t, a_t)$
    State s_t transfer to the next state s_{t+1}
    Update $Q(s_t, a_t)$ according to (20)
    Next time step $t \longleftarrow t + 1$
  end while
end for

5. Implementation

Using the above scenario, we conduct an evaluation by implementing an actual testbed that can generate and trade energy. In this testbed, we set two participants, the seller and buyer in the platform, and conduct several tests for the case in which an actual energy transaction could be made. The details of the tests are as follows: (1) Check the possibility that it is possible to transfer the values to the stepper motor, which is obtained through reinforcement learning. (2) Set the environment in which the stepper motor can operate the solar panel module. (3) Evaluate whether energy is transferred between two participants. (4) Check that the transmission is done, and ETHs are exchanged during the transaction process. Furthermore, we implement the proposed system as web pages and mobile applications so that users could perform transactions smoothly and obtain the necessary information they needed.

5.1. Testbed Hardware

The proposed testbed uses an 'SCM 5WA' solar cell module (solar panel), 'ESC 1206' charge controller, and 'KB 4.5 Ah 12 V' battery. The 42-angle stepper motor 'NEMA17' is used to move the solar panel. Additionally, the 'L298N 1 channel' is used to connect the motor to Raspberry Pi. At this time, the 'DC-DC BUCK/BOOST' converter is used to adjust the voltage because the voltage sizes are all different.

It moves the stepper motor and controls the solar cell module through the action values obtained through reinforcement learning. The controller stores the incoming energy in the battery and manages the energy stored in the battery for use. Using this function of the controller, the testbed checks that when the seller sends the specified amount of energy to the buyer, all that energy is sent to the buyer. Additionally, it connected the LED light bulb to the controller of the buyer to check whether the energy received from the seller can be used immediately. Figure 5 shows the blueprint of our energy transaction. Furthermore, Figure 6 is a real testbed that we implemented.

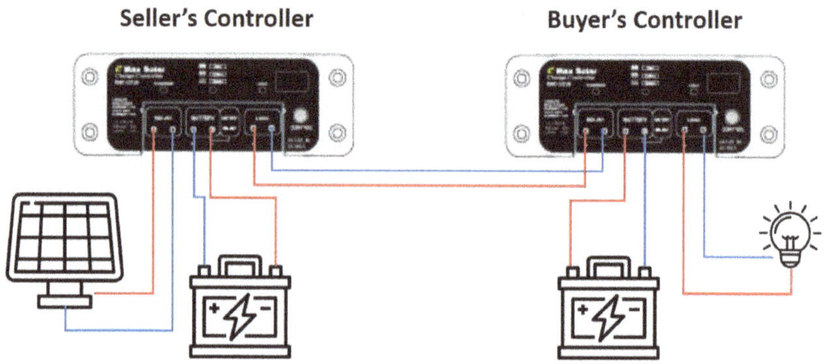

Figure 5. Blueprint of controllers in testbed.

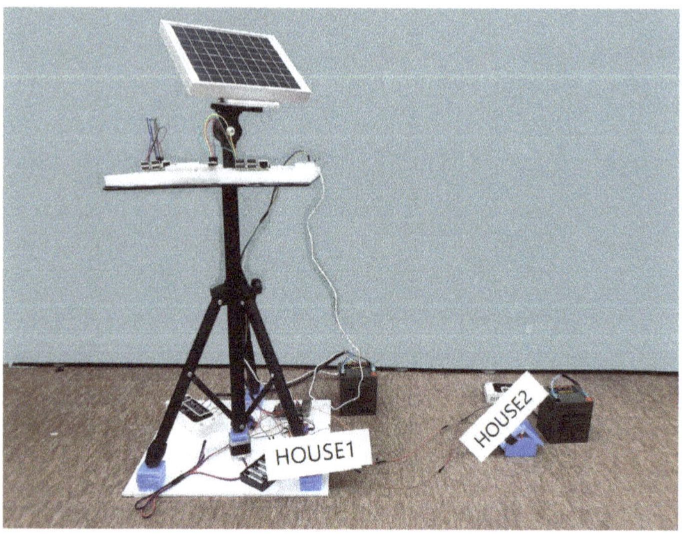

Figure 6. Testbed of platform.

5.2. Software

The proposed testbed uses Solidity to write the terms of the smart contract as represented in Algorithm 2 and proceeds with the transaction.

To proceed with Ethereum's deal, the user needs a personal wallet, and the user uses MetaMask [25], Google's extension program that allows users to safely manage Ethereum's personal wallet. MetaMask can only be used with Chrome, a PC browser; therefore, the proposed testbed creates a web page for energy trading to enable smooth trading. Energy generation is simulated using Python 3.7. We implement Algorithm 1 using Python and control the action value obtained from it.

A web page, which provides the real deal experience implemented in Figure 7a, gives the user the real-time price of Ethereum, the number of Ethereum and token users, and the address of the user's personal wallet. In Figure 7b, a compartment enters the amount of energy a user wants to purchase. If the order quantity is entered, the price will be automatically converted. When the user presses the "Purchase of Energy" button, a pop-up window appears informing them that the energy transaction will be performed at Figure 7c. When the deal is processed, it appears that MetaMask is executed as shown in Figure 7d and Ethereum will be sent to the seller's wallet. Upon the approval of the transaction, it may be confirmed that transaction approval is conducted as shown in Figure 7e,f and then completed. Following this process, the number of Ethereum and tokens currently in use are updated in real time.

(**a**)

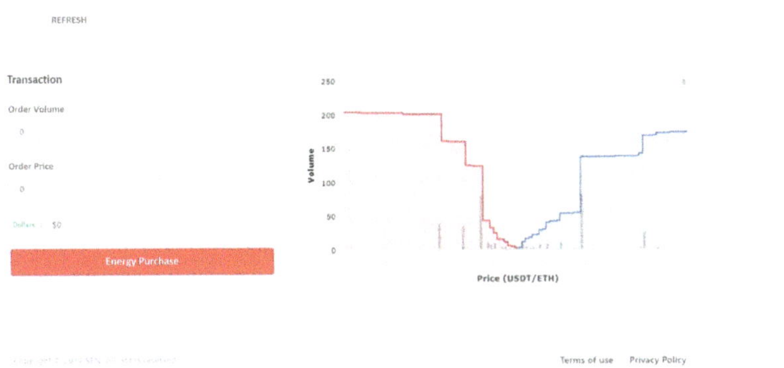

(**b**)

Figure 7. *Cont.*

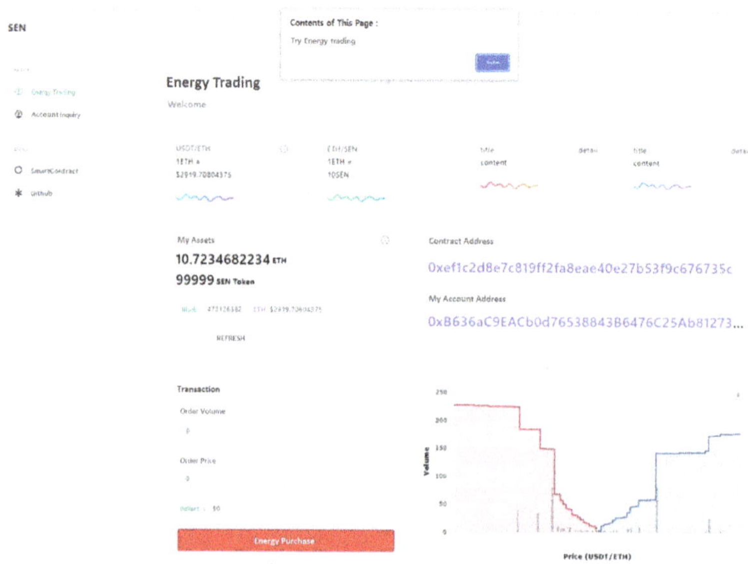

(**c**)

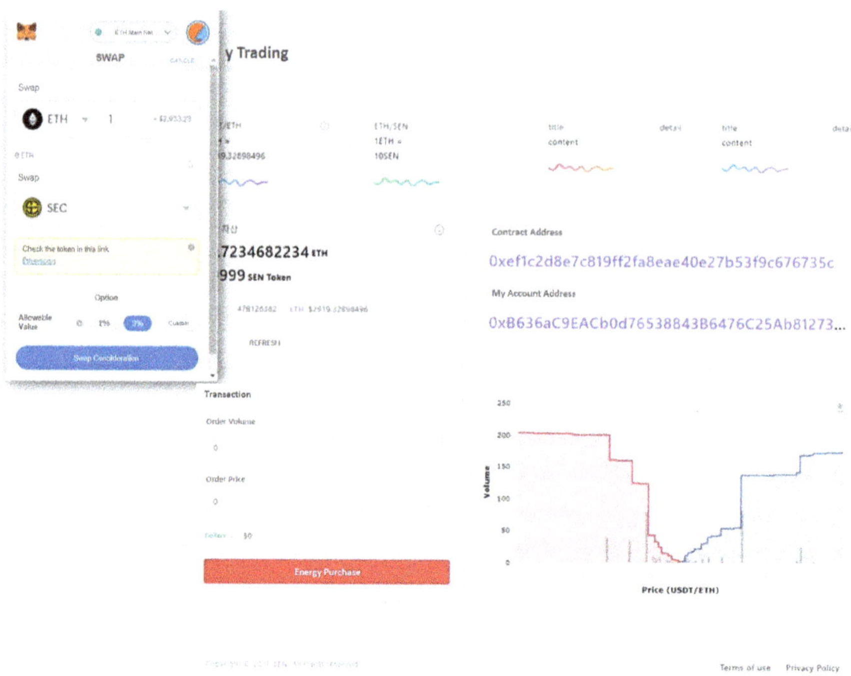

(**d**)

Figure 7. *Cont.*

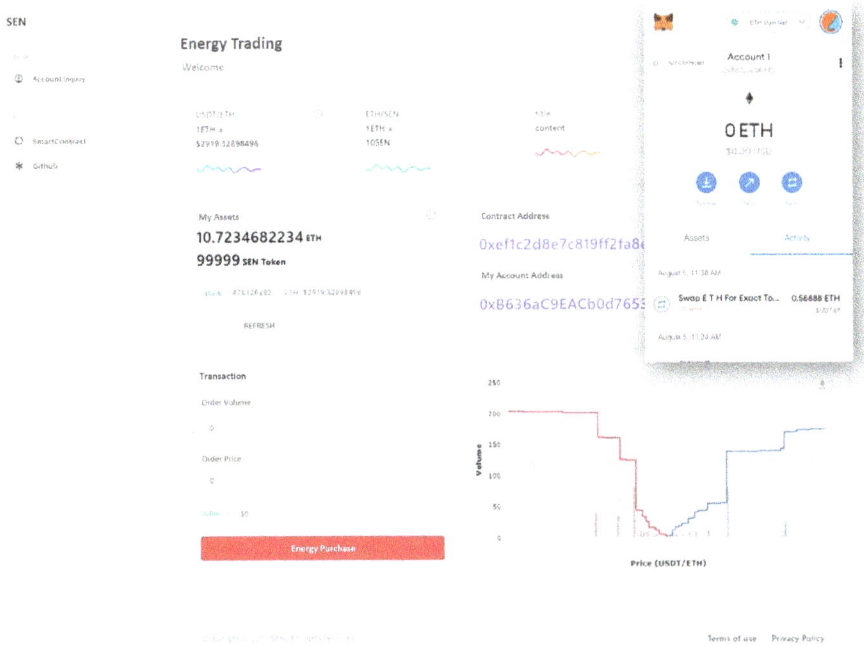

(e)

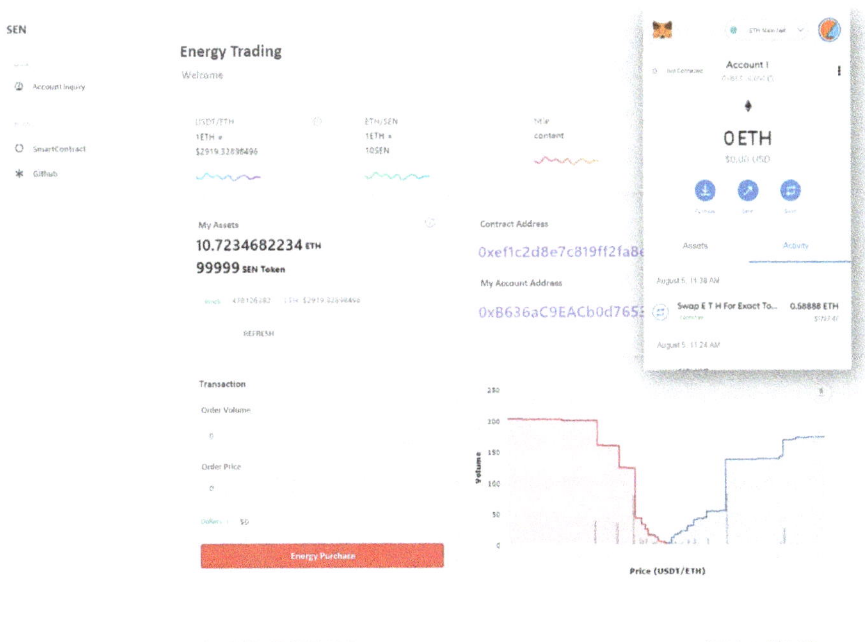

(f)

Figure 7. Process of energy trading in web page. (**a**) Real time user's information; (**b**) Purchase section; (**c**) Purchase progress pop-up window; (**d**) Ethereum transaction preview; (**e**) Transaction in progress; (**f**) Transaction completion.

Algorithm 2 Transaction Condition in Smart Contract.

```
if Seller's energy is bigger than Buyer's requirement then
    build smart contract
    while Exist Energy do
        while Exist token do
            if Check balance of token then
                if Check balance of Ether then
                    swap token and Ether
                    record transactions on the blockchain
                end if
            end if
        end while
    end while
end if
```

In the case of the proposed testbed, we implement an Android mobile application using Android Studio, Java, and Etherscan for the convenience of market participants. However, because MetaMask's characteristics make transactions impossible on mobile devices, the application mainly serves to provide information to the user. Figure 8a is the main page of the mobile application showing the current Ethereum's market price. Additionally, Figure 8b provides contract address information and does not provide the user's account address because MetaMask cannot be installed on a mobile device. In Figure 8c, the price can be verified by entering the desired quantity of purchases. Here, Figure 8d shows a graph of price and volume. Finally, by entering the user account address information shown in Figure 8e, we can check the user's transaction history as depicted in Figure 8f.

After conducting actual energy transaction simulations using the hardware testbed and web page, we could check that the energy is moved. Additionally, the LED light bulb is checked through the energy received and it is proved that the energy received could be used, and the Ethereum exchanged through the transaction is reflected and recorded in real time.

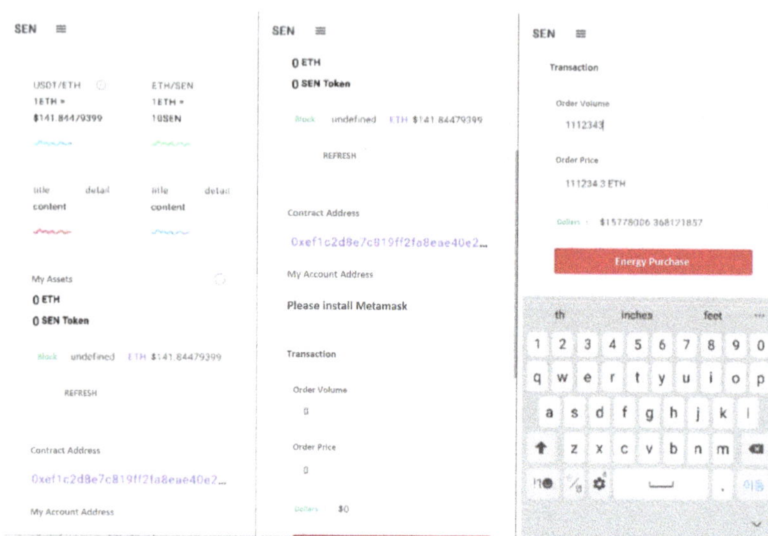

(a) (b) (c)

Figure 8. *Cont.*

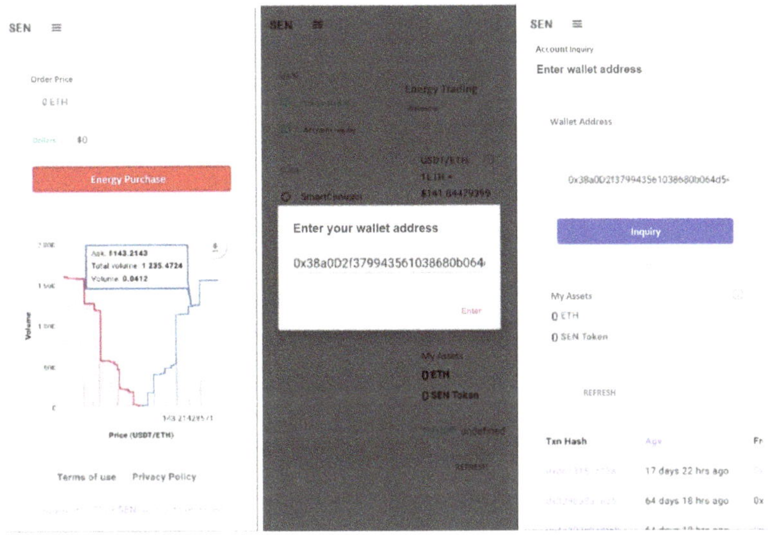

(d) (e) (f)

Figure 8. Provided information in mobile application. (**a**) Main page; (**b**) Contract address; (**c**) Market price; (**d**) Price with volume; (**e**) Account lookup; (**f**) Transaction history.

6. Evaluation

To analyze the performance of the proposed system, we conduct a simulation by adjusting various parameter variables. Furthermore, the proposed system is compared with other algorithms to determine whether it is effective. To proceed with the simulation, we generate solar energy from several buildings at KAIST over a long time and used this value as the actual training data.

Figure 9 is the result of ten tests each with various parameter values, averaging the amount of solar energy that can be generated per day as a result. In Figure 9a, we see that 8000 episodes are needed to achieve optimal learning effects. Additionally, Figure 9b shows that the learning rate α is the best parameter value of 0.1. However, Figure 9c shows that the discount factor γ_i does not have a significant impact on the training result.

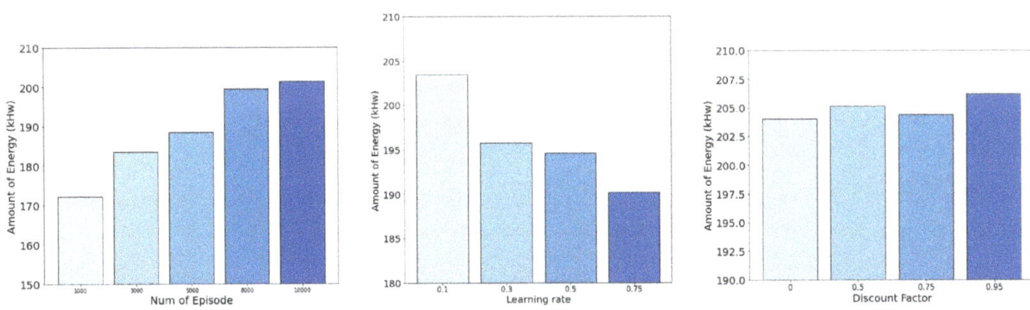

(a) (b) (c)

Figure 9. Adjusting parameters according to the amount of solar energy generated per day. (**a**) Adjusting number of episode; (**b**) Adjusting learning rate; (**c**) Adjusting discount factor.

Several algorithms are compared to analyze the performance of the proposed system. (1) A static panel that generates energy without moving the panel. (2) Regular panels move at the same angle every time and generate energy. (3) The heuristic panel moves in such a way that it can generate the most energy during that period. Finally, (4) the proposed system moves according to the completed Q-table and generates energy.

The simulation results compared to the various systems are shown in Figure 10. The proposed system generates 35% and 16% more energy than the fixed and regular panels, respectively. Additionally, the heuristic algorithm, which moves to the optimal location, collects 0.06% more energy than the proposed system, indicating that there is no significant difference in energy collection.

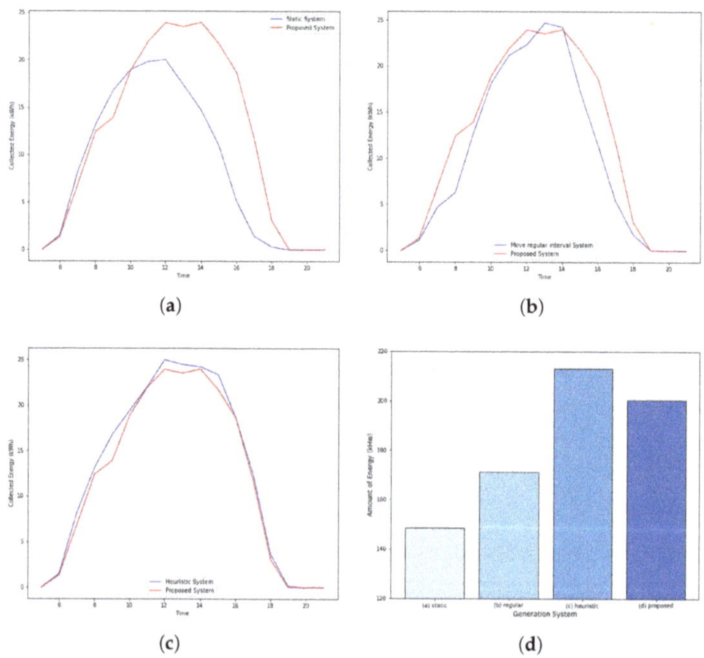

Figure 10. Simulation result. (**a**) Compared with static system; (**b**) Compared with regular system; (**c**) Compared with heuristic system; (**d**) Total amount of generated energy.

However, the proposed system no longer needs to train once the Q-table converges. Therefore, we believe that the proposed system will be more effective than a heuristic algorithm that should continue checking the energy collection. As a result, we have confirmed that the proposed system is effective in terms of solar panel generation compared with the other operation methods by comparing the energy generation rates and their effectiveness.

To analyze the profits of each market participant according to the auction progress, we pre-determine the value of variables and progress the numerical simulation results. For the sales, electricity rate C_{sell} and future expected cost C_{op} are set to KRW 0.1 and KRW 0.5. In addition, battery efficiency η is set to 0.9, and the amount of energy generated by each time period is set to be located between 1–10 kWh. In the case of the buyer, the price of electricity purchased through the conventional market is set to 0.08–0.24 per kWh considering the progressive billing system in the Republic of Korea. In addition, we set the average amount of electricity required for each time period to 5 kWh for the target of the small-scale power consumer. At this time, numerical simulation results are progressed to analyze the strategies of each market participant.

As depicted in Equation (5), a value of battery efficiency is related with the seller's decision. Here, low efficiency means a large energy loss in storing progress, so sellers take a strategy to sell energy rather than store it. According to Figure 11a, when the battery efficiency is less than 0.2, it can be confirmed that the seller is taking a strategy to sell the whole energy as the financial profit when storing progress is relatively low. In addition, the seller's strategy according to the relative electricity rates is shown in Figure 11b. Here, an increase in the value of the auxiliary variable α means that the current energy sale price C_{sell} increases or the future expected price C_{op} decreases. In this case, it can be seen that the seller takes a strategy to sell more energy at the present time and try to reduce the energy stored.

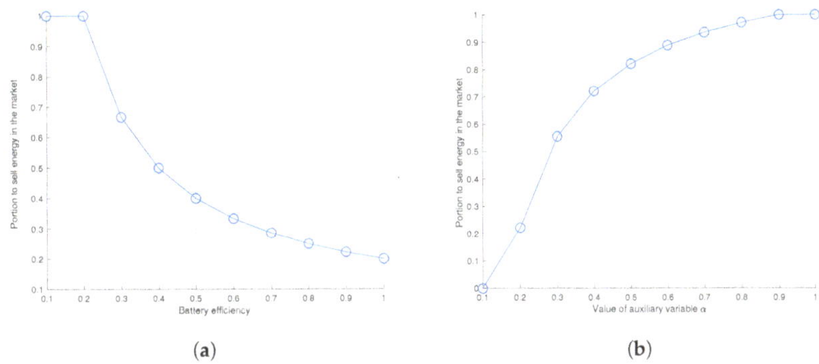

Figure 11. Determination of transaction ratio of seller according to environmental changes. (**a**) Compare with static system; (**b**) Compare with static system.

For the case of the buyer, their strategies are related with the amount of transaction energy, which is determined by the seller. When the transaction energy E is greater than a buyer i's required energy D_i, the buyer pays an unnecessary cost as mentioned in Equation (6). To prevent such a problem, the buyer tries to reduce the bidding price as the amount of energy in the auction increases, as depicted in Figure 12a. Here, the buyer's bidding price is closely related to the amount of energy in the auction, but also closely related to the success rate according to the bidding price. The fact that the higher value of the auxiliary variable γ_i means that the probability of successful bidding is increased even at the bidding price of the buyer i is low. Therefore, when γ_i increases, the bidding price tends to decrease as depicted in Figure 12b. In addition, buyer's bidding price is also affected by the electricity price as argued in Equations (10) and (12). From these equations, it is possible to estimate that the buyer's bidding price tends to increase proportionally according to the increment of market price.

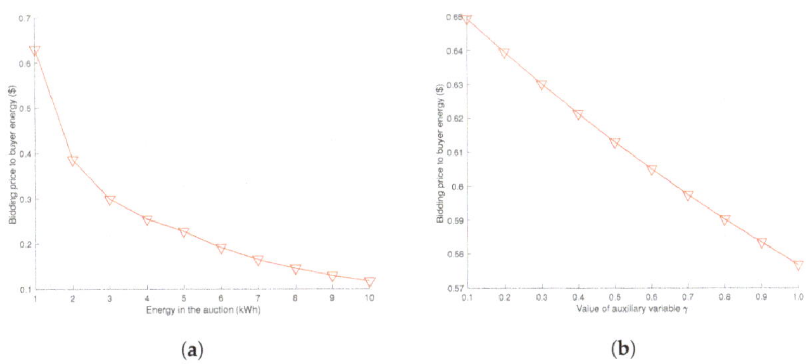

Figure 12. Determination of bidding cost of buyer i according to environmental changes. (**a**) Compared with static system; (**b**) Compared with static system.

Furthermore, we verify that energy trading is going smoothly with our testbed. On the implemented platform, when the buyer purchases energy from the seller, the energy moves from the seller to the buyer, as shown in Figure 13. To check whether the energy moved, we mount an LED sensor on the buyer's testbed. We confirm that energy trading occurs by checking that the LED sensor is lit up as much as the energy received when the buyer receives the energy.

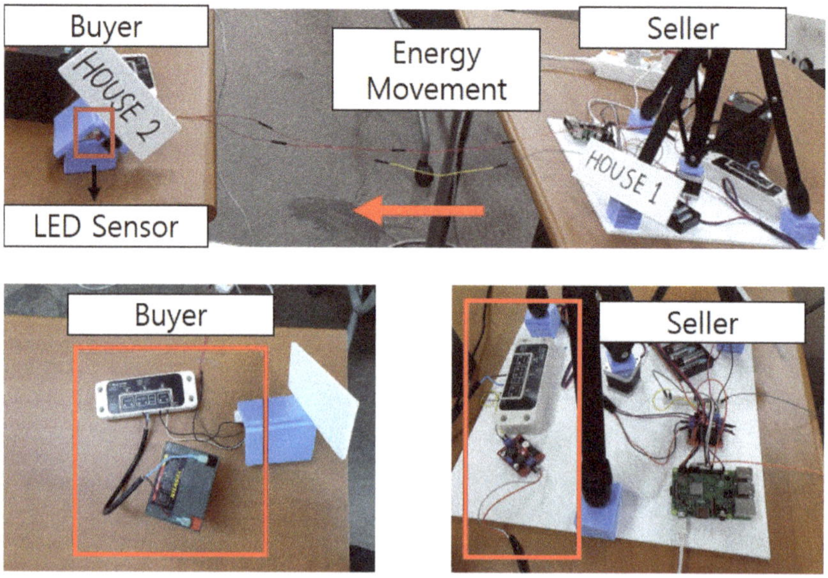

Figure 13. Energy trading test.

7. Conclusions

This study proposes a solar energy generation and transaction platform (EggBlock) using blockchain technique. The proposed EggBlock platform can maximize solar energy generation by controlling the angle of generator, and balance the power supplement by utilizing the proposed energy transaction scheme. Through the various simulation results, we showed that the dynamic panel control mechanism on the EggBlock converges well to the optimal directions with short iterations, which results in an average energy generation gain of 35%. In addition, in the energy trading, we also analyzed the strategies of each market participant according to the changes in the electricity price and environment of the actual power system. In our future study, the uncertainty of solar energy generation should be considered both in the energy trading and generation to address the practical concerns.

Author Contributions: Conceptualization, S.K., J.L. and J.K.; Data curation, H.O.; Formal analysis, J.L. and H.O.; Funding acquisition, H.O.; Project administration, J.L.; Software, J.K.; Supervision, J.K.; Visualization, S.K. and J.K.; Writing—original draft, S.K. and J.K.; Writing—review & editing, J.L. and H.O. All authors have read and agreed to the published version of the manuscript.

Funding: This research is supported in part by the National Research Foundation of Korea (NRF) grant funded by the Korea government (MSIT) (No. 2021R1F1A1048098) and in part by the Institute of Information and Communications Technology Planning and Evaluation (IITP) grant funded by the Korea Government [Ministry of Science and ICT (MSIT)] (A Study on 5G based Intelligent IoT Trust Enabler) under Grant 2020-0-00833.

Institutional Review Board Statement: Not applicable.

Informed Consent Statement: Not applicable.

Data Availability Statement: Not applicable.

Conflicts of Interest: The authors declare no conflict of interest.

References

1. SEIA. *Solar Energy Introduction*; Technical Report; SEIA: Washington, DC, USA, 2017.
2. Leeton, U.; Uthitsunthorn, D.; Kwannetr, U.; Sinsuphun, N.; Kulworawanichpong, T. Power loss minimization using optimal power flow based on particle swarm optimization. In Proceedings of the ECTI-CON2010: The 2010 ECTI International Confernce on Electrical Engineering/Electronics, Computer, Telecommunications and Information Technology, Chiang Mai, Thailand, 19–21 May 2010; pp. 440–444.
3. Detyniecki, M.; Marsala, C.; Krishnan, A.; Siegel, M. Weather-based solar energy prediction. In Proceedings of the 2012 IEEE International Conference on Fuzzy Systems, Brisbane, QLD, Australia, 10–15 June 2012; pp. 1–7.
4. Rodríguez, F.; Fleetwood, A.; Galarza, A.; Fontán, L. Predicting solar energy generation through artificial neural networks using weather forecasts for microgrid control. *Renew. Energy* **2018**, *126*, 855–864. [CrossRef]
5. Lee, J.; Guo, J.; Choi, J.K.; Zukerman, M. Distributed energy trading in microgrids: A game-theoretic model and its equilibrium analysis. *IEEE Trans. Ind. Electron.* **2015**, *62*, 3524–3533. [CrossRef]
6. Zhang, C.; Wu, J.; Zhou, Y.; Cheng, M.; Long, C. Peer-to-Peer energy trading in a Microgrid. *Appl. Energy* **2018**, *220*, 1–12. [CrossRef]
7. Pei, W.; Du, Y.; Deng, W.; Sheng, K.; Xiao, H.; Qu, H. Optimal bidding strategy and intramarket mechanism of microgrid aggregator in real-time balancing market. *IEEE Trans. Ind. Inform.* **2016**, *12*, 587–596. [CrossRef]
8. Mihaylov, M.; Jurado, S.; Moffaert, K. NRG-X-change. In Proceedings of the 3rd International Conference on Smart Grids and Green IT Systems, Barcelona, Spain, 3–4 April 2014; pp. 101–106.
9. Mengelkamp, E.; Notheisen, B.; Beer, C.; Dauer, D.; Weinhardt, C. A blockchain-based smart grid: Towards sustainable local energy markets. *Comput. Sci. Res. Dev.* **2018**, *33*, 207–214. [CrossRef]
10. Zhang, T.; Pota, H.; Chu, C.C.; Gadh, R. Real-time renewable energy incentive system for electric vehicles using prioritization and cryptocurrency. *Appl. Energy* **2018**, *226*, 582–594. [CrossRef]
11. Murdock, H.E.; Gibb, D.; André, T.; Appavou, F.; Brown, A.; Epp, B.; Kondev, B.; McCrone, A.; Musolino, E.; Ranalder, L.; et al. Renewables 2019 Global Status Report. 2019. Available online: https://repository.usp.ac.fj/11648/1/gsr_2019_full_report_en.pdf (accessed on 18 March 2022).
12. Kuik, O.; Branger, F.; Quirion, P. Competitive advantage in the renewable energy industry: Evidence from a gravity model. *Renew. Energy* **2019**, *131*, 472–481. [CrossRef]
13. Furukakoi, M.; Adewuyi, O.B.; Matayoshi, H.; Howlader, A.M.; Senjyu, T. Multi objective unit commitment with voltage stability and PV uncertainty. *Appl. Energy* **2018**, *228*, 618–623. [CrossRef]
14. Cole, W.J.; Frazier, A. *Cost Projections for Utility-Scale Battery Storage*; Technical Report; National Renewable Energy Lab. (NREL): Golden, CO, USA, 2019.
15. Hassija, V.; Chamola, V.; Garg, S.; Krishna, D.N.G.; Kaddoum, G.; Jayakody, D.N.K. A blockchain-based framework for lightweight data sharing and energy trading in V2G network. *IEEE Trans. Veh. Technol.* **2020**, *69*, 5799–5812. [CrossRef]
16. Chen, X.; Zhang, T.; Ye, W.; Wang, Z.; Iu, H.H.C. Blockchain-based electric vehicle incentive system for renewable energy consumption. *IEEE Trans. Circuits Syst. II Express Briefs* **2020**, *68*, 396–400. [CrossRef]
17. Hassan, M.U.; Rehmani, M.H.; Chen, J. DEAL: Differentially private auction for blockchain-based microgrids energy trading. *IEEE Trans. Serv. Comput.* **2019**, *13*, 263–275. [CrossRef]
18. Gramoli, V. From blockchain consensus back to Byzantine consensus. *Future Gener. Comput. Syst.* **2020**, *107*, 760–769. [CrossRef]
19. Zhu, S.; Song, M.; Lim, M.K.; Wang, J.; Zhao, J. The development of energy blockchain and its implications for China's energy sector. *Resour. Policy* **2020**, *66*, 101595. [CrossRef]
20. Watkins, C.J.; Dayan, P. Q-learning. *Mach. Learn.* **1992**, *8*, 279–292. [CrossRef]
21. Wang, Q.; Zhu, X.; Ni, Y.; Gu, L.; Zhu, H. Blockchain for the IoT and industrial IoT: A review. *Internet Things* **2020**, *10*, 100081. [CrossRef]
22. Levin, J. Auction Theory. 2004; Volume 20286. Available online: http://www.mohamedelafrit.com/education/CNAM/ESD208-Incitations-et-Design-Economique/Lectures/MOOCs/Stanford-Econ286/Stanford-Auction_Theory-Jonathan_Levin.pdf (accessed on 18 March 2022).
23. Kim, J.; Park, H.; Lee, G.H.; Choi, J.K.; Heo, Y. Seal-bid renewable energy certification trading in power system using blockchain technology. In Proceedings of the 2020 International Conference on Information and Communication Technology Convergence (ICTC), Jeju Islan, Korea, 21–23 October 2020; pp. 1752–1756.
24. Kim, J.; Lee, J.; Park, S.; Choi, J.K. Battery-wear-model-based energy trading in electric vehicles: A naive auction model and a market analysis. *IEEE Trans. Ind. Inform.* **2018**, *15*, 4140–4151. [CrossRef]
25. Google. *Introduction of Metamask*; Technical Report; Google Inc.: Mountain View, CA, USA, 2020.
26. Sutton, R.S.; Barto, A.G. *Reinforcement Learning: An Introduction*; MIT Press: Cambridge, MA, USA, 2018.

energies

MDPI

Article

LPSRS: Low-Power Multi-Hop Synchronization Based on Reference Node Scheduling for Internet of Things

Mahmoud Elsharief [1,2], Mohamed A. Abd El-Gawad [3], Haneul Ko [4] and Sangheon Pack [5,*]

1 Department of Electronic Engineering, Hanbat National University, Daejeon 34158, Korea; mahmoud@hanbat.ac.kr
2 Department of Electrical Engineering, Al-Azhar University, Cairo 11651, Egypt
3 National Telecommunication Institute, Cairo 11768, Egypt; mohamed.gawad@nti.sci.eg
4 Department of Computer Convergence Software, Korea University, Sejong 30019, Korea; heko@korea.ac.kr
5 School of Electrical Engineering, Korea University, Seoul 02841, Korea
* Correspondence: shpack@korea.ac.kr

Abstract: Time synchronization is one of the most fundamental problems on the internet of things (IoT). The IoT requires low power and an efficient synchronization protocol to minimize power consumption and conserve battery power. This paper introduces an efficient method for time synchronization in the IoT called low-power multi-hop synchronization (LPSRS). It employs a reference node scheduling mechanism to avoid packet collisions and minimize the communication overhead, which has a big impact on power consumption. The performance of LPSRS has been evaluated and compared to previous synchronization methods, HRTS and R-Sync, via real hardware networks and simulations. The results show that LPSRS achieves a better performance in terms of power consumption (transmitted messages). In particular, for a large network of 450 nodes, LPSRS reduced the total number of transmitted messages by 53% and 49% compared to HRTS and R-Sync, respectively.

Keywords: HRTS; IoT; LPSRS; low power; R-sync; scheduling; time synchronization

Citation: Elsharief, M.; El-Gawad, M.A.A.; Ko, H.; Pack, S. LPSRS: Low-Power Multi-Hop Synchronization Based on Reference Node Scheduling for Internet of Things. *Energies* **2022**, *15*, 2289. https://doi.org/10.3390/en15062289

Academic Editors: Antonio Cano-Ortega, Francisco Sánchez-Sutil, Marco Pau and Aurora Gil-de-Castro

Received: 9 February 2022
Accepted: 19 March 2022
Published: 21 March 2022

1. Introduction

The internet of things (IoT) plays a remarkable role and makes the quality of our life efficient. It has been extensively studied and developed in recent years. The IoT is generally applied to many areas, including healthcare, transportation, wireless networks, and industry. The wireless sensor nodes (WSNs) are one of the prime parts of IoT applications. They face many challenges to exchange information and cooperate in wireless applications [1,2]. One of these challenges is power consumption. Generally, sensor nodes are frequently battery-powered in most IoT applications. To reduce the battery power consumption and increase the battery lifetime, sensor nodes switch between active and sleep modes. To wake up sensor nodes at a predetermined time, most of them require time synchronization [3,4]. Sensor nodes, as a result, require trust and reliable synchronization [5].

Sensor nodes are frequently equipped with a relatively inexpensive hardware clock oscillator [6]. As a result, reliable time synchronization for such sensor nodes is more difficult [7]. Time synchronization of such wireless networks with such sensor nodes cannot be well-preserved for an extended period of time unless an accurate synchronization technique is performed on a constant schedule [7]. In time division multiple access (TDMA) networks, a node's wake-up time might be completely erroneous due to a lack of synchronization, resulting in severe network connectivity failure. Therefore, time synchronization is considered an essential and important aspect of the sensor nodes' operations. It offers a common reference time for all sensor nodes in the entire network [8,9].

Generally, to keep the whole network well synchronized and to minimize time offsets caused by clock drift, all sensor nodes should require continuously exchanging timing messages [10]. Exchanging timing messages between the nodes is commonly utilized for time synchronization in sensor node networks. Since the transmission of messages usually accounts for most of the power consumption, frequent synchronization significantly increases power consumption [5,8]. Generally, time synchronization in IoT sensor nodes has two main problems: (1) communication overhead, which can be represented by the number of transmitted messages, and (2) accuracy. Both of these problems lead to a significant rise in power consumption [4,9].

Several protocols for sensor node synchronization have been developed. These protocols have been developed to synchronize all nodes in large networks [11]. The majority of these protocols aimed to enhance synchronization accuracy at the expense of higher communication overhead [2,4]. On the other hand, there are a few proposed protocols that are interested in reducing power consumption by reducing overhead connections [4,9]. Moreover, due to frequent collisions, these protocols suffer from message loss. Collisions have an effect on power usage and may result in synchronization failure [5,7].

All of these issues and problems motivate us to establish the low-power synchronization based on reference node scheduling (LPSRS) for IoT. It can save an amount of energy by reducing the number of timing messages sent between sensor nodes. Furthermore, LPSRS implements a technique that fully and completely prevents node collisions during the time synchronization process. According to experimental results and simulations, the performance of LPSRS has been extensively investigated compared to HRTS [12] and R-Sync [9]. The results demonstrate that LPSRS reduces the total number of transmitted messages by 53%, 49%, and 13% compared to HRTS, R-Sync, and FADS, respectively, and, thus, LPSRS can minimize the power consumption for time synchronization in the IoT.

The remainder of the paper is structured as follows: We address the related work in Section 2, while the preliminaries and system model of LPSRS are described in Section 3. Section 4 elaborates on the operations of LPSRS. Section 5 presents the experimental and simulation setup and discusses the results. This is followed by the simulation results. Finally, in Section 6, we conclude our work.

2. Related Work

Sensor nodes' time synchronization has been extensively researched over many decades. Many methods for clock synchronization have been proposed in the literature, all of which seek to reduce power consumption by improving accuracy and lowering communication overhead.

The flooding time synchronization protocol (FTSP) [13] pursues to minimize the error of synchronization by using the least square linear regression (LSLR) to compensate for the clock drift. Lenzen et al. have reported that the global synchronization error of FTSP rises exponentially with the size of the network [14,15]. They, then, propose a scalable technique called PulseSync, which can quickly flood the timing information all over the network [14]. Yildirim et al. indicated possible drawbacks of quick flooding and suggested a clock speed agreement algorithm. He proposed flooding with clock speed agreement (FCSA) [6] and an adaptive value tracking synchronization (AVTS) [16]. FCSA and AVTS, however, need many messages which increase the chance of collision and result in excessive power consumption [11].

The reference broadcast synchronization (RBS) [17] is typically a receiver-to-receiver protocol (RRP). By excluding the transmitter's delays from the offset computations, it improves the accuracy of the time offset estimate. In RBS, however, it sends a large number of messages in each cycle of the synchronization [7]. This leads to two different problems: a high chance of collision and excessive power consumption. Gong et al. developed the energy-efficient coefficient exchange synchronization protocol (CESP) to solve these issues [18]. Compared to RBS, it makes use of the synchronization coefficient to relatively reduce communication costs. CESP, on the other hand, is based on a stationary reference node and, thus, absorbs a lot of energy [9].

In [19], the timing-sync protocol for sensor networks (TPSN) was introduced. Although TPSN has good scalability, it requires regular resynchronization [20]. Furthermore, because of the enormous number of message exchanges required by TPSN, it consumes a lot of power [7]. Hierarchy reference broadcast synchronization (HRTS) [12] contains three stages that are repeated at every hierarchical level all over a multi-hop network. In HRTS, it is presumed that each node is aware of its adjacent neighbors. Despite being a scalable and lightweight protocol, HRTS has two drawbacks: high communication costs and excessive collisions [11].

M. Elsharief et al. introduced a synchronization technique called FADS [7]. FADS uses a scheduling technique to organize message transmission among sensor nodes, which reduces the number of collisions. FADS reduces latency by speeding up the scheduling process. During the scheduling process, FADS, however, requires a relatively large number of message transmissions, which consumes a lot of power [21].

The robust time synchronization (R-sync) for the industrial internet of things (IIOT) was also introduced in [9]. It focuses on locating and reconnecting the network's isolated nodes that have lost synchronization [22]. Despite the reality that R-sync needs fewer messages than TPSN, it lacks a collision avoidance technique.

In contrast to previous approaches, average time synchronization (ATS) [23] is a fully distributed method. To calibrate its compensation parameters, ATS employs a cascade of two consensus methods, and it allows nodes to converge to a steady-state virtual clock [5]. However, ATS suffers from frequent collisions, which might result in synchronization method failure [7]. To address the aforementioned shortcoming of ATS, the selective average time synchronization (SATS) [24] protocol has been suggested. It employs a dynamic programming method to increase the speed and accuracy of convergence in sparse WSNs. It achieves faster convergence than other protocols (ATSP [25] and CCS [26]). SATS, on the other hand, increases the number of transmitted messages since each node in SATS must forward every received message [27]. Based on ATS, the multi-hop average consensus time synchronization protocol (MACTS) [28] has been proposed. To improve the algebraic connectedness of the large-scale network, it leverages the notion of virtual communication linkages among multi-hop nodes. MACTS, like SATS, must send additional messages throughout the synchronization process. Message forwarding in large-scale networks, such as MACTS, can be delayed due to congestion among surrounding nodes [27].

The notion of density-based scheduling was introduced in [29], which can minimize overhead by reducing the number of reference nodes. It may, nevertheless, experience reception failure as a result of frequent packet collisions. Furthermore, it frequently suffers from coverage issues since it does not ensure the synchronization of some nodes that are not within the wireless range of any chosen reference nodes [7]. In [21], the low-power scheduling for synchronization protocol (LPSS) was introduced. It is a free collision m method. LPSS reduces the number of messages and accelerates the scheduling process. It, however, does not provide a synchronization scheme. In [30], Geo-TimeSyc was proposed. Geo-TimeSyc partitions the network into separated regions to avoid concurrent transmission of neighboring nodes. It reduces the probability of collisions [31]. As Geo-TimeSyc is implemented with RBS, it needs a large number of message exchanges. Recently, we proposed an energy-efficient synchronization technique called EERS [4]. It is a free collision protocol. Although ERRS substantially reduces energy consumption, it is only suitable for networks in which the nodes' locations are known.

From a thorough analysis of prior studies, we found that most of the prior techniques suffer from a serious problem of high power consumption. Hence, it is highly challenging to provide low-power synchronization. This problem motivates us to find a new technique to resolve it. Our approach, LPSRS, uses a scheduling method of [21] that tries to cover all nodes with as few reference nodes as possible. Then, it applies the synchronization scheme to synchronize all nodes in the entire network. LPSRS can save a large amount of energy by minimizing the usage of messages were sent and preventing collisions. As a result, LPSRS's contributions can be summarized in the following points:

(1) LPSRS substantially reduces transmission overhead (number of transmitted messages), thereby lowering power consumption.
(2) LPSRS utilizes a scheduling approach to address the collision issue, which can reduce power consumption.
(3) LPSRS is constructed into a real wireless sensor network with different configurations and an extensive simulation with large-scale networks.

3. Preliminaries

3.1. System Model

A sensor network is designed as a graph $G = (V, E)$ consisting of a set of vertices, $V = \{1, 2, \ldots, \ldots, N\}$, representing the sensor nodes and a set of edges, $E \subseteq V \times V$, representing the two-way communication links among the nodes. We refer to the nodes located within the wireless range of a node $i \in V$ as the neighbors of node i. These neighbor nodes of node i are denoted by $\mathbb{V}(i) = \{j \in V | \{i, j\} \in E\}$. Each node in this article has its own unique ID number (≥ 1) as well as a hardware clock. Simply for the purpose of clarity, we assume that the network is fixed and the communication between the nodes is fully reliable. Besides, we also presume that each node is supplied with an omnidirectional antenna and has a steady wireless range of an r meter radius. Moreover, we assume that the sink node recognizes all neighbor nodes of every node in the network [21].

3.2. Message Format

LPSRS utilizes two types of messages: the first one is used for the scheduling process, while the other one is used for the synchronization process. The first scheduling message includes the following fields:

(1) *DestAddr*: a destination address of the message.
(2) *MesN*: a number of split messages of the scheduling message. Its default value is 1.
(3) *SrcID*: a unique identifier of the sender.
(4) *Type*: the type of message is Sch (i.e., scheduling message).
(5) *RefNodes*: list of reference nodes.
(6) *SchSlots*: list of scheduled time slots.
(7) *SeqN*: sequence number of the scheduling messages. Its default value is 0.

Meanwhile, the synchronization message has the following fields:

(1) *DestAddr*: the destination address of the message.
(2) *SrcID*: the unique identifier of the sender.
(3) *SID*: specified reference node. Its default value is -1.
(4) *Type*: the type of messages, including $Sync1$, $Sync2$, and $Sync3$.
(5) *SeqN*: sequence number of the synchronization messages. Its default value is 0.
(6) *RT*: the timestamp at the moment of receiving $Sync1$. Its default value is -1.
(7) *TS*: the timestamp at the moment of sending $Sync2$. Its default value is -1.
(8) *RefT*: the timestamp of the reference time when receiving $Sync3$. Its default value is -1.
(9) *OF*: the offset value when sending $Sync3$. Its default value is 0.

4. LPSRS Algorithm

LPSRS objects to decreasing power consumption while synchronizing the whole WSN. It actually uses a scheduling methodology that tries to cover as many nodes as possible with a small number of reference nodes. LPSRS consists of two processes: (1) the scheduling process and (2) the synchronization process. The scheduling process runs only during the configuration phase or when the network topologies are significantly altered. It offers a low-power approach to the aim of guaranteed synchronization for a tree-based network topology without collisions. Meanwhile, the synchronization step is repeated on a regular schedule with a $Psyn$ interval to keep the time offset of all nodes within an allowable threshold.

4.1. Scheduling Process

In this section, we explain in detail the scheduling algorithm, which aims to reduce communication costs while ensuring that all nodes in a tree-based network topology are completely covered. The suggested solution provides each reference node a distinct allocated time slot, ensuring that the synchronization operation is performed at a distinct time slot to avoid collisions.

We simply describe the core idea of the scheduling algorithm. Since finding the minimum list of connected reference nodes is an (NP)-hard problem [32], we simply use the breadth-first search (BFS) method to approximately find the minimum set of reference nodes. BFS is a method for traversing the graph layer. The total time complexity of BFS is $O(E+V)$ [33]. As the sink node (predefined node) has complete awareness of the system architecture, it simply uses BFS to identify the level of each node in the network. As shown in Figure 1 and Algorithm 1, the sink node begins from the sink node and finds out all the adjacent nodes at the current depth before moving on to the nodes at the next depth level (lines 8 to 12). The sink also determines all reference nodes, $RefNodes$, and their scheduled time slots, $SchSlots$ (lines 13 to 18). Here, $RefNodes$ are the nodes responsible to spread time messages to all nodes in the network. To prevent reusing the same reference node, we set $Parent_{flag[j]}$ to a *true* value (line 15). Next, all reference nodes and their related time slots are listed in the scheduling message. The scheduling message is then broadcast to all nodes in the network. To finish the synchronization process, each reference node uses its assigned time slot [21].

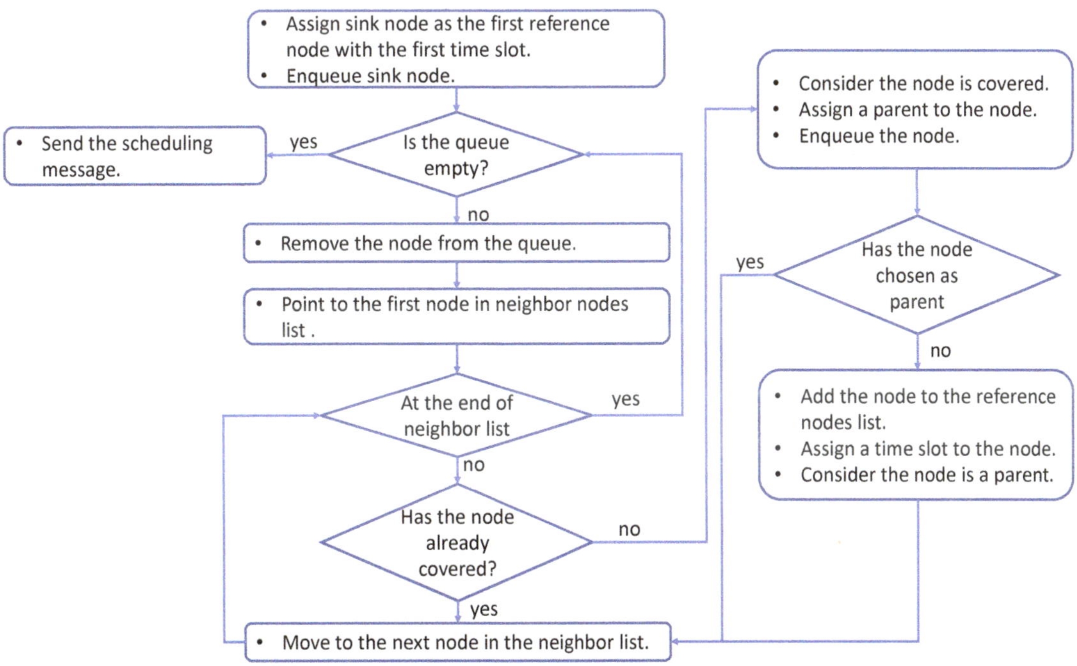

Figure 1. Scheduling process's flow chart.

Algorithm 1: LPSRS pseudo-code for the scheduling process (sink node only).

```
Initial : Q ← ∅, CoverNodes_flag[ ] ← false, cnt ← 0, Node_Parent[ ] ← ∅, Parent_flag[ ] ← false
s ← sink node
RefNodes(1) ← s
Schslot(1) ← 0
CoverNodes_flag[s] ← true
enqueue(Q, s)
while (Q! = Null) do
            current i ← dequeue(Q); // Remove i form queue
    for (every edge (i, j) ∈ E) do
        if (CoverNodes_flag[j] == false) Then
            CoverNodes_flag[j] ← true
          Node_Parent[j] ← i
          enqueue(Q, j)
          if (Parent_flag[i] == false) Then
            RefNodes(cnt) ← i
              Parent_flag[j] ← true
            SchSlots(cnt) ← cnt
            cnt ← cnt + 1
          end if
        end if
    end for
end while
send scheduling message
<0xFFFF, s, Sch, MesN, RefNodes, SchSlots, SeqN>
```

When a node receives a scheduling message, it applies Algorithm 2. Each node responds only if its ID has been stored in $RefNodes$ (lines 5 to 10). Then, it preserves the typically associated time slot, $mySchslot$. For each reference node, the waiting timer is initially started (WT) as soon as it receives the scheduling message. The value of WT can be computed as below:

$$WTvalue = mySchslot - Schslot_{SrcID} \tag{1}$$

Here, $WTvalue$ is the amount of time each reference should wait before transmitting the scheduling message. $Schslot_{SrcID}$ and $mySchslot$ are the sender's time slot and the current reference time slot, respectively (lines 12 to 15). The current reference transmits the scheduling message to its nearby nodes as soon as WT eventually expires (lines 17 and 18). Additionally, each reference node examines the sequence number field, $SeqN$, to avoid sending the scheduling message multiple times. If the $SeqN$ of the current message is less than or equal to the $SeqN$ of the previous message, every reference node does not forward the message. It only adds the $SrcID$ of the sender node to its reference nodes neighbor list (lines 2 to 4). The reference nodes neighbor list is used by the synchronization process as described in Section 4.2.

For a large network, if the size of $RefNodes$ and $SchSlots$ exceeds the maximum packet size, the scheduling message is divided into multiple small messages, $MesN$. Besides, the assigned time slot width of each reference node is increased to avoid collisions. It will be equal to $MesN$ times the time slot width. For example, if the maximum packet size is 50 bytes, the assigned time slot is 10 ms, and $RefNodes$ and $SchSlots$'s sizes are 80 bytes, the scheduling message is split into two messages, $MesN$ equals 2, and the time slot width is increased to be 20 ms.

In this paragraph, we clearly demonstrate an example of the suggested technique using the network represented in Figure 2. As illustrated in Figure 3, the sink node, S, employs Algorithm 1 to construct a tree. It actually finds that the reference nodes are S (sink node), B, C, I, and F and that their allocated time slots are 0, 1, 2, 3, and 4, respectively (lines 6~21). Then, the sink node builds and broadcasts the scheduling message to its neighbor nodes (line 22). As a result, S fills the scheduling message as follows.

Algorithm 2: LPSRS pseudo-code for the scheduling process (all nodes).

```
Initial : cnt1 ← 0, cnt2 ← 0, send_flag ← false, SeqN ← 0, mySchsSlot ← 0
■ Upon receiving Scheduling Message
if(Received SeqN ≤ SeqN) Then
   add SenderID to RefNeighbors list
end if
if (NodeID is in RefNodes) Then
    cnt1 ← position of NodeID in RefNodes
    mySchsSlot ← SchSlots(cnt1)
    cnt2 ← position of SenderID in RefNodes
    SchSlot_SrcID ← SchSlots(cnt2)
end if
WaitingTime ← (mySchSlot − SchSlot_SrcID)
if (WaitingTime > 0 and send_flag == false) Then
   send_flag ← true
   Setup waiting timer WT(WaitingTime)
end if
SeqN ← Received SeqN
■ Upon WT expires
send scheduling message
<0xFFFF, NodeID, Sch, MesN, RefNodes, SchSlos, SeqN>
```

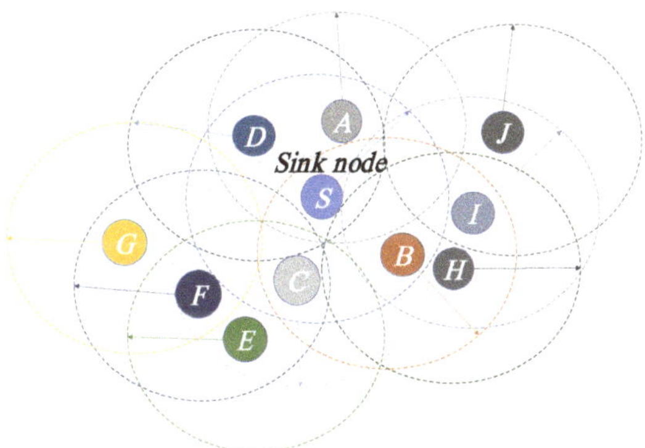

Figure 2. Sensor network example.

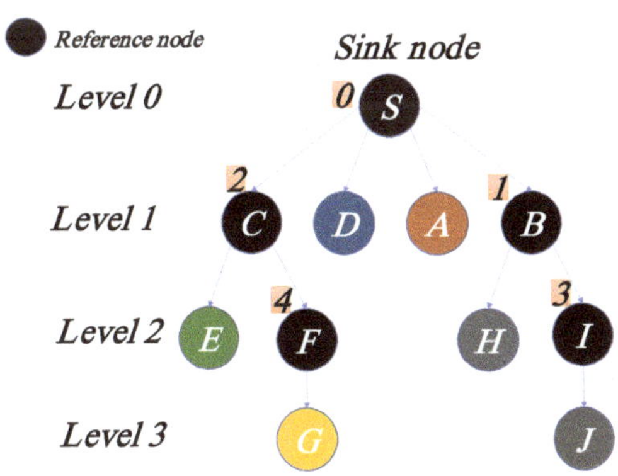

Figure 3. The tree created by the LPSRS scheduling process.

DestAdd	*SrcID*	*Type*	*MesN*	*RefNodes*	*SchSlots*	*SeqN*
0xFFFF	S	Sch	1	S, B, C, I, and F	0,1,2,3, and 4	1

Upon receiving the scheduling message, each node uses Algorithm 2 to check the $SeqN$ (line 2). Here, the $SeqN = 1$, which is greater than the previous $SeqN$ (default value = 0). As a result, each node examines whether or not it has been designated as a reference node. If it is a reference node, it sets its scheduled time slot to $mySchSlot$ and starts WT (lines 5~20). After the timeout expires, it sends the received packet. (Line 22). Node C, for example, is designated as a reference node, and the time slot is 2. C then determines its waiting time, which is equal to 2. (2 (C's $mySchSlot$)–0 ($Schslot_{SrcID}$(sender node))). Next, C delivers the received message to its adjacent nodes just as soon as the waiting timer expires.

4.2. Synchronization Process

In this section, we investigate and discuss the periodic synchronization process steps in detail. We conduct the synchronization process frequently, with a time period of $Psyn$ to conserve the time offset of all nodes within a suitable range (depending on the application). Once the scheduling process selects all reference nodes and their time slots, the synchronization process begins with the sink as the first reference node. The sink node's time is considered the reference time to all surrounding nodes. The synchronization process consists of three simple steps that are repeated in each hierarchical level over a multi-hop network. Since each reference node has its own time slot, *Schslot*, it should perform the synchronization process within the scheduled time to avoid collisions. The synchronization process of LPSRS occurs in a single broadcast domain. *R* is the reference node, *SIR* is the specified reference node, and *K* represents the other neighbor's nodes of *R*. Algorithm 3 and Figure 4 show how to estimate the total offset for frequent synchronization. In the first step, at time $T_{1,R}$, the $R =$ sink node broadcasts synchronization message number 1 ($Sync1$) to all its surrounding nodes. $Sync1$ includes the specified reference node, SIR, that has been randomly pre-selected (from the reference nodes neighbor list) by SIR to send back a response (line 1 to 6).

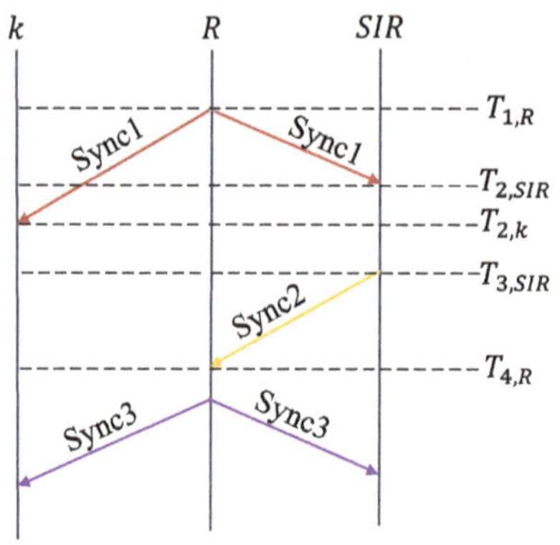

Figure 4. The synchronization process of LPSRS in a single broadcast domain. *R* is the reference node, SIR is the specified reference node, and *K* represents the other neighbor's nodes of *R*.

Every node $k \in \mathbb{U}(R)$ saves the time stamp, $T_{2,k}$ for the arrival time of $Sync1$ (line 9). Here, $\mathbb{U}(R)$ denotes a group of nearby nodes inside the wireless range of R. Next, SIR sends synchronization massage 2 ($Sync2$) back to R. $Sync2$ contains the arrival time of $Sync1$, $T_{2,SIR}$, as well as the time stamp for the transmit time, $T_{3,R}$, of $Sync2$ (line 10 to 13).

Once R gathers all time stamps $T_{1,R}$, $T_{2,SIR}$, $T_{3,SIR}$, and $T_{4,R}$ (the time right after receiving $Sync3$), R estimates the offset by

$$O_{R,SIR} = \frac{(T_{2,SIR} + T_{3,SIR}) - (T_{4,R} + T_{1,R})}{2} \tag{2}$$

Here, $O_{R,SIR}$ is the offset between R and SIR. R then broadcasts a synchronization message number ($Sync3$) to all its neighbor nodes, $k \in \mathbb{U}(R)$, which contains time $T_{2,SIR}$ and $O_{r,SIR}$ (line 18 to 23). Each node, k, in $\mathbb{U}(R)$ receives $Sync3$ and checks the sequence number, $SeqN$. It only reacts if the current $SeqN$ in the message is higher than the previous one. Then, (line 26) it estimates the offset difference between SIR and k, $O_{SIR,k}$, according to SIR by

$$O_{SIR,k} = T_{2,SIR} - T_{2,k} \tag{3}$$

Then, (line 27) every node, k, determines its total offset, $O_{R,k}$, as

$$O_{R,k} = O_{R,SIR} + O_{SIR,k} \tag{4}$$

Here, $O_{R,k}$ is the offset difference between R and k. Finally, each node, k, compensates its local time by subtracting the total offset, $O_{R,k}$, from its old local time, T_k^{old}, using

$$T_k^{new} = T_k^{old} - O_{R,k} \tag{5}$$

where T_k^{old} denotes the old local time, and T_k^{new} is the new local time of node k (line 28). After all, the reference nodes within the wireless range of the sink node become synchronized and each reference node initiates the synchronization timer (ST) to repeat the same procedure until no further node can be synchronized (line 30 to 34). Each reference uses its time slot, $Schslot$, to avoid message collision.

For more clarification, we show an example of the synchronization method considering the depicted network in Figure 4. In the beginning, all nodes use Algorithm 3 for the synchronization process, and S fills the synchronization message (Type: $Sync1$) as follows.

DesAdd	SrcID	SID	Type	SeqN	RT	ST	RefT	OF
0xFFFF	S	B	$Sync1$	1	−1	−1	−1	0

Next, it broadcasts $Sync1$ to all its neighbor nodes at the time $T_{1,S}$. Since it has a list of neighbor reference nodes (B and C), it selects node B to be the responder (lines 1~7).

All nodes (B, A, D, and C) receive $Sync1$ and keep the arrival time of $Sync1$ ($T_{2,B}$, $T_{2,A}$, $T_{2,D}$, and $T_{2,C}$) (line 9). Next, each reference node (B and C) starts its own ST (lines 14~17). Node B sends back $Sync2$ to node S. $Sync2$ consists of the arrival time of $Sync1$, $T_{2,B}$, as well as the time stamp for the transmit time, $T_{3,B}$, of $Sync2$ (lines 10~13). Node B sets the synchronization message as follows.

DesAdd	SrcID	SID	Type	SeqN	RT	ST	RefT	OF
S	B	−1	$Sync2$	1	$T_{2,B}$	$T_{3,B}$	−1	0

As soon as node S receives the $Sync2$, it records the arrival time of $Sync2$, $T_{4,S}$. Then, it calculates the offset using (2) and broadcasts $Sync3$ to all its neighbor nodes (lines 19–23). In addition, S fills the synchronization message (Type: $Sync3$) as follows.

DesAdd	SrcID	SID	Type	SeqN	RT	ST	RefT	OF
0xFFFF	S	−1	$Sync3$	1	−1	−1	$T_{2,B}$	$O_{S,B}$

Nodes (B, A, D, and C) receive $Sync3$ and check $SeqN$. They find the $SeqN = 1$ is higher than the previous one (default value = 0). Then, they estimate their offset and compensate their local time using (3), (4), and (5) (lines 25 to 29).

Nodes B, C, I, and F repeat the same process after their ST expires (lines 31 to 34). The synchronization process requires only three broadcast messages per reference node, regardless of the network size or density. Additionally, the scheduling process reduces the set of reference nodes. Thus, LPSRS can significantly reduce the number of message exchanges and is, therefore, well suited for low-power applications.

Algorithm 3: LPSRS pseudo-code for the synchronization process.

1. **if** $(NodeID = sink\ node)$ **Then**
2. $R \leftarrow sink\ node$
3. $SIR \leftarrow select\ rondlmly\ fromRefNeighbors\ list$
4. $SeqN \leftarrow 1$
5. $T_{1,R} \leftarrow ReadCurrent\ Time$
6. $send\ Sync1$ $< 0xFFFF, R, SIR, Sync1, SeqN, -1, -1, -1, 0 >$
7. **end if**
8. ■ **Upon receiving** $Sync1$
9. $T_{2,k} \leftarrow Read\ Current\ Time\ //$ k represents all neighbor nodes of R
10. **if** $(NodeID == SIR)$ **Then**
11. $T_{2,SIR} \leftarrow T_{2,k}$ $T_{3,SIR} \leftarrow Read\ Current\ Time$
12. $send\ Sync2$ $< R, NodeID, -1, Sync2, SeqN, T_{2,SIR}, T_{3,SIR}, -1, 0 >$
13. **end if**
14. **if** $(NodeID$ is in $RefNode)$ **Then**
15. $WaitingTime \leftarrow (mySchSlot - SchSlot_{SrcID})$
16. Setup Synchronization timer $ST(WaitingTime)$
17. **end if**
18. ■ **Upon receiving** $Sync2$
19. **if** $(NodeID == R)$ **Then**
20. $T_{4,R} \leftarrow Read\ Current\ Time$
21. $O_{R,SIR} \leftarrow ((T_{2,SIR} + T_{3,SIR}) - (T_{4,R} - T_{1,R}))/2$
22. $send\ Sync3 < 0xFFFF, R, -1, Sync3, SeqN, -1, -1, T_{2,SIR}, O_{R,SIR} >$
23. **end if**
24. ■ **Upon receiving** $Sync3$
25. **if** $(Received\ SeqN > SeqN)$ **Then**
26. $O_{k,SIR} \leftarrow T_{2,SIR} - T_{2,k}$
27. $O_{R,k} \leftarrow O_{R,SIR} + O_{NodeID,SIR}$
28. $T_k^{new} \leftarrow T_k^{old} - O_{R,k}$
29. **end if**
30. ■ **Upon ST expires**
31. $R \leftarrow NodeID$
32. $SIR \leftarrow select\ rondlmly\ from\ RefNeighbors\ list$
33. $T_{1,R} \leftarrow ReadCurrent\ Time$
34. $send\ Sync1$ $< 0xFFFF, R, SIR, Sync1, SeqN, -1, -1, -1, 0 >$

5. Experimental and Simulation Results

In this section, we present our achieved experimental results, followed by the simulation results. We have used the number of transmitted messages in each protocol during the synchronization process as the performance metric. Before discussing and explaining the experimental results, as described below, we show how the number of transmitted messages is the key performance indicator related to power consumption.

5.1. Power Consumption Analysis

In general, to analyze power consumption, we focus on the major power consumption caused by packet transmission and reception on sensor nodes. To save energy, wireless networks often cycle between sleep and active modes. As a result, power consumption is roughly proportional to the number of broadcasts and received packets, assuming a MAC-layer protocol that reduces idle listening time. Besides, wireless networks are broadcast domains [34]. As a result, for each broadcast, there are many receptions. Assume we have a wireless network with N nodes that is equally distributed, with each node surrounded by m nodes. The total power consumption, P_T, can be expressed by transmission power, P_{TX}, and reception power, P_{RX}, as defined by

$$P_T = A(P_{TX} + mP_{RX})\left(\frac{L}{D}\right) \tag{6}$$

where A indicates the number of transmitted messages, L indicates the trace length in seconds, while D indicates the duration of the message. The P_{RX} has different modes. For example, NXP's MCU with the RF transceiver chip KW01Z128 [35] has different modes. At the highest mode, the P_{RX} is almost equal to P_{TX}. On the other hand, at the lowest mode, P_{RX} is around 0.2 of P_{TX}. In this article, for sake of simplicity, we have chosen $P_{RX} \approx P_{TX}$. Since $P_{RX,}$ is almost equal to P_{TX} [35], we can, therefore, simplify (6) into

$$P_T = A\frac{P_{TX}L(m+1)}{D} \tag{7}$$

According to (7), power usage is dependent on the number of messages sent, since the number of transmitted messages has a big impact on power consumption. In addition, we analyze the required number of the transmitted messages, M, for each protocol [35].

In R-sync, each node transmits a message to help build the spanning tree. The number of sent messages in the network is three times greater than the number of referenced links at the start of the time synchronization procedure. Additionally, there are further required messages for pulling isolated nodes. Thus, the total number of transmitted messages of R-Sync is: $M_{R\text{-}Sync} = N + 3(R-1) + P$. Here N is the node number, R is the number of reference nodes, and P is the number of pulling messages [9]. We note that $M_{R\text{-}Sync}$ is directly proportional to the number of nodes and the number of selected reference nodes.

Meanwhile, HRTS constructs a multi-level hierarchy of the network. At every level, each node sends two messages and obtains a response from the selected reference node. Therefore, the total number of transmitted messages of HRTS is $M_{HRTS} = 2N + R$ [2]. M_{HRTS} is directly proportional to the number of nodes [5].

Lastly, LPSRS requires $M_{LPSRS} = 3(R-1)$ messages. However, LPSRS minimizes the number of reference nodes by using an efficient scheduling technique for the reference nodes. Besides, LPSRS is able to avoid collisions during its operations.

5.2. Experimental Results

The LPSRS protocol was built in C++ and implemented in an embedded CPU in the hardware sensors. We used the Arduino Nano RF (ANRF) as the hardware sensor. The ANRF utilizes an NRF24L01 RF transceiver and an ATmega328P CPU [36,37]. Moreover, it has a 2 kB memory and 32 KB Flash. The ANRF operated at a frequency of 2.4 GHz, and the bitrate was 50 kbps. We used 0 dBm as the transmitted power [38]. We created a testbed of 25 nodes, as shown in Figure 5, to build a wireless sensor network. For simplicity of the testing, with the help of the programming, we designed a wireless network of 3-way grid topology to test our protocol over a multi-hop network as depicted in Figure 6. In the 3-way grid topology, each node only receives incoming messages from nodes that have a common solid line. For example, in Figure 6, N12 only receives messages from N6, N7, N11, N13, N17, and N18. Therefore, we can evaluate LPSRS and other protocols under a realistic transmission contention and conditions of congestion.

Figure 5. The 25-node setup.

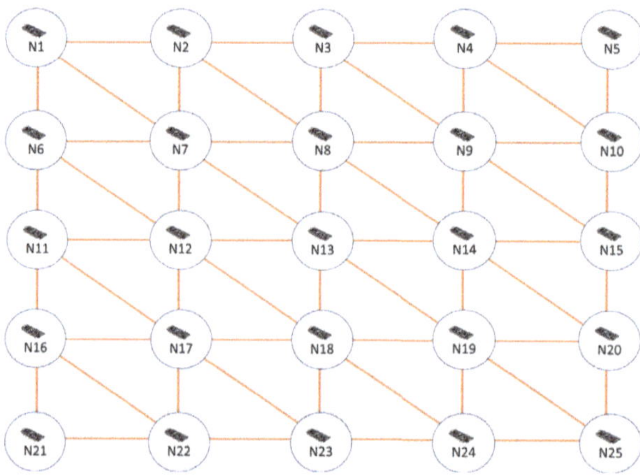

Figure 6. The 3-way grid network.

Table 1 shows the number of the transmitted message of LPSRS, R-sync, and HRTS for the network topology of Figure 6. It demonstrates that LPSRS, R-sync, and HRTS exchanged 42, 57, and 64 messages, respectively. LPSRS outperforms other technics (R-sync and HRTS). It can decrease the number of transmitted messages by around 26% and 35% compared to R-Sync and HRTS. Although the improvement in LPSRS looks moderate when compared to other protocols in the tiny test network shown in Figure 6, it develops fast as the network size increases. Due to the complexity of building large-scale actual hardware networks, we demonstrated this enhancement using simulations.

Table 1. Comparison of LPSRS, R-sync, and HRTS.

LPSRS	R-sync	HRTS
42	57	64

5.3. Simulation Results

To further test our proposed LPSRS protocol, we originally built a simulator that runs the LPSRS and HRTS algorithms. Meanwhile, the R-Sync simulation findings [9] have been adopted. In MATLAB, simulations were created employing wireless networks of varying sizes and ranges. We assumed the N sensor nodes were uniformly disrupted and scattered at random in a square area of 1000 m × 1000 m. Each node had a wireless communication range, denoted by *r*. We also presumed that all nodes were similar and self-contained. The sink node had two positions: (1) at the network's center and (2) at the network's corner. Due to the homogeneous distribution of the nodes, there was no significant change in the findings due to the position of the sink node. In this article, to prevent repetition, we only presented the findings if the sink node was placed in the network's center, as other locations yielded similar results. In addition, we tested with LPSRS in the simulation using a wide variety of 12 distinct scenario networks, ranging from sparse to congested. We could simulate several different sorts of sensor nodes with varying wireless ranges by varying the transmission range. Changing the number of nodes in the network, on the other hand, allowed us to show the behavior of our protocol from a sparse network (200 nodes) up to a dense network (450 nodes).

Through simulations, we measured the number of transmitted messages in each protocol during the synchronization process, which is the key performance indicator related to power consumption. As described in Section II, HRTS and R-Sync often incur a large number of collisions during the synchronization process, while LPSRS is guaranteed to have no collisions. For the sake of simplicity, we assumed that no collisions happen even in HRTS and R-Sync.

Figure 7 depicts the number of messages delivered by all three protocols (LPSRS, R-Sync, and HRTS) across a range of network different sizes from 200 to 450 nodes. For a large network with 450 nodes, it can be seen that R-Sync and HRTS generated 970 messages and 1071 messages, respectively. On the other hand, LPSRS generated around 500 messages. That is, LPSRS outperformed the other two protocols, as it decreased the number of transmitted messages to around 49% and 53% compared to R-Sync and HRTS, respectively.

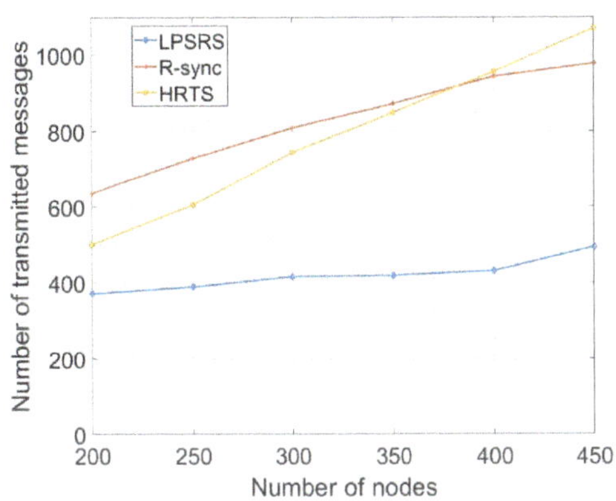

Figure 7. Required number of messages (*r* = 85 m, N = 240: 450).

Figure 8 shows the number of transmitted messages under different communication ranges from 85 m to 160 m. It can be found that LPSRS substantially outperformed the other methods. For a network with a communication range of 85m, LPSRS decreased the number of transmitted messages to around 55% and 34% compared to R-Sync and HRTS, respectively. On the other hand, for a network with a long communication range of 160 m, LPSRS generates around 175 messages. LPSRS outperformed HRTS and R-Sync by a factor of around 3 and 1.4, respectively.

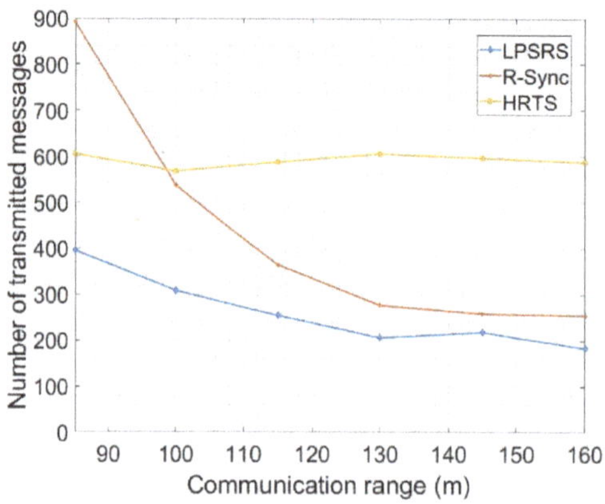

Figure 8. Required number of messages (r = 85 m: 160 m, N = 240).

Based on the results presented above, it is reasonable to conclude that LPSRS consumes less energy than the other techniques. LPSRS, on the other hand, does not compromise timing accuracy. The offset calculations described by HRTS [12] and DTSync [7] during the synchronization process were used, which is recognized to provide high timing accuracy. For more details, we compared DTSync and HRTS in our previous work with more protocols (see Table 11 [11]). Furthermore, LPSRS can manage high-density networks without losing its key features, making it a highly scalable and flexible protocol.

6. Conclusions

In this article, we suggested LPSRS as a time synchronization protocol for the IoT. It addressed the power consumption and collision issues in IoT networks. LPSRS significantly decreased the number of messages, resulting in significant power savings. We constructed LPSRS into a real wireless sensor network with different configurations and an extensive simulation with large-scale networks. According to the experimental results, LPSRS requires fewer messages than prior approaches such as R-Sync and HRTS. It can decrease the number of transmitted messages by around 26% and 35% compared to R-Sync, and this benefit, in turn, reduces LPSRS's energy use. In addition, we can consider LPSRS as a semi-centralized synchronization protocol, as some tasks are completed by the sink node and other tasks are completed locally by each node. Therefore, LPSRS is very convenient for those applications in which the network topology is recognized, e.g., smart metering.

Author Contributions: M.E., M.A.A.E.-G. and S.P. came up with the concept for this study; M.E., H.K. and S.P. developed the research technique; M.E., M.A.A.E.-G. and H.K. carried out the experiments and simulations; M.E., H.K. and S.P. evaluated the data and wrote the paper. All authors have read and agreed to the published version of the manuscript.

Funding: This research was supported in part by Korea University and in part by the MSIT (Ministry of Science and ICT), Korea, under the ITRC (Information Technology Research Center) support program (IITP-2021-0-01810) supervised by the IITP (Institute for Information and Communications Technology Planning and Evaluation).

Institutional Review Board Statement: Not applicable.

Informed Consent Statement: Not applicable.

Data Availability Statement: Not applicable.

Acknowledgments: We gratefully appreciate the anonymous reviewers' variable reviews and comments.

Conflicts of Interest: The authors declare no conflict of interest.

References

1. Farzad, A.; Najafi, K. BlueSync: Time Synchronization in Bluetooth Low Energy with Energy Efficient Calculations. *IEEE Internet Things J.* **2021**. [CrossRef]
2. Xintao, H.; Kim, K.S.; Lee, S.; Lim, E.G.; Marshall, A. Improving Multi-Hop Time Synchronization Performance in Wireless Sensor Networks Based on Packet-Relaying Gateways with Per-Hop Delay Compensation. *IEEE Trans. Commun.* **2021**, *69*, 6093–6105.
3. Phan, L.-A.; Kim, T.; Kim, T. Robust neighbor-aware time synchronization protocol for wireless sensor network in dynamic and hostile environments. *IEEE Internet Things J.* **2020**, *8*, 1934–1945. [CrossRef]
4. ElSharief, M.; El-Gawad, M.A.A.; Ko, H.; Pack, S. EERS: Energy-Efficient Reference Node Selection Algorithm for Synchronization in Industrial Wireless Sensor Networks. *Sensors* **2020**, *20*, 4095. [CrossRef] [PubMed]
5. Sarvghadi, M.A.; Wan, T.-C. Message passing based time synchronization in wireless sensor networks: A survey. *Int. J. Distrib. Sens. Netw.* **2016**, *12*, 1280904. [CrossRef]
6. Yildirim, K.S.; Kantarci, A. Time synchronization based on slow-flooding in wireless sensor networks. *IEEE Trans. Parallel Distrib. Syst.* **2013**, *25*, 244–253. [CrossRef]
7. ElSharief, M.; El-Gawad, M.A.A.; Kim, H.W. Density table-based synchronization for multi-hop wireless sensor networks. *IEEE Access* **2017**, *6*, 1940–1953. [CrossRef]
8. Choudhury, N.; Matam, R.; Mukherjee, M.; Shu, L. Beacon synchronization and duty-cycling in IEEE 802.15. 4 cluster-tree networks: A review. *IEEE Internet Things J.* **2018**, *5*, 1765–1788. [CrossRef]
9. Qiu, T.; Zhang, Y.; Qiao, D.; Zhang, X.; Wymore, M.L.; Sangaiah, A.K. A robust time synchronization scheme for industrial internet of things. *IEEE Trans. Ind. Inform.* **2018**, *14*, 3570–3580. [CrossRef]
10. Yildirim, K.S. Gradient descent algorithm inspired adaptive time synchronization in wireless sensor networks. *IEEE Sens. J.* **2016**, *16*, 5463–5470. [CrossRef]
11. EMahmoud, l.; El-Gawad, M.A.A.; Kim, H. Fads: Fast scheduling and accurate drift compensation for time synchronization of wireless sensor networks. *IEEE Access* **2018**, *6*, 65507–65520.
12. Hui, D.; Han, R. TSync: A lightweight bidirectional time synchronization service for wireless sensor networks. *ACM SIGMOBILE Mob. Comput. Commun. Rev.* **2004**, *8*, 125–139.
13. Miklós, M.; Kusy, B.; Simon, G.; Lédeczi, Á. The flooding time synchronization protocol. In Proceedings of the 2nd International Conference on Embedded Networked Sensor Systems, Baltimore, MD, USA, 3–5 November 2004; pp. 39–49.
14. Lenzen, C.; Sommer, P.; Wattenhofer, R. PulseSync: An efficient and scalable clock synchronization protocol. *IEEE/ACM Trans. Netw.* **2015**, *23*, 717–727. [CrossRef]
15. Lenzen, C.; Sommer, P.; Wattenhofer, R. Optimal clock synchronization in networks. In Proceedings of the 7th ACM Conference on Embedded Networked Sensor Systems, Berkeley, CA, USA, 4–6 November 2009; pp. 225–238.
16. Yildirim, K.S.; Gurcan, O.; Yildirim, K.S. Efficient time synchronization in a wireless sensor network by adaptive value tracking. *IEEE Trans. Wirel. Commun.* **2014**, *13*, 3650–3664. [CrossRef]
17. Elson, J.; Girod, L.; Estrin, D. Fine-grained network time synchronization using reference broadcasts. *ACM SIGOPS Oper. Syst. Rev.* **2002**, *36*, 147–163. [CrossRef]
18. Gong, F.; Sichitiu, M.L. CESP: A low-power high-accuracy time synchronization protocol. *IEEE Trans. Veh. Technol.* **2016**, *65*, 2387–2396. [CrossRef]
19. Saurabh, G.; Kumar, R.; Srivastava, M.B. Timing-sync protocol for sensor networks. In Proceedings of the 1st International Conference on Embedded Networked Sensor Systems, New York, NY, USA, 5–7 November 2003; pp. 138–149.
20. Leandro, T.B.; Heimfarth, T.; de Freitas, E.P. Enhancing time synchronization support in wireless sensor networks. *Sensors* **2017**, *17*, 2956.
21. ElSharief, M.A.; El-Gawad, M.A.A.; Kim, H.; El-Gawad, M.A. Low-Power Scheduling for Time Synchronization Protocols in a Wireless Sensor Networks. *IEEE Sens. Lett.* **2019**, *3*, 1–4. [CrossRef]
22. Atis, E.; Fafoutis, X.; Duquennoy, S.; Oikonomou, G.; Piechocki, R.; Craddock, I. Temperature-resilient time synchronization for the internet of things. *IEEE Trans. Ind. Inform.* **2017**, *14*, 2241–2250.
23. Luca, S.; Gamba, G. A distributed consensus protocol for clock synchronization in wireless sensor network. In Proceedings of the 2007 46th IEEE Conference on Decision and Control, New Orleans, LA, USA, 12–14 December 2007.
24. Panigrahi, N.; Khilar, P. Multi-hop consensus time synchronization algorithm for sparse wireless sensor network: A distributed constraint-based dynamic programming approach. *Ad Hoc Netw.* **2017**, *61*, 124–138. [CrossRef]
25. Jianshe, W.; Jiao, L.; Ding, R. Average time synchronization in wireless sensor networks by pairwise messages. *Comput. Commun.* **2012**, *35*, 221–233.
26. Maggs, M.K.; O'Keefe, S.G.; Thiel, D. Consensus clock synchronization for wireless sensor networks. *IEEE Sens. J.* **2012**, *12*, 2269–2277. [CrossRef]
27. Phan, L.-A.; Kim, T. Fast consensus-based time synchronization protocol using virtual topology for wireless sensor networks. *IEEE Internet Things J.* **2021**, *8*, 7485–7496. [CrossRef]
28. Shi, F.; Tuo, X.; Ran, L.; Ren, Z.; Yang, S.X. Fast convergence time synchronization in wireless sensor networks based on average consensus. *IEEE Trans. Ind. Inform.* **2020**, *16*, 1120–1129. [CrossRef]

29. Lim, H.; Kumar, S.; Kim, H. Density-driven scheduling of low power synchronization for wireless sensor networks. In Proceedings of the 2016 International Conference on Electronics, Information, and Communications (ICEIC), Danang, Vietnam, 27–30 January 2016.
30. Ranjan, R.; Varma, S. Collision-free time synchronization for multi-hop wireless sensor networks. *J. Comput. Intell. Electron. Syst.* **2012**, *1*, 200–206. [CrossRef]
31. Ranjan, R.; Varma, S. Challenges and implementation on cross layer design for wireless sensor networks. *Wirel. Pers. Commun.* **2016**, *86*, 1037–1060. [CrossRef]
32. Kim, D.; Wu, Y.; Li, Y.; Zou, F.; Du, D.-Z. Constructing Minimum Connected Dominating Sets with Bounded Diameters in Wireless Networks. *IEEE Trans. Parallel Distrib. Syst.* **2008**, *20*, 147–157.
33. Cormin, T.; Leiserson, C.E.; Rivest, R.L.; Stein, C. *Introduction to Algorithms*; MIT Press: Cambridge, MA, USA, 2009.
34. *IEEE Std 802.15.4-2015*; IEEE Standard for Low-Rate Wireless Networks. IEEE Standards Association: Piscataway, NJ, USA, 2016; Volume 802.
35. NXP. MKW01Z128 Highly-Integrated, Cost-Effective Single-Package Solution for Sub-1 GHz applications. Document Number: MKW01Z128, Rev. 5,03/2014. Available online: http://www.nxp.com (accessed on 5 December 2021).
36. Arduino Nano User Manual. Available online: https://www.arduino.cc/en/uploads/Main/ArduinoNanoManual23.pdf (accessed on 5 January 2022).
37. nRF24L01+ Preliminary Product Specification. Available online: https://www.sparkfun.com/datasheets/Components/SMD/nRF24L01Pluss_Preliminary_Product_Specification_v1_0.pdf (accessed on 5 January 2022).
38. ATmega328P [DATASHEET]. Available online: http://ww1.microchip.com/downloads/en/DeviceDoc/Atmel-7810-Automotive-Microcontrollers-ATmega328P_Datasheet.pdf (accessed on 5 January 2022).

electronics

MDPI

Article

A Monitoring System Based on NB-IoT and BDS/GPS Dual-Mode Positioning

Zhibo Xie *, Ruihua Zhang, Juanni Fang and Liyuan Zheng

Department of College of Information and Intelligence Engineering, Zhejiang Wanli University, Ningbo 315104, China

* Correspondence: xiezhibo@zwu.edu.cn

Abstract: Monitoring system is widely used to detect the environment parameters such as temperature, humidity and position information in cold chain logistic, modern agriculture, hospital and so on. Poor position precision, high communication cost, high packet loss rate are the main problems in current monitoring system. To solve these problems, the paper presents a new monitoring system based on Narrow Band Internet of Things (NB-IoT) and BeiDou system/Global System Position (BDS/GPS) dual-mode positioning. Considering the position precision, a dual-mode positioning circuit based on at6558 is designed, and the calculation formula of the positioning information of the monitored target has been derived. Subsequently, a communication network based on wh-nb75-ba NB-IoT module is designed after compared with the LoRa technology. According to the characteristics of high time correlation of sensor data, an adaptive optimal zero suppression (AOZS) compression algorithm is proposed to improve the efficiency of data transmission. Experiments prove the feasibility and effectiveness of the system from the aspects of measurement accuracy, positioning accuracy and communication performance. The temperature and humidity error are less than 1 °C and 5% RH respectively with the selected sensor chips. The position error is decided by several factors, including the number of satellites used for positioning, the monitored target moving speed and NB-IoT module lifetime period. When the monitored target is stationary, the positioning error is about 2 m, which is less than that of the single GPS or BDS mode. When the monitored target moves, the position error will increase. But the error is still less than that of the single GPS or BDS mode. Then the AOZS compression algorithm is used in actually experiment. The compression ratio (CR) of it is about 10% when the data amount increasing. In addition, the packet loss rate test experiment proves the high reliability of the proposed system.

Keywords: NB-IoT; BDS/GPS dual-mode positioning; data compression algorithm

Citation: Xie, Z.; Zhang, R.; Fang, J.; Zheng, L. A Monitoring System Based on NB-IoT and BDS/GPS Dual-Mode Positioning. *Electronics* **2022**, *11*, 2493. https://doi.org/10.3390/electronics11162493

Academic Editors: Antonio Cano-Ortega, Francisco Sánchez-Sutil and Aurora Gil-de-Castro

Received: 7 July 2022
Accepted: 7 August 2022
Published: 10 August 2022

1. Introduction

Monitoring system can detect environmental parameters such as temperature, humidity and location information, and send these information to the monitoring center, which greatly reduce the workload of staff and enhanced management efficiency. Hence, it has gotten more and more recognition and application in cold chain logistic monitoring, modern agriculture arrangement, hospital monitoring and so on. Due to hardware technology, communication network and other reasons, the system still has the following deficiencies:

(i) Poor positioning precision, especially for the moving object. Currently, the positioning error is about 50 m for the stationary object, while that is more than 100 m for moving object. (ii) High cost, especially for the communication cost. The hardware cost is a one-time investment, but the communication cost needs to be paid for every day. Actually, the transmission data amount of a monitoring system is small, but it needs to communicate with the monitoring center through 3 g/4 g network, so it has to pay for the expensive network resources, and the cost of communication is very high. (iii) Unreliable network performance, poor stability and high packet loss rate. The packet loss rate of existing monitoring systems is about 10% when the distance is about 1 km. When the communication network becomes worse, the packet loss rate will be higher. Therefore, it is essential to develop a monitoring system with high reliability and low cost.

Generally, the monitoring system mainly includes the three parts: communication network design, position function design and realization, and data transmission. The data transmission network of the monitoring system belongs to long-distance network. 4 g and low-power wide area network (LPWAN) are main long-distance communication. 4 g has high power consumption and high traffic cost, which is not suitable for non-real time communication. LoRa and NB-IoT are representative technologies of LPWAN [1]. LoRa is a physical layer technology which uses a proprietary spread spectrum technique to modulate signals in sub-GHz ISM bands. The bidirectional communication of LoRa is provided by the chirp spread spectrum (CSS) modulation that spreads a narrow-band signal over a wider channel bandwidth. The resulting signal has low noise levels, enabling high interference resilience and is difficult to detect or jam [1]. It provides long-range communication up to 10–40 km in rural areas and 1–5 km in urban areas and has very high energy efficiency [2–4]. Compared to LoRa, NB-IoT can use the current 3 g/4 g network to save the network cost and shortens the developing period with the license frequency band [5–7]. It uses a minimum system bandwidth of 180 kHz for downlink and uplink communication and can be deployed in three operating modes: (a) Stand-alone, (b) Guard band, and (c) In-band [8–10]. The physical channels and signals of NB-IoT are time-division multiplexed. The data rate for uplink is about 160 to 200 kHz and 160 to 250 kHz for downlink. The coverage is 18 km in cities and 25 km in suburbs. Because of the above advantages, NB-IoT has been widely used in many fields [11–13]. It is regarded to be a very important technology and a large step for 5 g IoT evolution [14,15]. Many famous companies have shown great interest in NB-IoT as part of 5 g systems, and spent lots of effort in the standardization of NB-IoT [16,17]. Shi proposed a smart parking system using NB-IoT communication technology, which can effectively improve the utilization rate of the existing parking facilities [18]. Anand presented a remote monitoring mechanism for the water level in a storage tank using NB-IoT [19]. Haibin studied NB-IoT in smart hospitals [20]. An infusion monitoring system was developed to monitor the real-time drop rate and the volume of remaining drug during the intravenous infusion. Srikanth put forward the utilization of onshore narrowband IoT infrastructure for tracking of containers transported by marine cargo vessels while operating near the coastline [21]. Xihai applied NB-IoT in an Information monitoring system to reduce the power consumption [22]. Cao applied the NB-IoT in intelligent traffic lights system for urban areas in Vietnan to reduce traffic congestion [23]. The above studies show that NB-IoT is oriented to applications that require high QoS and low latency and has strong links, high coverage, low power, and low cost [24]. Because of these, the paper proposed the monitoring system scheme based on NB-IoT.

BeiDou satellite System (BDS) is a global positioning system independently developed by China. Its space station consists of 5 geostationary orbit satellites and 30 non-geostationary orbit satellites, while the space station of Global Position System (GPS) consists of 24 satellites (21 working satellites and 3 standby satellites). The user terminal of BDS has double-direction message communication, and the user can transmit short-message information of 40–60 Chinese characters per time [25]. GPS does not have the function of short-message communication. Unlike GPS which uses dual-frequency signals, BeiDou-3 uses triple-frequency signals, which can better eliminate the effect of the ionosphere and improve the positioning reliability and accuracy [26–28]. With the initial service provided by the BDS foundation strengthening system, it can provide meter-level, sub-meter-level, decimeter-level and even centimeter -level service. In addition, BeiDou-3 satellite network has laid an "inter satellite link" in space. Thus, all satellites in the constellation can be connected without global stations, and the satellites can continue to provide services even if they are disconnected from the ground. Because of these advantages, BDS begins to be widely used to measure height [29], vehicle position [30], anomaly detection [31], and train position [32], etc. Some scholars propose to combine BDS with other positioning technologies to produce higher cost performance [33–35]. However, for both BDS or GPS, the number of observation satellites in a single satellite navigation system is limited, they will become extremely vulnerable in the case of severe environmental interference, and they cannot guarantee the positioning accuracy and availability of the receiver. Since GPS and BDS have common features in system design and positioning principle, the receiver can simultaneously receive the satellite signals of the two satellite navigation systems for dual-system integrated positioning to avoid the situation that a single satellite system cannot locate due to the lack of satellites. Therefore, in theory, BDS/GPS dual mode positioning can optimize the satellite position and improve the accuracy and availability of positioning results. This paper will use the BDS/GPS dual mode positioning system to improve the positioning accuracy.

In wireless sensor networks, there is a large amount of redundant information in the original data collected by sensor nodes, including the temporal redundancy collected by the same node at adjacent times and the spatial redundancy collected by adjacent nodes in geographical areas [36]. If the data carrying a large amount of redundant information is transmitted, the communication bandwidth will be wasted and increased network delay and node energy consumption, which will affect the stability and life of the whole sensor network system. Compressing redundant information before transmitting original data is a mechanism that can effectively reduce node energy consumption. In recent years, researchers have proposed many data compression algorithms for wireless sensor networks. The main algorithms for wireless sensor data are divided into compression based on time correlation and compression based on space correlation. Data compression algorithm based on time correlation is a kind of typical compression algorithm. It often focuses on mining data time correlation and removing data time redundancy with the help of some classical coding technologies, such as Huffman [37], LZW [38], S-LZW [39], LEC [40], RLE [41]. The data compression algorithm based on spatial correlation is also a typical compression algorithm, which is often combined with clustering mechanism [42–44], and strives to fully mine the spatial correlation of data and reduce and balance the energy consumption of each node of the network. Data compression algorithms based on temporal and spatial correlation have attracted more and more attention. For example, the algorithm proposed by Ciancio and Donoho [45,46] not only involves removing the temporal redundancy of data, but also discusses how to establish an optimal path, so that the spatial redundancy can be removed to the greatest extent when the data is transmitted along this path. Difference mechanism is often used in data compression [47–49]. The common point of the data compression algorithm based on the difference mechanism is that by selecting a reference data, a single sensor node only needs to transmit the difference between the original sensing data and the reference data, so as to remove the temporal redundancy, or the adjacent sensor nodes in the geographical region only need to transmit the difference between their original sensing data and the reference data, so as to remove the spatial redundancy. The difference between these algorithms is the choice of difference coding. Differential Code Compression Method (DCCM) is the typical algorithm. The disadvantages of DCCM algorithm are: (i) simply taking the average value of data as the reference value, which is lack of rationality; (ii) The correlation between data is not mined. So the paper will propose a new algorithm to compress the sensor data.

In view of the problems in the above literature, the paper proposes a monitoring system scheme with high positioning accuracy, low cost and high network reliability. Three main contributions of this paper can be summarized as follows:

(i) A new positioning system based on BeiDou System/Global Position System is proposed to improve the position accuracy. The hardware and software are introduced in detail in the paper. In the dual positioning system, more satellites can be obtained to calculate position information. The calculation formulas have also been derived.
(ii) A new data compression algorithm is proposed. The new algorithm removes data redundancy according to the time correlation between data, and the compression rate is about 90%, while the complexity of it is similar to that of the commonly used algorithms.
(iii) A transmission network system based on NB-IoT for the compressed data is proposed. Compared with LoRa technology, the system is more stable, more reliable and lower

packet loss rate through the experiment. The development period of the system is shorter and the cost is lower.

The remaining paper is organized as follows. Section 2 describes the monitoring system in detail, including hardware design and software design. Section 3 gives the results and test data. Section 4 presents the discussion and analysis. Finally, Section 5 presents conclusions.

2. Materials and Methods

According to the framework of IoT, the network frame of the proposed scheme is divided into four parts: data acquisition layer which includes sensor nodes and sink nodes, communication layer which is NB-IoT station, application layer which is IoT cloud platform and user layer which is the monitoring center. The sensor node is the detection terminal which is used to detect the information around the sensor and send these information to the sink node. The sink node is used to receive the information from the sensor nodes and compress the information and send it to the IoT cloud platform. The whole hardware frame of the monitoring system is shown in Figure 1. The monitoring system includes one monitoring center which is in PC, and two mobile carriages. Each carriage includes one sink node and three sensor nodes. The sensor node is mainly composed of a stm8 MCU, a temperature and humidity sensor, a RFID module and lithium battery. Stm8 controls the temperature and humidity sensor to collect the temperature and humidity information nearby, and then send the information to the sink node through the RFID. The sink node is mainly composed of a stm32 MCU, a TFT display panel, a RFID module, a BDS/GPS module, a NB-IoT module and battery. Stm32 receives the temperature and humidity information from the sensor node through RFID and gets the position information from BDS/GPS module, then compresses and sends these data to the OneNET cloud platform through the NB-IoT module. The monitoring center is in PC. The temperature and humidity of each sensor node, the sink node position can be acquired in monitoring center by accessing the cloud platform. My SQL database is used to manage current and historical data. Real time map of each mobile carriage can also be displayed.

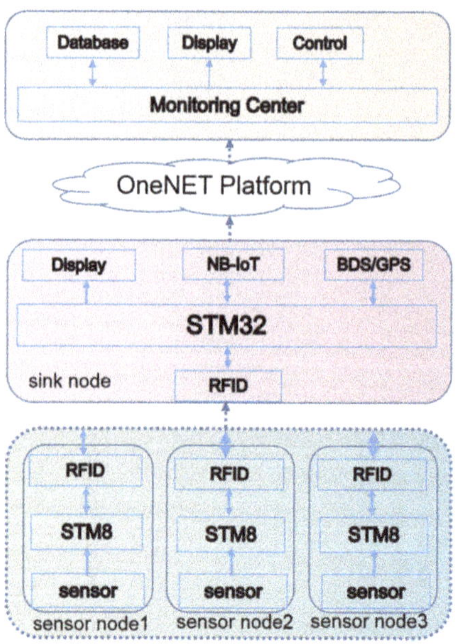

Figure 1. Hardware frame of the monitoring system.

2.1. Hardware Design

The hardware of monitoring system includes two parts, one is the sink node hardware, the other is the sensor node hardware. The hardware of sink node includes stm32f103c8t6, cc1101 RFID module, wh-nb75-ba NB-IoT module, at6558 BDS/GPS positioning module, TFTLCD display screen, key module, 3.3 V voltage regulator module and several LED lights. The system is powered by 5 V battery. The physical hardware is shown in Figure 2. The wh-nb75-ba can access mobile developer platform OneNET for free, communicate with MCU by UART and configure with AT instruction set [50]. The positioning data collected by BDS/GPS module and temperature and humidity data sent by sensor nodes are sent to NB-IoT module through UART port during UART interrupt, and finally uploaded to cloud platform. At6558 chip is used in the positioning module with BDS/GPS dual positioning mode to obtain higher positioning accuracy. It is a real six in one multi-mode satellite navigation and positioning chip, which contains 32 tracking channels and can receive global navigation satellite system (GNSS) signals of six satellite navigation systems at the same time, and realize joint positioning, navigation and timing. This chip has high sensitivity, low power consumption and low cost, which is suitable for vehicle navigation, hand-held positioning and wearable devices [51]. The chip communicates with MCU through UART serial port. The baud rate of UART serial port is set to 9600, and the data format is strictly in accordance with international NMEA0183 standard. It is a low power chip. The working current is less than 23 mA, the sleep current is less than 10uA. RFID module cc1101 is a kind of RF application which is lower than 1 GHz for ultra-low power consumption. It has high data transmission speed and long transmission distance. It is connected with MCU through 4-wire SPI interface and provides two universal digital output pins with configurable functions [52]. The 2.3-inch TFTLCD is a color LCD, which can exchange data with MCU through SPI interface. CH340 is used to down load the program in the PC to the stm32. In addition, some LED lights are used to indicate whether modules are connected successfully or not. Some keys are used to reset system or initialize modules.

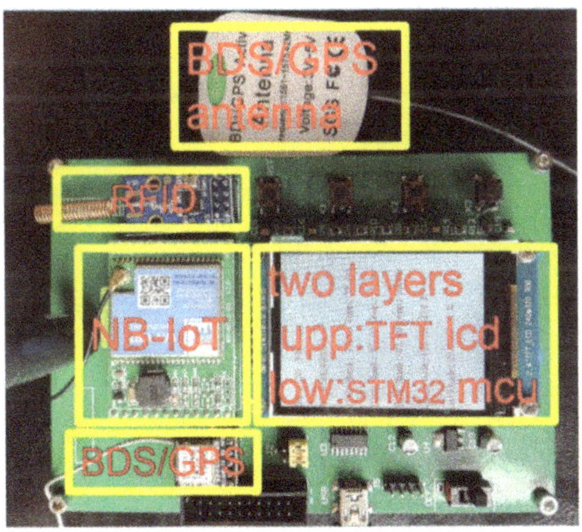

Figure 2. The Sink node physical diagram.

The hardware of each sensor node includes stm8s103f3p MCU, cc1101 module, dht11 temperature and humidity sensor chip, using dry battery for power supply. The hardware block diagram of sensor node is shown in Figure 3. The stm8s103f3p MCU of the sensor node which has various communication protocols such as I2C, SPI and UART are designed in the 20 pins. It has 8 KB flash program memory and 1 KB RAM space that is fully competent for the current temperature and humidity acquisition and subsequent data acquisition. There are 46 working state configuration registers and 32 command registers. The temperature and humidity sensor DHT11 has a single bus bidirectional serial communication interface, which can be directly connected with the serial port of MCU. It can measure temperature and humidity at the same time. The measurement accuracy of humidity is ±5% RH, that of temperature is ±1 °C. In view of the low precision requirement, and the focus of our research is low power consumption, so we use this chip. It should be emphasized that in order to reduce the power consumption, the MCU does not read the temperature and humidity value of DHT11 all the time, but only reads when receiving the reading request from the sink node. SWD is used to load down the program from the PC to stm8s103f3p.

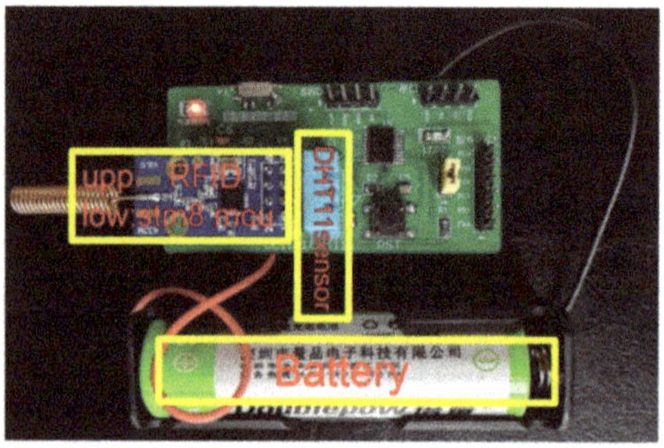

Figure 3. Sensor node physical diagram.

2.2. Software Design

According to the hardware frame in Figure 1, the software of monitoring system includes three parts: (i) the sink node software. Receiving the data sent by each sensor node, starting the positioning module to obtain the positioning information, configuring the cloud platform, and compressing and sending the received data and so on must be completed by the software of the sink node. Before the sink node sends data, the cloud platform must be configured according to the actual situation. Then the data sent by the sink node can be stored in the cloud platform. The cloud platforms of different NB-IoT companies are different. People need to refer to the user manual for configuration. (ii) the sensor node software. The main task of sensor node software is to start each sensor on the node to make them work normally, then collect their data and send them to the sink node. (iii) the monitoring center software. The software of the monitoring center needs to download temperature, humidity and positioning information from the cloud platform and store them in the database. In addition, it also needs to display the location information on the map and draw the temperature and humidity curve.

2.2.1. Software Design of Sink Nodes

The software design of sink node mainly focuses on four subprograms: NB-IoT subprogram, RFID subprogram, BDS/GPS subprogram and data compression subprogram. The main program flow chart of sink node is shown in Figure 4. After the initialization of each module, the main program will be looped in each module subprogram to deal with each module in real time. The watchdog is added to the main program to reset the program to prevent the program from getting stuck or running away.

OneNET platform is a NB-IoT cloud platform developed by China Mobile Communication Company. It can communicate with multiple sink nodes and can read data from multiple sink nodes at the same time by multithreading. We should login in the platform "https://open.iot.10086.cn (accessed on 20 January 2022)" to register the device name. Then add the objects for each device, as well as the number of points and properties of each object. In our experiment, there are three objects in each device, namely temperature, humidity and position information. Finally, the data type of each object should be described. For example, the data type of temperature is 31 bits floating point. According to the Internet Protocol for Smart Objects (IPSO) Alliance Technical Guideline, the longitude, the object ID of latitude, humidity and temperature is 3300, 3303 and 3304 separately. The instruction format is described detailly in the manual [53]. After the NB-IoT module is connected to the OneNET platform, the platform will record the life cycle of the sink device (the life cycle is configured 3600 s in the initialization). When the life cycle expires, the OneNET platform will issue a life cycle update request, and the sink node can update the life cycle. Or the sink node can actively update the life cycle before the life cycle expires. In this paper, the life cycle is automatically updated, and the life cycle update flag is activated by setting a certain time through the timer. During the life time, the NB-IoT and cloud platform can communicate. The NB-IoT subprogram is shown in Figure 5. At first, the timer will judge whether it exceeds 3600. If yes, the NB-IoT module will be initialized. If no, it continues to judge whether the receive flag bit is 1. If it is 1, it means that the reception is completed, and the data needs to be sent and the receive flag should be cleared and over. Otherwise, the life cycle flag will be judged, if it is 1, it means the time is over and the new request of update the life cycle should be sent. The two flags are active in timer interrupt and UART interrupt respectively.

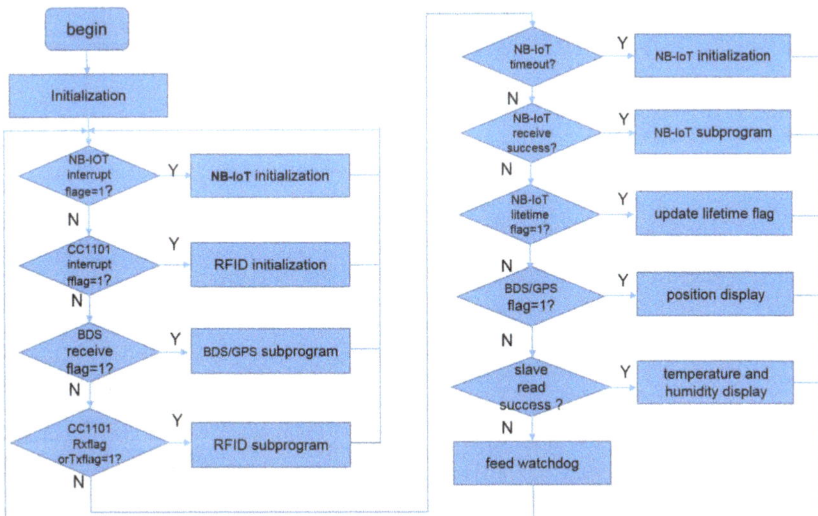

Figure 4. Main program flow chart of sink node.

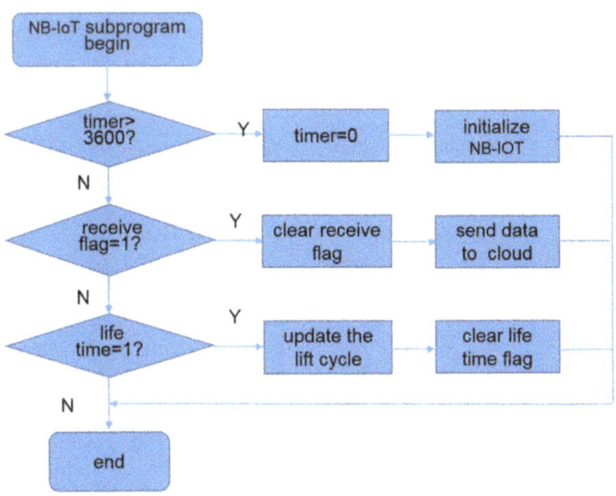

Figure 5. The Flowchart of NB-IoT subprogram.

One sink node should receive data from several sensor nodes. How to receive data from multiple sensor nodes efficiently and successfully? Here, a polling mechanism is proposed. As Figure 6 shown, the sink node sends 'read request' to sensor node1, sensor node2, and sensor node3 in turn. If sensor node1 send 'answer request' in time, then the data of sensor node1 will be allowed to send, the sink node will receive the data from sensor node1. Then next sensor node. The RFID module cc1101 in sink node accesses a sensor node every second in turn. The response timeout mechanism of the sub node is set up to avoid the data transmission errors due to the fact that the sub node is not in the transmission range and does not respond when the sink node sends a request.

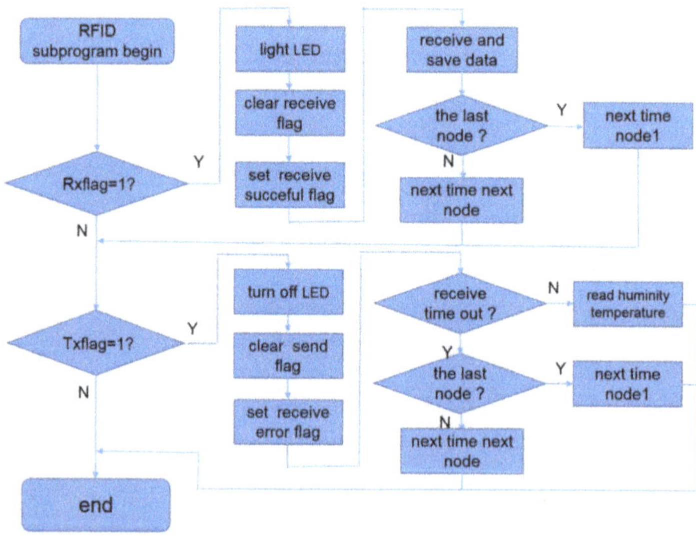

Figure 6. The flowchart of RFID subprogram of sink node.

RFID initialization mainly refers to the function configuration of cc1101 chip in RFID module. Specifically, through SPI communication mode, we can read and write the internal register of cc1101, so as to complete the setting of fundamental frequency, modulation and demodulation mode, baud rate, packet length and other related parameters. For one of the two general digital output pins with settable functions contained in cc1101, the level jump from low to high occurs when receiving data. The data can be received and read by configuring the rising edge of the I/O port of MCU to trigger interrupt. The flowchart of RFID subprogram is shown Figure 6.

BeiDou System is a positioning and navigation system independently developed by China. The positioning accuracy of BDS can reach 2.5 m in the Asia Pacific region and 10 m in the world. The test accuracy is 0.2 m/s; Timing accuracy is 10 ns. It also has its unique short message communication function. GPS is Global Positioning System. It is well known in the world and developed by the United States. We use BDS/GPS dual system for positioning to obtain higher positioning accuracy. The positioning principle is as follows:

$$\begin{aligned} \rho_B^{(m)} &= r_B^{(m)} + E_B^{(m)} + i * \delta_{t_B} + j * \delta_{t_G} - \tau_B^m + I_B^m + T_B^m + \varepsilon_B^m \\ \rho_G^{(n)} &= r_G^{(n)} + E_G^{(n)} + j * \delta_{t_G} + i * \delta_{t_B} - \tau_G^n + I_G^n + T_G^n + \varepsilon_G^n \end{aligned} \quad (1)$$

where, ρ, r, E, δ_t,τ, I, T respectively represent the pseudo range measurement value of the receiver to a star, the real distance, ephemeris error, receiver clock error, satellite clock error, ionospheric delay, tropospheric delay and pseudo range measurement noise. Superscript m, n denotes different satellites. Subscripts B and G indicate different satellite systems. B represents Beidou and G represents GPS. $i = 1$, $j = 0$. If (x, y, z) is used to represent the position coordinates of the unknown receiver, and $(x(n), y(n), z(n))$ is used to represent the position of satellite n, then $r^{(n)}$ is equal to the following expression.

$$r^{(n)} = \sqrt{\left(x - x^{(n)}\right)^2 + \left(y - y^{(n)}\right)^2 + \left(z - z^{(n)}\right)^2} \tag{2}$$

In Expressions (1) and (2), ρ, τ, I, T and the position of satellite can be calculated by original observation, navigation message and corresponding model. If the pseudo range measurement noise is ignored, five unknown parameters need to be solved, namely, the position of the receiver, the BDS clock difference of the receiver and the GPS clock difference $((x, y, z)> \delta_{t_B}, \delta_{t_G})$. Define error correction pseudo range measurements $\rho_c^{(n)}$ is as Expression (3) shown.

$$\rho_c^{(m)} = \rho^{(n)} + \tau^{(n)} - I^{(n)} - T^{(n)} \tag{3}$$

Then, the pseudo range observation equation of BDS/GPS dual system can be expressed as Expression (4):

$$\begin{aligned} \rho_{c,B}^{(m)} &= \sqrt{\left(x - x_B^{(m)}\right)^2 + \left(y - y_B^{(m)}\right)^2 + \left(z - z_B^{(m)}\right)^2} + i * \delta_{t_B} + j * \delta_{t_G} \\ \rho_{c,G}^{(n)} &= \sqrt{\left(x - x_G^{(n)}\right)^2 + \left(y - y_G^{(n)}\right)^2 + \left(z - z_G^{(n)}\right)^2} + j * \delta_{t_B} + i * \delta_{t_G} \end{aligned} \tag{4}$$

Linearize Expression (4) through the first-order Taylor expansion to obtain the linearized matrix equation as Expression (5) shown.

$$G \begin{bmatrix} \Delta x \\ \Delta y \\ \Delta z \\ \Delta \delta_{t_B} \\ \Delta \delta_{t_G} \end{bmatrix} \tag{5}$$

where,

$$G = \begin{bmatrix} -I_{B,k-1}^{(1)} & -p_{B,k-1}^{(1)} & -q_{B,k-1}^{(1)} & 1 & 0 \\ -I_{B,k-1}^{(2)} & -p_{B,k-1}^{(2)} & -q_{B,k-1}^{(2)} & 1 & 0 \\ \cdots & \cdots & \cdots & \cdots & \cdots \\ -I_{B,k-1}^{(m)} & -p_{B,k-1}^{(m)} & -q_{B,k-1}^{(m)} & 1 & 0 \\ -I_{G,k-1}^{(m+1)} & -p_{G,k-1}^{(m+1)} & -q_{G,k-1}^{(m+1)} & 0 & 1 \\ \cdots & \cdots & \cdots & \cdots & \cdots \\ -I_{G,k-1}^{(m+n)} & -p_{G,k-1}^{(m+n)} & -q_{G,k-1}^{(m+n)} & 0 & 1 \end{bmatrix} \tag{6}$$

$$b = \begin{bmatrix} \rho_{C,B}^{(1)} - r_{B,k-1}^{(1)} - i * \delta_{t_{B,k-1}} - j * \delta_{t_{G,k-1}} \\ \rho_{c,B}^{(2)} - r_{B,k-1}^{(2)} - i * \delta_{t_{B,k-1}} - j * \delta_{t_{G,k-1}} \\ \cdots \\ \rho_{c,B}^{(m)} - r_{B,k-1}^{(m)} - i * \delta_{t_{B,k-1}} - j * \delta_{t_{G,k-1}} \\ \rho_{c,G}^{(m+1)} - r_{G,k-1}^{(m+1)} - j * \delta_{t_{B,k-1}} - i * \delta_{t_{G,k-1}} \\ \cdots \\ \rho_{c,G}^{(m+n)} - r_{G,k-1}^{(m+n)} - j * \delta_{t_{B,k-1}} - i * \delta_{t_{G,k-1}} \end{bmatrix}_0 \tag{7}$$

$\left(-I_{B,k-1}^{(m)}, -p_{B,k-1}^{(m)}, -q_{B,k-1}^{(m)}\right)$ is the directional cosine of the observation vector from the receiver to satellite m. Using the principle of weighted least squares to solve Expression (5), continue to use Newton iterative algorithm to solve the location result.

In order to reduce the utilization rate of CPU, the positioning data of BDS/GPS is transmitted to memory of MCU by UART port through DMA mode every second. The

UART idle interrupt is triggered after all the positioning data are sent successfully, and the MCU responds to the interrupt to read the data in DMA for subsequent processing. The process of interrupt response program is similar to NB-IoT. It is not described in detail. The subprogram flowchart of BDS/GPS is given in Figure 7. After the BDS/GPS module parameters are set successfully, we can use the serial port assistant to read the module positioning information.

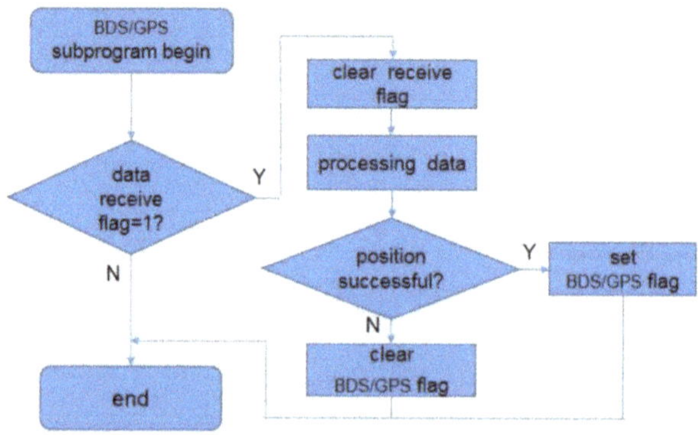

Figure 7. The flowchart of BDS/GPS subprogram.

Data compression subprogram is used to compress data redundancy, reduce power consumption of sink node and improve data transmission efficiency. According to the characteristics of small amount of data transmission and good local time correlation of the system studied in this paper, the wireless remote data compression system designed includes sequence correlation packet processing and Adaptive Optimal Zero Suppression (AOZS) compression. Correlation grouping processing refers to the grouping rearrangement of the original sequence according to the transmission data format to obtain n groups of incremental subsequences with good time correlation. Then AOZS compression is performed on each subsequence to eliminate the time redundancy in the sequence.

Adaptive optimal zero suppression (AOZS) compression algorithm is improved from differential code compression algorithm, which is suitable for the compression of incrementally sorted data sequences. AOZS algorithm reduces the number of coded data and removes the time redundancy in the original sequence by eliminating zeros and encoding the zero eliminational factor, and selects the best coding bits by calculating the compressed data length to shorten the length of the final coding. Assume that the data sequence collected by the sink node in an upload cycle can be expressed as D.

$$D = \begin{bmatrix} d_1 \\ d_2 \\ \vdots \\ d_m \end{bmatrix} = \begin{bmatrix} v_{11} & \cdots & v_{1n} \\ \vdots & \ddots & \vdots \\ v_{m1} & \cdots & v_{mn} \end{bmatrix} \tag{8}$$

where, m is the acquisition times of sensor terminal in an upload cycle; d_i is the data sequence value collected for the i-th time; n is the number of measurement parameters; v_{jk} is the value of the k-th measurement parameter in the j-th data acquisition. There are N data that appear only once, and the minimum data is recorded as α, the maximum difference between adjacent data is recorded as β. When in data compression, the relevant bit factor $\{r_1, r_2, \ldots, r_m\}$ of the original data sequence should be recorded at first:

$$r = \begin{cases} 1, \ d_i = d_{i-1}, i = 2, 3, \ldots, m \\ 0, d_i > d_{i-1}, i = 2, 3, \ldots, m \end{cases} \tag{9}$$

The relevant bit factor records the repetition of sequence adjacent data. By default, the relevant bit factor r_1 of $d_1 = 0$. The coding bit factor C_x is determined by the minimum binary coding bit factor C_α and the maximum difference binary coding bit factor C_β of the sequence as Expression (10).

$$C_x = \begin{cases} C_\beta, & C_\alpha \le C_\beta \\ C_\gamma, & C_\alpha > C_\beta \end{cases} \tag{10}$$

Here, $C_\gamma \in [C_\beta, C_\alpha]$. Then, the maximum value that can be represented by a set of C_x bit codes $d_x = 2^{C_x} - 1$. 3-bit binary is used to record the encoded bit information in AOZS algorithm. 000, 001, ..., 111 means using 2, 3, ..., 9 bits binary to coding d_i respectively. The relationship of C_x and d_i is shown in Table 1. The relationship of coding length L_x and coding bit factor C_x is shown as Expression (11).

Table 1. The relationship of bit numbers, code, C_x and d_i.

Bit Numbers	Code	C_x (x = 2, 3, ..., 9)	d_x (i = 2, 3, ..., 9)
2	000	$C_2 = 2$	$d_2 = 3$
3	001	$C_3 = 3$	$d_3 = 7$
4	010	$C_4 = 4$	$d_4 = 15$
5	011	$C_5 = 5$	$d_5 = 31$
6	100	$C_6 = 6$	$d_6 = 63$
7	101	$C_7 = 7$	$d_7 = 127$
8	110	$C_8 = 8$	$d_8 = 255$
9	111	$C_9 = 9$	$d_9 = 511$

$$L_x = \left(\left[\frac{\alpha}{d_x} \right] + N - 1 \right) C_x + M + 3 \tag{11}$$

where, [ ] means to take up as an integer. M is the numbers of $\{d_1, d_2, \ldots, d_m\}$, notes the minimum of L_x as L_{min}, and its corresponding code bit factor C_x is the best code numbers, noted as $C_{optimal}$. Zero elimination operation refers to subtracting an integer value (recorded as zero elimination factor) from all data of the sequence, and finally making all data of the original sequence become 0. The zeroing factor s_i of the i-th order is recorded as Expression (12).

$$s_i = min\left\{ d_{min}, d_{optimal} \right\} \tag{12}$$

where, d_{min} is the minimum value of sequence.

After the above parameters are determined, the algorithm records the sequence coding information and related information, and uses $C_{optimal}$ bit number binary to encode the all zero elimination factors. The flow chart is shown in Figure 8. Table 2 shows the implementation process of the AOZS algorithm.

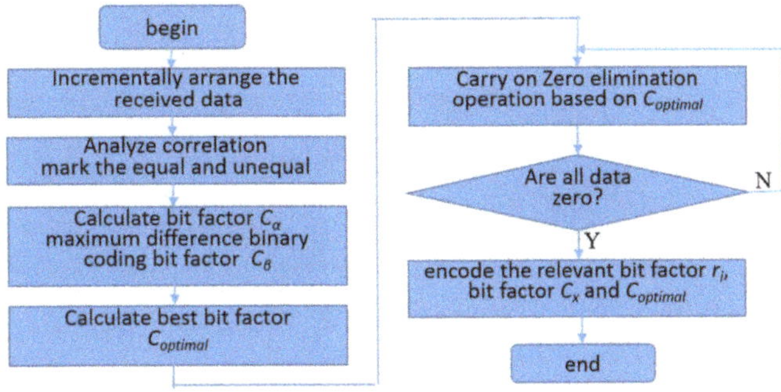

Figure 8. The flowchart of data compress algorithm.

Table 2. An example of AOZS algorithm.

Orignal Data	$d = 15$	$d = 4$	$d = 6$	$d = 2$	r
15	0	0	0	0	0
19	4	0	0	0	0
19	4	0	0	0	1
23	8	6	0	0	0
25	10	8	2	0	0
25	10	8	2	0	1

So, the compression code is 100,001001,1111,0100,0110,0010.

Compression Ratio (CR) is used to describe the efficiency of compression.

$$CR = \frac{S_{CP}}{S_{OR}} \tag{13}$$

Here, S_{CP} is the amount of compressed data, S_{OR} is the amount of original data. Obviously, the smaller the CR, the smaller the proportion of the compressed data to the original data, and the better the compression performance.

2.2.2. Software Design of Sensor Nodes and Monitoring Center

Software of each sensor node is simpler than that of sink node. It includes two parts. One is to read the humidity data and temperature data of the DHT11. The other is to drive the Bluetooth module to send these data to the sink node. The flowchart is given in Figure 9. It needs to complete the initialization of MCU peripherals and related modules. Clock initialization is used to set the working frequency of the system. Timer initialization is used to read temperature and humidity sensor data. Because the single line communication mode of DHT11 does not have a standard communication format, it is necessary to use a timer to simulate the communication sequence to realize the reception and transmission of data. Finally, the cc1101 module should be initialized.

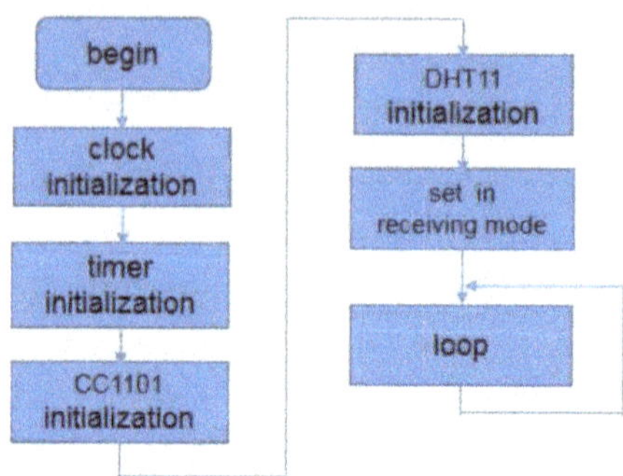

Figure 9. The flowchart of sensor nodes.

The monitoring center is developed with C++ language, which mainly realizes the following functions: (i) according to the longitude and latitude coordinates obtained from the cloud platform, it can display the location of the mobile carriage (sink node) in real time; (ii) it can display the real-time temperature and humidity in the carriage; (iii) it can dynamically draw the temperature and humidity change line chart; (iv) Using database to manage the collected data, it can save the historical data for data analysis. Figure 10 is the operation flowchart of monitoring center.

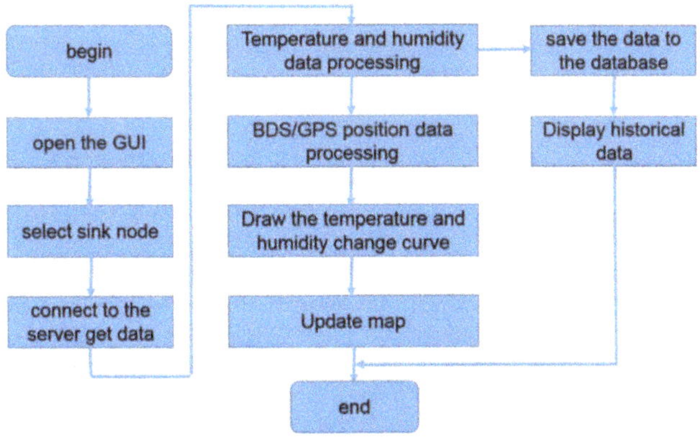

Figure 10. The operation flowchart of monitoring center.

3. Results

3.1. Measurement Accuracy

First, we do the measurement accuracy experiment. The experience was carried on the author's campus from 22 September to 21 October 2021. The campus is located at 120°55′ E and 28°51′ N. The weather was sunny during the test. The campus is spacious. The parameters of cc1101 are a carrier frequency of 433 MHz, a baud rate of 100 kbps, and a modulation mode of 2 FSK. The transmission power of the NB-IoT circuit is 13 dBm, the antenna gain is 3 dB, and the transmission rate is 3.9 Kbps. Each test point continuously sends and receives 1000 data packets. The packet loss rate of the whole network is less than 1% within 8 km, and the packet loss rate is 0% within 400 m. We read the humidity data and temperature data 10 times of 6 sensor nodes every day for a week and compare the data of thermometers and hygrometers which are put near the sensor node simultaneously. Then, the error was calculated and drawn in Figure 11a,b. The temperature error is less than 1 °C, as shown by the red line on Figure 11a. The average of temperature error is about 0.5 °C, as shown by the black line on the Figure 11a. The humidity error is less than 5% RH, as shown by the red line on Figure 11b. The average of humidity error is about 2% RH, as shown by the black line on the Figure 11b. The error of temperature and humidity are mainly decided by the DHT11 chip precision.

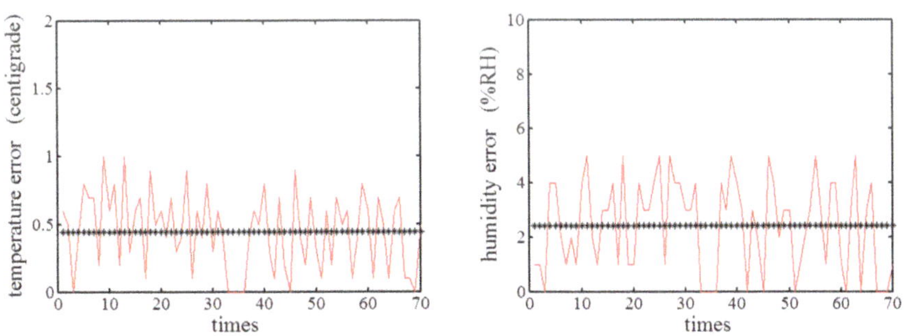

(**a**) (**b**)

Figure 11. The measurement error. (**a**) temperature error (**b**) humidity error.

Then, one sink node and its three sensor nodes were placed in a mobile car. We put a Leica GNSS (teaching edition) in the car, which is a professional position for the measuring instrument. The car was moved in different speed, and we read the latitude and longitude information of the monitoring center and Leica Receiver simultaneously. Let the longitude and latitude test by Leica GNSS are *LonA*, *LatA*. Let the longitude and latitude test by our system are *LonB*, *LatB*. Then the position error can be calculated by the Expression (14).

$$\begin{array}{c} \Delta Lon = (LonA - LonB) \times 1000 \times \big(111.413 \times cos\big(LatB \times \frac{\pi}{180}\big) \\ -0.094 \times cos\big(3 \times LatB \times \frac{\pi}{180}\big)\big) \\ \Delta Lat = (LatA - LatB) \times 1000 \times \big(111.133 - 0.59 \times cos\big(2 \times LatB \times \frac{\pi}{180}\big)\big) \\ Distance = \sqrt{\Delta Lon^2 + \Delta Lat^2} \end{array} \quad (14)$$

The relationship between positioning error and vehicle speed is shown in Figure 12.

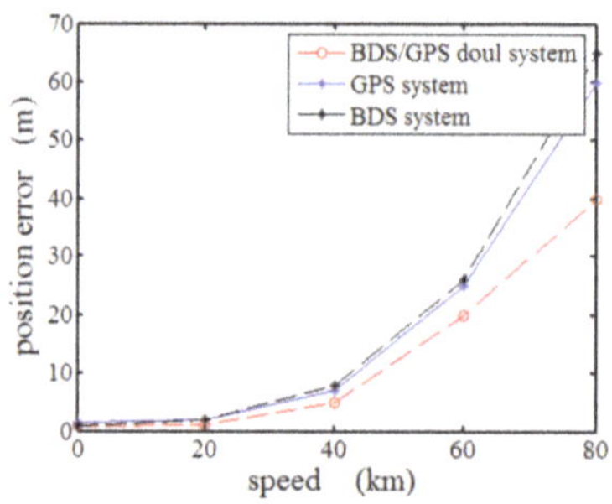

Figure 12. The relationship between positioning error and vehicle speed.

The positioning error is decided by several factors, such as the number of satellites used for positioning, vehicle speed, NB-IoT life cycle, etc. The average number of observable satellites under the GPS/BDS dual mode is 9, and the positioning performance is better than that of GPS single system and BDS single system. The main reason is that the number of available satellites increases, and the geometry configuration is enhanced. Under BDS/GPS dual positioning system, more positioning satellites can be obtained, so the accuracy is higher than that of GPS or BeiDou single positioning system. When the vehicle is static, the positioning error is about 2 m. When the car moves, the positioning error increases. The faster the car speed, the greater the positioning error. When the life period of NB-IoT is set as 3600 s, and the speed is less than 40 km/h, the positioning error is less than 10 m. When the speed is about 60 km/h, the positioning error is about 20 m. The larger the life cycle of NB-IoT, the greater the positioning error, because the larger the life cycle of NB-IoT, the greater the transmission latency.

3.2. Network Performance Test

Network performance test includes data compression rate and transmission packet loss rate. Limited by the experimental conditions, it is impossible to obtain a large number of test data of sensor nodes. Therefore, the experimental data on data compression rate is taken from the temperature data of Intel-Berkeley University Joint Research Laboratory in reference [40]. Compare the compression ratio between the AOZS algorithm proposed in this paper and the commonly used DCCM (Differential Code Compression Method) algorithm, as shown in Figure 13. Under the condition of the same amount of node data collection, the compression ratio of AOZS algorithm is lower than DCCM algorithm, and the compression performance is better, because AOZS algorithm makes full use of the correlation between data, and the coding based on the optimal bit factor removes the redundant information to the greatest extent. The more data the node collects, the higher the time correlation of the data. The coding factor of AOZS algorithm can describe more original data and fully mine the time correlation of data. Therefore, the compression ratio becomes smaller and smaller and tends to be stable gradually. With the increase of the number of sensor data, the compression rate of DCCM algorithm is maintained at about 50%, and that of AOZS algorithm is maintained at about 10%.

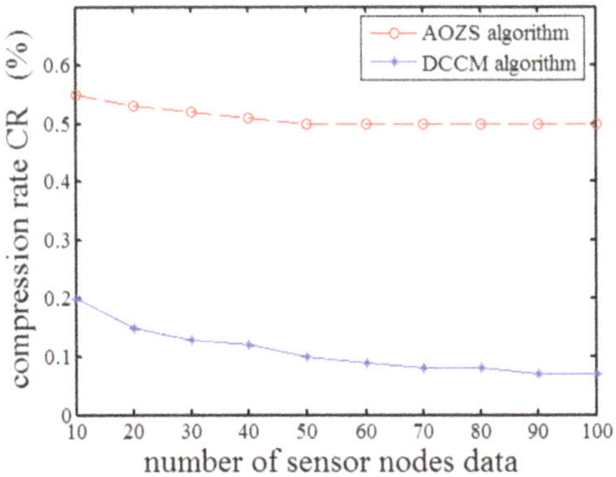

Figure 13. The Comparison of data compression ratio.

Another indicator of network communication reliability is packet loss rate. Sx1268 is a new generation 433 MHz LoRa half duplex transceiver chip produced by Semtech in 2018. It is also one of the commonly used Lora chips at present. So we compare the communication reliability between Sx1268 LoRa module and our wh-nb75-ba NB-IoT module. Figure 14 is the comparison of packet loss rate of our NB-IoT module and LoRa Sx1268 module under the same transmitting and receiving condition. When the distance is less than 250 m, the packet loss rate of both circuits is nearly 0. With the increase of distance, the packet loss rate of LoRa module increases significantly, while that of NB-IoT module increases little. When the distance is 400 m, the packet loss rate of LoRa module is about 1.5%, that of NB-IoT module is still nearly 0. when the distance is 600 m, the packet loss rate of LoRa module is about 2%, that of NB-IoT module is about 0.5%. when the distance is 800 m, the packet loss rate of LoRa module is about 5%, that of NB-IoT module is about 0.7%. when the distance is 800 m, the packet loss rate of LoRa module is about 5%, that of NB-IoT module is about 0.7%. When the distance is 1000 m, the packet loss rate of LoRa module is about 10%, that of NB-IoT module is about 1%. When the distance is 1200 m, the packet loss rate of LoRa module is about 15%, that of NB-IoT module is about 1.2%.

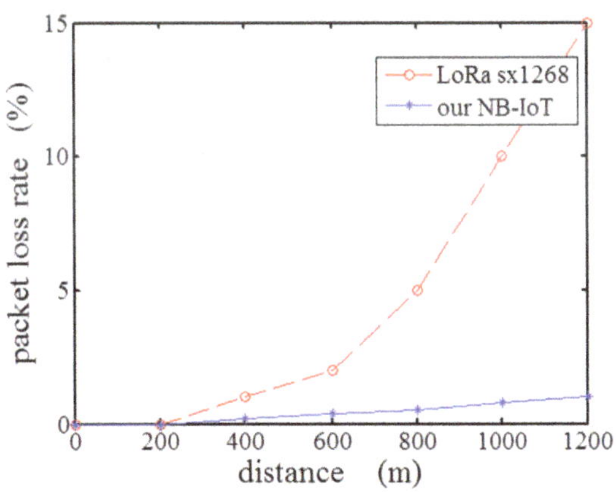

Figure 14. The packet loss rate of our NB-IoT module and LoRa sx1268.

4. Discussion

Through these tests aboved, the monitoring system realizes the higher position precision. It is shown that the BDS/GPS dual mode position have higher position precision than that of single BDS or GPS. When the monitored target is stationary, the positioning accuracy is only determined by the positioning module. The position calculation formula under the dual-mode positioning module is deduced as above. When the monitored target moves, the positioning accuracy is jointly determined by the positioning module, vehicle speed and life cycle. However, under the same vehicle speed and the same life cycle of NB-IoT, the monitoring system accuracy of dual-mode positioning is still higher than that of single BDS or GPS positioning mode. Considering the characteristics of sensor data in monitoring system, an adaptive optimal zero suppression (AOZS) algorithm based on time correlation is proposed in this paper. After testing and comparing with the commonly used differential code compression method (DCCM) algorithm, the data compression rate of the new algorithm can be as high as 90%, which greatly reduces the amount of data transmission in the communication network and improves the network performance and transmission efficiency. With the increase of the number of sensor data, the compression rate of DCCM algorithm is maintained at about 50%, and that of AOZS algorithm is maintained at about 10%. Packet loss rate is the main indicator of communication network performance. We tested and compared the packet loss rate of the monitoring system based on wh-nb75-ba NB-IoT module and the monitoring system based on sx1268 LoRa module which is mainly used. When the distance is less than 250 m, the packet loss rate of both circuits is nearly 0. With the increase of distance, the packet loss rate of LoRa module increases significantly, while that of NB-IoT module increases little. The greater the distance, the greater the difference between the packet loss rate data of the two circuits.

5. Conclusions

A new monitoring system is proposed in the paper, based on NB-IoT and BDS/GPS dual-mode positioning. The whole monitoring system includes three parts: sensor node, sink node and monitoring center. The sensor node which is based on cc1101 RFID circuit realizes the detection of surrounding temperature and humidity and data transmission. The sink node receives and compresses the temperature and humidity data from the sensor node, obtains the positioning information through at6558 BDS/GPS positioning module, and uploads these data to the cloud through wh-nb75-ba NB-IoT module. The monitoring center can download data from the cloud and save it to the local machine, and can analyze the historical data through an operation interface.

Experiments and analysis show that the proposed scheme has better positioning accuracy, better data compression ratio and transmission performance. The temperature and humidity error are less than 1 °C and 5% RH especially with the selected chip. The position error is decided by several factors, including the number of satellites used for positioning, the monitored target moving speed and NB-IoT module lifetime period. When the monitored target is stationary, the positioning error is about 2 m, which is less than that of the single GPS or BDS mode. The AOZS compression algorithm is used to improve compression ratio (CR). The CR is about 10% when the data amount increasing.

The scheme of this paper had encouraged experiments and was efficient and practicable in monitoring system. However, many aspects, still need to be further studied, such as transmission delay, multi-sensor nodes and low-power circuits. Furthermore, optimizing the network structure to reduce its consumption and accomplishing end-to-end network will be the main direction of our work.

Author Contributions: Conceptualization, Z.X. and R.Z.; methodology, Z.X.; software, Z.X. and R.Z.; validation, J.F. and L.Z.; formal analysis, J.F.; investigation, L.Z.; writing—original draft preparation, Z.X.; writing—review and editing, R.Z.; visualization, J.F.; funding acquisition, Z.X. All authors have read and agreed to the published version of the manuscript.

Funding: This research was supported by the Commonweal Projects of Zhejiang Province (Grant No. LGN20F010001) and General Project of Zhejiang Education Department (Grant No. Y201940951).

Conflicts of Interest: The authors declare no conflict of interest.

References

1. Reynders, B.; Meert, W.; Pollin, S. Range and coexistence analysis of long range unlicensed communication. In Proceedings of the 2016 23rd International conference on Telecommunications, Thessaloniki, Greece, 16–18 May 2016.
2. Centenaro, M.; Vangelista, L.; Zanella, A.; Zorzi, M. Long-range communications in unlicensed bands: The rising stars in the IoT and smart city scenarios. *IEEE Wirel. Commun.* **2016**, *23*, 60–67. [CrossRef]
3. Patel, D.; Won, M. Experimemtnal study on low power wide area networks(LPWAN) for mobile Internet of Things. In Proceedings of the 2017 IEEE 86th Vehicular Technology Conference, Toronto, ON, Canada, 24–27 September 2017.
4. Raza, U.; Kulkarni, P.; Sooriyabandara, M. Low power wide area networks: An overview. *IEEE Commun. Surv. Tutor.* **2017**, *19*, 855–873. [CrossRef]
5. Li, Y.; Cheng, X.; Xao, Y. Smart choice for the smart grid: Narrowband Internet of Things (NB-IoT). *IEEE Internet Things J.* **2018**, *5*, 1505–1515. [CrossRef]
6. Duangsuwan, S.; Takarn, A.; Nujankaew, R. A study of air pollution smart sensors LPWAN via NB-IoT for Thailand smart cities 4.2. In Proceedings of the 10th International Conference on Knowledge and Smart Technology, Chiang Mai, Thailand, 31 January–3 February 2018.
7. Lauridsen, M.; Krigslund, R.; Rohr, M. An empirical NB-IoT power consumption model for battery lifetime estimation. In Proceedings of the 2017 IEEE 86th Vehicular Technology Conference, Toronto, ON, Canada, 24–27 September 2017.
8. Sotirios, K.G.; Margot, D.; David, P.; Luc, M.; Kostas, E.P. A Novel Design Approach for 5G Massive MIMO and NB-IoT Green Networks Using a Hybrid Jaya-Differential Evolution Algorithm. *IEEE Access* **2019**, *7*, 105687–105700.
9. Li, H.; Dong, W.; Wang, Y.; Gao, Y.; Chen, C. Enhancing the Performance of 802.15.4-Based Wireless Sensor Networks with NB-IoT. *IEEE Internet Things J.* **2020**, *7*, 3523–3534. [CrossRef]
10. Pilar, A.M.; Pablo, A.; Jonathan, P.G. An Analytical Performance Evaluation Framework for NB-IoT. *IEEE Internet Things J.* **2019**, *6*, 7232–7240.
11. Ahmad, A. Toward Achieving a Balance between the User Satisfaction and the Power Conservation in the Internet of Things. *IEEE Internet J.* **2022**, *8*, 10998–11015.
12. Gayathri, N.; Rakesh, K.S.; Rizwan, P. Ensemble Classification and IoT-Based Pattern Recognition for Crop Disease Monitoring System. *IEEE Internet J.* **2021**, *8*, 12847–12854.
13. Joseph, H.A.; Tang, Y.; Celestine, I.; Desire, N. A Secure Multiuser Privacy Technique for Wireless IoT Networks Using Stochastic Privacy Optimization. *IEEE Internet J.* **2022**, *9*, 2566–2577.
14. Ayoub, W.; Samhat, A.E.; Nouvel, F.; Mroue, M.; Prévotet, J.C. Internet of Mobile Things: Overview of LoRaWAN, DASH7, and NB-IoT in LPWANs standards and Supported Mobility. *IEEE Commun. Surv. Tutor.* **2019**, *21*, 1561–1581. [CrossRef]
15. Chakrapani, A. NB-IoT Uplink Receiver Design and Performance Study. *IEEE Internet Things J.* **2020**, *7*, 2469–2482. [CrossRef]
16. Nancy, A.A.; Ravindran, D.; Raj Vincent, P.D.; Srinivasan, K.; Gutierrez Reina, D. IoT-Cloud-Based Smart Healthcare Monitoring System for Heart Disease Prediction via Deep Learning. *Electronics* **2022**, *11*, 2292. [CrossRef]
17. Bencak, P.; Hercog, D.; Lerher, T. Indoor Positioning System Based on Bluetooth Low Energy Technology and a Nature-Inspired Optimization Algorithm. *Electronics* **2022**, *11*, 308. [CrossRef]
18. Shi, J.; Jin, L.P.; Li, J.; Fang, Z.X. A smart parking system based on NB-IoT and third-party payment platform. In Proceedings of the International Symposium on Communications and Information Technologies, Cairns, Australia, 25–27 September 2017.
19. Mandza, Y.S.; Raji, A. IoTivity Cloud-Enabled Platform for Energy Management Applications. *IoT* **2022**, *3*, 73–90. [CrossRef]
20. Zhang, H.; Li, J.; Wen, B.; Xun, Y.; Liu, J. Connecting Intelligent Things in Smart Hospitals Using NB-IoT. *IEEE Internet Things J.* **2018**, *5*, 1550–1560. [CrossRef]
21. Srikanth, K.; Dmitri, M.; Aleksandr, O.; Sergey, A. Performance Analysis of Onshore NB-IoT for Container Tracking during Near-the-Shore Vessel Navigation. *IEEE Internet Things J.* **2020**, *7*, 2928–2943.
22. Zhang, X.; Zhang, M.; Meng, F.; Qiao, Y.; Xu, S.; Hour, S. A Low-Power Wide-Area Network Information Monitoring System by Combining NB-IoT and LoRa. *IEEE Internet Things J.* **2019**, *6*, 590–598. [CrossRef]
23. Cao, T.P.; Duy, D.P.; Hoang, V.T.; Trung, V.T.; Phat, N.H. Applying the IoT platform and green wave theory to control intelligent traffic lights system for urban areas in Vietnam. *KSII Trans. Internet Inf. Syst.* **2019**, *13*, 34–51.
24. Sinha, R.S.; Wei, Y.; Hwang, S.H. A survey on LPWAN technology: LoRa and NB-IoT. *ICT Express* **2017**, *3*, 14–21. [CrossRef]
25. Wu, Z.; Liu, R.; Cao, H. ECDSA-Based Message Authentication Scheme for BeiDou-II Navigation Satellite System. *IEEE Trans. Aerosp. Electron. Syst.* **2019**, *55*, 1666–1682. [CrossRef]
26. Wei, H.; Pascale, D. BeiDou Time Transfer with the Standard CGGTTS. *IEEE Trans. Ultrason. Ferroelectr. Freq. Control* **2016**, *63*, 1005–1012.
27. Su, M.; Yang, Y.; Jiang, J. BeiDou system satellite-induced pseudorange multipath bias mitigation based on different orbital characteristic for static applications. *IET Radar Sonar Navig.* **2020**, *14*, 242–251. [CrossRef]
28. Qin, H.; Liu, P.; Cong, L.; Ji, W. Triple-Frequency Combining Observation Models and Performance in Precise Point Positioning Using Real BDS Data. *IEEE Access* **2019**, *7*, 69826–69836. [CrossRef]
29. Zhang, Y.; Tian, L.; Meng, W.; Gu, Q.; Han, Y.; Hong, Z. Feasibility of Code-Level Altimetry Using Coastal BeiDou Reflection (BeiDou-R) Setups. *IEEE J. Sel. Top. Appl. Earth Obs. Remote Sens.* **2015**, *8*, 4130–4140. [CrossRef]
30. Wu, Y.H.; Zheng, M.H.; He, W.; Chen, Z.M.; Hua, B. Intelligent vehicle safety system based on BeiDou satellite navigation system. *IET Intell. Transp. Syst.* **2019**, *13*, 967–974. [CrossRef]

31. Fan, L.; Tu, R.; Zhang, R.; Zheng, Z.; Liu, J.; Hong, J.; Lu, X. Real-time BDS signal-in-space anomaly detection method considering receiver anomalies. *IET Radar Sonar Navig.* **2019**, *13*, 2220–2229. [CrossRef]
32. Wang, E.; Yang, D.; Wang, C.; Huang, Y.; Qu, P.; Pang, T. Optimized Fault Detection Algorithm Aided by BDS Baseband Signal for Train Positioning. *Chin. J. Electron.* **2020**, *29*, 34–40. [CrossRef]
33. Wang, L.; Li, Z.; Yuan, H.; Zhao, J.; Zhou, K.; Yuan, C. Influence of the time-delay of correction for BDS and GPS combined real-time differential positioning. *Electron. Lett.* **2016**, *52*, 1063–1065. [CrossRef]
34. Jiang, W.; Chen, S.; Cai, B.; Wang, J.; ShangGuan, W.; Rizos, C. A Multi-Sensor Positioning Method-Based Train Localization System for Low Density Line. *IEEE Trans. Veh. Technol.* **2018**, *67*, 10425–10437. [CrossRef]
35. Zhu, Y.; Liu, Y.; Wang, Z.; Li, Q. Evaluation of GBAS flight trials based on BDS and GPS. *IET Radar Sonar Navig.* **2020**, *14*, 233–241. [CrossRef]
36. Ketshabetswe, L.K.; Zungeru, A.M.; Mangwala, M.; Chuma, J.M.; Sigweni, B. Communication protocols for wireless sensor networks: A survey and comparison. *Heliyon* **2019**, *5*, e01591. [CrossRef]
37. Capo-Chichi, E.P.; Guyennet, H.; Friedt, J.M. K-RLE: A new data compression algorithm for wireless sensor network. In Proceedings of the IEEE 2009 Third International Conference on Sensor Technologies and Applications (SENSORCOMM), Athens, Glyfada, 18–23 July 2009.
38. Zhang, H.; Fan, X.P.; Liu, S.Q. Design and realization of improved LZW algorithm for wireless sensor networks. In Proceedings of the International Conference on Information Science and Technology(ICIST), Changsha, China, 26–28 May 2011.
39. Das, P. Implementing Dial-On-Demand Technique for Inter and Intra Cluster Communication in Energy Conserving Postbox Delay Tolerant Networks. In Proceedings of the International Conference on Computational Intelligence, Communications, and Business Analytics, Kalyani, India, 27–28 July 2018.
40. Ahmad, M.S.; Lata, S.; Mehfuz, S.; Ahmad, A. Lossless Compression Algorithm for Energy Efficient Wireless Sensor Network. In Proceedings of the 2019 International Conference on Power Electronics, Control and Automation (ICPECA), New Delhi, India, 16–17 November 2019.
41. Mo, Y.B.; Qiu, Y.B.; Liu, J.Z. A data compression algorithm based on adaptive huffman code for wireless sensor networks. In Proceedings of the 2011 International Conference on Intelligent Computation Technology and Automation (ICICTA), Nanjing, China, 3–6 September 2011.
42. Xie, L.; Chen, L.J.; Chen, D.X. Clustering-based approximate scheme for data aggregation over sensor networks. *J. Softw.* **2009**, *20*, 1023–1037.
43. Hu, J.; Shen, L.F. Novel clustering algorithm for wireless sensor networks. *J. Commun.* **2008**, *29*, 20–29.
44. Zhu, X.R.; Shen, L.F.; Tak-Shing, P.Y. Hausdorff clustering and minimum energy routing for wireless sensor networks. *IEEE Trans. Veh. Technol.* **2009**, *58*, 990–997. [CrossRef]
45. Ciancio, A.; Pattem, S.; Ortega, A. Energy-efficient data representation and routing for wireless sensor networks based on a distributed wavelet compression algorithm. In Proceedings of the Fifth International Conference on Information Processing in Sensor Networks, Palo Alto, CA, USA, 7–10 July 2006.
46. Donoho, D.L. Message passing algorithms for compressed sensing: I. motivation and construction. In Proceedings of the IEEE Information Theory Workshop 2010, Cairo, Egypt, 2–6 January 2010.
47. Marcelloni, F.; Vecchio, M. A simple algorithm for data compression in wireless sensor networks. *IEEE Commun. Lett.* **2008**, *12*, 411–413. [CrossRef]
48. Fan, X.H.; Li, S.N.; Du, P.L. Simple algorithm for self-adapting lossless data compression in WSN. *Comput. Meas. Control* **2010**, *18*, 463–465.
49. Xia, Y.C.; Chen, L.L.; Chen, X. Research on data compression in wireless sensor networks-with wavelet lifting algorithm and difference mechanism. *Comput. Eng. Appl.* **2010**, *46*, 109–112.
50. WH-NB75-BA Datasheet. Available online: http://www.mokuai.cn/products/67.html (accessed on 4 June 2022).
51. AT6558 Manual. Available online: http://www.icofchina.com/pro/dingwei/2016-07-29/5.html (accessed on 4 June 2022).
52. CC1101 Data Sheet. Available online: https://wenku.baidu.com/view/c2b0081b227916888486d769.html (accessed on 4 June 2022).
53. NMEA0183 Protocol. Available online: https://blog.csdn.net/st526403649/article/details/54946529 (accessed on 4 June 2022).

smart cities

MDPI

Article

A Prosumer-Oriented, Interoperable, Modular and Secure Smart Home Energy Management System Architecture

Pedro Gonzalez-Gil *, Juan Antonio Martinez and Antonio Skarmeta

Departamento de Ingeniería de la Información y las Comunicaciones, Facultad de Informática, Universidad de Murcia, 30100 Murcia, Spain
* Correspondence: pedrog@um.es

Abstract: As prices on renewable energy electricity generation and storage technologies decrease, previous standard home energy end-users are also becoming producers (prosumers). Together with the increase of Smart Home automation and the need to manage the energy-related interaction between home energy consumers and Smart Grid through different Demand Response approaches, home energy management becomes a complex and multi-faceted problem, calling for an extensible, interoperable and secure solution. This work proposes a modular architecture for building a Smart Home Energy Management System, integrable with existing Home Automation Systems, that considers the use of standard interfaces for data communication, the implementation of security measures for the integration of the different components, as well as the use of semantic web technologies to integrate knowledge and build on it. Our proposal is finally validated through implementation in one real smart home test-bed, evaluating the system from a functional standpoint to demonstrate its ability to support our goals.

Keywords: energy management systems; home automation; smart homes; Internet of Things

Citation: Gonzalez-Gil, P.; Martinez, J.A.; Skarmeta, A. A Prosumer-Oriented, Interoperable, Modular and Secure Smart Home Energy Management System Architecture. *Smart Cities* **2022**, *5*, 1054–1077. https://doi.org/10.3390/smartcities5030053

Academic Editors: Antonio Cano-Ortega, Francisco Sánchez-Sutil and Aurora Gil-de-Castro

Received: 23 July 2022
Accepted: 19 August 2022
Published: 24 August 2022

1. Introduction

In the wake of increasingly present climate change effects, the scientific community proposes a "decarbonization" of our society, from transportation and industry to energy sectors. As society shifts from fossil fuel usage for transportation [1] and heating to electricity, our total electrical energy usage and dependence increases, imposing an excessive load on our already-ageing power distribution grid, leading to the development of different Demand Response (DR) strategies and mechanisms for the Smart Grid (SG). At the same time, electricity production is shifting towards renewable energy sources. These new energy sources, though, are not as controllable and predictable as traditional ones, in which production could be more easily matched to demand, increasing the need for effective DR.

Traditional demand-side management measures have taken the approach of reducing total energy consumption by increasing appliances' energy efficiency and leveraging technology to reduce wastage. More recently, as more producer-consumers (*Prosumers*) join the grid; leveraging Distributed Energy Resources (DERs) both for local energy generation and storage [2,3] also come as interesting opportunities to help both: reduce global carbon emissions and make better use of the grid.

According to the International Energy Agency (IEA) [4], electricity wholesale prices in Spain, France, Germany and the United Kingdom, increased in 2021 from three to more than four times in respect to the 2016–2020 period. This steep increase in electricity prices, together with the cost reduction of renewable energy production technologies such as solar Photo-Voltaic (PV), as well as electrical energy accumulation through different battery technologies, have greatly promoted its application in residential installations, as a way to reduce the electricity bill.

In this present situation of increased electricity prices, reduced costs for the application of DERs in residential installation, the need for cooperation between energy consumers/prosumers and the grid and the variable nature of renewable energy production; new and innovative solutions are needed to deal with the complexity of our present and future energy system.

1.1. Smart Home Energy Management

The Internet of Things (IoT): the interconnection of everyday physical devices through Information and Communication Technology (ICT), presents itself as a great candidate to automate and manage the ever-increasing complexity of those places where us humans dwell, as well as helping to shape electricity demand, better match production and consumption and quickly react to variability, while reducing the electricity bill and keeping within user-defined comfort parameters (Figure 1).

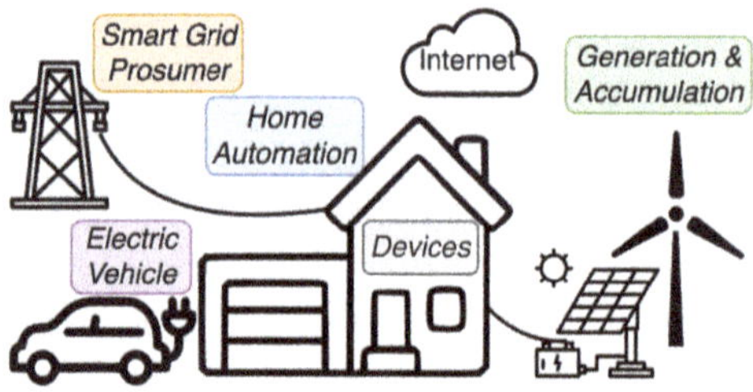

Figure 1. Smart Home energy management.

Many advances have come forth in the last years regarding the Smart Home (SH). More specifically, in the field of home automation, where a plethora of devices and appliances are available: smart bulbs, refrigerators, washing machines, tumble dryers, electric vehicles (and chargers), Home Ventilation and Air Conditioning (HVAC) and the likes. Several different technologies have coalesced and started to settle among the general public, designed to integrate many of those devices under a single point of control: the Smart Hubs. Among the most known commercial cloud-based solutions for Smart Hubs are those from Google, Amazon and Apple; however, there is also a growing community of Do It Yourself (DIY) enthusiasts, that have opted for open and often self-hosted solutions such as Home Assistant [5], Domoticz [6] and OpenHab [7] to name but a few. These open solutions come as great frameworks for the rapid development of new ideas beyond home automation, allowing us to leverage already existing integrations of different devices and platforms.

The present picture of a modern prosumer-oriented SH is thus a complex ecosystem in which energy management has to comply with a broad range of concerns (Figure 2). The interaction with external agents, like the SG, Smart Hubs and other external services, places high interoperability expectations both in data modelling and communication interfaces, as well as in the privacy concerns of the home inhabitants [8,9].

To this end, new and innovative solutions must be brought forth, such as the use of semantic web technologies to bridge the gap among the many different services and domains of home energy management, standard interfaces to facilitate communications with other agents and security mechanisms to facilitate setting boundaries to information access.

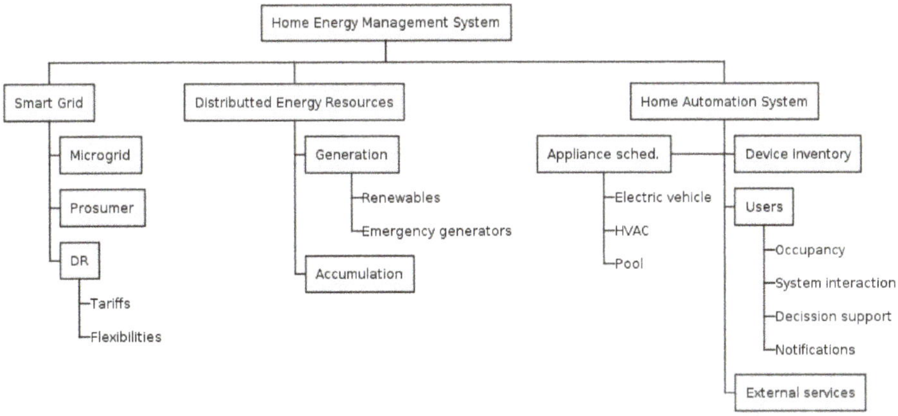

Figure 2. Home energy management concerns.

1.2. Contribution of This Paper

This work proposes a Smart Home Energy Management Systems (SHEMSs) [10–12] architecture design capable of integrating the many different facets of energy management of a modern home, in an interoperable, standard-based and secure way so that consumers/prosumers and grid are benefited. To that end, we have leveraged the use of semantic technologies for their potential to store and extract knowledge from heterogeneous sources, providing a formal common ground [13] as well as existing ontologies that capture the represented concepts. Additionally, to be able to offer a solution that stands to be integrated by different vendors and parties, we propose the use of standard-based secure technologies for information exchange.

The rest of this document is structured as follows: in Section 2, we reflect on the previously existing work on this topic. In Section 3, we propose a secure architecture for modular SHEMS, integrable with existing Home Automation Systems (HASs), which is later showcased by implementation in a real scenario in Section 4. Finally, we discuss the results obtained in Section 5, providing conclusions and possible future lines of work in Section 6.

2. State of the Art and Related Work

There is a broad and extensive bibliography related to the work on Home Energy Management Systems (HEMSs) architecture and strategies, semantic representation of SH and HEMS concepts, DR strategies from within the SH and interfacing with the SG. In this section, we will briefly cover some of the most relevant works from our proposal perspective, sorted in reverse chronological order and categorised into two families: existing architectures and ontological work related to the interoperability of the solution.

2.1. Architecture and Security

Machorro-Cano et al. present HEMS-IoT [14]; a machine learning-based HEMS for energy saving, ensuring comfort and security while reducing energy consumption. The proposed architecture consists of seven layers, from presentation to device, in which information security is considered between presentation, IoT services and management layers, contemplating both authentication and authorisation. It does not mention DR or DER management as part of the proposed HEMS, nor explores the possibility of its integration with existing HAS, although its management layer does apply a semantic approach to home management, utilising a self-developed ontology to represent the main domotic concepts. Overall it presents a monolithic structure that goes from device to presentation, where interoperability has not been approached.

Elshaafi et al. [15] explore an approach to decentralised automated DR and home energy management. The proposed architecture is implemented using a multi-agent system with three different levels: home, aggregator and distribution system operator (being the latter two SG-related). At the home level, they define only two agents: the device agent and the HEMS agent. Their combined objective is to reduce the energy bill while respecting user preferences and comfort. While device agents encapsulate the communications with the home devices, the HEMS agent is responsible for the grunt of the work: planning, optimising and communicating with SG agents. This architecture does not consider the existence of a previous HAS. The final solution is presented in the form of a home gateway device that will directly interact with smart devices, although it does consider DER management through the different device agents. It addresses interoperability by proposing the usage of standard communication interfaces (OSGi: Open Services Gateway initiative [16]) and REST interfaces) and the use of Web Ontology Language (OWL) for information modelling. Finally, the proposed architecture takes into consideration the privacy and security of home users by using XACML (eXtendible Access Control Markup Language [17]) as an attribute-based access control platform, controlling the visibility of home devices to the HEMS agent.

The MAS2TERING project [18] defines a multi-agent system consisting of the following agents: *Distribution system operator agent*, *Aggregator agent*, *Central home energy management agent*, *Microgeneration agent*, *Appliance agent* and *Battery agent* whose interactions aim at delivering demand-side management through the supply chain, from generation to consumer appliances. This architecture does not cover security aspects, nor the integration with existing HASs or home energy management.

Zhang et al. propose iHEMS [19], a publish-subscribe communications infrastructure, using Information-Centric Networking (specifically Content Centric Networking) as the communication backbone. It does not rely on securing the communication channels but on encrypting the data itself, using a secure-group communications scheme above the pub-sub layer. The architecture itself contemplates the different devices interconnected through the Information Centric Networking-based pub-sub substrate, as well as a *Directory Service*, which devices use to publicise their data and a *Group Controller* in charge of key management for encryption/decryption.

Digital Environment Home Energy Management System (DEHEMS) project [20] proposes a service-based architecture, consisting of a remote server where the knowledge base is deployed, which in turn is fed from a *Data Collector* located in the home, to which sensors, devices, appliances and display devices are connected using wireless interfaces.

Rosello-Busquet et al. [21] propose a home gateway for a HEMS system, to control the devices in a home network at the service level. Built over the OSGi framework, it uses DogOnt ontology as the base for its knowledge base data repository. The architecture is composed of six bundles: *Knowledge Base*, *Interface*, *Network n*, *Networks Manager*, *Manager* and *Network Emulator*.

ThinkHome by Reinisch et al. [22] describes a multi-agent system architecture with two main premises: ensuring energy efficiency at home and comfort optimisation. The control strategies realised by the multi-agent system are split into problem aspects which are directly mapped to the different agents of the framework: *Control*, *Users*, *Global Goals*, *Context Inference*, *Auxiliary Data*, *Knowledge Base Interface* and *Building Automation System Interface*.

After the bibliographical review performed in architecture and security, we can conclude that none of the reviewed works considers all the previously described concerns of a modern prosumer-oriented SH in their architecture. None addresses data model and communications interoperability simultaneously to enable successful interaction with other service providers, such as external device platforms, Smart Hubs and the SG. Finally, security was covered by some works, but mostly in the communications domain, not placing the focus on data privacy, which represents a major concern in the complex SH ecosystem.

2.2. Ontological Background

To model the information in SHEMSs, a wide range of topics had to be covered, which resulted in an extensive survey of the different existing related ontologies. The topics or areas covered include grid DR management, home DR management, DER management, energy metering and performance assessment, energy saving advice, (home) infrastructure, user preferences, weather and environmental and sensor data: to name but a few. It is needless to say that no single ontology covers all of the required areas and that they all differ in the level of detail with which the different concepts are captured.

To summarise the ontological state of the art, Table 1 categorises the most relevant surveyed works according to our specific use case. Later, in Section 3.4, the mapping of those ontologies to specific elements in our proposal will be presented.

Table 1. Ontology summary.

Ontology	Application Field
OSEIM [23,24]	Semantic reasoning for intelligent energy management.
SARGON [25]	Smart grid and building energy automation.
DABGEO * [26]	Integration of smart home energy-related ontologies.
EnergyUse [27]	Home energy saving advice.
MAS2TERING [18]	Supporting USEF implementation on smart grid.
SAREF4EE [28,29]	Interoperability of smart appliances.
DNAS [30]	Energy efficiency through occupant behaviour.
ProSGv3 [31]	Modelling prosumer-oriented smart grid.
MIRABEL [32]	Flexibility description for demand response.
BonSAI [33]	Modelling of service-oriented smart building.
ThinkHome [22,34]	Home energy assessment and device control.

* Grayed rows are imported in DABGEO.

Smart Energy Domain Ontology (SARGON) [25] extends SAREF (Smart Applications REFerence [35]) but, unlike SAREF4EE, focuses on the control and monitoring of distribution electrical grids, integrating it with building energy automation.

OSEIM and NewOSEIM [23,24] leverage semantic reasoning over an ontology presenting knowledge about the internal and external environment of a home, to achieve intelligent energy management.

DABGEO [26] is an ontology semantically equivalent to OEMA [36], a previous ontology by the same authors. It improves on the former by offering a modular ontology that can be imported into subsets, facilitating its adoption in custom use-case scenarios. As its predecessor, it links and extends concepts from other previous ontologies. The base of the ontology network is ThinkHome, to which SAREF4EE, EnergyUse and ProSGV3 have been added.

EnergyUse framework [27] for home energy-saving advice applications, enriches PowerONT [37] (the power consumption ontology) with other ontologies and maps it to JSON-LD.

MAS2TERING ontology [18,38], developed under the homonymous project, implements USEF (Universal Smart Energy Framework [39]) through multi-agent systems. The purpose of the ontology is the representation of the data of different SG domains and to provide interoperability between SG agents and stakeholders. It is based in Energy@Home [40] and CIM (International Electrotechnical Commission's Common Information Model [41]).

The DNAS framework presents an ontology [30] to represent energy-related occupant behaviour to understand total energy consumption in buildings.

The MIRABEL [32,42] ontology for modelling flexibility in SG energy management, allows actors to express their energy flexibility for a specific device. It also represents energy profiles for devices, as well as production and storage devices.

BOnSAI [33,43] presents an ontology for incorporating Ambient Intelligence in Smart Buildings that can be used for energy management and monitoring. Includes concepts about functionality, QoS, hardware, users and context.

3. Proposal

The general or conceptual architecture of our proposal (Figure 3) is composed of multiple Energy Management Components (EMCs) responsible for a specific domain within home energy management, which interact with each other to pursue their goals, cooperating toward a common objective of increasing energy efficiency and reducing energy costs of the home.

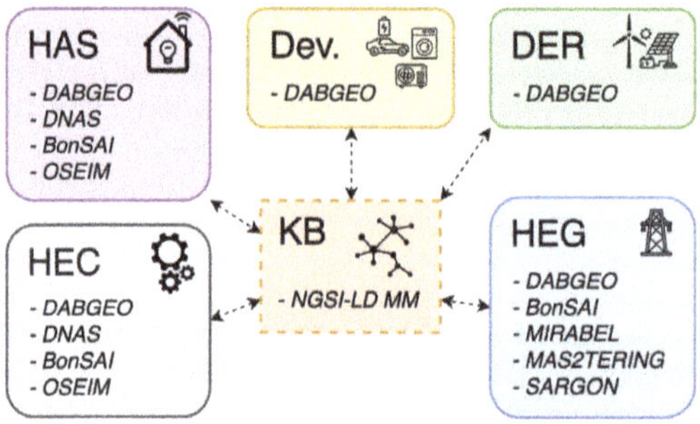

Figure 3. Knowledge Base-centred architecture of the Smart Home Energy Management System.

3.1. Energy Management Components

The following list describes the different EMCs that govern home energy management, as well as some of the relationships between them:

1. *Home Automation System (HAS)*: in charge of communication back and forth with the Home Automation System, keeps it updated regarding energy budgets, production and accumulation status and current consumption. High-level information produced by other components, can be used by the HAS to provide home occupants with energy-related decision support tools and notifications. It also updates the KB with information from the HAS, such as devices inventory, appliance status and scheduling, presence and occupancy information as well as information coming from external sources, such as real-time weather information and forecasts. This information can be used to build and feed different DER and energy consumption forecast models and allow the HEMS to plan and adapt to changing conditions.
2. *Distributed Energy Resource (DER)*: is responsible for dealing with energy resources, both for energy production and accumulation. Its purpose is to: (a) to produce high-level information, such as production and storage forecasts needed by the Home Energy Controller (HEC) component (5) and (b) to operate the different DERs safely, trying to minimise energy costs for the home and (c) cooperating with HEC for DR strategies implementation.
3. *Home Energy Gateway (HEG)*: updates the status of the grid-tie, whether we are injecting or consuming power from the grid, as well as pricing information and contractual parameters (e.g., maximum usable power). It can also relay energy information back to the utility company, such as energy usage schedules and forecasts, appliance inventory and usage patterns and, in general, any information that the homeowner is willing to share that can help the utility company to offer a better deal to the client. It is also responsible for communicating DR strategies, such as flexibilities or acting as a relay for Packetized Energy Management (PEM) brokering with the grid (or micro-grids) and keeping track of total grid energy costs.
4. *Device*: represents devices whose energy management can be dealt with directly by the HEC component (5). Examples of such devices would be big consumers, continuously-running appliances like HVAC or heating systems, swimming pool pumps, water heaters and electric vehicles, whose energy consumption schedule can impact to a great extent on home energy management and usually can be re-scheduled or curtailed under certain conditions with minimal to no impact for the home occupants.

Usually, these devices will be the most critical ones in DR strategies such as flexibility management.

5. *Home Energy Controller (HEC)*: in charge of control and optimisation. It can make use of different strategies and techniques, from semantic reasoning to the use of traditional optimisation methods or artificial intelligence. Its purpose is to deal with and mediate between components, ensuring that the energy budget is optimised and providing useful high-level information to be shared with others, such as HAS (1) and HEG (3).

It is worth mentioning at this point that EMCs are not single or individual instances. Different device components could be designed for a washing machine and an electric vehicle charger. Conversely, components do not have to be implemented in a single final unit; for example the HEC could be split into different software modules in charge of energy brokering, total energy consumption forecasting and deciding different strategies for DR.

3.2. Knowledge Base

As depicted in Figure 3, we propose an architecture centred around the KB. Components are capable of realising their goals as well as interacting and cooperating by using the Knowledge Base (KB), where all the information of the system is stored.

The information stored in the KB includes meta-information regarding different aspects, such as data typing and ontological links to the represented concepts. This "enriched" information is called *context* in our system.

The structure and semantic roots of context in our proposal and the mechanisms to interact with it, are further described in Section 3.5. The components responsible for its management are first described in Section 3.3. Finally, the KB is also used as a means for coordination and orchestration among EMCs in the system, leveraging its publish/subscribe nature and flexible data model. This is further described in Section 3.7.

3.3. Functional Architecture

The functional architecture of our proposed HEMS consists of three layers (Figure 4) and a transversal security layer, in which the different components are structured as follows:

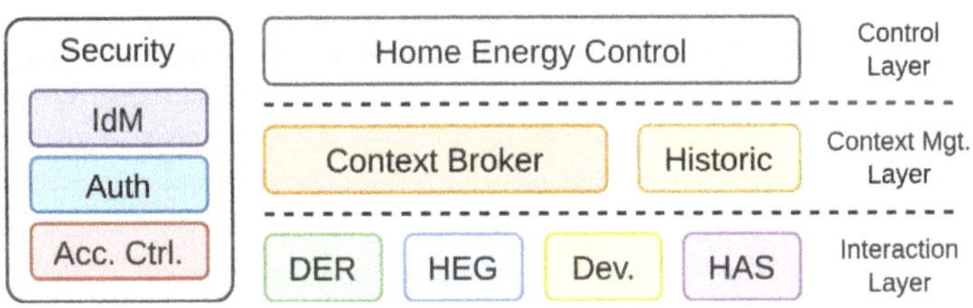

Figure 4. Layered architecture of the Smart Home Energy Management System.

1. *Control Layer*: this is the core of the management system, where all the modules related to the HEC component coexist. This layer is responsible for the strategy and the scheduling. It will try to realise the global system objectives by working with the information provided by the remaining components. To both: accrue data from the *Interaction Layer (3)* and send back directives regarding scheduling and any operational parameters, any modules in this layer will use the *Context Broker* component, placed in the *Context Management Layer (2)*. The orchestration of the system, by which the rest of the EMCs are controlled, is further described in Section 3.7.
2. *Context Management Layer*: this layer acts as the information backbone of the system, providing several mechanisms and characteristics that enable the interoperability and modularity of the system. The main feature of this layer is that it is based on the European Telecommunications Standards Institute (ETSI) NGSI-LD (Next Generation Service Interfaces for Linked Data) [44] standard: evolution of NGSI, both of them used as the communications standard of the FIWARE [45] project. The CB is the component responsible for making all the information (context) of the KB accessible,

as well as the provider of means to search, access and update that context. Its specific characteristics and details regarding the communication between components, the ontological foundation and the structure of the information model are described. This layer also holds the historic component, responsible for storing selected historical information (used for forecasting and data analysis) and making it available to EMCs. Both the CB and the historic component can be instantiated from the many already available implementations of FIWARE Generic Enablers, promoting re-usability.

3. *Interaction Layer*: this layer comprises all the components responsible for interacting with the devices and entities related to the HEMS. It is the boundary layer of the SHEMS with the rest of the SH and the world, acting as an adaptor between the internal NGSI-LD interfaces and the different external interfaces. It is through this layer that information from devices, the grid, DERs and the HAS, reach the KB. It is also in this layer that the management of the system crystallises in specific commands sent to devices or communications with external services. Finally, all semantic adaptation between the SHEMS and the devices, services and external agents will take place in this layer.
4. *Security & Privacy*: lastly this layer permeates the whole system. It is responsible for ensuring secure data access as well as privacy. It mediates the information exchange between the components and the KB, providing authentication and authorisation components for access control to the CB and other components in the architecture. The components that form the security and privacy layer and their operation are further described in Section 3.6.

3.4. Information Model

Semantic web technologies present a compelling solution to provide interoperability across vendors and providers of different services, platforms and devices. Under the semantic web, information has to be semantically annotated (linked) to ontological concepts. Information enriched with semantic tags is called context in NGSI-LD, the underlying technology on which the KB of this proposal is based.

According to NGSI-LD Information Model [44], context is structured in the form of entities. These entities have identity, type, properties and can be linked to one another through relationships. Entities are exchanged in the form of JSON-LD documents that follow a *Core MetaModel*. Bundled with those properties and relations, NGSI-LD introduces the use of special JSON-LD attributes (represented by the *Cross-Domain Ontology*) linking to semantic concepts from other *Domain-Specific Ontologies*, creating an "onion" model (Figure 5).

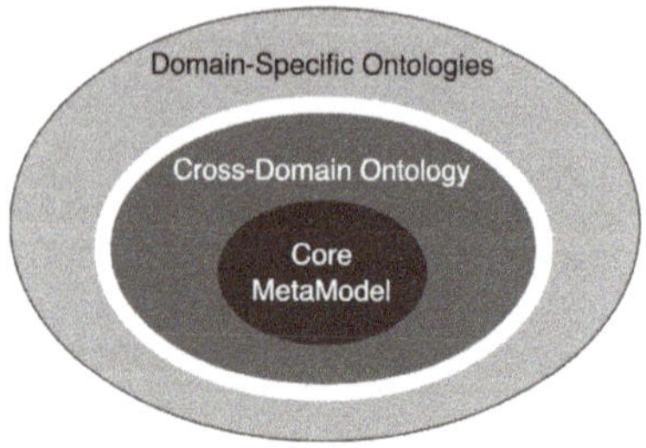

Figure 5. NGSI-LD Information Model.

A wide range of existing ontologies are already available for us to use in our proposal, closely related to the domain of SHEMSs and covering DR and SG interfacing; the most relevant ones have already been introduced in Section 2.2 and summarised in Table 1.

In our proposal, we have selected DABGEO as the main ontology, as it covers all of the most relevant concepts related to home energy management: appliances and devices description and power consumption, grid-related information (from tariffs to prosumer-related concepts), electrical energy generation and accumulation, as well as user related information (user preferences, occupancy and other related concepts). Specific examples of the mapping of DABGEO concepts to NGSI-LD will be provided in Section 4.3.

As already indicated, no ontology covers all possible aspects and scenarios in the domain of SHEMSs. Table 2 represents the subset of ontologies that can complement DABGEO in our proposal and the EMCs for whom they can be relevant (e.g., MAS2TERING could be applied in the HEG in DR scenarios where USEF is being implemented).

Table 2. Complementary ontologies.

Ontology	HEC	HEG	HAS
DNAS [30]	X		X
BonSAI [33]	X	X	X
MIRABEL [32]		X	
MAS2TERING [18]		X	
SARGON [25]		X	
OSEIM [23,24]	X		X

Finally, to the best of our knowledge, no previous mapping from NGSI-LD to the selected ontology has been previously proposed. The approach we have followed in this work has been to map OWL elements of DABGEO to NGSI-LD according to Table 3. In the case of OWL's object properties linking to individuals, the representation as NGSI-LD nested properties can be considered if the following criteria are met: (1) linked individuals are exclusively related to a single entity, (2) their existence depends on it, (3) have low number of properties and/or relationships and (4) will not be used as search keys in the KB.

Table 3. OWL and NGSI-LD mapping.

OWL	NGSI-LD
Individual	Entity
Class	Entity type
Object property	Relationship or nested property
Datatype property	Property

3.5. Context Management

In NGSI-LD, context (entities) can be created, updated, retrieved and deleted through REST API exchanging JSON-LD (JSON for Linking Data [46]) documents. This API also provides mechanisms to subscribe to changes in context, forming a publish/subscribe information management system. The CB is the single central point of the architecture where all information can be accessed.

The CB offers advanced mechanisms to search context information available in the system, offering different filters and query mechanisms to retrieve information. Some of those filters can also be used with the subscription mechanism, allowing components to receive updates on tailored sets of information of their interest.

Finally, the CB can also act as a relay and directory for other providers of information previously registered: in this way, EMCs can act as *Context Providers (CPs)*, responsible for sub-sets of the KB, capable of answering queries from the CB and other peer components (Figure 6). This mechanism is relevant in cases where information is calculated upon request or cases where information changes constantly but is requested with low frequency.

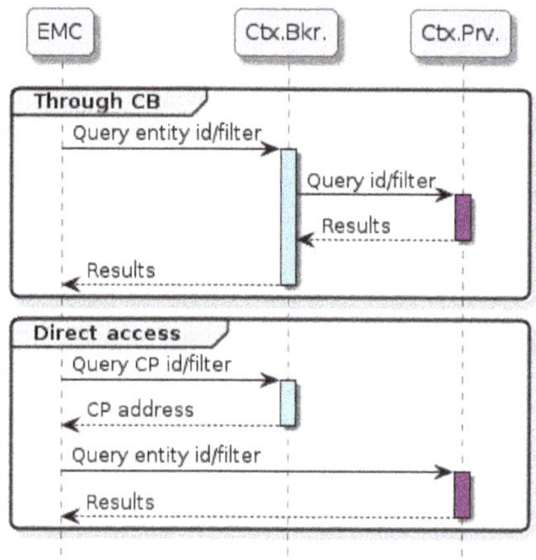

Figure 6. Context Provider information access sequence.

3.6. Security and Privacy

To provide secure and private access to context information, our proposal introduces the use of DCapBAC (Distributed Capability-Based Access Control) [47], a derivation of the Attribute-Based Access Control (ABAC) scheme in which the authorisation checking against the policy base and the enforcement of the actual policy are decoupled (Figure 7).

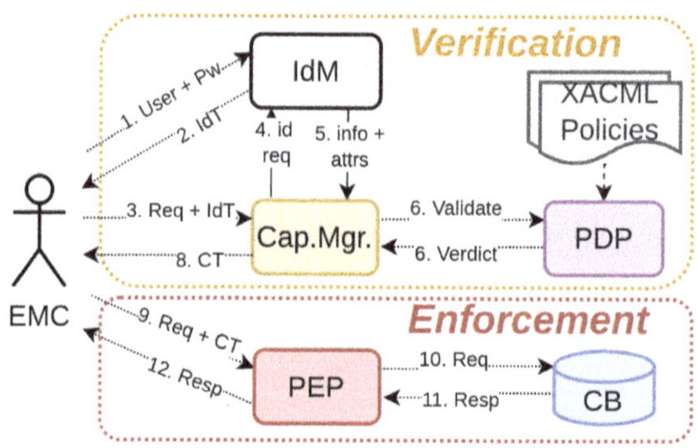

Figure 7. Context authorisation and access sequence.

In the following list, the security components of the architecture are described:

1. *Identity Management (IdM)* component: provides the authentication service to the architecture. It stores identity information together with attributes that will be later used by the authorisation component. In our architecture, we propose the use of standard interfaces, such as OAuth 2 [48] and OpenID Connect [49].
2. *Capability Manager (CM)* component: acts as the authorisation facade for components, granting or denying access to the requested resource. It receives requests from components and interacts with the IdM and the PDP to perform its task.
3. *Policy Decision Point (PDP)* component: validates requests using identity information against the set of XACML (eXtendible Access Control Markup Language [17]) policies stored in the system. Those policies are managed via the next component in the list: the PAP.
4. *Policy Administration Point (PAP)* component: offers a single administration point to manage the policies for the system. Policies are stored in XACML, defined as conditions that a subject must meet to act on a resource. In this case, the conditions

can be based on attributes from the IdM identity, the actions are HTTP verbs and the resources are HTTP resources, represented as URLs.

5. *Policy Enforcement Point (PEP)* component: act as a transparent reverse HTTPS proxy, which enforces the verdict issued by the PDP, without the need for further XACML policy evaluation.

The process by which EMCs access information begins with the authentication process (Figure 8) represents a simplified version of a typical OpenID interaction for authentication of an EMC against the IdM. As a result of this interaction, the component obtains an Identity Token (IdT) which will be used in the next phase.

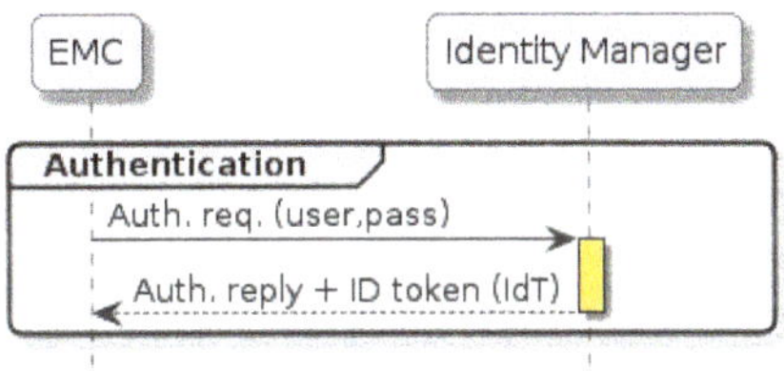

Figure 8. Authentication sequence.

Authorisation in DCapBAC begins with a request to get access to a resource, accompanied by the IdT previously obtained. This request follows the classical XACML structure and interface (Figure 9), in which the EMC's request is matched in the PDP against XACML policies to verify whether the request can be granted or not. The difference is that instead of immediately accessing the resource after a positive verdict, this interaction will result in the issuing of a Capability Token (CT).

The final phase (Figure 10), is the actual access to the context information. In this case, the CB is protected by a transparent PEP. This component will only grant requests that are valid according to the attached CT. Access can take place several times with the same CT for as long as it is valid, effectively reducing the authorisation delay of each request, as there is no further validation against XACML policies on each access.

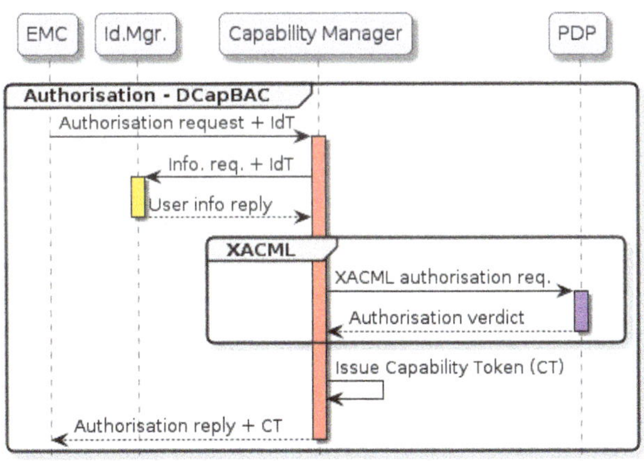

Figure 9. Authorisation sequence.

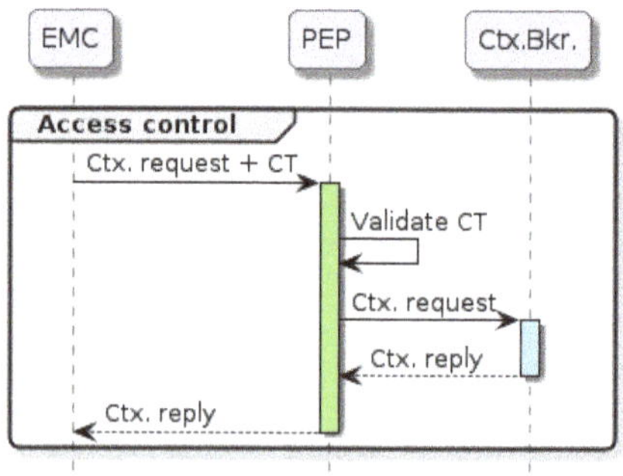

Figure 10. Authorisation enforcement sequence.

Communications between EMCs, security components and CB take place via REST API calls and are secured with standard web technologies, using HTTPS protocol and SSL certificates.

3.7. System Orchestration

EMCs not only share information through the KB but also communicate with each other in an asynchronous message-passing manner, using the publish/subscribe functionality defined in NGSI-LD. We have implemented a mechanism inspired by FogFlow [50,51] task orchestration, in which tasks receive input from other tasks and the orchestrator by using intermediary entities in the KB. Those entities hold input/output information for their tasks.

In our proposal, EMCs subscribe to specific entities by which the HEC sends and updates the desired outcome, schedule or management commands that the EMC requires as input for its operation (Figure 11). That same mechanism is used by the HEC to receive output from the rest of the EMCs.

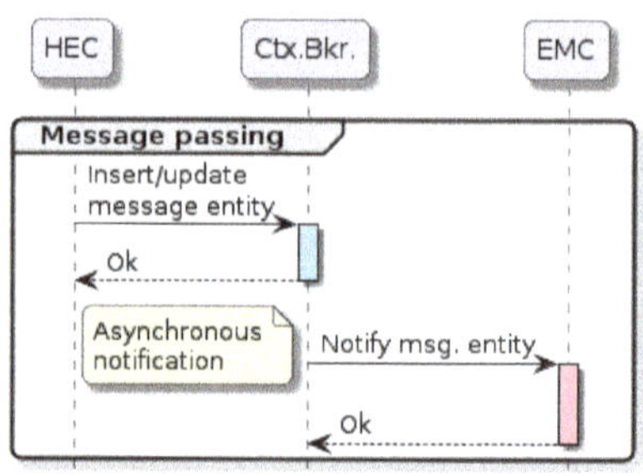

Figure 11. Orchestration asynchronous message passing.

This orchestration mechanism is also secured with the proposed security and privacy components, covered in Section 3.6. This way, XACML policies can be put in place to ensure that only the expected EMC will be able to access its input message entities and update its output ones.

4. Validation

To validate our proposed architecture, an implementation use case is being conducted for the SHEMS of a single home. This home has an already existing HAS installation, electricity production and storage facilities and some high-consumption devices.

Regarding the specific algorithms and methods used in the EMCs implemented, we have purposely omitted implementation details. The two main reasons behind this decision are: (a) the implementation of some of the EMCs is still under development, being bound to change in the near future as new approaches and algorithms are being tested and (b) this proposal's concern is the architecture by which different implementations of the EMCs can cooperate in an interoperable and secure way thus fully describing the algorithms and optimisations would only lead to possible confusion by the reader and an over-extended work.

Never the less, in this section we will offer some implementation details regarding EMCs, as well as specific results that support our goal of improving energy efficiency in our test-bed scenario, together with the specific objectives that guide our SHEMS and the core architecture components selected for our demo, as well as examples of communication between EMCs.

4.1. Test-Bed Description

The test-bed (Table 4) consists of a two-story house at ground level, located in Murcia, in the south-east of Spain, with a patio and an underground garage. A small swimming pool is located on the patio, whose filtration and chlorination systems can be controlled by the SHEMS.

Table 4. Test-bed summary.

Element	Description
Home automation system	Home Assistant, running in a Raspberry Pi 4, 4 GB RAM
Photo voltaic array	14 panels, 480 W rated
Inverter	Ingecon Sun Storage 1Play
Battery	32 cells in series, LiFePo4 chemistry, 280 Ah rated capacity
Battery management system	Batrium WatchMon-CORE, 2 Batrium CellMate K9 and ShuntMon 500 A
Grid power meter	Carlo Gavazzi EM112
Grid tariff	5.4 KW peak input, with constant pricing. PV surplus feed-in possible with constant price rebate
Appliance controller * (Coffee machine)	TP-Link HS110 Smart Plug (with inbuilt power meter), controlled by Home Assistant
Pool controller	Node-RED based, running in a Raspberry Pi zero. Power-meter integrated via Modbus.

* An espresso machine (ECM Synchronika), with 1.6 kW peak power consumption is connected to this smart plug.

The existing HAS is based on the Home Assistant software (Figure 12), running on a Raspberry Pi 4 and is already capable of controlling HVAC, central heating (underfloor heating), window blinds and some lights, smart TVs and the tumble-dryer among other appliances. Different sensors are integrated into the HAS, monitoring temperature and humidity in different rooms. It also provides weather information via external web services, as well as home occupancy status through presence detection of different inhabitants at home.

Figure 12. Test-bed's Home Assistant installation graphical interface.

On the roof, an array of PV panels (Figure 13), connected to an inverter, provides up to 6 kW of electric power. An accumulation system is currently in development, with 28 kWh of planned storage capacity, protected by a Battery Management System (BMS) and directly controlled by the inverter.

Figure 13. Aerial view of the test-bed location. Pool, PV panels, HVAC and heating units visible.

The PV and storage systems provide real-time information on production, consumption and State Of Charge (SOC) through their controllers. Moreover, some high-consumption appliances have dedicated power meters for a more granular energy consumption composition breakdown; such is the case of the pool filtration system and the espresso machine.

Finally, a grid tie provides energy. The contract with the electricity company establishes a constant price tariff with 5.4 kW peak power. It also allows grid feed-in from the PV installation, with a rebate proportional to the injected energy. To monitor import/export energy, a grid-tie power meter has also been installed and integrated.

4.2. Tasks

The general objective of the test-bed SHEMS is to achieve the most efficient and cost-effective operation of the system and to do so it takes the following tasks:

1. To manage battery accumulation parameters (maximum SOC and Depth Of Discharge (DOD), charging and discharging regimes and the likes), schedule and manage battery charging and manage stored energy usage.
2. To schedule and manage the central heating system, as well as HVAC.
3. To schedule and manage swimming pool filtration and chlorination system.
4. To react to inhabitants' actions that offset the expected energy profiles.
5. To inform the user about energy consumption patterns and energy management status, raise alerts and provide energy-related suggestions to improve efficiency and reduce consumption.

To fulfil its tasks, the SHEMS will use and generate information from its different EMCs, such as grid-tie parameters (maximum usable power, current and planned energy pricing and grid inject rebate depending on tariff), current and forecast energy consumption, production and storage, schedules of operation of different devices and even occupancy and weather information.

4.3. Knowledge Base and Security Components

For the KB and security components, we have selected some Generic Enablers from the FIWARE project: the CB implementation of choice is the *Orion-LD* [52] broker and on the IdM, *Keyrock Identity Management Generic Enabler* is being used. For the DCapBAC components, we have selected the open-source implementation of the PAP-PDP, PEP and CM provided by IoTCrawler project [53,54].

Information in the KB is structured in entities (Listing 1), according to NGSI-LD's *Core MetaModel*, and linked to DABGEO [26] ontology (Figure 14).

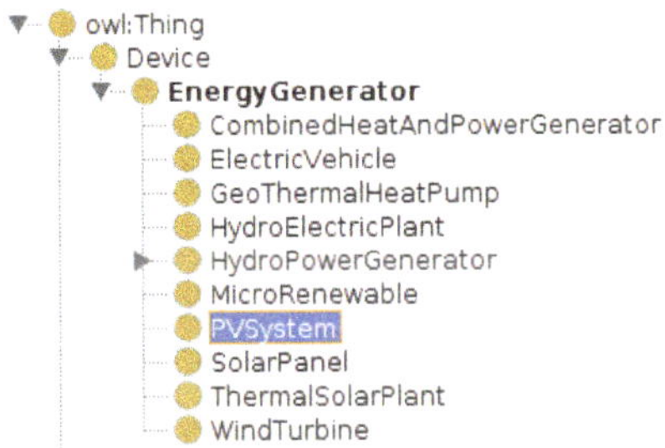

Figure 14. Protegé view of PVSystem in DABGEO.

Listing 1. JSON-LD representation of the photo-voltaic inverter.

```
{
  "id": "urn:ngsi-ld:PVSystem:Device:0042",
  "type": "PVSystem",
  "deviceName": {
    "type": "Property",
    "value": "INGECON SUN STORAGE 1Play TL M"
  },
  "maxProducesEnergy": {
    "type": "Property",
    "value": 6,
    "unitCode": "KW"
  },
  "@context": [
    {
      "deviceName": "http://www.purl.org/oema/enaeq/deviceName",
      "maxProducesEnergy": "https://www.auto.tuwien.ac.at/downloads/
          thinkhome/ontology/EnergyResourceOntology.owl#"
    },
    "https://uri.etsi.org/ngsi-ld/v1/ngsi-ld-core-context.jsonld"
  ]
}
```

Orchestration is performed via asynchronous message passing and requires the subscription of the EMCs to the input entities of their domain. One such example is the control of the swimming pool filtration and chlorination system (Listing 2). In case of exceeding the power constraints of the system (e.g., if the energy imported from the grid is close to, or exceeding, the maximum power defined by the energy provider contract), the HEC will issue a modification to the entity representing the *controllerDesiredStatus* of the filtration and chlorination system, asking the device controller to stop the system as a result of energy constraints. That modification will trigger the appropriate subscription to send a notification to the swimming pool device EMC (Listing 3) and it will stop operation until the HEC clears that status.

Listing 2. Subscription in NGSI-LD.

```
{
  "id": "urn:ngsi-ld:Subscription:PoolController:0001",
  "description": "Pool system dev.ctr. subscription",
  "type": "Subscription",
  "entities": [{
    "type": "PoolController",
    "id": "urn:ngsi-ld:PoolController:ID:0001"
    }],
  "watchedAttributes": ["controllerDesiredStatus"],
  "notification": {
    "attributes": ["controllerDesiredStatus"],
    "format": "normalized",
    "endpoint": {
      "uri": "http://pool:1880/notifications",
      "accept": "application/json"
    }
  }
}
```

Listing 3. Notification in NGSI-LD.

```
{
  "subscriptionId": "urn:ngsi-ld:Subscription:PoolController:0001",
  "data": [{
    "id": "urn:ngsi-ld:PoolController:ID:0001",
    "type": "PoolController",
    "controllerDesiredStatus": {
      "type": "Property",
      "value": "OFF",
        "observedAt": "2022-06-10T10:11:57.000Z"
      }
  }]
}
```

Security in the system begins with the definition of different identities for the different EMCs, that will be used to interact with the DCapBAC authorisation components. A set of XACML policies is set in place, governing which EMC can act over the different entities stored in the KB. In NGSI-LD, the different HTTP verbs are mapped to the create, retrieve, update and delete functionalities, offering good granularity on the different actions that can be taken on entities. The entities affected by the policy can be defined in terms of identifier (literal or pattern) and different query filters based on entity type or other properties.

When an EMC wants to interact with information in the KB, it first needs to authenticate with the IdM, obtaining an IdT (Figure 15). It then obtains a CT from the CM for the action to be performed and the information associated in the KB. Finally, it will perform that interaction against the PEP, attaching the CT in the request, which will grant (or deny) its access. The security process is explained in detail in Section 2.1.

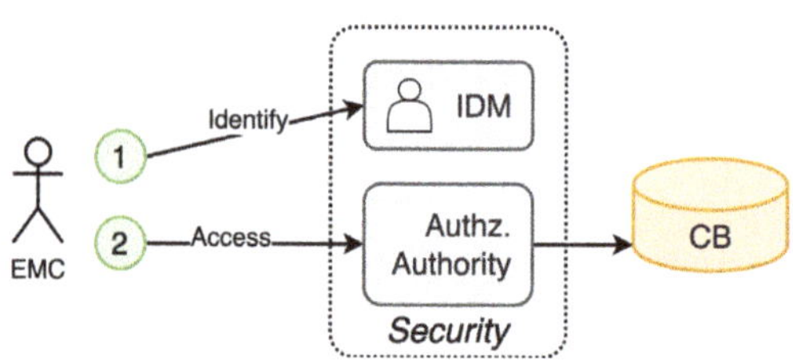

Figure 15. Secure access of Energy Management Components to the Context Broker.

The added benefit of DCapBAC is that further requests will not need to perform the authentication and authorisation phases again and that the access control performed by the PEP doesn't need to interact with the PDP, thus reducing the latency of the enforcement.

4.4. Energy Management Components

The EMCs of our test-bed (Figure 16) are at different development stages. We have used Node-RED [55] as the development tool of choice for its quick turnaround and simplicity. The enumeration that follows this text summarises the details and implementation status of each of the components:

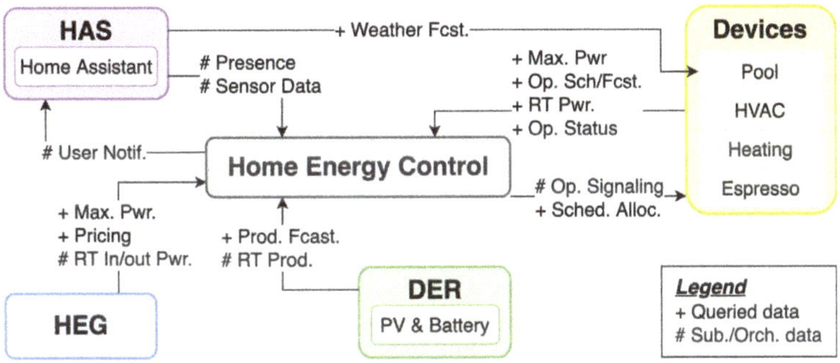

Figure 16. Test-bed Energy Management Components and interactions.

1. *HAS*: it communicates with the Home Assistant instance of the test-bed. It draws information from it regarding weather forecasts, sensor data and home occupation and updates it on the KB for other EMCs to use. It also receives messages from the HEC, to notify users about different conditions. The interaction between the HAS component and Home Assistant takes place via REST API, leveraging Home Assistant's user notification mechanisms (Figure 17).
2. *HVAC and Heating devices*: these devices were already integrated into Home Assistant, utilising custom ESP8266 devices and the ESPHome [56] integration in Home Assistant. These devices offer another REST API to interact with, which we have leveraged due to its simplicity and to avoid using Home Assistant as an intermediary to interact with them. Currently, their components only send status updates to the KB and support stopping temporarily the operation upon request from the HEC (via the orchestration mechanism) to avoid overloading the grid input. In the future, operation scheduling for the heating system could be implemented and integrated into the energy management strategy, as well as operation forecasting for the HVAC to predict when users want to use it.
3. *Espresso machine device*: this device is controlled via a smart plug (see Table 4), allowing users to remotely turn the machine on or off. This device is (also) integrated into Home Assistant [57]. This time we have opted to implement its component interacting with Home Assistant's REST API, instead of interacting directly with the smart plug. The reason behind this decision is merely to save effort by re-using Home Assistant's API. In the current implementation, it only updates real-time energy consumption and status (on/off) in the KB, but in the future, we expect to be able to better integrate it into the SHEMS by implementing other features such as operation forecasting and scheduling.
4. *Pool device*: the pool filtration EMC has been directly built into its controller (which is also based on Node-RED). It updates operation status and real-time power consumption in the KB. It also requests an operation schedule to the HEC for the number of hours of filtration that it deems appropriate, based on weather forecasts retrieved from the KB (which are updated by the HAS component). This compo-

nent also reacts to HEC's inputs (through orchestration mechanism) to temporarily stop/resume operation.

5. *PV and battery DER*: the component responsible for the PV array and battery interacts with the inverter via Modbus-TCP (Figure 18). The current implementation of the controller updates in the KB, real-time production and SOC of the batteries. Once the battery is finally installed, we plan on developing the mechanism to receive from the HEC messages to update its maximum SOC and DOD to optimise battery usage.
6. *HEG*: this component updates in the KB the real-time power input/output as measured in the grid tie power meter, to which it is connected via Modbus. It also updates the maximum power that the grid-tie can deliver as well as the maximum it can be fed from the PV installation. This component also retrieves the hourly prices of electricity, both consumed from the grid and fed-in, from ESIOS [58] via REST API, according to the Spanish PVPC tariff, although the current test-bed electricity contract is a constant-price one.
7. *HEC*: the current implementation of this component is capable of generating schedules for electricity consumption upon request, trying to maximise PV-generated electricity usage. It currently does so by using forecast cloud coverage information (coming from the HAS component) as well as expected sunset and sundown times. It also reacts to notifications regarding high-consumption devices and takes decisions based on the current energy budget (PV, battery and grid) to signal viable devices to temporarily stop operation. Finally, it also notifies users through the HAS component (Figure 17) when: (a) the grid energy import reaches 80%, (b) when high-consumption devices are active and the home is not occupied and (c) when contingency mechanisms (such as temporarily stopping operation of a device) have been implemented.

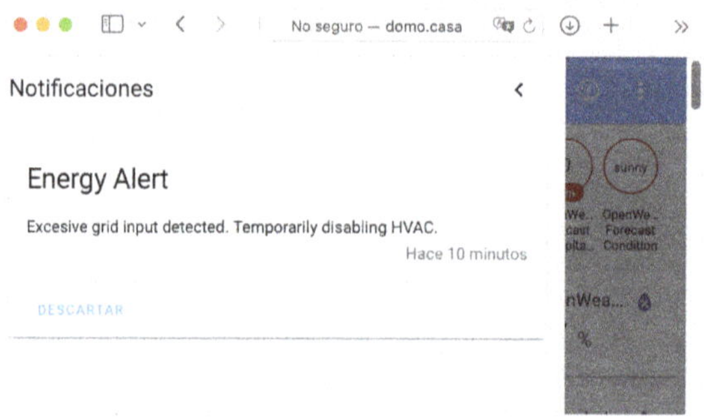

Figure 17. Alerting and notification through Home Assistant.

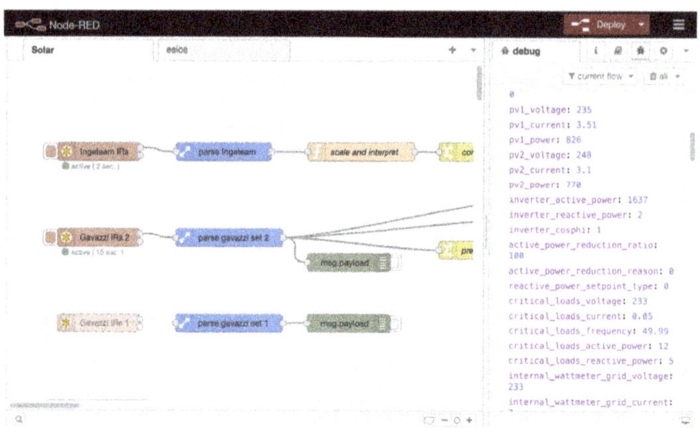

Figure 18. Node-RED development of the DER component for PV.

4.5. Results

Among the various sub-systems of the test bed, we have decided to show results obtained in the filtration and electrolysis chlorination system of the pool, as they present a good balance between simplicity, interaction with other EMCs and improvement in energy cost reduction.

Prior to the SHEMS implementation, this sub-system was controlled with an electric plug-in timer switch. Its programming had to be manually configured several times a year, to adapt to the differences in temperature as well as sun incidence. Figure 19 shows the peak power demand and production, on an hourly basis, for an average spring/summer work day.

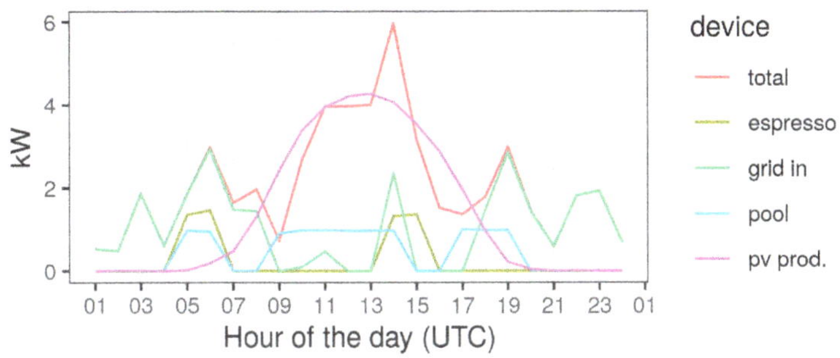

Figure 19. Power peaks prior to Smart Home Energy Management System.

From it, we can deduce the pool filtration and chlorination system schedule, in which we can notice two gaps (6 a.m. to 9 a.m. and 2 p.m. to 5 p.m.). Those gaps correspond to specific periods of the day in which users are specially energy-active at home on a common replacedweek dayweek day: the time of breakfast and lunch. At those times, users utilise high consumption devices such as the espresso machine, the electric stove or the microwave, that have shown in the past the possibility of triggering the grid input power breaker when used in conjunction with the pool filtration system. Thus a conservative approach was taken, avoiding those time ranges. It is worth noticing that, although the second gap coincides with a high PV production period, cloud coverage has occasionally produced dips in production, leading to power breaker trips.

Figure 20 represents the same variables of the previous example and similar conditions (average spring/summer work day in our test-bed home), but this time the pool has a device controller integrated into the SHEMS. The implementation of the pool's EMC decides filtration and chlorination time based on local weather information and forecasts (updated in the KB by the HAS module) and asks for a schedule from the HEC, which provides it based on energy cost, resulting in an allocation during PV production.

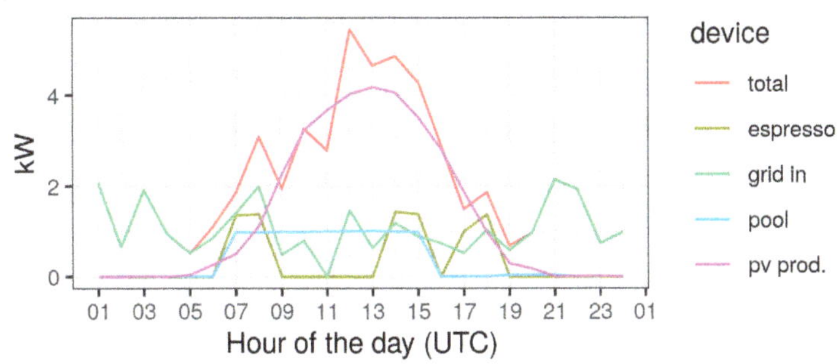

Figure 20. Power peaks with Smart Home Energy Management System.

This schedule would have incurred an increased risk of power breaker trips in the previous scenario, but the SHEMS-integrated pool controller receives commands from the HEC to temporarily stop operation when there is the risk of exceeding maximum grid power constraints, therefore, avoiding the need to schedule out of PV production hours. The HEC sends control commands to the pool filtration system to stop/resume operation, based on the power status of the whole system, as described in the previous Section 4.4. The orchestration of the control commands has also been showcased in Section 4.3 and the messages exchanged in Listings 2 and 3.

Finally, Figure 21 compares the total accumulated energy imported from the grid on an hourly basis for the two previous examples, showing that the conservative strategy followed by the timer implementation could be associated with less efficient use of the PV system by allocating filtration time outside of production hours, in turn producing an increase in grid energy import, compared to the SHEMS control.

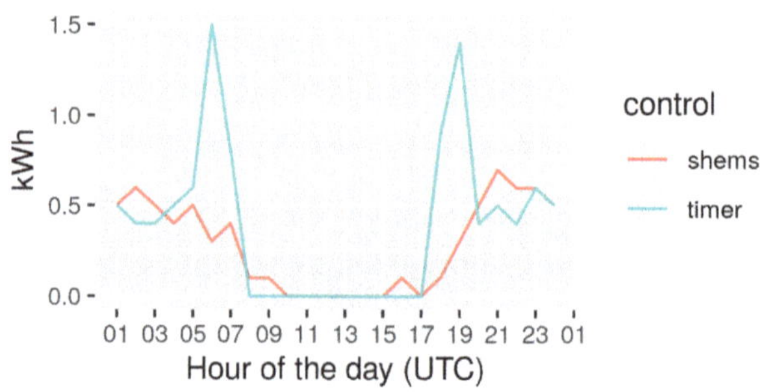

Figure 21. Grid energy import comparison.

5. Discussion

We want to open up the discussion by stating that the results shown in the previous subsection cannot be taken as proof that our test-bed SHEMS succeeds at obtaining better energy efficiency than the previous scenario, nor of the degree to which such benefit could be obtained. They have only been offered to support our claim that the system is capable of successfully integrating the many different actors present in the energy management of a Smart Home to take action depending on information coming from diverse sources.

With our demonstration, we show that we have integrated an existing Home Automation System (Home Assistant), making all of its information available to the rest of the SHEMS and leveraged it as a convenient way to reach users through notifications.

We have created a HEG proof of concept implementation, capable of integrating electricity pricing in our SHEMS, establishing good footing for the integration of future DR implementations. This HEG can act as an intermediary between the SHEMS and the grid (or micro-grids), opening the possibility of securely and privately sharing selected information from the system with the grid, which could be beneficial in DR and Demand Flexibility scenarios.

We have integrated DERs in the form of a PV installation and its attached accumulation battery system, as well as different high consumption devices, with different levels of functionality, using different communication technologies and integrated their information for other components of the SHEMS to use.

Finally, we have showcased an example of a simple energy management implementation capable of scheduling consumption for PV energy optimisation and reacting to high electricity demand by notifying users and disabling non-critical high-consumption devices.

This architecture brings to the field a framework for modular SHEMSs where different EMCs can be built by different parties and where DR, Home Automation, DER, devices and users, have been considered. As an added benefit, the core elements of the architecture can be readily deployed from COTS components, many of them coming from the FIWARE project, such as the security stack as well as the Context Broker.

In our proposal, all the information in the system can be accessed securely and privately way by any of the EMCs of the system. Moreover, this information is accessed using standard communication protocols specifically designed for interoperability. On top of that, information is formatted and structured following Semantic Web principles, leveraging existing ontologies representing all the necessary concepts, achieving data interoperability. Lastly, using the NGSI-LD communications standard, we implemented an orchestration mechanism for energy management, leveraging NGSI-LD's publish/subscribe functionality. These four aspects are the main contribution of our work.

6. Conclusions and Future Work

In this work, a modular, interoperable and secure architecture for building SHEMSs has been proposed, presenting a set of EMCs for the management of the energy of a smart home. The presented architecture also considers prosumer interactions with the grid, local generation and accumulation optimisation, the management of high-consumption devices and the integration with existing HASs, while considering data security and privacy through access-control mechanisms.

The proposal is based on the NGSI-LD standard, which is used both as a semantic KB and asynchronous message passing for orchestration. Using this standard opens the possibility of re-utilising existing implementations of different FIWARE components in our architecture, such as the Context Broker, used for the KB, as well as some security components, like IdM and PEP. Lastly, existing implementations of other projects that lay within the FIWARE ecosystem have also been re-utilised, such as the DCapBAC components from project IoTCrawler.

The ontological foundation of the information model has been established from a selection of existing ontologies, analysed in Section 2.2, from which DABGEO has been selected as the base ontology, together with an array of complementary ontologies for specific use cases.

The proposal has been validated through the presentation of a test-bed scenario consisting of a single prosumer home governed by an existing HAS, which has been integrated into the SHEMS. The central components for the architecture, in charge of context management and security, have been instantiated from existing implementations. Examples of the representation of information and orchestration of different tasks between components have been showcased.

Finally, results in the form of a SHEMS capable of scheduling high-consumption devices for better PV utilisation and reacting to high energy consumption by notifying users and disabling non-critical loads, have demonstrated the feasibility of the solution, successfully being able to schedule the pool chlorination system based on PV production while integrating information coming from the HAS to react to changes in consumption by the home users. This has been possible thanks to the capacity of the system to integrate diverse information sources and its ability to provide secure mechanisms to access information within the different components instantiated.

This work opens the door to future proposals on multi-faceted SHEMSs in which to optimise complex systems controlling accumulation and generation, DR strategies communicating with the SG and, at the same time, leveraging the existing HASs installation to retrieve information from it and even interact with it.

It also opens the door for the development and deployment of specific EMCs that can be readily plugged into any SHEMS following our architecture proposal. One such example could be that of specific HEG implementations by different energy providers that would allow communication between homes and the SG to implement elaborated DR strategies.

Finally, it presents the possibility of creating specific SHEMS frameworks and implementations, ready to be deployed and integrated with other existing solutions in a single-click fashion, that would allow the easy deployment of a SHEMS by layman users, which in turn, would become a pivotal factor for the widespread deployment of advanced DR strategies.

Author Contributions: Conceptualisation, P.G.-G. and A.S.; methodology, J.A.M.; software, P.G.-G.; validation, A.S. and J.A.M.; formal analysis and investigation, P.G.-G.; data curation, P.G.-G.; writing—original draft preparation, P.G.-G.; writing—review and editing, A.S. and J.A.M.; supervision, A.S.; project administration, A.S.; funding acquisition, A.S. All authors have read and agreed to the published version of the manuscript.

Funding: This work was supported by Grant PID2020-112675RB-C44 funded by MCIN/AEI/10.13039/501100011033.

Data Availability Statement: Not applicable.

Conflicts of Interest: The authors declare no conflict of interest. The funders had no role in the design of the study; in the collection, analyses, or interpretation of data; in the writing of the manuscript; or in the decision to publish the results.

Abbreviations

The following abbreviations are used in this manuscript:

CB	Context Broker
CM	Capability Manager
CT	Capability Token
DER	Distributed Energy Resource
DOD	Depth Of Disch
DR	Demand Response
EMC	Energy Management Component
HAS	Home Automation System
HEC	Home Energy Controller
HEG	Home Energy y Gateway
HEMS	Home Energy Management System
HVAC	Home Ventilation and Air Conditioning
IdM	Identity Management
IdT	Identity Token
IoT	Internet of Things
KB	Knowledge Base
OWL	Web Ontology Language
PAP	Policy Administration Point
PDP	Policy Decision Point
PEP	Policy Enforcement Point
PV	Photo-Voltaic
SG	Smart Grid
SH	Smart Home
SHEMS	Smart Home Energy Management System
SOC	State Of Charge

References

1. Almeida, J.; Soares, J. Integration of electric vehicles in local energy markets. In *Local Electricity Markets*; Elsevier: Amsterdam, The Netherlands, 2021; pp. 21–36. [CrossRef]
2. Gonçalves, I.; Gomes, Á.; Henggeler Antunes, C. Optimizing the management of smart home energy resources under different power cost scenarios. *Appl. Energy* **2019**, *242*, 351–363. [CrossRef]
3. Yang, Q.; Ehsan, A.; Jiang, L.; Fang, X. Optimal energy dispatch in residential community with renewable DGs and storage in the presence of real-time pricing. In *Smart Power Distribution Systems: Control, Communication, and Optimization*; Elsevier: Amsterdam, The Netherlands, 2018; pp. 447–465. [CrossRef]
4. IEA. *Electricity Market Report*; Technical Report; IEA: Paris, France, 2022.
5. Home Assistant Website. Available online: https://www.home-assistant.io (accessed on 22 July 2022).
6. Domoticz Website. Available online: https://domoticz.com (accessed on 22 July 2022).
7. OpenHab Website. Available online: https://www.openhab.org (accessed on 22 July 2022).

8. Singh, P.; Masud, M.; Hossain, M.S.; Kaur, A. Blockchain and homomorphic encryption-based privacy-preserving data aggregation model in smart grid. *Comput. Electr. Eng.* **2021**, *93*, 107209. [CrossRef]
9. Milchram, C.; van de Kaa, G.; Doorn, N.; Künneke, R. Moral Values as Factors for Social Acceptance of Smart Grid Technologies. *Sustainability* **2018**, *10*, 2703. [CrossRef]
10. Badar, A.Q.H.; Anvari-Moghaddam, A. Smart home energy management system—A review. *Adv. Build. Energy Res.* **2022**, *16*, 118–143. [CrossRef]
11. Lashkari, B.; Chen, Y.; Musilek, P. Energy management for smart homes-state of the art. *Appl. Sci.* **2019**, *9*, 3459. [CrossRef]
12. Mahapatra, B.; Nayyar, A. Home energy management system (HEMS): Concept, architecture, infrastructure, challenges and energy management schemes. *Energy Syst.* **2019**, *13*, 643–669. [CrossRef]
13. Cuenca, J.; Larrinaga, F.; Eciolaza, L.; Curry, E. Towards cognitive cities in the energy domain. In *Studies in Systems, Decision and Control*; Springer International Publishing: Berlin/Heidelberg, Germany, 2019; Volume 176, pp. 155–183. [CrossRef]
14. Machorro-Cano, I.; Alor-Hernández, G.; Paredes-Valverde, M.A.; Rodríguez-Mazahua, L.; Sánchez-Cervantes, J.L.; Olmedo-Aguirre, J.O. HEMS-IoT: A big data and machine learning-based smart home system for energy saving. *Energies* **2020**, *13*, 1097. [CrossRef]
15. Elshaafi, H.; Vinyals, M.; Grimaldi, I.; Davy, S. Secure Automated Home Energy Management in Multi-Agent Smart Grid Architecture. *Technol. Econ. Smart Grids Sustain. Energy* **2018**, 3, 4. [CrossRef]
16. Open Services Gateway Initiative Website. Available online: https://www.osgi.org (accessed on 22 July 2022).
17. Xtendible AccessControl Markup Language Website. Available online: http://docs.oasis-open.org/xacml/3.0/xacml-3.0-core-spec-os-en.html (accessed on 22 July 2022).
18. Hippolyte, J.L.; Howell, S.; Yuce, B.; Mourshed, M.; Sleiman, H.A.; Vinyals, M.; Vanhee, L. Ontology-based demand-side flexibility management in smart grids using a multi-agent system. In Proceedings of the IEEE 2nd International Smart Cities Conference: Improving the Citizens Quality of Life, ISC2 2016, Trento, Italy, 12–15 September 2016; pp. 1–7. [CrossRef]
19. Zhang, J.; Li, Q.; Schooler, E.M. IHEMS: An information-centric approach to secure home energy management. In Proceedings of the 2012 IEEE 3rd International Conference on Smart Grid Communications, SmartGridComm 2012, Tainan, Taiwa, 5–8 November 2012; pp. 217–222. [CrossRef]
20. Shah, N.; Chao, K.M.; Zlamaniec, T.; Matei, A. Ontology for home energy management domain. In *Communications in Computer and Information Science*; Springer: Berlin/Heidelberg, Germany, 2011; Volume 167, pp. 337–347. [CrossRef]
21. Rossello-Busquet, A.; Soler, J.; Dittmann, L. A novel home energy management system architecture. In Proceedings of the 2011 UKSim 13th International Conference on Modelling and Simulation, UKSim 2011, Cambridge, UK, 30 March–1 April 2011; pp. 387–392. [CrossRef]
22. Reinisch, C.; Kofler, M.J.; Iglesias, F.; Kastner, W. Thinkhome energy efficiency in future smart homes. *Eurasip J. Embed. Syst.* **2011**, *2011*, 104617. [CrossRef]
23. Saba, D.; Sahli, Y.; Hadidi, A. An ontology based energy management for smart home. *Sustain. Comput. Inform. Syst.* **2021**, *31*, 100591. [CrossRef]
24. Saba, D.; Sahli, Y.; Abanda, F.H.; Maouedj, R.; Tidjar, B. Development of new ontological solution for an energy intelligent management in Adrar city. *Sustain. Comput. Inform. Syst.* **2019**, *21*, 189–203. [CrossRef]
25. Haghgoo, M.; Sychev, I.; Monti, A.; Fitzek, F.H. SARGON—Smart energy domain ontology. *IET Smart Cities* **2020**, *2*, 191–198. [CrossRef]
26. Cuenca, J.; Larrinaga, F.; Curry, E. DABGEO: A reusable and usable global energy ontology for the energy domain. *J. Web Semant.* **2020**, *61–62*, 100550. [CrossRef]
27. Burel, G.; Piccolo, L.S.G.; Alani, H. EnergyUse—A Collective Semantic Platform for Monitoring and Discussing Energy Consumption. In *Lecture Notes in Computer Science (Including Subseries Lecture Notes in Artificial Intelligence and Lecture Notes in Bioinformatics)*; Springer International Publishing: Berlin/Heidelberg, Germany, 2016; Volume 9982, pp. 257–272. [CrossRef]
28. Daniele, L.; den Hartog, F.; Roes, J. Created in Close Interaction with the Industry: The Smart Appliances REFerence (SAREF) Ontology. In *Lecture Notes in Business Information Processing*; Springer International Publishing: Cham, Switzerland, 2015; Volume 225, pp. 100–112. [CrossRef]
29. Daniele, L.; Solanki, M.; Den Hartog, F.; Roes, J. Interoperability for smart appliances in the IoT world. In *Lecture Notes in Computer Science (Including Subseries Lecture Notes in Artificial Intelligence and Lecture Notes in Bioinformatics)*; Springer International Publishing: Cham, Switzerland, 2016; Volume 9982, pp. 21–29. [CrossRef]
30. Hong, T.; D'Oca, S.; Turner, W.J.N.; Taylor-Lange, S.C. An ontology to represent energy-related occupant behavior in buildings. Part I: Introduction to the DNAs framework. *Build. Environ.* **2015**, *92*, 764–777. [CrossRef]
31. Gillani, S.; Laforest, F.; Picard, G. A generic ontology for prosumer-oriented smart grid. In Proceedings of the CEUR Workshop Proceedings, Rome, Italy, 24–25 November 2014; Volume 1133, pp. 134–139.
32. Verhoosel, J.; Rothengatter, D.; Rumph, F.J.; Konsman, M. An ontology for modeling flexibility in smart grid energy management. In *eWork and eBusiness in Architecture, Engineering and Construction, Proceedings of the European Conference on Product and Process Modelling 2012, ECPPM 2012, Reykjavik, Iceland, 25–27 July 2012*; CRC Press: Boca Raton, FL, USA, 2012; pp. 931–938. [CrossRef]
33. Stavropoulos, T.G.; Vrakas, D.; Vlachava, D.; Bassiliades, N. BOnSAI: A smart building ontology for ambient intelligence. In Proceedings of the WIMS'12: 2nd International Conference on Web Intelligence, Mining and Semantics, Craiova, Romania, 13–15 June 2012; ACM Press: New York, NY, USA, 2012; pp. 1–12. [CrossRef]

34. Kofler, M.J.; Reinisch, C.; Kastner, W. A semantic representation of energy-related information in future smart homes. *Energy Build.* **2012**, *47*, 169–179. [CrossRef]
35. Smart Applications Reference Website. Available online: https://saref.etsi.org (accessed on 22 July 2022).
36. Cuenca, J.; Larrinaga, F.; Curry, E. A unified semantic ontology for energy management applications. In Proceedings of the CEUR Workshop Proceedings, Cagliari, Italy, 11–12 September 2017; Volume 1936, pp. 86–97.
37. Bonino, D.; Corno, F.; De Russis, L. Poweront: An ontology-based approach for power consumption estimation in smart homes. In *Lecture Notes of the Institute for Computer Sciences, Social-Informatics and Telecommunications Engineering, LNICST*; Giaffreda, R., Vieriu, R.L., Pasher, E., Bendersky, G., Jara, A.J., Rodrigues, J.J., Dekel, E., Mandler, B., Eds.; Springer International Publishing: Cham, Switzerland, 2015; Volume 150, pp. 3–8. [CrossRef]
38. MAS2TERING Project Website. Available online: http://www.mas2tering.eu (accessed on 22 July 2022).
39. Universal Smart Energy Framework Website. Available online: https://www.usef.energy (accessed on 22 July 2022).
40. Energy@Home Website. Available online: http://www.energy-home.it/SitePages/Home.aspx (accessed on 22 July 2022).
41. International Electrotechnical Commission's Common Information Model. Available online: https://webstore.iec.ch/publication/74467 (accessed on 22 July 2022).
42. Mirabel Ontology. Available online: https://sites.google.com/site/smartappliancesproject/ontologies/mirabel-ontology (accessed on 22 July 2022).
43. BOnSAI Ontology. Available online: http://lpis.csd.auth.gr/ontologies/ontolist.html#bonsai (accessed on 22 July 2022).
44. Cim, E.; Management, C.I. NGSI-LD Information Model. Available online: https://www.etsi.org/deliver/etsi_gs/CIM/001_099/009/01.01.01_60/gs_CIM009v010101p.pdf (accessed on 22 July 2022).
45. FIWARE Website. Available online: https://www.fiware.org (accessed on 22 July 2022).
46. JSON for Linking Data website. Available online: https://JSON-LD.org (accessed on 22 July 2022).
47. Truong, H.; Hernández-Ramos, J.L.; Martinez, J.A.; Bernal Bernabe, J.; Li, W.; Marin Frutos, A.; Skarmeta, A. Enabling Decentralized and Auditable Access Control for IoT through Blockchain and Smart Contracts. *Secur. Commun. Netw.* **2022**, *2022*, 1–14. [CrossRef]
48. OAuth 2.0 website. Available online: https://oauth.net/2/. (accessed on 22 July 2022).
49. OpenID Connect Website. Available online: https://openid.net/connect/ (accessed on 22 July 2022).
50. Cheng, B.; Solmaz, G.; Cirillo, F.; Kovacs, E.; Terasawa, K.; Kitazawa, A. FogFlow: Easy Programming of IoT Services Over Cloud and Edges for Smart Cities. *IEEE Internet Things J.* **2018**, *5*, 696–707. [CrossRef]
51. FogFlow Website. Available online: https://fogflow.readthedocs.io (accessed on 22 July 2022).
52. Orion-LD - Linked Data Context Broker Website. Available online: https://fiware-academy.readthedocs.io/en/latest/core/orion-ld/index.html (accessed on 22 July 2022).
53. IoTCrawler GitHub Repositories. Available online: https://github.com/orgs/IoTCrawler/repositories (accessed on 22 July 2022).
54. IoTCrawler Project Website. Available online: https://iotcrawler.eu/ (accessed on 22 July 2022).
55. Node-RED Website. Available online: https://nodered.org/ (accessed on 22 July 2022).
56. ESPHome Website. Available online: https://esphome.io (accessed on 22 July 2022).
57. Home Assistant's TP-Link Integration Website. Available online: https://www.home-assistant.io/integrations/tplink/ (accessed on 22 July 2022).
58. Sistema de Información del Operador del Sistema (ESIOS)—Red Eléctrica de España Website. Available online: https://www.esios.ree.es/es/pvpc (accessed on 22 July 2022).

smart cities

MDPI

Article

SDS: Scrumptious Dataflow Strategy for IoT Devices in Heterogeneous Network Environment

Zeeshan Rasheed [1], Shahzad Ashraf [2,*], Naeem Ahmed Ibupoto [1], Pinial Khan Butt [3] and Emad Hussen Sadiq [4]

1 Department of Computer Science, Mir Chakar Khan Rind University, Sibi 82000, Pakistan
2 College of Internet of Things Engineering, Hohai University, Changzhou 213031, China
3 Information Technology Center, SAU Tandojam, Hyderabad 70060, Pakistan
4 Department of Energy Engineering, Duhok Polytechnic University, Duhok 1006, Iraq
* Correspondence: nfc.iet@hotmail.com

Abstract: Communication technologies have drastically increased the number of wireless networks. Heterogeneous networks have now become an indispensable fact while designing the new networks and the way the data packet moves from device to device opens new challenges for transmitting the packet speedily, with maximum throughput and by consuming only confined energy. Therefore, the present study intends to provide a shrewd communication link among all IoT devices that becomes part of numerous heterogeneous networks. The scrumptious dataflow strategy (SDS) for IoT devices in the heterogeneous network environment is proposed and it would deal with all link selection and dataflow challenges. The SDS would accomplish the targeted output in five steps: Step 1 determines the utility rate of each heterogeneous link. Step 2 develops a link selection attribute (LSA) that gauges the loads of network features used for the link selection process. Step 3 calculates the scores of all heterogeneous networks. Step 4 takes the LSA table and computes the network preference for different scenarios, such as round trip time (RTTP), network throughput, and energy consumption. Step 5 sets the priority of heterogeneous networks based on the scores of network attributes. Performance of the proposed SDS mechanism with state of the art network protocols, such as high-speed packet access (HSPA), content-centric networking (CCN), and dynamic source routing (DSR), was determined by conducting a simulation with NS2 and, consequently, the SDS exhibited its shrewd performance. During comparative analysis, in terms of round trip time, the SDS proved that it utilized only 16.4 milliseconds to reach IoT device 50 and was first among all other protocols. Similarly, for network throughput, at IoT device 50, the throughputs of the SDS are recorded at 40% while the rest of other protocols were dead. Finally, while computing the energy consumption used to reach IoT device 50, the SDS was functional and possessed more than half of its energy compared to the other protocols. The SDS only utilized 302 joules while the rest of the protocols were about to die as they had consumed all of their energy.

Keywords: wireless communication; scrumptious; heterogeneous network; IoT device; routing

Citation: Rasheed, Z.; Ashraf, S.; Ibupoto, N.A.; Butt, P.K.; Sadiq, E.H. SDS: Scrumptious Dataflow Strategy for IoT Devices in Heterogeneous Network Environment. *Smart Cities* **2022**, *5*, 1115–1128. https://doi.org/10.3390/smartcities5030056

Academic Editor: Pierluigi Siano

Received: 13 June 2022
Accepted: 2 September 2022
Published: 5 September 2022

1. Introduction

The deployment process of traditional wireless cellular networks is inherited from past scenarios and requires time-to-time upgradation. The state of the art cellular systems are basically encompassed in base stations and user terminals adopting the same standards as followed by the cellular system in other regions. Currently, wireless networks are keystones, with several diverse application fields, such as wireless sensor networks, cloud facilities, cyber physical systems [1], infrastructures protection, and command control, with several other likely examples. The wireless workstation networking has taken the place of former customary technologies to provide better services with configurability, flexibility, and interoperability [2].

Contemporary network technologies are comprised of software and hardware components. The accessibility of faster and reliable hardware is altering the equilibrium between software and hardware with generating changes in the network of structured devices. Currently, several software-built nodes based on the Python language coexist with hardware-built nodes with the provision of many analogous functions in conjunction with different computing tools [3].

Heterogeneous networks [4] contain numerous existing radio access network (RAN) technologies, such as WiMAX, Wi-Fi, EUTRAN, etc., and have various architectures, different transmission mechanisms, and base stations with a volatile performance range. Configuration networks are used to improve the user experience and reduce RAN and core network (CN) bottlenecks. HetNet [5] can also help to route and manage intelligent IP traffic, and implement efficient load balancing and resource allocation. This is the aggregation of heterogeneous network radio resources and the traffic between selective or packet-switched or circuit-switched HetNet. 3GWLAN has been considered beyond other inter-technology options. Heterogeneous networks include interconnected nodes and different types of links. Such interconnected structures contain a wealth of information that can be used to mutually strengthen nodes and links and transfer knowledge from one type to another.

The heterogeneous wireless sensor network demands an abundant network resource. It attempts to deploy a replication connectivity mechanism in a fast mobile architecture. The real measurement output exhibits that the replication connection mechanism drastically reduces network uncertainty, controls the packet loss ratio, and proactively improves the network throughput. A small network fidelity significantly influences the outcomes of assessments when the network complexity is not inconsequential, so the obtainability of integrated simulation-centered tools to maintain the whole network course is an actual need to evade the underestimation and equivocation of network glitches [6].

Modular simulation is very useful in building an articulate network model with varying levels of applications. The accessibility of modular, programmable, extensible, open-source, community-driven, and community-supported simulation frameworks produces simulation events with desirable outcomes even in a heterogeneous network system using varying simulated nodes [7].

1.1. Smart Homes

Smart homes are getting popular due to two factors. First, sensing and actuation techniques as well as wireless sensor networks, have dramatically advanced. Second, nowadays, individuals rely on technology to answer their worries about their quality of life and home security. Intelligent and automated services are provided via a range of IoT-based sensors in smart homes, which help individuals who forget to automate daily duties and maintain routines and can save energy by automatically shutting off lights and electrical gadgets. Motion sensors are employed for this purpose and security can also be achieved. Sensors collect data from the environment by conserving energy (light, temperature, humidity, gas, fire events).

The data from the heterogeneous sensor is given to the context aggregator, which then transmits it to the context recognition service engine. This engine chooses services depending on their context. When the humidity rises, for example, an application can automatically switch on the air conditioner. If there is a gas leak, you can also switch off all of the lights. Smart home applications are highly beneficial to the elderly and the disabled. Health can be monitored and professionals can be notified immediately in the event of an emergency. The floor is outfitted with pressure sensors, which aid in tracking a person's activity in a smart home and detecting falls. CCTV cameras may be used in smart homes to record interesting occurrences. There are countless challenges and questions about smart home applications [8]. Security and privacy are of paramount importance, as all data about what is happening at home is recorded [9]. An intruder can attack the system and cause it to behave maliciously if its security and dependability are not ensured. When such abnormalities are noticed, smart home systems are meant to warn the owner.

In continuation to smart homes, there are IoT-based applications that govern with smart homes. The role of the most relevant applications to the proposed systems are discussed as follows.

1.1.1. IoT-Based Transport

Sensors and cognitive information processing systems can be used by IoT-based transportation apps to govern everyday traffic in the city. The primary aims of intelligent traffic systems are to reduce traffic congestion, make parking easier and stress-free, appropriately route traffic, and eliminate accidents by identifying intoxicated drivers. GPS sensors for position information, accelerometers for speed, gyroscopes for direction, RFID for vehicle identification, and infrared for counting passengers and cars are examples of IoT devices with sensor technologies for these sorts of applications, as well as sensors and cameras for documenting traffic and vehicle movements.

1.1.2. IoT-Based Water Systems

The current level of water shortages in most regions of the world urges critics to effectively manage the water supplies. As a result, most cities are opting for smart solutions that include the installation of a large number of meters on water supply pipes and storm drains. Smart water meters come in a variety of styles. These meters may be used to determine the amount of water entry and outflow as well as potential leaks. Water metering systems based on IoT are also employed in combination with data from meteorological satellites and river water sensors. They can also assist us in forecasting flooding.

1.1.3. IoT-Based Social Meetings

"Opportunistic IoT" [10] refers to information exchange between opportunistic devices (devices that seek communication with other devices) depending on mobility and availability of contacts in the neighborhood. Personal gadgets with sensing and short-range communication capabilities include tablets, wearables, and mobile phones. When there is a shared goal, people may find and interact with one another.

1.1.4. IoT-Based Supply Chain Management

IoT seeks to simplify the actual processes of business and information systems [11]. One can easily trace items in a supply chain from the point of manufacturing to the point of final distribution by using sensor technologies, such as RFID and NFC. Real-time data is recorded and processed for future reference. RFID tags connected to cargo can also record information regarding product quality and ease of use.

The proposed mechanism (SDS) achieves the required goals in five steps:

- Step 1: The service is separated into patterns and the attributes of each pattern are examined before using the utility function to determine the utility value for each network feature.
- Step 2: Network attribute weights are calculated using the link selection attribute (LSA). Based on this, signal inference is completed.
- Step 3: The network attribute score is calculated using the network attribute utility and weights.
- Step 4: Network settings for different scenarios are calculated using the LSA.
- Step 5: Based on the evaluation of network attributes, unpleasant networks are prioritized. This allows the user to select the network with the highest score.

The key contributions of this work are as follows:

- The link selection attribute (LSA) specifies the network selection criteria that match the predefined values from the data corpus. This selection strategy ensures that only the best network is selected. This mechanism has been explained in Section 3.2 with the help of Figure 3.
- The performance of IoT devices in a heterogeneous network was analyzed by calculating results in terms of round trip time, network throughput, and energy consumption.

- The results were obtained by simulation with the NS2 simulator.
- The proposed strategy allows users to choose the most appropriate network, improve interoperability between devices, and reduce unnecessary handovers between different networks.
- Finally, the results are compared to state of the art protocols, such as high-speed packet access (HSPA), content-centric networking (CCN), and dynamic source routing (DSR).

The rest of this paper is divided into further sections. The current literature on heterogeneous networks is placed in Section 2. Section 3 contains a description of the proposed model, which accomplishes the target output in five steps. Performance analysis is described in Section 4. The conclusions and future research directions are given in Section 5.

2. Literature Review

Traditional network resource selection approach usually chooses the network that is the best performer among all available networks. However, with diverse services, each service demands rigid features. Furthermore, various users have diverse preferences. As a result, the goal of the study introduced in this white paper was to create an access selection algorithm that takes network, service, and user demographics into consideration.

Considering the low-speed moving environment, Fung po et al. [12] investigated the performance of high-speed packet access (HSPA). Not only static scenarios were taken into account, but many mobile scenarios, including subways, trains, and city buses were also considered. However, limiting the deployment of commercial networks, it analyzed only 3G networks, and all measurements primarily looked at network and transport layer parameters. Therefore, the network access layer has great priority but SNR was fully ignored during the entire transmission.

Similarly, another work from Mahfuzur [13] developed a content-centric networking (CCN) mechanism in the 4G/5G network, where various heterogeneous networks are converged. They also offered a unique mobility management method to enable content and network variety by using the mobile network's rich computing resources. Rather than establishing a communication link to the information source, they promised to enable more efficient, quicker, and secure content delivery. Furthermore, they examined existing mobility options and assessed the efficacy of a seamless content delivery mechanism in terms of content transfer time, throughput, and the data transmission success ratio. Their suggested solution uses name-based routing rather than content or device addresses. For content transfer, this system primarily employs two basic messages: the Interest packet and the Data packet. The Interest packet provides a request for a requested material, which includes information such as content name, content type, and content version. The Data packet comprises the original data as well as the content name, security information, and numerous additional properties, such as hop distance and content source description. However, they disregarded some of the issues, such as excessive energy usage and data packet delay rationing, which significantly reduce network longevity.

Qin et al. [14] proposed the reactive DSR source routing protocol for cognitive radio ad hoc networks to send IoT data from the IoT gateway to non-constrained networks inside the cognitive radio ad hoc networks. DSR, in particular, falls to the reactive routing protocol group since it may find routes from source to destination only when requested and needed. DSR is a source-routing system that allows for on-demand routing. DSR broadcasts routes to its neighbors but does not overload them with data. It only follows routes by calculating total distance or counting the number of nodes between the source and destination nodes. Nodes in the DSR mechanism keep route cache information that contains the path sequence from the source. Route maintenance and route discovery are the two processes used in DSR.

The authors [15] claimed to improve end-to-end throughput while minimizing the latency. The gateway distributed routers also act as IoT gateway nodes, gathering and encapsulating the data into the distributed cognitive radio ad hoc network. For the IoT network to be simulated in the Cooja simulator, an assumption was made to get IoT data from the LLN node to the LNN gateway node (LBR), which also functions as a cognitive source node. When a packet reaches the CR source node, it is wrapped in IP-in-IP and delivered via the cognitive radio network simulator. Channel route identification and restoration delays via local or global channel route recovery approaches have an impact on the end-to-end cumulative network throughput inside CRAHNs. As a result, the chance of PU spectrum handoff is higher than in the absence of active PU transmitters. As a consequence, the performance of source routing with various PU transmitter nodes was being assessed in order to compute the aggregate network throughput of IoT data inside the ad hoc network. The performance of the cognitive source routing protocol was compared to the existing hybrid cognitive AODV routing protocols, licensed control channel-based AODV routing protocols, unlicensed AODV-based routing protocols, and traditional IEEE 802.11 DCF-based routing techniques. Overall, this method is only suitable for sparse networks and is impractical for dense networks; no alternate measure for dense settings was provided.

Z. Yang et al. [16] employed a directional antenna to enhance the amount of concurrent noninterfering broadcasts within the cognitive radio network. This raises the possible end-to-end throughput in multihop communication while simultaneously lowering node power consumption. In other words, by minimizing interference with directional antennas, directional cognitive control and IoT application transmission would help in obtaining greater end-to-end throughput. The authors omitted to compute power consumption, which seems to be a major flaw in this work.

Ashraf et al. [17] suggested a lower power listening (LPL) technique to monitor malfunctioning nodes and energy waste in a wireless network using ContikiMAC Cooja. In both the centralized and distributed models, energy usage is lowered. By presenting a stochastic model for wireless sensor networks, the author calculated energy consumption with end-to-end latency. The suggested model, however, incorporates cylindrical propagation but lacks common spherical propagation.

2.1. Advantage of Heterogeneous Network

It has been noted that network performance is poor in high-speed mobile scenarios, unable to fulfil user requests for network resource access. The elements influencing user network performance are examined layer by layer, and the benefits of heterogeneous networks are analyzed.

2.1.1. Transport Layer

Each user is familiar with the shifting patterns of TCP throughput at various speeds. In every case, the performance of TCP throughput [18] appears to be the weakest, even the average of TCP throughput is the lowest, and the volatility of TCP throughput is the most dramatic. Given the statistics from the cumulative distribution function (CDF) [19], it is challenging to meet user demand for network resource access over a single wireless network. Taking use of diverse networks may be able to meet the user's need for network access.

2.1.2. Network Layer

It was determined from the data travelling throughout the transport layer that the network performance of a single wireless network is poor. However, with heterogeneous networks, the overall network performance has a lot of opportunity for improvement. The benefits of heterogeneous networks with a network layer have been studied because when speed of the movement grows, the packets take longer to transport and are even lost. This might mean trouble for applications that are extremely sensitive to packet delays, such as real-time gaming. Using diverse networks may help to decrease transmission delays [20].

3. Proposed SDS Model

The propagation of wireless heterogeneous networks is a challenge that researchers have related to networking, to develop unique methodologies and techniques that can attain affordable, reproducible, and credible results for the experimentation and development of many network designs, and for the smooth flow of data in heterogeneous networks.

The proposed model developed a trust engine for the smooth data flow in heterogeneous networks for the provision of flexible and creative research experimentation by using an advanced simulation technique.

The proposed SDS layout is illustrated in Figure 1, with four heterogenous network environments and access points. The main router has been fixed while other dedicated routers are associated with each heterogeneous network and each is linked with a radio link.

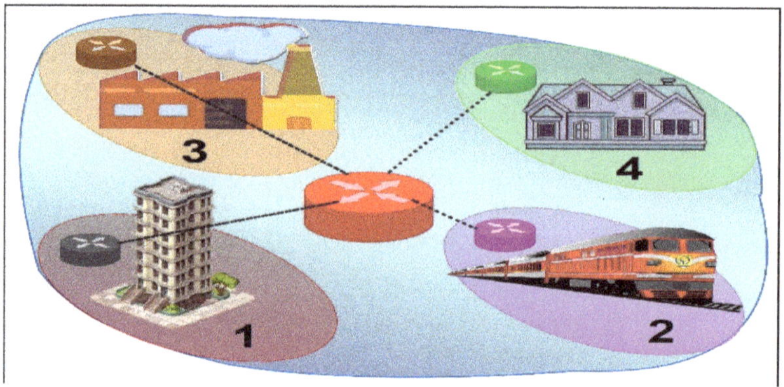

Figure 1. SDS heterogeneous network with router configuration.

Each heterogeneous network, such as Network 1, covers the dense residential population where a number of IoT devices are interconnected. Similarly, Networks 2, 3, and 4 represent the railway transport, industrial environment, and sparse residential area, respectively. Each heterogeneous network has different dynamics and challenges.

3.1. Selection of Prudent Network

The interconnectivity and relative performance of IoT-enabled devices have been examined using network simulator (NS2). The volatile nature of wireless network scan easily be analyzed when discrete data are feed. It possesses extensive libraries and a variety of communication protocols. The results are a hallmark for future investigation.

Initially, three hosts and two routers (one main router and another related to a particular heterogeneous network) were configured after creating the devices and linking routers and hosts in the network to perform basic tests. Four networks (10.0.100.0/24, 10.0.200.0/24, 10.300.0/24) and (10.0.1.0/24) were configured with dynamic OSPF routing protocol [1] running for the network, linking two routers to transfer network information from one router to a subsequent one.

As the execution begins, the IoT-enabled wireless devices initiate the packet broadcast mechanism illustrated in Figure 2, where a number of IoT devices broadcast the data packets in the heterogeneous environment.

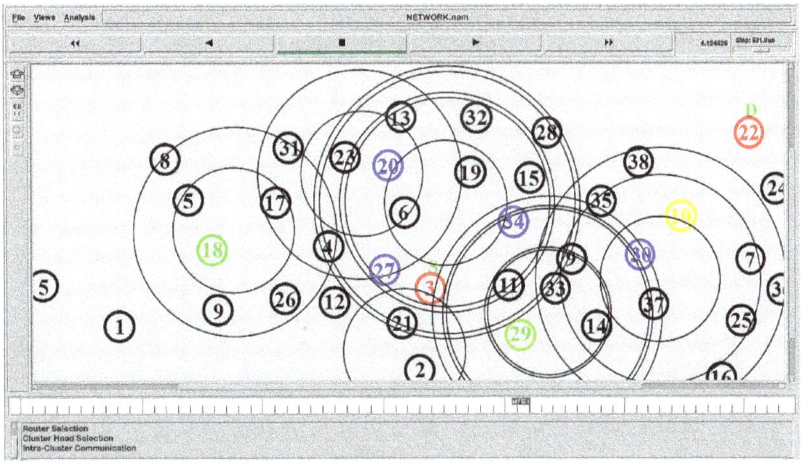

Figure 2. SDS heterogeneous network simulation in process.

The proposed scenario is designed to appropriately represent the deviating values of response metrics from their corresponding performance metric bounds in order to conserve network capacity and increase transmission ability. The values of the performance metrics are entered into the system to maximize and decrease the performance measure. Furthermore, the deviating value represents the balance of the trade-off between device demand and routing efficiency; both values are critical in balancing the weight of communication expenses. Using communication potential, the suggested SDS approach computes the optimal fitness value. The communication potential is associated with each device, which affects the behavior of particles to find the overall dataflow output value of a single parameter at a time. These values represent the new parameters upon which the performance of each heterogeneous network is analyzed in terms of round trip time (RTTP), network throughput, and energy consumption.

3.2. IoT Device to Device Link Selection Mechanism

The routing path between the IoT devices are represented by $d\Delta$, which, in fact, foretells the best quality path but not a legitimate scrumptious link. Therefore, a predefined link selection attribute (*LSA*) is considered and all records are maintained in a data table corpus. Equation (1) shows the entire process of the link selection mechanism, where four varieties of links are determined. Sometimes, it happens that the best quality link is achieved but it does not belong to a targeted destination; therefore, such links are not considered as legitimate links.

$$LSA = \begin{cases} Scrumptious\ link, \\ \quad Average\ link, \\ \quad Fair\ link, \\ \quad Uncouth\ link, \end{cases} \qquad LSA_{scrumptious} < d\Delta \tag{1}$$

Consider *pt* as an absolute data packet sent from the source device and *ps* to be the destination device's successfully accepted packet. The *pti* represents the data packet sent from device node i, and would determine the connection quality through calculating the total *LSA* and signal-to-noise ratio (*SNR*) of *pti* in relation to the accessible networks. Equation (2) may be used to compute the link factor estimator (*LFE*) parameter.

$$lfe = D(i_{pt}, S) - D(j_{pt}, S) \tag{2}$$

The $D(i_{pt}, S)$ and $D(j_{pt}, S)$ are taken from the Euclidean distance formula, i.e., from source device i to the destination device j. S represents the source region where nodes are found as $i = (i1, i2, i3, \ldots, in)$. The packet transmission from the i sensor node moves toward the destination region denoted by D and, therefore, respected packets are *pt_i1*, *pt_i2*, *pt_i3*.

The *LFE* parameters are calculated between the nodes (i) belonging to the source (S) region and the nodes (j) located at the destination (D) region, as illustrated in Figure 3.

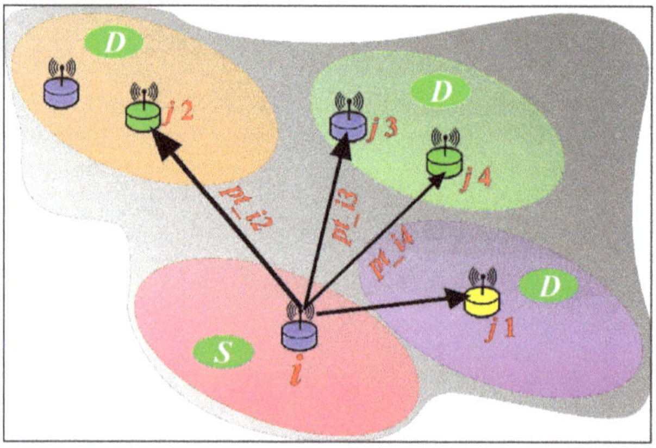

Figure 3. IoT-enabled device-to-device communication link selection mechanism.

The *pti* packet is created by the source device i, while *pt_i2*, *pt_i3*, and *pt_i4* represent the packet formed from source device i towards destination devices as a result of the displacement impact of nearby network overlapping. When the link quality reaches an acceptable level, the fixed device factor and link threshold parameter are verified according to the criteria given in Table 1.

Table 1. LSA link corpus.

Metric Type	SNR	*LFE*	Scalability
Scrumptious link	>32	>110	>150
Average link	18–32	100–110	90–150
Fair link	10–18	50–100	30–90
Uncouth link	0–10	0–50	0–30

The *SNR* computes the signal-to-noise ratio by combining the loudness of the received signal and the background noise. Obtaining the *LFE* mean, the *SNR* strongly suggests that higher *LFE* and *SNR* threshold parameters result in a delicious connection. The selection of a delectable link between the source device and the next device was tested by executing a thorough test paradigm and, therefore, Algorithm 1 ratifies the subject finding's conclusion.

Description of Algorithm 1.

The packet (packet*pt*) created by the source device (device*i*) is distributed over the full transmission zone I via the first communication connection (*pt_i1*), which then expands towards the remainder of the next device inside the transmission zone as (*pt_i2*) and (*pt_i3*) and continues as seen in Figure 3. Furthermore, the link factor estimator determines delectable links by employing an extensive link testing technique and yielding astute outcomes (line 5–23). It considers when the value of the link factor estimator of the transmitted packet from sensor node device i becomes greater or equal to the entire displaced route between source node i and the destination node j. This entire segment remains shorter to the link factor estimator of the transmitted packet by sensor node device i having the same parameters as that of the scrumptious link, where SNR and LFE is calculated between 32 to 100 and the overall condition is given as ($LFE.\ tc_{pti} \geq d\Delta$ and $<$ ($LFE.\ tc_{pti} =$ Scrumptious link)). Then, this link is considered to be an "Average link" stated as ($LFE.\ tc_{pti} =$ average link) on line #12. After establishing a strong link between the source and the next device, it advances to the next device, which requires a steady communication link with constant transmission power. Using an absolute transmitted packet (*pt*) and an acknowledgement packet (*ps*), the link factor estimator (*LFE*) and signal-to-noise ratio (*SNR*) are calculated (line 28–35). As a result, the link selection attribute corpus table is updated with the devices' current condition.

Algorithm 1 Link consistency estimation and packet forwarding mechanism

1: **Procedure** LinkFactorEstimator {(*LFE.tc*), device***i***, packet***pt***}//Link consistency estimation
2: ***F*(*i*) = {*pt_i1*, *pt_i2* . . . . *Pt_i*$_n$}**
3: ***Device i*** transmits packet ***pt*** over distance $d\Delta$
4: **Switch** $LFE.\ tc \leftarrow$ Types
5: **Case** 1: *Scrumptious_Link*
6: **if** $LFE.\ tc_{pti} < d\Delta$ **then**
7: $LFE.\ tc_{pti} =$ scrumptious link
8: **endif**
9: **EndCase**
10: **Case 2**: *Average_Link*
11: **if** $LFE.\ tc_{pti} \geq d\Delta$ & $<$ ($LFE.\ tc_{pti} =$ Scrumptious link) **then**
12: $LFE.\ tc_{pti} =$ average link
13: **endif**
14: **EndCase**
15: **Case 3**: *Fair_Link*
16: **if** $LFE.\ tc_{pti} \geq d\Delta$ & $<$ ($ALQ.\ tc_{pti} =$ average link) **then**
17: $LFE.\ tc_{pti} =$ fair link
18: **endif**
19: **EndCase**
20: **Case 4**: *Uncouth_Link*
21: **if** $d\Delta <$ $LFE.\ tc_{pti}$ **then**
22: $LFE.\ tc_{pti} =$ uncouth link
23: **endif**
24: **EndCase**
25: **end Procedure**
26:
27: **Procedure PacketTransmission**(device***i***, packet***pt***, d_{irj}, *lfe*, *snr*, (*TransmittingDevicei2,i3*), *lct*}//Devices transmit the packets
28: ***pt*** ←absolute transmitted packet
29: ***ps*** ←packet received and acknowledge by destination
30: **if** LinkFactorEstimator(*lfe*) = $LFE.\ tc_{pti}$ **then**
31: **goto** line 2
32: Debuts: *F*(*i*) = φ
33: for ($i_{2,3}$) = 1: Max
34: Max = T_{ip} //transmission impulses
35: **if** Signal-to-NoiseRatio (*snr*) = $LFE.\ tc_{pti}$ **then**
36: **goto** line 2
37: **endif**
38: **End**for
39: **Endif**
40: Compute $N\left(d_{irj}, f\right) \geq D_0$
41: update LFETable//Link Corpus Table
42: *F* (*i*) = *F* (*i*) +{$pt_{2,3}$}
43: **end Procedure**

3.3. Packet Transmission between IoT Devices

Usually, various applications are used to avail the IoT devices for packet movement. In fact, it builds a device-level heterogenous environment. The device *i* receives numerous data packets from other devices, such as *i*1 to *i*2. The colored lines between all devices illustrate the various applications. This scenario shows that even at the device level, the applications also communicate in a heterogenous environment.

4. Performance Analysis

After the completion of the simulation process, the output results were collected according to the given scenario, and simulation setting parameters are shown in Table 2. These parameters maintain the stability of the execution process, and the fixed values, for instance, number of networks, the specified area distance among the devices, transmission communication range, number of devices for each network, and the transmission rounds, etc., are indeed in real-time perception. The execution process directly depends on these parameters, which might change the output of the results. For instance, transmission starting energy has been fixed to 7 Joules for the entire round. If at some stage this energy level changes, it would definitely affect the results and could cause dysfunction. The NS2 simulator is the best approach for wireless communication when the real-time data transmission would be required in an efficient manager. The proposed SDS model was compared to three state of the art protocols: high-speed packet access (HSPA), content-centric networking (CCN), and dynamic source routing (DSR). The output performance is analyzed on the basis of round trip time (RTTP), network throughput, and energy consumption.

Table 2. Simulation setting values.

Parameter	Value
Number of networks	4
Area	500 × 500 m^3
Distance devices	10 m
Number of devices in each network	[10–30]
Communication range	500 m
Type of protocol	OSPF
Start energy	100 J
Medium	Wireless
Bandwidth capacity	100 Kbps
Packet generation rate	0.03 pkts/min
Energy consumption	2 W; 1.75 W; 8 mW
Data packet volume	64 bytes
Data packet interval (Hello)	99 s
Packet creation time	15 s
No. of runs	50

4.1. Computing Round Trip Time

The RTTP represents the amount of time it takes for a source device to send a request to the destination device and to receive acknowledgment. Usually, this technique is used to determine the health of a network connection. Analyzing the results shown in Figure 4, the overall response time of HSPA, CCN, DSR, and the proposed SDS protocols can be observed more scrumptiously. It can be seen that the network devices are responding quickly and remain active. If these devices take longer, in contrast to the standard time, it means that there are some anomalies that hinder the packet movement as well as the network lifespan.

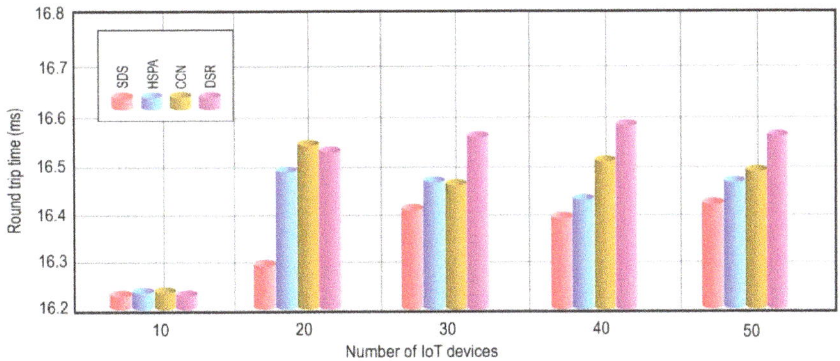

Figure 4. Round trip time computation.

When compared to the rest of the protocols, the SDS takes substantially less time. In reality, our suggested technique uses LSA to identify the best link between the source and the following device. The chosen communication link carried the most energy, but the identical devices coupled with other protocols carried far less energy and were judged inoperable. It produces tangle-free routing and seamless network functioning.

It can be observed that as the number of devices increased, i.e., to 20, the RTTP for CCN somehow became greater, but when approaching device 30, it decreased. This happened due to some unavoidable changes in the routing port, which sometimes blocks the traffic but soon releases. Although HSPA worked better than the other protocols in certain ways, since devices with modest distances were picked more frequently, the data reveal a fast delay when reaching device 50. For DSR, transmission appeared to be more crucial, reducing network lifetime and requiring additional resources to alter response time. It is worth noting that device responsiveness appears to be lower in CCN than in others. The source device continues to deliver data packets until the energy level falls below a certain threshold, putting a heavy pressure on the server. The SDS, on the other hand, changes the momentum during packet transmission and responds quickly. From device 10 to 50, the RTTP time remained less than all other protocols.

4.2. Network Throughput

Throughput is the rate at which a packet or information is successfully transmitted over a network and recognized by the destination device. Network throughput and the packet dissemination ratio both assess network strength, and throughput is directly proportional to the packet dissemination ratio in general. When the network became denser and the number of devices increased, the extravagant communication load immediately impacted the SDS's performance metrics. In this case, the LSA's judicious link selection considerably influences the throughput ratio. The result shown in Figure 5 vouches for the tremendous achievement made by the proposed SDS as compared to the HSPA, CCN, and DSR. At IoT device 10, the throughput of the SDS reached more than 75%, followed closely by HSPA. When SDS reaches device 50, the amazing results can be illustrated by its throughput maintaining high energy levels while the rest of the protocols are about to lose energy.

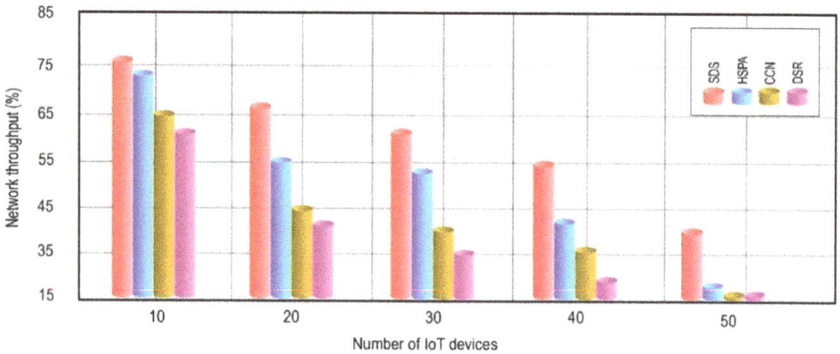

Figure 5. Proposed SDS network throughput.

4.3. *Energy Consumption*

The overall energy consumption by all IoT devices during packet transmission is known as system energy consumption. As the SDS establishes the communication link by considering the predefined values available in the LSA corpus table, only the prudent links are chosen, and the rest of the links are ignored, consequently, it prevents energy wastage and energy is only utilized for the prudent link. In Figure 6, it can be seen that the SDS has only utilized a confined energy level when the link was selected by the IoT device 10 whereas HSPA, CCN, and DSR utilized exorbitant energy. Similarly, at IoT device 50, the SDS still has a substantial energy level while the rest of the protocols' energy is almost empty. Considering the above discussion, it can be concluded that the proposed SDS mechanism has extraordinary performance as compared to the rest the protocols (HSPA, CCN, and DSR). This only became possible due to adopting a prudent LSA technique.

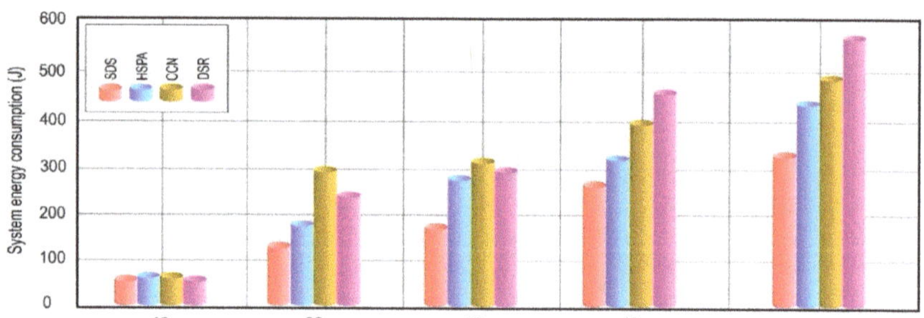

Figure 6. Overall system energy consumption.

5. Conclusions

This study was focused on streamlining the better dataflow mechanism by establishing the prudent communication link among IoT devices for a community based wireless network. The proposed scrumptious dataflow strategy (SDS) has achieved this by developing an LSA data corpus that possessed the pre-defined link parameters. The results were obtained through an NS2 simulator. The output performance of this system has vouched for the statement that was made in the methodology section about the performance. The results were obtained on the basis of a round trip time (RTTP), network throughput, and system energy consumption, and were compared with the results of HSPA, CCN, and DSR protocols. The comparison proved that the SDS performed much better than the rest of the protocols. To compute round trip time, the SDS utilized only 16.4 milliseconds to reach IoT device 50, and was first to do so. Similarly, for network throughput, at IoT device 50, the throughputs are recorded at 40% while the rest of the other protocols died. Finally, when the energy consumption used to reach IoT device 50 was computed, the proposed SDS was functional and possessed more than half of its energy compared to other protocols. The SDS only utilized 302 joules while the rest of the protocols were about to die as they had consumed all of their energy.

Author Contributions: S.A. implemented the idea and designed images and tables. Z.R. solved the topology design problem and wrote the algorithm; N.A.I. accepted the entire module. Comprehensive proofreading has been performed by P.K.B. and E.H.S. All authors have read and agreed to the published version of the manuscript.

Funding: This research received no external funding.

Informed Consent Statement: Informed consent was obtained from all subjects involved in the study.

Data Availability Statement: The associated data are available within the manuscript.

Conflicts of Interest: The authors declare no conflict of interest.

References

1. Ashraf, S. Culminate Coverage for Sensor Network through Bodacious-Instance Mechanism. *Manag. J. Wirel. Commun. Netw.* **2019**, *8*, 9.
2. Sethi, P.; Sarangi, S.R. Internet of Things: Architectures, Protocols, and Applications. *J. Electr. Comput. Eng.* **2017**, *2017*, 1–25. [CrossRef]
3. Guo, X.; Omar, M.H.; Zaini, K.M. Multiattribute Access Selection Algorithm Supporting Service Characteristics and User Preferences in Heterogeneous Wireless Networks. *Wirel. Commun. Mob. Comput.* **2020**, *2020*, 1–27. [CrossRef]
4. Ashraf, S.; Ahmed, T.; Aslam, Z.; Muhammad, D.; Yahya, A.; Shuaeeb, M. Depuration based Efficient Coverage Mechanism for Wireless Sensor Network. *J. Electr. Comput. Eng. Innov. JECEI* **2020**, *8*, 145–160. [CrossRef]
5. Ashraf, S.; Ahmed, T.; Raza, A.; Naeem, H. Design of Shrewd Underwater Routing Synergy Using Porous Energy Shells. *Smart Cities* **2020**, *3*, 74–92. [CrossRef]
6. Begum, S.; Nianmin, Y.; Shah, S.B.H.; Abdollahi, A.; Khan, I.U.; Nawaf, L. Source Routing for Distributed Big Data-Based Cognitive Internet of Things (CIoT). *Wirel. Commun. Mob. Comput.* **2021**, *2021*, 1–10. [CrossRef]
7. Qadeer, W.; Rosing, T.S.; Ankcorn, J.; Krishnan, V.; De Micheli, G. Heterogeneous wireless network management. In Proceedings of the Third International Conference on Power—Aware Computer Systems, San Diego, CA, USA, 1 December 2003; Springer: Berlin/Heidelberg, Germany, 2003; pp. 86–100. [CrossRef]
8. Subekti, A.; Perdana, D.; Muldina, N.R.; Arifl̃ndra, I. Dynamic Source Routing and optimized Link State Routing Performance in Multipath Fading Environment with Dynamic Network Topology. In Proceedings of the 2019 4th International Conference on Information Technology, Information Systems and Electrical Engineering (ICITISEE), Yogyakarta, Indonesia, 20–21 November 2019; pp. 373–378. [CrossRef]
9. A Design of Home Subscriber Server for IP Multimedia Service in All-IP UMTS Network | IEEE Conference Publication | IEEE Xplore. Available online: https://ieeexplore.ieee.org/document/1259079 (accessed on 28 May 2022).
10. Rahman, M.H.; Sejan, M.A.S.; Chung, W.-Y. Multilateration Approach for Wide Range Visible Light Indoor Positioning System Using Mobile CMOS Image Sensor. *Appl. Sci.* **2021**, *11*, 7308. [CrossRef]
11. Network Access Selection Algorithm Based on Balanced Profits between Users and Network. Available online: https://www.hindawi.com/journals/wcmc/2019/6981657/ (accessed on 28 May 2022).
12. Tso, F.P.; Teng, J.; Jia, W.; Xuan, D. Mobility: A Double-Edged Sword for HSPA Networks: A Large-Scale Test on Hong Kong Mobile HSPA Networks. *IEEE Trans. Parallel Distrib. Syst.* **2012**, *23*, 1895–1907. [CrossRef]
13. Bosunia, M.R.; Jeong, S.-H. Efficient Content Delivery for Mobile Communications in Converged Networks. *Wirel. Commun. Mob. Comput.* **2019**, *2019*, 1–12. [CrossRef]
14. Qin, L.; Kunz, T. Pro-active route maintenance in DSR. *ACM SIGMOBILE Mob. Comput. Commun. Rev.* **2002**, *6*, 79–89. [CrossRef]
15. Arfeen, Z.A.; Sheikh, U.U.; Azam, M.K.; Hassan, R.; Shehzad, H.M.F.; Ashraf, S.; Abdullah, P.; Aziz, L. A comprehensive review of modern trends in optimization techniques applied to hybrid microgrid systems. *Concurr. Comput. Pract. Exp.* **2021**, *33*, e6165. [CrossRef]
16. Zhang, Y.; Zheng, T.; Dong, P.; Luo, H.; Pang, Z. Comprehensive Analysis on Heterogeneous Wireless Network in High-Speed Scenarios. *Wirel. Commun. Mob. Comput.* **2018**, *2018*, 4259510. [CrossRef]
17. Ashraf, S.; Gao, M.; Chen, Z.; Kamran, S.; Raza, Z. Efficient Node Monitoring Mechanism in WSN using Contikimac Protocol. *Int. J. Adv. Comput. Sci. Appl.* **2017**, *8*. [CrossRef]
18. Saleem, S.; Ashraf, S.; Basit, M.K. CMBA—A Candid Multi-Purpose Biometric Approach. *ICTACT J. Image Video Process.* **2020**, *11*, 6. [CrossRef]

19. Ashraf, S.; Saleem, S.; Ahmed, T.; Aslam, Z.; Muhammad, D. Conversion of adverse data corpus to shrewd output using sampling metrics. *Vis. Comput. Ind. Biomed. Art* **2020**, *3*, 19. [CrossRef] [PubMed]
20. Ashraf, S.; Saleem, S.; Ahmed, T.; Arfeen, Z.A. Succulent link selection strategy for underwater sensor network. *Int. J. Comput. Sci. Math.* **2022**, *15*, 224. [CrossRef]

sensors

MDPI

Article

Internet-of-Things Based Hardware-in-the-Loop Framework for Model-Predictive-Control of Smart Building Ventilation

Abdelhak Kharbouch [1,2,*], Anass Berouine [1,3], Hamza Elkhoukhi [1,2], Soukayna Berrabah [1], Mohamed Bakhouya [1,*], Driss El Ouadghiri [2] and Jaafar Gaber [4]

1 LERMA Lab, College of Engineering, The International University of Rabat, Technopolis Rabat-Shore Rocade Rabat-Salé, Sala El Jadida 11100, Morocco
2 I&A Laboratory, Faculty of Science, Moulay Ismail University of Meknès, B.P. 11201 Zitoune, Meknès 50070, Morocco
3 ENSIAS, Mohamed V University, Rabat 10713, Morocco
4 Univ. Bourgogne Franche-Comte, UTBM, FEMTO-ST UMR CNRS 6174, 25000 Belfort, France
* Correspondence: abdelhak.kharbouch@uir.ac.ma (A.K.); mohamed.bakhouya@uir.ac.ma (M.B.)

Abstract: In this work, a Hardware-In-the-Loop (HIL) framework is introduced for the implementation and the assessment of predictive control approaches in smart buildings. The framework combines recent Internet of Things (IoT) and big data platforms together with machine-learning algorithms and MATLAB-based Model Predictive Control (MPC) programs in order to enable HIL simulations. As a case study, the MPC algorithm was deployed for control of a standalone ventilation system (VS). The objective is to maintain the indoor Carbon Dioxide (CO_2) concentration at the standard comfort range while enhancing energy efficiency in the building. The proposed framework has been tested and deployed in a real-case scenario of the EEBLab test site. The MPC controller has been implemented on MATLAB/Simulink and deployed in a Raspberry Pi (RPi) hardware. Contextual data are collected using the deployed IoT/big data platform and injected into the MPC and LSTM machine learning models. Occupants' numbers were first forecasted and then sent to the MPC to predict the optimal ventilation flow rates. The performance of the MPC control over the HIL framework has been assessed and compared to an ON/OFF strategy. Results show the usefulness of the proposed approach and its effectiveness in reducing energy consumption by approximately 16%, while maintaining good indoor air quality.

Keywords: Internet of Things; model predictive control; hardware in the loop; machine learning; energy efficiency; smart buildings

Citation: Kharbouch, A.; Berouine, A.; Elkhoukhi, H.; Berrabah, S.; Bakhouya, M.; El Ouadghiri, D.; Gaber, J. Internet-of-Things Based Hardware-in-the-Loop Framework for Model-Predictive-Control of Smart Building Ventilation. *Sensors* **2022**, *22*, 7978. https://doi.org/10.3390/s22207978

Academic Editors: Antonio Cano-Ortega and Francisco Sánchez-Sutil

Received: 29 September 2022
Accepted: 13 October 2022
Published: 19 October 2022

1. Introduction

Heating, ventilation, and air-conditioning (HVAC) systems are considered among the main building's energy consumers. They account for approximately 50% of the global energy usage in buildings and 36% of all energy-related CO_2 emissions worldwide [1,2]. Therefore, HVAC systems need to be efficiently designed and controlled, in reference to international standards, to ensure optimal trade-off between the occupants' comfort and energy efficiency in buildings [3,4]. On the other hand, to assess the energy performance in the design of HVAC management services in buildings, four main comfort metrics need to be considered, which are the visual comfort, acoustic comfort, thermal comfort, and the Indoor Air Quality (IAQ) [5]. This latter has been identified as one of the most important metrics influencing the indoor environmental comfort of the occupants as well as one of the main sources of energy consumption in buildings [4,6], which depends mainly on standalone ventilation management systems.

On the other hand, the indoor concentration of CO_2 is considered among the most important parameters for developing efficient control strategies of VSs [7]. The aim is to minimize their electrical energy consumption, while providing good IAQ to the occupants.

The objective is to keep the CO_2 concentrations within the comfort range by providing the required fresh air from outside to the inside of the building using optimal ventilation flow rates. Basically, the majority of conventional building's VSs are operated by simple rules-based controllers (e.g., intuitive ON/OFF controllers or simple PID controllers). Most of them are based on predefined operating parameters, which use normal ventilation rates, to provide the amount of outside air demanded by the building. However, their control mechanism is still inefficient regarding the performance decrease in frequent system context changes as well as dealing with the time-delay [8]. Typically, the ventilators acts automatically on behalf of many buildings' context-awareness parameters, such as the indoor temperature (based on the envelope characteristics), control modes of the VSs, and occupants' presence [9]. This can affect the energy operation flexibility, indoor environmental comfort, and occupants' productivity due to uncontrollable ventilation rates, resulting in wasted energy [10].

Recent studies highlighted that weather conditions and occupants' behavior are the most important information that can help to improve a building's services (e.g., VSs) [7]. Occupancy detection systems in buildings are mostly involved in extracting meaningful occupancy information, which could be used for setting up different control strategies. Different occupancy parameters can be collected from the building's environment including occupants' presence, number, activity, identity, location, and tracks. All of these metrics can be integrated in Building Energy Management Systems (BEMS) as a primary input for controlling active/passive systems, such as HVAC, standalone ventilation, and lighting [11–14]. Most recent research work investigated the development of intelligent methods by integrating machine learning, deep learning, and reinforcement learning [15,16]. In addition, advanced techniques from automation, system modeling and optimization, Internet of Things (IoT) for real-time systems monitoring, data processing and context-aware computing techniques could be combined for the development of BEMS [17].

As is commonly known, new innovative designs of equipment and system components need to be tested while going through extensive essays [18]. The aim is to validate and properly ensure their reliability before deploying them in real-sitting scenarios. The tests can either be run in a laboratory (small scale), using only pure simulations, or by combining both ways, resulting in HIL simulations [19]. Unlike conventional simulations, concrete testing in laboratories may be seen as the most accurate and is a sure indicator of performance. However, it has some limitations. First, it can generate high costs and is subject to many constraints. For instance, the number of tests that can be run over a period of time and under the same conditions are very limited. Hence, comparing different control approaches or different products providing the same function becomes challenging. On the other hand, numerical simulation is another used method among engineers and researchers to properly evaluate the performances of control methods of many applications, which can be deployed in buildings or other sectors. Numerical models could capture the dynamic of the equipment and the building while considering real weather conditions (if available), the combined internal loads (gains, lighting, occupancy, etc.), and other stimuli. Furthermore, once the numerical model is mature enough, it can be used repetitively to evaluate equipment's control at lower costs. For this, the model should prove its fidelity and accuracy in mimicking the real system's behavior and the related building. As can be noticed, a variety of validated models and toolkits are available for a variety of domains using different simulation tools [20]. It is worth mentioning that, during the recent decades, co-simulation capabilities expanded the modeling scope further to other domain systems at a very precise resolution [21]. However, some problems cannot be tackled easily through simulations, especially if numerical models cannot capture all necessary details [22].

In parallel, recent advances in IoT and big data technologies allow for real time data monitoring and processing, while enabling predictive analytics and advanced systems' control. In fact, IoT is considered the most important emerging technology, allowing for the development of advanced and smart connected solutions varying from eHealth, industry and transportation to energy management and smart control [23–31]. Any system or device having the capability to connect to a network and communicate over the Internet is considered a thing in an IoT infrastructure [32]. This latter provides required tools to manage, control, monitor, visualize and process the things' data (e.g., embedded devices, smartphones, smart actuator, sensors). In parallel to this progress, the integration of smart energy grids with IoT and big data techniques has recently emerged into what is named the Internet of Energy or Energy Internet [19,20]. In fact, with the emergence of smart power meters and smart electrical appliances, it is now possible for users to closely monitor energy consumption while having the ability to plan and manage their consumption. The IoT infrastructure makes it possible to capture and analyze sensor data in real time, allowing consumers to interact with data and decision making [33].

The main aim behind the framework proposed in this paper is to fill the gap between simulations and real case experimental validation of control approaches and mechanisms. The framework could be used not only to control buildings systems but also for other use cases in which the experimental validation of a developed control model is needed. A flexible architecture of the platform has been introduced and its components are detailed to provide an easy to implement solution for similar applications. The work presented in this paper focuses on the integration of IoT/big data techniques with simulation tools in order to enable HILS. The aim is to join both field testing and numerical modeling by combining hardware and software to form HILS frameworks. These latter make it easy to assess multiple tests under the same conditions and, eventually, to accommodate for dangerous operations. As a case study, to show the usefulness of the HILS framework, a Model Predictive Control approach (MPC) was deployed on standalone VS. The framework integrates recent IoT and big data platforms together with machine-learning algorithms and MATLAB-based MPC model.

In summary, the objective of the work is twofold, first to show the usefulness of the proposed IoT based HIL framework together with the integrated machine learning model and smart control technique of MPC, and second, to study the performance of the MPC model combined with forecasted occupancy number and real-time test site's contextual data. The goal of the experimentation is to maintain the indoor CO_2 concentration at the standard comfort range while enhancing the energy efficiency. Setting up a field operational testing predictive control techniques is a very challenging and time consuming task. This work could leverage the gap between simulations and real application of predictive control in smart buildings.

The remainder of this paper is structured as follows. Section 2 presents recent work related to advanced strategies for smart control and IoT-HIL based approaches. In Sections 3 and 4, the description of the used materials as well as the architectures of the proposed control strategies and the IoT-HIL platform will be presented. In Section 5, results are presented to demonstrate the accuracy of the proposed models as well as the developed framework. Conclusions and perspectives are presented in Section 6.

2. Related Work

Recent research work showed that reducing energy consumption in buildings, especially those related to HVAC systems, can be attained through the usage of advanced control strategies. In this regard, two main approaches of rule-based control algorithms have recently emerged in the field of advanced HVAC control: Learning based approaches (e.g., fuzzy logic, Artificial Neural Networks (ANNs), fuzzy and adaptive fuzzy neural networks and genetic algorithms) and MPC [34]. Among these control algorithms, MPC has been introduced as one of the most powerful control techniques used to manage complex processes, such as in HVAC systems [35] studies. This control technique can handle nonlinear processes and their dynamics according to different objectives functions, such as those related to indoor air quality and thermal comfort improvement [36,37].

On the other hand, one of the most efficient ways of conducting field operational testing appears to be the HIL simulations seeing its various advantages (low cost, accurate results, etc.) [38]. This new concept is becoming widely used in developing and testing complex real-time embedded systems [39]. This is mainly done by adding, through mathematical representations (also referred to as "plant simulation"), the complexity of the plant to be controlled into the test bed [40]. To perform HIL simulations, electrical emulation of sensors and actuators is used to interface between the "plant simulation" and the "system under test". In fact, the plant simulation controls the value of the emulated sensor, which is then read by the embedded system under test. In HIL simulations for system synthesis, major physical equipment and their associated controllers are integrated with simulated devices or building spaces to investigate behaviors under realistic dynamic conditions.

In the last decade, researchers focused their interest on the HIL approach and used it not only in automotive and spatial systems but also in buildings' equipment testing and control. For instance, Missaoui et al. [41] proposed new BEMS strategies to support demand side management and to validate them using a Power-Hardware-in-the Loop (PHIL) test bench. However, the proposed solution can be used to validate control algorithms in a reasonable time. Schneider et al. in [42] focused their work on investigating the interaction of a real circulating pump with the hydronic network of a virtual building energy and control system. The presented model, using Modelica for building simulation, is used to bridge the gap between the design and commissioning stage of a control algorithm for HVAC components. The used model is a single-family dwelling with limited complexity. The comparison between simulation results and measured data proved the accuracy of the model with a mean relative error less than 4%. De la Cruz et al. in [43] presented, in their paper, the implementation of an HIL real time simulation test bunch for Air-to-Water-Heat-Pumps (AWHP). This will allow HVAC manufacturers to optimize the control of their systems and to improve their efficiency. A real AWHP was tested under real climate conditions, as for the thermal loads, they were calculated through the connection of the AWHP and a virtual building, simulated using Modelica software, via HIL real time simulation. Seifried et al. in [44] proposed a new model, based on the interconnection of a prominent building automation protocol, namely BACnet, and the PowerDEVS simulator to facilitate HIL testability of new and existing building automation system components. Huang et al. in [22] presented an agent-based framework for HIL simulations, which could either be used for investigating the controller performance or HIL for system synthesis. In other words, it is possible to involve controllers as well as other major equipment in the test to ensure that their dynamic behavior is being correctly captured. Zahari et al. [45] developed a control algorithm to bring the HIBORO helicopter prototype into equilibrium. The developed algorithm is a combination of the MPC and the black box nonlinear autoregressive model. Using the Xpc Target rapid prototype under Simulink, HIL simulations have been run for different set points to evaluate the performances of the proposed model. This latter contains inertial measurement unit sensor software, the MPC, and C/T blocks for capturing and generating Pulse Width Modulation (PWM) signals. The controller proved its efficiency in terms of stabilizing the prototype under all disturbances.

Samano-Ortega et al. [46] developed a platform for the validation of photovoltaics (PVs) system controllers using IoT and HIL concept. The platform englobes five main parts: (i) a control emulator based on HIL, producing the behavior of PVs' arrays, a converter, and Alternating-Current (AC) loads, (ii) Cloud database, (iii) smart sensors for load monitoring, (iv) residential PVs (RPVs) connected to the Internet, and (v) a mobile application for tracking and monitoring. The main principle is that measured voltage and current of the AC loads (using smart sensors) and the production of RPVs are downloaded to the HIL, which reproduces the behavior of the PVs and loads in real-time. The platform proved its efficiency in emulating the behavior of the installed PVs with a mean relative error of 0.42% and the AC load with a mean absolute error of 10 mA. Conti et al. [47] showed the relevance of the dynamic coupling between an air-source heat pump and a building apartment, located in Pisa (Italy), in winter in terms of energy performances under three different operational modes. The adopted HIL extensive experimental campaign proved its potential in properly estimating the energy consumption as well as developing advanced operational strategies. Frison et al. [48] developed a simple low cost MPC controller, which has been evaluated using HIL experiments, for assessing, under realistic conditions, the energy performances of a heat pump system.

Furthermore, due to the large availability of smart low-cost embedded devices (e.g., Arduinos, Raspberry pi, NVidia Nano, actuators, and distributed sensors), and data streaming processing tools, such as Storm/SAMOA and Kaa applications [49], and generally, the advances of information and communication of IoT technologies [50], the implementation of optimal control strategies for improving the energy efficiency as well as indoor air quality and thermal comfort is becoming immediate and more viable [51]. Their application has been widely studied for the development and deployment of intelligent context-aware services and applications, such as occupancy prediction [52], healthcare [31], transportation and logistics [53], smart grids [54,55], and smart homes [56]. For example, Huchuk et al. [16] evaluated numerous classification machine learning algorithms and models for predicting occupants' presence in smart buildings using thermal data. Further, Zhang et al. [57] presented a literature review about the integration of machine learning for predicting occupancy patterns to improve indoor air quality, while optimizing energy use. In addition, online machine learning techniques (e.g., vertical Hoeffding tree and self-adjusting memory for KNN) can be included for predicting occupants' number and presence using environmental data, such as CO_2 temperature and humidity [58,59]. IoT and HIL concepts could provide an integrated solution to cover the important aspects of BEMS by enabling the collection, monitoring, and processing of stream data together with machine-learning techniques. These latter are, for instance, used to compute accurate forecasts, which are required for the MPC to compute accurate predictions, i.e., forecast optimal actions for real-time control of a building's services.

In this work, a case study that focuses on a standalone VS, is worked out to assess the usefulness and effectiveness of the proposed HIL framework. In fact, data that has been collected from a set of sensors, such as temperature, motion, and CO_2 concentration, is used to predict occupancy patterns [30]. These latter are then fed to the MPC to control indoor CO_2 dynamics by forecasting the optimal ventilation rates.

3. Materials and Methods

In this section, an HIL experiment of a closed-loop VS driven by the MPC is performed. The MPC ventilation controller model, which has been previously designed, developed, and validated using simulations [28,29], has been physically deployed to the Raspberry Pi (RPi) located at the EEBLab test site. The RPi-in-the-loop experiment has been run under realistic conditions to dynamically actuate the fans of the VS and to assess the controller's performance in terms of energy efficiency and indoor CO_2 improvement. In fact, the deployed VS is made of two standalone controlled fans, which are respectively responsible for bringing the fresh outdoor air to the indoor and draining the CO_2 out of the building. More precisely, these two fans are installed in both side walls and operate instantaneously under the same control signals. The VS can provide a maximum airflow rate of 440 m^3/h, which is equivalent to a rated speed of 3800 rpm and is powered by a photovoltaic solar system. The occupancy information was used as a disturbance as well as a forecast input for the MPC.

3.1. Description of the Case Study Building: EEBLab

The considered specimen, named Energy Efficient Building Laboratory (EEBLab), is a rectangular cavity, which is part of a set of two identical prefabricated structures (Figure 1), located at the International University of Rabat. Each test bed is 12 m^2 of occupied surface and 30 m^3 of volume. Additionally, each prefabricated has one single glazed window on its south façade. The laboratory has been made essentially for implementing and testing different scenarios related to eHealth, Energy efficiency, ICT, and renewable energies integration and control. The main aim is to investigate the integration of recent IoT, big Data technologies, and advance real-time machine learning algorithms for developing context-aware services and applications.

Figure 1. (**a**) Energy Efficient Building Lab (EEBLab) test site; (**b**) Interior side wall of EEBLab with ventilation fan and a set of sensors and other equipment.

3.2. The HOLSYS Internet of Things Platform Setup

The HOLSYS platform has been in development and passed from many stages namely, the use of Kaa project IoT platform and the upgrade to the ThingsBoard open source IoT platform. It allows configuring, supervising, and acquiring connected sensing and actuating nodes. Its aim is to allow the development and deployment of IoT based scenarios related to smart energy efficient buildings as well as eHealth and Smart Mobility. The HOLSYS platform follows the general architecture presented in Figure 2. The four layers define the different general sections/aspects of an IoT platform, namely, Sensing/actuating, Data Acquisition, Processing, and Visualization.

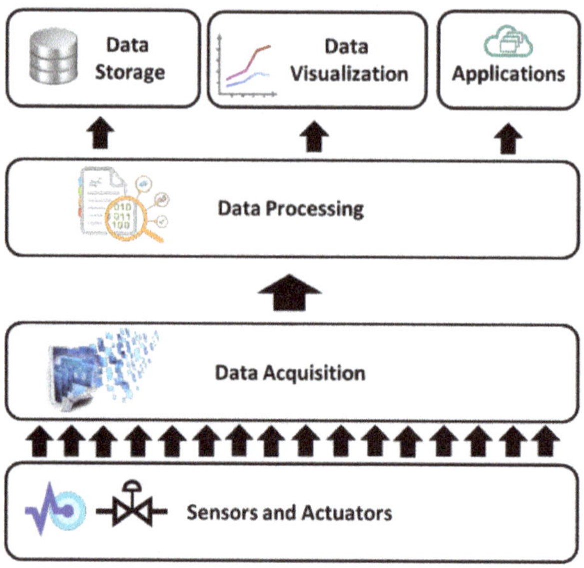

Figure 2. General architecture of an Internet of Things platform.

3.2.1. Sensing and Actuation Layer

Sensors and actuators represent all embedded sensors and actuators together with control units considered as one device and presented to the platform as an IoT node. This latter is capable of receiving and sending stream data while ensuring the execution of all control strategies sent by the platform. The communication between deployed nodes and the HOLSYS platform may pass through wired or wireless protocols. In this paper, only MQTT, REQUEST and REST have been used as the EEBLab is accessible in the campus network either wirelessly or through an ethernet connection.

The deployed IoT devices (nodes) are built using low-cost microcontrollers (e.g., NodeMCU, Arduino, STM32S) and presented to the platform using Raspberry Pi 3 and 4 B+ (RPi). A set of nodes can be connected in serial mode (USB), Serial Peripheral Interface (SPI), Inter-Integrated Circuit (I2C) or Bluetooth to the RPi gateway where NodeRed controls the inputs and the outputs of the set. NodeRed is a flow-based platform, developed originally by IBM, for facilitating the process of wiring hardware devices together with Application Programming Interfaces (APIs) and online services as part of the Internet of Things. A NodeRed flow is a set of connected logical NodeRed nodes linked together to form a processing logic with inputs and outputs. The resulting logic flow ensures gathering input data from wired or wireless IoT nodes, pre-processing and aggregating them to finally output structured data into local or remote storage systems.

Figure 3 presents the deployed sensors and actuators used in this study. (a) represents the Indoor air quality node connected to an RPi via a USB cable. It gathers the indoor CO_2 concentration in Part Per Million (PPM) using an MH-Z14A sensor with an accuracy of ±50 PPM +3% reading value; (b) shows the control node of both inlet and outlet fans used to ventilate the EEBLab. The 12 V and 440 m^3/h fans are controlled with PWM generated from an Arduino nano. Speed can be controlled from 0, for OFF mode, to 255 PWM for max speed (ON mode). However, a relay has been added to completely turn OFF the fans if 0 PWM has been triggered to save energy; (c) presents the deployed RPi based weather station with wind speed and direction, solar irradiance, ambient temperature and relative humidity sensors for outdoor environmental data. Used sensors are, respectively, an analog anemometer and magnetic direction sensors, SR20 pyranometer and DHT22 together with DS18B20; (d) depicts gate door motion sensors to determine the true occupants' number inside the EEBLab. They are based on infrared emitters and receivers, which are aligned together at the door entrance, to detect the exact occupants' number.

Figure 3. The deployed IoT devices; (**a**) Indoor CO_2, temperature and humidity node; (**b**) Inlet and outlet ventilator speed control node; (**c**) Weather station node for outdoor air quality; (**d**) Occupants' number node.

3.2.2. Data Acquisition Layer

Data collected at the first layer (Sensing and Actuation), using the deployed sensors, are sent via HTTP requests and MQTT by the RPi gateways to the HOLSYS platform, which is deployed in the remote server, as depicted in Figure 4. The HOLSYS platform is based on the open source Thingsboard IoT platform in its community edition. Services and packages together with all connectors enabling the acquisition of all the deployed IoT devices are installed and configured in the cluster composed of one performant master and three slaves. They are HP computers with Intel core i3 and i5 with 4 Gbytes of RAM and 500 Gbytes of storage each.

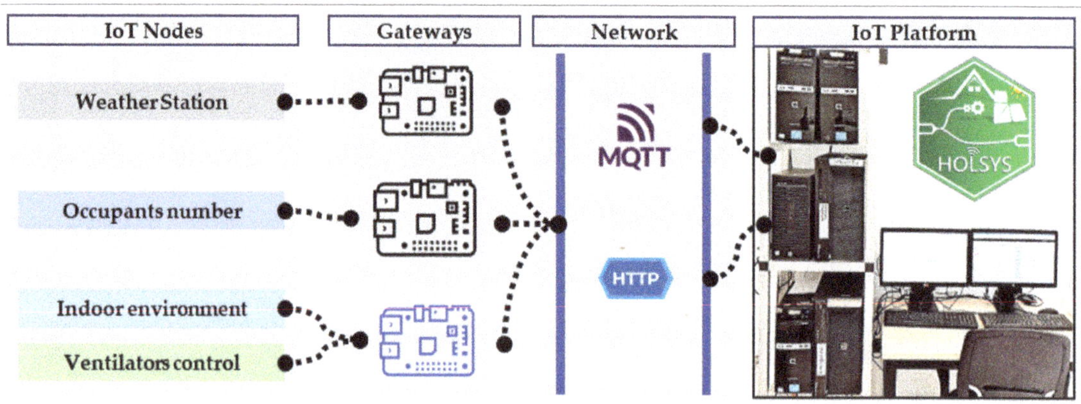

Figure 4. Data transfer from IoT nodes to the HOLSYS platform via RPi gateways over MQTT and HTTP.

The HOLSYS acquisition layer is a set of RPis representing the deployed nodes to the platform. A RPi is a tiny credit-card sized computer using the Raspbian operating system, a free version based on Debian and optimized for its limited power. The three gateways that have been deployed are a 4 Gbyte RAM RPi 4 B+ and two 1 Gbyte RAM RPi 3 B+. NodeRed is used in these RPis to acquire data from the serial connected nodes to pre-process and store them locally in files for a backup. Each node is represented to the platform by a unique token and identifier, which is used to secure the communication and to identify the streamed data. HTTP requests have been used as a backup transfer protocol in case the MQTT broker becomes unfunctional. However, the HTTP protocol is not suitable for IoT architectures as it is more energy consuming and has a bigger data packet size. On the other hand, MQTT has been designed to provide a lightweight messaging technique enabling small packet size for faster transfer. The MQTT is an IoT data transfer protocol having publish/subscribe architecture. It is based on a broker to which all clients, either subscribers, publishers, or both at the same time, should be connected to (see Figure 5).

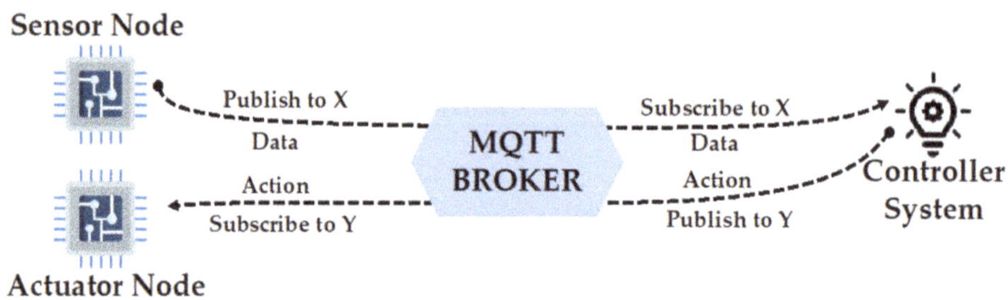

Figure 5. MQTT stream data flow from/to sensor/actuator/controller.

The installed broker is the central communication point where data is exchanged between clients based on a topic. This latter is a category in which a given client has published data. From the other side, the subscriber will get the data from the given topic.

3.2.3. Data Processing Layer

The processing of data can be performed in real-time or batch manners in the HOLSYS platform servers or by third-party applications deployed elsewhere. The local processing can be a simple aggregation of each received tuple of data or a complex processing using the rule-based engine. It is a customizable and configurable system for complex event processing. It allows for filtering, enriching, and transforming incoming data and triggering various actions, for example, notifications or communication with external systems. In the current study, a rule chain has been implemented to save received data to the HOLSYS database (NoSQL Cassandra) after applying filters on them according to each data source (e.g., Temperatures, Humidity and CO_2 ranges; null values, missing data) and forward, via MQTT and Kafka, the filtered data to external sources for other applications. This latter plays a key role for integrating machine learning applications. For instance, the occupancy forecast, an important input to the MPC algorithm, is performed as an external application while using the filtered data coming from the platform. In fact, Apache Kafka, a high-performance real-time data streaming technology capable of handling large amounts of events, is used as a pipeline to transmit stream data between the platform and other applications (e.g., machine learning algorithms). This part shall be detailed in the next section. Furthermore, MQTT has been used to send data to the consumers in the MATLAB MPC control as it is supported by default as a communication method implemented by MATLAB. Figure 6 depicts the communication between the platform and the processing layer. Kafka and MQTT are the main tools used to allow the processing layer to receive data for occupancy forecasts and MPC control. For instance, the forecast model receives the indoor CO_2 concentration and the real occupancy number, respectively over MQTT and Kafka, to forecast 10 steps ahead. Real-time forecasted values are fed as an input to the MPC controller in order to predict the required control actions, which are sent back to the HOLSYS platform for execution.

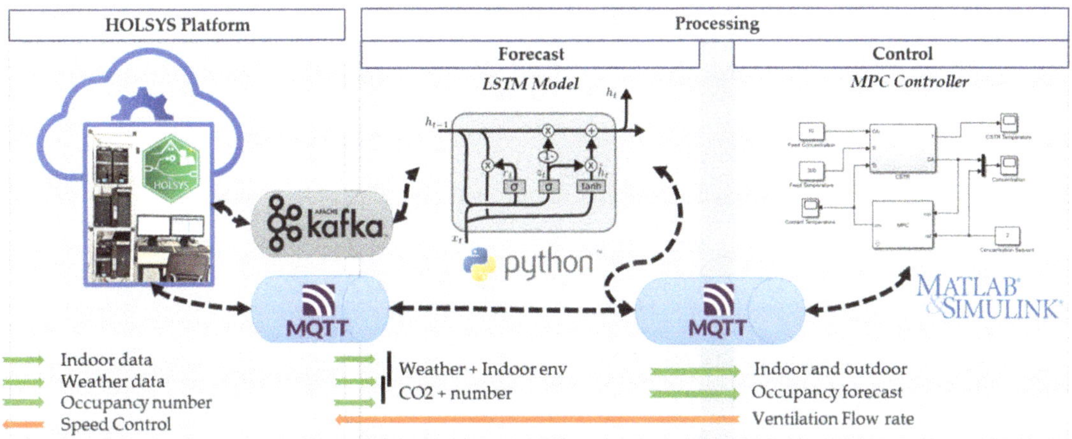

Figure 6. General architecture platform for controlling ventilation system based on occupancy forecast and CO_2 measurement.

3.2.4. Data Storage, Visualization, and Applications Layer

This layer presents all external services that can be connected to the platform by means of supported data transfer protocols and technologies. Mainly, Kafka, MQTT, and HTTP requests are used to allow external third-party applications to connect, produce data, and consume available resources of the HOLSYS platform. In addition to the local NoSQL Cassandra database, which is deployed by default for storing data at the platform level, Mongodb is also used to store backup data for batch processing, serving other applications, such as training the forecast model of occupancy. The main reason behind using Mongodb is its architecture based on storing data as JSON format but with a special syntax called BSON. Each set of data is stored as a document into a collection while its content can be unstructured and different from other documents, resulting in the NoSQL principle, which does not need a tabular and relational concept. Furthermore, the collected data is broadcasted on MQTT using adequate topics for all other applications, mainly those requiring shared resources, such as weather data. As for data visualization, Grafana tool is being integrated with Cassandra and the Mongodb database for real-time visualization of data streams.

3.3. Data Measurement/Preparation for Occupancy Prediction

Occupancy information, as stated above, is one of the most important inputs to context-driven control approaches for efficient control of building equipment. Several studies show the effectiveness of integrating occupancy in MPC for HVAC, ventilation and lighting systems control [60,61]. In order to collect accurate occupancy prediction, different techniques can be used, including PIR sensors, cameras, RFID, Wi-Fi, Bluetooth-low-energy (BLE), and environmental data (e.g., CO_2, temperature, and humidity) [62–64].

In this work a data set containing almost 28,000 instances for one day from 8:30 until 19:00 has been used. Data are collected from EEBLab using the occupancy number node and stored into a Mongodb data (see Figure 7). The occupancy profile varied between one and seven occupants during the day.

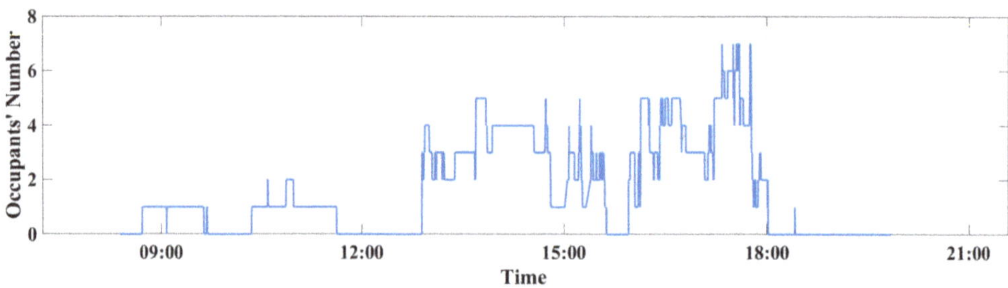

Figure 7. Occupants' numbers over the day in EEBLab.

Deep learning-based occupancy forecasting techniques have been investigated [65]. The first two recurrent neural network (RNN) based methods, long short-term memory (LSTM) and gated recurrent unit (GRU), have been evaluated and compared in terms of accuracy and root mean square error. These algorithms are classified as extensions of RNN by integrating internal gates which help in deciding whether to keep or throw out the past relevant information compared to traditional RNN. The idea is to evaluate the first generated model and then decide which could be deployed in the EEBLab. Therefore, in this study, LSTM model performs well and has been selected to be exploited in the experiment, due to its effectiveness in terms of accuracy (LSTM 98.7%, GRU 97.5%) and root mean square error (RMSE) (LTSM 3.34, GRU 3.73) parameters. In fact, Apache Kafka has been used, in this case study, to consume the actual number of the occupancy, coming from the HOLSYS platform, to forecast the next 10 steps ahead, serving as a real time input for the MPC controller model.

3.4. MPC for Predective Control

The work presented in [29], presents the developed dynamic model for CO_2-based MPC of a building's VS. The model, describing this system, is a state-space model, which is based on the relationship between the input/output airflow rates and indoor CO_2 concentrations. The MPC controller model was tuned to be deployed in a real case scenario, considering the real context of the EEBLab, in particular, the building space and occupancy number profile as well as the characteristics of the VS. Simulations have been conducted to the following highlights during the tuning of the MPC controller input parameters:

- The controller output provides a slow response to CO_2 set point and occupancy changes if the prediction horizon is short (i.e., $2 \leq P \leq 5$ steps ahead);
- The output of the controller acts faster on changes, which means that the controller's prediction ability increases if the prediction horizon is long (i.e., $P \geq 10$ steps ahead);

- A longer control horizon (i.e., M $\geq$ 10 steps ahead), the response of the controller output becomes too aggressive and therefore overshoots, which does not occur when using a small control horizon.

In fact, the control horizon M and the prediction horizon P inputs are the key design parameters of the MPC. They have a significant impact on its performance (i.e., settling/rise time and stability), especially in the presence of disturbances.

In this experimental study, as schematized in Figure 8, the optimal control problem (OCP) of the MPC is solved for every time interval (30 s) in which its optimal control output is calculated for the entire horizon P. The inputs to the OCP are the forecasted occupancy number, the outdoor CO_2 concentration, and the previous measurement of indoor CO_2 concentration (k−1), along with the system constraints (i.e., indoor CO_2 set point and airflow limits).

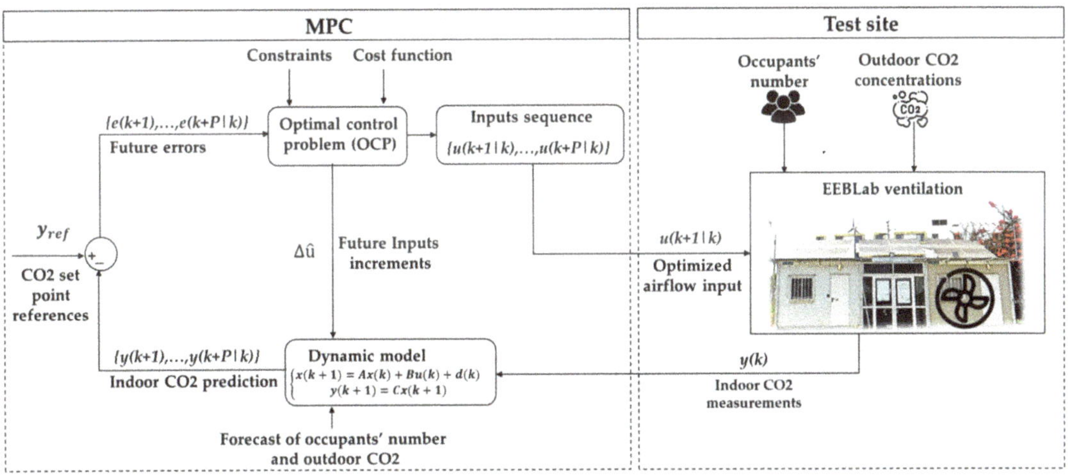

Figure 8. The general structure of the MPC framework for the EEBLab ventilation control system.

The prediction and control horizons used in the MPC framework are respectively P = 10 (i.e., 300 s) and M = 5 (i.e., 150 s) steps ahead. For occupancy, the forecasted number is used to control the indoor CO_2 dynamics, including the CO_2 generated by the occupants over the horizon P. The forecast of the occupants' number and measurements of the indoor/outdoor CO_2 are forwarded to the OCP. The optimized control output (i.e., minimal required airflow) is fed back to the dynamic model to calculate the future predictions of indoor CO_2 concentrations for the entire prediction horizon P. This calculation is repeated every time interval. At each time interval, the future occupancy and prediction of the indoor CO_2 concentrations along with the constraints are updated and passed to the OCP to plan the next sequence of control inputs to be applied at that time. Only the first optimal input of the control sequence is implemented, and the remaining input values are discarded.

To solve the OCP, the following quadratic cost function is used, which reduces the future error $\hat{e}$ between the CO_2 set point references y_{ref} and predicted indoor CO_2 concentration $\hat{y}$ through the prediction Horizon P. This is mainly achieved by applying the optimal control increment action $\Delta\hat{u}$ in which the minimum of airflow u is delivered and the indoor CO_2 concentration y is maintained within comfort bounds. Q and R represent weighting matrices. The set point of indoor CO_2 concentration is defined at 550 PPM, whereas the outdoor CO_2 is kept at a constant value of 400 PPM.

$$\underset{\Delta u}{Minimize} J = \frac{1}{2}\sum_{K=0}^{k=P}\left[\left(\hat{y} - y_{ref}\right)Q\left(\hat{y} - y_{ref}\right)^T + (\Delta\hat{u})R(\Delta\hat{u})^T\right],$$

Subject to, $y < 550$ PPM and $0 < u < 440\ \mathrm{m^3/h}$ $(\sim 0.12\ \mathrm{m^3/s}\ \sim\ 122.22\ \mathrm{L/s}\ \sim\ 259\ \mathrm{scfm})$.

4. Real-Time Implementation

This section presents the implementation of the MPC framework for controlling the VS with the aim to improve both the indoor air quality and energy saving. An HIL experiment in which the MPC model controller is physically implemented in an RPi development board is conducted. Figure 9 shows the blocks that have been integrated to enable the communication between the MPC VS model, which is carried out with the MPC toolbox of MATLAB/Simulink environment and the HOLSYS platform.

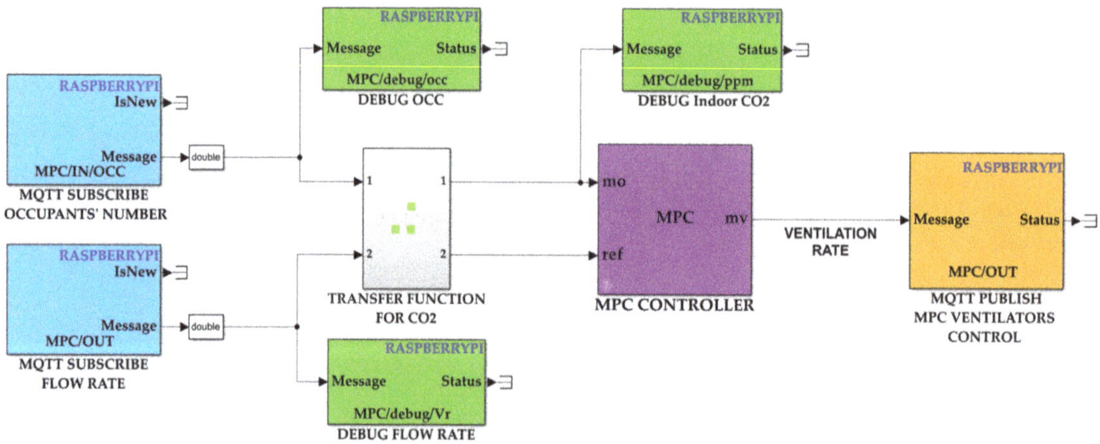

Figure 9. MATLAB/Simulink model for ventilation system's control using MPC.

To enable the use of MQTT in the MATLAB/Simulink model, the Raspberry Pi support package for MATLAB and Simulink have been installed using MATLAB Add-ons.

Many blocks are available in the Simulink libraries under "Simulink Support Package for Raspberry Pi Hardware". The two used blocks are "MQTT subscribe" and "MQTT publish". The former subscribes to the topics "MPC/IN/OCC" and "MPC/IN/CO_2" to get, respectively, the forecasted occupant's number and CO_2 measurements from the forecast model and the HOLSYS platform. The latter publishes the required flow rate into the "MPC/OUT" topic to control the ventilation speed. As shown in Figure 3b, the ventilation control node is wired to the inlet and outlet fans and controls them using PWM. In fact, the node receives the required flow rate from the platform through MQTT and transforms it to the corresponding PWM signal. The general architecture of the experimental setup from sensing till the control execution is illustrated in Figure 10.

The MPC controller, which is run from MATLAB/Simulink simulator, is emulated and embedded into the RPi as an independent hardware in the network.

However, Simulink is able to keep monitoring the simulation run time as well as the inputs and outputs of the MPC model. Using an MQTT publisher and subscriber tool, it is possible to inject test data into the model or monitor any variable from all over the experimental setup and its system entities.

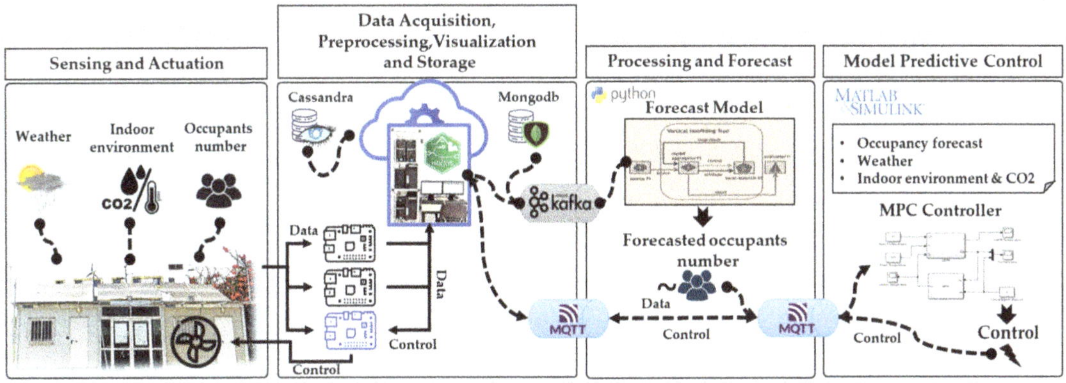

Figure 10. The experimental setup architecture of HIL implementation of the MPC for EEBLab ventilation system control enabled by the HOLSYS IoT platform.

5. Results and Discussions

In this section, the results obtained from the experimental setup of the different deployed systems are presented. In fact, after connecting everything together, the simulations have been run and data are collected during the experiment's time. The experimentation started at 13:36 at the EEBLab test site. All the windows were closed and the only source of fresh air was the ventilation inlet. The HVAC was set to off. The weather station readings at the time were 23 °C for the ambient temperature and 56% for the relative humidity. The behavior of two employees at an office of 12 m^2 was simulated. Other staff joined the team from time to time. While the occupants were performing moderate activities (e.g., using their personal computers and reading several articles while conversing), CO_2 was changing its levels and increasing with more people inside the test site. At each new visit, the door was opened and closed in approximately 5 to 8 s. However, the influence of the door openings and the air exchanged during this time has not been taken into consideration. In fact, the objective behind the experimental setup is to show the usefulness of the proposed framework with all its components. It is a proof of concept of the intercommunication of different entities that form the entire concept. The idea is to integrate control strategies and modern technologies into a holistic framework for enabling real time monitoring and control of buildings' systems.

First, advanced methods for forecasting indoor occupancy are implemented. Real occupancy data is sent to the server to be processed and exploited by the deployed forecasting model. It is implemented to read 10 instances and forecast 10 values ahead. The model gets new data every 1 min. It means that the model is able to forecast 10 min ahead. the forecasted occupancy data is sent to the MPC model to measure the flow rate needed to adjust optimal operation of the VS. The calculated accuracy and root mean square error (RMSE) parameters of the LSTM forecasting are respectively, 3.34% and 98.7%. Secondly, an MPC control strategy is integrated for VS's control. All together, these systems have been inter-communicated via the deployed IoT platform. The simulation model of the MPC controller has been compiled via MATLAB/Simulink and embedded into the RPi 4 B+ installed into the test site. Afterwards, the installed sensors began to inject input data to the forecasting and MPC models and outputs (controls) were executed by the ventilators.

In order to assess the performance of the MPC against the ON/OFF, three metrics have been generated: (i) the regulation of indoor CO_2 concentration, (ii) the ventilation flow rate evolution, and (iii) the instantaneous power consumption, which are calculated using the smart metering platform [26]. Experiments have been conducted using the above-mentioned set-up and the three metrics have been evaluated for the ON/OFF and MPC controllers during five hours and a half from 13:30 to 19:00. The occupants' number together with the behavior of indoor CO_2 concentration, ventilation flow rate, and power consumption of two controllers can be observed in Figures 11–14.

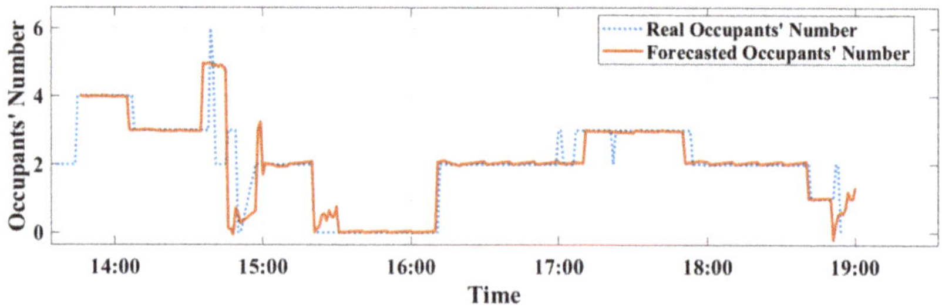

Figure 11. Occupancy forecasting results using LSTM.

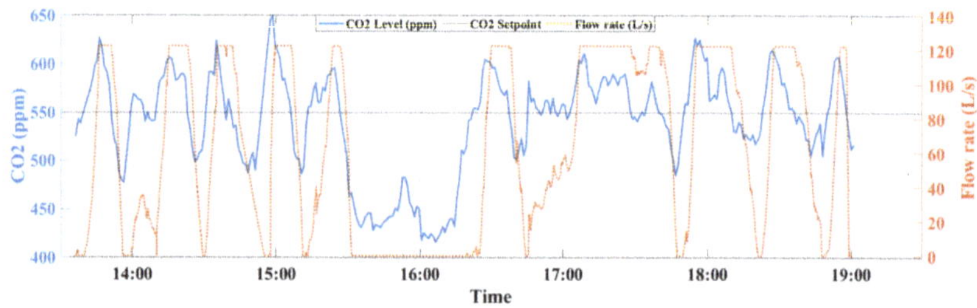

Figure 12. The MPC flow rate output together with the CO_2 concentration variation.

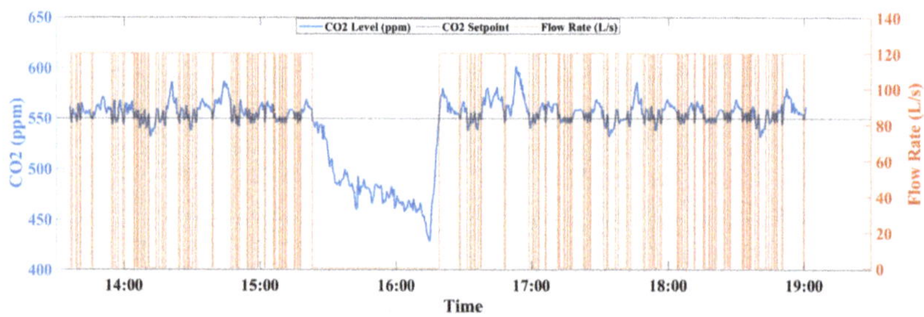

Figure 13. The ON/OFF flow rate output together with the CO_2 concentration variation.

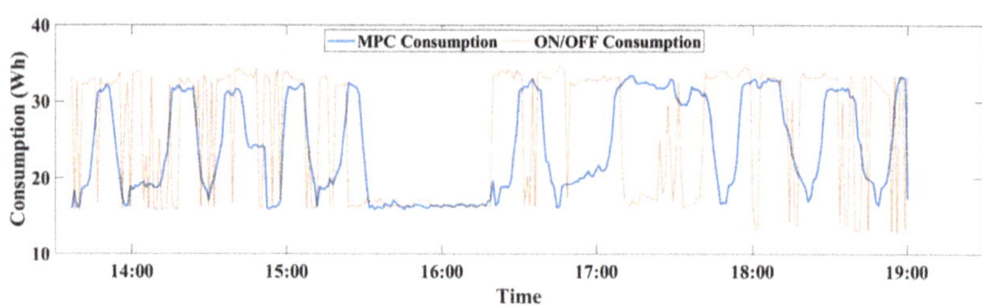

Figure 14. Energy consumption variation of the ventilation system during control for both ON/OFF and MPC controllers.

As can be seen from Figure 11, the forecasted occupants' number seems to be close and to correlate well with the collected real occupants' number. A minor difference is seen at the peak points of the real occupancy.

The ON/OFF approach has been chosen for comparison as it is the most used control approach in VSs. It is a simple control mechanism which triggers full On or full Off in case of CO_2 variation from the fixed setpoint. The ON/OFF control was deployed using the above-mentioned approach the next day.

In terms of CO_2 regulation, both controllers provide good performance in maintaining the CO_2 concentration with faster settling/rise responses for the ON/OFF to achieve and maintain the desired level, which is fixed to 550 PPM setpoint. Unlike the ON/OFF, the MPC was able to provide a better transient response to refresh the air inside the EEBLab using the optimal ventilation rate, as can be observed from Figures 12 and 13.

For energy consumption, the obtained results presented in Figure 14, showed that the MPC outperforms the ON/OFF and allowed higher performance in improving energy savings. This performance can be explained by the predictive mechanism of the MPC, which includes the optimized criterion $\Delta\hat{u}$ that predict the effective ventilation flow rate according to the indoor CO_2 dynamics, including the CO_2 generated by occupants.

Regarding the total energy consumption of the VS during this experiment time period, the MPC outperformed the ON/OFF control and allowed a significant energy reduction by 16.44%. It can be noticed from Figure 14, that the peak energy consumed by the ventilators is reached only a few times by the MPC control unlike the ON/OFF method. The total energy consumed by MPC control is 119.4 Wh while the ON/OFF consumed a total of 142.88 Wh.

6. Conclusions and Perspectives

In this work, an HIL based framework was introduced for standalone VSs using MPC control method. The objective was to assess the effectiveness of the proposed framework in terms of indoor air quality improvement and energy efficiency in real-setting scenario. In fact, a Simulink based HIL model was proposed and implemented in the EEBLab to assess the effectiveness of MPC control. Contextual data are collected using the HOLSYS IoT platform and LSTM machine learning models have been integrated for real time occupants' number forecasting. Resulting forecast data have been exploited by the MPC for optimal regulation of the ventilation flow rate. The performance of the MPC over the HIL framework has been assessed and compared to the ON/OFF strategy. Experimental results showed that both controllers provide acceptable performance in regulating the indoor CO_2 concentration. However, the MPC allowed significant energy reduction by approximately 16% compared to ON/OFF.

As a perspective of this work, the framework will be applied and experimented for the HVAC system's control using MPC. This latter has already been validated by simulations [66]. Furthermore, additional experiments will be conducted to shed more light on the integration of IoT and machine learning algorithms for setting up context-driven control approaches of different building services, including lighting, shading, and HVAC systems. Additionally, the perspective includes integrating the proposed framework for developing other buildings services, such as renewable energy production forecasting and predictive control of power systems [24].

Author Contributions: Conceptualization, A.K., A.B. and H.E.; methodology, A.K.; software, A.K., A.B. and H.E.; validation, A.K., A.B., H.E., S.B., M.B. and D.E.O.; formal analysis, A.K., A.B. and H.E.; investigation, M.B.; resources, M.B. and J.G.; data curation, A.K.; writing—original draft preparation, A.K., A.B., H.E. and S.B.; writing—review and editing, M.B., J.G. and D.E.O.; visualization, D.E.O.; supervision, M.B. and D.E.O.; project administration, M.B.; funding acquisition, M.B. and J.G. All authors have read and agreed to the published version of the manuscript.

Funding: This research was supported by IRESEN under the HOLSYS project (2020–2022), Green Inno-PROJECT-2018.

Institutional Review Board Statement: Not applicable.

Data Availability Statement: Not applicable.

Conflicts of Interest: The authors declare no conflict of interest. The funders had no role in the design of the study; in the collection, analyses, or interpretation of data; in the writing of the manuscript; or in the decision to publish the results.

References

1. Nejat, P.; Jomehzadeh, F.; Taheri, M.M.; Gohari, M.; Majid, M.Z.A. A global review of energy consumption, CO_2 emissions and policy in the residential sector (with an overview of the top ten CO_2 emitting countries). *Renew. Sustain. Energy Rev.* **2015**, *43*, 843–862. [CrossRef]
2. Del Mar Castilla, M.; Álvarez, J.D.; Normey-Rico, J.E.; Rodríguez, F.; Berenguel, M. A multivariable nonlinear MPC control strategy for thermal comfort and indoor-air quality. In Proceedings of the IECON 2013—39th Annual Conference of the IEEE Industrial Electronics Society, Vienna, Austria, 10–13 November 2013; pp. 7908–7913. [CrossRef]
3. Addendum n to ANSI/ASHRAE Standard 62-2001, Ventilation for Acceptable Indoor Air Quality. Available online: https://www.ashrae.org/File%20Library/Technical%20Resources/Standards%20and%20Guidelines/Standards%20Addenda/62-2001/62-2001_Addendum-n.pdf (accessed on 25 September 2022).
4. *EN 13779:2007*; Ventilation for non-Residential Buildings—Performance Requirements for Ventilation and room-Conditioning Systems. 2007. Available online: https://standards.iteh.ai/catalog/standards/cen/c8d82246-bdec-4100-abad-7dc911c420f9/en-13779-2007 (accessed on 23 September 2022).
5. Marzouk, M.; Atef, M. Assessment of Indoor Air Quality in Academic Buildings Using IoT and Deep Learning. *Sustainability* **2022**, *14*, 7015. [CrossRef]
6. *EN 15251:2007*; Indoor Environmental Input Parameters for Design and Assessment of Energy Performance of Buildings Addressing Indoor Air Quality, Thermal Environment, Lighting and Acoustics. 2007. Available online: https://standards.iteh.ai/catalog/standards/cen/92485123-bf64-40e3-9387-9724a642eae8/en-15251-2007 (accessed on 23 September 2022).
7. Scislo, L.; Szczepanik-Scislo, N. Air Quality Sensor Data Collection and Analytics with IoT for an Apartment with Mechanical Ventilation. In Proceedings of the 2021 11th IEEE International Conference on Intelligent Data Acquisition and Advanced Computing Systems: Technology and Applications (IDAACS), Cracow, Poland, 22–25 September 2021; Volume 2, pp. 932–936. [CrossRef]
8. Afram, A.; Janabi-Sharifi, F. Theory and applications of HVAC control systems—A review of model predictive control (MPC). *Build. Environ.* **2014**, *72*, 343–355. [CrossRef]
9. Kukadia, V.; Palmer, J. The effect of external atmospheric pollution on indoor air quality: A pilot study. *Energy Build.* **1998**, *27*, 223–230. [CrossRef]
10. Redlich, C.A.; Sparer, J.; Cullen, M.R. Sick-building syndrome. *Lancet* **1997**, *349*, 1013–1016. [CrossRef]
11. Rueda, L.; Agbossou, K.; Cardenas, A.; Henao, N.; Kelouwani, S. A comprehensive review of approaches to building occupancy detection. *Build. Environ.* **2020**, *180*, 106966. [CrossRef]
12. Tekler, Z.D.; Low, R.; Yuen, C.; Blessing, L. Plug-Mate: An IoT-based occupancy-driven plug load management system in smart buildings. *Build. Environ.* **2022**, *223*, 109472. [CrossRef]
13. Zou, H.; Zhou, Y.; Jiang, H.; Chien, S.-C.; Xie, L.; Spanos, C.J. WinLight: A WiFi-based occupancy-driven lighting control system for smart building. *Energy Build.* **2018**, *158*, 924–938. [CrossRef]
14. Balaji, B.; Xu, J.; Nwokafor, A.; Gupta, R.; Agarwal, Y. Sentinel: Occupancy based HVAC actuation using existing WiFi infrastructure within commercial buildings. In Proceedings of the 11th ACM Conference on Embedded Networked Sensor Systems, Roma, Italy, 11–15 November 2013; Association for Computing Machinery: New York, NY, USA, 2013; pp. 1–14. [CrossRef]
15. Zanon, M.; Gros, S.; Bemporad, A. Practical Reinforcement Learning of Stabilizing Economic MPC. In Proceedings of the 2019 18th European Control Conference (ECC), Naples, Italy, 25–28 June 2019; pp. 2258–2263. [CrossRef]
16. Huchuk, B.; Sanner, S.; O'Brien, W. Comparison of machine learning models for occupancy prediction in residential buildings using connected thermostat data. *Build. Environ.* **2019**, *160*, 106177. [CrossRef]
17. Lohani, D.; Acharya, D. SmartVent: A Context Aware IoT System to Measure Indoor Air Quality and Ventilation Rate. In Proceedings of the 2016 17th IEEE International Conference on Mobile Data Management (MDM), Porto, Portugal, 13–16 June 2016; Volume 2, pp. 64–69. [CrossRef]
18. Ma, Z.; Wang, S.; Xiao, F. Online performance evaluation of alternative control strategies for building cooling water systems prior to in situ implementation. *Appl. Energy* **2009**, *86*, 712–721. [CrossRef]
19. Fathy, H.K.; Filipi, Z.S.; Hagena, J.; Stein, J.L. Review of hardware-in-the-loop simulation and its prospects in the automotive area. *Proc. SPIE* **2006**, *6228*, 117–136. [CrossRef]
20. Crawley, D.B.; Hand, J.W.; Kummert, M.; Griffith, B.T. Contrasting the capabilities of building energy performance simulation programs. *Build. Environ.* **2008**, *43*, 661–673. [CrossRef]
21. Wetter, M. Co-simulation of building energy and control systems with the Building Controls Virtual Test Bed. *J. Build. Perform. Simul.* **2011**, *4*, 185–203. [CrossRef]

22. Huang, S.; Wang, W.; Brambley, M.R.; Goyal, S.; Zuo, W. An agent-based hardware-in-the-loop simulation framework for building controls. *Energy Build.* **2018**, *181*, 26–37. [CrossRef]
23. Malek, Y.N.; Kharbouch, A.; Khoukhi, H.E.; Bakhouya, M.; Florio, V.D.; Ouadghiri, D.E.; Latre, S.; Blondia, C. On the use of IoT and Big Data Technologies for Real-time Monitoring and Data Processing. *Procedia Comput. Sci.* **2017**, *113*, 429–434. [CrossRef]
24. Elmouatamid, A.; Ouladsine, R.; Bakhouya, M.; Elkamoun, N.; Zine-Dine, K.; Khaidar, M. MAPCAST: An Adaptive Control Approach using Predictive Analytics for Energy Balance in Micro-Grid Systems. *Int. J. Renew. Energy Res.* **2020**, *10*, 945–954.
25. Bakhouya, M.; NaitMalek, Y.; Elmouatamid, A.; Lachhab, F.; Berouine, A.; Boulmrharj, S.; Ouladsine, R.; Felix, V.; Zinedine, K.; Khaidar, M.; et al. Towards a context-driven platform using IoT and big data technologies for energy efficient buildings. In Proceedings of the 2017 3rd International Conference of Cloud Computing Technologies and Applications (CloudTech), Rabat, Morocco, 24–26 October 2017; pp. 1–5. [CrossRef]
26. Berouine, A.; Lachhab, F.; Malek, Y.N.; Bakhouya, M.; Ouladsine, R. A smart metering platform using big data and IoT technologies. In Proceedings of the 2017 3rd International Conference of Cloud Computing Technologies and Applications (CloudTech), Rabat, Morocco, 24–26 October 2017; pp. 1–6. [CrossRef]
27. Berouine, A.; Akssas, E.; Naitmalek, Y.; Lachhab, F.; Bakhouya, M.; Ouladsine, R.; Essaaidi, M. A fuzzy logic-based approach for HVAC systems control. In Proceedings of the 2019 6th international conference on control, decision and information technologies (CoDIT), Paris, France, 23–26 April 2019; pp. 1510–1515.
28. Berouine, A.; Ouladsine, R.; Bakhouya, M.; Lachhab, F.; Essaaidi, M. A Model Predictive Approach for Ventilation System Control in Energy Efficient Buildings. In Proceedings of the 2019 4th World Conference on Complex Systems (WCCS), Ouarzazate, Morocco, 22–25 April 2019; pp. 1–6. [CrossRef]
29. Berouine, A.; Ouladsine, R.; Bakhouya, M.; Essaaidi, M. Towards a Real-Time Predictive Management Approach of Indoor Air Quality in Energy-Efficient Buildings. *Energies* **2020**, *13*, 3246. [CrossRef]
30. Elkhoukhi, H.; NaitMalek, Y.; Berouine, A.; Bakhouya, M.; Elouadghiri, D.; Essaaidi, M. Towards a Real-time Occupancy Detection Approach for Smart Buildings. *Procedia Comput. Sci.* **2018**, *134*, 114–120. [CrossRef]
31. Kharbouch, A.; Naitmalek, Y.; Elkhoukhi, H.; Bakhouya, M.; Florio, V.D.; Ouadghiri, M.D.E.; Latre, S.; Blondia, C. IoT and Big Data Technologies for Monitoring and Processing Real-Time Healthcare Data. *IJDST* **2019**, *10*, 17–30. [CrossRef]
32. Nord, J.H.; Koohang, A.; Paliszkiewicz, J. The Internet of Things: Review and theoretical framework. *Expert Syst. Appl.* **2019**, *133*, 97–108. [CrossRef]
33. Khan, N.S.; Ghani, S.; Haider, S. Real-Time Analysis of a Sensor's Data for Automated Decision Making in an IoT-Based Smart Home. *Sensors* **2018**, *18*, 1711. [CrossRef] [PubMed]
34. Behrooz, F.; Mariun, N.; Marhaban, M.H.; Mohd Radzi, M.A.; Ramli, A.R. Review of Control Techniques for HVAC Systems—Nonlinearity Approaches Based on Fuzzy Cognitive Maps. *Energies* **2018**, *11*, 495. [CrossRef]
35. Godina, R.; Rodrigues, E.M.G.; Pouresmaeil, E.; Matias, J.C.O.; Catalão, J.P.S. Model Predictive Control Home Energy Management and Optimization Strategy with Demand Response. *Appl. Sci.* **2018**, *8*, 408. [CrossRef]
36. Prívara, S.; Široký, J.; Ferkl, L.; Cigler, J. Model predictive control of a building heating system: The first experience. *Energy Build.* **2011**, *43*, 564–572. [CrossRef]
37. Yuan, S.; Perez, R. Multiple-zone ventilation and temperature control of a single-duct VAV system using model predictive strategy. *Energy Build.* **2006**, *38*, 1248–1261. [CrossRef]
38. Ellis, G. Chapter 13—Model Development and Verification. In *Control System Design Guide*, 4th ed.; Ellis, G., Ed.; Butterworth-Heinemann: Boston, MA, USA, 2012; pp. 261–282. ISBN 978-0-12-385920-4.
39. Bacic, M. On hardware-in-the-loop simulation. In Proceedings of the 44th IEEE Conference on Decision and Control, Seville, Spain, 15 December 2005; pp. 3194–3198. [CrossRef]
40. HIL Simulation|OPAL-RT. OPAL-RT. 2022. Available online: https://www.opal-rt.com/hardware-in-the-loop/ (accessed on 10 October 2022).
41. Missaoui, R.; Warkozek, G.; Bacha, S.; Ploix, S. Real time validation of an optimization Building Energy Management strategy based on Power-Hardware-in-the-loop tool. In Proceedings of the 2012 3rd IEEE PES Innovative Smart Grid Technologies Europe (ISGT Europe), Berlin, Germany, 14–17 October 2012; pp. 1–7. [CrossRef]
42. Schneider, G.F.; Oppermann, J.; Constantin, A.; Streblow, R.; Müller, D. Hardware-in-the-Loop-Simulation of a Building Energy and Control System to Investigate Circulating Pump Control Using Modelica. In Proceedings of the 11th International Modelica Conference, Versailles, France, 21–23 September 2015; Available online: https://ep.liu.se/en/conference-article.aspx?series=ecp&issue=118&Article_No=24 (accessed on 25 September 2022).
43. Tejeda De La Cruz, A.; Riviere, P.; Marchio, D.; Cauret, O.; Milu, A. Hardware in the loop test bench using Modelica: A platform to test and improve the control of heating systems. *Appl. Energy* **2017**, *188*, 107–120. [CrossRef]
44. Seifried, S.; Preyser, F.J.; Kastner, W. Enabling hardware-in-the-loop for building automation networks: A case study for BACnet and PowerDEVS. In Proceedings of the IECON 2017—43rd Annual Conference of the IEEE Industrial Electronics Society, Beijing, China, 29 October 2017–1 November 2017; pp. 8119–8125. [CrossRef]
45. Taha, Z.; Deboucha, A.; Kinsheel, A.; Ghazilla, R.A.B.R. Development of Real-Time Hardware in the Loop Based MPC for Small-Scale Helicopter. In *Frontiers of Model Predictive Control*; IntechOpen: London, UK, 2012; Available online: https://www.intechopen.com/chapters/29697 (accessed on 10 October 2022).

46. Sámano-Ortega, V.; Méndez-Guzmán, H.M.-G.; Padilla-Medina, J.; Aguilera-álvarez, J.; Martínez-Nolasco, C.; Martínez-Nolasco, J. Control Hardware in the Loop and IoT Integration: A Testbed for Residential Photovoltaic System Evaluation. *IEEE Access* **2022**, *10*, 71814–71829. [CrossRef]
47. Conti, P.; Bartoli, C.; Franco, A.; Testi, D. Experimental Analysis of an Air Heat Pump for Heating Service Using a "Hardware-In-The-Loop" System. *Energies* **2020**, *13*, 4498. [CrossRef]
48. Frison, L.; Kleinstück, M.; Engelmann, P. Model-predictive control for testing energy flexible heat pump operation within a Hardware-in-the-Loop setting. *J. Phys. Conf. Ser.* **2019**, *1343*, 012068. [CrossRef]
49. Apache Storm. Available online: https://storm.apache.org/ (accessed on 25 September 2022).
50. Wen, J.T.; Mishra, S. (Eds.) *Intelligent Building Control Systems*; Advances in Industrial Control; Springer International Publishing: Cham, Switzerland, 2018; ISBN 978-3-319-68461-1.
51. Serale, G.; Fiorentini, M.; Capozzoli, A.; Bernardini, D.; Bemporad, A. Model Predictive Control (MPC) for Enhancing Building and HVAC System Energy Efficiency: Problem Formulation, Applications and Opportunities. *Energies* **2018**, *11*, 631. [CrossRef]
52. Jeon, Y.; Cho, C.; Seo, J.; Kwon, K.; Park, H.; Oh, S.; Chung, I.-J. IoT-based occupancy detection system in indoor residential environments. *Build. Environ.* **2018**, *132*, 181–204. [CrossRef]
53. Zantalis, F.; Koulouras, G.; Karabetsos, S.; Kandris, D. A Review of Machine Learning and IoT in Smart Transportation. *Future Internet* **2019**, *11*, 94. [CrossRef]
54. Ghasempour, A. Internet of Things in Smart Grid: Architecture, Applications, Services, Key Technologies, and Challenges. *Inventions* **2019**, *4*, 22. [CrossRef]
55. Elmouatamid, A.; NaitMalek, Y.; Bakhouya, M.; Ouladsine, R.; Elkamoun, N.; Zine-Dine, K.; Khaidar, M. An energy management platform for micro-grid systems using Internet of Things and Big-data technologies. *Proc. Inst. Mech. Eng. Part I J. Syst. Control Eng.* **2019**, *233*, 904–917. [CrossRef]
56. Alaa, M.; Zaidan, A.A.; Zaidan, B.B.; Talal, M.; Kiah, M.L.M. A review of smart home applications based on Internet of Things. *J. Netw. Comput. Appl.* **2017**, *97*, 48–65. [CrossRef]
57. Zhang, W.; Wu, Y.; Calautit, J.K. A review on occupancy prediction through machine learning for enhancing energy efficiency, air quality and thermal comfort in the built environment. *Renew. Sustain. Energy Rev.* **2022**, *167*, 112704. [CrossRef]
58. Elkhoukhi, H.; Bakhouya, M.; El Ouadghiri, D.; Hanifi, M. Using Stream Data Processing for Real-Time Occupancy Detection in Smart Buildings. *Sensors* **2022**, *22*, 2371. [CrossRef]
59. Elkhoukhi, H.; NaitMalek, Y.; Bakhouya, M.; Berouine, A.; Kharbouch, A.; Lachhab, F.; Hanifi, M.; El Ouadghiri, D.; Essaaidi, M. A platform architecture for occupancy detection using stream processing and machine learning approaches. *Concurr. Comput. Pract. Exp.* **2020**, *32*, e5651. [CrossRef]
60. Jin, Y.; Yan, D.; Zhang, X.; An, J.; Han, M. A data-driven model predictive control for lighting system based on historical occupancy in an office building: Methodology development. *Build. Simul.* **2021**, *14*, 219–235. [CrossRef]
61. Esrafilian-Najafabadi, M.; Haghighat, F. Occupancy-based HVAC control systems in buildings: A state-of-the-art review. *Build. Environ.* **2021**, *197*, 107810. [CrossRef]
62. Depatla, S.; Muralidharan, A.; Mostofi, Y. Occupancy Estimation Using Only WiFi Power Measurements. *IEEE J. Sel. Areas Commun.* **2015**, *33*, 1381–1393. [CrossRef]
63. Tekler, Z.D.; Low, R.; Gunay, B.; Andersen, R.K.; Blessing, L. A scalable Bluetooth Low Energy approach to identify occupancy patterns and profiles in office spaces. *Build. Environ.* **2020**, *171*, 106681. [CrossRef]
64. Hahnel, D.; Burgard, W.; Fox, D.; Fishkin, K.; Philipose, M. Mapping and localization with RFID technology. In Proceedings of the IEEE International Conference on Robotics and Automation, 2004. Proceedings. ICRA '04. 2004, New Orleans, LA, USA, 26 April 2004–1 May 2004; Volume 1, pp. 1015–1020. [CrossRef]
65. Elkhoukhi, H.; Bakhouya, M.; Hanifi, M.; El Ouadghiri, D. On the use of Deep Learning Approaches for Occupancy prediction in Energy Efficient Buildings. In Proceedings of the 2019 7th International Renewable and Sustainable Energy Conference (IRSEC), Agadir, Morocco, 27–30 November 2019; pp. 1–6. [CrossRef]
66. Berouine, A.; Ouladsine, R.; Bakhouya, M.; Essaaidi, M. A predictive control approach for thermal energy management in buildings. *Energy Rep.* **2022**, *8*, 9127–9141. [CrossRef]

electronics

Article

Design and Implementation: An IoT-Framework-Based Automated Wastewater Irrigation System

Shabana Habib [1,*], Saleh Alyahya [2], Muhammad Islam [2], Abdullah M. Alnajim [1], Abdulatif Alabdulatif [3] and Abdullah Alabdulatif [4]

1 Department of Information Technology, College of Computer, Qassim University, 6633, Buraydah 51452, Saudi Arabia
2 Department of Electrical Engineering, College of Engineering and Information Technology, Onaizah 56447, Saudi Arabia
3 Department of Computer Science, College of Computer, Qassim University, Buraydah 51452, Saudi Arabia
4 Department of Computer, College of Sciences and Arts in Al-Rass, Qassim University, Al-Rass 51451, Saudi Arabia
* Correspondence: s.habibullah@qu.edu.sa

Abstract: Automation is being fueled by a multifaceted approach to technological advancements, which includes advances in artificial intelligence, robotics, sensors, and cloud computing. The use of automated, as opposed to conventional, systems, has become more popular in recent years. Modern agricultural technology has played an important role in the development of Saudi Arabia in addition to upgrading infrastructure and plans. Agriculture in Saudi Arabia is dependent upon wells, which are insufficient in terms of water supplies. Thus, irrigation is used for agricultural fields, depending on the soil type, and water is provided to the plants. Two essential elements are necessary for farming, the first is the ability to determine the soil's fertility, and the second is the use of different technologies to reduce the dependence of water on electrical power and on/off schedules. The purpose of this study is to propose a system in which moisture sensors are placed under trees or plants. The gateway unit transmits sensor information to the controller, which then turns on the pump and recycles the water flow. A farmland's water pump can be remotely controlled and parameters such as moisture and flow rate can be monitored using an HTTP dashboard. In order to evaluate the applicability of IOT-based automatic wastewater irrigation systems, a pilot test was conducted using the developed framework. Theoretically, such a system could be expanded by including any pre-defined selection parameters.

Keywords: sensors; microcontroller; mobile networking; data processing; visualizations; Internet of Thing (IoT); GUI

Citation: Habib, S.; Alyahya, S.; Islam, M.; Alnajim, A.M.; Alabdulatif, A.; Alabdulatif, A. Design and Implementation: An IoT-Framework-Based Automated Wastewater Irrigation System. *Electronics* **2023**, *12*, 28. https://doi.org/10.3390/electronics12010028

Academic Editors: Antonio Cano-Ortega and Francisco Sánchez-Sutil

Received: 10 November 2022
Revised: 8 December 2022
Accepted: 11 December 2022
Published: 22 December 2022

1. Introduction

The Gulf Cooperation Council (GCC) nations and Saudi Arabia are already classified as water-scarce countries, with only Oman above the extreme scarcity threshold of 500 cubic meters per capita per year [1]. The growth of agricultural production is greatly affected by climate change, urban population concentration, crop diseases, and greenhouse emissions, all of which highlight the need to meet the growing demand for food and energy. With water resources already scarce, the Middle East and North Africa regions will be particularly vulnerable to climate change [2]. Saudi Arabia is one of the largest arid countries without permanent rivers or lakes in the Arabian Peninsula [3,4]. In some parts of the region, temperatures can reach more than 50° Celsius (C) (122° Fahrenheit (F)), resulting in extremely hot and dry conditions. The country has one of the poorest natural renewable water resources [5]. Renewable natural resources can be affected by a range of variables, including water demand in different sectors and the effects of global warming. There is a recurring theme in research papers on water resources and agriculture related to climate change. Because of global warming, water adaptation measures have been considered to ensure food production and people's access to water, and the preservation of ecosystems. Further, the worsening drought conditions in the peninsular countries have created a need for increasing the water supply from alternative sources and the implementation of water conservation. Moreover, humans and the environment should be assured of their safety as a result of water scarcity or water quality and supply in the face of intensifying climate-related impacts. There are various risks associated with climate change, including water shortages, water quality reductions, soil salinity increases, biodiversity loss, irrigation requirements, and the cost of emergency and remedial action to make sure that farmers plant a crop with a guarantee of success and generating income. Agriculture is the biggest consumer of water in Saudi Arabia, amounting to up the 70% of the total water use, creating a demand for optimization strategies. The number of studies focusing on reducing irrigation water consumption has increased because of the introduction of smart water management. In addition to social, economic, and climate change policies, technological innovations can improve water management, according to some studies [6].

Because of the ever-increasing demand for food necessities and the diminishing supply of those necessities, there is always room for improvement in the production of agricultural produce through the adoption of an agro-tech-based system. Only through cultivation can such things be provided. There is no doubt that this is an imperative factor in human societies when it comes to increasing and maintaining the demand for food production. However, the scarcity of land and water has resulted in a decreasing volume of water available on farms. Therefore, farmers are using irrigation to supplement their water supply system through using high-tech-based systems. It is possible to define irrigation as the science of artificially applying water to land or soil. The methods unfolding in the agriculture sector are related to what is termed precision agriculture capable of estimating the chemistry of the soil to determine the water needs to raise a given crop for productive yield. This means that plants are to be provided with water according to the type of soil to be irrigated. The virgin fertility of the soil is also being investigated using sensor technology. This is in conjunction with the collection and study of the characteristics of the soil moisture dynamics profile. The purpose is to use this data as feedback to improve farming efficiency [7]. The use of renewable energy-powered technology in Seawater Reverse Osmosis (SWRO) desalination plants has had an impact on the preparation for embracing Industry 4.0 in every sector with promising results benefitting humans and agriculture [8,9].

An IoT-enabled framework for automated farming is essentially a network consisting of sensory electronics installed and distributed on the farm at specific locations and connected through a bi-directional communication channel with the management center equipped with related hardware devices harnessed through purposefully developed programming modules. This kind of agriculture is growing in quality, sustainability, transparency, and cost-effectiveness to maximize crop production with lower labor costs. Herein, we propose a system that uses WSN to retrieve environmental data in real time in order to optimize irrigation timing implementation.

The rest of the paper consists of a discussion of related work in Section 2, and an outline of the intelligent communication system in Section 3, the monitoring flowchart in Section 4, Data collection in Section 5, with design dashboard in Section 6, while implementation of test scenarios for real-time implementation are described in Section 7, the results and discussion in Section 8, and conclusions and further work are given in Section 9.

2. Related Work

The Kingdom of Saudi Arabia, with a total area of 2.25 million km^2, is a large country, bordered by Jordon and Iraq and Kuwait in the north, on the east by the Gulf, Bahrain and the United Arab Emirates, on the south by the Sultanate of Oman and Yemen, and in the west by the Red Sea, giving it a coastline of 1750 km. Saudi Arabia is on its way to achieving self-sufficiency in poultry and dairy products [10,11].

Automatic farming systems are based on the collection of data related to soil moisture content, the temperature profile of an indoor covered agriculture farm shed, rainfall data, and intra-field wind speed profiles [12,13]. The related data archives are stored in a computer server that is capable of providing climate prediction, soil condition analysis, and disturbance analysis [14,15].

A fuzzy controller was used by Jamroen et al. to adjust irrigation taking into account the moisture index and water stress [16] by continuously monitoring the soil moisture content and crop water variability. Lloret et al. have used flooding methods to carry out irrigation automatically by observing remotely crop variables and climatic conditions using an application [17] that allows the farmer to remotely control water gates in the water canals. Based on the difference in environmental leaf temperatures and light intensity in hydroponic crops under temperature control between 28 and 32 degrees Celsius, Puengsungwan et al. propose a method for determining root stress in soilless cultivation. According to their proposal, the obtained patterns are divided into regions representing natural roots, stress roots, and rotten roots. Using IoT technologies, they were able to reduce the system's response time from 5 min to less than 60 s. In addition, they increased detection efficiency from 85% to 95% compared to a method that uses the Easy Leaf Area application to determine the drop in leaf area [18].

A variety of irrigation systems have been used in soil and soilless crops. Different approaches exist to estimate irrigation requirements in soil cultivation, highlighting the importance of monitoring and controlling the temperature, soil moisture, evapotranspiration, and water stress index used for operating micro-sprinklers, drip irrigation, and ventilation fans. Several authors, including González-Amarillo [19] and Fernández-Ahumada [20], use humidity or temperature measurements to properly maintain conditions within an agricultural shed by turning on all related irrigation, ventilation, and heating systems. By measuring daily evapotranspiration, Mohamed [21] and Poyen [22] estimate irrigation requirements. Mohamed employs the Penman–Monteith method, whereas Poyen includes the capability of automatically selecting between Hargreaves and Samni, Kharufa, and Penman–Monteith methods according to the type of climate and geographical factors by employing modern technologies and intelligent systems for improving water productivity and conservation. These are some examples (among many others) of how the word "smart" is used in works and prototypes that show different capabilities and features. In an effort to solve the problem of one-size-fits-all qualification, the research community has proposed different taxonomies that classify smart artifacts based on their features [23]. Bin Ahmadon et. al. consists of two major phases in their paper. Phase one is design, and phase two is implementation. Here are the scenarios for two major phases in which a service must be implemented in a different [24].

Agro-IoT was an agricultural IoT project that ran from 2015 to 2020 and featured an architecture that consists of six layers, including sensors, actuators, wireless nodes, etc. In addition, there is a network layer that includes communication protocols, an intermediate layer for software, an application layer for data analysis and prediction, and a user layer where results are communicated to farmers, experts in the field, and the supply chain. Among the goals of the project were real-time monitoring of low-scale greenhouses, early disease detection, identification of crop varieties, optimizing irrigation facilities, and pesticide use and fertilizers efficiency. Similarly, solar precision agriculture is seen as a four-layer IoT architecture, consisting of the sensory layer, the network layer that takes into account IoT nodes and base stations, the decision layer that involves server services, workstations, business and knowledge bases, and the application availability layer that provides information for researchers, experts, and farmers [25,26].

The implemented irrigation solution developed in this research is not restricted to the control of irrigation systems. It also analyzes every component of the system, from water pipes to sprinklers, to detect leaks and malfunctions, while improving efficiency and reducing costs. The proposed solution uses Wireless Sensor Networks (WSN) to collect real-time information data directly in the field to verify the existing conditions. By combining this information with weather forecasts, evapotranspiration from soil and other surfaces and plants, and farm specifications, artificial intelligence algorithms are able to determine how much water is needed for a particular section of the farm by adjusting the irrigation controller accordingly.

3. Intelligent Communication Systems

As shown in Figure 1, the IoT-based irrigation system consists of six main elements: Data Acquisition, Data Monitoring, Communication, Data Processing, Storage, and Visualization.

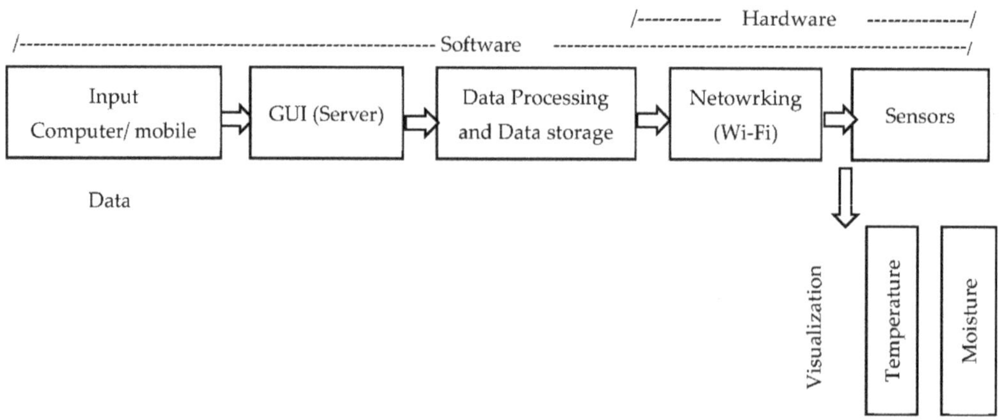

Figure 1. Block diagram of the IoT-based irrigation system.

3.1. *Data Acquisition*

An array of sensors must be used to obtain data from soil moisture sensors, rainfall sensors, ambient humidity sensors, and temperature sensors, among other devices. The various sensors used to monitor crops are configured with a variety of threshold values. This corresponds to the type of crop being monitored and depends on the type of sensor. Since the sensors sense and acquire data at intervals, the measurements will be sent through Wi-Fi modules to be processed according to the parameters of each of the measurements.

3.2. *Data Monitoring*

A supervisory system is required to manage and monitor the information gathered by the sensors (temperature, rainfall, and moisture), as well as on crops and their cycles. A Graphical User Interface (GUI) is used to display the information that is stored in the database. A statistical analysis of the amount of water consumed by crops with respect to their dates and sections has been carried out as a result of monitoring. Finally, all data are requested through the use of web services, which are made available on the Internet for access by authorized personnel.

3.3. *Communication Networking*

The sensors are connected to Arduino microcontroller development boards that use ESP8266 Wi-Fi modules to communicate with the database, and everything is done via the cloud since the entire solution relies on the cloud for connectivity. The database is not directly connected to the sensor because it is not secure. Data is entered into the database using a Web Service (WS). The Wi-Fi module transmits two fields: (1) the identification number (ID), and (2) the value of the sensor. Each type of sensor is assigned a unique ID based on its IP address and a unique number. As an identifier, it serves the purpose of identifying the source of information. The data is processed to determine if irrigation is required. The watering schedule is updated when it is determined that watering is necessary based on the interpretation of the information. An electronic valve is incorporated into the irrigation system and is controlled by the Arduino via the Wi-Fi module. Since the Wi-Fi Module does not have a public IP address, it cannot be identified from the cloud, and therefore requests cannot be sent from the cloud to the module. In the same manner, the Wi-Fi module continuously queries the web service for information. The valve is activated in response to an update in the WS with the time the system was powered on. The Wi-Fi module is assigned a static IP address whenever it is connected to an Internet gateway, making it easier to access from the Internet. In order to monitor data in a timely manner, data visualization along with database information is required. A web application for data visualization has been used in this case study, but other methods may also be employed.

Previous projects used the data collected in order to increase the productivity of the garden or farm, as well as to reduce the amount of water used. Although these ideas are useful in theory, they may not be able to be implemented in a practical scenario. In our system, the algorithm enables the system to understand how much water should be applied to the field based on the moisture in the field; however, the field is watered only when moisture levels are low. Using this approach, we were able to save approximately 25% of our budget. Despite this, our previous systems lack a certain level of reliability, which can be attributed mainly to the selected sensors. In addition, they need to be tested in a larger scenario. In order to improve results and to make sure that the solution works in multiple environments, the wireless network will play a crucial role in ensuring its effectiveness.

3.4. Data Processing

Soil moisture, temperature, and rainfall are received as sensor data in the form of three integer values. Decisions regarding the crop can be made based on the information provided by this data. It is necessary to note that the valve will only open if the soil moisture is below the level required for the crop. In the case of sprinkler irrigation, the water temperature will determine whether irrigation is required. The system will not permit the watering of the plants in the event of rain. The valve can either be opened or no action taken depending on the values collected by the sensors.

3.5. Storage

Databases are necessary to store information about crop types, sensors, moisture, and statistical data. The database used in the application is schematically as shown in Figure 2. The purpose of this structure is to store information about crops, their cycles, and the order in which each irrigation should be performed. Furthermore, information about the monitored sections is saved, as well as that of the sensors associated with each section. The system uses the IoT to determine the amount of water needed for every crop in every cycle using a "smart irrigation" system. Instead of creating another smart irrigation controller, in this work, we have created a smart sustainable irrigation solution. This solution also analyzes every component of the system, from the pipes that supply the water to the sprinklers. This is because it analyzes their interaction. During this analysis, the aim is to detect potential leaks or malfunctions, improve efficiency and reduce costs in a sustainable way in line with the requirements of a circular economy.

3.6. Visualization

The administrator will be able to control the variables (sensor data), as well as irrigate the crop in order to ensure that the resulting data are presented in an appropriate manner. There are different irrigation configurations for each type of crop when it comes to visualization and interaction with users. When this is not possible, the possibility of adjusting the configuration of crops is available based on a predetermined composition.

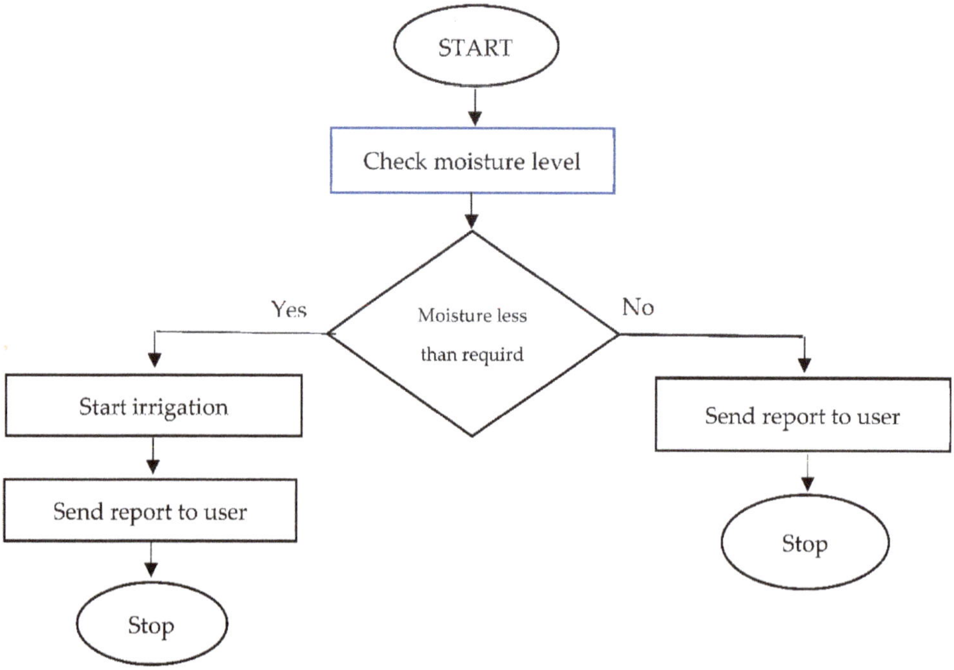

Figure 2. Flowchart of the IoT-based irrigation system.

4. Flowchart

Initially, the user interacts with the irrigation system using a mobile application or any other connected interface to extract data regarding the current soil conditions and moisture in order to change the state of the irrigation system. The sensor node/mote measures the soil moisture, temperature, and humidity of the surrounding environment of—in this case of this study—each date palm tree.

The information can be received using an automated system as shown in Figure 2. Users can make a decision with respect to watering based on the data available/provided on the application or system. The method will lead to accurate agricultural techniques to closely monitor the conditions of the field and use real-time data to realize the most efficient irrigation.

5. Data Collection and Connection

All sensor nodes have been installed and are connected to the network master node. The node will send information packets to every sensor hub bearing the sensor hub ID if communication is present in the node. After acquiring the data packets from the sensors, the measured values of soil parameters are analyzed—for example: Soil moisture assessment (30%). A pH value is used to determine whether the soil is acidic or alkaline. The sensors assess the quality of the equipment, and these quality values are stored in the memory of the Arduino microcontroller unit (MCU). In response to these stored values, the MCU responds by performing specific actions according to the listed conditions. When data packets are received from the sensor node, the MCU transmits them wirelessly to the master node, thereby participating in the Internet of Things (IoT) environment activities.

6. Designed Monitoring Dashboard

The dashboard can be thought of as a server that can be designed using a mobile device, as shown in Figure 3. Using the dashboard, the user will be able to monitor the amount of water flow, and parameters such as temperature and humidity. In addition, users are also able to turn on/off the water pump just by operating the monitoring dashboard. There is a control button (On/Off) shown in Figure 4 that shows the installed applications with the dashboard.

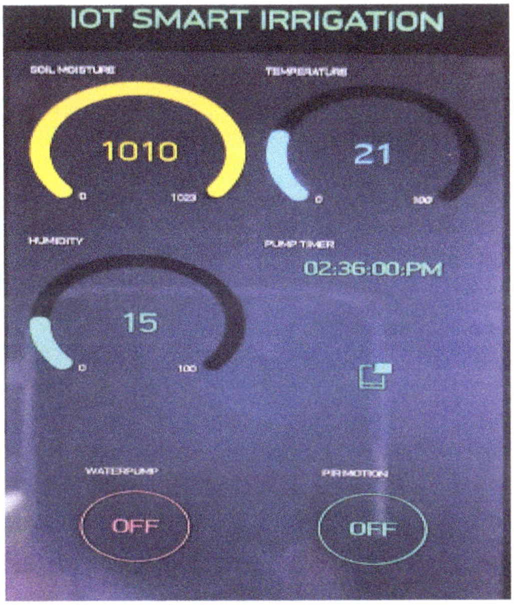

Figure 3. Dashboard display of the mobile device application.

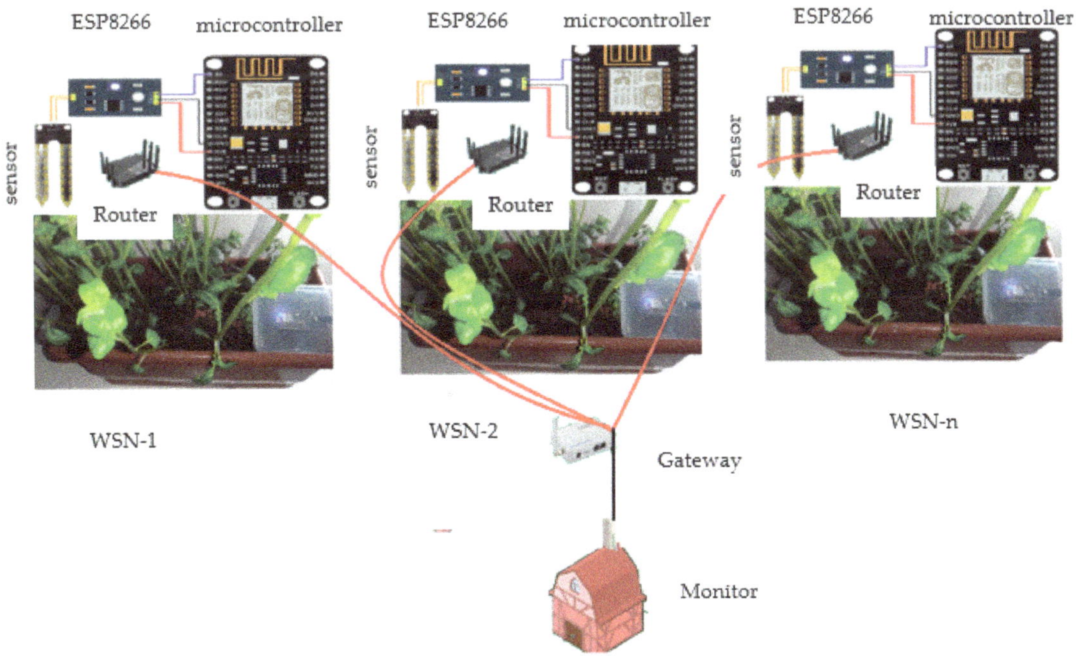

Figure 4. Implementation of IOT-based systems at medium and large scales.

The sensor and electronics are powered by DC power derived from solar panels, and the amount of power generated is determined by the rating of the solar panels. As shown in Figure 3, the solar system produces DC power for charging the battery via the charging kit. This power can be directed to the inverter by the inverter circuit. The inverter is needed to convert the DC power to AC because older water pumps run on AC power. In order to implement this system, we have used a number of hardware components, such as: (a) Solar panels, (b) Charging kits, (c) Batteries, and (d) Inverters. It has become possible to implement the software on a mobile phone. However, the oversized bloated network model is not conducive to the deployment of the device's software.

7. Real-Time Implementation

A schematic diagram of the automated irrigation system is shown in Figure 1. The system is designed with devices capable of controlling water delivery, a management unit, a soil moisture sensing unit, and an on-spot processing unit, as shown in Figure 4.

The remote database is an installation to store the contiguous farming data structure. Climatic parameters, including soil moisture level, are obtained from the sector of each sensory mote. Through the personal interface of the internet utility, the user can control the water valve by converting the positioning monitor to real-time climate situations within the field.

There are many factors that affect actual water usage, including the type of plants, the evapotranspiration rate, the amount of field area, the type of valves and the pipes, the distance between the valves, and the number of waterings per day. Taking all these parameters into consideration, and using Equation (1), we can calculate the actual amount of water that needs to be applied to the field.

$$\text{Time} = \frac{\text{Area} \times (\text{K}_\text{c} + \text{ET}_\text{r}) \times 60}{\text{F} \times \text{N} \times 100} / \text{Period} \tag{1}$$

where Time is the irrigation time in minutes, K_c is the crop coefficient (here, the coefficient ranges between approximately 7.04 mm/day and 11.7 mm/day), F is the water flow per valve in m^3/h, N is the number of valves, and Area is the field area in m^2.

Based on Equation (1) and how much data can be input to achieve accurate results, three analyses have been done and compared to the performance of a normal irrigation system:

1. Calculation of irrigation times using Equation (1) and forecast data;
2. The irrigation system is controlled by Equation (1) and sensor data;
3. Controlled irrigation system using Equation (1) and sensor data to estimate irrigation times.

There is no doubt that the concept and applications of the IoT (Internet of Things) are becoming more and more relevant in our world. This can be attributed to the fact that there is the possibility of interconnecting any object or device through the network. This can be done by using various types of communication technologies and protocols. This interconnection can be achieved by using a variety of different types of networks, some of the best-known and widely disseminated ones being WiFi (Wireless Fidelity), LTE (Long-Term Evolution), Bluetooth, and so on. Based on the proposed system, we intend to validate the implementation of a WSN environment in medium-scale crops for a specific crop variation and an area specific to the crop. The main objective of this study was to minimize the cost of the sensor devices whilst optimizing the technical characteristics that the deployed devices should have. As a result of this information, it will be possible in the future to produce on a large scale monitored and sustainable crops, as shown in Figure 5.

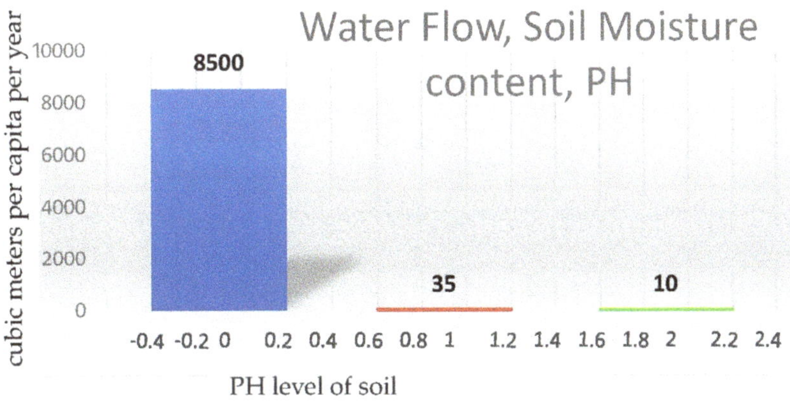

Figure 5. An overview of the output in a graphical format—soil moisture.

8. Results and Discussion

IoT applications have developed a variety of specialized communication protocols that have the potential to become long-term standards with global acceptance and that are associated with significant benefits at the time of their implementation as well. It is still critical to know the specifications and technical capabilities of each protocol in order to choose the one that is appropriate for your application and environment. Therefore, the macro project that led to the publication of this article proposed the creation of a model that could be used to analyze and simulate the behavior of the key performance indicators of the system. The information provided in this paper is structured in such a way that it becomes a very comprehensive decision-making tool that contributes to the research and development of relevant knowledge in this field as well as the implementation of it in environments.

As a result of providing the sensors with data in the form of voltage, this data is presented to the microcontroller. The microcontroller, after calibrating these values, gives us the moisture content of the soil in percentage values. The data is sent continuously from the IoT device to the online site through the wireless network to be displayed on the computer screen being used. The dashboard display is also refreshed every three seconds to make sure the real-time data availability, and just in case of possible Internet disruption, reloading page will let us know about it.

This model is designed to manage water for irrigating date palm trees. The development and improvement of date palms require a lot of water in order to grow and improve the quality of the dates at harvest time. However, it should be noted that consistent dry weather leads to insufficient water as a result of dispersal, waste, and penetration. Precise irrigation will help in achieving remarkable yields from the date palm. However, at the same time, the quality of the dates will still depend on the decision of when to harvest. Figure 5 shows the output of the proposed model in graphical form. There is no better way to display data than to use a graphical representation in the form of a diagram. Also, visual representations are much more memorable. As we not only display the moisture content (data) as text but also in visual form, this helps the user to quickly and easily notice the sudden changes in the data or information that has been displayed.

Data were collected from each sensor node of the WSN containing real-time information about the environmental conditions of the farm. The set of data collected that represents the year as a whole is shown in Figure 6. The winter season in the study area starts at the end of October and lasts until the beginning of February. November and December are the months when the weather is at its most pleasant. As a result, less water is needed for irrigation, which is in contrast with summertime, when it is mostly dry and needs a lot more water.

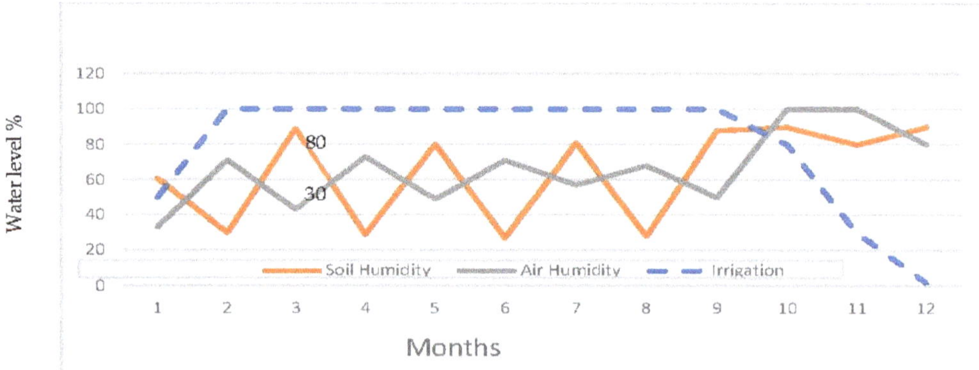

Figure 6. Data collected during the whole year in Saudi Arabia.

The first thing to note is that, in the testing period, there are no rain periods in Saudi Arabia. This means that watering is mainly done when soil moisture is less than 30%. It is also possible to note that humidity levels in Saudi Arabia are high at that time of year, which helps the soil moisture to decline more slowly. Therefore, the water used for irrigation can be less. The measurement of humidity parameters in the soil and the subsequent processing of this information constitutes a fundamental element for decision-making in the technical environments of smart agriculture. This leads, for example, to establishing the appropriate sowing and harvesting periods to improve the productivity of dates in Saudi Arabia. Smart farms are an especially significant aspect that must be addressed by systems engineering, electronic engineering, and data science. The analysis of large volumes of information using time series, big data, data mining, and other analysis methodologies allows the finding of patterns that allow intelligent decision-making. This allows us to improve and optimize different plants in the Agro sector. Thanks to the integration of the IoT and different sensors in cloud platforms, it is possible to improve many aspects related to plant growth. This includes saving on inputs and pest control.

The novelty of this work lies in the implementation of the idea through the use of components and devices available in the market as shown in Figure 6. However, it is through the development of the user interface shown in Figure 4 that the authors realized their search for real-time data acquisition. The quantifiable characterization of the soil through the use of spectral characteristics for each pixel in an image of a scene is another attractive area to be pursued for cloud-based non-intrusive irrigation systems, soil data, or underground water quality collection [27].

9. Conclusions

This proposed irrigation system can measure soil moisture and temperature as well as the atmospheric temperature of the field and can transmit the data in real-time to the user using Wi-Fi and an IoT server. The IoT-based irrigation system is superior to other irrigation systems recently proposed and developed. This is because, in the past, the process of implementing automatic irrigation was done in a traditional, luxurious, and inefficient manner, which results in a small profit margin and misfortunes in production. This paper proposes the utilization of IoT communication to develop an automated irrigation system for agricultural monitoring.

This analysis has allowed the authors to confirm that the implementation model has the capability of being more effective, accurate, and responsive in a very short period of time. Additionally, the proposed model is better than existing models based on the Internet of Things due to the advantages it offers compared to those models: by creating a dashboard based on the HTTP protocol, users will be able to change parameters such as moisture and water flow rates through IoT devices, as well as turn on/off water pumps. Future work may include the integration of sensor grids in order to be able to determine important parameters such as pH, CEC, SAR, organic constituents of the soil, and other crucial parameters. The paper is produced with a critical review of contemporary literature and the use of an algorithm to support the title. It makes a useful contribution to the body of engineering knowledge through the use of simple electronics and simple program code for application development.

Author Contributions: Conceptualization, S.A. and S.H.; methodology, M.I.; software, S.H. and M.I.; validation, A.M.A. and Abdullah. Alabdulatif formal analysis, S.H. and A.A. (Alabdulatif Alabdulatif); investigation, S.A. and M.I.; resources, A.A. (Abdullah Alabdulatif) and A. Alabdulatif; data curation, M.I.; writing—original draft preparation, S.A., S.H. and M.I.; writing—review and editing, S.A., A.M.A., A.A. (Alabdulatif Alabdulatif) and A.A. (Abdullah Alabdulatif) visualization, A.M.A.; A.A. (Alabdulatif Alabdulatif) and A.A. (Abdullah Alabdulatif).; supervision, S.A.; project administration, A.M.A.; funding acquisition, S.H.; All authors have read and agreed to the published version of the manuscript.

Funding: This research was funded by the Deputyship for Research & Innovation, Ministry of Education, Saudi Arabia for funding this research work through project number (QU-IF-2-4-5-26275).

Institutional Review Board Statement: Not applicable.

Informed Consent Statement: Not applicable.

Data Availability Statement: Not applicable.

Acknowledgments: The authors extend their appreciation to the Deputyship for Research & Innovation, Ministry of Education, Saudi Arabia for funding this research work through project number (QU-IF-2-4-5-26275). The authors also thank Qassim University for technical support.

Conflicts of Interest: The authors declare that there is no conflict of interest regarding the publication of this paper.

References

1. Hati, A.J.; Singh, R.R. Smart Indoor Farms: Leveraging Technological Advancements to Power a Sustainable Agricultural Revolution. *AgriEngineering* **2021**, *3*, 728–767. [CrossRef]
2. Sowers, J.; Vengosh, A.; Weinthal, E. Climate change, water resources, and the politics of adaptation in the Middle East and North Africa. *Clim. Chang.* **2011**, *104*, 599–627. [CrossRef]
3. Ibrahim, K.; Alnajim, A.M.; Naveed Malik, A.; Waseem, A.; Alyahya, S.; Islam, M.; Khan, S. Entice to Trap: Enhanced Protection against a Rate-Aware Intelligent Jammer in Cognitive Radio Networks. *Sustainability* **2022**, *14*, 2957. [CrossRef]
4. Kamienski, C.; Soininen, J.P.; Taumberger, M.; Dantas, R.; Toscano, A.; Salmon Cinotti, T.; Filev Maia, R.; Torre Neto, A. Smart Water Management Platform: IoT-Based Precision Irrigation for Agriculture. *Sensors* **2019**, *19*, 276. [CrossRef]
5. Albattah, W.; Habib, S.; Alsharekh, M.F.; Islam, M.; Albahli, S.; Dewi, D.A. An Overview of the Current Challenges, Trends, and Protocols in the Field of Vehicular Communication. *Electronics* **2022**, *11*, 3581. [CrossRef]
6. Iglesias, A.; Santillán, D.; Garrote, L. On the barriers to adaption to less water under climate change: Policy choices in mediterranean countries. *Water Resour. Manag.* **2018**, *32*, 4819–4832. [CrossRef]
7. Castellanos, G.; Deruyck, M.; Martens, L.; Joseph, W. System Assessment of WUSN Using NB-IoT UAV-Aided Networks in Potato Crops. *IEEE Access* **2020**, *8*, 56823–56836. [CrossRef]
8. Karavas, C.S.; Dimitriou, E.; Balafoutis, A.T.; Manolakos, D.; Papadakis, G. Development of a computational tool for the design of seawater reverse osmosis desalination systems powered by photovoltaics for crop irrigation. *J. Green Energy Sustain.* **2022**, *2*, 1–22. [CrossRef]
9. Karavas, C.S.; Arvanitis, K.G.; Papadakis, G. Optimal technical and economic configuration of photovoltaic powered reverse osmosis desalination systems operating in autonomous mode. *Desalination* **2019**, *466*, 97–106. [CrossRef]
10. Zuhaib, M.; Shaikh, F.A.; Tanweer, W.; Alnajim, A.M.; Alyahya, S.; Khan, S.; Usman, M.; Islam, M.; Hasan, M.K. Faults Feature Extraction Using Discrete Wavelet Transform and Artificial Neural Network for Induction Motor Availability Monitoring—Internet of Things Enabled Environment. *Energies* **2022**, *15*, 7888. [CrossRef]
11. Baig, M.B.; Straquadine, G.S. Sustainable agriculture and rural development in the Kingdom of Saudi Arabia: Implications for agricultural extension and education. In *Vulnerability of Agriculture, Water and Fisheries to Climate Change*; Springer: Dordrecht, The Netherlands, 2014; pp. 101–116.

12. Ratnakumari, K.; Koteswari, S. Design & implementation of innovative IoT based smart agriculture management system for efficient crop growth. *J. Eng. Sci.* **2020**, *11*, 607–616.
13. Boursianis, A.D.; Papadopoulou, M.S.; Shaouha, A.G.; Sarigiannidis, P.; Nikolaidis, S.; Goudos, S.K. Smart Irrigation System for Precision Agriculture—The AREThOU5A IoT Platform. *IEEE Sens. J.* **2021**, *21*, 17539–17547. [CrossRef]
14. Susmitha, A.; Alakananda, T.; Apoorva, M.L.; Ramesh, T.K. Automated Irrigation System using Weather Prediction for Efficient Usage of Water Resources. *IOP Conf. Ser. Mater. Sci. Eng.* **2017**, *225*, 012232. [CrossRef]
15. Ayaz, M.; Ammad-Uddin, M.; Sharif, Z.; Mansour, A.; Aggoune, E.H.M. Internet-of-Things (IoT)-based smart agriculture: Toward making the fields talk. *IEEE Access* **2019**, *7*, 129551–129583. [CrossRef]
16. Jamroen, C.; Komkum, P.; Fongkerd, C.; Krongpha, W. An Intelligent Irrigation Scheduling System Using Low-Cost Wireless Sensor Network toward Sustainable and Precision Agriculture. *IEEE Access* **2020**, *8*, 172756–172769. [CrossRef]
17. Lloret, J.; Sendra, S.; García-Fernández, J.; García, L.; Jimenez, J.M. A WiFi-Based Sensor Network for Flood Irrigation Control in Agriculture. *Electronics* **2021**, *10*, 2454. [CrossRef]
18. Fahlgren, N.; Gehan, M.A.; Baxter, I. Lights, camera, action: High-throughput plant phenotyping is ready for a close-up. *Curr. Opin. Plant Biol.* **2015**, *24*, 93–99. [CrossRef]
19. González-Amarillo, C.A.; Corrales-Muñoz, J.C.; Mendoza-Moreno, M.A.; González-Amarillo, A.M.; Hussein, A.F.; Arunkumar, N.; Ramírez-González, G. An IoT-Based Traceability System for Greenhouse Seedling Crops. *IEEE Access* **2018**, *6*, 67528–67535. [CrossRef]
20. Fernández-Ahumada, L.M.; Ramírez-Faz, J.; Torres-Romero, M.; López-Luque, R. Proposal for the Design of Monitoring and Operating Irrigation Networks Based on IoT, Cloud Computing and Free Hardware Technologies. *Sensors* **2019**, *19*, 2318. [CrossRef]
21. Mohammed, M.; Riad, K.; Alqahtani, N. Efficient IoT-Based Control for a Smart Subsurface Irrigation System to Enhance Irrigation Management of Date Palm. *Sensors* **2021**, *21*, 3942. [CrossRef]
22. Poyen, F.B.; Ghosh, A.; Kundu, P.; Hazra, S.; Sengupta, N. Prototype Model Design of Automatic Irrigation Controller. *IEEE Trans. Instrum. Meas.* **2021**, *70*, 1–17. [CrossRef]
23. Costa, N.; Rodrigues, N.; Seco, M.A.; Pereira, A. SL: A Reference Smartness Level Scale for Smart Artifacts. *Information* **2022**, *13*, 371. [CrossRef]
24. Bin Ahmadon, M.A.; Yamaguchi, S.; Mahamad, A.K.; Saon, S. Physical Device Compatibility Support for Implementation of IoT Services with Design Once, Provide Anywhere Concept. *Information* **2021**, *12*, 30. [CrossRef]
25. Kour, V.P.; Arora, S. Recent Developments of the Internet of Things in Agriculture: A Survey. *IEEE Access* **2020**, *8*, 129924–129957. [CrossRef]
26. Sivaiah, N.; Sowmya, K.P.S.; Susmitha, K.; Sai, N.A.; Suma, N. Internet of Things (IoT) Enabled Water Monitoring System. *Iconic Res. Eng. J.* **2018**, *1*, 40–43.
27. Singh, S.; Kasana, S.S. Estimation of soil properties from the EU spectral library using long short-term memory networks. *Geoderma Reg.* **2019**, *18*, e00233. [CrossRef]

electronics

Article

Evaluation of IoT-Based Smart Home Assistance for Elderly People Using Robot

Abdulrahman A. Alshdadi

Department of Information Systems and Technology, College of Computer Science and Engineering, University of Jeddah, Jeddah 23890, Saudi Arabia; alshdadi@uj.edu.sa

Abstract: In the development of Internet-of-things (IoT)-based technology, there is a pre-programmed robot called Cyborg which is used for assisting elderly people. It moves around the home and observes the surrounding conditions. The Cyborg is developed and used in the smart home system. The features of a smart home system with IoT technology include temperature control, lighting control, surveillance, security, smart electricity, and water sensors. Nowadays, elderly people may not be able to maintain their homes appropriately and may feel uncomfortable performing day-to-day activities. Therefore, Cyborg can be used to carry out the activities of elderly people. Such activities include switching off unnecessary lights, watering plants, gas control, monitoring intruders or unknown persons, alerting elderly people in emergency situations, etc. These activities are controlled by web-based platforms as well as smartphone applications. The issues with the existing algorithms include that they are inefficient, require a long time for implementation, and have high storage space requirements. This paper proposes the k-nearest neighbors (KNN) with an artificial bee colony (ABC) algorithm (KNN-ABC). In this proposed work, KNN-ABC is used with wireless sensor devices for perceiving the surroundings of the smart home. This work implements the automatic control of electronic appliances, alert signal processors, intruder detection, and performs day-to-day activities automatically in an efficient way. GNB for intruder detection in the smart home environment system using the Cyborg human intervention robot achieved an accuracy rate of 88.12%, the Artificial Bee Colony algorithm (ABC) achieved 90.12% accuracy on the task of power saving in smart home electronic appliances, the KNN technique achieved 91.45% accuracy on the task of providing a safer pace to the elderly in the smart home environment system, and our proposed KNN-ABC system achieved 93.72%.

Keywords: smart home; security; KNN; CYBORG; sensor devices; ABC; elderly people; Gaussian Naïve Bayes

Citation: Alshdadi, A.A. Evaluation of IoT-Based Smart Home Assistance for Elderly People Using Robot. *Electronics* **2023**, *12*, 2627. https://doi.org/10.3390/electronics12122627

Academic Editors: Antonio Cano-Ortega and Francisco Sánchez-Sutil

Received: 10 April 2023
Revised: 4 June 2023
Accepted: 5 June 2023
Published: 11 June 2023

1. Introduction

Due to the development of technology, dynamic changes have occurred in the automation and application of robotics and related systems. Nowadays, robotics plays a vital role in various applications, reducing the workload of human beings as well as errors made by humans. Robots are used in different surveillance processes such as detection of gas leaks and minimizing the risk of disaster through leakage in the chemical industry. Surveillance is the process of closely monitoring an industry, person, or group in the same and different situations. Surveillance is mainly needed in monitoring public places, border areas, companies, and industries in which the intervention of humans is difficult. This surveillance takes place with the help of an embedded system of robots. A robot is a pre-programmed electronic machine that replaces human work through automation and provides accurate results while minimizing error and improving time efficiency [1]. IoT-based devices are linked with one another by a network that connects electronic home appliances, vehicle-based electronic devices, actuators, and software, allows the exchange of information between one device and another. IoT devices can interact with other devices via Wi-Fi communication module by using the wireless sensor networks (WSN) in smart home electronic appliances and by Low Power Wireless Personal Area Networks (LoW-PAN) using RFID (Radio-Frequency Identification). An IoT-based smart home environment operates sensor-based devices remotely using mobile applications [2,3].

Human Interaction Robots (HIR) are mainly used in activities with a social component, such as medicine, neuroscience, cognitive science, and robotics. In order to provide security, the need for human intervention can be replaced with Cyborg. This robot can assist elderly people who are home alone, helping them to avoid crime due to home invasion or theft. In this case, it is necessary to provide security to elderly people by implementing a smart home environment system that contains the required sensor devices and it can transmit sensor signals through a communication module in order to alert the user and allow them to take precautionary steps [4,5]. The smart home secure environment enhances the lifestyle of human beings by providing security, detecting gas leakage in the kitchen, monitoring temperature and humidity in the home, detecting intruders, and more. This can be achieved by monitoring the surroundings of the smart home using a Raspberry Pi-based wireless camera, capturing images with related information, and sending it to the server. The main components of Cyborg are DC motors, a battery, and a wheel chassis, and it can be implemented in either automatic or manual mode [6].

Many research works have been implemented in smart home environments. The main issues are that they are inaccurate and inefficient, consumption time is high, and large amounts of storage space are required. This paper proposes a smart home environment for assisting elderly people using the KNN-ABC technique. It uses sensor-based electronic home appliances to monitor the surroundings of the smart home, detect intruders, and generate an alert notification to a registered mobile device or through a mobile app.

The contributions of this work are as follows:

1. To implement smart home assistance for elderly people by using the KNN algorithm to monitor the status of electronic home appliances and provide an ON/OFF state using Cyborg.
2. To save energy in smart home systems using the Artificial Bee Colony (ABC) algorithm.
3. Analysis of the proposed KNN-ABC using the metric measures of precision, recall, F1-score, and accuracy.

This paper is written in five sections including this introduction. In the remainder of this paper, Section 2 discusses relevant previous works on smart home systems, Section 3 describes the proposed methodology, Section 4 describes the results and evaluates the outcomes, and Section 5 concludes the paper.

2. Related Work

Smart home electronic appliances based on IoT technology require automatic ON/OFF operation using a remote control-based application, voice-based technology, or fixed-time scheduling. A notification can be sent to the user by the server. This control is completely based on the activities of the user and passing the commands which can be triggered the activities through the mobile phone [7–9]. C. Victor et al. [10] proposed an IoT-based sensor system for monitoring the temperature in the environment. Using a temperature sensor, the system can collect sensor signals and store them in the server. Gladence et al. [11] proposed a client–server-based machine learning algorithm implemented for establishing an automated smart home environment control system able to interact with humans who send commands or triggering the smart appliances. M. Wendy et al. [12] presented a review of effective smart home technology to support elderly people in aspects related to health and security issues. Mehmood et al. [13] proposed an innovative concept involving managing a cloud storage platform, detecting hindrances, activating IoT devices by passing commands, executing those commands, and then transmitting the information to the registered users via mobile notification. To monitor health-related issues for elderly people in smart homes, various machine learning algorithms (LSTM, SVM, and RNN) can be used. IoT devices can closely observe health conditions of elderly people, analyze their symptoms, and make predictions related to disease, as well as helping patients to consult their physicians and alert them to take medicine at the proper time [14]. Sensor devices are used with wireless networks, software, and computers to detect threats which affect the smart home environment. The implementation of the CNN model produces efficient detection of threats [15]. The Cyborg system can be used to save power, as it is able to automatically switch unnecessary electronic devices into the OFF state. In addition, it can detect the presence of human beings in the external surroundings of the smart home. At the same time, it can send a notification to the resident to perform important activities such as taking medicine, watering plants, etc. The proposed smart home system interfaces with sensor devices and assists elderly people in the smart home environment based on the generated sensor signals [16]. Table 1 enumerates related works on smart home environment systems along with the technology and sensor measurements employed by the respective systems.

Many earlier works demonstrated the use of IoT technology for energy efficiency, monitoring, and activity detection in a smart home environment. Below, we present selected works, which are tabulated in Table 1 along with their prominent features.

In [17], the author presented a smart home remote control system based on wireless sensor networks that collect positioning information and use actuators to control electrical appliances and operate alarms. In [18], X. Gengyi applied support vector machine (SVM) in a smart-home energy monitoring system using a cloud computing-based platform. The proposed solution improves energy efficiency and makes it easier for human interaction. In [19], C. Zhou et al. proposed a design for a smart home system based on virtual reality. Virtual reality was used to improve control interaction in the smart home. Their experimental results indicated that control methods could be simplified and costs reduced by as much as twenty percent through the use of virtual reality. In [20], P. Sharma et al. proposed a design for an IoT system using NodeMCU for real-time supervision of sensor measurements, allowing the user to control electrical loads in a smart home. O. Taiwo et al. [21] proposed a smart home automation mobile application that uses an Arduino microcontroller and personal area communication technologies such as Zigbee and Bluetooth. The practicality of the system was demonstrated through a simulation of the smart home environment.

In [22], M. S. Soliman et al. proposed a smart home automation system based on Arduino and Labview that allows the user to control temperature, save energy, and detect intruders. M. Naing et al. (2019) [23] demonstrated a proposed smart home automation system through a prototype implementation employing two Arduino Nano sensors. Sensors for measuring temperature, smoke, and motion were interfaced with these microcontrollers, which in turn interfaced with actuators to control and secure the home. R. D. Manu et al. (2019) [24] proposed a smart home system able to measure and respond to human activities using long-short term memory (LSTM) deep learning-based decision-making. S. K. Saravanan et al. (2019) [25] proposed a smart home controller using Arduino and Android. A smart door actuator was secured using a multi-factor authentication mechanism. L. D. Liao et al. (2019) [26] proposed the design of a smart home system using Arduino–Uno that provides user control and monitoring through a mobile application. Temperature and motion sensors were connected and controlled by the system to demonstrate its application in a smart home environment.

D. Popa et al. (2019) [27] demonstrate a smart home application where measurements of energy consumption and other sensor data could be stored on a cloud and later analyzed using machine learning methods for improved environmental sustainability and energy efficiency.

The authors of [28] applied linear discriminant analysis to classify power quality disturbances and carry out a performance analysis using KNN, naive Bayes, support vector machine (SVM), and random forest (RF) classifiers. Their results showed that higher classification accuracy was obtained in the presence of noise. In [29], Moraes et al. used a naive Bayes algorithm to propose a structured data mining model that can predict whether a smaller enterprise can join a business association with given attributes. The proposed approach can be utilized as a decision assistance tool for business associations to choose member enterprises. In [30], the authors used four different classifiers, i.e., KNN, naive Bayes, decision tree, and random forest approaches, to distinguish between defective and non-defective metal parts using laser-induced-breakdown spectroscopy. The above-mentioned works show that machine learning algorithms can be used to make accurate predictions and to inform decisions in many situations and for a variety of data formats.

To make the literature survey more comprehensive, below we include several recent optimization methods for feature selection and classification. The authors of [31] proposed a hybrid feature selection method using a combination of the Butterfly optimization algorithm and the Ant Lion optimizer for breast cancer prediction. The proposed hybrid method outperforms both component methods for breast cancer diagnosis in terms of accuracy, sensitivity, specificity, and area under the receiver operating characteristic curve.

Chakraborty et al. [32] proposed an improved whale optimization method for segmentation of chest X-Rays from patients with symptoms of COVID-19. During the global search phase, a random initialization is used to exploit after exploration. The proposed method outperformed the original method in terms of segmentation accuracy.

Sayed et al. [33] adopted a hybrid approach combining a convolutional neural network with Bald Eagle optimization to improve detection performance in melanoma skin cancer prediction. The robustness and accuracy of the proposed approach were verified as being superior through a comparison with state-of-the-art methods.

Xing et al. [34] proposed a modified whale optimization method using a quasi-Gaussian "bare bones" method. The modified method was able to promote diversity and expand the scope of the solution space.

Piri et al. [35] proposed a modified optimization method based on the Harris Hawk optimizer. This method, called multi-objective quadratic binary Harris Hawk optimization, uses a KNN classifier to extract the optimal feature subsets. The proposed methodology proved superior thanks to its better combination of fitness assessment criteria.

Table 1. List of related works on smart home environment systems and on the technologies and sensor measurement approaches employed by the respective systems.

Author	Technology	Monitoring Function
Ruili Zheng (2022) [17]	IoT	Indoor smart monitoring and modern lifestyle.
X. Gengyi (2021) [18]	Machine Learning algorithm SVM	Energy monitoring in a smart home system
C. Zhou et al. (2021) [19]	Virtual Reality	Classifying human activity with R&D
P. Sharma et al. (2020) [20]	Cloud server-based NodeMCU	Electricity measurements in a smart home
O. Taiwo et al. (2020) [21]	Zigbee, Bluetooth, and Arduino technologies	Health care monitoring system
M. S. Soliman et al. (2020) [22]	LabVIEW	PIR Motion Sensor, indoor detection in a smart home
M. Naing et al. (2019) [23]	Arduino-based smart home control	Temperature, humidity sensor
R. D. Manu et al. (2019) [24]	IoT-based Deep Learning algorithm	Motion sensor, PIR sensor
S. K. Saravanan et al. (2019) [25]	Android-based smart home control	Multiple authentication processes, privacy preservation using key generation
L. D. Liao et al. (2019) [26]	Android-based smart home control	health care monitoring, multifunctional operating systems
D. Popa et al. (2019) [27]	Deep Learning	Energy reduction and power saving

3. Proposed Methodology

This section describes the proposed IoT-based Smart Home Assistance system for elderly people using Cyborg.

3.1. Cyborg in Smart Home Assistance Technology

The proposed model KNN-ABC with the Cyborg is designed for applications in monitoring the surroundings of the smart home and assisting elderly people. This robotic model is specifically designed to switch off unnecessary lights, water plants, detect gas leaks, monitor for intruders or unknown persons, alert elderly residents in emergency situations, etc. These tasks are handled by sensor devices connected via Wi-Fi and control interfaces. This control system interface in the smart home allows elderly people to easily access activities in an efficient manner. It satisfies the basic requirements of elderly people by cleaning the home, detecting gas leaks in the kitchen, setting alarms and reminders about important work, and more. The KNN-ABC system uses PIR (Passive Infrared Sensor), a type of electronic sensor used to set alerts for security and automatic ON/OFF operation of fans and lights. The presence or absence of a human in the room is detected using the ZIGBEE communication protocol. For the detection of gas leaks, an LPG sensor is used to produce an alert signal. To clean the house using a vacuum cleaner, the Cyborg robot is moved around the smart home surroundings using the follower technique. The IR sensor (Infrared Sensor) is used to detect obstacles along the movement path of the robot, and it produces the buzzer that makes a beeping sound. For reminders about medication scheduling and important activities, a real-time clock (RTC) is used to set alarms and produce reminder messages on the LCD screen. Figure 1 shows the overall diagram of the proposed smart home assistance system for elderly people using KNN-ABC.

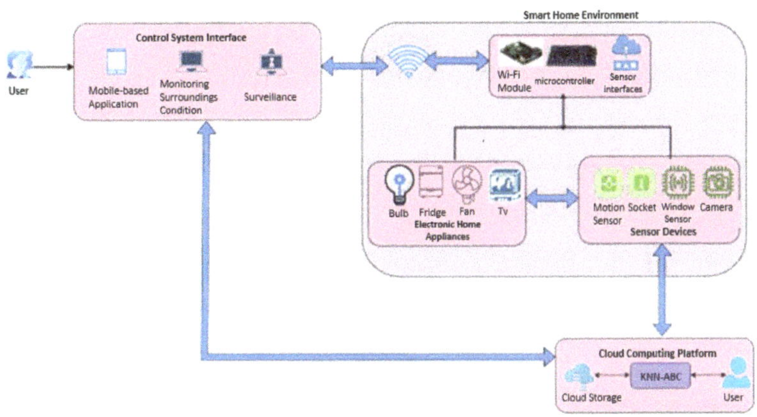

Figure 1. Framework of the proposed smart home assistance system for elderly people.

Figure 1 shows the smart home assistance control for elderly people that controls the electronic appliances, sensor devices, electronic home appliances, control system interfaces, and cloud-based computing platform modules. The KNN-ABC system module contains the home environment, control system interface for the user, sensor-based electronic home appliances, and cloud computing platform. The user can communicate with the home environment via Wi-Fi using either a mobile-based or web-based application. The home environment contains sensor devices and communication modules, and can be linked with electronic home appliances. The received sensor signals and their information are stored in the cloud storage platform back-end of the KNN-ABC system model.

3.1.1. Control System Interface

In the KNN-ABC model, the interaction between the user and the home environment at the front-end takes place in the control system interface via Wi-Fi through the mobile-based or web-based application. The surveillance camera live stream monitors the surroundings of the home environment. Using the customized Android application, users can easily access the smart home surroundings status (e.g., humidity, temperature, ON/OFF status of electronic home appliances such as fans and lights, presence of intruders). Data generated by the various sensors are stored in the cloud computing platform for future reference. The control application of the smart home is uses Android. The live stream of the surveillance camera can be displayed on a laptop, mobile device, or desktop computer.

3.1.2. Smart Home Environment

The KNN-ABC-based smart home environment comprises three modules: the communication interface module, electronic home appliances module, and sensor device module. The sensor devices are incorporated with IoT-based electronic home appliances and linked together using a microcontroller to provide communication with the smart home's outer surrounding environment via the wireless network. The microcontroller uses the ESP8266 module and Wi-Fi with HTTPS/IP and TCP/IP as the communication protocol. In order to function, the microcontroller requires a power supply and the ESP32 camera module in an Arduino board. The in-built function of the Wi-Fi chip has an ESP32 camera board with wireless connectivity. Therefore, ESP8266 and ESP32 are used for communication.

3.1.3. Cloud Computing Platform

This is an intelligent module that sends commands to the system to carry out various activities on behalf of elderly residents. It comprises a cloud storage system that stores the signals received from sensor devices for future reference. The smart home assistance system for elderly people is enhanced using the KNN-ABC algorithm, and is used for monitoring the surroundings of the smart home environment, detecting gas leaks, monitoring the surveillance camera, automatically turning electronic home appliances on and off, generating alert signals for medication scheduling, and providing other alert notifications about any unusual activities that may be occurring inside the smart home environment. The steps involved in the Smart Home Assistance System for Elderly People using Cyborg is shown in Figure 2.

3.2. *K-Nearest Neighbour Algorithm (KNN)*

The smart home environment uses a robot named Cyborg, which is used to assist elderly people by switching off unnecessary lights, watering plants, gas control, monitoring for intruders, providing alerts in emergency situations, etc. Sensor signals are collected from various electronic home appliances

$$s = \{ea_1, ea_2, ea_3, \ldots, ea_n\}, \tag{1}$$

where s represents the sensor signal values and $ea_1, ea_2, ea_3, \ldots, ea_n$ represent the various electronic home appliances.

The sensor signals of electronic home appliances are used to detect the ON/OFF states of appliances perform other functions such as intruder detection using the KNN algorithm. If an intruder is detected, the sensor signal values of electronic home appliances and coordinate value of the safer place are collected and stored in the system as a dataset. When the KNN-ABC system model is activated, it analyzes and detects the nearest coordinate value from the stored is dataset and predicts a safe place for the residents to go. The steps involved in the implementation of KNN algorithm are provided below:

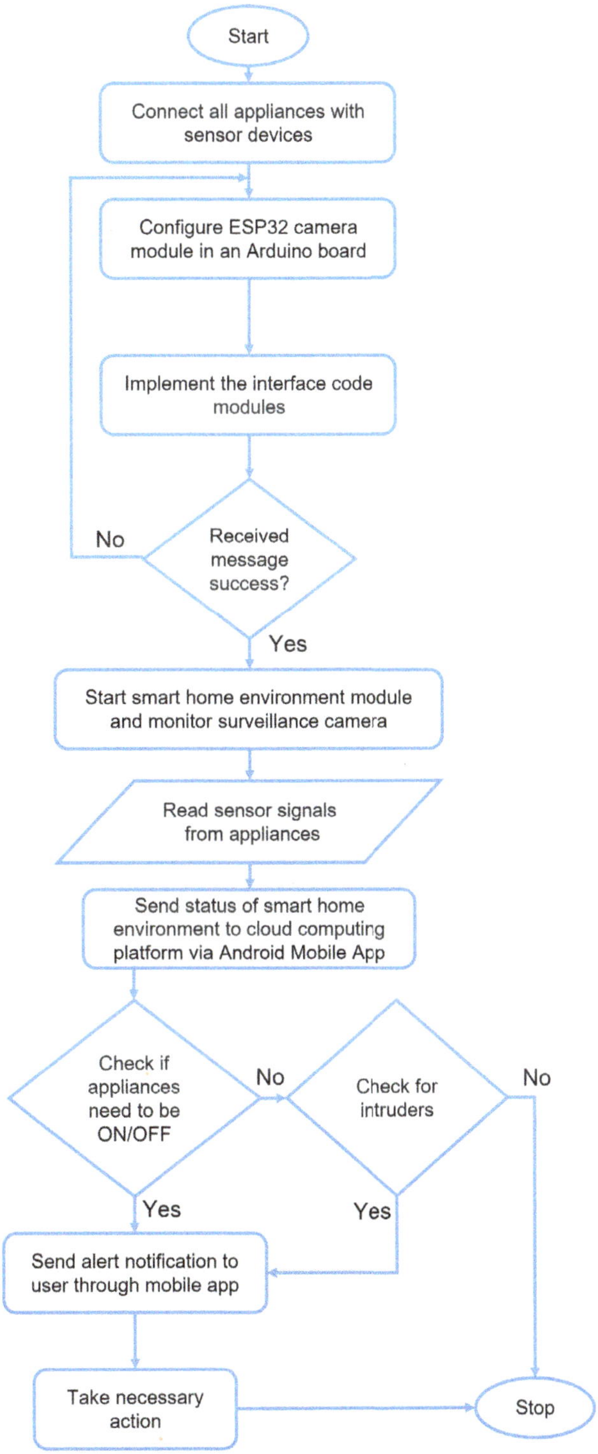

Figure 2. Working framework of smart home system for assisting elderly people.

- KNN-1: Used to train the model of KNN-ABC model using dataset.
- KNN-2: Used to evaluate the distance between the user's location and the coordinate position of value in the dataset, as follows:

$$dist(p_1, p_2) = \sqrt{\sum_{i=1}^{n}(p1_i - p1_t)^2 + (p2_i - p2_t)^2} \quad (2)$$

Here, $dist(p1, p2)$ is the Euclidean distance between the current location and the targeted location in the dataset.
- KNN-3: Sorts all distances in ascending order to select the nearest point.
- KNN-4: The value of $dist(p1, p2)$ is processed by the ABC algorithm as a control input for ON/OFF control of electronic home appliances in the smart home system. It can monitor the sensor signal values using a mobile-based or web-based application.

3.3. *Artificial Bee Colony Algorithm for Power Saving in Smart Home System*

In the smart home, the environment uses a robot named Cyborg to assist elderly people by switching off unnecessary lights, watering plants, gas control, monitoring for intruders, providing alerts during emergency situations, etc. Using the ABC algorithm, it detects the nearest safe place for residents to move to during an emergency situation. Similarly, the collected sensor signals from the electronic home appliances are stored in the dataset. to obtain more accurate and efficient detection of the ON/OFF state, the Artificial Bee Colony algorithm (ABC) is implemented. The ABC algorithm functions by connecting a socket system with the camera sensor and reading the sensor signals from the various electronic home appliances, gas sensor, camera, and PoseNet human positional sensor. It collects all real time information, including the position of human beings in the smart home, at regular intervals of time and stores it in the cloud storage platform in a dataset using the communication module. The locations of human beings can then be retrieved from cloud storage and compared with the location of electronic appliances to check the distance between the human being and the electronic appliances. If the human being is far away from electronic appliances, then the system receives instructions to turn off the unnecessary smart electronic appliances; otherwise, it can use the information to operate the smart electronic appliances in safety mode. The ABC procedure is explained below.

- ABC 1: Initialize the population size; M is the initial nectar coordinate value and maximum number of iterations.
- ABC 2: Search for a new nectar source by selecting bees from the total number of number of bees N; the bee group size is M, and the spatial dimensionality for bees searching for new nectar sources is S. From the current nectar source, a bee starts its searching process within its neighborhood. The newly created nectar source contains:
  - ABC 2.1: The spatial dimension space of the current nectar source is split into regular intervals based on the following formula:

$$R_{k,l}^{h} = Q_k^l + \frac{(2h - H)}{H}(Q_k^l - Q_n^l), h \in [0, H], \quad (3)$$

  where R denotes the h-th interval of point from division of the current nectar source, Q denotes k-th current honey source generated in the l-th dimension space, and Q denotes the n-th current honey source generated in the l-th dimension space.
  - ABC 2.2: In each interval of R, the interval is divided into several sub-intervals Y based on the formula

$$R_{k,l}^{h,z} = R_{k,l}^{h} + \sin\left(\frac{y - rand(0,1)}{2y}\pi\right)\left(R_{k,l}^{h} - R_{k,l}^{h+1}\right), y \in [1, Y] \quad (4)$$

where R denotes the y-th sub-interval of the current nectar source, $rand(0,1)$ denotes the random distribution of values between 0 and 1 at a uniform rate, and $R+1$ denotes the $R+1$-th interval point produced by division of the next current nectar source.

- ABC 2.3: For every sub-interval of the current nectar source, its fitness function is calculated, then the sub-interval point is selected based on the largest fitness value.
- ABC 2.4: The difference between Q and the fitness value representing the nectar source is evaluated using the following formula:

$$F_k^l = min\{fit(F_{k,l}^h) - fit(R_k^l)\}, fit(F_{k,l}^h) - fit(R_k^l) > 0, h \in [1, H] \quad (5)$$

where F_k denotes the difference between Q and the fitness value of the nectar source, $fit(v)$ represents the fitness value of thee nectar source in the regular interval of R, and $fit(q)$ denotes the fitness value of R.

- ABC 2.5: F_i is selected as the nectar source, and it is treated as a new source of nectar.

- ABC 3: Compute the fitness value for newly created nectar source compared to the current nectar source.
- ABC 4: Compare the fitness value of the newly created nectar source with the current nectar source.
- ABC 5: Discard the nectar source with the lower fitness value.
- ABC 6: Based on the probability of the nectar source value, select pickers for following the bees.
- ABC 7: For the selected pickers, update the nectar source values.
- ABC 8: For the current nectar source value, search the closest nectar source.
- ABC 9: Repeat steps 2–5 and retain the nectar source with the largest fitness value.
- ABC 10: Increment the iteration.
- ABC 11: Stop the search process when it reaches the maximum iteration and choose the highest fitness value as the coordinate value of the target node in the nectar source.

The ABC algorithm can achieve the an effective search process in terms of locating the target node; it has good accuracy for determining the position of human beings in a smart home environment and can provide power savings for electronic home appliances in the smart home control system. Despite these advantages, it is inefficient in remote control of smart electronic home appliances. Therefore, the ABC algorithm is modified by implementing the following steps.

- KNN-ABC 1: Randomly generate the initial nectar source from the values of M based on the target nectar source of the bees. Based on the target nectar source with the maximum fitness value, randomly generate the initial nectar source values N using

$$Q_l^{I0} = Q_{min}^{I0} + rand(0,1)(Q_{max}^{I0} - Q_{min}^{I0}), \quad (6)$$

where Q_l is the l-th initial honey nectar source generated in the k-th dimension spatial space, Q_{min} is the minimum nectar source value of the k-th source of honey, Q_{max} represents the k-th source of honey generated in the spatial dimensional space, and $rand(0,1)$ generates random numbers between 0 and 1 and is uniformly distributed in the system.

- KNN-ABC 2: For every nectar source of honey, compute the reverse honey nectar source using

$$Q_l^{I0'} = rand(0,1)(Q_{max}^{I0} - Q_{min}^{I0}) - Q_l^{I0}, \quad (7)$$

where Q_l is the the reverse nectar source of honey for the l-th initial generation of honey in the k-th spatial dimensional space.

- KNN-ABC 3: Compute the fitness values for all initial nectar sources as well as for the reverse nectar sources. Based on their fitness values, arrange them in descending order to generate the nectar source value set. The first N nectar honey sources are chosen as the target nodes.
- KNN-ABC 4: To improve the quality of the initial nectar honey source by ensuring an even distribution of nodes in the system, in the smart home environment control system the position of the target node is based on the terms of its speed and stability. This allows for more accurate remote control of all smart electronic home appliances. The fitness value is evaluated using

$$fit(F) = \min\left(\sqrt{((a - q_l)^2 + (b - q_l)^2 - D\lambda_{pi})}\right), k = 1, 2, \ldots, \phi \tag{8}$$

where $fit(F)$ denotes the fitness value of the nectar honey source F at the position of the l-th beacon node, D denotes the average hop distance which is sent by the first beacon node of nectar Q, q_l denotes the total number of hops between the l-th beacon node and the source of honey, and ϕ is the number of beacon nodes.

Using Equation (8) and the modified artificial bee colony algorithm (ABC) improves the accuracy of finding the target node while minimizing the error rate, and is able to provide both effective remote control of smart home electronic appliances and position monitoring of the electronic home appliances via exact angle measurement.

4. Result and Discussion

The proposed KNN-ABC system was implemented in Python 3.6, and was compared with Gaussian Naive Bayes (GNB) [36], Artificial Bee Colony algorithm (ABC) [17], and KNN [37]. Table 2 shows the measures of Precision, Recall and Sensitivity for the different algorithms used in the smart home system. Data were collected from the Kaggle website [38].

Precision: Precision quantifies the number of true positive predictions provided by a given technique. It is calculated as follows:

$$\text{Precision} = \frac{TP}{TP + FP} \times 100. \tag{9}$$

Recall: The percentage of correctly classified true positive predictions is evaluated by calculating the recall, as follows:

$$\text{Recall} = \frac{TP}{TP + FN}. \tag{10}$$

F1-Score: The F1-Score is a measure of accuracy based on precision and recall values. It is calculated as follows:

$$\text{F1-Score} = \frac{2 \times \text{Precision} \times \text{Recall}}{\text{Precision} + \text{Recall}}. \tag{11}$$

Specificity: Specificity is used to measure the proportion of actual negative cases that a technique rightly predicts. Specificity is calculated as follows:

$$\text{Specificity} = \frac{TN}{TN + FP} \times 100. \tag{12}$$

Accuracy:

$$\text{Accuracy} = \frac{TP + TN}{TP + TN + FP + FN} \times 100. \tag{13}$$

MSE: The mean squared error (MSE) calculates the average of the squares of the differences between the predicted values and actual values.

$$\text{MSE} = \frac{1}{n}\sum_{i=1}^{n}(y_{pi} - y_{ai})^2. \tag{14}$$

MAE: The mean absolute error (MAE) calculates the average of the squares of the differences between the predicted values and actual values.

$$\text{MAE} = \frac{1}{n}\sum_{i=1}^{n}|y_{pi} - y_{ai}|. \tag{15}$$

Table 2. Metrics of Precision and Recall in Smart Home System.

Algorithm	Precision	Recall	Sensitivity
GNB	65.16%	55.35%	68.76%
ABC	72.11%	67.65%	70.37%
KNN	66.78%	68.46%	71.11%
KNN-ABC	88.32%	78.54%	83.65%

Table 2 shows a performance comparison between the proposed KNN-ABC technique and existing algorithms. Here, GNB using Smart Home assistance for elderly people using the Cyborg robot reached a sensitivity of 68.76%, precision of 65.16%, and recall of 55.35%, the Artificial Bee Colony algorithm (ABC) for power saving in smart home electronic appliances reached a sensitivity of 70.37%, precision of 72.11%, and recall of 67.65%, and the KNN technique for detecting the a nearest safer place in an emergency situation reached a sensitivity of 71.11%, precision of 66.78%, and recall of 68.46%. The proposed KNN-ABC technique for detecting intruders, finding the nearest safe place in an emergency, and saving power on smart home electronic appliances attained a sensitivity of 83.65%, a precision of 88.32%, and a recall of 78.54%. Figure 3 shows the F1 scores of the different algorithms tested for smart home assistance for elderly people using the Cyborg system. The F1 score is the weighted harmonic mean of the precision and recall, with 0.0 being the worst and 1.0 being the best.

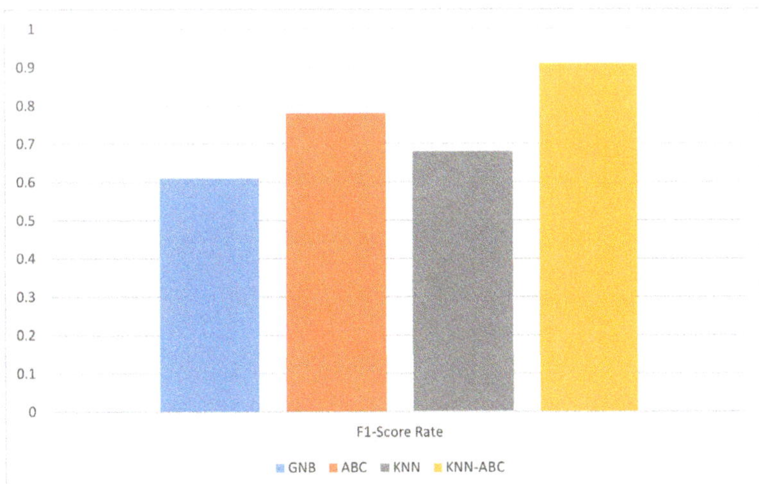

Figure 3. F1-Score.

Figure 3 shows the F1 scores of the techniques used in the comparative analysis. Our proposed work produced the best result at 0.91, while the GNB algorithm produced the worst result at 0.61. The error rate of the smart home assistance system with the different algorithms is shown in Table 3.

Table 3. Error rates.

Algorithm	MAE	MSE
GNB	0.295	0.088
ABC	0.321	0.379
KNN	0.273	0.218
KNN-ABC	0.154	0.056

From Table 3, it can be seen that our proposed KNN-ABC approach had the lowest error rate and produced better outcomes than the other algorithms. Figure 4 shows the correlation matrix for predicting intruders in the smart home environment.

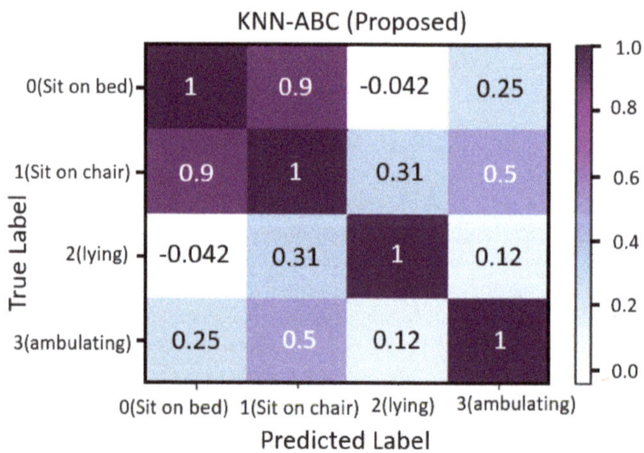

Figure 4. Confusion matrix.

In Figure 4, the diagonal values are not meaningful as they are self-correlated, i.e., with the variable itself. The values shown to the left and right of diagonal are considered mirror images of each other. The highly correlated variables are shown as darker boxes. Here, standing activities of intruders are highly correlated with one another. Therefore, detection of intruder with activity is predicted as sitting on the bed. Figure 5 shows the accuracy rate of the different tested techniques.

From Figure 5 shows the accuracy rates of the different techniques in the smart home environment system: for detection of intruders, GNB reached 88.12%; for power saving, Artificial Bee Colony algorithm (ABC) reached 90.12%; for determining the safest place to go in an emergency situation, the KNN technique reached an accuracy rate of 91.45%; finally, our proposed KNN-ABC approach reached 93.72%. Figure 6 shows the computation times for the various techniques tested in the smart home system.

Figure 6, shows that of the various techniques, our proposed KNN-ABC approach requires the lowest computation time.

Limitations and Future Work: This work considers a new hybrid approach, i.e., KNN-ABC, which combines the virtues of the KNN and ABC methods to improve on these methods as well as on the GNN method in terms of recall rate, precision, accuracy, F1 score, and computational complexity. However, the performance evaluation in this paper was restricted to smart home environment applications, and warrants further investigation to determine its improvement over other machine learning methods. Such an investigation and comparison that considers more advanced machine learning classifiers and data from other applications is recommended as a future extension of this work.

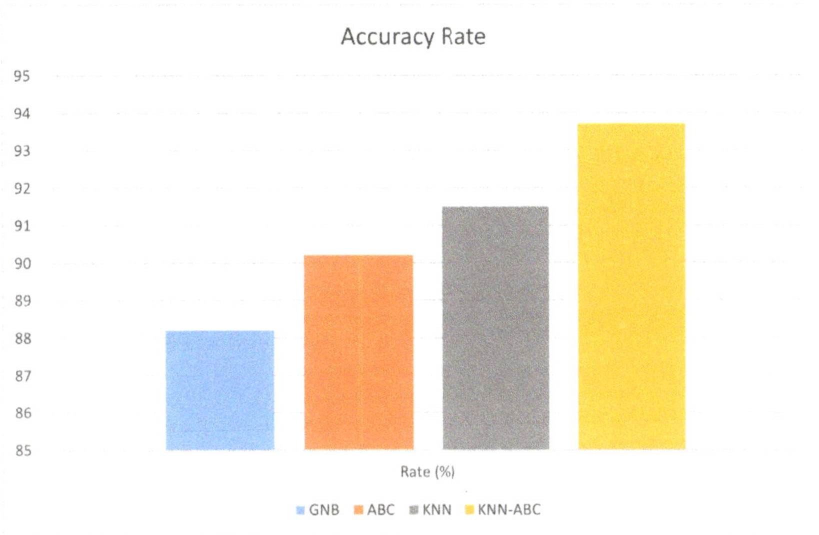

Figure 5. Accuracy rates.

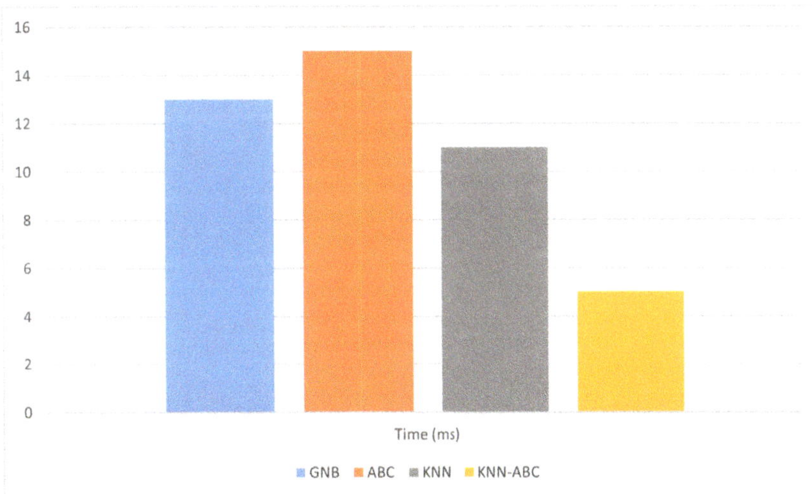

Figure 6. Computation times.

5. Conclusions

This paper proposes the implementation of a smart home environment control system through various sensor modules, control modules, and a human intervention robot named Cyborg; the system is used for control of the automatic ON/OFF state of various electronic appliances, detection of intruders in the smart home, and sending various alert notifications to the user. This system is intended to assist elderly people in an effective and efficient way with a fast response time. The accuracy rate of the various techniques was tested; GNB for the detection of intruders in the smart home environment system using the human-intervention robot named Cyborg reached 88.12%, the Artificial Bee Colony algorithm (ABC) for power saving in smart home electronic appliances reached 90.12%, the KNN technique for predicting safe locations in an emergency situation reached an accuracy rate of 91.45%, and our proposed KNN-ABC algorithm reached 93.72%. In the future, this work could be extended by providing the Cyborg robot system with more security features, including biometric concepts such as facial recognition and fingerprint identification. In summary, the proposed KNN-ABC algorithm shows better accuracy, precision, and recall rate than the GNB, KNN, and ABC algorithms. Better performance means more accurate prediction of activities and can lead to user satisfaction, a greater sense of security and safety, and improved quality of life.

Funding: This research received no external funding.

Data Availability Statement: Not applicable.

Conflicts of Interest: The authors declare no conflict of interest.

References

1. Sumathi, S.; Aditya, S.; Archanaa, B.; Priya, G.L. Cyborg—A Surveillance Droid Using Raspberry Pi and Internet of Things. *Int. Res. J. Eng. Technol.* **2020**, *7*, 529.
2. Wang, C.; Liu, Q.; Xing, L.; Guan, Q.; Yang, C.; Yu, M. Reliability analysis of smart home sensor systems subject to competing failures. *Reliab. Eng. Syst. Saf.* **2022**, *221*, 108327. [CrossRef]
3. Alghayadh, F.; Debnath, D. Hid-smart: Hybrid intrusion detection model for smart home. In Proceedings of the 2020 10th Annual Computing and Communication Workshop and Conference (CCWC), Las Vegas, NV, USA, 6–8 January 2020; IEEE: Piscataway, NJ, USA, 2020; pp. 0384–0389.
4. Alghayadh, F.; Debnath, D. A hybrid intrusion detection system for smart home security based on machine learning and user behavior. *Adv. Internet Things* **2021**, *11*, 10–25. [CrossRef]
5. Cele, B. Quarter One Crime Statistics, South African Government. 2022. Available online: https://www.gov.za/speeches/minister-bheki-cele-quarter-one-crime-statistics-20222023-19-aug-2022-0000 (accessed on 7 April 2023).
6. Taiwo, O.; Ezugwu, A.E. Internet of things-based intelligent smart home control system. *Secur. Commun. Netw.* **2021**, *2021*, 1–17. [CrossRef]
7. Dhanusha, C.; Kumar, A.S. Deep recurrent Q reinforcement learning model to predict the Alzheimer disease using smart home sensor data. In *Proceedings of the IOP Conference Series: Materials Science and Engineering*; IOP Publishing: Bristol, UK, 2021; Volume 1074, p. 012014.
8. Gupta, P.; McClatchey, R.; Caleb-Solly, P. Tracking changes in user activity from unlabelled smart home sensor data using unsupervised learning methods. *Neural Comput. Appl.* **2020**, *32*, 12351–12362. [CrossRef]
9. Pattamaset, S.; Choi, J.S. Irrelevant data elimination based on a k-means clustering algorithm for efficient data aggregation and human activity classification in smart home sensor networks. *Int. J. Distrib. Sens. Netw.* **2020**, *16*, 1550147720929828. [CrossRef]
10. Chang, V.; Martin, C. An industrial IoT sensor system for high-temperature measurement. *Comput. Electr. Eng.* **2021**, *95*, 107439. [CrossRef]
11. Gladence, L.M.; Anu, V.M.; Rathna, R.; Brumancia, E. Recommender system for home automation using IoT and artificial intelligence. *J. Ambient. Intell. Humaniz. Comput.* **2020** , 1–9. Available online: https://link.springer.com/article/10.1007/s12652-020-01968-2 (accessed on 7 April 2023).
12. Moyle, W.; Murfield, J.; Lion, K. The effectiveness of smart home technologies to support the health outcomes of community-dwelling older adults living with dementia: A scoping review. *Int. J. Med Inform.* **2021**, *153*, 104513. [CrossRef]
13. Mehmood, F.; Ullah, I.; Ahmad, S.; Kim, D. Object detection mechanism based on deep learning algorithm using embedded IoT devices for smart home appliances control in CoT. *J. Ambient. Intell. Humaniz. Comput.* **2019**, 1–17. Available online: https://link.springer.com/article/10.1007/s12652-019-01272-8 (accessed on 7 April 2023).
14. Mshali, H.; Lemlouma, T.; Moloney, M.; Magoni, D. A survey on health monitoring systems for health smart homes. *Int. J. Ind. Ergon.* **2018**, *66*, 26–56. [CrossRef]
15. Alghayadh, F.; Debnath, D. A hybrid intrusion detection system for smart home security. In Proceedings of the 2020 IEEE International Conference on Electro Information Technology (EIT), Chicago, IL, USA, 31 July–1 August 2020 ; IEEE: Bristol, UK , 2020; pp. 319–323.
16. Thomas, A.; Joseph, G.B.; Augustine, M. CYBORG-The Smart Home Assistance Robot. *Int. Adv. Res. J. Sci. Eng. Technol.* **2017**, *4*, 118–120. [CrossRef]
17. Zheng, R. Indoor smart design algorithm based on smart home sensor. *J. Sens.* **2022**, *2022*, 2251046 [CrossRef]
18. Xiao, G. Machine learning in smart home energy monitoring system. In *Proceedings of the IOP Conference Series: Earth and Environmental Science*; IOP Publishing: Bristol, UK, 2021; Volume 769, p. 042035.

19. Zhou, C.; Huang, T.; Liang, S. Smart home R&D system based on virtual reality. *J. Intell. Fuzzy Syst.* **2021**, *40*, 3045–3054.
20. Sharma, P.; Kantha, P. 'Blynk' cloud server based monitoring and control using 'NodeMCU'. *Int. Res. J. Eng. Technol.* **2020**, *7*, 1362–1366.
21. Taiwo, O.; Ezugwu, A.E.; Rana, N.; Abdulhamid, S.M. Smart home automation system using zigbee, bluetooth and arduino technologies. In Proceedings of the Computational Science and Its Applications–ICCSA 2020: 20th International Conference, Cagliari, Italy, 1–4 July 2020; Proceedings, Part VI 20; Springer: Berlin/Heidelberg, Germany, 2020; pp. 587–597.
22. Soliman, M.S.; Alahmadi, A.A.; Maash, A.A.; Elhabib, M.O. Design and implementation of a real-time smart home automation system based on arduino microcontroller kit and labview platform. *Int. J. Appl. Eng. Res.* **2017**, *12*, 7259–7264.
23. Naing, M.; Hlaing, N.N.S. Arduino based smart home automation system. *Int. J. Trend Sci. Res. Dev.* **2019**, *3*, 276–280. [CrossRef]
24. Manu, R.D.; Kumar, S.; Snehashish, S.; Rekha, K. Smart home automation using IoT and deep learning. *Int. Res. J. Eng. Technol.* **2019**, *6*, 1–4.
25. Saravanan, S.; Nainar, A.; Marichamy, S. Android based smart automation system using multiple authentications. *IRE J.* **2019**, *3*, 60–65.
26. Liao, L.D.; Wang, Y.; Tsao, Y.C.; Wang, I.J.; Jhang, D.F.; Chu, T.S.; Tsao, C.H.; Tsai, C.N.; Chen, S.F.; Chuang, C.C.; et al. Design and validation of a multifunctional android-based smart home control and monitoring system. *IEEE Access* **2019**, *7*, 163313–163322. [CrossRef]
27. Popa, D.; Pop, F.; Serbanescu, C.; Castiglione, A. Deep learning model for home automation and energy reduction in a smart home environment platform. *Neural Comput. Appl.* **2019**, *31*, 1317–1337. [CrossRef]
28. Singh, G.; Pal, Y.; Dahiya, A.K. Classification of Power Quality Disturbances using Linear Discriminant Analysis. *Appl. Soft Comput.* **2023**, *138*, 110181. [CrossRef]
29. Moraes, J.d.; Schaefer, J.L.; Schreiber, J.N.C.; Thomas, J.D.; Nara, E.O.B. Algorithm applied: Attracting MSEs to business associations. *J. Bus. Ind. Mark.* **2020**, *35*, 13–22. [CrossRef]
30. Lin, J.; Yang, J.; Huang, Y.; Lin, X. Defect identification of metal additive manufacturing parts based on laser-induced breakdown spectroscopy and machine learning. *Appl. Phys. B* **2021**, *127*, 1–10. [CrossRef]
31. Thawkar, S.; Sharma, S.; Khanna, M.; kumar Singh, L. Breast cancer prediction using a hybrid method based on Butterfly Optimization Algorithm and Ant Lion Optimizer. *Comput. Biol. Med.* **2021**, *139*, 104968. [CrossRef] [PubMed]
32. Chakraborty, S.; Saha, A.K.; Nama, S.; Debnath, S. COVID-19 X-ray image segmentation by modified whale optimization algorithm with population reduction. *Comput. Biol. Med.* **2021**, *139*, 104984. [CrossRef]
33. Sayed, G.I.; Soliman, M.M.; Hassanien, A.E. A novel melanoma prediction model for imbalanced data using optimized SqueezeNet by bald eagle search optimization. *Comput. Biol. Med.* **2021**, *136*, 104712. [CrossRef] [PubMed]
34. Xing, J.; Zhao, H.; Chen, H.; Deng, R.; Xiao, L. Boosting whale optimizer with quasi-oppositional learning and Gaussian barebone for feature selection and COVID-19 image segmentation. *J. Bionic Eng.* **2023**, *20*, 797–818. [CrossRef]
35. Piri, J.; Mohapatra, P. An analytical study of modified multi-objective Harris Hawk Optimizer towards medical data feature selection. *Comput. Biol. Med.* **2021**, *135*, 104558. [CrossRef] [PubMed
36. Shen, J.; Fang, H. Human activity recognition using gaussian naive bayes algorithm in smart home. In *Proceedings of the Journal of Physics: Conference Series*; IOP Publishing: Bristol, UK , 2020; Volume 1631, p. 012059.
37. Siddiq, M.; Wibawa, I.; Kallista, M. Integrated Internet of Things (IoT) technology device on smart home system with human posture recognition using kNN method. In *Proceedings of the IOP Conference Series: Materials Science and Engineering*; IOP Publishing: Bristol, UK , 2021; Volume 1098, p. 042065.
38. Kaghazgarian, M. Activity Recognition with Healthy Older People. 2018. Available online: https://www.kaggle.com/datasets/marklvl/activity-recognition-with-healthy-older-people (accessed on 7 April 2023).

sensors

MDPI

Article

Unified Environment for Real Time Control of Hybrid Energy System Using Digital Twin and IoT Approach

Lamine Chalal [1,2,*], Allal Saadane [1] and Ahmed Rachid [2]

1 Icam School of Engineering, Lille Campus, 6 Rue Auber, B.P 10079, CEDEX, 59016 Lille, France; allal.saadane@icam.fr

2 Laboratory of Innovative Technologies, University of Picardie Jules Verne, 80000 Amiens, France; rachid@u-picardie.fr

* Correspondence: lamine.chalal@icam.fr; Tel.: +33-202263943

Abstract: Today, climate change combined with the energy crisis is accelerating the worldwide adoption of renewable energies through incentive policies. However, due to their intermittent and unpredictable behavior, renewable energy sources need EMS (energy management systems) as well as storage infrastructure. In addition, their complexity requires the implementation of software and hardware means for data acquisition and optimization. The technologies used in these systems are constantly evolving but their current maturity level already makes it possible to design innovative approaches and tools for the operation of renewable energy systems. This work focuses on the use of Internet of Things (IoT) and Digital Twin (DT) technologies for standalone photovoltaic systems. Based on Energetic Macroscopic Representation (EMR) formalism and the Digital Twin (DT) paradigm, we propose a framework to improve energy management in real time. In this article, the digital twin is defined as the combination of the physical system and its digital model, communicating data bi-directionally. Additionally, the digital replica and IoT devices are coupled via MATLAB Simulink as a unified software environment. Experimental tests are carried out to validate the efficiency of the digital twin developed for an autonomous photovoltaic system demonstrator.

Keywords: digital twin; internet of things; real-time monitoring; rule-based control; battery; standalone photovoltaic system

Citation: Chalal, L.; Saadane, A.; Rachid, A. Unified Environment for Real Time Control of Hybrid Energy System Using Digital Twin and IoT Approach. *Sensors* **2023**, *23*, 5646. https://doi.org/10.3390/s23125646

Academic Editors: Antonio Puliafito, Antonio Cano-Ortega and Francisco Sánchez-Sutil

Received: 3 May 2023
Revised: 3 June 2023
Accepted: 12 June 2023
Published: 16 June 2023

1. Introduction

Renewable energy systems, such as solar panels and wind turbines generate power from natural resources that are available intermittently. Since they produce variable power, their effective dissemination can be accelerated by better control and monitoring. Traditionally, data acquisition systems (DAQs), usually centralized, are used for collecting all system data [1]. However, the cost of commercial DAQs is the most significant barrier for greater diffusion. IoT (Internet of things) based smart meters have recently gained substantial popularity for control and measurement data. Indeed, their ability to communicate data over networks offers a wide range of applications. Therefore, IoT devices can potentially be useful for real-time energy management. Accordingly, energy production efficiency and reliability can be significantly improved ([2–4]). This can help reduce dependence on traditional fossil fuel-based energy sources and promote renewable energy.

However, implementing IoT in renewable energy systems faces several challenges due to the variety of protocols and devices available in the market [5]. This can make it difficult to integrate IoT devices while ensuring their compatibility with the existing infrastructure [6].

On the other hand, defined as the combination of the physical system and its digital model, the Digital Twin (DT) paradigm can be used to predict energy production and consumption. It also enables predictive maintenance [7]. Therefore, it is crucial to implement techniques that provide a wide range of operational data about the actual system [8]. IoT technology can be associated with DT thanks to their ability of actuation and sensing.

In this work, our objective is to build a DT of a standalone PV system to deal with real-time energy management challenges. Figure 1 depicts the diagram of the digital twin as developed in this work:

- Physical system: made up of two PV panels, batteries, DC loads, a solar emulator, and a weather station.
- Smart sensors and actuators: including devices required for control and monitoring purposes.
- Digital counterpart implemented in MATLAB: including mainly EMR-based model, Real-Time monitoring, and control systems.

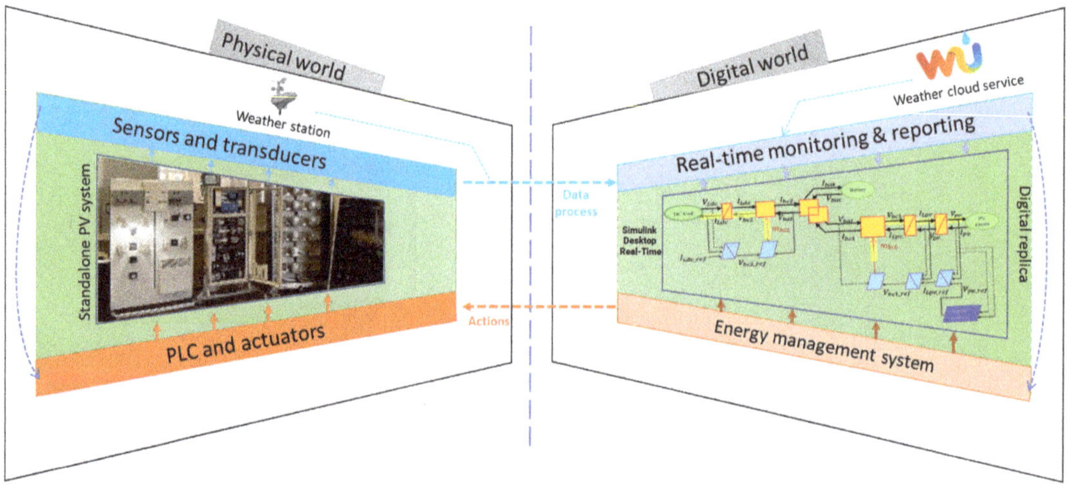

Figure 1. Diagram of the digital twin of the standalone PV system.

To experimentally validate the proposed framework, we have developed a lab demonstrator according to IoT-based architecture for embedded and distributed instrumentation. Furthermore, to cope with the devices' heterogeneity, we use MATLAB Simulink. Indeed, this is a comprehensive software environment that can communicate with sensors and Programmable Logic Controllers through client/server applications.

This document is structured as follows. First, we present a review of the literature related to our work. In the second part, we detail our test bench which constitutes the physical part of the proposed Digital Twin. In the third part, we present an approach based on the Energetic Macroscopic Representation (EMR) formalism as a numerical counterpart of our system. Finally, the experimental validation of the resulting digital twin will be discussed.

2. Literature Review

Several studies have considered photovoltaic systems monitoring. In [9], the authors used the IoT and MQTT (MQTT: Message Queuing Telemetry Transport) in web-based monitoring. They implemented this approach to monitor the performance of the solar panels and the battery system, as well as the energy consumption of a living laboratory. Similar work is reported in [10]. When compared to other protocols, MQTT has a small footprint, making it much more suitable for resource-constrained environments. Despite several benefits, it is important to note that MQTT brokers do not provide the same level of entity authentication or encryption capabilities [11]. Moreover, IoT devices are often not interoperable, and it is difficult to integrate external sources of information and cloud computing to use energy more efficiently. Indeed, it requires the design and implementation of hierarchical architectures and standardized solutions to facilitate interoperability. Till today, no standard solution is established yet [12]. However, most providers share IoT middlewares, which has fostered the emergence of cross-domain applications.

In [13], the authors developed a Smart Home monitoring system using Power Line Communication (PLC) which has the advantage of not needing additional communication cables [14]. This article also demonstrates the potential of using PLCs to monitor individual photovoltaic panels.

Moreover, the review work [1] provides an overview on the importance of monitoring systems for photovoltaic plants (electrical and meteorological data). The article reviews different types of monitoring systems that are currently available for PV plants, including hardware and software aspects. The authors discuss the advantages and disadvantages of each type of system and provide examples of commonly used components.

There exist several commercial software for monitoring and simulation of PV systems such as LabVIEW ([15–18]) and MATLAB Simulink ([19,20]).

Furthermore, studies in [20–24], present an energy management system by using Programmable Logic Controller. Compared to other hardware control systems, PLCs have specific advantages as ruggedness, noise immunity, modularity, low cost, and small footprints [25].

We also reviewed several papers dealing with Digital Twins. This concept is particularly popular in the context of industry 4.0, where it is mainly implemented for manufacturing systems [26]. Nowadays, there is a significant trend to apply this concept to the electrical energy field [27]. That said, there are few concrete applications. Moreover, as there are several misconceptions about digital twin definition [28], it is important to distinguish the digital twin, the digital model, and the digital shadow. In fact, the digital model is defined as a digital copy of a physical system without any data exchange and is generally used for simulation and design purposes. Likewise, the Digital Shadow is a combination of a physical system and its digital model with a one-way data exchange.

Although some authors claim to use the concept of the Digital Twin as just defined above, most of the articles deal with the "digital model" or "digital shadow". For instance, authors in [27] study the digital twin possibilities for fault diagnosis purpose of PV system. However, they use a digital shadow instead of digital twin. We find the same confusion in the articles [29–33].

Relatively to our contribution, Table 1 summarizes the state of art of the current literature dealing with digital twin applications.

In this work, we propose a Digital Twin of a complete photovoltaic system using MATLAB software as a unified environment. This framework is suitable to address the following topics:

- Real-time access to multi-protocol data for monitoring purposes.
- Modelling, simulation, and real-time control of PV systems
- Implementing of innovative energy management systems
- Reporting

In other words, this experimental platform can be used to compare simulation results and monitored data in real time. Indeed, it could be used to develop new approaches for fault detection and prediction issues. This integrated environment allows on the one hand to have a large panel of toolboxes (modelling, code generation, machine learning, advanced control, cloud computing . . .). Therefore, it could be easily used for advanced control and optimization purposes.

Table 1. Literature review related to IoT and DT usage for energy management.

Reference	IoT Capability	Multi-Protocols	Energy Management	Monitoring	Digital Model	Digital Shadow	Digital Twin
[34]	✓	✓	✗	✓	✗	✗	✗
[29]	✓	-	✗	✓	✓	✓	✗
[35]	✗	✗	✗	✓	✓	✓	✗
[9]	✓	✓	✓	✓	✗	✗	✗
[36]	✓	✓	✗	✓	✗	✗	✗
[37]	✓	✓	✗	✓	✗	✗	✗
[16]	✗	✓	✗	✓	✗	✗	✗
[30]	✓	-	✗	✓	✓	✓	✗
[38]	✓	✓	✗	✓	✗	✗	✗
[39]	✗	-	✓	✓	✗	✗	✗
[10]	✓	-	✓	✓	-	-	✗
Proposed	✓	Unified environment	✓	✓	✓	✓	✓

3. Materials and Methods

The concept of the digital twin, object of this work, is implemented on a test rig. In this section we describe the structure of the demonstrator, its instrumentation and control system.

3.1. PV System Description

The stand-alone system is composed of the following elements (Figure 2):

- Solar emulator as artificial light source,
- 2 × 215 Wp photovoltaic panels (SunPower Co., San José, CA, USA),
- 28 Ah batteries as storage system (Victron Energy B.V., Almere, The Netherlands),
- DC loads.
- Power converters (Victron Energy B.V., Almere, The Netherlands).

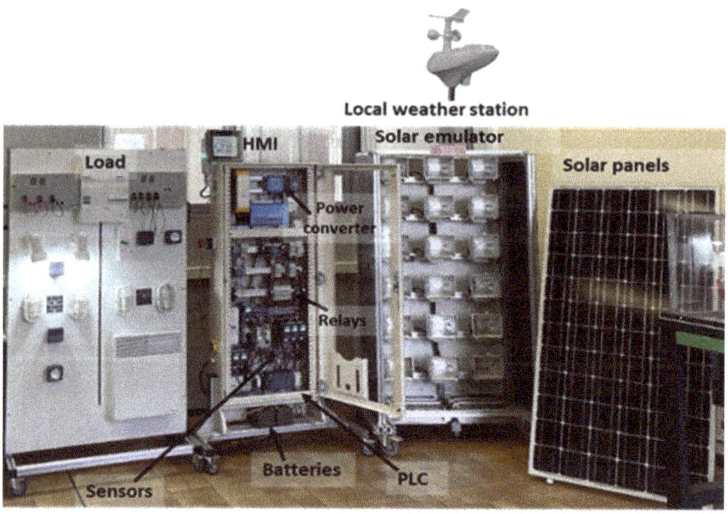

Figure 2. PV system demonstrator.

3.2. Sensors and Data Aquisition

Developing a digital twin for renewable energy systems requires constant data collection and monitoring [40]. Therefore, it is necessary to implement techniques that provide a wide range of data:

- Weather data (temperature, irradiance, wind speed ...);
- Real time electrical data (energy, currents, voltages, batteries' state of charge ...).

The experimental platform is equipped with sensors of various technologies that do not use the same communication protocol. To transmit measurement data to a unified software environment, we have developed a hardware and software architecture based on the following components:

- An IoT architecture combining smart sensors for electrical data [41] and a weather station. As IoT, these devices use heterogeneous communication protocols [42], including Modbus TCP/IP, HTTP, and PROFINET;
- MATLAB Simulink, as an integrated environment concentrating all the operational data of the system, constitutes the core of the digital twin.

Figure 3 depicts the demonstrator overview and its control system architecture.

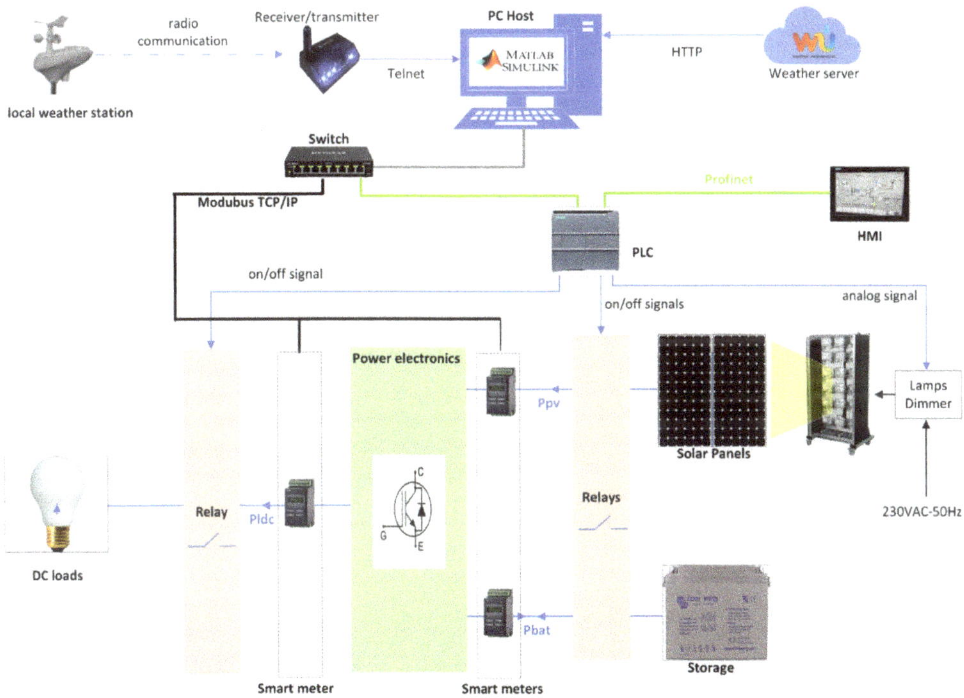

Figure 3. Control system architecture of the PV plant.

3.3. Control System

At first level, the embedded control system is based on a S7-1200 PLC which controls solar emulator and relays. This allows experimental tests to be carried out according to real-time weather data or with historical data.

At the second level, the overall control of the PV system is performed within MATLAB Simulink programs using the embedded PLC as a slave. Data communication is carried out using OPC-UA and Industrial Communication Toolbox™.

Figure 4 describes the unified architecture proposed in this work.

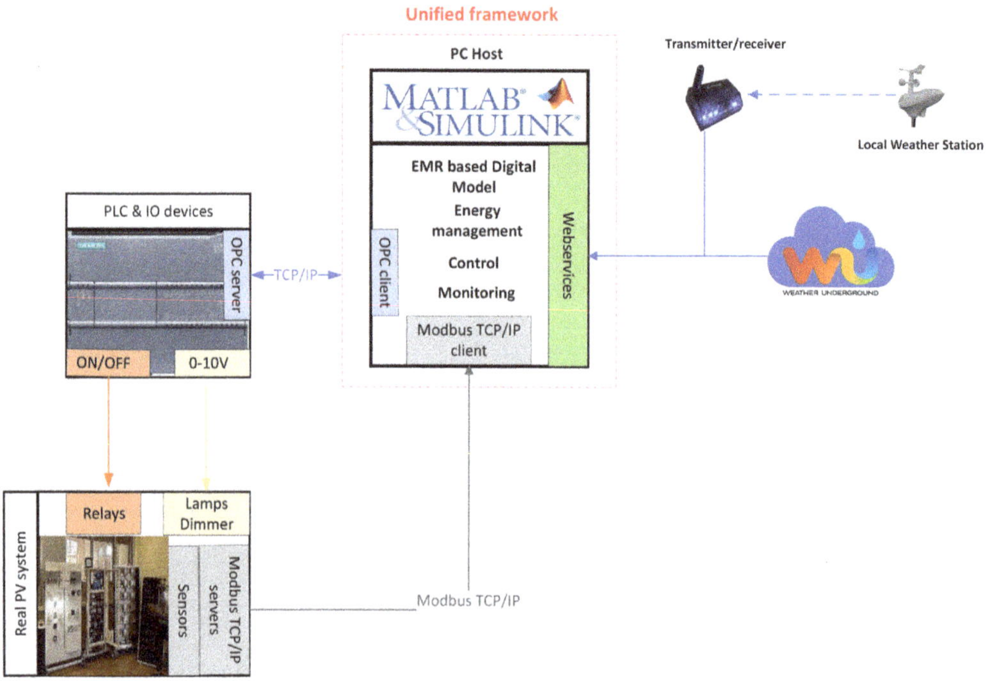

Figure 4. Digital Twin architecture overview.

4. PV System Modeling

This section details our approach to modeling the PV system. The aim is to develop the digital replica counterpart of the Digital Twin.

4.1. Solar Panels

A photovoltaic (PV) power system consists of two solar panels (SPR-215-WHT). PV parameters of each panel are given in Table 2.

Table 2. Parameters of the SPR-215-WHT solar panel at STC form datasheet.

Parameter	Value
Maximum power P_{max}	215 Wp
Voltage at maximum power point V_{mp}	39.8 V
Current at maximum power point I_{mp}	5.40 A
Open circuit voltage V_{oc}	48.3 V
Short Circuit Current I_{sc}	5.80 A
Voltage temperature coefficient β_{oc}	−136.8 mV/°C
Current temperature coefficient α_{sc}	3.5 mA/°C
Number of cells per module n_s	72

PV panel is composed of several cells. Each PV cell is made up of semiconductor materials which can convert solar irradiance into electrical energy. Based on the electronics theory of semiconductor p-n junction, it can be described by a current source. The studied panel model in this work is represented by an equivalent circuit. It consists of a single diode for the cell polarization function and two resistors (series and shunt) for the losses (Figure 5). The equivalent circuit is composed of an ideal current source I_{ph} in parallel, reverse diode, series resistance R_s and parallel resistance R_{sh}.

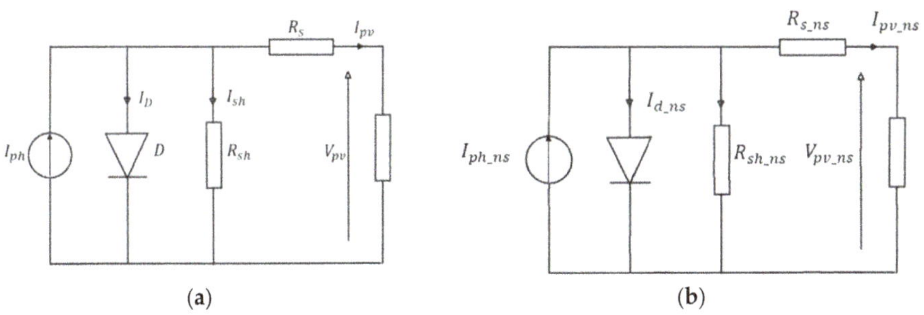

Figure 5. Equivalent circuit model. (**a**) PV cell; (**b**) PV panel composed of n_s cells in series.

The $I_{pv} = f(V_{pv})$ characteristic of this model is given by the following equation:

$$I_{pv} = I_{ph} - I_D - \left(\frac{V_{pv} + R_s I_{pv}}{R_{sh}}\right) \quad (1)$$

where:

- I_{pv} is the current generated by solar panel;
- I_{ph} is the photocurrent, which is linearly proportional to irradiance and depends on the temperature as shown in the following equation:

$$I_{ph} = (I_{phn} + \alpha_{sc}\Delta T)\frac{G}{G_n}$$

  - where $\Delta T = T - T_n$ ($T_n = 25$°C), G is the incident of irradiation on the solar panel, and G_n (1000 W/m^2) at standard conditions (STC);

- I_D is the diode current:

$$I_D = I_0[\exp\left(\frac{qV_{pv}}{AKT}\right) - 1] \quad (2)$$

  - where: I_0 is the PV cell reverse saturation current that mainly depends on the temperature, q is the electronic charge of an electron (1.6×10^{19} C), T is the temperature of the PV cell, k is Boltzmann's constant (1.38×10^{23} J/K), and A is the diode ideality factor.

A PV panel is made up of numerous identical PV cells connected in series to provide a higher voltage. A PV module composed of n_s identical cells in series can also be represented by the equivalent circuit shown Figure 4a but the circuit needs to be modified [43] as follows (Figure 5b):

$$R_{s\_ns} = n_s R_s,\ R_{sh\_ns} = n_s R_{sh},\ A_{ns} = n_s n,\ I_{0\_ns} = I_0,\ I_{ph\_ns} = I_{ph} \quad (3)$$

The solar panel datasheet does not include some important parameters, such as R_s and R_{sh}. To obtain them, we used the PV array tool of Matlab by setting the PV parameters of the studied solar panel. Table 3 shows the extracted parameters.

Table 3. Parameters of the SPR-215-WHT solar panel at STC form datasheet.

Parameters	Values
R_{s_ns}	0.72262 Ω
R_{sh_ns}	198.7727 Ω
I_0	3.8896×10^{-15} A
A_{ns}	0.74816

The mathematical model of the studied solar panel is then implemented under MATLAB Simulink and the results are compared to datasheet data as shown in Figure 6.

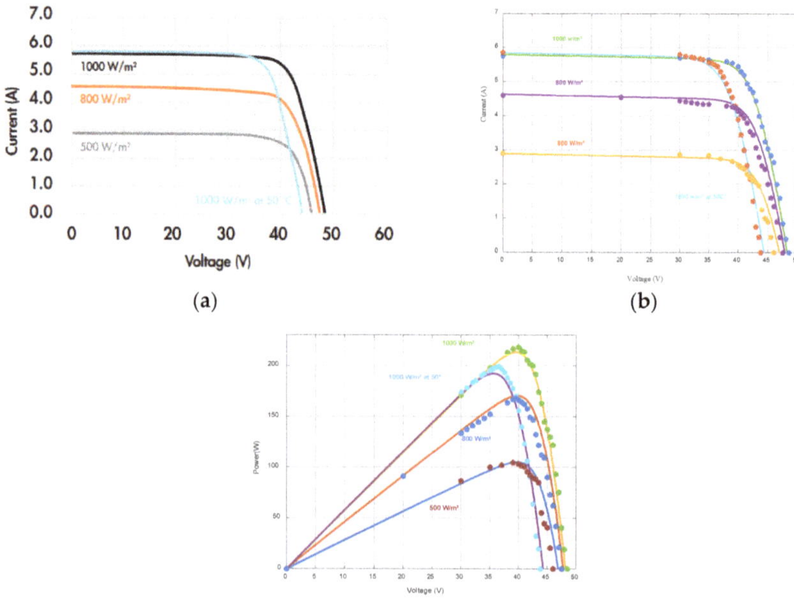

(**a**) (**b**)

(**c**)

Figure 6. Actual data vs. data computed by the model: (**a**) I-V characteristics from datasheet; (**b**) actual I-V characteristics vs. computed I-V characteristics; (**c**) actual P-V characteristics vs. computed P-V characteristics.

The dotted curves in Figure 5 represent data from the datasheet and the continuous ones come from the mathematical model. These results show a good correlation between the model and the solar panel manufacturer's data.

To assess the model's effectiveness, we conducted indoor tests of the PV panel. The PV panel was illuminated using an artificial light source, while its output was connected to a rheostat. To obtain the I-V curves, measuring instruments were integrated, as depicted in Figure 7a.

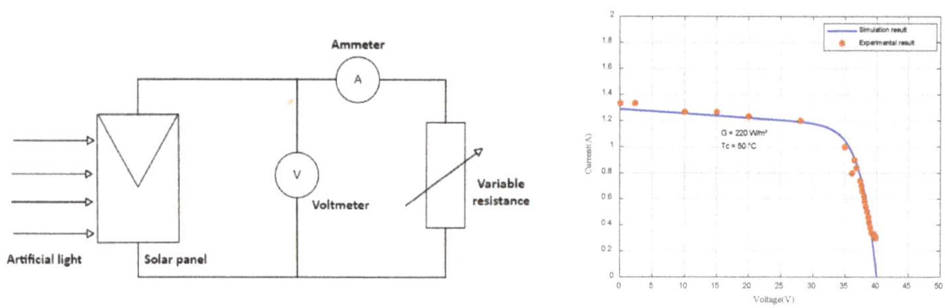

(**a**) (**b**)

Figure 7. Experimental test and comparison of I-V curves: (**a**) experimental components of the tested PV panel; (**b**) comparison between real test and the mathematical model.

Figure 7b shows the simulation and experimental results, confirming the accuracy of the implemented mathematical model.

4.2. Batteries Bank

Due to the inherent variability of PV power, the battery plays a crucial role in standalone PV systems. In this system, two 12 V AGM batteries are installed and connected in a series configuration. The utilization of AGM batteries provides additional benefits, as they facilitate recombination and effectively mitigate gas emissions during overcharge. Consequently, the demand for room ventilation is reduced, and the batteries do not emit any acid fumes during normal discharge operations [44].

The battery model in Figure 8 considers a constant internal resistance. This resistance is connected in series with the internal voltage source which depends on various parameters [45].

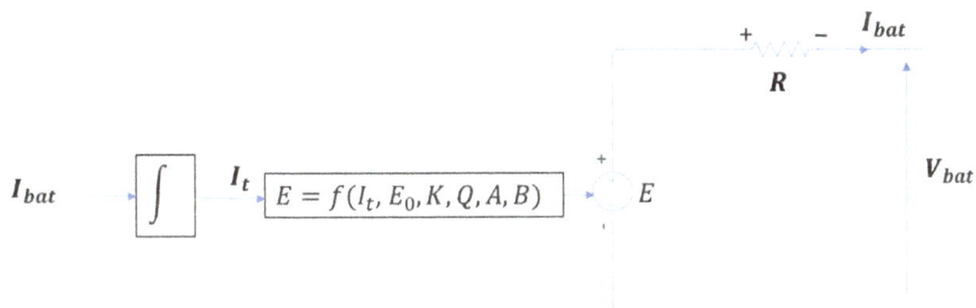

Figure 8. Nonlinear Battery model.

The terminal voltage of the battery is given by:

$$V_{bat} = E - RI_{bat}(t) \tag{4}$$

where:

- R: is the internal resistance;
- I_{bat}: is the battery current.

The controlled voltage source E is described by the following equation:

$$E = E_0 - K\frac{Q}{Q - \int I_{bat}(t)dt} + Aexp(-B\int I_{bat}(t)dt) \tag{5}$$

where:

- E_0 : Fully charged voltage;
- K: polarization voltage (V);
- Q : battery capacity (Ah);
- A: exponential zone amplitude (V);
- B: exponential zone time constant inverse (Ah).

The parameters of this equivalent circuit can be identified by considering the discharge characteristics with a nominal current.

We use a lead-acid battery, which features a nominal voltage of 12 volts (E_0), a capacity of 14 Ah (Q), a rated current of 0.7 A, and an internal resistance of 14 mΩ (Table 4).

Table 4. Parameters of an AGM battery from datasheet.

Capacity	Nominal Voltage	Internal Resistance	Weight
14 Ah/0.7 A	12 V	14 mΩ	4.05 kg

The model parameters (K = 0.12, A = 0.66, B = 7) are obtained through graphical estimation based on the rated current discharge characteristic curve. The simulation results are depicted in Figure 9. The Q-V discharge curve is composed of three sections. The first section represents the exponential voltage drop if the battery is initially fully charged. The second phase represents the charge that can be extracted from the battery until the voltage drops below the battery nominal voltage. Finally, the third section represents the total discharge of the battery, when the voltage drops quickly. The width of these sections depends on the battery type.

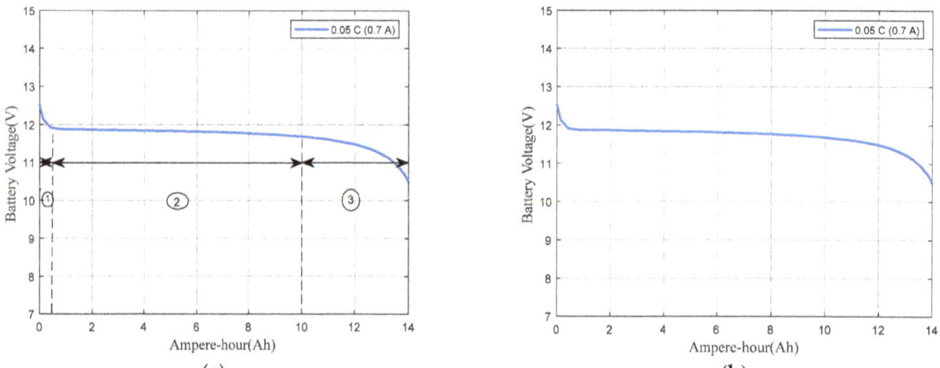

Figure 9. (**a**) Discharge curve (Q-V); (**b**) Discharge curve (Hours-V).

To estimate the state of charge of the battery (SoC), the well-known Coulomb counting method is used due to its simplicity. It relies on measuring the current and the estimation of the initial state of charge of the battery.

$$SoC(t) = SoC_0 - \frac{1}{C_{nom}} \int_{t_0}^{t} i_{batt}(t)dt * 100 \tag{6}$$

4.3. Power Electronics

Figure 10 illustrates the simplified PV system architecture, comprising a solar panel, a battery for energy backup, and power converters (MPPT regulator) that connect these components to the DC load.

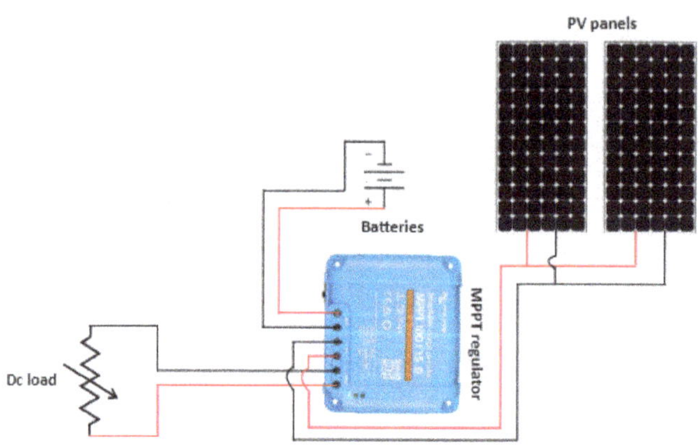

Figure 10. Simplified PV system architecture.

We determined the converter topology based on the current state of the art since it was not specified by the manufacturer. Photovoltaic energy harvesting relies primarily on irradiance and solar panel temperature, resulting in variable PV voltage. Therefore, implementing Maximum Power Point Tracking (MPPT) is essential to maximize power extraction regardless of weather conditions. A DC-DC converter is used to enable voltage adjustment from the photovoltaic panels to match the battery voltage. This converter performs step-up and step-down functions by controlling the input-to-output voltage ratio through the variation of the converter switching device's on-off duty cycle. Another converter is required to supply and regulate the voltage for the DC load. In the literature, this configuration is commonly referred to as a two-stage DC-DC converter (Figure 11), as reported in [46].

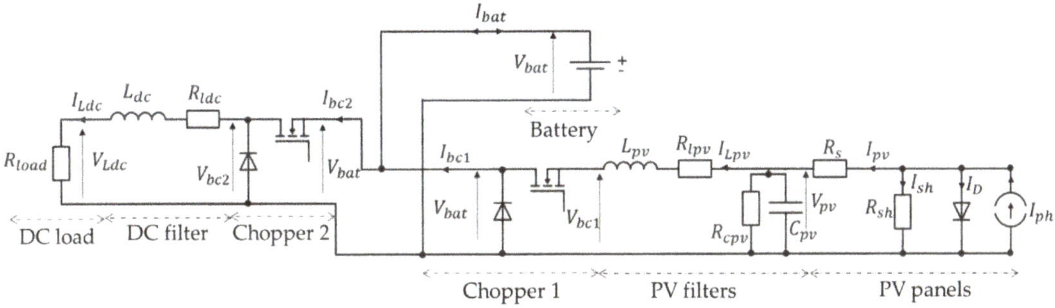

Figure 11. Topology of stand-alone hybrid PV system.

4.4. EMR Representation of the Digital Replica

In the proposed system, the battery imposes a constant voltage, and the PVs relates to the battery through the buck converter. It enables effective Maximum Power Point Tracking (MPPT) control for the PV panels. Moreover, RC (R_{cpv}, C_{pv}) and RL (R_{lpv}, L_{pv}) filters are inserted between the PV and the buck converter to reduce voltage and current ripple. Finally, the battery is connected with the DC Load via a chopper and a RL (R_{ldc}, L_{dc}) filter.

EMR (Energetic Macroscopic Representation) is a graphical formalism for organizing models and control of subsystems within a complete system [47]. The advantage of the EMR formalism lies in its ability to provide a comprehensive and systematic approach for modeling and analyzing complex energy systems. Firstly, each component is translated into EMR elements, and their inputs and outputs are defined according to the causality principle (action/reaction). Moreover, the system is decomposed into basic subsystems with interactions using colored pictograms (orange and green). Furthermore, the control blocks are depicted as blue parallelograms. Table 5 depicts the main EMR elements.

Table 5. Some elements of EMR and of control pictograms [48].

Pictrograms	Pictograms Significance
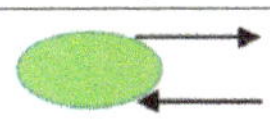	Source of energy
	Element with energy accumulation
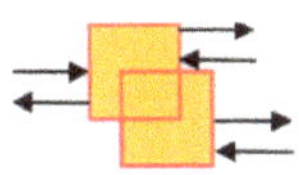	Coupling devices for energy distribution

Table 5. *Cont.*

Pictograms	Pictograms Significance
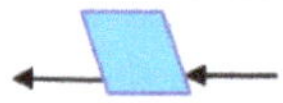	Control bloc without energy accumulation
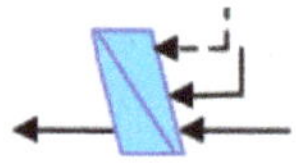	Control block with controller
	Strategy block

Based on the information on all parts of the demonstrator and hypothesis, the EMR (Energetic Macroscopic Representation) of the studied system is designed. The modeling methodology and the EMR organization method are provided in [49]. In Figure 12, the entire EMR organization of the Digital model is depicted.

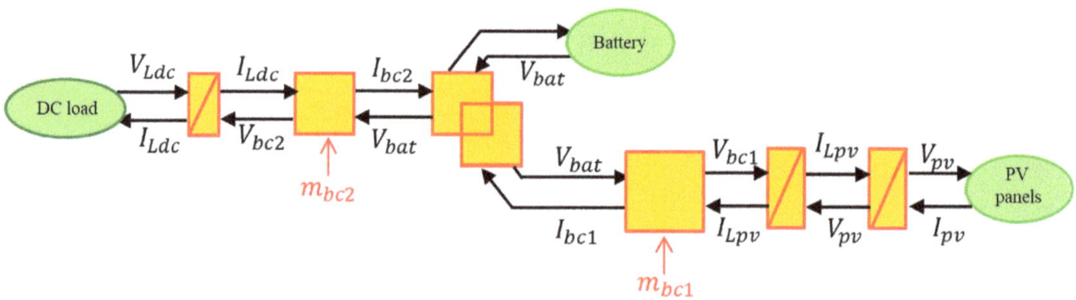

Figure 12. EMR of the studied hybrid system.

The control chains are deduced using inversion control theory [50] where the control structure of a system is considered as an inverse model of the system (Figure 13).

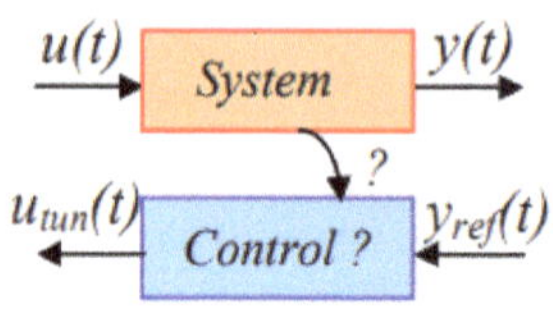

Figure 13. Inversion-based control principle.

The tuning paths (yellow lines in Figure 14) are defined according to the following objectives:

1. Extract the maximum of the solar power by acting on chopper 1 to find the optimal solar panels voltage;
2. Satisfy the DC load demand by acting on chopper 2.

The control scheme of the hybrid system is obtained by inverting the EMR element by element according to the tuning chains (see lower part of Figure 14).

Conversion elements are inverted directly as they have no time-dependence behavior. The accumulation elements (rectangle with forward slash) cannot be inverted physically to avoid derivation. Thus, an indirect inversion is made by using IP controller. Table 6 shows three examples of direct and indirect inversion.

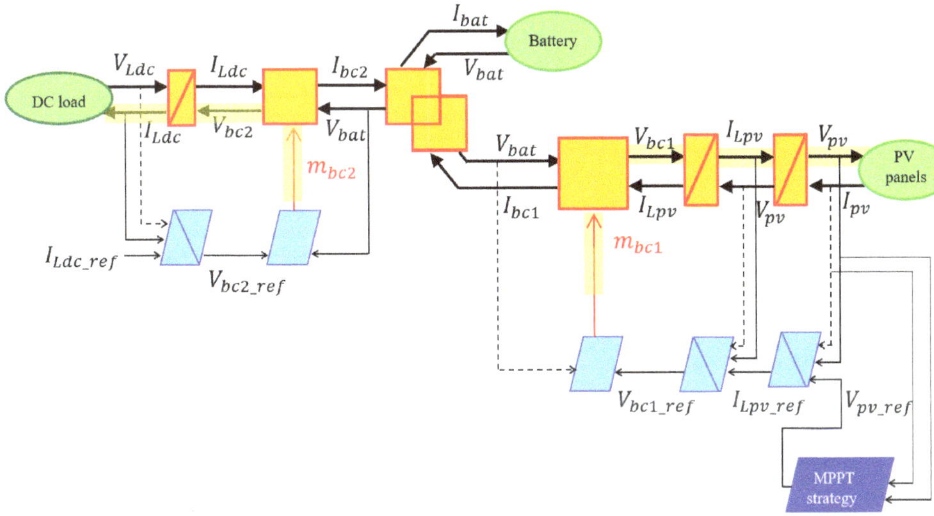

Figure 14. EMR and deduced control of the studied hybrid system.

Table 6. Direct and indirect inversion of EMR blocks.

	System	Equations	Block Diagrams	EMR and Based Inversed Based Control
Chopper1	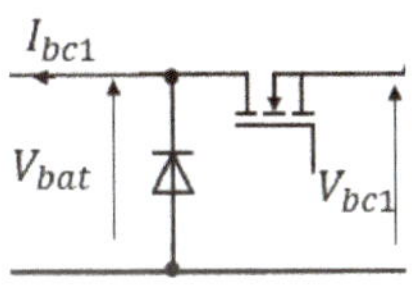	$m_{bc1} = \frac{V_{bat\_meas}}{V_{bc1\_ref}}$ $m_{bc1} = \frac{I_{Lpv}}{I_{bc1}}$	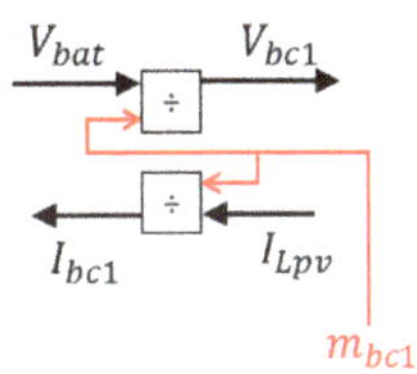	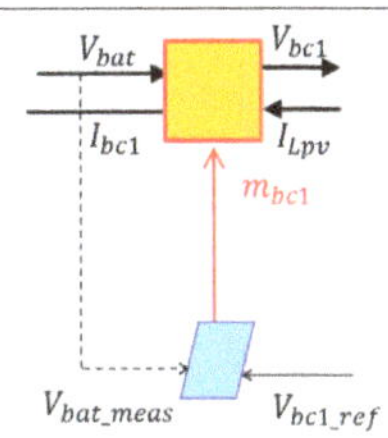
RL filter	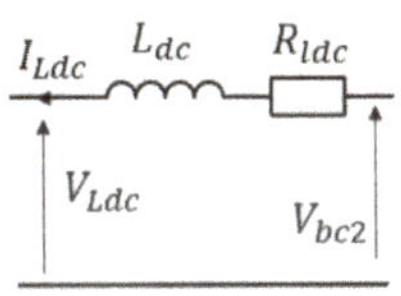	$H(s) = \frac{(\frac{1}{R_{ldc}})}{1+\frac{L_{dc}}{R_{ldc}}s}$	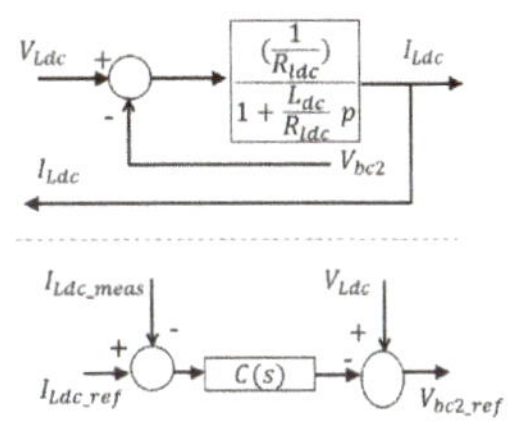	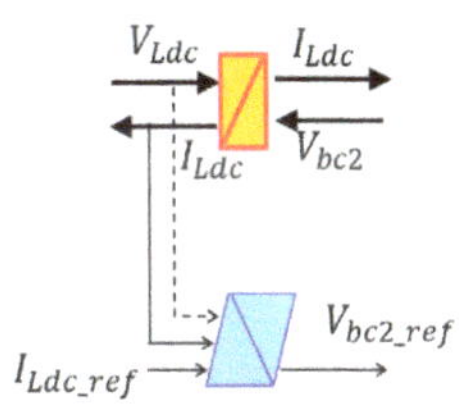
RC filter	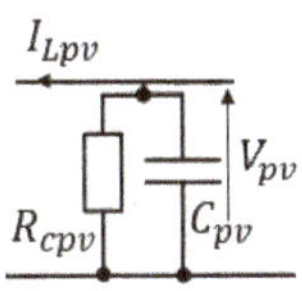	$H(s) = \frac{R_{cpv}}{1+R_{cpv}C_{pv}s}$	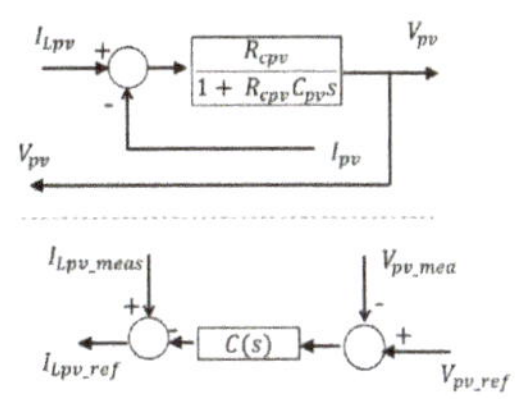	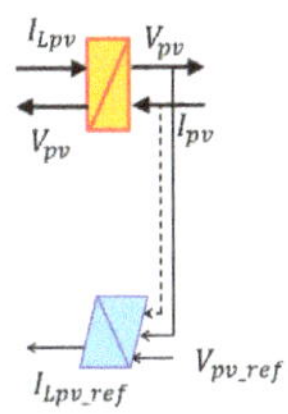

The MPPT (Maximum Power Point Tracking) algorithm is widely used in PV systems. In this study, the Perturb and Observe (P&O) method is implemented. As shown in Figure 15a, the power curves versus the PV panels output voltage present maximum power points (empty circles). A Perturb and Observe Maximum Power Point Tracking strategy [51] is implemented to define the reference voltage imposed on PV panels to obtain the maximal

PV power whatever the irradiance and temperature are (Figure 15b). The P&O algorithm begins by sensing PV voltage and current voltage. The value of the current power (V(k) X I(k)) is then compared to the previous power measurements. If the difference between the two measurements is equal to zero, then the value of the voltage is used as a reference to control the PV voltage thanks to the chopper. If the value of the difference is not equal to zero and if an increase in PV voltage generates an increase in power, this means that there is a convergence to MPP (Maximum Power Point). However, if the power decreases, the PV voltage reference must be reduced to converge to the MPP. The developed algorithm is implemented by the MPPT strategy block (dark blue block) as shown in Figure 15a.

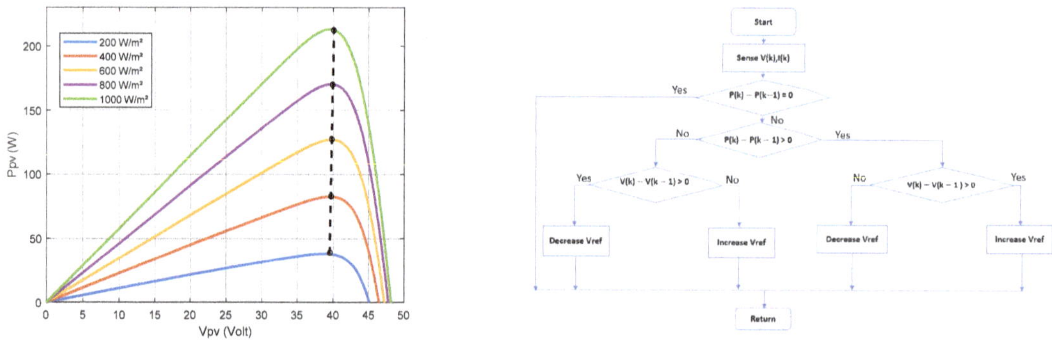

(a) (b)

Figure 15. (**a**) PV Power versus solar panel voltage for different irradiance (T = 25°) (**b**) Flowchart of perturb and observe algorithm.

The following figure (Figure 16) presents the implemented digital model under MATLAB Simulink thanks to a Simulink library containing the EMR basic pictograms.

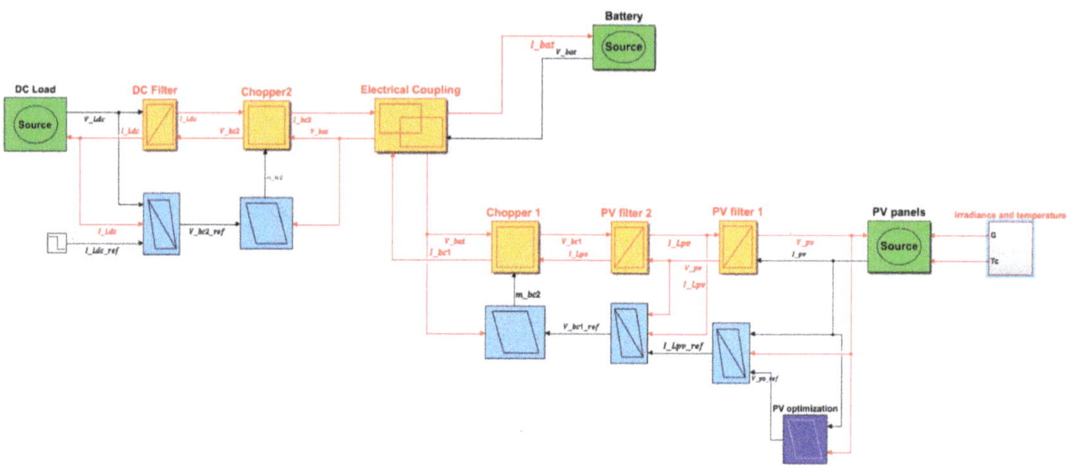

Figure 16. MATLAB Simulink model of the studied hybrid system and its control.

To evaluate the effectiveness of the developed digital model, simulations have been carried out with real-time solar irradiance and temperature data (Figure 17a) for a typical day (30 June 2022). The digital model can be run and connected to real time weather conditions thanks to Simulink Desktop Real Time. This latter provides a real-time kernel for executing Simulink models on a laptop or desktop running Windows or Mac OS X.

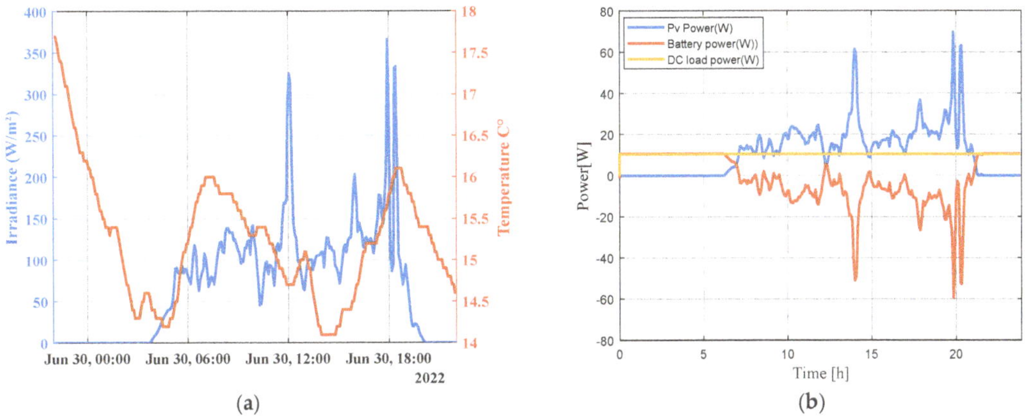

Figure 17. (**a**) Monitored irradiance and temperature profiles; (**b**) PV, battery, and load power curves.

Figure 17b shows simulation results of P_{pv}, P_{bat}, P_{DCl}. PV panels cannot provide enough energy to feed the DC load ($P_{pv} < P_{DCl} = 10$ W) before the time 0.1 s, 6:30 a.m. Thus, the battery provides the DC load during this period. After 6:30 a.m., the photovoltaic panels begin to produce electricity and the battery provides the difference to satisfy the battery. When PV power exceeds the load demand, the battery is charging.

Figure 18a shows simulation results for I_{pv}, I_{bat}, I_{DCl} while Figure 18b shows the state of charge of the battery. The battery is discharging when PV panels output is insufficient, and it is charging when PV power is higher than DC load demand.

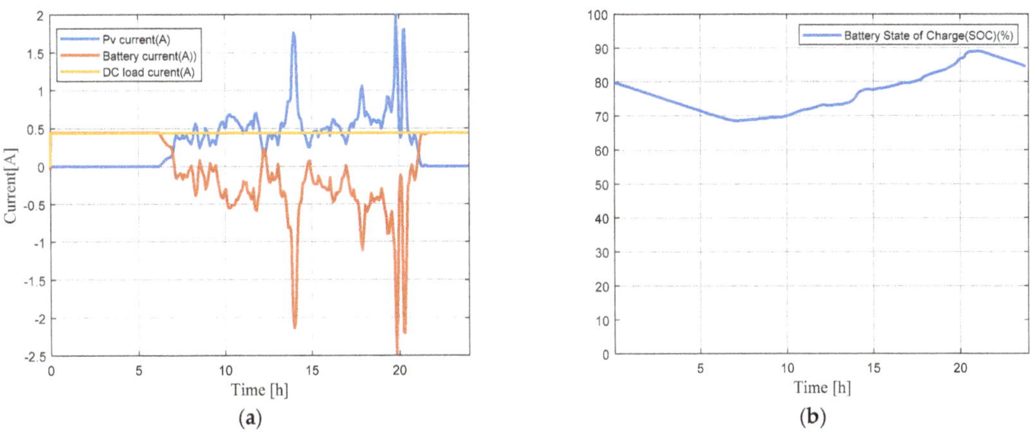

Figure 18. (**a**) PV, Battery, and DC load currents; (**b**) Battery state of charge.

5. Experimental Results and Discussion

Real-time measurements were carried out to show the effectiveness of the proposed unified framework, where the three electromagnetic relays are activated. These measurements are subsequently compared in real time with the digital model data, as outlined in Figure 1.

The applied solar irradiance is presented in Figure 19. From the figure, we can see that the artificial source causes a slight delay and behaves like a first-order system with non-linear behavior depending on the increase or decrease in light. This is due to the light dimmer.

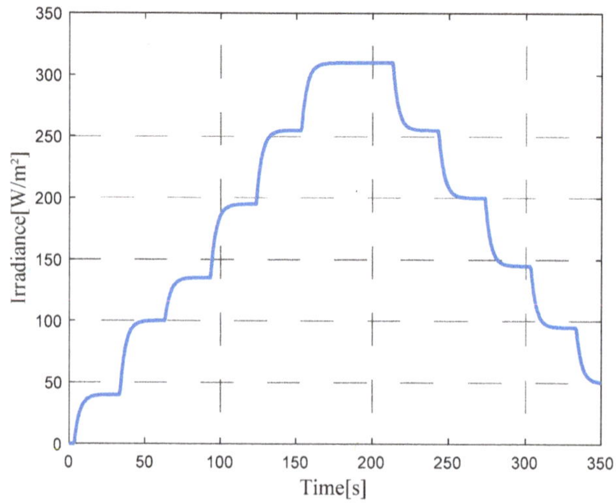

Figure 19. Applied solar irradiance (artificial light source).

Figure 20 represents the measured PV current and the PV numerical model, highlighting a slight deviation observed in the illumination levels around 300 W/m^2. This difference is due to both the approximation of the mathematical model and the non-uniform illumination of the PV panels. Furthermore, it should be noted that the maximum relative error is approximately 6%, and the maximum absolute error is around 0.1, which is considered low.

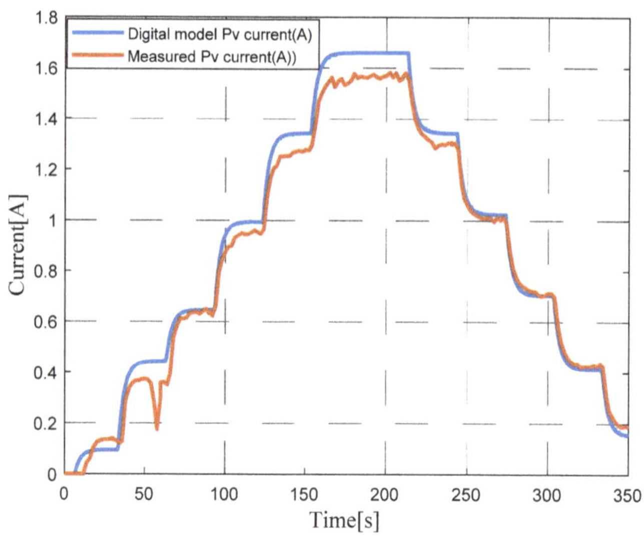

Figure 20. Digital model PV current vs measured PV current.

In Figure 21, the measured and digital model load currents are depicted. The tests were initiated by activating a single lamp and subsequently both lamps. This led to an increase in the DC current from 0.23 A to 0.46 A.

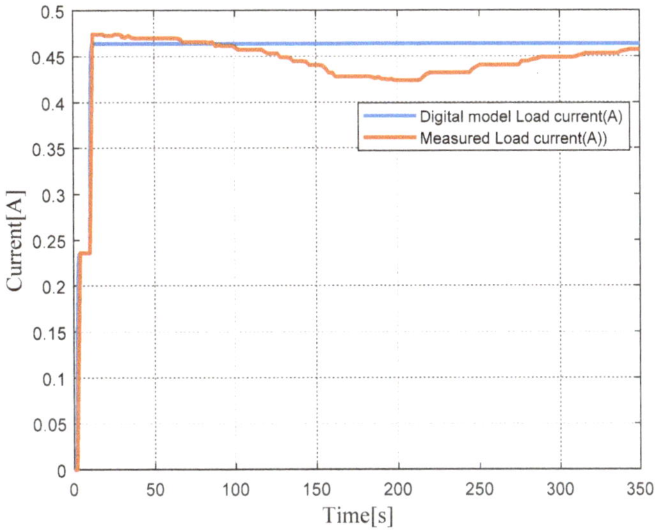

Figure 21. Digital model vs measured load currents.

Measured and digital model battery currents are depicted in Figure 22. A positive current refers to a battery discharge, while a negative current refers to the charging phase.

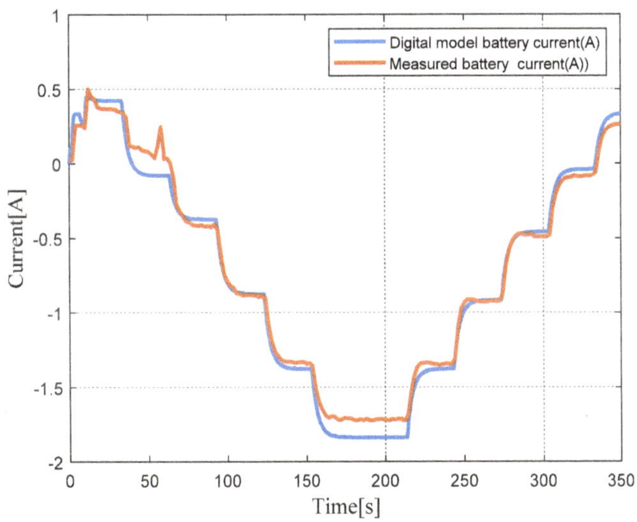

Figure 22. Digital model battery current vs measured battery current.

Figure 23 represents PV power, battery power and DC load power, respectively. From 0 to 50 s, the demanded power is higher than the PV power. The battery is then in discharging mode. However, when the PV power is higher than the demanded DC power, the battery is charging. The comparison between real data and the digital model data shows a good correlation between them. Indeed, the maximum relative deviations between the real and digital model data are 8% for PV power, 12% for battery power, and 2% for load power.

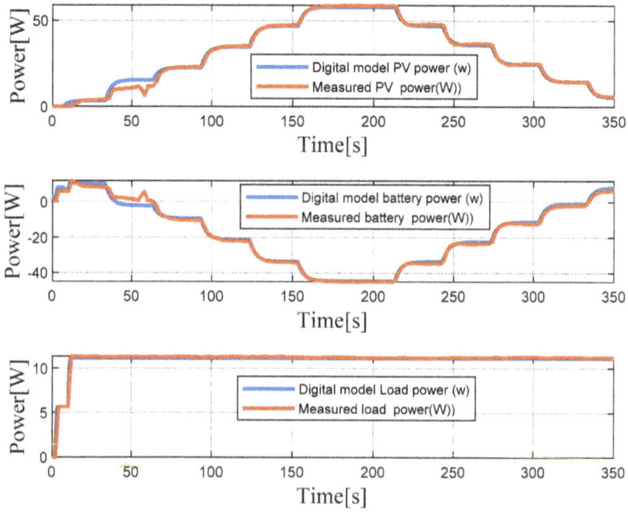

Figure 23. Digital model data vs measured PV, battery, and load power.

To establish a feedback loop between the virtual and real word of the digital twin, we implement an energy management algorithm in MATLAB Simulink. The rule-based energy management algorithm sends control signals to the PLC (s7 1200). The advantage of using a rule-based control approach is its ease of implementation in real-time [52]. The flowchart of the energy management algorithm is illustrated in Figure 24. It receives measurement data as inputs and generates relay control signals as outputs, which are subsequently transmitted to the PLC.

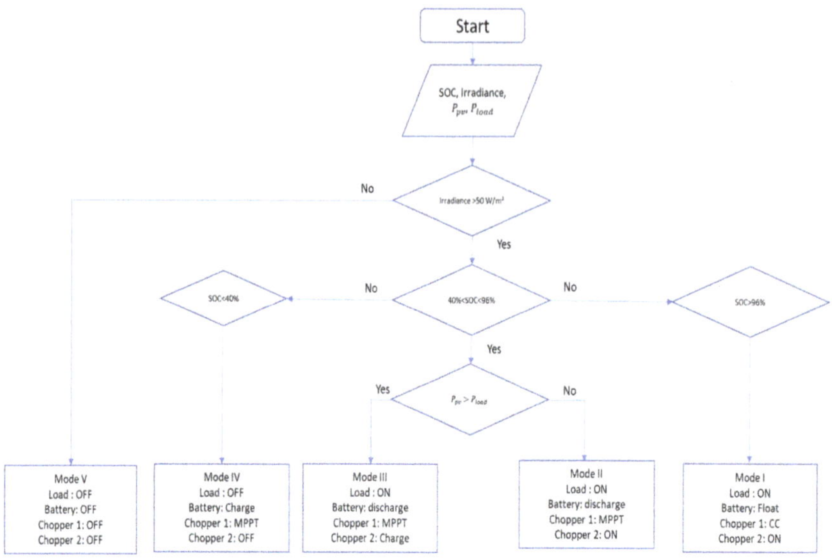

Figure 24. Rule-based algorithm flowchart.

The energy management system generates five different modes. They are described in Table 7.

Table 7. Different modes of the energy management system.

Modes	Load	Chopper 1	Chopper 2	Battery
Mode I: When the battery SOC reaches 96%, which indicates a nearly full charge, the system is operating in float mode. At this point, the PV power output is greater than the load demand. Therefore, the battery should not continue to charge, and the Victron regulator automatically switches to a lower charge voltage mode called float mode to maintain the battery charge level and prevent overcharging.	ON	FLOAT	ON	Float
Mode II: This mode is activated when the battery's SOC is within the normal range of 40 to 96 % and the power generated by the PV panels is lower than the load demand. In such a scenario, the PV panels alone are unable to meet the load requirement, and hence the battery is used to supplement the power supply. The battery operates in the discharge mode, while the PV converter operates in the MPPT mode, and the load remains connected.	ON	MPPT	ON	Discharge
Mode III: This mode is activated when the battery's SOC is within the normal range of 40 to 90 percent and the power generated by the PV panels is higher than the load demand. In such a scenario, the load is powered solely by the PV panels, and any surplus power is utilized to charge the battery. The battery operates in the charging mode, while the PV converter remains in the MPPT mode.	ON	MPPT	ON	Charge
Mode VI: This mode is activated when the battery's SOC drops below 40%, and the power generated by the PV panels is higher than the minimum power required, which is a fixed small value. In such a scenario, the load demand exceeds the PV power output, and the fully discharged battery cannot supplement the power supply. However, the PV panels can still generate power, which can be used to charge the battery after the load is disconnected. In this mode, the PV converter operates in the MPPT mode, and the battery is charging, while the load remains disconnected.	OFF	MPPT	OFF	Charge
Mode V: This mode occurs when irradiance is very low or non-existent. In this case, the system goes into complete off mode until solar radiation starts again. The battery charges during the day and is ready to supply the load in case of absence of solar radiation.	OFF	OFF	OFF	OFF

To assess the effectiveness of the energy management system, the initial load is set to 112 W. Figure 25 displays the recorded external solar irradiance for a one-hour period on a cloudy day. Subsequently, these data are transmitted to the dimmer, enabling the generation of adjustable artificial lighting.

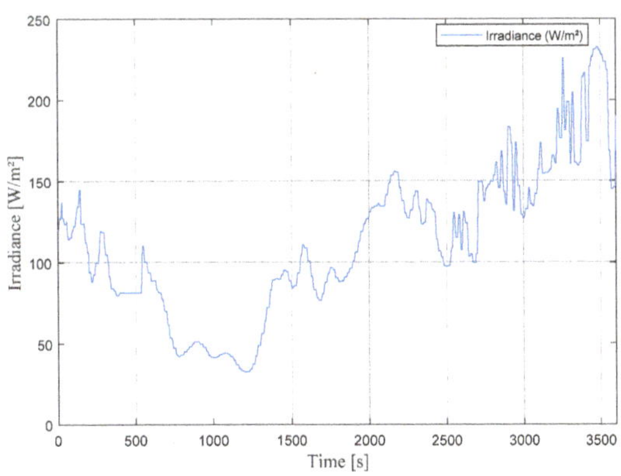

Figure 25. Measured irradiance.

Initially, the total power required by the load is provided by both the battery and solar power sources, as the PV alone cannot fully support the load (Figure 26). During the period from 0 to 757 s, the system operates in mode II.

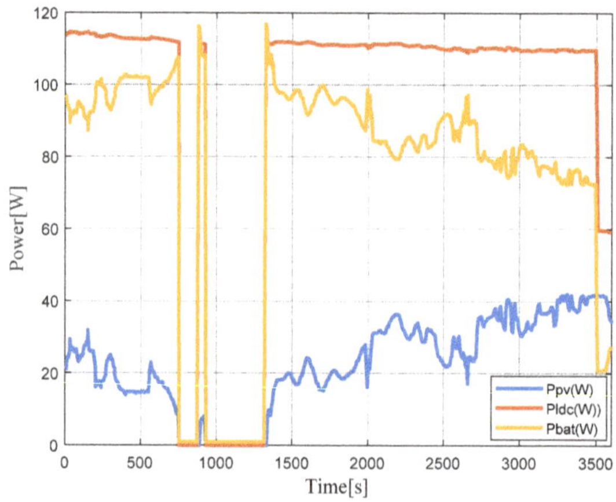

Figure 26. Measured PV, battery, and load power.

Between 757 s and 876 s, as well as from 932 s to 1317 s, the solar irradiance remains below 50 W/m^2, leading the system to switch to mode V (system OFF). From 1317 s to 3508 s, the system once again operates in mode II. During the period from 3508 s to 3600 s, the load demand gradually decreases from 111 W to 60 W, causing the battery to generate less power to meet the load requirements.

Figure 27a shows the battery voltage waveform, while Figure 27b displays the battery State of Charge. The battery primarily operates in discharging mode, except during periods when the system is not in operation (mode I).

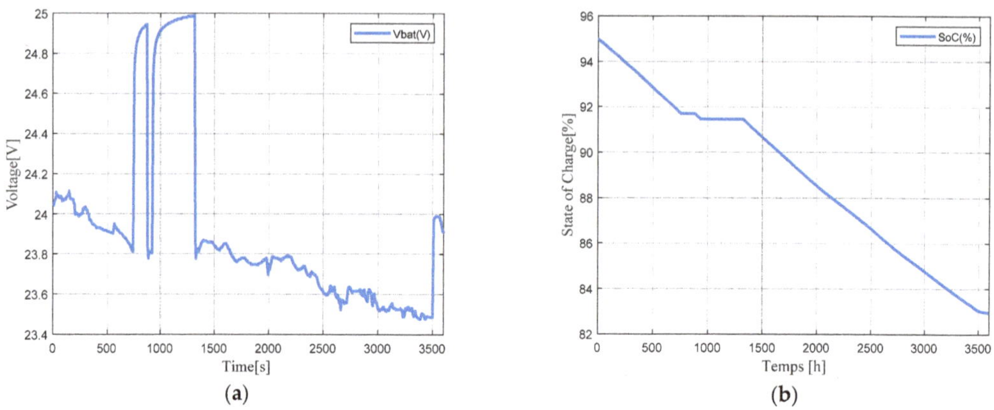

Figure 27. (**a**) Battery voltage; (**b**) Battery state of charge.

6. Conclusions

In this work, we discussed the challenges of implementing IoT in renewable energy systems and the potential of digital twins for predicting energy production and consumption. We proposed a framework that combines the use of Internet of Things (IoT) and Digital Twin (DT) technologies for standalone photovoltaic systems. The digital twin is defined as the combination of the physical system and its digital model, allowing for bidirectional data communication. Furthermore, we designed a digital model of the PV system by using Energetic Macroscopic Representation formalism. Experiments performed in real time and their analysis demonstrate the effectiveness of the proposed framework. As future work, we plan to use this platform to explore machine learning's potential to enhance energy management of PV systems.

Author Contributions: Conceptualization, L.C. and A.S.; Formal analysis, L.C., A.S. and A.R.; Methodology, L.C.; Investigation, L.C., A.S. and A.R.; Software, L.C. and A.S.; Supervision, L.C.; Writing—original draft, L.C.; Experimental tests, L.C., Writing—review and editing, L.C., A.S. and A.R. All authors have read and agreed to the published version of the manuscript.

Funding: This research received no external funding.

Institutional Review Board Statement: Not applicable.

Informed Consent Statement: Not applicable.

Conflicts of Interest: The authors declare no conflict of interest.

References

1. Madeti, S.R.; Singh, S. Monitoring system for photovoltaic plants: A review. *Renew. Sustain. Energy Rev.* **2017**, *67*, 1180–1207. [CrossRef]
2. Motlagh, N.H.; Mohammadrezaei, M.; Hunt, J.; Zakeri, B. Internet of Things (IoT) and the Energy Sector. *Energies* **2020**, *13*, 494. [CrossRef]
3. Shrouf, F.; Miragliotta, G. Energy management based on Internet of Things: Practices and framework for adoption in production management. *J. Clean. Prod.* **2015**, *100*, 235–246. [CrossRef]
4. Adhya, S.; Saha, D.; Das, A.; Jana, J.; Saha, H. An IoT based smart solar photovoltaic remote monitoring and control unit. In Proceedings of the 2016 2nd International Conference on Control, Instrumentation, Energy & Communication (CIEC), Kolkata, India, 28–30 January 2016; pp. 432–436. [CrossRef]
5. Baidya, S.; Potdar, V.; Ray, P.P.; Nandi, C. Reviewing the opportunities, challenges, and future directions for the digitalization of energy. *Energy Res. Soc. Sci.* **2021**, *81*, 102243. [CrossRef]
6. Mukhopadhyay, S.C.; Suryadevara, N.K. *Internet of Things: Challenges and Opportunities*; Springer: Berlin/Heidelberg, Germany, 2014; Volume 9, pp. 1–17. [CrossRef]
7. Yu, W.; Patros, P.; Young, B.; Klinac, E.; Walmsley, T.G. Energy digital twin technology for industrial energy management: Classification, challenges and future. *Renew. Sustain. Energy Rev.* **2022**, *161*, 112407. [CrossRef]
8. Agostinelli, S.; Cumo, F.; Guidi, G.; Tomazzoli, C. Cyber-Physical Systems Improving Building Energy Management: Digital Twin and Artificial Intelligence. *Energies* **2021**, *14*, 2338. [CrossRef]
9. Rachid, A.; Djedjig, A. IoT and MQTT based web monitoring of a solar living laboratory. In Proceedings of the 2022 2nd International Conference on Digital Futures and Transformative Technologies (IcoDT2), Awalpindi, Pakistan, 24–26 May 2022. [CrossRef]
10. Voutsinas, S.; Karolidis, D.; Voyiatzis, I.; Samarakou, M. Development of an IoT power management system for photovoltaic power plants. In Proceedings of the 2022 11th International Conference on Modern Circuits and Systems Technologies (MOCAST), Bremen, Germany, 8–10 June 2022. [CrossRef]
11. Dinculeană, D.; Cheng, X. Vulnerabilities and Limitations of MQTT Protocol Used between IoT Devices. *Appl. Sci.* **2019**, *9*, 848. [CrossRef]
12. Martín-Lopo, M.M.; Boal, J.; Sánchez-Miralles, A. A literature review of IoT energy platforms aimed at end users. *Comput. Netw.* **2020**, *171*, 107101. [CrossRef]
13. Han, J.; Choi, C.-S.; Park, W.-K.; Lee, I.; Kim, S.-H. PLC-based photovoltaic system management for smart home energy management system. *IEEE Trans. Consum. Electron.* **2014**, *60*, 184–189. [CrossRef]
14. Miller, D.; Mirzaeva, G.; Townsend, C.D.; Goodwin, G.C. The Use of Power Line Communication in Standalone Microgrids. *IEEE Trans. Ind. Appl.* **2021**, *57*, 3029–3037. [CrossRef]
15. Chouder, A.; Silvestre, S.; Taghezouit, B.; Karatepe, E. Monitoring, modelling and simulation of PV systems using LabVIEW. *Sol. Energy* **2013**, *91*, 337–349. [CrossRef]
16. Dabou, R.; Bouraiou, A.; Ziane, A.; Necaibia, A.; Sahouane, N.; Blal, M.; Khelifi, S.; Rouabhia, A.; Slimani, A. Development of autonomous monitoring and performance evaluation system of grid-tied photovoltaic station. *Int. J. Hydrogen Energy* **2021**, *46*, 30267–30287. [CrossRef]
17. Atia, D.M.; El-Madany, H.T.; Atia, Y.; Zahran, M.B. Real-Time Implementation of Energy Management for Photovoltaic/Battery/Diesel Hybrid System Based on LabVIEW. *Int. J. Renew. Energy Res.* **2022**, *12*, 1105–1116. [CrossRef]
18. Vergura, S.; Natangelo, E. Labview-Matlab Integration for Analyzing Energy Data of PV Pants. *Renew. Energy Power Qual. J.* **2010**, *1*, 1248–1252. [CrossRef]

19. Dhillon, J.; Unni, A.; Singh, N. Design and Simulation of a PV System with Battery Storage Using Bidirectional DC-DC Converter Using Matlab Simulink. In Proceedings of the 2022 1st International Conference on Sustainable Technology for Power and Energy Systems (STPES), Srinagar, India, 5–6 July 2022. [CrossRef]
20. Manolakos, D.; Papadakis, G.; Papantonis, D.; Kyritsis, S. A stand-alone photovoltaic power system for remote villages using pumped water energy storage. *Energy* **2004**, *29*, 57–69. [CrossRef]
21. Gouda, O.; Amer, G.; Elkhodary, T.; Awaad, M. Optimum Design and Implementation of Stand-alone Tracking Photovoltaic Power System based on PLC and Microcontroller. In Proceedings of the 14th International Middle East Power Systems Conference (MEPCON'10), Cairo, Egypt, 19–21 December 2010; pp. 119–124.
22. Kowsalya, M.; Elakya, A.; Pradeep, R. Solar Operated PLC Based Automated Irrigation System with Fault Preventer. In Proceedings of the 7th International Conference on Advanced Computing and Communication Systems (ICACCS), Coimbatore, India, 19–20 March 2021. [CrossRef]
23. Lee, J.; Lee, S.; Cho, H.; Ham, K.S.; Hong, J. Supervisory control and data acquisition for Standalone Hybrid Power Generation Systems. *Sustain. Comput. Inform. Syst.* **2018**, *20*, 141–154. [CrossRef]
24. Mohammed, N.; Danapalasingam, K.A.; Majed, A. Design, Control and Monitoring of an Offline Mobile Battery Energy Storage System for a Typical Malaysian Household Load Using PLC. *Int. J. Power Electron. Drive Syst.* **2018**, *9*, 180–188. [CrossRef]
25. Alphonsus, E.R.; Abdullah, M.O. A review on the applications of programmable logic controllers (PLCs). *Renew. Sustain. Energy Rev.* **2016**, *60*, 1185–1205. [CrossRef]
26. Tao, F.; Zhang, H.; Liu, A.; Nee, A.Y.C. Digital Twin in Industry: State-of-the-Art. *IEEE Trans. Ind. Inform.* **2018**, *15*, 2405–2415. [CrossRef]
27. Pan, H.; Dou, Z.; Cai, Y.; Li, W.; Lei, X.; Han, D. Digital twin and its application in power system. In Proceedings of the 2020 5th International Conference on Power and Renewable Energy (ICPRE), Shanghai, China, 12–14 September 2020; pp. 21–26.
28. Fuller, A.; Fan, Z.; Day, C.; Barlow, C. Digital Twin: Enabling Technologies, Challenges and Open Research. *IEEE Access* **2020**, *8*, 108952–108971. [CrossRef]
29. Fahim, M.; Sharma, V.; Cao, T.-V.; Canberk, B.; Duong, T.Q. Machine Learning-Based Digital Twin for Predictive Modeling in Wind Turbines. *IEEE Access* **2022**, *10*, 14184–14194. [CrossRef]
30. Agostinelli, S.; Cumo, F.; Nezhad, M.M.; Orsini, G.; Piras, G. Renewable Energy System Controlled by Open-Source Tools and Digital Twin Model: Zero Energy Port Area in Italy. *Energies* **2022**, *15*, 1817. [CrossRef]
31. Zaballos, A.; Briones, A.; Massa, A.; Centelles, P.; Caballero, V. A Smart Campus' Digital Twin for Sustainable Comfort Monitoring. *Sustainability* **2020**, *12*, 9196. [CrossRef]
32. Moi, T.; Cibicik, A.; Rølvåg, T. Digital twin based condition monitoring of a knuckle boom crane: An experimental study. *Eng. Fail. Anal.* **2020**, *112*, 104517. [CrossRef]
33. Livera, A.; Paphitis, G.; Pikolos, L.; Papadopoulos, I.; Montes-Romero, J.; Lopez-Lorente, J.; Makrides, G.; Sutterlueti, J.; Georghiou, G.E. Intelligent Cloud-Based Monitoring and Control Digital Twin for Photovoltaic Power Plants. In Proceedings of the 2022 IEEE 49th Photovoltaics Specialists Conference (PVSC), Philadeplhia, PA, USA, 5–10 June 2022; pp. 0267–0274. [CrossRef]
34. Al-Ali, A.; Zualkernan, I.A.; Rashid, M.; Gupta, R.; Alikarar, M. A smart home energy management system using IoT and big data analytics approach. *IEEE Trans. Consum. Electron.* **2017**, *63*, 426–434. [CrossRef]
35. Jain, P.; Poon, J.; Singh, J.P.; Spanos, C.; Sanders, S.R.; Panda, S.K. A Digital Twin Approach for Fault Diagnosis in Distributed Photovoltaic Systems. *IEEE Trans. Power Electron.* **2019**, *35*, 940–956. [CrossRef]
36. Cheddadi, Y.; Cheddadi, H.; Cheddadi, F.; Errahimi, F.; Es-Sbai, N. Design and implementation of an intelligent low-cost IoT solution for energy monitoring of photovoltaic stations. *SN Appl. Sci.* **2020**, *2*, 1–11. [CrossRef]
37. de Arquer Fernández, P.; Fernández, M.A.F.; Candás, J.L.C.; Arboleya, A.P. An IoT open source platform for photovoltaic plants supervision. *Int. J. Electr. Power Energy Syst.* **2021**, *125*, 106540. [CrossRef]
38. Pereira, R.I.; Dupont, I.M.; Carvalho, P.C.; Jucá, S.C. IoT embedded linux system based on Raspberry Pi applied to real-time cloud monitoring of a decentralized photovoltaic plant. *Measurement* **2018**, *114*, 286–297. [CrossRef]
39. Abdallah, E.L.; Ouassaid, M.; Zidani, Y. Development of a Real-Time Framework between MATLAB and PLC through OPC-UA: A Case Study of a Microgrid Energy Management System. 2022, p. 13. Available online: https://www.researchsquare.com/article/rs-1265626/v1 (accessed on 22 March 2022).
40. Qian, C.; Liu, X.; Ripley, C.; Qian, M.; Liang, F.; Yu, W. Digital Twin—Cyber Replica of Physical Things: Architecture, Applications and Future Research Directions. *Future Internet* **2022**, *14*, 64. [CrossRef]
41. Transducer of, D.C. Circuit Parameters with Ethernet | Lumel. Available online: https://www.lumel.com.pl/en/catalogue/product/transducer-of-d-c-circuit-parameters-with-ethernet-p30h (accessed on 17 June 2022).
42. Jaloudi, S. Communication Protocols of an Industrial Internet of Things Environment: A Comparative Study. *Future Internet* **2019**, *11*, 66. [CrossRef]
43. Wang, Y.-J.; Hsu, P.-C. Analytical modelling of partial shading and different orientation of photovoltaic modules. *IET Renew. Power Gener.* **2010**, *4*, 272–282. [CrossRef]
44. Guo, S. The Application of Genetic Algorithms to Parameter Estimation in Lead-Acid Battery Equivalent Circuit Models. Master's Thesis, University of Birmingham, Birmingham, UK, 2010; p. 137.
45. Tremblay, O.; Dessaint, L.-A.; Dekkiche, A.-I. A Generic Battery Model for the Dynamic Simulation of Hybrid Electric Vehicles. In Proceedings of the IEEE Vehicle Power and Propulsion Conference, Arlington, TX, USA, 9–12 September 2007; pp. 284–289.

46. Jamshidpour, E.; Jovanović, S.; Poure, P. Equivalent Two Switches and Single Switch Buck/Buck-Boost Circuits for Solar Energy Harvesting Systems. *Energies* **2020**, *13*, 583. [CrossRef]
47. Bouscayrol, A.; Guillaud, X.; Delarue, P.; Lemaire-Semail, B. Energetic Macroscopic Representation and Inversion-Based Control Illustrated on a Wind-Energy-Conversion System Using Hardware-in-the-Loop Simulation. *IEEE Trans. Ind. Electron.* **2009**, *56*, 4826–4835. [CrossRef]
48. Mokhena, T.; Mochane, M.; Tshwafo, M.; Linganiso, L.; Thekisoe, O.; Songca, S. *Powerful Multilevel Simulation Tool for HiL Analysis of Urban Electric vehicle's Propulsion Systems Raul*; Intech Open: London, UK, 2016; pp. 225–240.
49. Lhomme, W.; Delarue, P.; Giraud, F.; Lemaire-Semail, B.; Bouscayrol, A. Simulation of a Photovoltaic Conversion System using Energetic Macroscopic Representation. In Proceedings of the 15th International Power Electronics and Motion Control Conference (EPE/PEMC), Novi Sad, Serbia, 4–6 September 2012. [CrossRef]
50. Gomand, J. Analyse de Systèmes Multi-Actionneurs Parallèles par une Approche Graphique Causale—Application a un Processus Électromécanique de Positionnement Rapide. Ph.D. Thesis, Ecole Doctorale no. 432: Sciences des Métiers de l'Ingénieur, l'École Nationale Supérieure d'Arts et Métiers, Paris, France, 2009.
51. Femia, N.; Petrone, G.; Spagnuolo, G.; Vitelli, M. Optimization of Perturb and Observe Maximum Power Point Tracking Method. *IEEE Trans. Power Electron.* **2005**, *20*, 963–973. [CrossRef]
52. Mirzaei, A.; Forooghi, M.; Ghadimi, A.A.; Abolmasoumi, A.H.; Riahi, M.R. Design and construction of a charge controller for stand-alone PV/battery hybrid system by using a new control strategy and power management. *Sol. Energy* **2017**, *149*, 132–144. [CrossRef]

IoT MDPI

Article

An IoT- and Cloud-Based E-Waste Management System for Resource Reclamation with a Data-Driven Decision-Making Process

Mithila Farjana , Abu Bakar Fahad, Syed Eftasum Alam and Md. Motaharul Islam *

Deptement of Computer Science and Engineering, United International University, Dhaka 1212, Bangladesh; mfarjana201127@bscse.uiu.ac.bd (M.F.); afahad201119@bscse.uiu.ac.bd (A.B.F.); salam201133@bscse.uiu.ac.bd (S.E.A.)

* Correspondence: motaharul@cse.uiu.ac.bd

Abstract: IoT-based smart e-waste management is an emerging field that combines technology and environmental sustainability. E-waste is a growing problem worldwide, as discarded electronics can have negative impacts on the environment and public health. In this paper, we have proposed a smart e-waste management system. This system uses IoT devices and sensors to monitor and manage the collection, sorting, and disposal of e-waste. The IoT devices in this system are typically embedded with sensors that can detect and monitor the amount of e-waste in a given area. These sensors can provide real-time data on e-waste, which can then be used to optimize collection and disposal processes. E-waste is like an asset to us in most cases; as it is recyclable, using it in an efficient manner would be a perk. By employing machine learning to distinguish e-waste, we can contribute to separating metallic and plastic components, the utilization of pyrolysis to transform plastic waste into bio-fuel, coupled with the generation of bio-char as a by-product, and the repurposing of metallic portions for the development of solar batteries. We can optimize its use and also minimize its environmental impact; it presents a promising avenue for sustainable waste management and resource recovery. Our proposed system also uses cloud-based platforms to help analyze patterns and trends in the data. The Autoregressive Integrated Moving Average, a statistical method used in the cloud, can provide insights into future garbage levels, which can be useful for optimizing waste collection schedules and improving the overall process.

Keywords: IoT; cloud; e-waste; pyrolysis; Generative Adversarial Networks; bio-fuel; recycling

Citation: Farjana, M.; Fahad, A.B.; Alam, S.E.; Islam, M.M. An IoT- and Cloud-Based E-Waste Management System for Resource Reclamation with a Data-Driven Decision-Making Process . *IoT* **2023**, *4*, 202–220. https://doi.org/10.3390/iot4030011

Academic Editor: Antonio Cano-Ortega and Francisco Sánchez-Sutil

Received: 13 May 2023
Revised: 24 June 2023
Accepted: 26 June 2023
Published: 6 July 2023

1. Introduction

E-waste refers to repudiated electronic devices, such as computers, mobile phones and other electronic equipment, that are at the verge of their efficacious use. Owing to the unrelenting momentum of technological innovation, a growing multitude of individuals are procuring electronic devices with regularity; thus, this begets roughly 54 to 60 million metric tons of e-waste every year, averaging some 7 kg of e-waste per capita. Pursuant to the Global E-waste Statistics Partnership, this is expected to increase to 74.7 Mt by 2030. By 2025, it is estimated that Asia will generate the highest volume of e-waste, at 24.4 million metric tons, followed by the Americas (13.4 million metric tons) and Europe (12.8 million metric tons). Scarcely around 15 percent of global e-waste was collected and recycled in 2014, with the remaining 85 percent being discarded in landfills or incinerated [1].

This situation gives rise to a profound disquietude and engenders a palpable sense of apprehension. It is incumbent upon us to take substantive action. The deleterious effects of electronic waste on the environment are manifold and unequivocal. It has been empirically demonstrated that the materials utilized in the construction of these devices, when containing high concentrations of lead and mercury, are capable of perniciously poisoning the surrounding soil in landfills. Once discarded, the components of e-waste become veritable toxins for the ecosystem, gradually seeping into the earth and causing damage to the atmosphere [2]. This process releases noxious chemicals into the air, thereby exacerbating air pollution. Furthermore, as these toxic materials are carried by rainwater or groundwater, they can affect both terrestrial and aquatic wildlife, rendering e-waste an omnipresent threat to environmental health. The identification and separation of e-waste from municipal solid waste (MSW) is a challenging task that requires significant resources. Moreover, the recycling of e-waste involves substantial costs and requires specialized techniques for sorting and processing [3]. Our study focuses on the separation and sorting of e-waste using machine learning and the recycling of plastic using pyrolysis, as well as the potential uses for the resulting bio-char by-product, and using metals to produce solar batteries. E-waste metals can be converted to solar batteries

for achieving sustainable and renewable energy sources, and we propose the use of time series data [4] for the continuous monitoring of the garbage level in the cloud, employing the Auto-regressive Integrated Moving Average (ARIMA) to forecast and analyze the life cycle [5].

Our system for collecting and sorting waste employs a combination of machine learning, cloud computing, and IoT technology, which streamlines the waste-to-asset process and centralizes it under a single sector. Our strengths in developing this system are convenience and efficiency in waste management; sustainability—by improving waste management and reducing the likelihood of overflowing bins, this system could help promote a more sustainable approach to waste disposal; and data collection and analysis—the system's ability to continuously update the trash level in the cloud and store data can provide insights into waste patterns. This helps inform waste management strategies. Our limitation for making this system is that difficulties, such as mode collapse, training instability, time series data and evaluating generated images may limit the GAN performance, while the quality of solar batteries and bio-fuels can vary due to impurities and chemical reactions, posing challenges for implementation in developing countries where establishing processes may be difficult.

This research revolves around addressing improvements in the efficiency of e-waste management. The primary objective is to explore and evaluate the feasibility and benefits of implementing IoT- and cloud-based smart systems in e-waste management processes, enabling seamless connectivity and communication between various devices and stakeholders involved in the e-waste management system. One of the focuses of our research is the utilization of machine learning algorithms for sorting e-waste. By using machine learning, our system will automatically identify e-waste. This not only saves labor in the sorting process but also enhances the accuracy and efficiency of recycling operations. The data-driven approach ensures the rapid collection of the e-waste, optimizes the utilization of available resources, enhances operational efficiency, and facilitates continuous improvement in e-waste management practices. By analyzing and interpreting relevant data, stakeholders can make informed decisions regarding waste collection, recycling methods, and resource allocation. In addition, our study describes how we can efficiently turn e-waste plastic into bio-fuel and bio-char. Over and above that, our research delves into the repurposing of e-waste metals for the production of solar batteries. With the ever-increasing demand for renewable energy sources, the conversion of e-waste metals into solar batteries offers a sustainable solution for both waste management and energy production. By utilizing these metals, it becomes possible to transform a potential environmental hazard into a valuable resource. Ultimately, the goal of this research is to contribute to a more sustainable and efficient e-waste management framework. By exploring the potential of IoT- and cloud-based systems, integrating machine learning techniques, investigating pyrolysis for recycling, repurposing e-waste metals for solar batteries, developing sustainable strategies, and promoting data-driven decision making, we can pave the way for a greener future and mitigate the environmental and health risks associated with e-waste.

The major contributions of this paper are summarized below:

- We have potentially enhanced the efficiency and accuracy of e-waste recycling processes, enabling a more effective sorting and separation of valuable components by leveraging these cutting-edge technologies.
- We have trained a digital image processing camera to recognize the perfect waste parts, and the interconnection between the sensors and cloud processes the data and recycles the wastes in a perfect manner.
- We have built a small prototype of an IoT-based waste management system with the help of the cloud, which will make this whole system automated.
- We are turning waste into assets by developing our waste management and recycling system.

This whole paper is organized in the following order: Section 2 provides the details of the related works; Section 3 provides information about our proposed system; Section 3.1 contains the proposed solution; Section 3.2 outlines the system architecture; Section 3.3 explains the methodology; Section 3.4 contains a flowchart; Section 3.5 shows the algorithm used; Section 4 contains performance analysis; Section 4.1 contains graphical analysis of e-waste level updates in the cloud; Section 4.2 shows an accuracy chart of the GAN algorithm; Section 4.3 contains graphical analysis of the pyrolysis method, Section 4.4 contains graphical analysis of solar battery production and the reduction in CO_2; Section 5 outlines limitations and future works; and Section 6 is our conclusion.

2. Related Works

The recycling industry faces a significant challenge in managing e-waste, as there is a pressing need to raise awareness among general people about the environmental and energy-saving advantages of recycling electronic devices. In Table 1 summary of related works are depicted. Addressing this challenge requires a comprehensive effort to educate users about the benefits of e-waste recycling [6]. Though it is a challenging task, it also presents notable opportunities to effectively navigate this complex field [7]. It requires a multifaceted approach that includes developing environmentally friendly products, effective waste collection, and safe and responsible recycling and disposal. Utilizing a magnetic field to segregate electronic waste into its constituent plastic and metal components represents a sophisticated approach. This method involves applying a magnetic field to the electronic waste, which causes the metallic components to be attracted to the magnet while the non-metallic plastic components remain unaffected [8]. Pyrolysis is an advanced technique that provides a sustainable and efficient solution for plastic parts of e-waste while also reducing the environmental impact of plastic waste [9]. The focus of the research paper [10] was to create an IoT-based monitoring system for e-waste, where they used microcontrollers and sensors to monitor e-waste. One effective strategy for reducing e-waste involves designing products with reusability in mind, inspiring creative reuse across different e-waste sources. Effective intervention strategies should aim to minimize exposure to toxic components in e-waste [11].

Bansod et al. [12] proposed a project that focuses on developing an IoT-based e-waste monitoring system. It utilizes ultrasonic sensors, an Arduino Mega 2560 microcontroller, and GSM communication to detect and monitor e-waste levels in real-time. The main benefits are efficient waste management, reduced overflowing bins, and improved planning of waste pick-ups. The limitations are a reliance on a 12 v source for the GSM module. This project's future work includes potential enhancements like incorporating a line follower robot for automated waste disposal. Bošnjakovic et al. [13] proposed a paper that examines the production of liquid fuel from plastic waste, focusing on technological, ecological, and economic aspects. Pyrolysis with a catalyst, particularly zeolite-based catalysts, is a well-established and mature technology for obtaining fuel from plastic waste. While up to 800 L of fuel can be obtained from one ton of waste plastic, real plants typically yield around 450 L. Disposing of waste plastics through fuel production offers significant environmental benefits, including reduced greenhouse gas emissions. However, waste separation, complex technical systems, and proximity to landfills for cost-effective transportation are limitations.The analysis underscores the large amount of plastic waste in Croatia and the potential for economically viable bio-fuel production with improved waste collection.

Table 1. Summary of related works.

Author and Year	Study Description	Limitations	Method Adopted
Bošnjaković et al. (2022) [13]	Technological and ecological dimensions of converting plastic waste into bio-fuel.	Sorting plastic from e-waste and the cloud; IoT use was not mentioned.	Pyrolysis to turn plastic waste into bio-fuel.
Devi et al. (2021) [9]	Emphasis on generating bio-fuels from plastic waste.	No discussion on IoT, the cloud, e-waste collection, or plastic-to-bio fuel conversion.	Process of proselytizing plastic waste into diesel fuel.
Shamsudin et al. (2022) [10]	IoT-based monitoring system using microcontrollers and sensors.	No discussion regarding the next steps after e-waste collection.	IoT-based project with microcontrollers and sensors.
Bansod et al. (2022) [12]	IoT-based system to detect e-waste.	Yet to implement a plan for utilizing the collected waste.	Monitoring garbage levels and communicating them through a GSM system.
Balakrishnan et al. (2015) [14]	Investigate the formation of bio-fuels from plastic scrap.	Generating bio-fuels; the methods for plastic collection are missing.	Pyrolysis to convert plastics to bio-fuels.
M H, Dinesh. (2020) [15]	Generate bio-oil using pyrolysis.	No mention of collecting plastic from e-waste or another place.	Thermal pyrolysis and catalytic pyrolysis.
kazi Shawpnil et al. (2023) [16]	QFD study conducted; combined efficient e-waste management methods.	No mention of the cloud, pyrolysis, bio-char, bio-fuel, or solar batteries.	Physical recycling for metallic parts, the biological method of mycoremediation, phytoremediation.
Abdullah Al Mamun et al. (2023) [17]	YOLOv5 to separate e-waste.	Pyrolysis, bio-char, bio-fuel, and solar batteries were not mentioned.	Pixy camera to recognize e-waste.

In Table 2, a more in-depth analysis of related works is mentioned, where Sankeerth et al. [18] proposed a smart waste management system, which utilizes ultrasonic sensors in bins to measure garbage levels, which are then transmitted to a server via Wi-Fi. The server monitors the bins across the city, notifying the garbage truck driver when the amount of waste in a bin exceeds 70% and it needs to be emptied. SMS notifications are sent to the driver, providing optimized routes based on collected data. Thaseen Ikram et al. [19] proposes a waste management model for smart cities using a hybrid genetic algorithm (GA)–fuzzy inference engine. The system uses IoT components—RFIDs and sensors—to collect and process waste information. The model combines a GA with fuzzy logic to optimize the fuzzy inference system (FIS) and improve waste management accuracy. The system employs cost-effective sensors and ensures reproducibility. Their experimental results show high accuracy and precision of 95.44% in waste management and recyclable item classification. The proposed model reduces errors and minimizes manual interpretation costs compared to traditional approaches, but there could be potential privacy and security risks. Their future work includes integrating advanced technologies and addressing scalability and interoperability challenges. The smart dustbin proposed by Pavithra M. et al. [20] automatically opens upon detecting a clap or foot tap and closes once garbage is disposed of. An ultrasonic sensor monitors the garbage level and sends alerts to the main garbage collector when it reaches capacity. H. Cai et al. [21] proposed a garbage monitoring system where they use a NodeMCU chip integrated with ultrasonic sensors to measure waste levels in bins, which are transmitted to a cloud server through the Ali-cloud IoT platform. To observe the real-time bin status, they have used a web page. The average number of cleanings before establishing this was 3 and afterwards it was 2.28; the average number of bin overflows before was 0.67 and afterwards it was 0.11, which added improvements in waste collection. The paper by Artang Sara et al. [22] introduces a tracking and tracing platform, which offers a user interface for users and administrators, providing essential information to users for the disposal of their e-waste. Integrating block-chain technology and circular economy approaches into the tracking platform and conducting comparative studies among different countries is their proposed future work.

Table 2. In-depth analysis of related works in relation to the benefits and risks of each approach.

Reference Paper	Limitations	Method Adopted	Benefits	Risk	Future Work
[18]	Focuses on data collection and monitoring without data analysis for process optimization.	Bins with ultrasonic sensors measure garbage levels, send data to a server via Wi-Fi, and optimize collection routes using SMS.	Direct message sending reduces the costs and maintenance for the embedded bins, enhancing independence and transparency.	The reliance on Wi-Fi and server stability poses a risk of data loss and potential failure in communication.	Incorporating a database and utilizing data analytics to optimize waste management processes and improve efficiency
[19]	Limited applicability may impact its practicality in different waste management contexts.	An IoT and fuzzy inference system with a genetic algorithm to create a waste disposal system	Enhanced waste management efficiency, cost reduction, and resource optimization	Reliability, accuracy, privacy, and security	The integration of additional advanced technologies, scalability, and interoperability
[20]	Foul odors emanating from the bins and the manual control of the dustbins can restrict mobility and flexibility in waste collection.	Uses sensors for gesture detection and garbage level monitoring, enabling automatic bin operation and timely emptying through an IoT web interface	The automated waste management system reduces labor costs and enables the timely disposal of garbage to the correct location.	The system relies on the accuracy and reliability of the sensors to detect the garbage levels. False readings could impact the efficiency.	The system contributes to waste reduction, resource conservation, and sustainable waste management by handling both metal and non-metal waste.
[21]	Alerts cleaners based on threshold parameters but does not differentiate the recycling of e-waste	A sensor-based device detects and monitors the garbage status and sends notifications to cleaners when thresholds are exceeded.	The notification system eliminates the need for continuous monitoring, as it alerts the cleaner when the dustbin requires cleaning.	The device might get damaged while using the dustbin as it is set in quite an unprotected manner.	An Android app will be developed for this in future and a better algorithm will be implemented here.
[22]	The limited waste registration and complex interface may impede tracking and user adoption in the application.	E-waste registration, QR code tracking, and Google API integration enable effective monitoring and proper disposal.	The stakeholders can track e-waste and its location effectively through unique identification, facilitating easy monitoring.	The only risk is the security issue as the users are giving certain mobile access via this application.	Create a sustainable system with robotics and blockchain for enhanced security and tracking capabilities.

3. Proposed System

3.1. Proposed Solution

Our main goal is to collect e-waste and send it for recycling in an efficient and automated manner. We are using the combination of the IoT and machine learning for gathering e-waste for recycling purposes. We will be placing the processing part of our system in a dumpster with the help of a Field-Programmable Gate Array (FPGA) using the GAN algorithm to distinguish the e-waste from other wastes. Our proposed solution entails the deployment of a smart bin to collect waste, which utilizes cloud-based technology to monitor and update the garbage level automatically. If the bin reaches its maximum capacity, the SIM900A module generates a message alerting the collectors. Upon collection, we implement a process to separate the metallic and plastic components of the waste. The plastic components undergo a pyrolysis process to yield bio-fuel, while the metallic components are repurposed for solar panel and battery production.

3.2. System Architecture

The proposed system architecture is depicted in Figure 1. The system includes several steps aimed at effectively managing e-waste. The initial phase (step-1) involves the classification of wastes based on their type, which will be conducted by machine learning. Subsequently, e-waste is collected and deposited into a smart bin, and based on the trash level data, a data-driven decision-making process is implemented in Figure 2 to determine whether a notification should be sent to the trash collector. This process involves evaluating the trash -level data against predetermined thresholds, and if the data exceeds these thresholds, a notification is triggered and sent to the trash collector in step-2 and step-3.

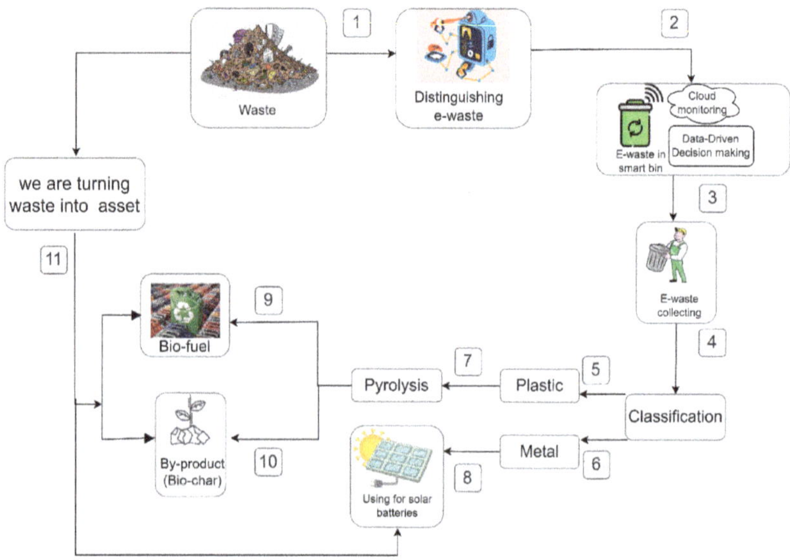

Figure 1. System architecture of our proposed solution.

The cloud-based system is continuously monitoring the level of trash in the background. In step-4, the e-wastes are separated into two categories, plastic and metal. The metal waste is processed for solar batteries in step-8, while plastic waste is converted into bio-fuel using the pyrolysis process and we obtain bio-char as a by-product in step-9 and step-10. In the final step of the process, the repurposed and transformed wastes are converted into valuable assets.

Figure 2 illustrates a data-driven decision-making process for e-waste collection. Trash level data are continuously monitored using an ultrasonic sensor in the trash bin. This data are collected, enabling real-time analysis of e-waste levels. Based on the analysis, notifications are sent to e-waste collectors, prompting them to collect e-waste from specific bins. The collectors follow the notifications, collect the e-waste, and ensure proper recycling methods are employed. This data-driven process optimizes the collection efficiency and helps in the timely and targeted collection of e-waste, contributing to environmental sustainability and proper e-waste management.

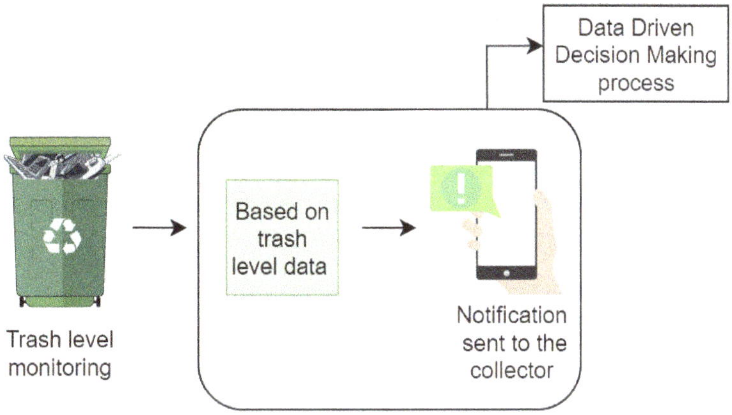

Figure 2. Data-driven decision making process.

Figure 3 is depicting three layers [23]. The sensor layer consists of a smart dustbin with an ultrasonic sensor that detects the level of trash inside the bin. The data collected from the sensors are sent to the cloud layer using the ESP-8266 Wi-Fi Module. The cloud layer receives the data from the sensor layer and stores it in a time series database. A time series database is designed to handle data that are collected over time, such as the trash level in the bin. The data stored in the time series database can be queried and analyzed to generate insights and predictions about the future. The Auto-regressive Integrated Moving Average algorithm is applied to the database to forecast the trash level for the future. The cloud layer also provides an interface for the user to view the trash level and other information in real time and send the value in the microcontroller. The user can access this interface through a web or mobile application. If the level of trash in the bin reaches a certain threshold, the microcontroller sends the notification using the GSM module to the application layer. The application layer receives the notification, and the collector collects the data after receiving it. The smart dustbin system uses a combination of sensors, cloud computing, and predictive algorithms to collect and analyze data about the trash level in the bin. These data are used to provide real-time notifications to the user and improve waste management processes.

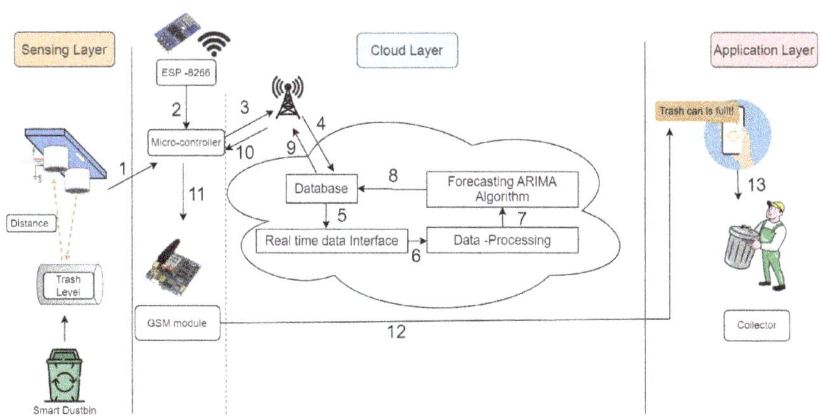

Figure 3. System architecture of collecting and monitoring trash using cloud and IoT.

3.3. Methodology

Our working process entails a dustbin for e-waste. The waste will move along a conveyor belt, and in the initial section of the belt our processing part will be incorporated, featuring a trained camera with machine learning and a Generative Adversarial Network (GAN) algorithm. The GAN requires high computation power and memory resources [24]. A powerful, dedicated hardware platform such as FPGA provides the high computation power. There are two sections to the GAN, namely the discriminator and the generator.

We provide the dataset to the discriminator, and the generator monitors the waste and creates an image according to it. The generator and discriminator images are then compared, and if they match the waste will be thrown into the e-waste smart bin, while other waste will be deposited in a different pile. The smart bin contains an ultrasonic sensor, which sends waste-level data to the database in the cloud through an ESP8266 Wi-Fi module; each data point that is sent from the sensor to the time series database should include a timestamp and the trash level reading in Figures 6–8. The time series database will store these data and make them available for querying and analysis. To forecast the trash level, we are using the ARIMA model [25]. The forecasting can be used to optimize the system by predicting when the trash level will reach the threshold and scheduling pick-ups accordingly. This can help reduce costs and improve efficiency. If the garbage level exceeds a threshold value, a notification will be sent to the collector via the SIM900A GSM module, and it will be sent for recycling. In the recycling process, the e-waste is churned through a robust blade and separated into plastic and metallic parts using a magnetic field. The plastic parts will be sent for the pyrolysis process. In pyrolysis there are several steps:

As the plastic is already shredded, it will increase the surface area for improved pyrolysis. Shredded plastic is fed into a pyrolysis reactor. Pyrolysis is a type of thermal treatment that breaks down complex organic materials (plastic) into simpler compounds using heat in the absence of oxygen. The end products of pyrolysis are typically a liquid fuel known as pyrolysis oil or bio-oil and a gaseous mixture known as syngas, which can be used for various applications, such as energy generation and chemical production [26]. Since pyrolysis takes place in the absence of oxygen, the reactor is sealed to prevent air ingress. The reactor is heated to an elevated temperature. The temperature and pressure inside the reactor are carefully controlled to ensure that the plastic is efficiently converted into bio-fuel. When plastic is heated, it begins to decompose into components, such as gas, liquid, and char. Gases and liquids are condensed and collected as bio-fuel.

Once the pyrolysis process is complete, the reactor is cooled and the bio-fuel is recovered from the condenser. The collected bio-fuel may require further purification to remove impurities such as water and acids. This can be performed by methods such as filtration or distillation. Finished bio-fuel products are stored in tanks or containers until use. If any organic material is mixed with the plastic waste, such as bio-solids, the by-products of pyrolysis, bio-char, can be recycled and used in various applications, such as soil amendment, carbon sequestration, and energy production [15]. This product has shown notable advantages in eliminating pollutants from wastewater [27] and enhancing soil quality [28]. Moreover, we are using scrubber and electrostatic precipitators to control pollution [29]. Metallic waste can be used for making solar batteries. The process of converting metal churns from e-waste into solar batteries involves several steps: The shredded metal is treated with acid or other chemicals to extract impurities and separate the pure metals. The pure metals are then processed using electrolysis, which involves passing an electric current through a solution containing the metal ions. This process causes the metal ions to gain or lose electrons, resulting in the formation of metal deposits on electrodes. The metal deposits are then used to produce various components of a solar battery, including the anode, cathode, and electrolyte. These components are assembled to create a functional solar battery that can store and release energy. The exact process of converting metal churns into solar batteries can vary depending on the specific type of metal and the desired end product. However, in general, the process involves a combination of chemical and electro-chemical techniques to purify and refine the metal and then assemble it into battery components.

The technical aspects of our methodology are as follows:

Data collection method:

Camera and machine learning: trained camera system with machine learning capabilities was used to monitor and capture images of the waste on the conveyor belt.

Ultrasonic sensor: the e-waste smart bin was equipped with an ultrasonic sensor to measure the waste level.

ESP8266 Wi-Fi module: the waste level data from the ultrasonic sensor were transmitted to a cloud database using an ESP8266 Wi-Fi module.

SIM900A GSM Module: when the garbage level exceeded the threshold, a notification was sent to the collector via a SIM900A GSM module.

Data analysis techniques: the ARIMA model was employed for forecasting the trash level.

3.4. Flowchart

In Figure 4, the flowchart describes the process of our IoT- and cloud-based e-waste management, starting with the aggregation of various types of waste. We utilize a trained camera, which has been trained with a GAN algorithm, for the classification of e-waste. Through image processing, it determines whether the waste is e-waste or not. If it is not e-waste, it is dumped in a different pile; otherwise, it is deposited in the smart bin. As e-waste is being disposed of, the waste level continues to increase; this increased level data are then updated in the cloud, and the system checks if the bin is full through an ultrasonic sensor. If it is not full, the process continues, or else a notification is sent to the collector. After collecting the e-waste, the recycling steps begin. It starts with churning the e-waste, followed by magnetic field separation. From separation, there are two parts—plastic and metallic churns. Plastic goes through pyrolysis and becomes bio-fuel, while metallic churns are processed for solar batteries. The process ends with the production of bio-fuel, with bio-char as a by-product, and solar batteries, representing our system's effective transformation and the recycling of e-waste into sustainable and eco-friendly materials.

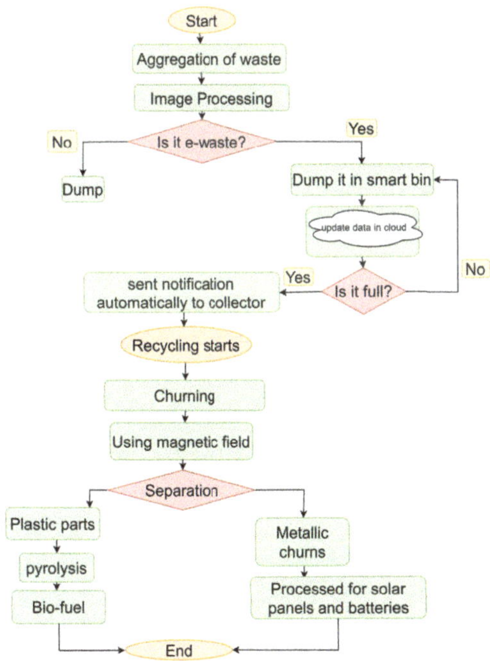

Figure 4. Flowchart of Proposed System.

3.5. Algorithm

For image processing, we are using the GAN, which is a very high-level algorithm. The accuracy of GAN algorithms for image processing is highly dependent on the specific use case and the techniques employed to train and optimize the model. Here, using GAN, the machine will be trained with real-life e-waste images and with the help of those images it will provide its decision. The pseudo code of the GAN algorithm is given below:

In Algorithm 1, the generator network G and discriminator network D are initialized with random weights. The hyper-parameters α, β, and γ are initialized. For a specified number of training iterations, the following steps are executed. For a specified number of discriminator updates per generator update, the following steps are executed. A mini-batch of m real images from the dataset is sampled. A mini-batch of m noise samples from a noise distribution is sampled. Fake images are generated using the generator network G. The discriminator network D is updated by minimizing the binary cross-entropy loss between the real images and the fake images, with the gradients computed using back-propagation. A mini-batch of noise samples from a noise distribution is sampled. The generator network G is updated by taking a gradient step on the loss function that maximizes the binary cross-entropy loss between the generated images and the real images, with the gradients computed using back-propagation. The hyper-parameters α and β are updated using a decay factor γ. The trained generator network G is returned.

Algorithm 1 Image classification using Generative Adversarial Networks

```
Initialize the generator network G with random weights
Initialize the discriminator network D with random weights
Initialize the hyper-parameters α, β, and γ
for number of training iterations do
    for number of discriminator updates per generator update do
        Sample a mini-batch of m real images from the data-set
        Sample a mini-batch of m noise samples from a noise distribution
        Generate fake images using the generator network
        Update the discriminator network.
    end for
    Sample a mini-batch of noise samples from a noise distribution
    Update the generator network by taking a gradient step on the loss function
    Update the hyper-parameters
        α ← γα
        β ← γβ
end for
return the trained generator network G
```

We use generator and discriminator neural networks to train on a dataset of real e-waste images. The goal is to train the generator network to produce images that are indistinguishable from real images, while the discriminator network learns to distinguish between real and generated images. During training, the generator produces images to try and fool the discriminator, and the discriminator tries to become better at distinguishing between real and generated images. Once trained, the generator can generate new images, which can be evaluated by comparing them to real images. If they are similar, the waste can be disposed of in the appropriate destination dustbin.

Algorithm 2 is used to detect the level of e-waste in a smart dustbin and send a notification to the garbage collector when the dustbin is almost full. The input variables for this algorithm are n (the number of iterations), x (the echoP input), and y (the trigP input). The algorithm starts by initializing x and y with the echoP and trigP inputs, respectively. The threshold distance is set to 4, which is the maximum distance at which the smart dustbin can detect e-waste. The algorithm then puts e-waste in the smart dustbin and enters a loop that runs as long as n is not equal to 0. Inside the loop, the value of y is set to

0 or Low initially. Then, the algorithm runs for 10 iterations, and the value of y is set to 1 or High. After 10 iterations, y is again set to 0. The value of x is set to 1 or High.

The algorithm then calculates the distance between the smart dustbin and the e-waste using time and the speed of sound. The garbage level is calculated by subtracting the distance from the total dustbin distance. This information is sent to the cloud using ESP8266. If the distance is greater than or equal to the threshold distance, the algorithm sends a notification using the SIM900A GSM module to the garbage collector. Otherwise, the smart dustbin collects the e-waste.

Algorithm 2 An algorithm for calculating the e-waste level in a smart dustbin

Require: $n \geq 0$
  $x = echoPin$
  $y = trigPin$
  $n = iteration$
  $thresholdDistance = 4$
  Put e-waste in Smart dustbin
  while $n \neq 0$ **do**
    $y \leftarrow$ 0 or Low
    for number of 10 iterations **do**
      $y \leftarrow$ 1 or High
    end for
    $y \leftarrow 0$
    $x \leftarrow$ 1 or High
    $distance \leftarrow \frac{time \times 0.034}{2}$
    $garbageLevel = totalDustbinDistance - distance$
                ▷ Send the distance and garbage level in cloud
    if $distance \geq thresholdDistance$ **then**
      Sent notification to garbage collector
    else
      Smart dustbin collects the e-waste
    end if
  end while

4. Performance Analysis

4.1. Graphical Analysis of E-Waste Level Updates in Cloud

Figure 5 depicts the updates of the garbage level in the cloud of a certain time period, where the initial level (at time = 1) was recorded as 28 cm, indicating an empty smart dustbin. As observed from the time axis (y-axis), the garbage level gradually decreased until it reached 9 cm at time = 8. At time = 9, the garbage level reached a threshold distance of 4 cm, after which it remained constant until time = 12. During this period, a notification was sent to the collector, who subsequently collected and emptied the e-waste from the smart bin. Following the trash collection, the garbage level increased and was recorded as 28 cm at time = 13.

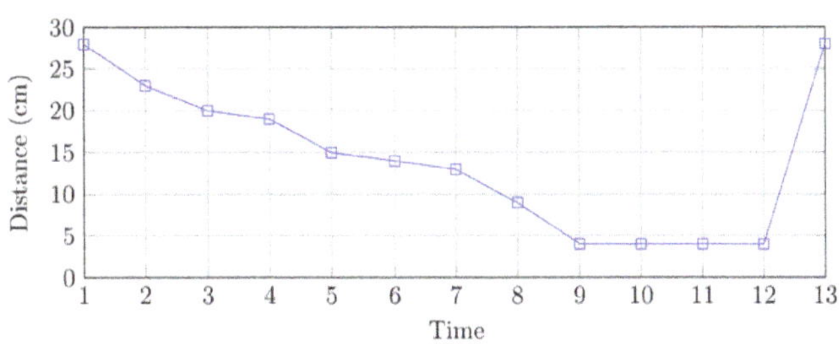

Figure 5. Graphical analysis of e-waste level update.

Figures 6–8 illustrate the empty space available in the smart trash bin and the process of updating the corresponding values in the cloud for a certain period of time. The distance between the contents and the top of the trash bin indicates the level of empty space available. A greater distance corresponds to a higher amount of empty space, while a lesser distance corresponds to a lower amount of empty space.

Figure 6. E-waste level update information in cloud.

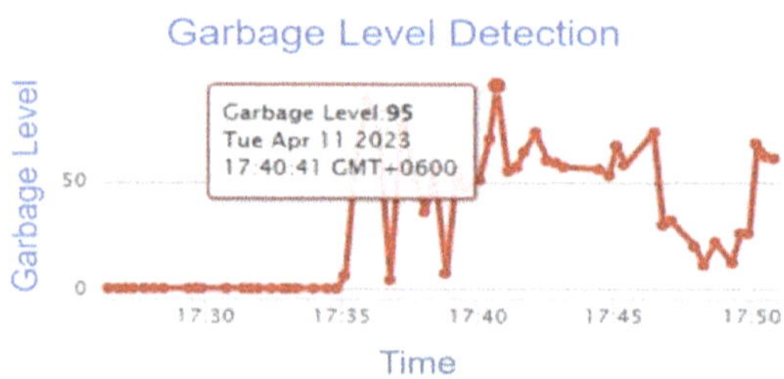

Figure 7. E-waste level update information in cloud with timestamp.

```
Content-Length: 30 Field-1: 37 data send
Waiting
Distance =29 cm
Send data to thingspeak:
Content-Length: 30 Field-1: 29 data send
Waiting
Distance =20 cm
Send data to thingspeak:
Content-Length: 30 Field-1: 20 data send
Waiting
Distance =10 cm
Send data to thingspeak:
Content-Length: 30 Field-1: 10 data send
Waiting
Distance =6 cm
```

Figure 8. E-waste level update information in cloud showing in serial monitor.

4.2. *Accuracy Chart of GAN Algorithm*

We are using the GAN algorithm where we are dividing our dataset into training, validation, and testing sets. The training set is used to train the model, the validation set is used to tune the model's hyper-parameters, and the testing set is used to evaluate the model's performance on unseen data.

This accuracy chart in Table 3 shows the performance of an e-waste recognition system that uses the GAN algorithm. The chart shows the precision, recall, and F1-score for each category of e-waste that the system is designed to recognize: smartphones, laptops, televisions, monitors, and other, which includes all other types of e-waste.

Table 3. Accuracy Chart for E-waste Recognition System using GAN Algorithm.

Category	Precision (%)	Recall (%)	F1-Score (%)
Phone	95	97	96
Laptop	90	85	87
TV	85	91	88
Monitor	92	89	90
Other	80	75	77
Overall	90	88	89

In Figure 9, precision is a performance metric that measures the accuracy of a system in identifying relevant items. It quantifies the proportion of correctly identified items among the total items identified by the system. The precision is calculated as the ratio of true positives (items correctly identified) to the sum of true positives and false positives (items incorrectly identified). A high precision value indicates that the system is effective in accurately identifying relevant items. It signifies that the system has a low rate of falsely identifying unrelated items as the target item. In our example, a precision of 95% implies that the system has a relatively low rate of falsely identifying non-smartphone items as smartphones.

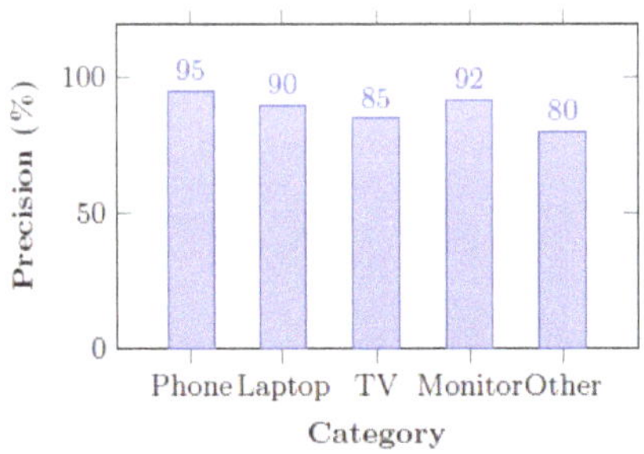

Figure 9. Precision of each category.

However, it is important to note that precision alone may not provide a complete picture of the system's performance. It should be considered in conjunction with other performance metrics, such as recall and the F1-score, to have a comprehensive evaluation of the system's effectiveness in identifying relevant items. This was considered in Figure 12. For instance, let us consider the example of a system that identifies smartphones among various items. The precision of the system in identifying smartphones is 95%, which means that out of all items identified as smartphones, 95% were actually phones. The remaining 5% could be items incorrectly classified as smartphones.

In Figure 10, recall serves as a performance metric that quantifies the completeness or comprehensiveness of a system in identifying relevant items. It is calculated by dividing the number of true positives (correctly identified items) by the sum of true positives and false negatives (items that were not identified as belonging to a particular category but should have been). Recall is particularly significant in situations where the consequences of false negatives are critical. By achieving a higher recall value, the system minimizes the chances of overlooking relevant items and provides a more comprehensive identification process.

For instance, consider a system designed to identify laptops among various objects. If the system achieves a recall of 85%, it indicates that out of all the actual laptops in the sample, 85% of them were correctly identified by the system. The remaining 15% represents the instances where the system failed to recognize laptops that were present.

A higher recall value suggests that the system is effective in capturing a larger proportion of the relevant items. It indicates a lower rate of false negatives, meaning that fewer items belonging to the target category are missed by the system. In our example, a recall of

91% signifies that the system has a relatively high ability to detect and include televisions in its identification process.

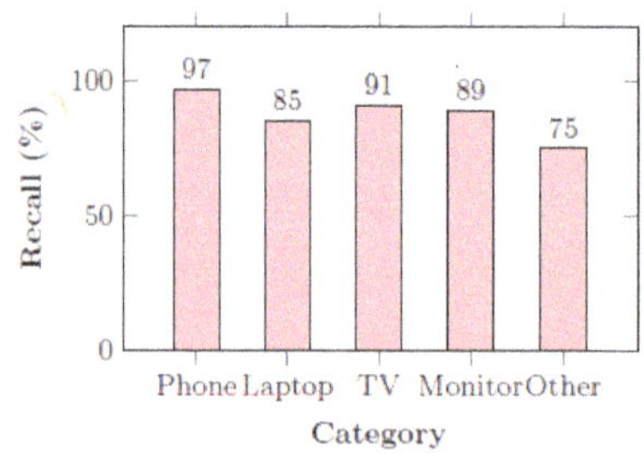

Figure 10. Recall of each category.

In Figure 11, the F1-score is a metric that encompasses both precision and recall to provide a comprehensive evaluation of the performance of a classification model. Precision and recall are both crucial aspects in assessing the effectiveness of such models. The F1-score offers a balanced measure by taking the harmonic mean of precision and recall.

$$F1\text{-}score = \frac{2 \cdot (precision \cdot recall)}{precision + recall}$$

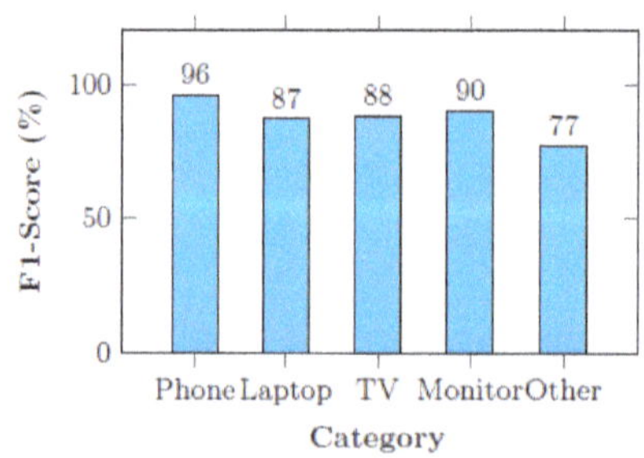

Figure 11. F1-Score (%) of each category.

This choice is made because the harmonic mean assigns more weight to smaller values, ensuring that both precision and recall are given equal consideration. By considering both precision and recall in the F1-score, it provides a unified indicator of overall performance. It strikes a balance between the two metrics, giving equal importance to correctly identifying relevant items (precision) and capturing the full extent of relevant items (recall). The F1-score is particularly valuable when the class distribution is imbalanced or when both precision and recall are of equal importance. It offers a single value that represents the overall effectiveness of the classification model, allowing for easier comparison and decision making.

In Figure 12, the overall performance of the system is represented by the "Overall" row of Table 3. Here, P represents precision, R represents recall, and F represents the F1-Score. This row displays the key performance metrics, including the precision, recall, and F1-score. According to the table, the system achieves an overall precision of 90%, recall of 88%, and F1-score of 89%. These metrics provide a comprehensive assessment of the system's performance across all categories. With a precision of 90%, the system demonstrates a high level of accuracy in correctly identifying e-waste items. Similarly, the recall of 88% indicates that the system is effective in capturing a significant portion of the actual e-waste items present in the sample.

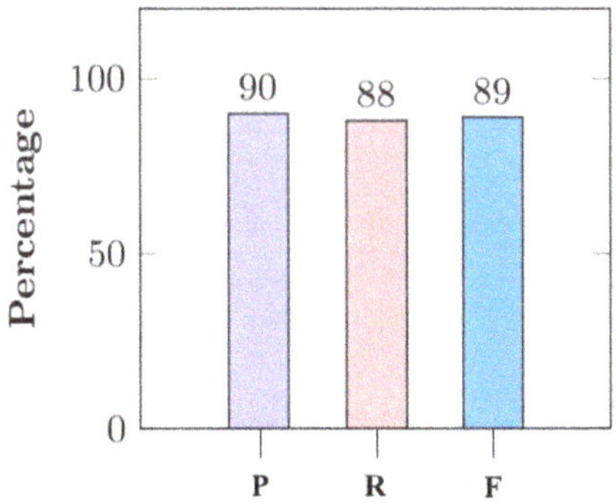

Figure 12. Overall performance of each category.

The F1-score of 89% is a balanced measure that combines precision and recall. It considers both metrics to provide an overall evaluation of the system's performance. This score indicates that the system maintains a good balance between precision and recall, achieving a harmonious trade-off between accurately identifying e-waste items, but may have some minor errors in specific categories.

4.3. Graphical Analysis of Pyrolysis Method

Figure 13 shows the yield of bio-fuel from plastic waste using the pyrolysis method. The x-axis represents the temperature in degrees Celsius, while the y-axis represents the yield of bio-fuel as a percentage. The blue line shows the relationship between the temperature and bio-fuel yield. As the temperature increases, the yield of bio-fuel also increases. At a temperature of 300 °C, the yield is 20%, which increases to 50% at a temperature of 500 °C. This graph suggests that the pyrolysis method can be an effective way of producing bio-fuel from plastic waste and that higher temperatures can result in a higher yield of bio-fuel. The legend indicates that the red line represents the bio-fuel yield.

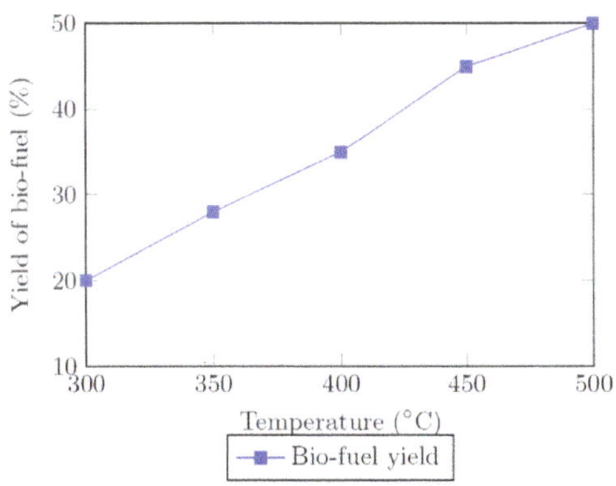

Figure 13. Yield of bio-fuel from plastic waste using pyrolysis method.

Table 4 [30] displays the results of the elemental analysis of mixed waste plastic pyrolysis liquid samples obtained from both thermal pyrolysis and catalyzed pyrolysis processes. The table shows the weight percentages of carbon (C), hydrogen (H), nitrogen (N), and sulfur (S) in the samples. The results indicate that the catalyzed pyrolysis process had a higher percentage of carbon and a lower percentage of hydrogen compared to the thermal pyrolysis process. Additionally, both processes showed similar percentages of nitrogen and sulfur in the pyrolysis liquid samples.

Table 4. Elemental analysis of mixed waste plastic pyrolysis liquid samples.

Weight (%)	Thermal Pyrolysis	Catalysed Pyrolysis
C	94.24	97.11
H	11.73	10.12
N	0.61	0.28
S	4.8	4.36

4.4. Graphical Analysis of Solar Battery Production and Reduction in CO_2

According to a study conducted in the Bangladesh University of Engineering and Technology, the projected growth of e-waste in Bangladesh from 2010 to 2035 is expected to increase from 0.13 million tons in 2010 to 4.62 million tons by 2035, indicating a significant rise in electronic waste generation over the given time period. The recycling rate of e-waste in the country stands at a mere 3%, with the remaining majority being indiscriminately deposited in landfills and rivers [31]. For each ton of e-waste that is collected and recycled, an impressive 1.44 tons of CO_2 emissions are effectively circumvented, as per the findings of an in-depth analysis conducted by the esteemed Belgian CO_2 environmental consulting firm, CO_2 logic [32].

Figure 14, shows the amount of CO_2 emissions in million tons through over the years if we recycle the metal from e-waste perfectly and reuse it for making solar batteries. The red line represents CO_2 emissions without recycling e-waste at all in Bangladesh and the blue line represents them after recycling only 3% of e-waste [31]. The green line represents the amount of CO_2 reduction if we recycle 60% of e-waste in Bangladesh. It is clearly visible that if we recycle the e-waste and produce solar batteries we will be able to reduce CO_2 emission to a large extent.

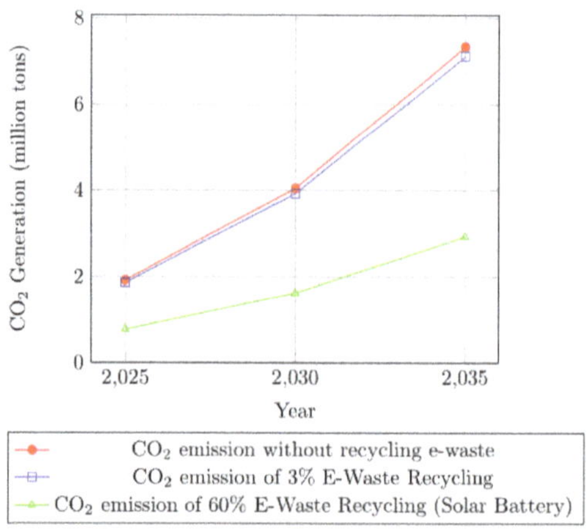

Figure 14. Reduction in CO_2 emissions with metal recycling from e-waste for solar batteries.

5. Limitations and Future Works

We have added the limitations and future works of our research:

The pyrolysis plant structure may vary according to its feedstock, requirements, products, and specific requirements. In conformity with HUAYIN, a manufacturer of waste tyre/plastic pyrolysis plants [33] typical a pyrolysis plant consists of six primary systems. The emission control system purifies the gas and confirms the emission of clean air. It also offers various de-dusters as per customer necessity while maintaining the standard of qualified emission. Pyrolysis is considered the future for plastic recycling techniques. We will implement preventive measures to mitigate and minimize the impacts of pyrolysis, such as the following:

- **Emissions and potential health/environment risks:** Pyrolysis can release gas, volatile organic compounds, and toxic substances, but our proposed process includes the implementation of a proper emission control system.
- **Energy inputs:** In some cases, traditional pyrolysis may use fossil fuels; our research focuses on using renewable energy sources as this will minimize the carbon footprint.
- **Contaminant release:** Plastic parts of e-waste can contain contaminants. Our proposed pyrolysis process involves control measures to ensure the safe handling, advanced sorting, and treatment of the e-waste plastic.

The proposed system will take every measure to reduce all the possible detrimental effects of pyrolysis. The major limitations of our system might be as follows:

- **Feedstock variability:** E-waste plastics may contain a different range of materials with different compositions. The consistency of the pyrolysis process can face some hurdles due to this.
- **Contaminants and impurities:** Despite thorough sorting and taking advance measures, some contaminants may still be present in the feedstock, which can affect the quality of the whole process and demean the standard of the by-products.
- **Pollutant emissions:** Though we are using an emission control system, making efforts to minimize the emissions, a comprehensive understanding about all kinds of pollutants and their potential impacts can impose limitations.

Continuous experimentation, development, and further research are necessary to enhance the characterization and monitoring of emissions, ensuring the safety and environmental sustainability of e-waste pyrolysis. For the future work of our research paper, we want to include some aspects: the further optimization of the pyrolysis process; the enhancement of data-driven decision making by leveraging advanced technologies; waste stream analysis for an effortless recycling process; solar batteries management and control to optimize their performance and prolong their lifespan; and the optimization of the recycling workflow to train the recyclers for different recycling processes.

6. Conclusions

The IoT- and cloud-based waste management and recycling system we have implemented successfully addresses the pressing issue of e-waste. Our study focused on the efficient separation and quick disposal of e-waste using the IoT, cloud computing, and machine learning. Our research results showcased numerous advantages, including enhanced efficiency, cost reductions, improved monitoring capabilities, and increased sustainability. Real-time data collection and analysis facilitated optimized waste collection routes, minimized the environmental impact, and successfully produced bio-fuel and solar batteries. Our research objectives were achieved through the implementation and evaluation of an IoT- and cloud-based waste management system, resulting in improved waste monitoring, optimized collection routes, and turning waste into assets by producing bio-fuel through pyrolysis and converting e-waste metal into solar batteries. Our study's outcomes align seamlessly with our initial research objectives, demonstrating the system's ability to overcome challenges associated with traditional waste management practices. However, there are some limitations, such as security and privacy concerns related to IoT devices and cloud infrastructure that must be addressed with robust measures to ensure data protection. In addition, the performance of the GAN algorithm can be affected by issues such as mode collapse, where the generator produces limited varieties of output, instability during training, and difficulty in evaluating the generated images. To sum up, the implementation of our IoT- and cloud-based waste management system has immense potential to revolutionize waste management practices. Its real-time data gathering, operational optimization, resource allocation, and production of recycled products offer substantial cost savings, a reduced environmental impact, and improved sustainability. However, addressing security concerns and conducting further research to ensure widespread adoption are necessary tasks for the successful implementation of such systems in the future.

Author Contributions: Conceptualization, M.F. and A.B.F.; methodology, M.F., A.B.F. and S.E.A.; software, M.F., A.B.F., S.E.A. and M.M.I.; validation, M.F., A.B.F., S.E.A. and M.M.I.; formal analysis, M.F. and A.B.F.; investigation, M.F., A.B.F., S.E.A. and M.M.I.; resources, M.F., A.B.F., S.E.A. and M.M.I.; data curation, M.F., A.B.F., S.E.A. and M.M.I.; writing—original draft preparation, M.F. and A.B.F.; writing—review and editing, M.F., A.B.F., S.E.A. and M.M.I.; visualization, M.F., A.B.F. and S.E.A.; supervision, M.F. and M.M.I.; project administration, M.F., A.B.F. and M.M.I. All authors have read and agreed to the published version of the manuscript.

Funding: This research received no external funding.

Data Availability Statement: Publicly available data-set were analyzed in this study. This data-set can be found here: [https://universe.roboflow.com/new-workspace-f7og7/e-waste-mx8fq/dataset/1].

Conflicts of Interest: The authors declare no conflict of interest.

References

1. Jafari, O.H.; Yang, M.Y. Real-time RGB-D based template matching pedestrian detection. In Proceedings of the 2016 IEEE International Conference on Robotics and Automation (ICRA), Stockholm, Sweden, 16–21 May 2016.
2. Mahipal, S.S.; Mayuri, K.; Manisha, N.; Shriyash, M.; Gaurav, S.; Bhaskar, C.; Rajeev, K. Effect of Electronic waste on Environmental & Human health-A Review. *IOSR J. Environ. Sci. Toxicol. Food Technol. (IOSR-JESTFT)* **2016**, *10*, 98–104.
3. Tanvir, A.; Rabeeh, G.; Fariborz, F.; Shahabuddin, M. E-Waste Recycling Technologies: An Overview, Challenges and Future Perspectives. In *Paradigm Shift in E-waste Management*; CRC Press: Boca Raton, FL, USA, 2022; pp. 143–176. [CrossRef]
4. Wayne, V.; Joseph, F. Time Series Analysis. In *Handbook of Psychology*; Research Methods in Psychology; John Wiley & Sons: New York, NY, USA, 2003; Volume 2, pp. 581–606, ISBN 9780471264385. [CrossRef]
5. Dubey, A.; Kumar, A.; García Díaz, V.; Sharma, A.; Kanhaiya, K. Study and analysis of SARIMA and LSTM in forecasting time series data. *Sustain. Energy Technol. Assess.* **2021**, *47*, 101474.
6. Tanskanen, P. Management and recycling of electronic waste. *Acta Mater.* **2013**, *61*, 1001–1011. [CrossRef]
7. Bazargan, A.; Lam, K.F.; Mckay, G. Challenges and Opportunities in E-Waste Management. In *E-Waste: Management, Types, and Challenges*; Li, Y.C., Wang, B.L., Eds.; Nova Science Publishers: Hauppauge, NY, USA, 2012; Chapter 2, ISBN 978-1-61942-217-9.
8. Zhao, P.; Xie, J.; Gu, F.; Sharmin, N.; Hall, P.; Fu, J. Separation of mixed waste plastics via magnetic levitation. *Waste Manag.* **2018**, *76*, 46–54. [CrossRef] [PubMed]
9. Devi, D.; Kumar, R.; Rajak, U. Plastic waste as a biofuel feedstock-A conceptual study. *IOP Conf. Ser. Mater. Sci. Eng.* **2021**, *1116*, 012029. [CrossRef]
10. Shamsudin, M.; Tong, A.; Firdhaus, M. *IOT Based E-Waste Monitoring System*; 2022.
11. Heacock, M.; Trottier, B.; Adhikary, S.; Asante, K.; Basu, N.; Brune, M.N.; Caravanos, J.; Carpenter, D.; Cazabon, D.; Chakraborty, P.; et al. Prevention-intervention strategies to reduce exposure to e-waste. *Rev. Environ. Health* **2018**, *33*, 219–228. [CrossRef] [PubMed]
12. Bansod, P. IoT Based Smart E-Waste Management System. *Int. J. Res. Appl. Sci. Eng. Technol.* **2022**, *10*, 881–890. [CrossRef]
13. Bošnjaković, M.; Galović, M.; Lackovic, I. Biofuel from plastic waste. In Proceedings of the 10th International Scientific and Expert Conference TEAM 2022, Slavonski Brod, Croatia, 21–22 September 2022.
14. Balakrishnan, B.K.; Shinde, A.; Kumar, A.; Kumar, A. Formation of Bio-Fuel from Waste Plastic Scrap. *Appl. Mech. Mater.* **2015**, *766–767*, 551–556.
15. Dinesh, M.H. Design and Fabrication Model is Used to Extract Waste Plastic to Bio Oil. *Int. J. Eng. Manag. Res.* **2020**, *6*, 68–71.
16. Shawpnil, K.; Nayeem, S.; Hossain, F.; Dayan, A.; Islam, M.M. EasyE-waste: A Novel Approach Towards Efficient and Sustainable E-waste Management. In Proceedings of the 2023 6th Springer International Conference on Intelligent Sustainable Systems (ICISS), Edinburgh, UK, 11–13 August 2023; SCAD College of Engineering and Technology: Tirunelveli, India, 2023.
17. Abdullah Al, M.; Sadman, S.; Motaharul, I. E-Waste Separation Using YOLOv5 and IoT. In Proceedings of the 6th International Conference on Sustainable Development (ICSD), Dhaka, Bangldesh, 19–20 February 2023.
18. Sankeerth, V.; Markandeya, V.; Ranga, E.; Bhavana, V. Smart Waste Management System Using IoT. *Int. J. Eng. Tech. Res.* **2020**, *9*, 4.
19. Thaseen Ikram, S.; Mohanraj, V.; Ramachandran, S.; Balakrishnan, A. An Intelligent Waste Management Application Using IoT and a Genetic Algorithm–Fuzzy Inference System. *Appl. Sci.* **2023**, *13*, 3943. [CrossRef]
20. Pavithra, M.; Alagu Esakkiammal, N.; Angel Melbha, A.; Aruleeswaran, R.; Balaji, N. IoT Based Automated Smart Waste Management System. *Int. J. Sci. Res. Sci. Eng. Technol. (IJSRSET)*. **2023**, *10*, 446–455. [CrossRef]
21. Cai, H.; Hu, J.; Li, Z.; Lim, W.H.; Mokayef, M.; Wong, C.H. An IoT Garbage Monitoring System for Effective Garbage Management. In Proceedings of the 2022 International Conference on Computer Engineering, Network, and Intelligent Multimedia (CENIM), Surabaya, Indonesia, 22–23 November 2022.
22. Artang, S.; Ghasemian, A.; Hameed, I. E-Waste Tracker: A Platform to Monitor e-Waste from Collection to Recycling. In Proceedings of the 37th ECMS International Conference on Modelling and Simulation (ECMS2023), Florence, Italy, 20–23 June 2023; Volume 37, ISBN 978-3-937436-80-7/978-3-937436-79-1 (CD).

23. Motaharul, I.; Bhuiyan, Z.A. An Integrated Scalable Framework for Cloud and IoT Based Green Healthcare System. *IEEE Access* **2023**, *11*, 22266–22282.
24. Zahran, M.; Wassal, A. A survey on GAN acceleration using memory compression techniques. *J. Eng. Appl. Sci.* **2021**, *68*, 47.
25. Fattah, J.; Ezzine, L.; Aman, Z.; Moussami, H.; Lachhab, A. Forecasting of demand using ARIMA model. *Int. J. Eng. Bus. Manag.* **2018**, *10*, 184797901880867. [CrossRef]
26. Miandad, R.; Barakat, M.; Aburizaiza, A.; Rehan, M.; Nizami, A.-S. Catalytic Pyrolysis of Plastic Waste: A Review. *Process. Saf. Environ. Prot.* **2016**, *102*, 822–838. [CrossRef]
27. Smith, K.M.; Fowler, G.; Pullket, S. Sewage Sludge-based Adsorbents: A Review of Their Production, Properties and Use in Water Treatment Applications. *Water Res.* **2009**, *43*, 2569–2594. [CrossRef] [PubMed]
28. Steiner, C.; Glaser, B.; Teixeira, W.G.; Lehmann, J.; Blum, W.E.H.; Zech, W. Nitrogen retention and plantuptake on a highly weathered central Amazonian Ferralsol amended with compost and charcoal. *J. Plant Nutr. Soil Sci.* **2008**, *171*, 893–899. [CrossRef]
29. Jaworek, A.; Krupa, A.; Czech, T. Modern electrostatic devices and methods for exhaust gas cleaning: A brief review. *J. Electrost.* **2007**, *65*, 133–155. [CrossRef]
30. Eze, W.; Madufor, I.; Onyeagoro, G.; Obasi, H.; Ugbaja, M. Study on the effect of Kankara zeolite-Y-based catalyst on the chemical properties of liquid fuel from mixed waste plastics (MWPs) pyrolysis. *Polym. Bull.* **2021**, *78*, 377–398. [CrossRef]
31. Assessment of Generation of E-Waste, Its Impacts on Environment and Resource Recovery Potential in Bangladesh. 28 June 2022. Available online: https://2022-06-28-11-42-330b838e623795f97fbe7501137d41ac.pdf (accessed on 6 June 2023).
32. Bo2W Impact on CO_2 Emissions. Available online: https://worldloop.org/e-waste/bo2w-impact-on-CO$_{2}$-emissions/ (accessed on 6 June 2023).
33. Pyrolysis Plant For Solid Wastes Recycling (16 August 2022). Available online: https://huayinre.com/pyrolysis-plant/ (accessed on 23 June 2023).

IoT

Article

IoT-Enabled Smart Drip Irrigation System Using ESP32

Gilroy P. Pereira *,†, Mohamed Z. Chaari † and Fawwad Daroge †

FAB-LAB, Qatar Scientific Club, Doha P.O. Box 9769, Qatar; chaari_zied@ieee.org (M.Z.C.); f.daroge@qsc.org.qa (F.D.)
* Correspondence: g.pereira@qsc.org.qa
† These authors contributed equally to this work.

Abstract: Agriculture, or farming, is the science of cultivating the soil, growing crops, and raising livestock. Ever since the days of the first plow from sticks over ten thousand years ago, agriculture has always depended on technology. As technology and science improved, so did the scale at which farming was possible. With the popularity and growth of the Internet of Things (IoT) in recent years, there are even more avenues for technology to make agriculture more efficient and help farmers in every nation. In this paper, we designed a smart IoT-enabled drip irrigation system using ESP32 to automate the irrigation process, and we tested it. The ESP32 communicates with the Blynk app, which is used to collect irrigation data, manually water the plants, switch off the automatic watering function, and plot graphs based on the readings of the sensors. We connected the ESP32 to a soil moisture sensor, temperature sensor, air humidity sensor, and water flow sensor. The ESP32 regularly checks if the soil is dry. If the soil is dry and the soil temperature is appropriate for watering, the ESP32 opens a solenoid valve and waters the plants. The amount of time to run the drip irrigation system is determined based on the flow rate measured by the water flow sensor. The ESP32 reads the humidity sensor values and notifies the user when the humidity is too high or too low. The user can switch off the automatic watering system according to the humidity value. In both primary and field tests, we found that the system ran well and was able to grow green onions.

Keywords: agriculture; agricultural technology; ESP32; Internet of Things

Citation: Pereira, G.P.; Chaari, M.Z.; Daroge, F. IoT-Enabled Smart Drip Irrigation System Using ESP32. *IoT* **2023**, *4*, 221–243. https://doi.org/10.3390/iot4030012

Academic Editors: Antonio Cano-Ortega and Francisco Sánchez-Sutil

Received: 15 May 2023
Revised: 14 June 2023
Accepted: 1 July 2023
Published: 7 July 2023

1. Introduction

Agriculture, which involves the cultivation of crops and animals, is one of the most essential practices for maintaining and growing the human population. Not only does it provide nourishment to human beings, but it is also helpful in eliminating extreme poverty and boosting the economy of a country. Agriculture accounts for 4% of the global gross domestic product (GDP) and is projected to feed about 9.7 billion people by 2050 [1].

Agriculture has always depended on technology in one way or another. People used extremely simple tools for farming more than 12,000 years ago. The farm tools were often made of wood or animal bones [2]. As time went on, humans developed better tools for farming. By the second agricultural revolution in the U.S., tractors were a common sight in farmlands [3].

Water is essential for plant growth and the distribution of mineral nutrients. Irrigation involves the application of water to the soil through a system of pumps, tubes, and sprays. It is commonly used in areas where rainfall is low [4]. There are many different types of irrigation systems. For sustainable agriculture in desert countries, where efficient water use is necessary, drip irrigation systems are a great fit [5]. With drip irrigation, water is directly applied to the soil (close to the roots of the plants) in the form of droplets over time. The most significant advantage of drip irrigation systems compared to other systems is the amount of water saved [6,7].

We can use the Internet of Things (IoT) in any application that requires data collection, automation, or control. With the growing popularity of IoT, there has been an increase in the ideas surrounding smart agricultural technology [8,9]. In this paper, we designed a smart IoT-enabled drip irrigation system using an ESP32 microcontroller. The system comprises an ESP32, solenoid valve, soil moisture sensor, temperature sensor, air humidity sensor, and water flow sensor. We used the Blynk IoT mobile app and web dashboard to collect irrigation data, turn the automatic irrigation feature on or off, manually open the valve if needed, and plot temperature, soil moisture content, and air humidity graphs. Opening the valve allows water to reach the roots of the plants. The soil moisture sensor constantly checks if the soil is dry. If the soil is dry and the temperature is ideal, the ESP32 can automatically open the valve and irrigate the field. Based on the humidity sensor readings, the user can turn off the automatic irrigation feature or turn it back on.

We set out to grow green onions from onion bulbs. After a week of running the drip irrigation system, spring onions began to grow from a few onion bulbs. The irrigation for the week was conducted automatically by the system with no interference from the users. Each spring onion received 0.676 gallons of water over the week, which is a satisfactory outcome. We also plotted the weekly temperature, soil moisture, and air humidity graphs.

The remainder of the paper is organized as follows: Section 2 illustrates various relevant works and how we build upon them. Section 3 covers the overview of the entire system. Section 4 highlights the hardware used, with further details provided in Sections 4.1–4.3. Section 5 covers the results of the implementation and tests. Finally, Section 6 provides the conclusion and future scope of our project.

2. Related Works

There have been prior works on smart irrigation systems and smart drip irrigation systems [10,11]. Refs. [12,13] provide overviews of smart irrigation systems. They talk about wireless communications, irrigation methods, sensors applicable to smart irrigation systems, and types of monitoring in this field. Likewise, ref. [14] provides a detailed breakdown of irrigation monitoring, control, and the scheduling system, while [15] investigates the use cases, challenges, and issues of IoT in agriculture.

In ref. [16], a smart irrigation system was designed using a resistive soil moisture sensor, temperature sensor, water flow meter, and Arduino UNO single board computer (SBC). The system monitors temperature and soil moisture level, and if the soil becomes dry or the temperature exceeds 30 °C, the field is irrigated. Ref. [17] details a smart system that monitors and controls agricultural production using IoT. It monitors the data and provides it to the farmer, who can use the data to control the system remotely when needed, reducing the workload.

In ref. [18], a smart irrigation system was designed using a resistive soil moisture sensor, temperature sensor, air humidity sensor, and Arduino UNO SBC. The system monitors and displays the temperature and humidity readings. If the soil is too dry, the motor is powered on so that the soil receives water. Ref. [19] proposed an innovative design of a solar-powered smart drip irrigation system using a node microcontroller unit (MCU) that monitors temperature and humidity via a DHT11 sensor, and the soil moisture value determines when the pump turns on. Ref. [20] proposes a smart farm using a long-range wireless area network (LoRaWAN).

We built on the prior works in five crucial ways. Firstly, we used an ESP32 microcontroller. The ESP32 is cheap, has built-in Wi-Fi, and Blynk IoT officially supports the ESP32. Secondly, we offer improved automatic irrigation. The ESP32 takes into account the actual time of the day, soil moisture content, and soil temperature before opening the valve and watering the plants. The temperature readings are not available just for monitoring. We used the temperature readings to ensure that we watered the plants at the best temperature for maximum water absorption. Thirdly, we improved the monitoring and control features. Along with soil moisture and temperature, the ESP32 also monitors the humidity and notifies the user when the humidity is too low or too high. The user can then decide to switch off the automatic irrigation feature or manually open/close the valve based on the monitored values.

Fourthly, we set the duration of watering the plants through the drip irrigation system with the help of the flow sensor. Lastly, the ESP32 collected real-time data. We used these data to track the date and time of irrigation, and to help ensure that we did not accidentally water the plants multiple times in a single day, with assistance from the flow sensor.

3. System Overview

The overview of the IoT-enabled smart drip irrigation system is shown in Figure 1. The ESP32 is the brain of the system. We connected the ESP32 to different sensors and a relay. The temperature sensor probe and soil moisture sensor probe were inserted into the soil and monitored the soil temperature and moisture levels, respectively. The water flow sensor provides data on the water flow rate, and the humidity sensor measures the humidity of the air. The system opens the solenoid valve to water the plants using a relay. The ESP32 uses Wi-Fi to communicate with the mobile app or web dashboard via Blynk cloud. We used the Blynk app to collect irrigation data, manually control the valve, and plot the soil temperature fluctuation graph.

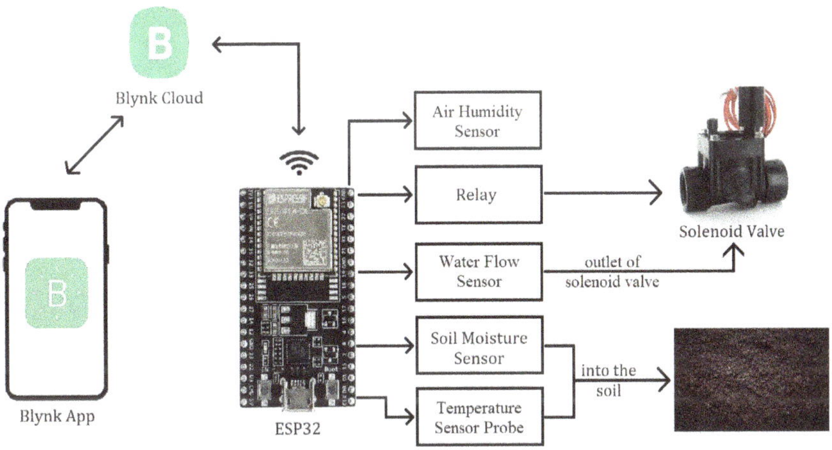

Figure 1. Overview of the IoT-enabled smart drip irrigation system [21].

It is best to water the plants in the morning or early in the evening. Watering the crops in the afternoon can lead to the water becoming hot and burning the plants. Watering the crops late in the evening may lead to water stagnation and encourage rot, fungal growth, and insects [22,23]. We used the hourly weather reports in Qatar [24,25] to set our morning irrigation window from 5 a.m. to 8 a.m. and evening irrigation window from 6 p.m. to 8 p.m. In these time windows, the weather is usually warm, and the temperature is between 24 °C and 30 °C, as shown in Figure 2. Using real-time data, the ESP32 will check the moisture and temperature of the soil within these time windows and water the plants if necessary.

We used an air humidity sensor to gather humidity data. If the temperature is very warm and the humidity is low, too much water will evaporate through transpiration. The water loss will lead to the plants attempting to absorb more water, and as they consume more water, they will consume more nutrients. Excess nutrients will cause the tips of the leaves to burn, and the leaves will wilt [26]. Hence, watering the soil when the humidity is too low may not be a good idea. The ESP32 will notify us if the humidity is too high or low. Based on the humidity readings, along with the other sensor data, we can turn the automatic irrigation feature on or off.

We can open the valve manually by using the app if there is a need to do so. The irrigation data are saved to the app as they help to recognize trends and eventually improve the system. The irrigation data consist of the date and time of irrigation, the temperature of the soil at the time of irrigation, and the rate of the flow of water in the drip line.

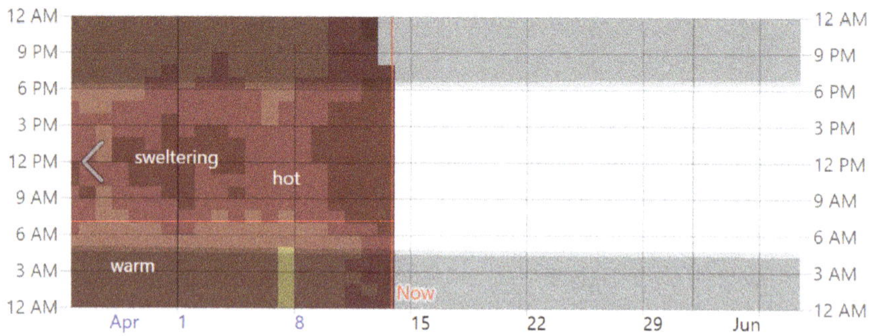

Figure 2. The hourly reported temperature in Qatar for May 2023, color-coded into bands. The shaded overlays indicate night and civil twilight (source: www.weatherspark.com (accessed on 13 May 2023)) [27].

We interfaced the ESP32 with a moisture sensor, temperature sensor, air humidity sensor, water flow sensor, and solenoid valve. Using the data from these sensors, the ESP32 determines when to open the solenoid valve. The solenoid valve controls the flow of water into the pipes of the drip irrigation system. The flowchart highlighting the logic programmed into the ESP32 is shown in Figure 3.

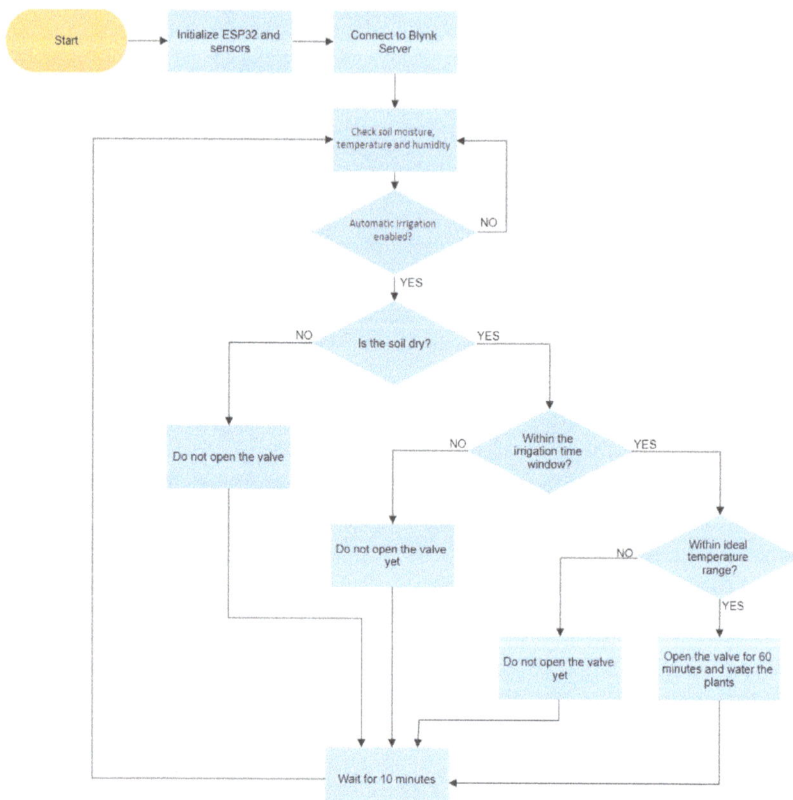

Figure 3. Flow chart of the IoT-enabled smart drip irrigation system.

When we turn on the system, the ESP32 initializes its non-volatile storage (NVS) flash, Wi-Fi, real-time operating system (RTOS), soil moisture sensor, temperature sensor, air humidity sensor, and flow sensor. The ESP32 then connects to the Blynk servers and checks the moisture content of the soil, temperature, and humidity. If the irrigation system is currently set to automatic irrigation and if the soil is dry, it then obtains the current time of the day and compares it to the morning and evening irrigation time windows. If the time is within the irrigation time window, the ESP32 will check if the soil temperature is within the ideal range. If the temperature is within the ideal range for maximum water absorption, the ESP32 will open the valve for an hour and water the plants. The ESP32 will then wait ten minutes before checking the soil moisture level.

If the soil is dry but the current time is not within the irrigation time windows, the ESP32 will not open the valve. Similarly, if the temperature is not in the ideal range, the ESP32 will not open the valve. If the soil is humid during the moisture check, the ESP32 will keep the valve closed. An improved flowchart based on the results of our tests is presented in Section 5.

4. Materials Used

4.1. Hardware

The system's main hardware components are a microcontroller, moisture sensor, temperature sensor, air humidity sensor, water flow sensor, solenoid valve, relay, and a step-down transformer.

4.1.1. Microcontroller—ESP32

The ESP32 is a low-cost, 32-bit microcontroller. It has built-in Bluetooth and Wi-Fi, making it useful for IoT applications. It can accommodate multiple sensors and devices with 48 general purpose input–output (GPIO) pins. We used the inbuilt Wi-Fi of the ESP32 to communicate with the Blynk mobile app or web dashboard. ESP32 sends irrigation information to Blynk cloud. We can control the valve or set the irrigation time using the mobile app.

4.1.2. Moisture Sensor—DFRobot

We used a SEN0308 DFRobot soil humidity sensor, which detects soil humidity and sends analog signals to the ESP32. The sensor is shown in Figure 4.

Figure 4. DFRobot moisture sensor.

The SEN0308 is a capacitive moisture sensor that offers improved waterproof performance, increased length, and high corrosion resistance [28]. It solves a critical issue encountered with commonly used resistive moisture sensor probes, which is probe corrosion. The SEN0308 has excellent corrosion resistance and can be inserted into the soil for long periods.

We inserted the sensor probe into the soil. The sensor measures changes in capacitance that are caused by alterations in the dielectric due to humidity [29]. It does not measure moisture directly but instead measures the moisture's ions. The sensor sends analog signals to the ESP32 based on the measurement, which is converted to a digital signal by the ESP32.

4.1.3. Temperature Sensor—DS18B20

We used a DS18B20 one-wire bus temperature sensor probe, as shown in Figure 5. These sensors provide up to 12-bit temperature measurements in Celsius and have an alarm function with non-volatile user-programmable lower and upper trigger points. Each sensor has a unique 64-bit ID burned in at the factory to differentiate them, which allows us to control multiple sensors with a single GPIO pin of a microcontroller. This sensor's significant advantages are its high accuracy and waterproofing [30].

Figure 5. DS18B20 waterproof temperature sensor.

4.1.4. Air Humidity Sensor—DHT22

We used a DHT22 humidity–temperature sensor, as shown in Figure 6. It is low-cost and uses a capacitive humidity sensor to measure the humidity in the air. It also uses a thermistor to measure the temperature. The data can be obtained from the data pin of the DHT22. The DHT22 is good for 0–99.9% humidity readings wit +/−2% accuracy [31].

Figure 6. DHT22 air humidity sensor.

4.1.5. Water Flow Sensor—FS300A G3/4 Inch

The FS300A consists of a water rotor, a hall-effect sensor, and a plastic valve body, as shown in Figure 7. The water flows in through the inlet and out through the outlet due to the flow of water and the wheel rolls, and so does the magnet. The rotation of the magnet triggers the hall-effect sensor, which outputs high- and low-square waves. We calculate the water flow by counting the square waves [32].

Figure 7. FS300A G3/4 inch water flow sensor.

4.1.6. Solenoid Valve—Hunter PGV-100G (24VAC)

We used a hunter PGV one-inch solenoid valve, which is an electrically controlled valve. This valve is shown in Figure 8. A solenoid is an electric coil with a movable magnetic core. Applying an electric current to this coil creates a magnetic field, which moves the core and allows water to flow. If the current is cut off, the valve closes, and the water flow stops [33].

Figure 8. Hunter PGV-100G solenoid valve.

4.1.7. Relay

A relay is a simple electrically controlled switch. By sending a signal from the ESP32, we can turn the switch on and supply a 24 V AC to the solenoid valve and open it.

4.1.8. Step-Down Voltage Regulator

The smart drip irrigation system is powered using a 12 V DC adapter. We used a step-down voltage regulator to supply the ESP32 with the 5 V needed for operation.

4.2. Drip Irrigation Setup

4.2.1. Acrylic Container

To make our irrigation system somewhat portable, we made an acrylic container. We first cut out the base and walls of the container from an 18 mm thick acrylic sheet using a computerized numerical control (CNC) machine and then glued them together and applied silicone, as shown in Figure 9a. Once the glue and silicone were dry, we added some screws to the container for strength and drilled drainage holes in the base of the acrylic container to allow for percolation, as shown in Figure 9b. The diameter of the holes ranged from 10 to 12 mm.

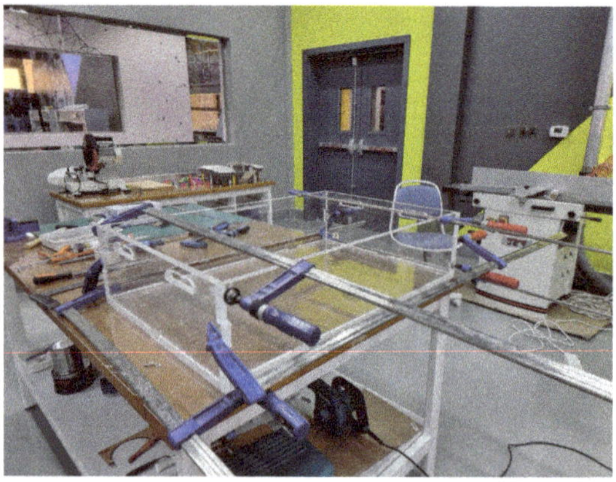

(a)

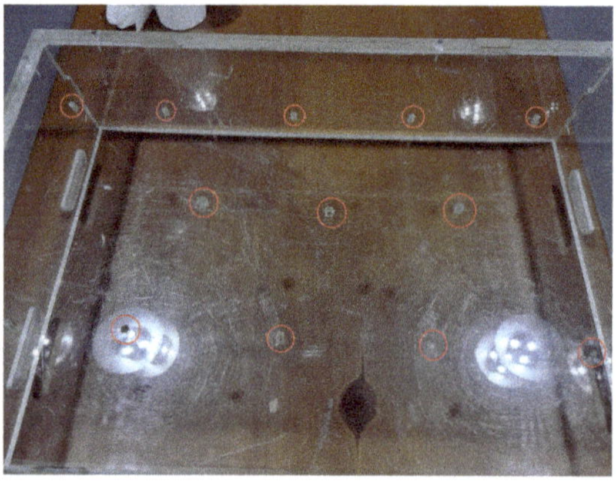

(b)

Figure 9. Preparing the acrylic container. (**a**) Curing the super glue and silicone. (**b**) Drainage holes drilled in the base of the container.

4.2.2. Soil

After preparing the container, we poured in one and three-fourths of a 50 L bag of all-purpose potting soil [34], as shown in Figure 10. This translates to 6.280 inches of soil, which is sufficient for growing green onions as they require 6–8 inches of soil [35]. The soil is a mixture of organic compost and moisture-retaining coir residues that prolong hydration and reduce drought stress, making the soil quite useful for the subtropical desert climate of Qatar. The elemental analysis of the soil is shown in Table 1.

Table 1. Test results of the elemental analysis of the soil [34].

Elemental Analysis	Results
Total organic carbon	45.5%
Total organic nitrogen	1.40%
Total nitrogen	1.40%
Sodium chloride	0.76%
Potential of hydrogen (pH)	6.7
Electrical conductivity (EC)	1.51 dS/m
Cadmium (Cd)	<0.01%
Chromium (Cr)	<0.01%
Selenium (Se)	<0.01%

Table 1. Cont.

Elemental Analysis	Results
Carbon nitrogen ratio	34:1
Copper (Cu)	<0.01%
Lead (Pb)	<0.01%
Molybdenum (Mo)	<0.01%
Nickel (Ni)	<0.01%
Potassium oxide (K_2O)	0.03%
Phosphorus pentoxide (P_2O_5)	<0.01%
Zinc (Zn)	<0.01%
Arsenic (As)	<0.01%
Mercury (Hg)	<0.01%
Organic matter	78.9%
Moisture content	42.2%

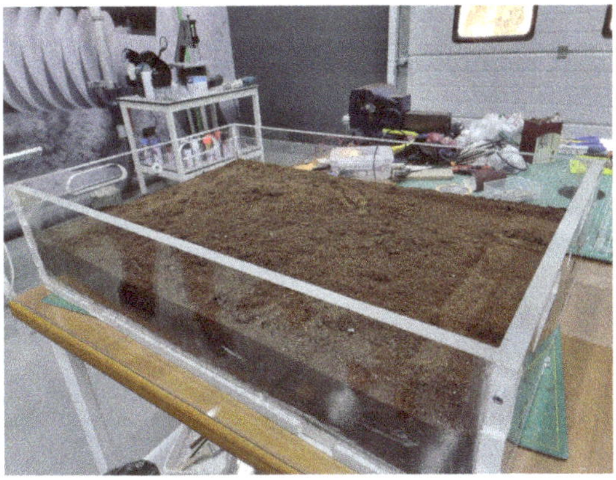

Figure 10. Adding soil to the container.

4.3. Drip Irrigation Piping

After the addition of soil, we drilled holes in two opposing walls and added the mainline, dripline, and drippers, as shown in Figure 11. We will connect the rest of the smart irrigation system to this basic drip irrigation setup later.

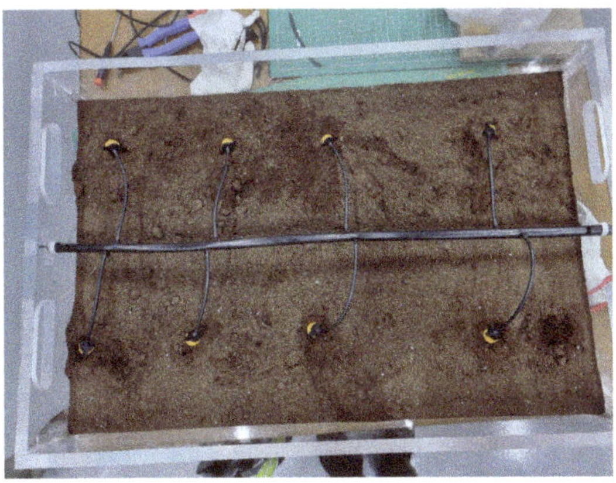

Figure 11. The addition of drip irrigation pipelines.

5. Experiments and Discussion

5.1. Primary Tests in the Laboratory

The initial tests were conducted in the lab to test the moisture sensor, temperature sensor, air humidity sensor, water flow sensor, solenoid valve, and firmware. The system was powered using a power supply. Using the relay, the ESP32 was able to control the solenoid valve. A loud click noise let us know when the valve was opened or closed. We calibrated the soil moisture sensor by first reading the value of the sensor in the air and then placing the probe in a glass of water and re-reading the value. After calibrating the moisture sensor, we proceeded to confirm that the ESP32, moisture sensor, and valve worked well together. To do so, we programmed the valve to open if the ESP32 was not in the cup of water. We then replaced the cup with a potted plant, re-calibrated the moisture sensor with the soil [36,37], and repeated the experiments.

We connected the DS18B20 temperature sensor to the ESP32 using the GPIO. As ESP32 does not have a dedicated 1-wire bus interface GPIO pin, we had to perform bit-banging on the GPIO to use the DS18B20. We followed the timing diagrams of the DS18B20 in order to write and read 1 s and 0 s. Lastly, we performed simple calibration by measuring the known temperatures [38]. We measured the temperatures of different objects with DS18B20 and compared them to the readings from a FLIR C3-X thermal camera. Figure 12a shows the temperature measured by the DS18B20 for a cup of hot water, and Figure 12b shows the thermal capture using the camera for the same cup. We repeated the temperature measurement with other test scenarios, and the results are shown in Table 2.

```
The temperature is 70.500000
Outside ideal watering temperature range
Enters next loop
The temperature is 70.000000
Outside ideal watering temperature range
Enters next loop
The temperature is 70.000000
Outside ideal watering temperature range
```

(a)

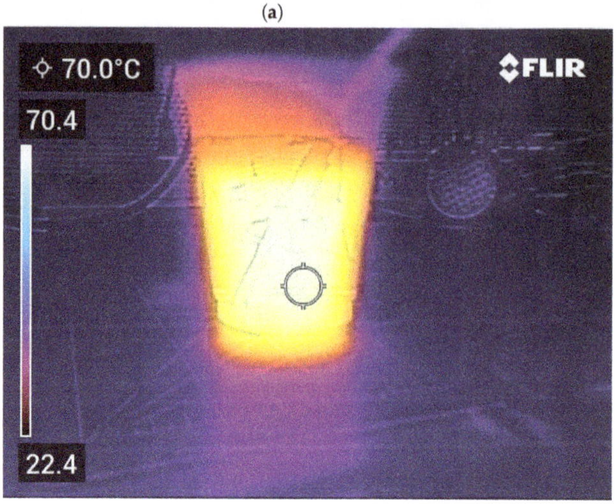

(b)

Figure 12. Measuring the temperature of a hot cup of water using the DS18B20 sensor and a thermal camera. (**a**) Temperature measured by the DS18B20 sensor. (**b**) Temperature measured by the thermal camera (FLIR C3-X, manufactured by Teledyne FLIR LLC, Wilsonville, OR, USA).

Table 2. Comparison of the DS18B20 temperature sensor with a thermal camera.

Test	Subject of Temperature Measurement	Temperature Measured by DS18B20 (°C)	Temperature Measured by Thermal Camera (°C)
1	Cup of hot water	70.0	70.0
2	Cup of room temperature water	25.0	24.3
3	Cup of cold water	11.5	12.0
4	Dry soil	21.5	21.2
5	Wet soil	24.0	23.4

As the measured temperature difference was less than $+/-1$ °C, we then proceeded to connect the air humidity sensor and water flow sensor. We set up an interrupt on the ESP32 GPIO pin to read the pulses from the flow meter. To calibrate the sensor, we poured a known amount of water through the flow meter and checked if the sensor could calculate the amount of water that flowed through [39]. We observed that the error was less than one percent.

5.2. Comparing the Sensor Readings to Weather Forecasts

To test the accuracy of the sensor measurements in an outdoor environment, we took the readings from the DS18B20 soil temperature sensor and DHT22 air temperature and humidity sensor and compared them to the data gathered from timeanddate.com [40]. The test was conducted over a period of 4 h on 16 June 2023. Table 3 compares the temperature measured by the soil temperature sensor (DS18B20), air temperature sensor (DHT22), and the temperature data obtained from timeanddate.com (accessed on 16 June 2023).

Table 3. Comparison of temperatures.

Time (GMT+3)	Temperature Obtained from timeanddate.com (accessed on 16 June 2023) (°C)	Soil Temperature Measured by DS18B20 (°C)	Air Temperature Measured by Thermal Camera (°C)
12:00	43	44	43
13:00	43	44.5	43
14:00	42	44.5	42
15:00	42	43.5	42

As seen in the table, the air temperature measured by the DHT22 and the data obtained from timeanddate.com (accessed on 16 June 2023) match. The soil temperature measured by the DS18B20 was always slightly more than the air temperature. A comparison of the air humidity is shown in Table 4.

Table 4. Comparison of Temperature.

Time (GMT+3)	Relative Humidity Obtained from timeanddate.com (accessed on 16 June 2023) (%)	Relative Humidity Measured by DHT22 (%)
12:00	18	16.9
13:00	15	16.4
14:00	14	14.6
15:00	15	17.1

The measured and observed humidity slightly differ. The differences observed may be due to the differences in measurement locations. timeanddate.com (accessed on 16 June 2023) has a weather station set up at Doha International Airport; our tests were done in Abu Hamour on a windy day, with wind speeds reaching 37 km/h. Despite the slight differences, all the sensors performed well and are fit for the smart irrigation system.

5.3. Testing the Solenoid Valve Outdoors

We used a large 400 US gallon water tank for the irrigation system, as seen in Figure 13a. We also installed a pump motor to ensure the water flowed with enough pressure to open the solenoid valve. We placed the motor wiring in a container with a gasket lid to protect the connections from water and put an acrylic container sealed with silicone over the motor to provide some protection from the rain, as shown in Figure 13b. Lastly, we made holes in one of the walls of the container to provide air circulation for the motor.

Once the water tank was ready, we connected the solenoid valve to the pump motor. The electronic components were placed in a plastic container with a rubber gasket to protect them from water. The printed circuit boards (PCBs) were fastened into place using screws and a raiser. We designed a case for the barrel jack of the adapter and 3D-printed it. Next, we drilled holes into the walls of the container to allow wires to pass through and applied silicone to prevent moisture from entering the container. The field-ready kit can be seen in Figure 14a,b.

(a)

Figure 13. *Cont.*

(b)

Figure 13. Water tank and pump motor for the irrigation system; (**a**) 400 US gallon water container; (**b**) water pump motor covered for water resistance.

(a)

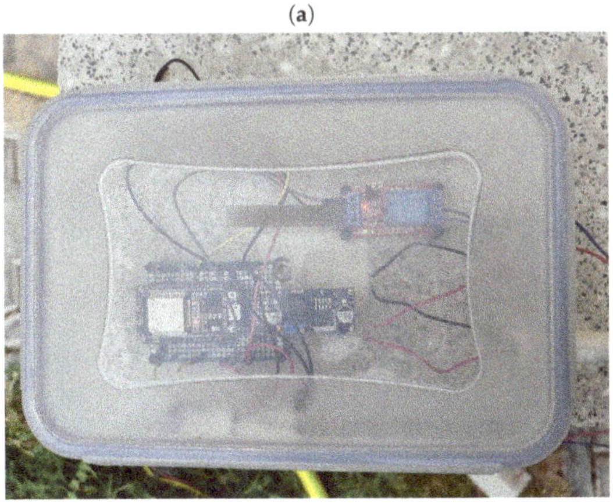

(b)

Figure 14. Plastic container for the smart drip irrigation system. (**a**) Components of the irrigation system placed in the container. (**b**) Sealed container ready for outdoor use.

We tested the system and found that the solenoid valve worked well and would let water flow through as directed by the ESP32. We connected the solenoid valve outlet to the inlet of the water flow sensor. We made more acrylic containers sealed with silicone to improve the irrigation system's overall dust and water resistance. We placed one container over the extension board and the other electronics. We put the other container over the solenoid valve and flow sensor. The containers are shown in Figure 15a,b. To keep the wiring between the PCB and solenoid valves free of water, we 3D-printed a simple case and sealed it using glue and silicone. This case is shown in Figure 15c. We also placed the DHT22 humidity sensor in a plastic enclosure and sealed it with silicone, as shown in Figure 15d.

We found the irrigation system capable of withstanding dust and light rain due to all the containers and silicone.

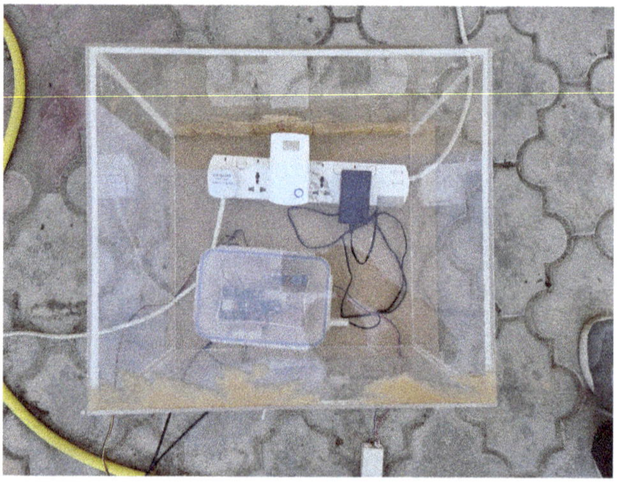

(a)

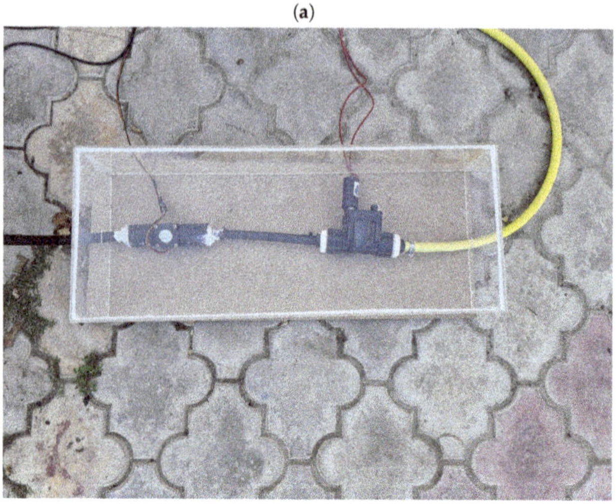

(b)

Figure 15. *Cont.*

(c)

(d)

Figure 15. Making the irrigation system water- and dust-resistant. (**a**) Protecting the electronics with an acrylic container. (**b**) Protecting the solenoid valve and water flow sensor with an acrylic container. (**c**) Protecting the solenoid wires with a 3D-printed case. (**d**) Protecting the DHT22 with a plastic case.

5.4. Testing the Entire Smart Drip Irrigation System in the Field

After confirming that the irrigation system was satisfactorily water- and dust-resistant, we connected the outlet of the water flow sensor to the drip irrigation system's mainline. We placed the drippers at different locations and positioned the moisture sensor and temperature sensor in the soil near one of the drippers. We placed the DHT22 on top of the acrylic case covering the valve. The entire smart drip irrigation system is shown in Figure 16.

We tested the system without any plants for the first few days and opened the valve for just ten minutes at a time. The water flow rate through the mainline was 10 L per hour, as per the flow meter sensor. We collected the water from the drippers using a bottle for ten minutes and measured it in a graduated cylinder. We repeated the tests multiple times and present our findings in Table 5. The average flow rate of 0.64 L per hour falls within the range specified by the manufacturer of the dripper [41].

Table 5. Measured flow rate of the mainline and drippers.

Test Number	Mainline Flow Rate (lph)	Flow Rate of Each Dripper (lph)	Total Flow Rate of Eight Drippers (lph)
1	16	0.71	5.68
2	10	0.59	4.72
3	12	0.66	5.28
4	10	0.60	4.8
5	12	0.66	5.28

Figure 16. Smart drip irrigation system working in the field.

We programmed the ideal temperature range to be between 27–32 °C or 80.6–89.6 °F. While maximum water is used at a soil temperature of 59 °F [42], it is not a temperature that is easy to reach in the summer months. Hence, we continued our tests with an attainable temperature range. We also programmed the acceptable relative humidity range to be between 25% and 90%; 25% was set as the lower limit as most plants grow best with a relative humidity of over 50%. Although many plants will tolerate lower levels, only those native to arid regions will tolerate humidity below 25% [43]. If the humidity reading crossed the lower or upper limit, we received a notification on the Blynk app. For the most part, the system worked as we expected. The ESP32 opens the valve if the soil is dry and the current time and temperature are within the programmed ideal range. If the soil is moist, the ESP32 will not open the valve. However, we discovered some issues during our tests.

At times, even though the ESP32 had opened the valve once, the soil around the sensor was not moist. This delay led to the ESP32 opening the valve a second time and over-watering the soil. To combat this, we modified the firmware to block the valve from being opened twice on the same day if the water flow sensor was already triggered earlier the same day. Hence, the flow sensor confirms that the plants received water. It was possible to manually open the valve via the app if we needed to do so.

Additionally, if the tank was empty during the irrigation window, the ESP32 still opened the valve and considered that it had done its job, but the soil was dry. If there is a water shortage, the ESP32 might miss the irrigation window and not water the plants. We handled this issue by modifying the firmware and setting up a notification. We received an app notification if the irrigation window had passed, the flowmeter was never triggered, and the soil was dry even though the ESP32 opened the valve. The Blynk app notification allowed us to inspect the tank and soil and decide between manually opening the valve or waiting for the next irrigation time window. The modified flow chart accommodating the above changes is shown in Figure 17.

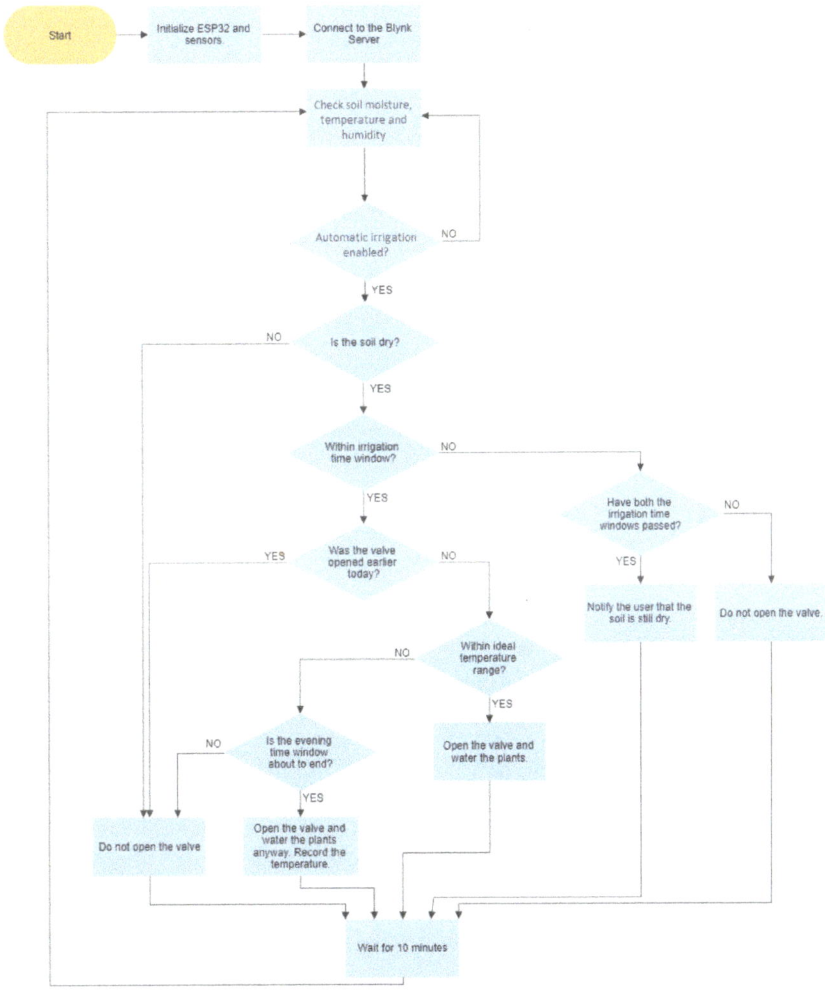

Figure 17. Final flow chart of the smart drip irrigation system firmware.

The most significant differences in the final flowchart occurred after confirming that the soil was dry. If the ESP32 conducts the moisture check within one of the irrigation time windows, it then checks if the valve was opened earlier today. If the system had opened the valve before, it would not open the valve again to prevent over-watering the plants. If the ESP32 has yet to open the valve and the temperature is in the ideal range, it will open the valve and water the plants.

Due to the extreme weather changes in Qatar this year, there were days when we were not even close to the ideal soil temperature for irrigation. We added a safety feature to ensure that the plants did not remain thirsty just because the soil was not at the ideal temperature. If the soil is dry, the ESP32 has not opened the valve as yet, and the evening time window is about to end, the ESP32 will open the valve to water the plants regardless and note down the temperature at the time of irrigation.

If the valve is opened once, the morning and evening irrigation time windows have passed, and the soil is still dry, the admin user will receive a notification about the dry soil. We can then decide if the valve needs to be opened manually through the app.

After we updated the ESP32 firmware according to the latest flowchart, we planted onion bulbs into the soil to grow spring onions. We placed the bulbs near the drippers. Later, we positioned the moisture sensor in the soil near one of the bulbs and drippers. We then opened the solenoid valve for sixty minutes compared to the ten minutes during the early testing phase. We then watered the onion bulbs.

6. Results

We checked back on the smart drip irrigation system in one week. The spring onions had begun to grow atop a few onion bulbs, and the plants were healthy and received plenty of water. The growing spring onions are shown in Figure 18.

Figure 18. Smart irrigation system used for growing spring onions.

Further details about the irrigation system are available on the Blynk IoT dashboard. The web dashboard is shown in Figure 19a,b.

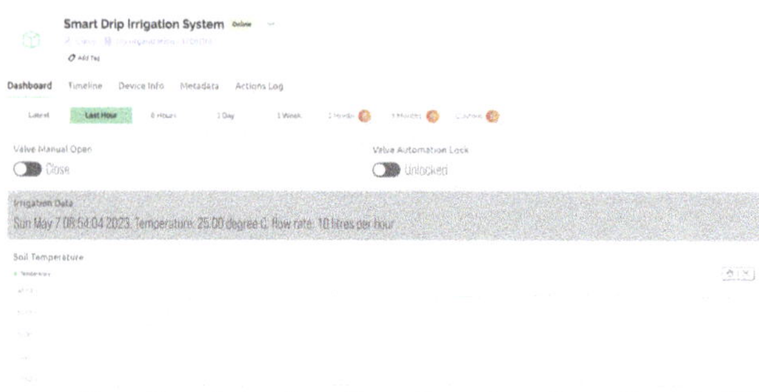

(a)

Figure 19. *Cont.*

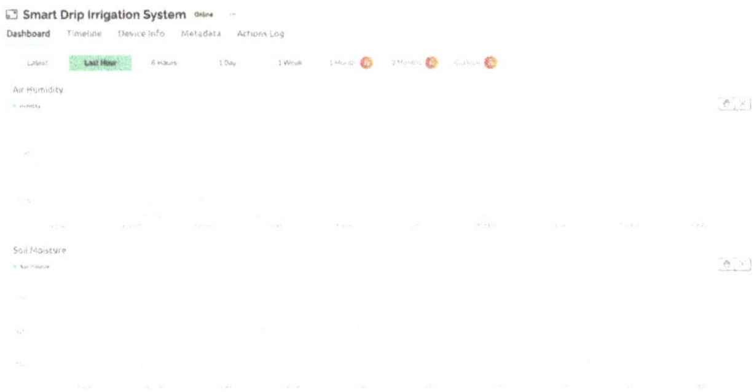

(**b**)

Figure 19. Web dashboard for the smart drip irrigation system. (**a**) Switches, irrigation data, and temperature graph. (**b**) Humidity and moisture content graph.

The "Valve Manual Open" button allows us to open the solenoid valve as needed and the "Valve Automation Lock" button allows us to block the automatic irrigation function. The "Irrigation Data" widget provides the exact date, time, soil temperature, and flow rate of when the ESP32 last opened the solenoid valve. The "Temperature" graph shows the fluctuation of temperature throughout the day, the "Air humidity" graph shows the variation of humidity, and the "Soil Moisture" graph shows the alteration of moisture in the soil.

We collected the irrigation data over one week and plotted the soil moisture content graph (Figure 20). The soil is dry if the moisture reading crosses 380. If the moisture reading is above 380 and the temperature and irrigation window are in the proper range, the ESP32 opens the valve to irrigate the field. The soil is considered wet if the reading is between 190 and 380, and very wet if the reading is between 0 and 190. As seen in the graph, the onions received water four times this week.

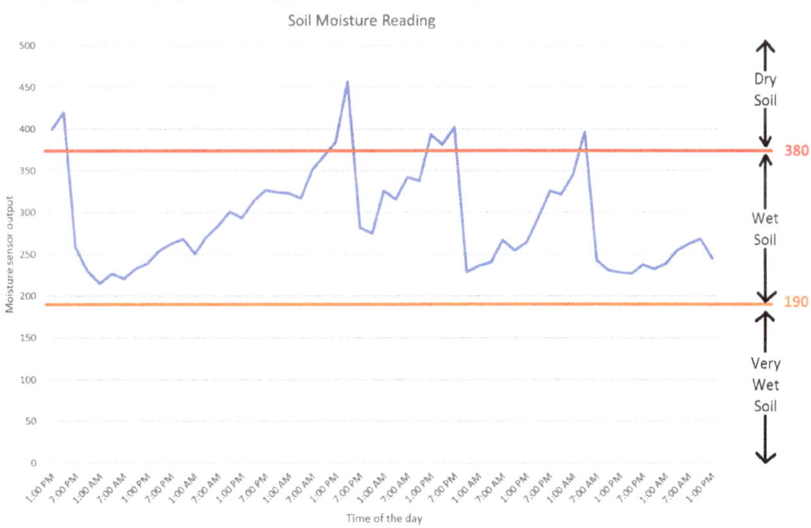

Figure 20. Soil moisture variation for one week.

Each dripper had a flow rate of 0.64 L per hour and watered an onion bulb for an hour a day for four days a week. In other words, the onion received 0.64 L of water a day. The total water provided to an onion over the week was as follows:

$$0.64 \times 4 = 2.56 \text{ L} \tag{1}$$

To convert the value into gallons, we divide Equation (1) by 3.785, as shown:

$$2.56/3.785 = 0.676 \text{ gallons} \tag{2}$$

Spring onions or shallots grown from the bulb need about an inch of water per week to grow well [44]. "One inch of water" refers to the amount of water necessary to cover one square foot of soil with one inch of water [45]. As there are 12 inches in a foot, the square inches of water needed per square foot of soil is as follows:

$$12 \times 12 = 144 \text{ sq. inch} \tag{3}$$

Moreover, 1 gallon is 231 cubic inches. Thus, when we divide 144 sq. inches by 231, we obtain the amount of water needed by onion bulbs per square foot as follows:

$$144/231 = 0.623 \text{ gallons} \tag{4}$$

As seen above, the actual amount of water provided to the onion bulbs in a week using the smart irrigation system in Equation (2) and the recommended amount of weekly water in Equation (4) are very close; hence, the smart irrigation system succeeds in providing adequate moisture to the plants.

We plotted the humidity variation graph for one week, as shown in Figure 21. The humidity was mostly above the lower limit of 25%. However, corresponding to some of the hottest hours of the day, it was not uncommon for the humidity to fall well below the 25% mark.

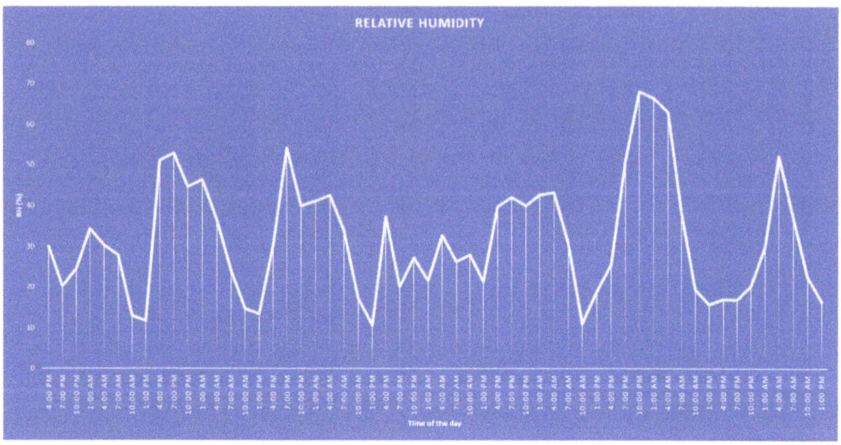

Figure 21. Humidity fluctuation throughout the week.

We plotted the temperature fluctuation graph for one week, as shown in Figure 22. The temperatures were the highest between 1 p.m. and 3 p.m. This timing also coincides with the lowest humidity readings of the day. The highest observed temperature was 46 °C. The temperature was comparatively low during the morning and evening irrigation time windows.

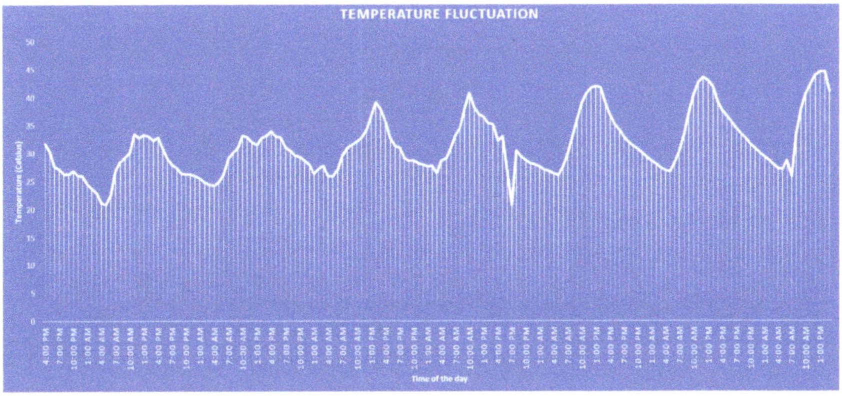

Figure 22. Temperature fluctuation throughout the week.

The moisture content, humidity, and temperature data collected are valuable for observing the weekly and monthly trends, making changes to the firmware, and are especially useful for building a greenhouse around the smart irrigation system.

7. Conclusions

We were successful in building an IoT-enabled smart drip irrigation system. It provides an enhanced automation feature, where if the soil is dry, the temperature is in the ideal range for maximum water absorption, the time falls within the designated morning or evening irrigation windows, and the ESP32 will open the solenoid valve and water the plants. We added safety features to prevent scenarios such as over-irrigation, missing the irrigation time, or leaving the plants thirsty.

Using the Blynk IoT dashboard, we can also monitor soil moisture, temperature, and air humidity. If the humidity is too low or too high, the admin user receives a notification on the Blynk app. We can use the Blynk dashboard to stop the automation function or manually open the valve based on the monitored data.

The smart drip irrigation system is currently being used to grow green onions from onion bulbs. While the system has been performing well, there are still some areas of improvement to explore, such as:

- Exploring a companion app where we can select the crop planted; the watering time would be changed accordingly to meet the crop needs.
- Expanding the system to control multiple sensors and valves.
- Exploring the impact of watering the plants at the ideal temperature for maximum water absorption make.
- Using Bluetooth or Wi-Fi mesh to control multiple smart drip irrigation systems.
- Exploring some portability options and solar energy systems.
- Integrating pH sensors, wind speed sensors, rain sensors, and more.

Author Contributions: Conceptualization, M.Z.C. and G.P.P.; methodology, G.P.P. and M.Z.C.; software, G.P.P.; validation, G.P.P. and F.D.; formal analysis, G.P.P.; resources, M.Z.C. and F.D.; data curation, G.P.P. and F.D.; writing—original draft preparation, G.P.P.; writing—review and editing, G.P.P. and M.Z.C.; supervision, M.Z.C. All authors have read and agreed to the published version of the manuscript.

Funding: This research received no external funding.

Data Availability Statement: Not applicable.

Acknowledgments: We thank Faisal Kuttikadavath (Qatar Scientific Club) for providing us with the space, tools, knowledge, and help needed to set up the garden for the smart drip irrigation system.

Conflicts of Interest: The authors declare no conflict of interest.

Abbreviations

The following abbreviations are used in this manuscript:

IoT	Internet of Things
SBC	single board computer
MCU	microcontroller unit
LoRaWAN	long-range wireless area network
NVS	non-volatile storage
RTOS	real-time operating system
GPIO	general purpose input–output
CNC	computerized numerical control
pH	potential of hydrogen
EC	electrical conductivity
PCB	printed circuit board

References

1. Agriculture Overview: Development News, Research, Data | World Bank. Available online: https://www.worldbank.org/en/topic/agriculture/overview#1 (accessed on 12 December 2022).
2. Woods, M.; Woods, M.B. *Ancient Agricultural Technology: From Sickles to Plows*; Twenty-First Century Books: Minneapolis, MO, USA, 2011; pp. 98–99. ISBN 978-0-7613-7269-1.
3. Bellis, M. Farm Machinery and Technology Changes from 1776 to 1990. Available online: https://www.thoughtco.com/american-farm-tech-development-4083328 (accessed on 13 December 2022).
4. Types of Agricultural Water Use | Other Uses of Water | Healthy Water | CDC. Available online: https://www.cdc.gov/healthywater/other/agricultural/types.html (accessed on 13 December 2022).
5. Ranjan, S.; Sow, S. Drip Irrigation System for Sustainable Agriculture. *Agric. Food* **2020**, *2*, 67–69.
6. Mangi, N. Performance and evaluation of drip irrigation system, and its future advantages. *Sch. Rep.* **2020**, *4*, 27–41.
7. Sathyapriya, E.; Naveenkumar, M.R.; Dhivya, V. An Empirical Study on Drip Irrigation. In Proceedings of the National Conference on Micro Irrigation, TNAU, Coimbatore, India, 1–3 March 2017.
8. Verdouw, C.; Wolfert, S.; Tekinerdogan, B. Internet of Things in Agriculture. In *CABI Reviews*; CABI International: Wallingford, UK, 2016; pp. 1–12. [CrossRef]
9. Verma, D.K.; Mishra, A.; Mishra, K. Role of IOT in Introducing Smart Agriculture. *Int. Res. J. Eng. Tech.* **2022**, *9*, 883–887.
10. Ghosh, S.; Sayyed, S.; Wani, K.; Mhatre, M.; Hingoliwala, H.A. Smart irrigation: A smart drip irrigation system using cloud, android and data mining. In Proceedings of the 2016 IEEE International Conference on Advances in Electronics, Communication and Computer Technology (ICAECCT), Rajarshi Shahu College of Engineering, Pune, India, 2–3 December 2016.
11. Swetha, R.N.; Nikitha, J.; Pavitra, B. Smart Drip Irrigation System for Corparate Farming-Using Internet of Things. *Int. J. Creat. Res. Thoughts* **2017**, *5*, 1846–1851.
12. Obaideen, K.; Yousef, B.A.; AlMallahi, M.N.; Tan, Y.C.; Mahmoud, M.; Jaber, H.; Ramadan, M. An overview of smart irrigation systems using IoT. *Energy Nexus* **2022**, *7*, 100124. [CrossRef]
13. Gamal, Y.; Soltan, A.; Said, L.A.; Madian, A.H.; Radwan, A.G. Smart Irrigation Systems: Overview. *IEEE Access* **2023**, *4*. [CrossRef]
14. Bwambale, E.; Abagale, F.K. Smart Irrigation Monitoring and Control. In *Encyclopedia of Smart Agriculture Technologies*; Zhang, Q., Ed.; Springer: Cham, Switzerland, 2022; pp. 1–7. [CrossRef]
15. Mohy-eddine, M.; Guezzaz, A.; Benkirane, S.; Azrour, M. IoT-Enabled Smart Agriculture: Security Issues and Applications. In *Artificial Intelligence and Smart Environment. ICAISE 2022. Lecture Notes in Networks and Systems*; Farhaoui, Y., Rocha, A., Brahmia, Z., Bhushab, B., Eds.; Springer: Cham, Switzerland, 2023; Volume 635, pp. 566–571. [CrossRef]
16. El Mezouari, A.; El Fazziki, A.; Sadgal, M. Smart Irrigation System. In Proceedings of the 10th IFAC Conference on Manufacturing Modelling, Management and Control MIM 2022, Nantes, France, 22–24 June 2022; pp. 3298–3303. [CrossRef]
17. Mabrouki, J.; Azoulay, K.; Elfanssi, S.; Bouhachlaf, L.; Mousli, F.; Azrour, M.; El Hajjaji, S. Smart System for Monitoring and Controlling of Agricultural Production by the IoT. In *IoT and Smart Devices for Sustainable Environment. EAI/Springer Innovations in Communication and Computing*; Azrour, M., Irshad, A., Chaganti, R., Eds.; Springer: Cham, Switzerland, 2022; pp. 103–115. [CrossRef]
18. Gomathy, C.K.; Vamsikumar, A.; Karthik, B.; Purushothamreddy, A. The smart irrigation system using IoT. *Int. Res. J. Eng. Tech.* **2021**, *8*, 462–468.
19. Yusuf, S.D.; Comfort, S.L.D.; Umar, I.; Loko, A.Z. Simulation and Construction of a Solar Powered Smart Irrigation System Using Internet of Things (IoT), Blynk Mobile App. *Asian J. Agric. Hort. Res.* **2022**, *9*, 136–147. [CrossRef]
20. Saban, M.; Bekkour, M.; Amdaouch, I.; El Gueri, J.; Ait Ahmed, B.; Chaari, M.Z.; Ruiz-Alzola, J.; Rosado-Muñoz, A.; Aghzout, O.A. Smart Agricultural System Based on PLC and a Cloud Computing Web Application Using LoRa and LoRaWan. *Sensors* **2023**, *23*, 2575. [CrossRef] [PubMed]

21. Pereira, G.P.; Chaari, M.Z. Comparison of Blynk IoT and ESP Rainmaker on ESP32 as beginner-friendly IoT solutions. In Proceedings of the 2022 International Conference on Internet of Things (ICIOT 2022), Honolulu, HI, USA, 10–14 December 2022. [CrossRef]
22. Carberry, A. How to Choose the Best Time for Watering a Garden: 7 Steps. Available online: https://www.wikihow.com/Choose-the-Best-Time-for-Watering-a-Garden (accessed on 14 December 2022).
23. When to Water—Southern Living Plants. Available online: https://southernlivingplants.com/planting-care/when-to-water/ (accessed on 14 December 2022).
24. Hourly Forecast for Doha, Qatar. Available online: https://www.timeanddate.com/weather/qatar/doha/hourly (accessed on 4 March 2023).
25. Qatar Historical Past Weather | Weather25.com. Available online: https://www.weather25.com/asia/qatar?page=past-weather#day=2&month=3 (accessed on 4 March 2023).
26. What Humidity Do Houseplants Need?—Be.Green. Available online: https://be.green/en/blog/what-humidity-houseplants-need (accessed on 13 June 2023).
27. Doha March 2023 Historical Weather Data (Qatar)—Weather Spark. Available online: https://weatherspark.com/h/m/150272/2023/5/Historical-Weather-in-May-2023-in-Qatar#Figures-ColorTemperature (accessed on 13 May 2023).
28. Waterproof Capacitive Soil Moisture Sensor SKU SEN0308-DFRobot. Available online: https://wiki.dfrobot.com/Waterproof_Capacitive_Soil_Moisture_Sensor_SKU_SEN0308 (accessed on 23 February 2023).
29. Koyuncu, H.; Gunduz, B.; Koyuncu, B. Construction of 3D Soil Moisture Maps in Agricultural Fields by Using Wireless Sensor Networks. *Gazi Univ. J. Sci.* **2021**, *34*, 84–98. [CrossRef]
30. Yudhana, A.; Kusuma, A.C. Water quality monitoring at paddies farming based on android. *IOP Conf. Ser. Mater. Sci. Eng.* **2018**, *403*, 012042. [CrossRef]
31. DHT22 Temperature-Humidity Sensor—Waveshare Wiki. Available online: https://www.waveshare.com/wiki/DHT22_Temperature-Humidity_Sensor (accessed on 13 June 2023).
32. G3-4 Water Flow Sensor | Seeed Studio Wiki. Available online: https://wiki.seeedstudio.com/G3-4_Water_Flow_sensor/ (accessed on 23 February 2023).
33. Odiagbe, M.; Eronu, E.M.; Shaibu, F.E. An Effective Water Management Framework Based on Internet of Things (IoT) Technology. *Eur. J. Eng. Res. Sci.* **2019**, *4*, 102–108. [CrossRef]
34. Mitras Trading Co. WLL | All Purpose Potting Soil. Available online: https://www.mitraswll.com/all-purpose-potting-soil-OF.html# (accessed on 6 June 2023).
35. Koehler, K. How to Grow Green Onions: Your Complete Guide. Available online: https://a-z-animals.com/blog/how-to-grow-green-onions-your-complete-guide/ (accessed on 6 June 2023).
36. Nagahage, E.A.A.D.; Nagahage, I.S.P.; Fujino, T. Calibration and Validation of a Low-Cost Capacitive Moisture Sensor to Integrate the Automated Soil Moisture Monitoring System. *Agriculture* **2019**, *9*, 141. [CrossRef]
37. Muzdrikah, F.S.; Nuha, M.S.; Rizqi, F.A. Calibration of Capacitive Soil Moisture Sensor (SKU:SEN0193). In Proceedings of the 2018 4th International Conference on Science and Technology (ICST), Yogyakarta, Indonesia, 7–8 August 2018.
38. DS18B20—Programmable Resolution 1-Wire Digital Thermometer. Available online: https://www.analog.com/media/en/technical-documentation/data-sheets/DS18B20.pdf (accessed on 23 February 2023).
39. Husni, N.L.; Dampito; Abdurrahman; Evelina; Handayani, A.S.; Rasyad, S.; Anisah, M. Modified Design of Water Metering System. *J. Phys. Conf. Ser.* **2020**, *1500*, 012018. [CrossRef]
40. Maia, R.F.; Netto, I.; Tran, A.L.H. Precision agriculture using remote monitoring systems in Brazil. In Proceedings of the 2017 IEEE Global Humanitarian Technology Conference (GHTC), San Jose, CA, USA, 19–22 October 2017; pp. 1–6. [CrossRef]
41. J-Turbo Key Plus Dripper. Available online: https://www.jains.com/PDF/Catalogue/Drip/Dripper/Turbo%20Key%20Plus.pdf (accessed on 6 April 2023).
42. Dwivedi, A.; Kumar, P.; Kumar, P.; Kumar, Y.; Sharma, Y.K.; Kayastha, A.M. Soil sensors: Detailed insight into research updates, significance, and future prospects. In *New Pesticides and Soil Sensors*; Grumezescu, A.M., Ed.; Academic Press: Cambridge, MA, USA, 2017; pp. 561–594. ISBN 978-0-12-804299-1.
43. Humidity, Temperature and Watering | Space for Life. Available online: https://espacepourlavie.ca/en/humidity-temperature-and-watering (accessed on 13 June 2023).
44. Iannotti, M. How to Grow Shallots. Available online: https://www.thespruce.com/growing-shallots-in-the-vegetable-garden-1403464 (accessed on 25 April 2023).
45. Russel, K. Water Inches—The Daily Garden. Available online: https://www.thedailygarden.us/garden-word-of-the-day/water-inches (accessed on 25 April 2023).

IoT

Review

A Tutorial on Agricultural IoT: Fundamental Concepts, Architectures, Routing, and Optimization

Emmanuel Effah [1,*,†], Ousmane Thiare [2,†] and Alexander M. Wyglinski [3,†]

1 Computer Science and Engineering Department, University of Mines and Technology, Tarkwa P.O. Box 237, Ghana
2 Department of Informatics, Gaston Berger University, Saint-Louis PB 234, Senegal; ousmane.thiare@ugb.edu.sn
3 Robotics Engineering Department, Worcester Polytechnic Institute, Worcester, MA 01609, USA; alexw@wpi.edu
* Correspondence: eeffah@umat.edu.gh
† These authors contributed equally to this work.

Abstract: This paper presents an in-depth contextualized tutorial on Agricultural IoT (Agri-IoT), covering the fundamental concepts, assessment of routing architectures and protocols, and performance optimization techniques via a systematic survey and synthesis of the related literature. The negative impacts of climate change and the increasing global population on food security and unemployment threats have motivated the adoption of the wireless sensor network (WSN)-based Agri-IoT as an indispensable underlying technology in precision agriculture and greenhouses to improve food production capacities and quality. However, most related Agri-IoT testbed solutions have failed to achieve their performance expectations due to the lack of an in-depth and contextualized reference tutorial that provides a holistic overview of communication technologies, routing architectures, and performance optimization modalities based on users' expectations. Thus, although IoT applications are founded on a common idea, each use case (e.g., Agri-IoT) varies based on the specific performance and user expectations as well as technological, architectural, and deployment requirements. Likewise, the agricultural setting is a unique and hostile area where conventional IoT technologies do not apply, hence the need for this tutorial. Consequently, this tutorial addresses these via the following contributions: (1) a systematic overview of the fundamental concepts, technologies, and architectural standards of WSN-based Agri-IoT, (2) an evaluation of the technical design requirements of a robust, location-independent, and affordable Agri-IoT, (3) a comprehensive survey of the benchmarking fault-tolerance techniques, communication standards, routing and medium access control (MAC) protocols, and WSN-based Agri-IoT testbed solutions, and (4) an in-depth case study on how to design a self-healing, energy-efficient, affordable, adaptive, stable, autonomous, and cluster-based WSN-specific Agri-IoT from a proposed taxonomy of multi-objective optimization (MOO) metrics that can guarantee an optimized network performance. Furthermore, this tutorial established new taxonomies of faults, architectural layers, and MOO metrics for cluster-based Agri-IoT (CA-IoT) networks and a three-tier objective framework with remedial measures for designing an efficient associated supervisory protocol for cluster-based Agri-IoT networks.

Keywords: Bluetooth Low-Energy (BLE); cluster-based Agricultural IoT (CA-IoT); fault management (FM); multi-objective optimization (MOO); wireless sensor network-based Agricultural IoT (WSN-based Agri-IoT)

Citation: Effah, E.; Thiare, O.; Wyglinski, A.M. A Tutorial on Agricultural IoT: Fundamental Concepts, Architectures, Routing, and Optimization. *IoT* **2023**, *4*, 265–318. https://doi.org/10.3390/iot4030014

Academic Editors: Antonio Cano-Ortega and Francisco Sánchez-Sutil

Received: 15 June 2023
Revised: 12 July 2023
Accepted: 17 July 2023
Published: 27 July 2023

1. Introduction and Tutorial Contributions

Currently, agriculture is the world's largest business, employing over one-third of the economically active global population and over 70% of the economically active population in Africa [1,2]. The impacts of high population growth rates and climate change-induced drought (according to Figure 1) on food security, unemployment threats and reduced crop quantity/quality make smart Agricultural Internet-of-Things technology (Agri-IoT) via precision farming and greenhouses the most promising remedy. However, the existing benchmarking Agri-IoT solutions can only be acquired, deployed, and managed by farmers with sufficient financial resources, an electricity grid, Wi-Fi/cellular coverage, and technical expertise in IoT, which is generally not the case in Ghana and Sub-Saharan Africa. These call for a paradigm shift in farming techniques, and the most promising game-changers are precision farming and greenhouses whose underlying technology is a robust, affordable, autonomous, and optimized, innovative WSN-based Agri-IoT [3] that satisfies the critical design expectations presented in Figure 2.

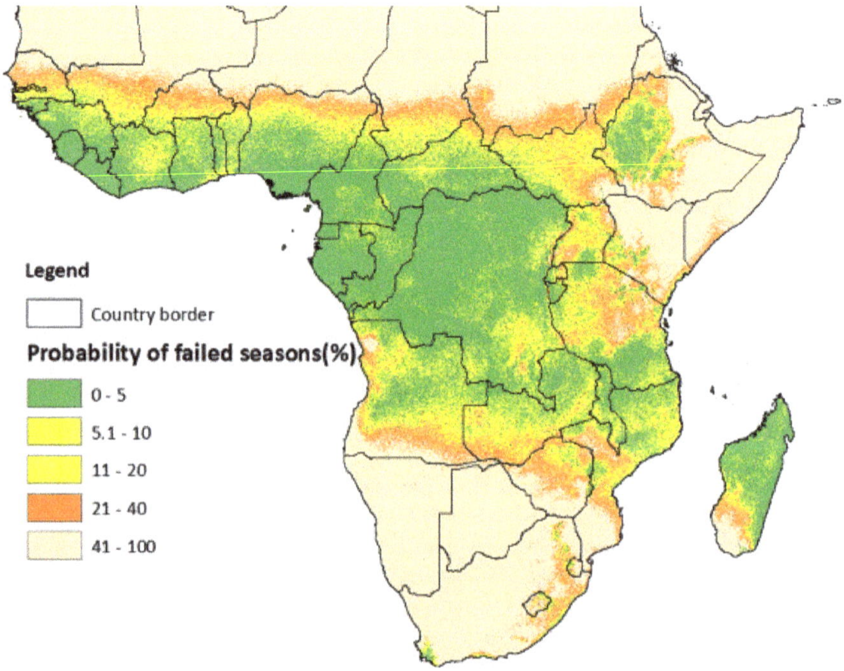

Figure 1. Seasonal failure probability-2014 [4] depicting the extent of climate change impact on Africa's farmlands.

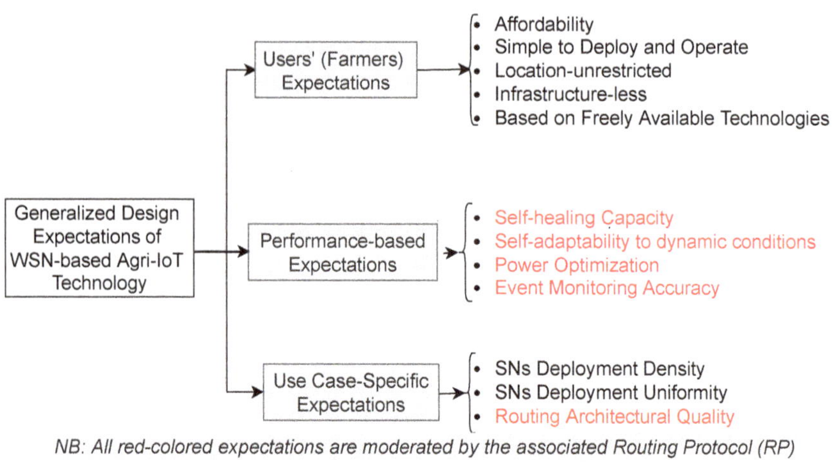

Figure 2. Generalized design expectations of WSN-based Agri-IoT technology.

Although few surveys and tutorials have been authored on this subject, they present mere classifications of communications trends on classical IoT [2,5–8] without any context-specific technical considerations of the critical design expectations in Figure 2. For instance, the authors in [2,6,7] examined IoT's communication infrastructure, platforms, standards, development trends, and possible network solutions in agriculture. Similarly, the roles of industrial IoT (thus, identification-based IoT (example, RFID [6], WSN [9], QR codes [5], barcodes) and communication-based IoT (example, ZigBee [5], Z-wave [6], MQTT [5,6], LoRa [10], SigFox [11], BLE [12], Li-Fi [5], Wi-Fi [13], Near-Field Communication (NFC) [5], and power line area network) were reviewed in terms of current research trends, applications, and main challenges in [5]. Although RFID tags and WSNs have similar data acquisition capacities, the authors concluded that WSN technology is more energy-efficient and suitable for Agri-IoT than the costly RFID technologies [5]. Overall, Agri-IoT technology has not yielded its intended paradigm transformation in the agricultural sector due to several technical challenges that have not received adequate contextual research considerations [14]:

1. **The agricultural setting is a unique area where conventional IoT technologies do not apply.** Existing Agri-IoT solutions are location-restricted because they are mostly based on Wi-Fi or cellular communication technologies and electricity grids with constrained coverages in Africa. A typical African agricultural setting lacks access to reliable electricity and the Internet for cellular/Wi-Fi-based technologies, and the intended users (farmers) of Agri-IoT technology are low-income earners with limited technological expertise. Common Agri-IoT applications mainly utilize architecture-restricted, high-resource-demanding routing techniques (e.g., routing over low-power and lossy networks protocol (RPL)) and communication standards (e.g., 4G, 5G, ZigBee, LoRa, Wi-FI, and long-term evolution (LTE)) [15], which are difficult to access in typical African farms. Consequently, Agri-IoT users in Africa expect a context-relevant solution that is affordable, simple to deploy and operate by non-experts, location-unrestricted, supportive of large-scale farm management, and based on freely available technologies that do not require licensing. Thus, they are unlike popular IoT use cases such as medical, vehicular, and industrial IoT, whose designs are mainly affected by critical factors including security, stable connectivity, and interference, respectively, Agri-IoT is compelled to drive on affordable battery-powered SNs, which make architecture, low-power communication technology, power optimization, cost, fault tolerance, multihop routing, scalability, and environmental impact critical design factors in order to address its resource or deployment-induced challenges [12,16,17].
2. **High susceptibility to faults and failures:** Agri-IoT networks are vulnerable to faults and failures since the resource-constrained SNs are densely deployed in hostile environments to autonomously operate via a network supervisory protocol with limited post-deployment maintenance services. This supervisory protocol must incorporate sufficient power optimization, auto-fault management (FM), and self-adaptability techniques in order to achieve the desired performance expectation. Due to the lack of an in-depth and context-relevant tutorial that bridges the gap between theoretical taxonomies and real-world designs, most canon Agri-IoT testbed solutions, such as those authored in [1,10,11,17–20], suffered abrupt failures during outdoor deployments.
3. **Agri-IoT technology lacks comprehensive context-based synthesis from SN design to field deployment.** The power- and resource-constrained SNs that form the WSN-based Agri-IoT network in the aforementioned context require limited data transmission rates, computational capabilities, memory capacities, communication distance, and operational stability. Consequently, the associated routing protocol [9,12,17,21], communication technology, and routing architecture [22–24] must support mechanisms that ensure packet size and communication distance moderation [16], efficient channel access management (CAM), and SN's tasks management. It is not a mere application of conventional IoT to a farm, as many authors attempted [1,10–12,17–20,23,25,26], which lacked application-specific requirements such as dense network inter-connectivity, higher information perceptibility, compre-

hensive intelligence services, remote monitoring, smart decision making, and the execution of precise control/actuation actions on the farm.

4. **Superficial consideration of desired communication technologies of Agri-IoT without considering the cluster-based architecture:** To date, Agri-IoT-related surveys and tutorials focused on high-power-demanding communication technologies (Wi-Fi and cellular-based technologies), the centralized architecture-constrained ZigBee standard, and the operation principles of conventional IoT as authored in [1,10,11,14,18,19] without an in-depth consideration of the unique case of Agri-IoT. It is well established that the cluster-based architecture is the best candidate for Agri-IoT application [12,16,17,24]; however, there are no systematic evaluations to cement this fact. For instance, most benchmarking WSN-based IoT testbed solutions are founded on the ZigBee IEEE 802.15.4 communication standard and high-resource-demanding Wi-Fi, cellular-based, and 6LoWPAN/IPv6 routing standards. These standards also thrive on wired or fixed IP-based infrastructural backbones, total Internet/electricity coverage, and highly complex graph-based and centralized routing protocols [1,10,11,14,18,19], leading to a lack of global significance because Africa, which is the focus of this study, has less than 50% electricity/Internet coverage [27]. Also, ZigBee, Wi-Fi and cellular-based communication technologies with centralized or flooding-based routing architecture [1,10,11,14,18,19] are capital-intensive, complex to manage, location-restricted, energy-inefficient, and over-reliant on fixed supporting infrastructure. Therefore, an in-depth contextual assessment of how low-power communication standards such as LoRa, SigFox, and Bluetooth Low-Energy (BLE) evolve in cluster-based Agri-IoT (CA-IoT) networks can be of immeasurable benefits to the IoT community and farmers.
5. **The role of Agri-IoT in eliminating food insecurity, improving crop quality, alleviating global poverty, and increasing agricultural production volumes has been underestimated [2,7,8,10,16,28,29].** The agricultural sector, which has been hindered by climate change, is the largest global employer [3]. To revitalize this sector, CA-IoT has emerged with the most promising opportunities to address food and employment insecurity issues and improve crop quality and economic conditions for the farmers. However, these benefits have not been fully realized due to insufficient research publicity.

To the best of our knowledge, no survey or tutorial articles have sufficiently considered these technical issues and provided sufficient technical guidelines for the designers of Agri-IoT systems to make well-informed decisions in order to achieve satisfactory network performance. Additional realistic research is needed regarding the contextual evaluation of SN design and deployment factors, fundamental network design concepts and requirements, multi-objective optimization (MOO) analysis of the parameters for designing the associated routing protocol, and efficient operational metrics of the WSN sublayer of the Agri-IoT using the cluster-based architecture. In addition, the assessment of the possibility of using low-power and accessible wireless communication technologies such as BLE via cluster-based architecture to achieve a complete infrastructure-less, cheaper, energy-efficient, self-healing, adaptive, and robust Agri-IoT network is imperative. Furthermore, a broader contextual overview covering all vital aspects such as the fundamental concepts of Agri-IoT, technical design requirements of SNs and WSN-based Agri-IoT, surveys of the benchmarking communication standards, routing protocols, and testbed solutions, and an in-depth case study on how to design a self-healing, energy-efficient, adaptive, and CA-IoT based on the performance and users expectations are illustrated in Figure 2. Such a reference document can help support researchers when they attempt to accurately model and optimize the performance of Agri-IoT [14] so that the performance gap between the simulated networks and the realized Agri-IoT testbed solutions [1] can be addressed. By way of addressing these technical challenges, this tutorial presents the following contributions:

- Perform an in-depth synthesis and review (1) the basic concepts of Agri-IoT, (2) the comprehensive design considerations of these networks, (3) the technical design re-

quirements of Agri-IoT, and (4) the up-to-date research progress on routing techniques, communication standards, and testbed solutions of WSN-based Agri-IoT.
- Systematically survey the benchmarking of WSN-based IoT networks' communication standards, FM techniques, routing and MAC protocols, and realization testbeds to respectively uncover the appropriate communication requirements for Agri-IoT, unveil the root faults and possible remedies in the WSN sublayer, derive a generalized taxonomy of routing architectures, and define appropriate routing paradigms for WSN-based Agri-IoT using the core PHY layer design metrics: affordability, self-healing capacity, energy-efficiency, location independence, and network adaptability.
- Systematic synthesis of canon cluster-based routing protocols to uncover the plethora of possible research gaps, derive a realistic taxonomy of MOO metrics and propose possible MOO remedies that can be implemented using CA-IoT routing architecture freely available low-power communication standards.
- Proposition of MOO-induced guidelines in the form of open issues that can help Agri-IoT designers to build adaptive, robust, fault-tolerant, energy-efficient, affordable, and optimized CA-IoT networks in both simulation and real-world implementations.

Overall, this tutorial is motivated to provide a contextualized, in-depth understanding of this technology and assist the reader in designing robust, affordable, and optimized Agri-IoT networks that can act as reliable game-changers to avert the stipulated challenges. Also, the critical design, deployment, and QoS requirements of WSN-based Agri-IoT networks from theoretical modeling to real-world deployment are unveiled in order to bridge the existing gap between the theory and practice of this technology [1,14].

The remainder of this paper is organized into the following sections: Section 2 provides a brief background comparative overview of WSN, IoT, and Agri-IoT technologies, while Section 3 focuses on their components, protocols, architectural layers, and proposed architectural layers for WSN-based Agri-IoT technology. Section 4 presents the detailed contextual design and implementation requirements of Agri-IoT networks, while Section V deduces the unique characteristics, challenges, and proposed performance expectations of the associated routing protocols for the WSN sublayer of Agri-IoT. Sections 6–8 present systematic surveys on routing protocols, FM techniques, and the canon real-world testbed implementations of WSN-based Agri-IoT solutions. Section 9 examines how the above discussions have evolved using a case study of cluster-based Agri-IoT (CA-IoT) for precision irrigation.Section 10 unveils open issues and future works, while Section 11 concludes the paper.

1.1. Comparative Overview of WSN, IoT, and Agri-IoT Technologies

A comparative overview of the underlying technologies (i.e., WSN, IoT, and Agri-IoT) forming the WSN-based Agri-IoT are compared from the perspective of architectural variations, users' expectations, and design and implementational differences in Table 1.

As depicted in Figure 3, WSNs are formed by spatially distributed, autonomous, resource-constrained SNs that wirelessly interconnect to communicate their sampled data to a BS for further monitoring or event tracking purposes without necessarily requiring the Internet. The main components of the WSN are the SNs, the BS/gateway, and the event sampling/routing software that supervises the entire network process. A node may route data directly or via relay SNs to the BS based on its location and assigned tasks. The BS locally takes actionable decisions and execution of the actuation actions. Although the WSNs are resource-constrained and fault-vulnerable, they constitute the inevitable part of this technology [2] and the underlying innovation of the WSN-based Agri-IoT framework. In contrast, classic IoT consists of IoT devices that sense and transmit their sampled information directly or via telemetry to the Internet for monitoring or event-tracking purposes, mostly via the centralized routing architecture. Like BS in WSNs, IoT devices can connect to the Internet/IoT cloud via fixed-line (thus, for a factory), 5G/4G/LTE cellular/mobile networks, or Wi-Fi for further processing, storage, and decisions/actions.

Table 1. Comparison of WSN, IoT, and Agri-IoT technologies.

Characteristics	WSN Technology	IoT Technology	Agri-IoT Technology
Internet Connectivity	SNs have no direct connection to the Internet, always via a BS/router/gateway if necessary	Nodes directly send sampled data to the Internet	SNs' Internet connectivity can be either direct or via a BS
Critical Design Factors/ Expectations	Application-specific	Security, interference, linking fleet	Power optimization, routing architectural support, fault tolerance, on-site auto-actuation demand, and self-adaptability to network dynamisms
Deployment Density	Application-specific	Moderate	High
Power Supply Constraints	Application-dependent	Application-specific	Compelled to drive on battery power
On-Site Electricity and Internet Coverage	May be possible	Required	Mostly inaccessible
Implementational Routing Architecture	Centralized or flooded	Mostly centralized	Contextualized cluster-based but inadequately researched
Communication Technology	Application-specific	May use high-power standards such as Wi-Fi, cellular-based, satelite, fixed-line, etc.	Requires low-cost low-power standards such as BLE, LoRa, SigFox, ZigBee, etc. that support cluster-based architecture
Users' Expectations	Performance stability	Performance stability	Affordability, autonomous performance stability, location-independence, simple to deploy and operate by non-experts, supportive of large-scale farm management, and based on freely available communication technologies that do not require licensing.
Network Type	Data-centric	Use information network directly	Mostly data-centric
Basic Components	Resource-constrained SNs, BS or Sink Node	May include smartphones, PCs, WSN, BS, Internet, IoT cloud with data analytic tools, and the user interface app.	WSN, BS, IoT-cloud with application-defined user apps and data analytical engines
Security and Privacy	Medium	High	Low
On-Site Actuation Required?	Not always	No	Yes
Network Participant Mobility during Operation	Usually static	Mobile	Application-specific

As presented in Figure 4, WSN-based Agri-IoT is an information- and knowledge-intensive intelligent feedback control system for farm monitoring, data sampling/computing, resource optimization, automation of farm operation (e.g., precision irrigation, chemical application, livestock monitoring, and disease management [16]), and actionable decision making via a variety of battery-powered and wirelessly connected SNs with sensing, processing, and communication capacities [2,29,30]. Unlike the WSN, Agri-IoT and IoT sample data to an Internet-based cloud. The SNs that form the WSN sublayer are spatially distributed and self-configured to achieve a myriad of remote sensing, surveillance/monitoring, and control applications via automated sensing, wireless communication, and computing, making informed decisions and performing actuation control [31] using precise, accurate, and timely sampled information about a real-world phenomenon [32].

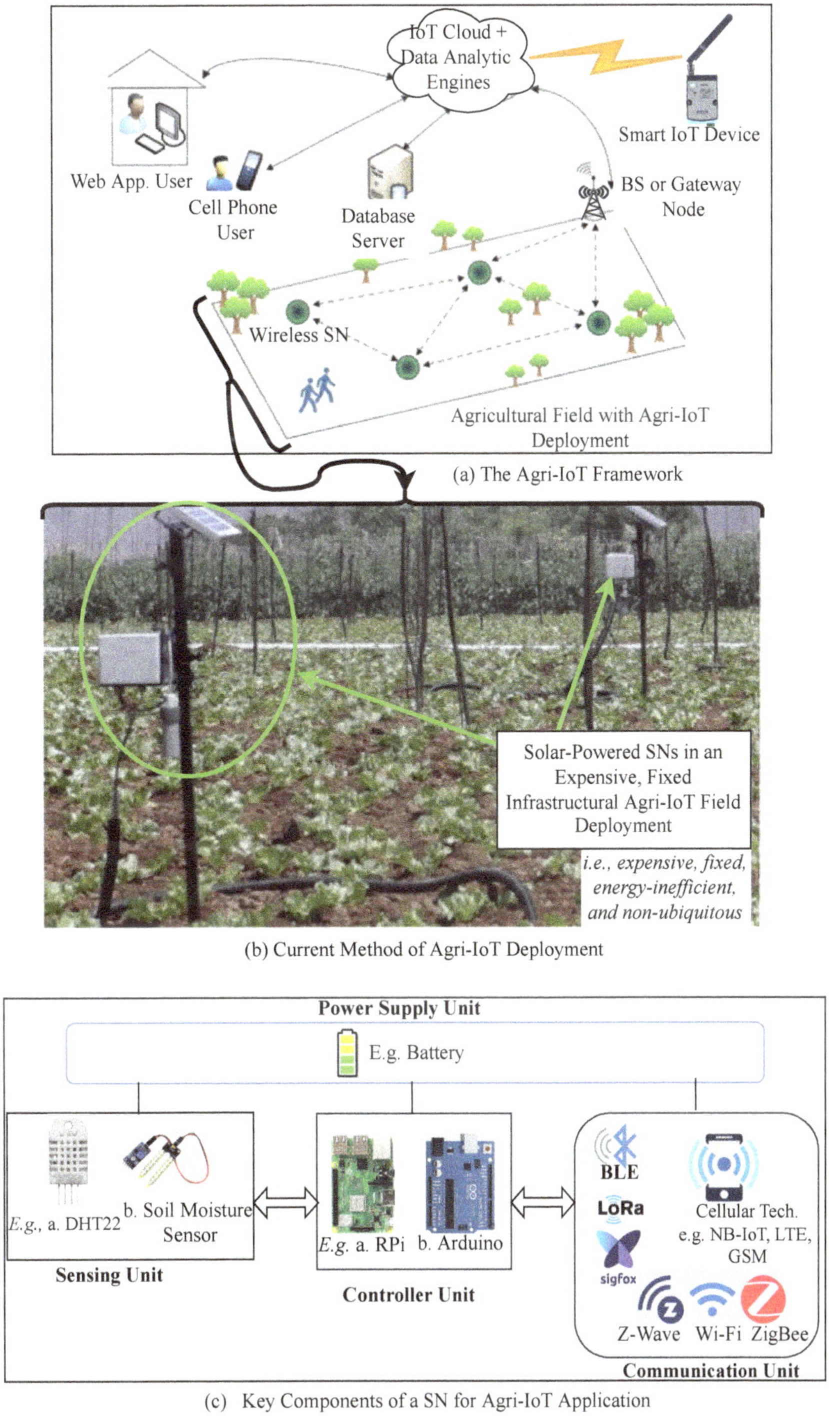

Figure 3. Generalized Agri-IoT framework consisting of: field layout overview of Agri-IoT framework (**a**), sample of classic Agri-IoT in the state of the art (**b**), and key components of an SN or a BS (**c**).

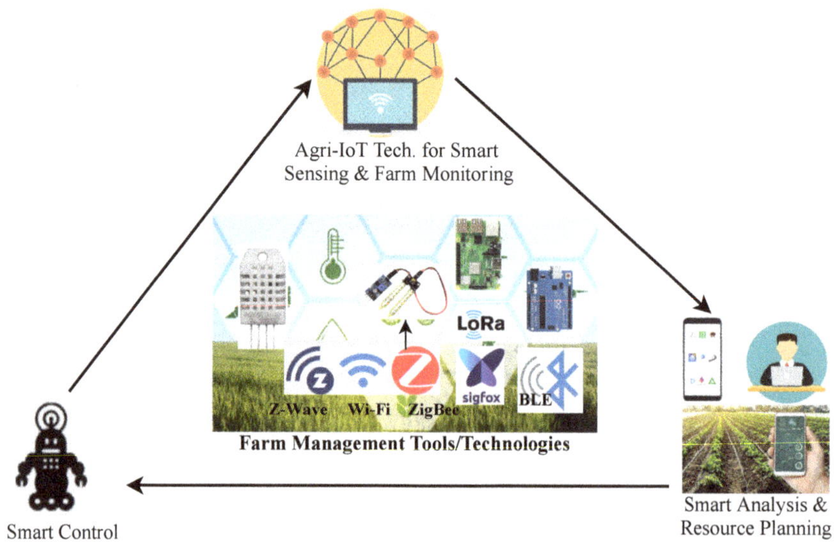

Figure 4. Conceptual framework: Agri-IoT-based farm monitoring and control cycle.

The main hardware components of an Agri-IoT framework, as presented in Figure 3 and Table 2, include the WSN (i.e., comprising the field-deployed SNs or IoT devices), a base station (BS) or gateway or actuator controller, cloud servers, and the user's monitoring/control devices. The on-farm participants (e.g., SNs and BS) in Agri-IoT are mostly battery-powered and must be equipped with sensing, computing, and communication abilities to form infrastructure-less, robust, self-healing, and self-configured WSNs for data collection and event management [33]. The core units of the SNs in Figure 3c and the BS are compared and contrasted in Table 2. As the framework in Figure 3a depicts, the IoT devices can sense, process, and transmit their sampled data directly to the Internet or IoT cloud without a gateway, whereas the SNs in WSN-based Agri-IoT perform likewise via a BS. This resource-sufficient BS interfaces between the IoT cloud/user and the WSN or actuator control system. It can also process the received data and locally execute actionable decisions via the actuator of the farm event being monitored. The received data can also be relayed to the analytical data engines in the IoT cloud via a wired and wireless medium for further processing and actions [13]. The resource-constrained WSN sublayer mainly uses data-centric protocols due to the SNs' high deployment densities, high network dynamics, and limited power supply of SNs. Although data-centric protocols are fragile and not standardized, they are more suitable than the high resource-demanding ID-based IPv4 or IPv6 protocols in the addressing space of the WSN-based Agri-IoT.

Table 2. Comparison of SN and BS.

Network Participant	Power Source	Communication Technologies	Controller Type	Processor/Memory Requirements	Requires Sensors
SN	Mostly battery-powered	Mostly relies on low-power, short-ranged standards such as BLE, LoRa, SigFox, and ZigBee for on-field communication	Can be Arduino-based, Raspberry Pi (RPi)-based, etc.	Low processing and storage powers but based on SN roles	Yes
BS	Can be battery-powered but mostly use a more reliable power supply	Mostly communicate with IoT cloud via fixed line, Wi-Fi, cellular technologies, and the WSN via the low-power standards, e.g., BLE, LoRa, SigFox, ZigBee, LoRa-based Satellite, etc.	Can be RPi or Arduino-based or a PC.	Requires high memory and processing powers	No

Agri-IoT combines WSN and IoT technologies into contextualized intelligent farm management systems to achieve higher event data quality and offer remote monitoring and control. WSN-based Agri-IoT consists of the WSN sublayer, the gateways, the cloud servers, and the remote interface application, as illustrated in Figures 3a and 5. Uniquely, the current trends of Agri-IoT mandate that both intra-SN and BS–cloud communication are based on low-power, ubiquitous, and freely available wireless standards [2]. Also, most Agri-IoT solutions support bidirectional communications between the BS/gateway and the cloud/users, whereby the BS updates the cloud/user database and receives actionable/control remote messages from the user or cloud analytical decision results for actuation purposes. The WSN-based Agri-IoT is the most dominant technology in the global smart farming use cases in the agricultural sector. The core tasks of SNs in a WSN-based Agri-IoT application, which are frequently supervised by the associated routing protocol, include network construction/management, data sensing, data processing/aggregation, fault tolerance, and communication [9,12]. Also, the routing architecture must be supported by the associated communication platform and the application-specific requirements of the network.

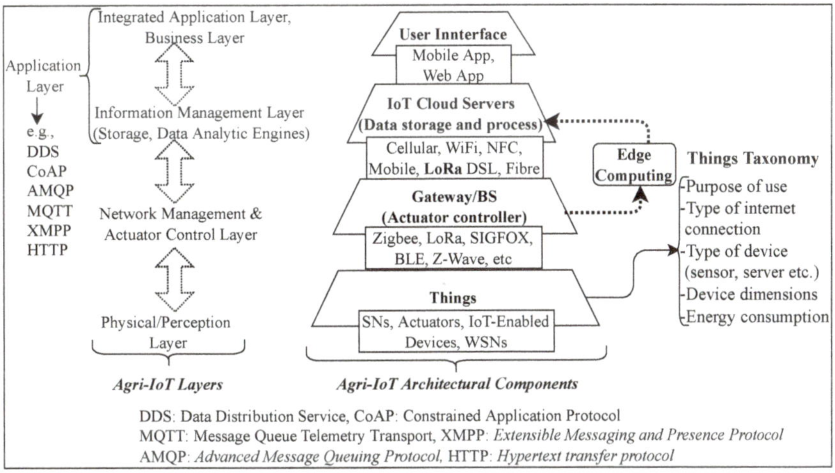

Figure 5. Proposed Agri-IoT architectural layers with core components of Agri-IoT ecosystem and the "things" taxonomy.

Unlike IoT and WSN whose design expectations are application-specific, WSN-based Agri-IoT requires holistic integrations of the expectations in Figure 2.

1.2. Classifications of IoT Applications and Specific Roles of Agri-IoT

Generally, IoT technology is application-specific. However, it has limitless applications and roles in the smart world agenda. Based on their intended purpose, WSN-based IoT systems can be broadly classified into condition monitoring and event-tracking categories [34], as illustrated in Figure 6.

The monitoring-based applications involve real-time event data collection and analysis, supervision, and operational control of systems. In contrast, tracking-based applications track changes in the phenomenon of interest, such as the locations of objects, persons, transported goods, animals, and vehicles. Both application domains can be subdivided into industrial, environmental, and societal IoT applications in Figure 6, where specific examples are provided for each application domain. For instance, monitoring-based applications may include indoor/outdoor environmental monitoring [6], industrial process monitoring [5,29], process control [2], greenhouse automation [7], precision agriculture (e.g., irrigation management, crop disease prediction, prediction of production quality, and pest and disease control) [2,8], biomedical or health monitoring [8], electrical grid network monitoring/control [12,29], military location monitoring [9], and so forth. Conversely, specific examples of tracking-based applications may include habitat tracking, traffic tracking, plant/animal condition tracking, and military target tracking, as outlined in Figure 6.

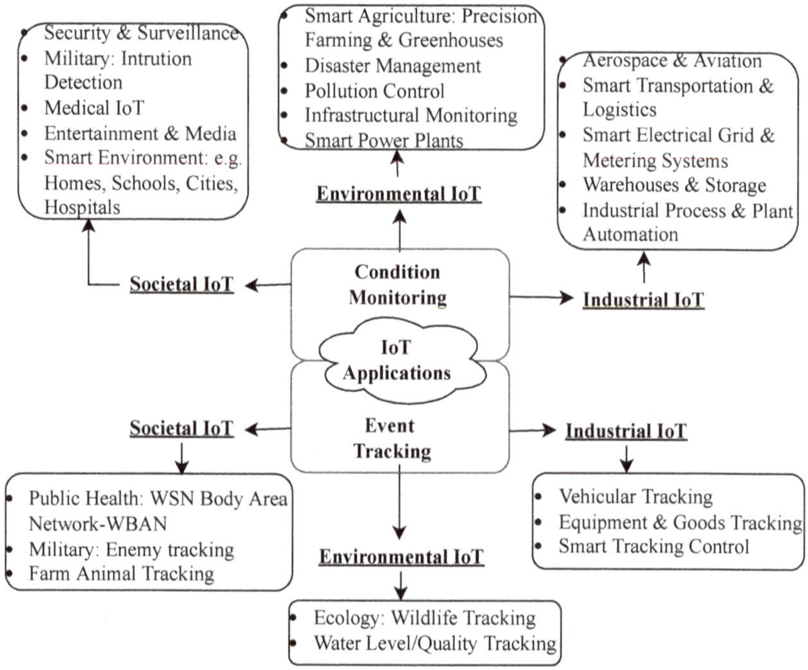

Figure 6. Generalized taxonomy of IoT applications.

1.3. *Agri-IoT Roles and Use-Cases*

The concept of intelligent farming involves data acquisition, data processing/planning, and smart control using the WSN and IoT technologies, big data, and cloud computing techniques to provide profitable solutions, as presented in Figure 7. These principal roles in Figure 7 define their use cases. For instance, monitoring the state of crops or the climate of the field using Agri-IoT technology can allow farmers to know precisely the amount of pesticides, water, and fertilizers required to attain optimal crop quality and production volume. However, the QoS requirements, the routing techniques, architectural requirements, and the operational dynamics differ from one use case to another. This tutorial focused on the critical and unique design requirements of WSN-based Agri-IoT, which is the backbone of the smart agricultural initiative [35]. The resulting use-cases in Figure 7 can be explained as follows:

1. *Agri-IoT for Climate Condition or Agronomical Monitoring:* This Agri-IoT system mostly comprises BS (i.e., weather stations) and a deployed WSN. The analytical data engines mine the sampled climate or crop condition data in the cloud to predict future climate conditions and farm automation plans. The most suitable crop and precise farming practices can then be predefined to improve agriculture production capacity and quality.
2. *Agri-IoT for Precision Farming:* This is the most famous application of Agri-IoT, whereby farming practices (e.g., irrigation, fertilizer application, etc.) are precisely and accurately controlled to optimize these resources. Here, the SNs are mostly fitted with soil sensors to collect a vast array of microclimatic data (e.g., soil moisture, temperature, and salinity) that can enable farmers to estimate optimal amounts of water, fertiliz-

ers, and pesticides needed by the crops to minimize resources' costs and produce healthier crops. Additionally, the BS controls the event actuation system via accurate data-driven real-time decisions on the crops using climate data, crop growth data, and disease infection data.

3. *Agri-IoT for Greenhouse Automation:* The Agri-IoT-based approach provides more accurate real-time information on greenhouse conditions, such as lighting, temperature, soil condition, and humidity, unlike manual greenhouse management. This allows precise remote monitoring and control or automation of all farming practices.
4. *Agri-IoT for Livestock Monitoring and Management:* In this system, SNs are attached to livestock to monitor their real-time health, track their physical location, and log their performance. This helps the farmer identify and isolate sick animals to avoid contamination and reduce staffing expenses.
5. *Agri-IoT for Predictive Analytics:* This Agri-IoT system provides highly relevant real-time data that can be analyzed to make essential predictions, such as crop harvesting time, risk of disease infection, yield volume, yield quality, and yield vulnerability, for proper planning.
6. *Agricultural Drones (Agri-Drones):* Agri-Drones, such as DroneSeed, are fitted with mobile SNs and farming tools to collect agricultural data or perform activities such as field surveillance, crop planting, pest control, farm spraying, crop monitoring, etc. For example, for Agri-Drones, all the above use cases utilize the WSN-based Agri-IoT framework.

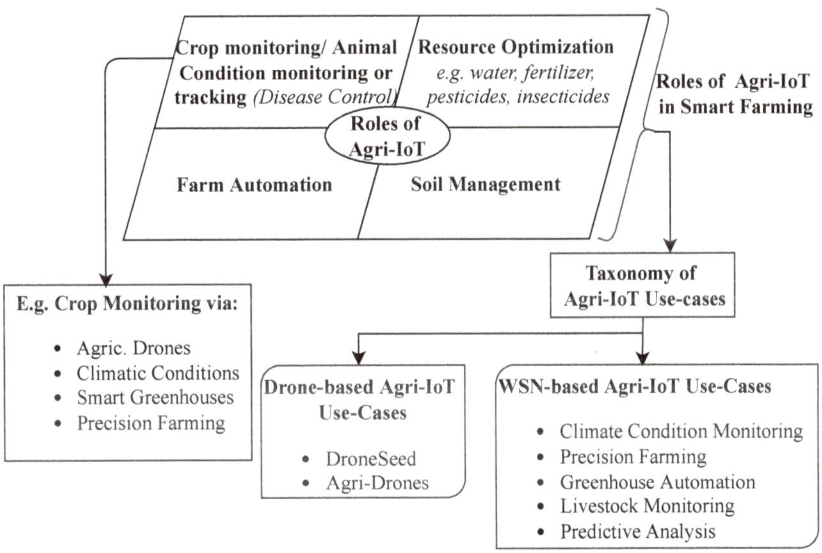

Figure 7. The roles of Agri-IoT in smart farming with specific use cases.

2. The Agri-IoT Ecosystem

The authors in [1,14] established that the existing real-world attempts of Agri-IoT could not meet both performance and user users' expectations because they are founded on the fundamental concepts and the operational principles of classic IoT and WSN technologies. To effectively achieve the expectations in Figure 2, it is imperative to conduct a systematic assessment of the related architectural layers in classic IoT and propose a suitable option for the WSN-based Agri-IoT ecosystem. Generally, the conventional IoT ecosystem consists of the network architectural layers and the data management platforms [2,7,8], which are further grouped into devices (sensors, actuators, and gateways/BS), network (BS to cloud), platforms/applications' cloud, and agents/users. Due to the domain-specific requirements of IoT applications and the incorporation of numerous heterogeneous devices with application-specific requirements, there are generally no unified or standardized IoT architectural layers. Therefore, most application-defined layers are frequently adapted from the canon architectural layers, which include the three-layer [5], the cloud-based [7], the service-oriented architecture (SOA) [2,7], and the fog-based [2,7,29], as illustrated in Figure 8.

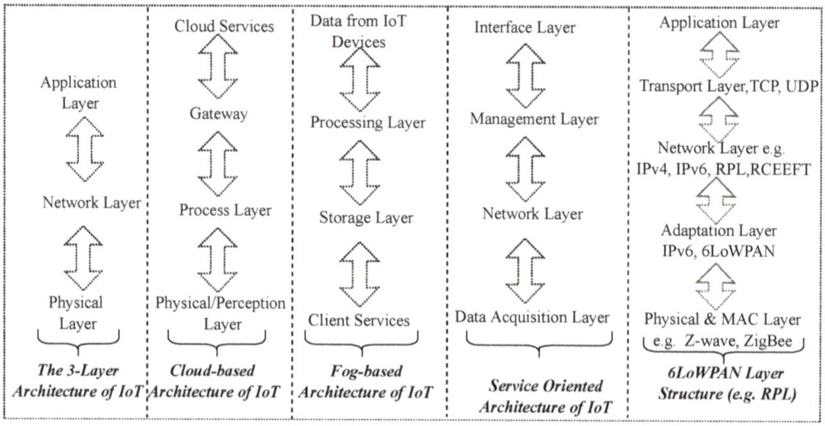

Figure 8. Different architectural layers in the state of the art of IoT ecosystem.

The fog-based architecture was adapted from the three-layer parent architecture to include cloud computing by offering computing, storage, and network information between the clients and the cloud services [29] in a decentralized manner. Here, cloud computing and fog/edge computing architectures only differ in where data computing occurs. These layers are not unified because the respective network layers do not cover all underlying technologies that transfer data to all IoT platforms [5]. Additionally, they are based on complicated centralized and flooding-based routing architectures, high-resource-demanding and capital-intensive Wi-Fi/cellular-based communication technologies. As well, they require wired infrastructural support in the farm, which is too complex, location-restricted, and capital-intensive for most low-income and non-expert farmers to implement and manage. Consequently, they are unsuitable candidates for the resource-constrained SNs in WSN-based Agri-IoT. By implication, there are no reference guidelines for designing Agri-IoT participants and supervisory protocols, controlling the speed of packet delivery, smoothing out SN's integration, unifying technology, and creating standardized Agri-IoT reference models, among other considerations. In contrast, an Agri-IoT ecosystem, depicted in Figure 3, consists of:

1. Agri-IoT network architectural layers: This shows how the physical network elements, network operation principles, and operational techniques interact throughout the entire ecosystem.
2. Network supervisory software/routing protocol and routing architectures: This contains the virtual arrangement of multiple network elements [8] and the event sampling/routing protocol that constructs the routing architecture, supervises sampling and moderates all communications in the PHY layer.
3. Data management platform: It hosts all high-resource-demanding data analytic engines, event databases, and remote control algorithms in a cloud model.

2.1. Proposed Architectural Layers for WSN-Based Agri-IoT

In designing an efficient Agri-IoT system of global significance, it is imperative to propose suitable architectural layers and evaluate how the various components interact in these layers. With the emerging advances in low-power, freely available, and boundless communication standards (e.g., BLE) and unfulfilled potentials of CA-IoT network [12,16], a new framework of cluster-based architectural layers for the WSN-based Agri-IoT ecosystem is proposed in the left side of Figure 5. The center portion of Figure 5 presents the key components/technologies required in each layer, while the Things taxonomies of hardware components from the related literature [4,8,29] are depicted on the right portion of Figure 5. The underlying layers in our four-tier layers in Figure 5 can be elaborated on as follows:

1. *Integrated Application and Management Layer:* This operates all agriculture-related applications that interface between the user (for example, farmer) and the Agri-IoT system to make decisions and execute remote actions to keep their crops or animals healthy. This layer manages the entire Agri-IoT system and its application-specific functionality, high-resource-demanding applications, and core business model in the cloud. This layer's security requirements are crucial to the next sublayer; however, these are beyond the scope of this research. The business or management sublayer maintains end-to-end data integrity and security by ensuring that data are transferred to the correct user. It also ensures that the correct user executes the actuation.
2. *Information Management Layer:* This handles data processing, storage, and other specialized cloud services and functionality that make precise, actionable decisions. In Agri-IoT, the sensory data are preprocessed locally to optimize communication power but can be further processed using analytic engines in the cloud for better decision making and remote monitoring and control. This layer can be embedded in the above application layers and hosted in the cloud in a typical Agri-IoT ecosystem.
3. *Network Management Layer:* This layer discovers, connects, and translates devices over a network, and it coordinates with the above application layers. It also contains the BS, which interfaces the resource-constrained WSN and cloud information network. By convention, the WSN sublayer must utilize low-power communication standards such as Zigbee, SigFox, LoRa, BLE, Z-Wave, SigFox, and IEEE P802.11ah (low-power Wi-Fi), while the BS-to-Cloud connectivity can be achieved via the traditional cellular networks, satellite networks, Wi-Fi, LAN, WAN, and LoRa, among others. Unlike classic IoT, Agri-IoT requires that the BS-to-Cloud connectivity utilize low-power communication standards. Also, since every communication standard for the resource-limited WSN sublayer comes with unique resource specifications and design tradeoffs between power consumption, routing architectural constraints, and bandwidth [4,14,17], the best connectivity option must be selected to achieve the desired application goals. Consequently, the stated WSN-based connectivity technologies can be classified using several distinct parameters, such as energy consumption rates, uplink/downlink data rates, packet size, SN-count per BS (gateway), network routing topology, the SNs' sensing range, the SNs' transmitter/receiver power, frequency bandwidth, channel width, etc. (refer to the right portion of Figure 5).
4. *Physical/Perception/Things Layer:* This layer refers to the field and all devices such as SNs, actuators, RFID tags, sensors, and edge devices that interact with the environment. This layer senses and collects the necessary information from the connected devices in the WSN sublayer to the BS. In Agri-IoT networks, the sampled microclimatic data can be processed and stored on the local BS, the cloud, or both. The activities in the cloud or application layers are beyond the scope of this tutorial.

2.2. Associated Hardware Components and Technologies Required in the Proposed Architectural Layers

To precisely model and design an Agri-IoT network of desired expectations (refer to Figure 2) using the proposed architectural layers shown in Figure 5, the knowledge of the principal components and technologies used in each of these layers and how they interact and adapt for their intended functions is imperative. As depicted in the middle of Figure 5, the Agri-IoT ecosystem is composed of the following core components/technologies:

1. **Things:** The Things unit is the physical interface between the tracked/monitored asset and the BS or actuator controller, which aligns with the physical or perception layer. It comprises the monitored/tracked asset (for example, field, crop, or animal), the SNs, or the entire IoT devices making up the WSN (for example, SNs, actuators, IoT-enabled

devices, WSNs, and other smart devices), the event sampling, and routing technology in the WSN. Since the SNs constituting this unit are resource-constrained, freely available communication standards such as Zigbee, BLE, Z-Wave, and IEEE P802.11ah (low-power Wi-Fi) are the most suitable for both SN–SN and SN–BS communications. The Things unit accesses the cloud/Internet via gateways (BS).
2. **Gateway (BS):** The BS interfaces the WSN out in the field and the applications situated in the cloud servers. This unit aligns with the network management and actuator control layer shown in the middle of Figure 5. The WSN sublayer may have more than one BS(s), each with the capacity to handle most resource-demanding computational tasks besides actuation execution, network construction, scheduling of event sampling, and network supervision services. They may also allow bidirectional communication with the cloud/user and WSN. Similar to standalone IoT devices, the BS can be equipped with 4G/5G/LTE/NB-IoT, cellular-based, Wi-Fi, LoRaWAN, or wired ethernet communication technologies to interact with the cloud, and low-power communication standards such as LoRa, low-power Wi-Fi, SIGFOX, UMTS, BLE, and Zigbee (Figure 5) to communicate with the sensor field. However, Agri-IoT networks require that both upper-layer and lower-layer communication technologies of the BS should be low-power, freely available, easy to deploy and manage, and platform-independent. The BS may preprocess or relay the raw data to the cloud for remote data processing. The BS(s) locations are strategically chosen to optimize network communication costs.
3. **IoT Cloud:** The Cloud unit aligns with the applications layer. It consists of an on-premises or remote server farm that hosts the applications layer, event data analytic engines, security protocols, robust IoT applications, user interface, and event database. The high resource-demanding data-processing tasks are mostly executed by well-equipped cloud-hosted applications to manage and store huge amounts of data, provide monitoring and data analytical services, enable communication with devices, and manage information access. The merits of edge computing can be exploited to ensure that large amounts of data are post-processed off-device to reduce the response times of the cloud.
4. **User Interface:** With the aid of a web or mobile app, the user or farmer can live-monitor the farm's conditions and execute control actions. Additionally, a presentation or business intelligence layer may be added to coordinate the activities of non-technical business users through dashboards and reports rather than with the application layer itself.

2.3. *Quality Expectations of Agri-IoT's Architectural Layers*

Although there is no unified, certified, and flexible Agri-IoT architecture layer, any suitable options deduced from the benchmarking architectures in Figure 8 must satisfy certain quality requirements, including:

1. Simultaneous data acquisition, analysis, and control from many sensors or actuators.
2. Minimization of huge raw data transmissions via data aggregation techniques to maximize actionable information quality.
3. Provision of reliable network architecture that supports energy-efficient routing, stable connectivity, self-adaptability, fault tolerance, operational simplicity/flexibility, platform independence, affordability, and location independence of Agri-IoT designs.
4. Support for automated/remote device management and updates.
5. Easy integration of each layer with existing applications and other IoT solutions via specified APIs.
6. Utilization of freely available, location-unrestricted, cheap, energy-efficient, and simple to deploy and manage by non-experts [4,29] underlying communication technologies in the PHY and network layers as well as based on open standards to guarantee interoperability.

3. Design and Implementation of Agri-IoT Networks

Despite the technical challenges associated with the WSN-based Agri-IoT, its potential contributions in the agricultural sector largely surpass the least complex, capital-intensive, pure IoT-based solutions, as illustrated in Figures 3b and 7. Due to the broader applicability and higher significance of the WSN-based Agri-IoT networks relative to the classic IoT networks, this study focuses on the former technology whose design and implementation involve four crucial phases, namely:

1. Custom-building of robust, affordable, energy-efficient, location-independent, and adaptive SNs and a BS that can form an infrastructure-less and easily manageable WSN. The SNs and the BS must consist of cost-effective, architecture-defined, and context-defined components so that the system operates stably and efficiently, becomes affordable to farmers, and easily integrates to any real-world scenario without any expensive, fixed/wired backbone connections. The low-power capabilities of the SNs help to easily integrate them into any precision farms and greenhouses to operate over the entire crop season without many technical hindrances.
2. Physical deployment of the SNs in the field, selection of the WSN's communication technology, and design of a suitable supervisory protocol to coordinate the construction of appropriate event routing architecture, the duty-cycle schedule of event sampling to the BS, fault management, data management, and network maintenance. Additionally, a range of techniques such as network participant mobility, cross-layer design, MAC techniques, data aggregation, self-healing techniques, nodes' duty-cycle schedule, security measures, localization, and communication specifications of the SNs can also be exploited in the associated routing protocols.
3. Selection of appropriate BS/gateway communication technology and design of a suitable higher protocol to update the cloud database and execute the actuation actions based on users' requests or decisions on processed event data.
4. Design of data analytical engines and applications in the cloud and users' remote monitoring and control interface app, which is beyond the scope of this tutorial.

These call for a systematic application-specific assessment of the hardware components selected for every use case.

3.1. Sensor Nodes Design Considerations

As illustrated at the bottom of Figure 3, a node for the WSN-based Agri-IoT network consists of four main units, which include the following:

1. Sensing Unit: This unit interfaces with the physical environment and records the physical phenomenon of interest. The type of sensor is application-specific and can be contact-based or non-contact-based. For instance, the STEMMA soil moisture sensor and the DHT22 sensor can be used to sample environmental temperature and humidity (refer to Figure 3c).
2. Controller Unit: This unit hosts the processor, storage, and connection pins for the other units and all auxiliary peripherals. The suitable controllers for building Agri-IoT SNs are Arduino-based and Raspberry-Pi-based (refer to the bottom of Figure 3) due to their ability to withstand extreme weather conditions. However, other off-the-shelf, application-specific controllers such as the ProPlant Seed Rate Controller, John Deere GreenStar Rate Controller, Viper Pro multi-function field computer, Radion 8140, Trimble Field-IQ, etc. are also available.
3. Communication Unit: This unit is the principal determinant of the node's power consumption, operational stability, and affordability, as well as the routing architecture in the associated supervisory protocol. The bottom of Figure 3 shows the available communication technologies, but an Agri-IoT-based SN demands an energy-efficient, affordable, freely available, simple, and reliable communication standard. Consequently, LoRa, BLE, ZigBee, LoRaWan, and SigFox are the best candidates based on

the support of the routing architecture of the resulting WSN, but the selection must be justified from the technology requirement metrics via a decision matrix.

4. Power Unit: Since the SNs are mostly battery-powered, the appropriate battery size and probable energy-harvesting techniques must be determined during the SNs' design according to the intended network lifespan and stability requirements. Modern trends in battery power banks with integrated solar-based energy-harvesting systems and power ratings above 30,000 Ah are available.

When selecting hardware components, adequate caution should be taken to avoid unit incompatibility, high operational complexities, unsuitable operational thresholds, and high energy consumption, among others. This implies that high component survivability and operational stability under different environmental conditions and the application specificities are vital to monitor.

3.2. *Wireless Spectrum and Core Communication Platforms of WSN-Based Agri-IoT*

The wireless electromagnetic (EM) spectrum, which has invisible, finite radio frequencies for wireless communication, can be licensed and sold exclusively by specific providers or unlicensed for free usage. For instance, the Industrial, Scientific, and Medical (ISM) frequency band (e.g., Bluetooth classic, BLE, Wi-Fi, ZigBee, and LoRaWAN) is an unlicensed microwave frequency band clustered around 2.4 GHz and globally reserved for applications such as Agri-IoT. Table 3 presents the various bands and their applications. A suitable candidate for a given Agri-IoT application is based on several factors, such as communication and the route architectural requirements, power consumption, cost, and environmental adaptability impacts.

Table 3. Wireless spectrum with the core communication platforms/applications.

Frequency Band	Applications
Licensed Band	
0–20 MHz	AM radio
86–108 MHz	FM radio
470–800 MHz	TV band
850–1900 MHz	Cellular-based: GSM/3G/4G/5G/LTE
Around 3.5 GHz	Satelite comm.
Unlicensed Band	
863–928 MHz	LoRa, LoRaWAN, SigFox *Legality location-dependent: e.g.,* 915 MHz (*Australia & North America*), 865 MHz *to* 867 MHz (*India*), 923 MHz (*Asia*)
Around 2.4 GHz	Wi-Fi, BLE, ZigBee, Classic Bluetooth
Around 5 GHz	Wi-Fi

3.3. *Factors to Consider When Deploying SNs and Designing the Supervisory Sampling/Routing Protocol*

After custom-building or selecting off-the-shelf SNs, the next activity is to deploy the SNs on the field and design a contextualized supervisory protocol to coordinate the aforementioned network's activities. The SNs' deployment in the field can be either random or deterministic. Both options require different methods to optimize the resulting network's performance. For instance, under the deterministic approach, the optimal parameters such as node uniformity and density must be predefined based on the distance thresholds of the associated communication technology (i.e., connectivity/distance range), the SNs' resource optimization mechanisms, the type of routing architecture, and the sensing range of the physical parameter to be measured. Since communication is the principal power consumer, the best ways to conserve power are to minimize communication distance and data sizes as well as operate the SNs in the appropriate sleep–active duty cycles using a cluster-based routing architecture [9,24,26].

Beyond the physical installation of the SNs at their most suitable in-range locations, the remaining activities, such as network construction, event sensing, data management, FM, network maintenance, sleep–active duty-cycle scheduling of SNs for sampling, network adaptability to turbulent and scalable conditions, power-optimization mechanisms, and network reconfiguration, among others, are controlled by the associated routing protocol [12,16,17,26,36]. This places crucial merits on the physical locations of the SNs in the field, thorough synthesis of network design factors, and assessment of available routing architectures/techniques, since this protocol manages all post-deployment tasks. This can be summarized into the core objectives of the routing protocol and its architecture, which include power optimization, self-healing of any faults without the obstruction of its normal operation, and self-adaptability to all turbulent and scalable conditions. From the analysis above, we can derive the critical primary factors to consider when designing a routing protocol for Agri-IoT networks, which are presented in Figure 9 and grouped into the following categories: SNs specifications, security issues, application-specific factors, communication standard compatibility and capacities, and other auxiliary factors. At the PHY layer level, which is the focus of this tutorial, these critical factors can translate into the stipulated core design objectives, which can be addressed via phase-based multi-objective optimization (MOO) formulation frameworks [12,23,24,37].

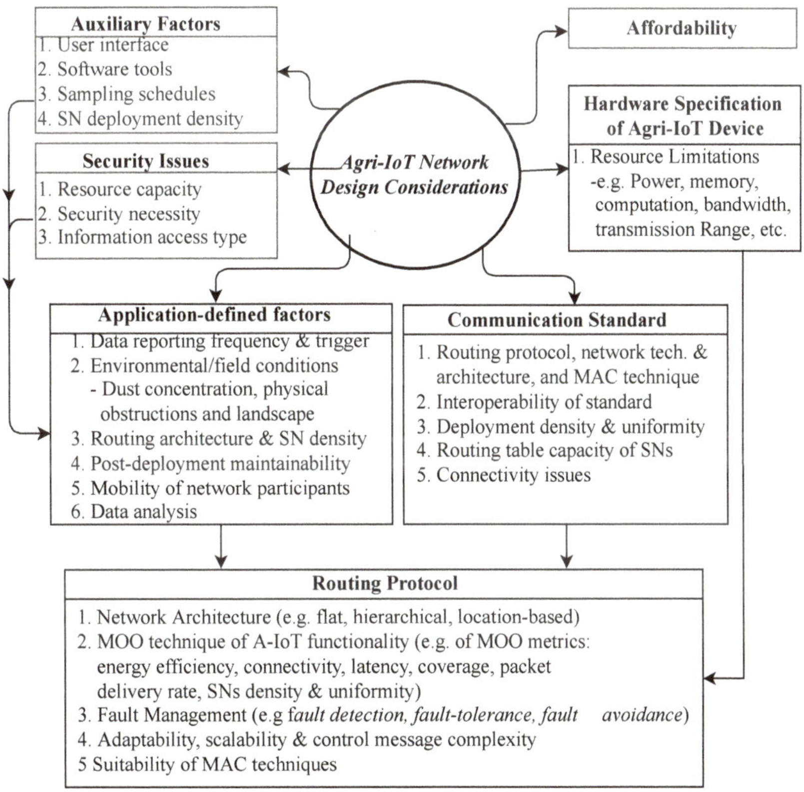

Figure 9. Principal design factors for Agri-IoT networks.

Hardware Specifications of SNs and BS Agri-IoT Device: The functional and resource capacities of participants' hardware units must be considered before their respective tasks in the protocol are assigned. For instance, the selected sensors' quality must suit the type of event information and its accuracy, the available communication platforms, and the general purpose of the Agri-IoT solution. Also, the communication standard must support the routing architecture and SNs' resource- and deployment-induced limitations. The crucial communication-based parameters of the SNs are illustrated in Table 4.

Table 4. Comparison of common communication platforms of the WSN sublayer of Agri-IoT.

Standard/	Network Size	P_{Tx}/mW	$P_{Rx-elec}/mW$	Range	Freq. Band	Data Rate/ Latency	Network Type	Energy	Topology
BLE/ IEEE 802.15.1 [6]	Application-definned	3–10	2×10^{-14}	10–50 m	2.4 GHz	1 Mbps/6 ms	PAN, WSN	Very Low	Star, mesh
Bluetooth Classic/ IEEE 802.15.1 [5]	7	215	200×10^{-14}	10–100 m	2.4 GHz	1–3 Mbps/100 ms	PAN	High	Scatternet
WiFi/ IEEE 802.11 a/c/b/d/g/n [7]	255	800–835	162	100 m	5–60 GHz	1 Mb/s–7 Gbps/50 ms	LAN	High	Point-to-hub
LoRaWAN/ LoRaWAN R1.0 [6,8]	10^4	25–100	2×10^{-14}	5–10 km	868/900 MHz	0.4–100 Kbps/NA	WAN	Very Low	Star
SigFox [2,6]	Undefined	122	10^6	15 miles	200 kHz	100–600 bps	PAN	Low	Star
ZigBee/IEEE 802.15.4 [2,23]	64,000+	36.9–100	77	10–20 m	2.4 GHz	20–250 Kbps/(20–30) ms	PAN, WSN	Low	P2P, tree, star, mesh
NB-IoT,LTE/2G-GSM, 4G-LTE [2,4]	1000	200–560	80	10–15 km	2.4 GHz	200 Kb/s–1 Gbps/1 s	WAN	Medium	Cellular system
*cc*2420//IEEE 802.15.4 [23]	64,000+	8.9–36.9	35.28	580 m	2.4 GHz	20–250 Kbps/40 ms	PAN	Low	P2P, tree, star, mesh
XBee PRO [24]	64,000+	36.9–63	6.31×10^{-11}	90 m–1.6 km	900 MHz	20–250 Kbps/40 ms	PAN, WSN	Low	P2P, tree, star, mesh
Jennic JN5121/IEEE 802.15.4	64,000+	100	45×10^{-9}	0.4 km	2.4 GHz	20–250 Kbps/30 ms	PAN, WSN	Low	P2P, tree, star, mesh
RFID/ISO 18000-6C [4,29]	Undefined	3000	unspecified	1–5 m	860–960 MHz	40–160 Kbps/ 45 ms	PAN	Low	Star

Cost or Affordability of the Resulting Agri-IoT System: In addition to being infrastructure-less, flexible, self-healing, adaptive, and energy-efficient, a WSN-based Agri-IoT must consist of cost-effective hardware and software components so that the system is affordable for farmers, since existing real-world solutions are too expensive and complicated [1,14]. Additionally, the installation, operational, and maintenance costs of the resulting WSN-based Agri-IoT network must be kept to a minimum so that it can be easily acquired.

Security Issues in Agri-IoT: Security is still a challenge in classic IoT systems that handle sensitive information, especially during cloud communications. Although Agri-IoT networks lack the requisite resource capacities in most large-scale, broadcast-based, distributed, and infrastructure-less WSN systems to achieve adequate data confidentiality, authenticity, integrity, and other security requirements, the security of the agricultural data is rarely a priority [2,4]. Nevertheless, the associated routing architecture, such as the clustering architecture, has an embedded capacity to resolve on-site security issues. In addition, both on-site and remote information access types (e.g., via a smartphone or desktop computer) must be selected based on solid internal infrastructure and security precautions to secure unwanted access to sensitive information.

The Application-Specific Factors: As indicated in Figure 9, the application-defined factors vary based on the Agri-IoT application, the field settings, network maintenance practices, intended event routing architecture, and network participants' mobility, among other factors. However, the routing protocol must incorporate all relevant operational efficiency factors of the routing software design objectives. Since the collected field data itself cannot make sense without using analytic data engines and predictive algorithms in machine learning, the BS or the application layer in the cloud should define appropriate data-processing frameworks to obtain accurate, actionable decisions from the collected data.

Communication Standards of Agri-IoT Devices: The power-constrained WSN sublayer of Agri-IoT network places hard restrictions on operational states of SNs' radio transceivers, code space, and processing cycles as well as memory capacities of SNs to enhance power savings [9,12,23]. The type of communication technology selected for a typical Agri-IoT is the principal predictor of its routing architecture, affordability, simplicity, adaptability, power-saving capacity, location independence, self-healing capacity, and event data quality [12,16]. Consequently, power and routing architectural limitations constrain the network design requirements. Despite the aforementioned technical challenges on the network's operational efficiency, interconnected SNs that form the WSN are expected to withstand extra operational disruptions caused by unfavorable weather conditions in the field [2,4]. Consequently, the de facto PHY-layer communication standards for this low-power, low bandwidth, and distance-limited communication Agri-IoT devices/SNs have been the energy-efficient platforms such as BLE, LoRa, Sigfox, and NB-IoT. Also, a suitable MAC technique is imperative in the routing architecture to curb all channel access challenges. For instance, the ZigBee/IEEE 802.15.4 standard focuses on the physical and the MAC layer specifications for WSNs, and it also supports the sleep–active or duty-cycle scheduled operation modes of SNs to enhance energy savings in centralized or mesh-based architectures. BLE does likewise in the highly endowed cluster-based routing architecture. Consequently, Agri-IoT network designers must make the most appropriate and critical decisions regarding the network's communication requirements when designing the routing protocol. Using Table 4, WSN-based Agri-IoT designers can make realistic design decisions regarding energy-efficient multihop routing, architectural requirements of routing protocol, bandwidth, routing table capacities, total communication cost, and the desired MAC technique. Additionally, the physical conditions within the agricultural environment such as atmospheric dust concentration, physical obstruction to wireless signal transmissions, and the terrain need to be considered.

Auxiliary Factors and Available Software Tool: Finally, the auxiliary factors can be non-exhaustive depending on the designer's financial capacity, user interface, information requisition model, cloud activities, operational expectations, and the available software tools. Additionally, an assortment of PHY-Layer design software tools for Agri-IoT experiments (thus, in both simulations and real-world testbed deployments) that can be used include NS-3 [9,38], OMNeT++, MATLAB/Simulink [9,12,39], Python [16], PAWiS [39], GloMoSim/QualNet [39,40], OPNET [12,39], SENSE [37,39], J-Sim [39], Ptolemy II [39], Shawn [9,39], and PiccSIM [12,39,41], among others. The key features that are frequently considered when selecting any of these software platforms include Python or MATLAB/Simulink compatibility for software model and hardware prototype integration during real-world operation, compatibility with low-power communication standards (e.g., BLE, LoRa, ZigBee, and SIGFOX), operating system support, programming language implementation, the density of simultaneously simulated or field-deployed SNs, co-simulation with other hardware, documentation, easy access to upgraded versions, and installation challenges [39]. MATLAB/Simulink and Python are the most commonly used experimental tools, since these software tools are well-equipped with the stipulated features.

4. Unique Characteristics and Challenges of WSN Sublayer of Agri-IoT

Unlike the traditional IoT, which generally relies on fixed hardware to route network traffic, a WSN sublayer of Agri-IoT combines automated sensing, computation, actuation, and wireless communication tasks into the SNs that are spatially distributed across the farm to autonomously form an infrastructure-less WSN [31]. A node may perform additional tasks such as local data processing (data aggregation), network construction, data redundancy, error control, data routing (e.g., in multihop networks), and network maintenance practices based on the network size, application specificity, and associated routing techniques. Also, the WSN can be equipped to observe heterogeneous conditions such as temperature, humidity, sound, color, location, light, vibration, and motion, using a wide variety of sensors contained within a task-scalable SN. Therefore, assuming that the accuracy and precision of event data in upper layers are preserved, the Agri-IoT's lifespan and its operational efficiency are rooted in the WSN's robustness. Thus, a deeper contextual exegesis into the design and maintenance of this sublayer is imperative. As opposed to conventional IoT and wireless ad hoc communication networks, the operational efficiency of the WSN sublayer, as well as Agri-IoT, hinge upon some application-specific characteristics and resource-constrained factors such as:

- *Higher SN Deployment Densities:* Generally, SNs are densely deployed in either a deterministic or random manner to provide the desired redundancies, spatial variability of soil, topography, distributed monitoring and processing, accurate and precise event reporting, and fault tolerance. However, this mostly leads to undesirable transmission overlaps, data redundancies from the simultaneous reporting of the same data, routing interferences, and packet collisions due to connectivity issues and the coexistence of common standards in the ISM band [42].
- *Limited Power Supply:* The SNs are frequently battery-powered, which does not only constrain their data transmission rate, computational capabilities, and communication distance but also subjects Agri-IoT to possible SN-out-of-service and data outlier faults due to rapid power depletion beyond certain thresholds [26,43]. Consequently, network power management through data-management-related, architectural-related, and communication-related parameters has been one of the principal research focuses in WSN-based IoT applications to improve network lifetime.
- *Fault Management (FM) (i.e., fault detection, fault tolerance, or fault avoidance):* The resource-constrained WSN is highly vulnerable to faults and failures due to high deployment densities and a lack of post-deployment maintenance services [25]. Although faults are inevitable in Agri-IoT for the stipulated reasons, their occurrence rates and effects on the network's functionality can be minimized, avoided, or tolerated without hindering the normal functionality of the network if the associated WSN's routing protocol is well-equipped with efficient self-healing and fault-avoidance (power-saving) mechanisms [12].

- *Self-Adaptability and Scalability:* Although WSNs are application-specific, the topological dynamism is inevitable due to node failures, node mobility, and scalable conditions. Therefore, the associated routing protocol and network architecture must adapt to these dynamic conditions using apt auto-reconfiguration and reactive multihop event routing techniques [44,45].
- *Network Architecture:* The underlying routing protocol of the WSN sublayer constructs a network architecture that can be flat, hierarchical (e.g., clustering, chain-based, and tree architectures) or location-based. This routing architecture prescribes the possible measures to achieve efficient local data processing, network maintenance, scalability, minimized communication overhead, prolonged network lifespan, and reduced network management complexities [25,36]. Therefore, a suitable network topology indirectly determines the resulting network's flexibility, scalability, reliability, communication strategy/costs, and the quality of the reported event data [12].
- *Mostly Requires On-site Actuation:* Regardless of where data are managed in a typical WSN-based Agri-IoT, the actionable decision signal must be sent to execute on-farm actuation.

Proposed Design Objectives of WSN-Based Routing Protocols for Agri-IoT and Realization Mechanisms

From the systematic evaluation of the unique characteristics and challenges of the WSN sublayer, a three-tier cluster-based framework that constitutes the condensed expected core design objectives and their corresponding remedial strategies of WSN-based routing protocols for Agri-IoT applications is demonstrated in Figure 10. Suppose the corresponding remedies in Figure 10 are implemented in the associated routing protocol. In that case, the desired power optimization, self-healing, and auto-adaptability expectations can transitively yield the desired event data quality and operational stability requirements or the global performance expectations of the resulting network.

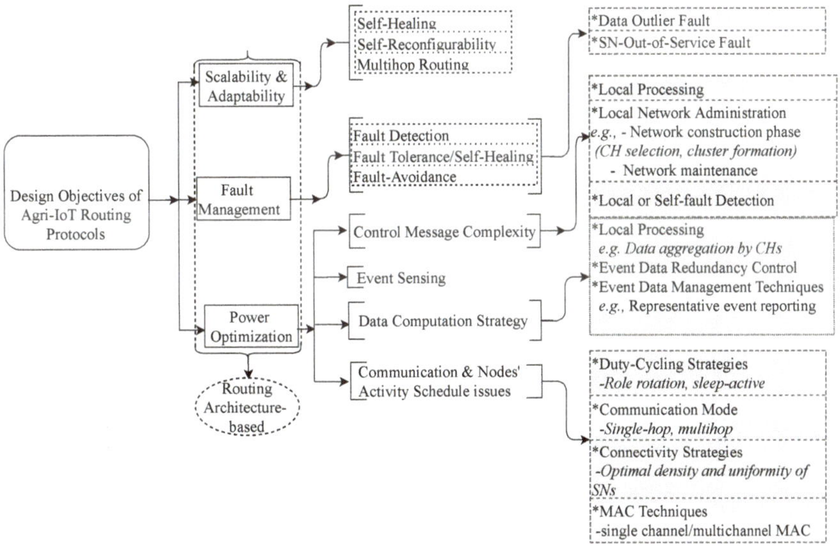

Figure 10. Proposed design objectives and strategies of WSN-based Agri-IoT routing protocols.

The importance of this three-tier framework can be expanded on as follows:

- An adaptive and scalable WSN-based routing protocol, as proposed in Figure 10, normally constructs a routing architecture that supports multihop routing, self-reconfiguration, self-healing, and local network administration at a minimal routing table size, communication cost, and and control message complexity requirement. Since communication is the principal power consumer, the operation of the routing protocol must invlove fewer control messages. Also, it must adapt to network turbulence due to SN failures. The cluster-based architecture exhibits the highest potential compared to related architectures [9,16,17,26]. The cluster heads (CHs) efficiently coordinate these activities by registering and tolerating all dynamism resulting from SN-out-of-service faults, increasing the network size and SN density.

- Due to the high vulnerability of SNs to faults and failures, it is imperative to deploy suitable FM techniques that can detect, tolerate, or avoid possible root faults such as SN-out-of-service and data outliers [25]. The adaptive clustering approach can effectively resolve SN-out-of-service faults, while the threshold-based decision theory at the local nodes and global levels can be suitable candidates for event data outlier detection and correction in the PHY layer. Since power mismanagement is the root cause of most faults and failures, the best fault-avoidance techniques optimize the nodes' power consumption rates.
- Figure 10 also outlines the suitable measures for power optimization in the WSN sublayer of Agri-IoT. In clustering approaches, power consumption in the constrained WSN can be managed via message complexity control, connectivity-related metrics, and communication-related parameters by exploiting the clustering architecture [46]. In addition to local data processing (data aggregation, data redundancy, and error checks) and local network administration (FM, adaptability to network dynamics), suitable MOO and multihop routing frameworks can be derived using the clustering architecture, total communication cost, and optimal cluster quality metrics to serve as a design optimization guide for the simulation and real-world implementations of the WSN phase of Agri-IoT.

To achieve the expectations in Figure 10, there is a need for an architecture-specific multi-objective assessment of the WSN's design cycle; from this, the associated parameters and theoretical models can be derived and then theoretically optimized and validated experimentally. A novel holistic MOO framework can help realize these expected goals in both simulation and real-world Agri-IoT implementations. Consequently, there exists the need to carry out a systematic survey and assessment on existing routing architectures, FM schemes, and routing protocols, and how these evolved in existing real-world realization testbeds of Agri-IoT. Such an in-depth literature synthesis can help assess these qualitative performance indicators constituting the root QoS metrics in Figure 10 as well as deduce application-specific guidelines for improving CA-IoT networks using a precision irrigation system as a case study.

5. State of the Art on Routing Protocols for WSN-Based Agri-IoT Applications

In Agri-IoT, it is not simply a matter of applying IoT to a farm; contextual due diligence on architecture, communication standard, cost, actuator, performance stability, control, and environmental impacts augment the routing protocol requirements. This section presents a systematic synthesis of WSN-applicable routing protocols under network architecture, the route discovery process, and protocol operation as illustrated in Figure 11. To help Agri-IoT designers make well-informed decisions concerning architectural selection, we classified the canon protocols based on routing architecture, route-discovery process, and operations in order to uncover their strengths, weaknesses, and contextual reasons why they can be adopted for Agri-IoT applications. Generally, event routing in every protocol can either be source-initiated or destination-initiated, and the optimal path selection from the constructed routing architecture can also be broadcast-based, probabilistic, cluster-based, or parameter-determined using location-related, weight-based, and content-based metrics [13]. Also, routing protocols must commonly resist link failures using mechanisms that ensure balanced network-wide power depletion rates, energy-efficient multihop routing, and effective implementation of the indispensable QoS metrics presented in Figure 10. The related routing protocols can be classified as illustrated in Figure 11.

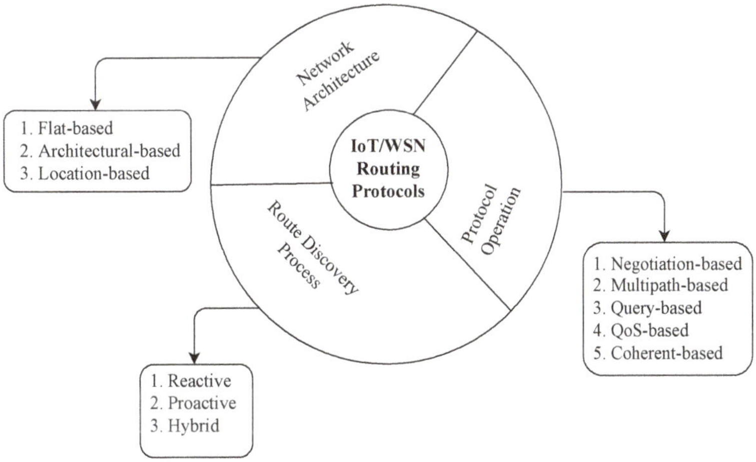

Figure 11. Taxonomy of WSN-based routing protocols of Agri-IoT.

5.1. Architectural-Based Routing Protocols

This class of protocols presented in Figures 11 and 12 can be sub-grouped into flat-based centralized or direct communication and decentralized [47] (e.g., flooding/peer-to-peer/graphical/mesh-like architectures), hierarchical/cluster-based/tree architectures, and the location-based protocols [37].

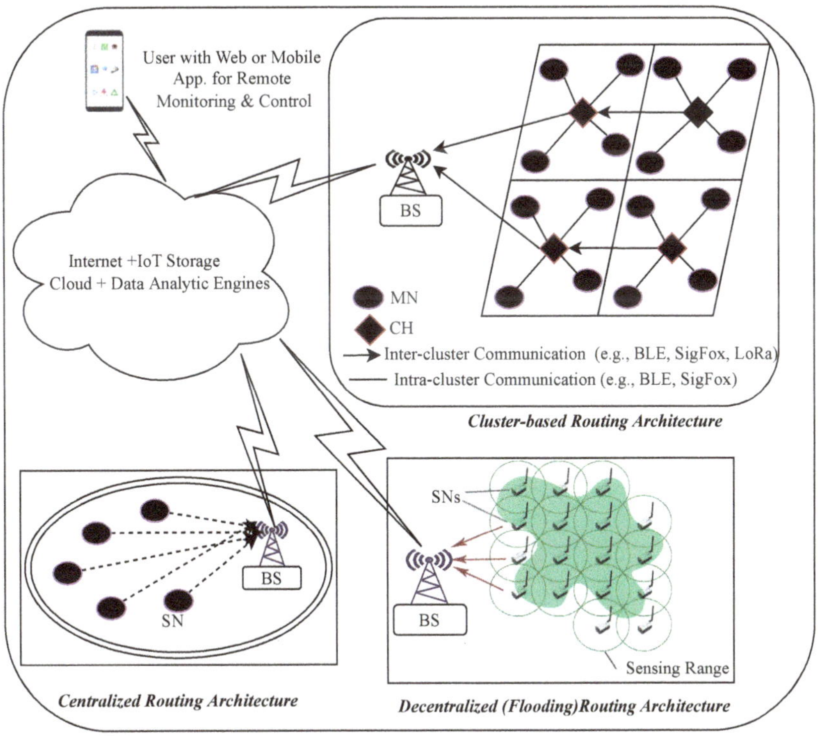

Figure 12. Sample network architectures: centralized-data-centric, cluster-based, and graph/flooding-based architectural frameworks of WSN sublayer.

The centralized protocols route data to the BS via single-hop routing, while the flooding and graph-based protocols flood data through multihop routing. The graph-based routing protocols construct a reactive or proactive graphical routing architecture with $G(V,E)$ where a node and path represent the vertex and edges, respectively. This method relies on resource-intensive routing techniques from graph theory used in classic IoT and ad hoc networks to transmit event data to the BS. In contrast, the clustering/tree topology depicted in Figure 12 groups the SNs into either static or dynamic clusters, each with an optimally selected CH to minimize the communication distances of the cluster's member nodes (MN). The CH is then tasked with aggregating the received readings from its MNs, executing error and measurement redundancy checks, and communicating directly (single-hop routing) or via a relay CH (RCH) using a multihop routing technique to the sink node or BS. However, the RCHs must be assigned fewer MNs to balance the network's power depletion rates, since aggregated packet forwarding inflicts extra energy burden on the RCHs [37]. Additionally, the CH can be equipped to perform extra roles such as FM, coordination of the reclustering process, network maintenance, relaying of aggregated packets in large-scale networks, and management of network dynamism [12]. In general, cluster-based routing protocols differ in terms of CH selection methods and coincide in terms of intra-cluster and inter-cluster multihop routing, local data processing by the CHs, and CH role rotation [47], which ensure balanced network-wide power depletion, prevent abrupt power exhaustion, and lead to exponential energy savings [37].

Although the flat-based architectures, such as centralized and flooding (see Figure 12), can be easily implemented in real-world small-scale Agri-IoT networks, they suffer severe packet collisions, communication bottlenecking at the BS, and high inaptness for scalable or turbulent large-scale WSNs where energy efficiency is a priority. Again, an optimized clustering approach can provide an ideal topology for addressing the proposed expectations in Figure 10, and it can also offer extra benefits such as minimized communication cost, stabilized network topology, efficient load management, improved network maintenance, and improved network traffic and channel access management [37,48]. The main challenge of the clustering method is how to achieve the desired cluster quality (e.g., optimal cluster count and cluster size) so that the computational, bandwidth, memory, and routing table capacities of the resource-constrained CHs are not exceeded. Typical examples of clustering protocols are the LEACH family of protocols, which include RCEEFT, ESAA, DEEC, SEP, and PEGASIS in [12].

In location-based routing architectures, routing decisions are made either reactively (e.g., Ad hoc On-demand Distance Vector—AODV) or proactively (e.g., RPL—Routing over Low-Power and Lossy Networks protocol), using the SNs' location information. This normally results in a decentralized, graphical architecture. Since the SNs that form the WSN are spatially deployed in the field without any IP-addressing schemes, location information is needed in order to establish communication between the nodes in a location-based architecture. The location information helps eliminate unwanted transmissions by collecting data from a specific region of interest. This architecture suffers from routing delays, high infrastructural cost, extreme difficulties in deployment and management, and high energy waste due to SNs' long idling durations. However, they are the most commonly used protocol in existing ZigBee-based Agri-IoT testbed solutions [1,10,14,17]. Since this approach yields non-energy-aware architectures, it is not suitable for Agri-IoT applications [12].

It is evident from the above discussions that Agri-IoT-based network architectures must be defined by the associated routing protocol using the design requirements in Figure 9 as well as the application-defined requirements [49] in order to enhance the performance expectations in Figure 10. In addition, the routing architecture must not compromise on the quality, precision, and accuracy of the event information. It must be in unison with the application-specific requirements to address possible deployments- and network-induced challenges, such as network turbulence and SN mobility.

5.2. Route Discovery-Based Protocols

As shown in Figure 11, route discovery-based protocols focus on when the route for data transmission is built and can be grouped into proactive, reactive, and hybrid protocols.

In proactive routing protocols, the routes are pre-created before they are needed. These protocols are table-driven, since every node stores a large routing table containing a list of all possible destinations, next-hop neighbors to those destinations, and the associated costs of all next-hop options. Proactive protocols such as the RPL and the APTEEN family of protocols [15] make local routing decisions using the routing table's content. For instance, the RPL operates as a distant-vector protocol for IPv6 low-power devices, utilizes the ZigBee/IEEE 802.15.4 standard on established IP infrastructure, and also supports the 6LoWPAN adaptation layer. RPL creates a multihop tree routing hierarchy of SNs, such that nodes can send data through their respective parent nodes to the BS/sink node in a flooded manner (Figure 12). Similarly, the BS or sink node can send a unicast message to a specific SN in order to complete a bidirectional operational framework of RPL. The optimal communication costs and routes are estimated by ranking the associated objective function (OF) metrics, which can be single-objective optimization, SOO metrics, or MOO metrics. This routing over LLNs (RoLL) restricts densely deployed and resource-limited SNs to communicate using peer-to-peer or extended star network topologies [13]. Technically, RPL builds a directed acyclic graph (DAG) with no outgoing edges from the root element (e.g., BS) to eliminate loops. RPL is the primary underlying routing protocol in most failed Agri-IoT testbed attempts. Although the proactive or RPL-based family of protocols are robust, reliable, scalable, and can relatively operate at minimized control messages with the help of timers, they are not suitable for Agri-IoT networks due to these technical challenges:

- The core of RPL/proactive protocols still suffers from key challenges such as energy wastage, a lack of adaptability/scalability, reliability, congestion, and security issues. Specifically, the energy expended by RPL-inherited protocols to create routes (e.g., establish and maintain routing tables) and transmit data can be too high for resource-constrained SNs in recent Agri-IoT applications.
- The underlying technology of RPL (e.g., ZigBee, 6LoWPAN, or IPv6) was designed for energy-sufficient devices with high processing and memory capacities. Therefore, RPL is inapt for typical resourced-constrained Agri-IoT networks (refer to Table 5).
- They require costly fixed IP infrastructural supports and utilize the centralized routing architecture, which becomes practically impossible to manage as the network scales.

Conversely, the source-initiated reactive or on-demand routing protocols only create the routes on-demand by a source to send data to a receiver. Reactive protocols (e.g., Ad hoc On-demand Distance Vector, AODV Protocol [13]) have no specific procedures for creating and updating routing tables with route information at regular intervals. For instance, the AODV is a loop-free, self-starting, and reactive routing protocol meant for LLNs (e.g., WSN-based IoT) that are characterized by node mobility, link failures, and packet losses. AODV mainly consists of the route discovery process (RREQ and RREP messages) and route maintenance (RERR and HELLO messages). Although reactive or AODV-based protocols can adapt to network dynamics and eliminate periodic updates, the associated flooding-based route–search process incurs severe overheads resulting in high control message complexity, high route acquisition latency, and high energy wastages due to longer SN idling periods. Consequently, these protocols are unsuitable for power-constrained WSN-based Agri-IoT applications.

The hybrid-based routing protocols merge the features of both reactive and proactive routing processes. However, hybrid protocols such as APTEEN [13] also require expensive fixed infrastructural support, which renders them unsuitable for Agri-IoT, even if the combined merits of reactive and proactive protocols are exploited.

A comparative assessment of the strengths and weaknesses of the parent WSN-based routing protocols for Agri-IoT applications is illustrated in Table 5.

Table 5. Comparison of some cardinal hierarchical WSN-based routing protocols for Agri-IoT in state of the art.

Protocol	Topology	Strength	Weakness	Suitability: Low-Power WSN Sublayer of Agri-IoT
LEACH and LEACH-inherited [9,12,21]	Tree or Cluster-based	• High power savings, • FM and adaptability, • Load balancing, • Less resource demanding than RPL, AODV	• Difficult to attain desired cluster quality	Suitable (optimal cluster quality yet required)
RPL and RPL-Inherited [15]	Graphical	• High adaptability, • High robustness, • Minimized control messages, • Suitable for small-scaled, power-sufficient networks	• High energy wastages, • High storage requirements, • Low reliability, • High congestion rates, • Unsuitable for large-scale turbulent networks, • More resource-demanding than AODV and LEACH-based methods [50,51]	Unsuitable (high resource demanding underlying technology, 6LoWPAN, and routing tables)
AODV and AODV-inherited [13]	Mostly graphical	• High adaptability, • Suitable for small-scaled, power-sufficient networks	• High control messages, • High energy wastages [28], • High-resource-demanding	Unsuitable (extremely high control message complexities during route construction and maintenance)

5.3. Operation-Based Routing Protocols

Finally, routing protocols can be classified based on the operation or communication model employed, which may include:

- *Negotiation-Based Protocols:* These protocols exchange negotiation messages or use meta-data negotiations between neighboring SNs before the actual data transfers to reduce redundant transmissions in the network. A typical example is the SPIN family of protocols [13].
- *Multipath-Based Protocols:* These use multiple routes simultaneously to accomplish higher resilience to route failure (i.e., fault tolerance) and load balancing.
- *Query-Based Routing Protocols:* These are receiver-initiated protocols whereby a destination node broadcasts a query to initiate a data-sensing task from a node through the network. A node having the data being queried sends it in response to the query.
- *Coherent and Non-Coherent Protocols:* The coherent routing method forwards data for aggregation after a minimum local pre-processing. However, in non-coherent routing, the nodes locally process the raw data before routing to the BS for further processing.
- *QoS-Based Routing Protocols:* These protocols' purpose is to satisfy a specific QoS metric or multiple QoS metrics such as low latency, energy efficiency, or low packet loss. These protocols ensure a balance between energy consumption and data quality in every event-reporting task.

In addition to route architectural construction and data transmission, efficient MAC must be embedded in the routing protocol to manage the wireless medium access and the duty-cycle/sampling schedules of the deployed SNs in Agri-IoT networks. As opposed to classic IoT, the MAC techniques in Agri-IoT are architecture-defined by the associated routing protocol to meet the energy efficiency requirements of the network via channel access management (CAM) and the moderation of the active–sleep duty cycles of the deployed SNs to save extra energy. The next subsection presents a concise overview of MAC techniques and their roles in WSN-based Agri-IoT networks.

5.4. MAC Techniques and Requirements for Agri-IoT

Next to node deployment, the routing protocol defines the network architecture and selects a suitable MAC technique and a communication pattern for the routing architecture. Unlike classic IoT, requirements for Agri-IoT applications include a low control message complexity and low latency MAC technique that moderates sampling schedules, access to a shared medium, transceiver operation modes, (e.g., packet transmission and reception, retransmission, collision, over-hearing, overhead handling, and idle listening) active–sleep duty cycles of the deployed SNs, and transceiver channels. Thus, an MAC protocol for WSN-based Agri-IoT applications must be architecture-specific and adaptive to network dynamics such as data transmission errors, interferences/packet collisions, and regular interfacing of the active–sleep duty-cycled schedules of the SNs' transceiver states (e.g., transmitting state, receiving state, idle state, and sleep state [52]) during packet transmission and reception in order to improve network throughput, energy efficiency, latency, and other QoS metrics.

Unlike MAC protocols for classic IoT, an efficient MAC technique for Agri-IoT must ensure exponential energy savings via channel assignment management (CAM) and active–sleep duty-cycle coordination in both time and channel perspectives. Based on these common dual tasks of Agri-IoT-based MAC (thus, duty-cycle optimization—DCO and channel access management—CAM), existing IoT-based MAC techniques can be classified as illustrated in Figure 13 and the state of the art in Table 6.

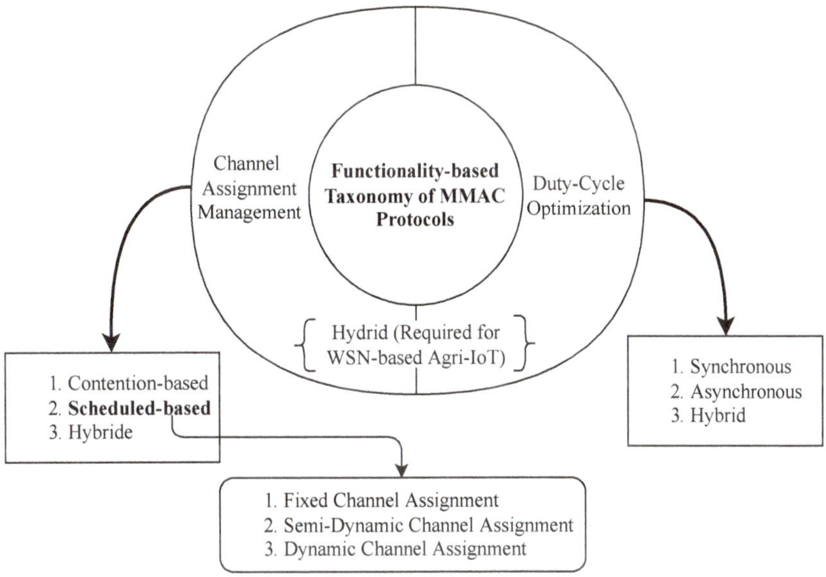

Figure 13. Proposed functionality-based MAC classification framework.

The CAM role eliminates packet collisions, overhearing, and over-emitting to ensure the desired functional balance, while the DCO task minimizes idle listening. A comparative assessment of related MAC methods used in recent WSN-based Agri-IoT applications in Table 6 affirms the need for further research on the functionality balance between DCO and CAM as well as a context-based MMAC approach for the LEACH family of protocols used in Agri-IoT applications.

Table 6. Summary of state of the art on duty-cycle and CAM MMAC protocols.

Name	Main Task	Application	Weakness	Approach	Overhead	Sync/Async
S-MAC, T-MAC, DS-MAC [53,54]	DCO	Event-driven with long idle listening times, collision-prone	High PC, complexity, latency	Contention-based, distributed MAC	RTS, CTS, ACK, SYNC	Sync
X-MAC [55]	DCO	High energy savings, throughput, collisions, delays	High complexity, higher PC, high collisions	Contention-based, distributed MAC	Preamble	Async
LA-MAC [56] Inherits X-MAC [55]	DCO	More energy savings than X-MAC, throughput, scalability collisions, low delays	High complexity, weak collision control measures	Contention-based, distributed MAC	Preamble	Async
B-MAC [57]	DCO	Delay-tolerant, high energy savings, throughput, DDR more than S-MAC,	High complexity, weak collision control measures, low throughput	Contention-based, distributed MAC (CSMA)	Preamble length	Async
(PEDAMACS) [58]	DCO with collision avoidance	Event-driven, energy-saving	High computational complexity, impracticable	Schedule-based, centralized MAC	RTS, CTS, ACK, SYNC, learning	Tight Sync
PW-MAC [59]	DCO	Low delay, long idle time	High complexity	Contention-based, distributed MAC	Beacon	Async
Cluster-based time synchronization [60]	DCO	High energy savings	High computational complexity	Schedule-based, cluster-based, distributed MAC	Schedule, CHs' formation	Tight Sync
LEACH [61]	DCO and CAM	Periodic sampling surveillance, energy balance, savings	High complexity, weak collision control measures	Schedule-based, cluster-based, distributed MAC	Schedule, CHs' selection	Tight Sync
PRIMA [62]	DCO and CAM	Periodic sampling/surveillance, balanced energy savings	High complexity, weak collision control measures	Schedule-based, cluster-based, distributed MAC	Schedule, CHs' selection	Tight Sync
WiseMAC [63]	DCO	High energy savings, collision, hidden terminal problem, poor duty schedule	High complexity, weak collision control measures, high PC	Hybrid, distributed MAC	Long wake-up preamble	Sync
Advanced WiseMAC [64]	DCO	Higher energy savings than WiseMAC, collision, hidden terminal problem	High complexity, weak collision control measures, poor duty schedule	Hybrid, distributed MAC	Shorter wake-up preamble than WiseMAC	Sync
WideMAC [65]	DCO	Wider duty-cycle ranges, aperiodic or periodic Tx, higher energy savings, low memory requirements	Weak collision control measures	Hybrid, distributed MAC	Preamble but short	Sync

Table 6. *Cont.*

Name	Main Task	Application	Weakness	Approach	Overhead	Sync/Async
EM-MAC [66]	CAM	Heavy traffic, delay-tolerant, hidden terminal problem	Prediction accuracy depends on the accuracy pseudorandom function	Schedule-based, predictive-based, dynamic CAM, distributive MAC	Initial preamble	Async
MCAS-MAC [67]	CAM	High energy savings, latency, low idle listening	Energy efficiency decreases with high traffic densities (high DDR)	Schedule-based, distributed MAC	Preamble	Async
AMMAC [68]	CAM and DCO	High energy savings, DDR	Time drift will affect accuracy	Contention-based, distributed MAC	Requires asynchronous modifications of duty cycles	Async.
LL-MCLMAC [69]	CAM	Improved end-to-end delay and throughput, low traffic with two time-slots	Data Tx on same control channel, susceptible to co-channel or adjacent channel interference	Semi-dynamic schedule-based, distributed MAC	Common control channel notification	Async
MC-LMAC [70]	CAM	Scalable WSNs, collision avoidance	High delays due to dynamic channel switching	Dedicated channel control, dynamic channels switching, schedule-based, distribute d MAC	Common control channel notification	Async

5.5. Overall Perspective

This section systematically surveyed core Agri-IoT-based routing protocols and evaluated the parent protocols (i.e., RPL, AODV, and LEACH/cluster-based families of protocols) for classic WSN-based IoT networks, of which LEACH-based methods are the best candidates for the resource-limited WSN-based Agri-IoT. However, the RPL and AODV have received more research considerations in terms of realizations in both simulations and practice [9,12,21]. Although the cluster-based architecture has unique endowments for realizing the proposed expectations in Figures 2 and 10, it lacks an in-depth design synthesis in the current state of the art that can uncover its contextualized performance optimization modalities for real-world Agri-IoT applications. In addition, the deployment requirements with trending technologies such as BLE, LoRaWAN, SigFox, 5G, LoRa via Satellite, and NB-IoT under both simulation and real-world operational conditions is imperative. Consequently, the following sections present in-depth overviews on FM, the benchmarking of WSN-based Agri-IoT testbed solutions, clustering methods in the existing state of the art, and how the possible deductions from these syntheses can evolve in a typical case-study such as a WSN-specific Agri-IoT routing protocol for precision irrigation.

6. State of the Art on FM Techniques for Classic WSN Sublayer of IoT

Since faults and failures are inevitable in the WSN sublayer of Agri-IoT networks (refer to Figure 10), it is imperative to reevaluate the faults, causes, types, strengths/weaknesses of existing FM (i.e., fault detection—FD, fault tolerance—FT, and fault-avoidance—FA) schemes, revisit their founding assumptions [71], and make appropriate recommendations for Agri-IoT network designers. In this section, we establish the root source/cause(s) of faults in the WSN sublayer by assessing the behaviors of the different fault types, examining the extent to which the existing FM schemes address these root faults, and exploring how these schemes will evolve in realistic WSN-based Agri-IoT networks based on their core assumptions, control message overheads/complexities, and energy-saving capacities. From this thorough assessment, this section proposes practical fault-avoidance-based FM techniques for the next generation of WSN-based Agri-IoT.

6.1. Systematic Overview of Faults, Sources, and Taxonomy of Faults in Agri-IoT

According to the fault–error–failure cycle depicted in Figure 14, a fault can be defined as any impairment that causes a system to produce erroneous results or leads to the failure of the entire system or specific components [72]. The prevalence of faults in WSN-based Agri-IoT is primarily due to the SN component malfunction, lack of post-deployment maintenance, or resource exhaustion [73], which can lead to either impaired event data quality (thus, sensory data error/outlier) or SN-out-of-service (thus, the shortened lifespan of SNs) [25].

Due to the high susceptibility of WSNs to faults, the supervisory routing protocol is expected to incorporate efficient FM mechanisms that can guarantee optimum event data quality and network availability. By implication, FM algorithms for WSNs must not be stand-alone as currently seen in the state of the art [73]; instead, they must be an integral aspect of the routing protocol that agrees with the core participants of the PHY layer, such as the SN, wireless communication medium, and the BS. As illustrated on the left of Figure 15, the WSN sublayer is the most prevalent source of faults in the Agri-IoT ecosystem, in which the SNs are the central origin of faults that can propagate to the upper layers [25,43,73]. This is because the BS is resource-sufficient mainly, and the link's reliability also hinges upon the SNs' availability, as indicated in Figure 15. At the local SN's level, each unit depicted at the bottom of Figure 3 is a potential source of fault/failure, but the degree of prevalence is frequently accelerated whenever power consumption is mismanaged through the disregard of any of the network design requirements and deployment conditions presented in later sections.

The different taxonomies of faults in the state of the art of the WSN sublayer [44,71,73–77], as illustrated on the left side of Figure 16, can be compared as follows:

- *Hard or permanent fault* refers to the inability of a node to stay active and communicate due to resource exhaustion or component malfunction, while in *soft or static faults*, nodes continue to work and communicate with other nodes, but they sense, process, or transmit erroneous data [44,74].
- The authors in [75,78] categorized faults as permanent (refers to SN-out-of-service faults), transient (caused by temporary conditions), intermittent (shows sporadic manifestations due to unstable behavior of hardware and software), and potential (due to depletion of hardware resources [78]).
- *Data inconsistency faults* can also result from faulty sensing, processing, and communication, which is frequently caused by power depletion below a certain threshold, while power failure occurs when a node exhausts its battery power completely [43,77,79].
- The authors in [73] classified faults into software and hardware faults based on software and hardware impairments, respectively.
- According to [71], faults can be either time-based, due to the depreciation of hardware components with time, or behavioral-based, due to SNs' inability to cope with harsh environmental and operating conditions.

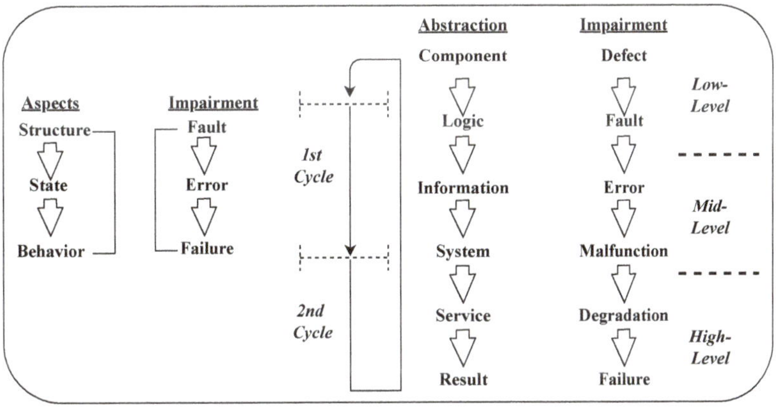

Figure 14. Fault–error–failure cycle [72].

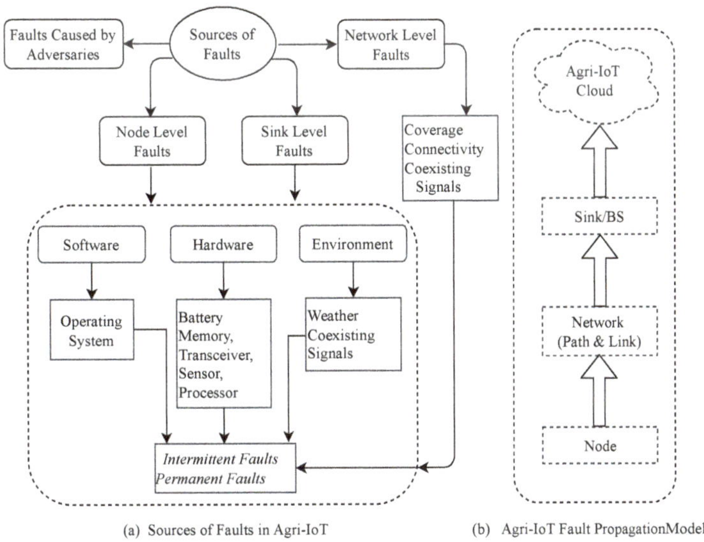

Figure 15. Faults in the WSN sublayer of Agri-IoT: sources and fault propagation model.

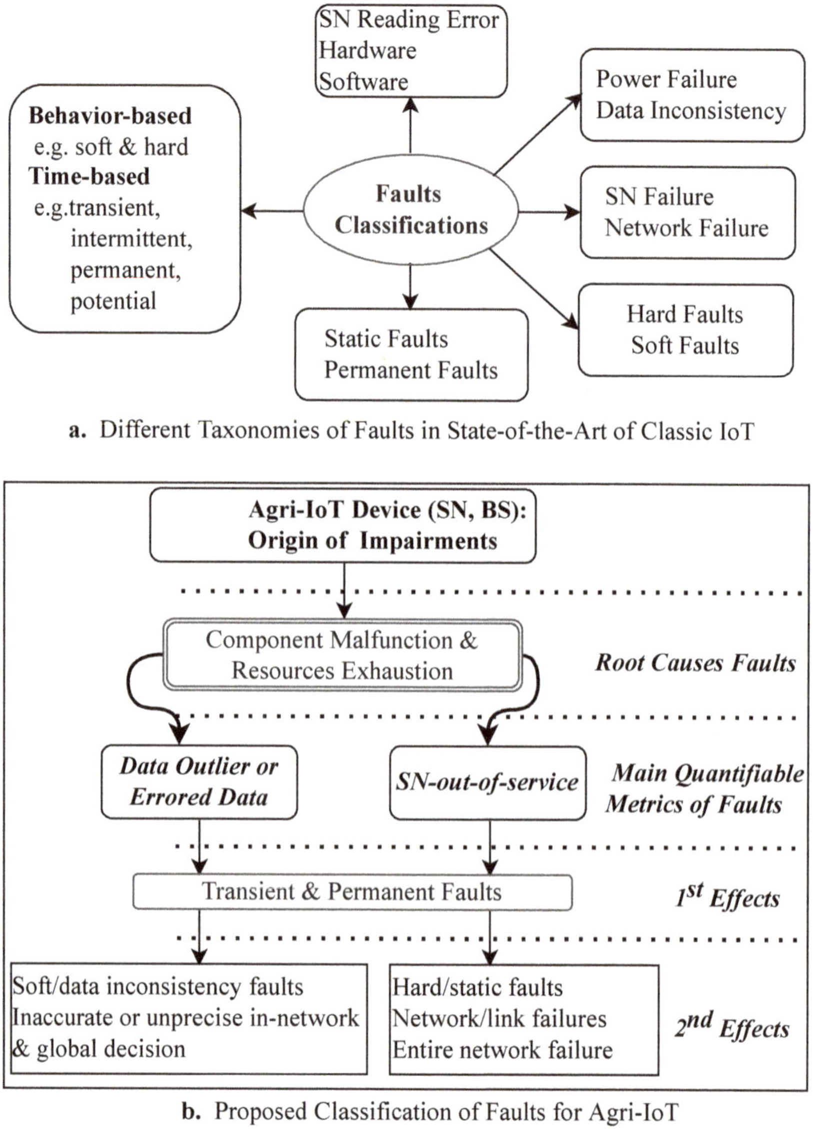

Figure 16. Classification of faults in the state of the art and proposed fault taxonomies for WSN-based Agri-IoT.

From the above definitions and the fault taxonomies on the left side of Figure 16, it can be deduced that hard, permanent, and static faults are practically manifested as SN-out-of-service, while soft, dynamic, and data-inconsistency faults can be observed as data outliers. Both SN-out-of-service and data outliers are consequences of unit malfunction or resource exhaustion and can be permanent or intermittent in behavior. Both conditions can impair the quality of event data and the global actionable decisions of the network. Therefore, the quality of FM schemes can be evaluated based on their capacities to effectively detect, tolerate, or avoid SN-out-of-service and data outlier faults. In summary, most FM schemes in the state of the art focus on their effects, instead of the root faults, which are the flaws in existing FM schemes [25]. Additionally, since the SN is the sole network device responsible for event sensing, data computation, packet forwarding, and communication in the WSN sublayer of Agri-IoT, it is the principal source of faults in Agri-IoT networks. A new fault classification framework shown in Figure 16 can be deduced from the above analysis.

Secondly, it is discernible that SNs' power mismanagement is the most prevalent origin of faults [43,80,81], which then propagate to the backend or application level (refer to the right side of Figure 15). For instance, communication, sensing, and computational accuracies of a node can be impaired when the battery energy falls below certain thresholds [43]. Also, network faults can be traced to power exhaustion and node failures, which create holes in the topology that divide the network into multiple disjointed segments [43]. On that account, faults can be avoided in WSN-based Agri-IoT if the energy-saving strategies presented in Figures 9 and 10 are effectively implemented.

Additionally, any FM scheme or fault-monitoring mechanism, be it proactive, reactive, passive, or active, must incorporate the following underlying qualities: thresholds that represent the probable fault conditions without false alarms, fault discovery, minimized message/time complexities, and self-healing and self-reconfiguration to neutralize the effects of the faults [43].

FM Framework and Architectures in WSN Sublayer of Agri-IoT

As illustrated in Figure 17, every FM scheme consists of three main steps, which include fault detection (FD), fault diagnosis, and fault recovery/tolerance (FT) [82,83], which always require input information. These steps are implemented in a decision-making framework that involves four major processes: data/information collection, FD model formulation, FD decision and fault classification, and tolerance of its effects using any of the FT mechanisms shown in Figure 17. Thus, the FD model detects the fault, the fault discovery technique distinguishes that fault from false alarms, while the FT mechanism helps to auto-heal and recover from the faults or failures [84]. Mainly, SN-out-of-service faults are detected and tolerated using self-reconfiguration techniques, whereas data outlier faults must strictly follow Figure 17.

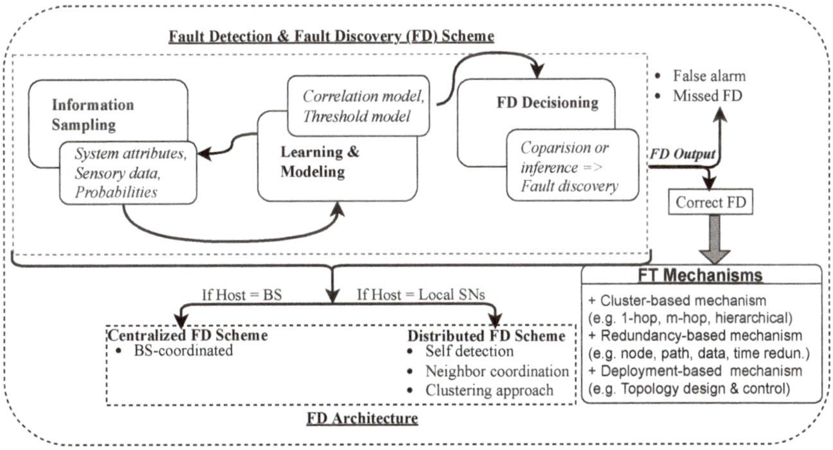

Figure 17. FM framework in WSN sublayer of Agri-IoT.

In addition, FM schemes can be implemented using either a centralized or distributed architecture [44,85,86]. In a centralized scheme, the FD/FT protocol is hosted and managed on the BS, whereas the distributed scheme hosts and manages this algorithm on the local SNs [87,88] (see Figure 17). The centralized approach is simpler for small-scaled networks but suffers many technical challenges, such as common point failure due to heavy message traffic at the BS and high SN energy waste. In contrast, the distributed approach saves power and controls message traffic on the BS because it allows local decision and self-

FD/FT with or without neighboring. According to Figure 17, the distributed architecture can be implemented in three major ways [43,89–91], which include self-detection, neighbor coordination, and the clustering approach. Since the basic design requirement of a WSN-based Agri-IoT is to maintain the healthy functionality and longevity of the SNs and the BS, any post-deployment impairments that cannot be self-fixed must be tolerated to not interfere with the core function of the network. Therefore, any automated FT mechanism that can be achieved through the self-reconfiguration and self-management for enhanced network availability, reliability, and dependability is encouraged in the WSN sublayer [92]. According to Figure 17, an efficient WSN-based Agri-IoT, therefore, requires a calculated mix of FT mechanisms based on the intended application.

6.2. *Systematic Survey of Fault Management Schemes in WSN-Based IoT*

FM in Agri-IoT networks has not received adequate conceptualized research considerations. As a result, existing Agri-IoT solutions inherit the FM propositions from the traditional WSN-based IoT networks, which have proven to be unsuitable [14]. This subsection presents a concise overview of these FM schemes, including their strengths, weaknesses, and underlying theories/concepts. It then proposes a more suitable remedy for WSN-based Agri-IoT technology. In canon centralized FM schemes (see references in [93–97]), the underlying FM algorithm is hosted and managed on the BS, while the local SNs host and manage the FM algorithm in distributed architectures [87,88]. Although the centralized approach is simpler for small-scale networks, it suffers many technical challenges, such as common point failure due to heavy message traffic at the BS, management difficulties, and high energy wastages on distant routing. This clearly explains why most outdoor Agri-IoT testbed experiments in [1,10,11,14,18,19] experienced severe FM complications to the extent that the networks became infeasible to operate or manage at higher scalability levels. However, the distributed approach (see references in [74,76,77,91,98–103]) saves power and controls message traffic and workload on the BS because it allows local decisions as well as local-FD/FT with or without neighboring nodes. The distributed FD/FT scheme can also be self-executed, neighbor-coordinated, or clustering-aided [89–91]. Although the clustering-based FM architecture has promising potential to improve energy conservation, network adaptability, and ease of implementation, it has not been extensively researched and exploited.

Again, distributed FD schemes are mainly established on the assumption that the failure of SNs is spatially uncorrelated, while event information is spatially correlated. Therefore, the FD's decision framework is frequently modeled using sensory data or statistical properties of the spatial or temporally correlated SNs [79,104–106] from the immediate neighborhood of a node [74,103] or data from farther SNs [107]. To date, the applicability of these solutions to the Agri-IoT context has attracted several technical challenges. Consequently, the strengths and weaknesses of the main results of the benchmarking FM schemes, their underlying assumptions, and how they addressed the critical fault-affinity factors such as energy conservation, FT/FA, control message complexity, and processor burden of the SNs, are presented in the comparative evaluation summary of Table 7.

Table 7. Comparative summary of FM schemes for WSN-based IoT networks.

Author/Year	Root Faults? (i.e., Data Outliers and SN-Out-of-Service)	FM Architecture	Unrealistic Assumptions	Energy Saving (FA)?	FT?	High Control Message Complexity	Stand-Alone?
[77] (2013)	Yes, both	Cluster-based	All SNs have the same lifetime; SNs record the same sensory data regardless of location	×	✓	High	×
[93] (2016)	Partial: data outliers	Centralized	SNs have binary sensing outputs	✓	×	Low	✓
[79] (2015)	Partial: data outliers	Distributed	All fault-free sensors measure the same physical value at any instant of time, while the faulty sensors measure different physical values	✓	×	Moderate	✓
[103] (2006)	Partial: SN-out-of-service	Distributed	All SNs must have enough neighbors	×	✓	High	✓
[74] (2009)	Partial: SN-out-of-service	Distributed	SNs must have unvarying detected initial status	×	✓	High	✓
[76] (2016)	Partial: SN-out-of-service	Distributed	SNs must have the same initial status and a predefined number of neighbors	×	✓	High	✓
[91,98,99] (2004, 2005, 2005)	Partial: SN-out-of-service	Distributed	All SNs have the same error detection probability, all neighboring nodes of an SN have identical levels of accuracy regardless of distance	×	×	High	✓
[100] (2009)	Partial: data outliers	Distributed	SNs have binary sensing outputs	×	✓	High	✓
[104] (2014)	Partial: data outliers & SN-out-of-service	Centralized	Silent on assumptions	×	No	Low	✓
[101] (2008)	Yes: transient faults	Distributed	All neighboring nodes have the same transmission range and reading values	×	×	High	✓
[105] (2016)	Partial: SN-out-of-service	Distributed	Silent on assumptions	×	×	Moderate	✓
[108] (2015)	Partial: data outliers	Distributed	Silent on assumptions	✓	✓	Low	✓
[109] (2009)	Partial: SN-out-of-service	Distributed	All nodes must have identical measurements, a quadrant must have the same number of SNs	×	✓	High	✓
[94] (2014)	Partial: SN-out-of-service	Centralized	Based on historical network data: assumes all SNs are healthy initially to obtain training data	×	×	Moderate	✓
[95] (2016)	Partial: SN-out-of-service	Centralized	Based on historical network data: assumes all SNs are healthy initially to obtain training data	×	×	Moderate	✓

Table 7. *Cont.*

Author/Year	Root Faults? (i.e., Data Outliers and SN-Out-of-Service)	FM Architecture	Unrealistic Assumptions	Energy Saving (FA)?	FT?	High Control Message Complexity	Stand-Alone?
[96] (2015)	Partial: SN-out-of-service	Centralized	Silent on assumptions	×	×	Moderate	✓
[110] (2018)	FT protocol	Distributed	Assumed centralized BS	✓	✓	High	✓
[111] (2017)	Effects: network failure	Distributed	All SNs are homogeneous in terms of energy, communication, and processing capabilities	×	×	High	✓
[97] (2018)	Partial: SN-out-of-service	Centralized	All faulty SNs must have at least a sleeping node in its proximity	×	✓	Moderate	✓
[112] (2018)	Partial: SN-out-of-service	Distributed	Silent on assumptions	✓	✓	High	✓
[113] (2016)	Partial: SN-out-of-service	Distributed	All faulty SNs must have at least a sleeping node in its proximity	×	✓	Moderate	✓
[114] (2013)	Partial: SN-out-of-service	Distributed	Silent on assumptions	×	×	Moderate	✓
[115] (2016)	Partial: data outliers	Distributed	Silent on assumptions	✓	✓	Moderate	✓

✓: Present or YES, ×: Absent or NO.

6.3. Theories/Concepts of Benchmarking FM Schemes and Their Shortcomings

The conceptual models/theories of the canon FD decision frameworks and the associated shortcomings can be expressed as follows:

- Statistical approaches such as Neyman–Pearson formulation [116], Bayesian statistics [77,103], and normal distribution test types (e.g., Thompson Tau statistical test [105]) are high-resource-demanding techniques that may apply to classic IoT. Still, they are unsuitable for power-constrained Agri-IoT devices or SNs. In addition to being stand-alone and without application specificity, these methods operate at high computational and control message complexities. Their operational efficiencies increase with increasing data dimensionality and also require a priori knowledge of data distribution, which is not possible in many real-life applications of Agri-IoT networks. Additionally, they rely on predefined thresholds to make local and global FD decisions. Therefore, regardless of the extensive research considerations of these methods, they are generally not suitable for low-power IoT applications, of which Agri-IoT is no exception.
- Graph-based FM techniques lack precise criteria for outlier detections [83,109], suffer higher computational complexities, and also make unrealistic assumptions about the data distribution. In addition, these approaches (e.g., De Bruijn graph theory [109] and depth-based techniques) are unsuitable for multidimensional and huge datasets.
- Machine learning decision concepts such as the k-out-of-n and majority decision rule concepts [93], naive Bayes, iterative algorithms [107], and neural network-based techniques, among others, are susceptible to high dimensional datasets, suffer high computational cost, and rely on sensitive model parameters.

In addition to the stipulated shortcomings, these benchmarking FM methods usually ignore the sensory data correlation (i.e., attribute correlation, spatial correlation, and temporal correlation) properties of SNs, require high communication overhead with high FD delays [83], and normally operate in an offline manner, which is inconsistent with the modus operandi of typical Agri-IoT. Hence, they are unsuitable for the recent low-power Agri-IoT applications.

6.4. Open Issues on Existing FM Solutions for Classic WSN-Based IoT Networks and Recommended Design Guidelines for Achieving Efficient FM in WSN-Based Agri-IoT

A fault in the WSN sublayer of Agri-IoT networks can be manifested as a data outlier and SN-out-of-service or node failure, both of which must be detected and resolved locally or globally using the spatially correlated event information and efficient threshold-based decision frameworks. Although there has been extensive research concerning FM schemes for the WSN sublayer, several technical challenges that require urgent contextual research considerations exist. They include the following:

1. Most faults in the PHY layer of Agri-IoT originate from the SNs' power exhaustion, which implies that the best fault-avoidance techniques are those that optimize power consumption. However, most FM schemes waste more energy and make the network prone to more faults/failures via high control messages and computational complexities.
2. Most FM schemes exist as stand-alone frameworks without architectural considerations and are founded on unrealistic assumptions, which make them difficult to incorporate into existing routing protocols.
3. The cluster-based routing architecture is endowed with many untapped local/global FM potentials and fault-avoidance capacities for the next-generation Agri-IoT. However, these promising potentials have received the least contextualized research considerations.

Existing FM solutions are meant for resource-sufficient and expensive classic WSN-based IoT, not resource-constrained, context-specific use cases like Agri-IoT networks. Regarding these technical challenges, this tutorial presents the following design guidelines for building efficient and realistic FM schemes for WSN-based Agri-IoT:

1. FM schemes must rely on realistic and contextual assumptions in order to detect and auto-tolerate sensory data outliers and SN-out-of-service faults in real-time routing protocols with minimal message, computational, and memory complexities. Such FM schemes will be suitable for all power-constrained WSN-based Agri-IoT applications.
2. Future works on FM schemes must be embedded into specific routing protocols so that their adaptability to topological dynamism and scalability in terms of network sizes and node densities can be assessed in an unsupervised manner. Therefore, fault detection and fault-tolerance schemes based on simple threshold-based theories are the best candidates for this context, since the threshold boundaries of agronomical metrics can be accurately computed from the historical data of the location.
3. FM schemes must incorporate redundancy check mechanisms by exploiting spatial and temporal correlations among sensory data.
4. FM schemes should maintain a good balance between local and global FDs as well as a reasonable detection rate and false alarm rate.

7. State of the Art on Real-World, Canon WSN-Based Agri-IoT Testbed Solutions

It is well documented that WSN-based Agri-IoT is the most reliable remedy for mitigating the negative impacts climate change has had on agricultural production, for which many architectural designs and testbed prototypes have been proposed [12,36]. In addition, since the autonomous, resource-constrained SNs in Agri-IoT are expected to operate without post-deployment maintenance checks, the issues of FM, power optimization, and self-organization during SN design and network deployment cannot be ignored in existing testbed solutions [12,117]. Essentially, the results from most research projects on Agri-IoT relied on simulation experiments [1,10,14], which have retained the gap between the philosophy of this technology and the comprehension of its real-world behavior for a more accurate performance assessment. This section presents a systematic performance assessment of the few real-world WSN-based Agri-IoT testbed solutions currently based on the classic WSN-based IoT principles. To understand how the benchmarking realization testbeds of Agri-IoT in [1,10,11,14,18,19] fared in real-world operational conditions, the results from their respective performances are systematically evaluated and summarized in Table 8. It was discovered that the current benchmarking testbed solutions in [1,10,11,14,18,19] are capital-intensive because they are reliant on fixed/location-restricted backbone infrastructures (see the middle of Figure 3), too complicated to deploy and manage by even expert users, based on unrealistic indoor conditions which do not commensurate real-world environmental conditions, and based on the high- power-demanding centralized or flooding architectures which further complicate network manageability when up-scaled. A concise and systematic survey of these benchmarking real-world Agri-IoT networks and their flaws in the state of the art is summarized in Table 8.

Additionally, it can be established from the comparative assessment of the benchmarking Agri-IoT testbeds in Table 8 [10,11,18,19] that the embedded communication technology, event routing architecture, and the SNs' power management are the core factors that made them capital-intensive and complicated to both experts and low-income farmers. Additionally, self-healing, reconfigurability, and adaptability mechanisms to faults were not deployed [1,14,17]; hence, faulty and turbulent conditions could not be tolerated. Furthermore, since the battery-powered SNs rely on expensive Wi-Fi and cellular communication technologies that are not freely accessible at all locations, the SNs exhausted their battery supply a few days after deployment. Similarly, those that relied on ZigBee/IEEE 802.15. 4 communication technologies with power-intensive 6LoWPAN or IPv6 protocols restricted the resulting network to drive on the problematic centralized or flooding architectures without any efficient FM techniques. As a result, these solutions used costly fixed IP infrastructural supports and the centralized routing architecture, making them practically impossible to manage as the networks scaled. This is why the SNs unstably exhausted their battery power and abruptly abridged network lifespans [1,10,11,14,18,19].

Table 8. Comparative analysis of WSN-based Agri-IoT testbed solutions.

Author/Deployment Type	Testbed Objective	Comm. Tech & Architecture	Weaknesses
[10] (Outdoor)	Disease control	IEEE 802.15.4/centralized, flooding	Relied on a fixed support system, expensive, power-inefficient, location-restricted
[11] (Outdoor)	Precision farming, to gather real-world experiences	ZigBee, Mica2 clones hardware and TinyOS software/centralized, flooding	Relied on a fixed support system, expensive, power-inefficient, location-restricted, no single measurement was achieved due to high network complexity
[18] (Indoor)	Data outlier detection and decision support system for precision irrigation testbeds	ZigBee/flooding-based	Results based on 3 SNs under unrealistic indoor conditions
[19] (Indoor)	Latency improvement	Fog computing, 6LoWPAN, 6LBR, and WiFi-based/centralized, flooding	Capital-intensive, energy-inefficient, high complexity, location-restricted
[1] (Indoor and Outdoor)	Gather real-world deployment experiences	ZigBee/centralized, flooding	Result focused on mere observation, not real-world deployment scenarios.

Therefore, the freely available low-power wireless technologies (e.g., LoRa, BLE, 5G, Z-wave, NB-IoT, and SigFox) that are founded on a suitable routing topology are the best candidates for making this ubiquitous application [1,16] cheap [1,20] and simple for all users. Thus, the cluster-based topology is more pivotal to addressing the above challenges of Agri-IoT [10,17] than the traditional cellular and WiFi technologies that are inaccessible in many farms, depending on their locations [10,20]. However, besides distance-power constraints, architectural support, and network manageability challenges, these freely accessible wireless communication technologies have specific limitations, which include:

1. ZigBee technology achieves the desired power savings only when deployed in star or centralized topology [14], and it operates at its low-power distance range (10–100 m) in line-of-sight mode depending on the environmental characteristics.
2. LoRa is limited to low-density and fixed network sizes (non-scalable), a low data rate, and a low message capacity [14]. It may require registration and expensive antennae, depending on its operation location.
3. SigFox supports a very low data rate and requires registration. LoRa and SigFox possess complex implementations because they both require specific modules to function and gateways.
4. WiFi, GPRS, cellular technologies, and NB-IoT are high power consumption standards and location-/architecture-restricted.
5. BLE has a short communication range but supports clustering architecture, which is the most optimal architecture for ensuring the best operational efficiency of WSN-based Agri-IoT deployments, since this architecture allows cluster isolation and management.

Therefore, a research opportunity exists for a flexible, ubiquitous, realistic, energy-efficient, self-healing, simple, low-cost, cluster-based, and wireless outdoor-based testbed that consists of infrastructure-less, task-scalable, and wirelessly configurable experimental SNs and a BS. It should also be deployed, re-deployed, monitored, controlled, and managed by non-experts to operate stably throughout the entire crop-growing season.

8. Case Study: Cluster-Based Agri-IoT (CA-IoT) for Precision Irrigation

As earlier established in Figure 2, the design and implementation of Agri-IoT networks are driven by unique critical factors, which are mainly determined by the associated routing architecture, communication technology, actuation management mechanisms, and environmental impacts. In the operation phase, these factors constitute the specific objectives in Figure 10, which the supervisory routing protocol must address in order to optimize performance efficiency and stability. As systematically established above, the LEACH-inherited cluster-based architecture has the most promising potential to address

these technical challenges. It helps to attain high power optimization via communication distance and packet minimization, efficient network administration/adaptability, high event data quality through auto-FM, and local data quality management, as indicated in Figure 10. So, this section presents a systematic analysis of how the merits of this architecture evolve in CA-IoT for precision irrigation use cases. Using the framework in Figure 12, the cluster-based architecture was pre-examined to uncover how the fundamental Agri-IoT design requirements and goals presented in the reference frameworks in Figures 2, 9 and 10 can unfold into realistic multi-parametric optimization metrics.

The conceptual architectural framework of the proposed network, as illustrated in Figure 18, can be implemented using Arduino-based or Raspberry Pi(RPi)-based micro-controllers, BLE and LoRa for intra-cluster, inter-cluster, and BS–cloud communications, respectively, and DHT22/STEMMA soil moisture sensors for measuring the respective ambient and soil microclimatic parameters. Also, a unit cluster from Figure 18 detailing the key network components of MNs, CH, BS, and the field-deployed precision irrigation system is shown in Figure 19. It is assumed that the core units constituting the MNs, CH, and BS, as illustrated in Figure 19, are optimally selected and designed after Figure 2. Using Figures 18 and 19 as the reference architectural frameworks for achieving our contextualized objectives, this section presents an in-depth systematic assessment and characterization of the scores of canon cluster-based routing protocols of conventional WSN-based IoT applications so that the desired MOO metrics can be appropriately deduced and adapted for the design of the associated routing for our case study.

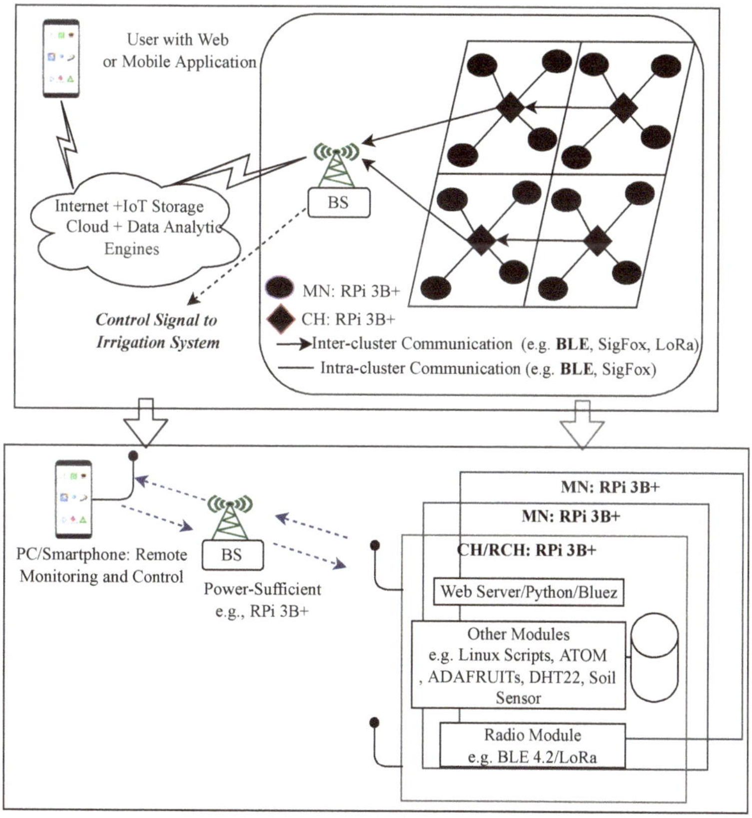

Figure 18. Conceptual architectural framework of the proposed CA-IoT for precision irrigation management.

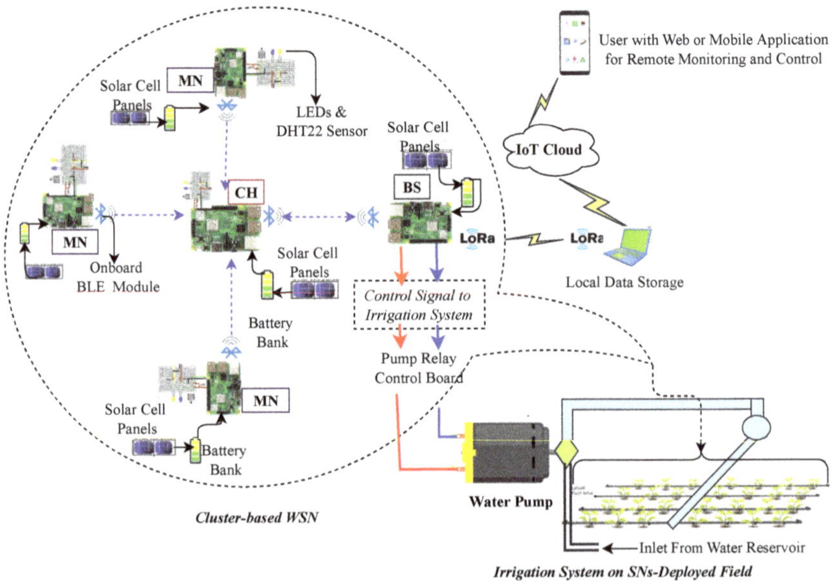

Figure 19. CA-IoT use case cluster illustrating the key network components: MNs, CH, and BS.

8.1. Characterization of Canon Clustering-Based Routing Protocols and Deduction of MOO Metrics

A systematic survey (refer to Table 9) and characterization of LEACH-based routing protocols were conducted using the clustering process, CH features, and cluster features, as indicated in Figure 20, in order to conceive the core MOO metrics for the proposed CA-IoT network framework. The clustering process, CH features, and cluster features define the performance optimalities and the quality of the sampled data of the resulting architecture.

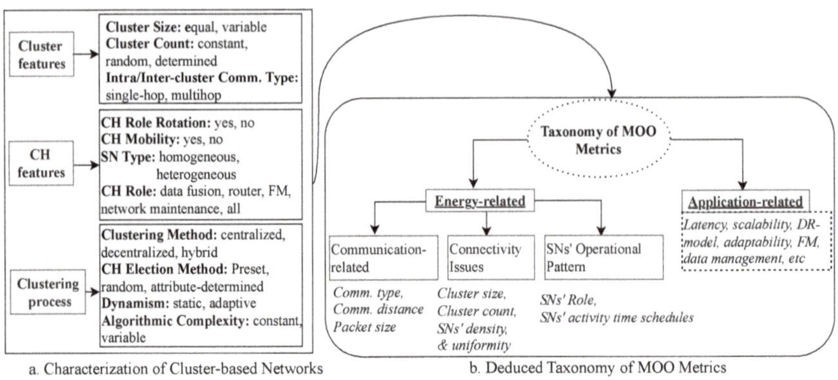

Figure 20. Characterization of cluster-based networks and deduced taxonomy of MOO metrics for optimizing Agri-IoT networks.

Table 9. Comparative summary of Agri-IoT-applicable clustering-based routing protocols using characterization parameters.

Protocol/Year	Hierarchy	DR-Model	Clustering Method	Comm. Type	Objective	CH Selection Method	Cluster Size	SN Mobility	SN Type	CH Role Rotation	Constant Time Complexity
LEACH, 2002 [21,47]	2-level	Time-driven	Decentralized	Intra: Single-hop Inter: Single-hop	Max. WSN lifespan	Random	uncontrolled	Static	Homogeneous	✓	×
SEP, 2004 [118]	2-level	Time-driven	Decentralized	Intra: Single-hop Inter: Single-hop	WSN stability pan	Random	uncontrolled	Static	Heterogeneous	✓	×
TL-LEACH, 2007 [119]	3-level	Time-driven	Decentralized	Intra: Single-hop Inter: Multihop	Data aggregation	Attribute-based	uncontrolled	Static	Homogeneous	✓	×
PECRP, 2009 [120]	multilevel	Time-driven	Hybrid	Intra: Single-hop Inter: Multihop	Max. WSN lifespan	Random Attribute-based	controlled	Static	Homogeneous	✓	×
LEACH-DT, 2012 [121]	3-level	Time-driven	Decentralized	Intra: Single-hop Inter: Single-hop	Max. WSN lifespan	Random Attribute-based	uncontrolled	Static	Homogeneous	✓	×
EESAA, 2012 [9]	2-level	Time-driven	Decentralized	Intra: Single-hop Inter: Single-hop	Max. WSN lifespan	Random Attribute-based	uncontrolled	Static	Homogeneous	✓	✓
DEEC, 2014 [122]	2-level	Time-driven	Decentralized	Intra: Single-hop Inter: Single-hop	WSN stability pan	Random	uncontrolled	Static	Heterogeneous	✓	×
DHCR, 2015 [123]	multilevel	Time-driven	Decentralized	Intra: Single-hop Inter: Multihop	Min. control messages Max. WSN lifespan	Random Attribute-based	controlled	Static	Homogeneous	✓	✓
HEER, 2016 [124]	multilevel	Time-driven	Decentralized	Intra: Multihop Inter: Single-hop	Max. WSN lifespan	Random Attribute-based	controlled	Static	Homogeneous	×	×
S-BEEM, 2017 [33]	2-level	Time-driven	Decentralized	Intra: Single-hop Inter: Multihop	Load balancing	Random	controlled	Mobile BS	Homogeneous	✓	✓
EAMR, 2018 [125]	multilevel	Time-driven	Decentralized	Intra: Single-hop Inter: Multihop	Min. control messages Max. WSN Lifespan	Random Attribute-based	controlled	Static	Homogeneous	✓	✓

✓: YES or Present, ×: NO or Absent.

As depicted in Figure 20a, the cluster features define the underlying connectivity issues, such as cluster quality indices (thus, cluster count, cluster size) and intra-cluster and inter-cluster communication types (thus, single-hop or multihop or both) [23,24]. From the network design viewpoint, the cluster quality depends on the optimality of the CH count and cluster sizes, which in turn rely on the core design parameters, such as the spatial density and uniformity of the deployed nodes, the specification of the wireless communication standard, the routing architecture, and the size of the network [47]. Since the deployment of SNs in a typical Agri-IoT can be controlled, the stipulated cluster quality properties can be optimized to resolve connectivity issues in Figure 20b. In a randomly deployed field, these cluster quality parameters can be optimized using a pairing-based SN duty-scheduling mechanisms [9,12].

Secondly, the CH features can be static, mobile, or role-rotated in both homogeneous or heterogeneous networks [9,12] based on the SNs' resource hierarchy. Additionally, the CHs can be assigned different tasks, such as data aggregation, FM, coordinating network reconfiguration or duty cycling, and network maintenance, depending on their resource capacities and network requirements. This case study is based on static SNs and the distributed network construction approach (see references in [9,12,33,126–132]), where the SNs locally manage the entire clustering process, and a CH is elected without the entire network's information.

As shown in Figure 20a, the clustering process can be characterized by the clustering method/network type (thus, centralized or distributed), the CH selection method, reclustering or network adaptability to topological or scalable conditions, and the complexities (thus, control message and computational complexities) of the entire network operation cycle. Unlike the static approach with fixed CH, the adaptive clustering approach selects CH based on the current network conditions and rotates this role. However, both approaches can incorporate self-reclustering techniques to self-heal SN-out-of-service faults. Data outlier faults can be best detected and corrected using threshold-based decision theory or spatial correlation methods with the least complexities. Due to the large-scale and high deployment densities of WSN-based Agri-IoT, the distributed clustering process is more suitable for enhancing local FM, scalability, network management, and power optimization than the centralized approaches [37,47].

Generally, the CA-IoT network can be optimized by formulating the deduced optimal decision metrics in Figure 20a into a MOO framework and multihop routing model in order to provide the guidelines for the design of the WSN sublayer of Agri-IoT. From the comparative evaluations of Figures 10 and 20a, a taxonomy of MOO metrics for designing an efficient WSN-based CA-IoT network is proposed in Figure 20b. To enhance the clarity of the state of the art on cluster-based protocols and justify the need for the proposed MOO metrics, a comparative summary based on the characterization parameters is presented in Table 9.

8.2. CH Election Techniques

A CH selection process is very critical to the resulting network's performance efficiency. In addition to centralized networks and the computationally expensive fuzzy-based clustering approaches [133,134], the efficiencies of all LEACH-inherited protocols are mainly dependent on their CH selection techniques [47,49]. Therefore, the correct estimation of the cluster quality metrics (i.e., CHs count and cluster size) is pivotal in attaining the objectives in Figure 10. With the aid of nodes' residual energy and location metrics, the optimal CH count and cluster size can be preset before network deployment. Currently, these metrics are randomly selected using a probabilistic approach in LEACH-inherited protocol [9,21] or derived using a deterministic or an attribute-based method [47,135]. However, the probabilistic clustering, such as the LEACH-inherited protocols, is expected to perform better in terms of network lifespan, minimal clustering overhead, improved connectivity, network/coverage stability, low latency, collision-free routing, load balancing, high network stability span, and algorithmic simplicity if the optimal CH count was

predefined correctly [136]. However, the CH count is randomly predefined in these protocols [9,21], which undermines the CH's stability and the resulting architecture's optimality. This challenge can be addressed via common CH selection metrics including Euclidean distance, intra-cluster/inter-cluster communication costs, energy-harvesting capacities, and probabilistic factors. To date, the related attempts in [49,126,137–139] only relied on the SNs' residual energy and location information to re-elect CHs after the initial CH count is defined, which cannot be ideal for WSN-based Agri-IoT.

For instance, an active SN in a particular round decides whether or not to become a CH by choosing a random number (r_n) ranging between 0 and 1 and comparing the number with a specified threshold Th. A node, therefore, becomes a CH for that round if $r_n < Th$, where Th is expressed as:

$$Th = \begin{cases} \frac{p_d}{1-p_d \times ((first-round) mod \frac{1}{p_d}}, & \text{if } n \in G \\ 0, & \text{otherwise} \end{cases} \quad (1)$$

where p_d is the desired percentage of CHs or CH count per round, and G is the number of SNs that have not been a CH in the previous $1/p$ rounds.

The authors in [119] proposed a three-layered LEACH (TL-LEACH) that operates in three functional phases—CH election, MN recruitment, and data transfer—to enhance the energy efficiency of LEACH. Their first-level CH election approach modified Equation (1) into an enhanced threshold $T(i)$, which is expressed as:

$$T(i) = \begin{cases} (r+1) \times mod(\frac{1}{p} \times p), & \text{if } i \in G \\ 0, & \text{otherwise} \end{cases} \quad (2)$$

where p is the CH count, r is the current round number, and G is the number of SNs that have not been a CH in the previous $1/p$ rounds. The second-level CHs are selected from the first-level CHs based on the shortest distance to the BS to function as aggregated packet forwarders or relay CHs (RCHs).

The authors in [120] introduced energy ($E(i)$) and distance ($D(i)$) attributes into Equation (1) to improve the load-balancing merit of LEACH. The resulting Th is expressed as:

$$Th = \begin{cases} \frac{p_d}{1-p_d[r \times mod \frac{1}{p_d}]} \times [E(i) + (1 - D(i))], & \text{if } n \in G \\ 0, & \text{otherwise} \end{cases} \quad (3)$$

Multihop routing via relay CHs (RCHs) was recommended for distant CHs in the future scope of [120].

In the LEACH presented with a distance-based threshold (LEACH-DT) algorithm in [121], the probability of becoming a CH depends on the relative distance between a node and the BS. This algorithm differs from the LEACH algorithm because the desired percentage of CHs (p_i) is predefined using Equation (5), while the threshold $T(I,r)$ is expressed as:

$$T(i,r) = \begin{cases} \frac{p_i}{1-p_i \times [r \times mod \frac{1}{p_i}]}, & \text{if } G_i(r) = 0 \\ 0, & \text{if } G_i(r) = 1 \end{cases} \quad (4)$$

Note that the terms retain their usual definitions, namely:

$$p(i) = k \frac{\xi_i}{\sum_{j=1}^{N} \xi_j}, 0 \leq p_i \leq 1, \quad (5)$$

where

$$\xi_i = 1/\overline{E_{CH}} \times d_i - \overline{E_{non-CH}}, \quad (6)$$

The variable d_i depicts the distance between node i and the BS, and E_{CH} and E_{non-CH} are the average residual energies in CHs and non-CHs, respectively. The authors further

established the need for a multihop routing approach in simulations and real-world WSNs to validate the countless theoretical propositions and benefits.

In the decentralized energy-efficient hierarchical cluster-based routing algorithm (DHCR) [123], SNs compete to become CHs. First, the BS broadcasts a trigger message at a specific range. The receiving nodes then compete to become a CH by disseminating a new message containing their residual energies and distances from the BS. Using this information, a neighboring node i within the target range receives the message and calculates its CHS_{nfun_i} as:

$$CHS_{nfun_i} = a \times \frac{Ere_i}{E_{max}} + b \times \frac{1}{Dis - To - BS_i}, \tag{7}$$

where Ere_i and E_{max} are the residual and initial energy levels of node i, respectively; $Dis - To - BS_i$ is the distance between node i and the BS, and a and b are real random values between 0 and 1 such that $a + b = 1$. The values of $Dis - To - BS_i$ of node i and its neighbors are compared, and the node with the highest $Dis - To - BS_i$ value is selected as the CH. A first-level CH broadcasts its residual energy, neighboring node count, and distance from the BS via a specific route. The next-level CHs receive the information and similarly repeat the procedure to ensure that every node determines a redistributor CH to the BS at the same time. A redistributor CH has more energy and fewer neighbors (neighboring degrees).

However, the Hamilton energy-efficient routing protocol (HEER) [124] creates an entire cluster of nodes, aggregates data, and transmits them to the BS via a Hamiltonian path that has been created by the entire cluster of nodes and controls the cluster size by selecting one node as the CH using the probability function p, which can be expressed as:

$$p = \frac{L_{message}}{F_{max}} \tag{8}$$

where $L_{message}$ is the size of every node, and F_{max} is the maximum size of a frame. The HEER protocol creates the clusters only once in the first round based on LEACH, and it role-rotates the CHs per the energy on the Hamiltonian path after a determined period.

Similarly, the two-phased EAMR protocol [125] randomly preselects the CH. A CH also selects its closest CH as its redistributor CH. The clusters are static over the entire network lifetime, and the CH role rotates randomly within the clusters according to a cluster replacement threshold. The new CH inherits the redistributor role if the old CH had one. Overall, since the node location, residual energy, and sleep schedule are indispensable in the CH selection process, the CH selection methods proposed by the authors in [9,12,36,120,140] are recommended WSN-based Agri-IoT applications.

8.3. Challenges of Existing MOO Frameworks and Recommended Future Works

As Figures 9 and 10 illustrate, the performance efficiency of an infrastructure-less WSN-based Agri-IoT mainly depends on the embedded MOO remedies in the associated supervisory routing protocol [12]. Several MOO frameworks have been researched since Agri-IoT networks are subjected to multiple design and operational constraints. A MOO framework is expected to formulate multiple objective functions from a set of MOO metrics to simultaneously optimize these multiple objectives, such as the maximal energy savings, highest connectivity, best latency, highest reliability, and balanced SN power depletion rates across the network. Although the MOO methods are the best candidates for Agri-IoT, the existing MOO solutions used in Agri-IoT are adopted from traditional WSN-based IoT without any contextual evaluation [12,16,26]. Consequently, they have not fulfilled their intended purposes due to several technical challenges, including the following:

1. They are limited to non-cluster-based network architectures, which implies that the promising potentials of the clustering architecture are not adequately exploited [9,12,50,51].
2. They are frequently implemented in the operational phase of the network, which makes it challenging to find global optimal solutions with a balanced tradeoff among

conflicting objective functions. The performance optimality of the Agri-IoT network starts from the SN design.

3. They rely on high-resource-demanding algorithms, such as mathematical programming-based scalarization methods, multi-objective genetic algorithms (MOGAs), heuristics/ metaheuristics-based optimization algorithms, and other advanced optimization techniques [23,26,48], making them unsuitable for the battery-powered SNs in Agri-IoT.
4. There are no contextual MOO guidelines based on Figure 20 to govern the PHY-layer design of Agri-IoT to achieve global optimal solutions with a balanced tradeoff among conflicting objectives. Consequently, there are conflicting scenarios in existing MOO solutions [50].

Therefore, there is an urgent demand for a realistic low-power MOO framework for CA-IoT networks that is founded on the core WSN design metrics and MOO taxonomy metrics in Figure 10 and the top of Figure 20, respectively. The following section assesses how evaluations and deductions evolve in a typical event sampling and routing protocol in a CA-IoT network for precision irrigation system management.

9. Design of WSN-Specific CA-IoT Routing Protocol

This section proposes a CA-IoT-based supervisory routing protocol that supports static SNs, rotatable/fixed CH roles, and enhanced SN resource management under the deterministic deployment approach. This can improve energy savings, connectivity, distance moderation, and multihop inter-cluster communication in the resulting network. The operational cycle and the embedded activities of our WSN-based CA-IoT protocol for precision irrigation application, as illustrated in Figure 21, include the following:

1. *Network Construction or Setup Phase:* This phase involves network modeling, CH election, and cluster formation, which is explained in Figure 21. The active–sleep duty-cycle scheduling ensures the SNs only switch to active mode during their scheduled sampling durations. In randomly deployed WSNs, redundant event reporting can be avoided using a correlated pairing-based active–sleep duty-cycle scheduling approach in [12]. The optimal CH count and cluster size must be predefined from the resource capacities of the SNs. After the initial CH election, the MNs are recruited and assigned their respective sampling and intra-cluster communication timeslots.
2. *Sampling, Data Management, and Transmission Phase:* The tasks executed in this phase include event sampling, intra-cluster and inter-cluster data transmissions, data outlier FM, and event data redundancy management. Since microclimatic soil parameters do not change swiftly [1,14], sampling can only be scheduled during the day at 3-hourly time intervals. In addition to power optimization, the clustering approach provides superb potential for both local and global FM using threshold-based FM theory and spatial correlation techniques. Based on the architecture in Figure 19 and the resource limitations of the SNs, it is recommended that the communication beyond the BS or gateway can utilize LoRa or Wi-Fi AirBox, whereas the intra-cluster and inter-cluster communications must be the freely available low-power BLE technology, since it is the most suitable for the clustering architecture.
3. *Network Maintenance and Reclustering Phase:* This phase resolves all unforeseen topological dynamics caused by the SNs' failures, network scalability, node mobility, and unexpected operational flaws, without interfering with the normal network functionality via adaptive reclustering, self-healing, and multihop routing techniques [12,23,24]. Here, a parent CH coordinates the election of child CHs (CCHs). While all non-CCHs switch to sleep mode, the CCHs recruit new MNs using location and residual energy parameters, assign them their respective sampling timeslots, and repeat Phase 2 afterward, as shown in Figure 21. SN-out-of-service faults are auto-detected and tolerated in this phase.

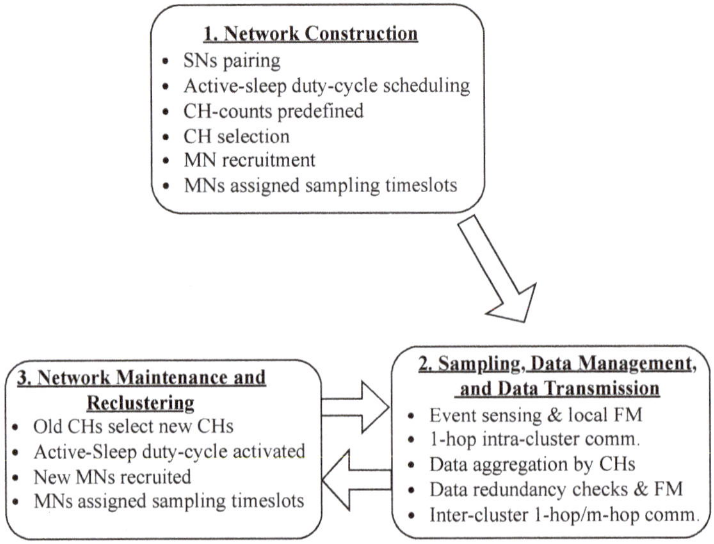

Figure 21. Proposed operation cycle for designing our CA-IoT network's routing protocol.

Additionally, Figure 21 uniquely incorporates correlated pairing-based duty cycling, constant control message complexity FM/data redundancy scheme, network construction/maintenance, and cluster quality measures that can ensure unprecedented energy savings and event data quality. This clustering approach can further minimize energy wastage via a suitable MAC method, a low-power wireless communication standard, data aggregation with data redundancy checks, and CH role rotation, among other factors. Although the various sections of the deduced MOO metrics have been implemented in our CA-IoT operational cycle, the most desired performance can be optimally attained when the MOO metrics are modeled into their respective objective functions, and their optimal values are determined and implemented in both simulation and testbed experiments in future works. Also, a realistic multihop routing framework can also be inculcated into this protocol for large-scale applications.

10. Open Issues and Future Works: Cluster-Based WSN-Specific Agri-IoT Networks

This tutorial has firmly established that the WSN-based Agri-IoT is an indispensable component of smart or precision farming and greenhouses, despite its resource- and deployment-induced challenges [12,26,141]. Unlike the conventional IoT, Agri-IoT is compelled to drive on batteries and affordable task-scalable SNs. However, it must meet the expectations in Figure 2 to guarantee a stable performance. The cluster-based routing technique has emerged with promising potential to mitigate these challenges. However, results from existing testbed solutions in this study show otherwise due to the absence of a contextualized in-depth overview of Agri-IoT as well as the following open issues which have not received extensive contextual research considerations in Agri-IoT applications:

1. The cluster-based routing architecture for WSN-based Agri-IoT has not received holistic and practical research considerations as far as FM, power optimization, and network adaptability are concerned. Therefore, there is a demand for multi-parametric optimization frameworks and guidelines for designing and implementing the WSN sublayer.
2. Concerning FM, most proposed schemes in the canon state of the art are standalone, have high control message and computational complexities (energy-inefficient), and are mostly incompatible with the clustering architecture [25,52]. The desired FM schemes for CA-IoT applications should be equipped with fault-avoidance mech-

anisms and the capacity to detect and self-heal root faults (SN-out-of-service and sensory data outliers [25]), not their effects.

3. Multihop routing, which is a requirement to attain the desired energy savings and network adaptability in large-scale CA-IoT networks, is asserted to be more energy-efficient only in simulation experiments [33,120,121,123–125,128–130] but not in real-world implementation [22–24]. This imbalance is due to a lack of a comprehensive and reliable theoretical multihop routing framework that is based on the total communication costs of multihop routing.
4. There is a demand for a more realistic and holistic MOO framework that can optimize the operational efficiency metrics such as cluster size, cluster counts, density/uniformity of nodes, communication distance, and activity schedule/duration, right from the network design phase to the operational phase of Agri-IoT networks.
5. Although the current literature supports adaptive clustering with CH role rotation ideology, there exists the need for an optimal initial CH-count estimator in order to improve the stability of CH elections and the architecture. Thus, the cluster quality indices (e.g., optimal cluster count and size) must be predetermined before defining them in the associated CH election method, since CH stability is compromised in most clustering methods [9,21,119–121,123,124].
6. Most protocols in the state of the art rely on perfect homogeneous networks, which is unrealistic due to variations in modular specifications and resource utilization and the fact that different SNs may have different communication and data computational roles. Therefore, a more realistic, contextualized, and adaptive clustering approach that leverages the gap between the philosophy and practice of Agri-IoT applications is needed.
7. In addition to the parent LEACH protocol [21,61] which is a complete suite application comprising routing, MAC, and physical characteristics for wireless communication in WSNs, most benchmarking MAC protocols purposed for traditional IoT applications are shelved, since they are developed in solitude without application specificity and network architectural considerations. A custom-built and holistic protocol suite for Agri-IoT remains a research opportunity.

11. Conclusion and Future Works

This tutorial presented: (1) a systematic overview of the fundamental concepts, technologies, and architectural standards of WSN-based Agri-IoT; (2) an evaluation of the technical design requirements of a robust, ubiquitous, self-healing, energy-efficient, adaptive, and affordable Agri-IoT; (3) a comprehensive survey of the benchmarking FM techniques, communication standards, routing protocols, MMAC protocols, and WSN-based testbed solutions; and (4) an in-depth case study on how to design a self-healing, energy-efficient, affordable, adaptive, stable, and cluster-based WSN-specific Agri-IoT from a proposed taxonomy of MOO metrics that can guarantee optimized network performance. Furthermore, this tutorial established new taxonomies of faults, architectural layers, and MOO metrics for CA-IoT networks. Using the open technical issues, it recommended application-specific requirements of Agri-IoT, general design expectations, and remedial measures, and it evaluated them in CA-IoT for precision irrigation in order to optimally exploit the proposed MOO metrics in a typical CA-IoT design in both simulation and real-world deployment scenarios. Overall, this tutorial can serve as a new reference document for the IoT community and Agri-IoT designers, since it adequately examined all critical aspects of WSN-based Agri-IoT networks from theoretical modeling to real-world implementation.

Author Contributions: The First Author contributed 60% while the second the third Authors contributed 20% each. Conceptualization, E.E.; methodology, A.M.W.; writing, review and editing, E.E., O.T. and A.M.W.; supervision, O.T. and A.M.W.; project administration, E.E.; and funding acquisition, E.E., and O.T. All authors have read and agreed to the published version of the manuscript.

Funding: The work was carried out with the financial support of icipe- World Bank Financing Agreement No D347-3A and the World Bank Korea Trust Fund Agreement No TF0A8639 for the

PASET Regional Scholarship and Innovation Fund. The views expressed herein do not necessarily reflect the official opinions of the donors.

Data Availability Statement: This research has no such data.

Conflicts of Interest: The authors declare no conflict of interest.

Abbreviations

SN	Sensor Node
WSN	Wireless Sensor Network
IoT	Internet of Things
Agri-IoT	Agricultural Internet-of-Things
CA-IoT	Cluster-based Agricultural Internet of Things
FD/FT	Fault Detection and Fault Tolerance
FA	Fault Avoidance
FM	Fault Management
MOO	Multi-Objective Optimization
BS	Base Station
MMAC	Multichannel Medium Access Control
MAC	Medium Access Control
BLE	Bluetooth Low-Energy
CH	Cluster Head
RCH	Relay Cluster Head
MN	Member Node
AODV	Ad hoc On-demand Distance Vector
RPL	Routing over Low-Power and Lossy Networks protocol
CAM	Channel Access Management
DCO	Duty-Cycle Optimization

References

1. Kumar, P.; Reddy, S.R.N. Lessons Learned From the Deployment of Test-Bed for Precision Agriculture. In Proceedings of the International Conference on Sustainable Computing in Science, Technology & Management (SUSCOM-2019), Jaipur, India, 26–28 February 2019; pp. 25686–25697. [CrossRef]
2. Abbasi, M.; Yaghmaee, M.H.; Rahnama, F. Internet of Things in agriculture: A survey. In Proceedings of the 2019 3rd International Conference on Internet of Things and Applications (IoT), Isfahan, Iran, 17–18 April 2019; pp. 1–12.
3. Gennari, P.; Moncayo, J.R. *World Food and Agriculture Statistical Pocketbook*; Food and Agriculture Organization of the United Nations: Rome, Italy, 2018; Volume 1, pp. 1–248.
4. Shiferaw, B.; Tesfaye, K.; Kassie, M.; Abate, T.; Prasanna, B.M.; Menkir, A. Managing vulnerability to drought and enhancing livelihood resilience in Sub-Saharan Africa: Technological, institutional and policy options. *Weather. Clim. Extrem.* **2014**, *3*, 67–79. [CrossRef]
5. Devi, K.H.; Gupta, M.V. IoT Application, A Survey. *Int. J. Eng. Technol.* **2018**, *7*, 891–896. [CrossRef]
6. Stoces, M.; Vaněk, J.; Masner, J.; Pavlík, J. Internet of Things (IoT) in Agriculture—Selected Aspects. *AGRIS On-Line Pap. Econ. Inform.* **2016**, *8*, 83–88. [CrossRef]
7. Lova, R.; Vijayaraghavan, V. IoT Technologies in Agricultural Environment: A Survey. *Wireless Pers. Commun.* **2020**, *113*, 2415–2446. [CrossRef]
8. Farooq, M.S.; Riaz, S.; Abid, A.; Abid, K.; Naeem, M.A. A Survey on the Role of IoT in Agriculture for the Implementation of Smart Farming. *IEEE Access* **2019**, *7*, 56237–156271. [CrossRef]
9. Tauseef, S.; Nadeem, J.; Talha, Q. Energy Efficient Sleep Awake Aware (EESAA) intelligent Sensor Network routing protocol. In Proceedings of the 15th International Multitopic Conference (INMIC), Islamabad, Pakistan, 13–15 December 2012; pp. 317–322. [CrossRef]
10. Hartung, R.; Kulau, U.; Gernert, B.; Rottmann, S.; Wolf, L. On the Experiences with Testbeds and Applications in Precision Farming. In Proceedings of the 15th ACM Conference on Embedded Network Sensor Systems, Delft, The Netherlands, 5 November 2017; pp. 54–61.
11. Langendoen, K.; Baggio, A.; Visser, O. Murphy loves potatoes: Experiences from a pilot sensor network deployment in precision agriculture. In Proceedings of the 20th IEEE International Parallel & Distributed Processing Symposium, Rhodes, Greece, 25–29 April 2006; Volume 51, pp. 8–13. [CrossRef]
12. Effah, E.; Thiare, O. Realistic Cluster-Based Energy-Efficient and Fault-Tolerant (RCEEFT) Routing Protocol for Wireless Sensor Networks (WSNs). In *Advances in Information and Communication*; Springer: Cham, Switzerland, 2020; pp. 320–337.

13. Nasser, N.; Karim, L.; Ali, A.; Anan, M.; Khelifi, N. Routing in the Internet of Things. In Proceedings of the GLOBECOM 2017—2017 IEEE Global Communications Conference, Singapore, 4–8 December 2017; pp. 1–6.
14. Jawad, H.M.; Nordin, R.; Gharghan, S.K.; Jawad, A.M.; Ismail, M. Energy-Efficient Wireless Sensor Networks for Precision Agriculture: A Review. *Sensors* **2017**, *8*, 1781. [CrossRef]
15. Clausen, T.; Herberg, U.; Philipp, M. A critical evaluation of the IPv6 Routing Protocol for Low Power and Lossy Networks (RPL). In Proceedings of the 2011 IEEE 7th International Conference on Wireless and Mobile Computing, Networking and Communications (WiMob), Shanghai, China, 10–12 October 2011; pp. 365–372.
16. Effah, E.; Thiare, O.; Wyglinski, A.M. Multi-Objective Modeling of Clustering-Based Agricultural Internet of Things. In Proceedings of the 2020 IEEE 92nd Vehicular Technology Conference (VTC2020-Fall), Victoria, BC, Canada, 18 November–16 December 2020. [CrossRef]
17. Effah, E.; Thiare, O.; Wyglinski, A.M. Energy-Efficient Multihop Routing Framework for Cluster-Based Agricultural Internet of Things (CA-IoT). In Proceedings of the 2020 IEEE 92nd Vehicular Technology Conference (VTC2020-Fall), Victoria, BC, Canada, 18 November–16 December 2020. [CrossRef]
18. Khan, R.; Ali, I.; Zakarya, M.; Ahmad, M.; Imran, M.; Shoaib, M. Technology-Assisted Decision Support System for Efficient Water Utilization: A Real-Time Testbed for Irrigation Using Wireless Sensor Networks. *IEEE Access* **2018**, *6*, 25686–25697. [CrossRef]
19. Ahmed, N.; De, D.; Hussain, I. Internet of Things (IoT) for Smart Precision Agriculture and Farming in Rural Areas. *IEEE Internet Things J.* **2018**, *5*, 4890–4899. [CrossRef]
20. Effah, E.; Dorgloh, W. GSM-Controlled Irrigation System (GSMCIS) for Vegetable Farmers in Ghana. *Ghana J. Technol.* **2016**, *1*, 21–24.
21. Mehmood, A.; Mauri, J.L.; Noman, M.; Song, H. Improvement of the Wireless Sensor Network Lifetime Using LEACH with Vice-Cluster Head. *Ad Hoc Sens. Wirel. Netw.* **2015**, *28*, 1–17.
22. Haenggi, M.; Puccinelli, D. Routing in Ad Hoc Networks: A Case for Long Hops. *IEEE Commun. Mag.* **2005**, *43*, 93–101.
23. Pešović, U.M.; Mohorko, J.J.; Benkič, K.; Čučej, Ž.F. Single-hop vs. multi-hop—Energy efficiency analysis in wireless sensor networks. In Proceedings of the 18th Telekomunikacioni forum TELFOR 2010, Belgrade, Serbia, 23–25 November 2010; pp. 471–474.
24. Haenggi, M. Twelve Reasons not to Route over Many Short Hops. In Proceedings of the IEEE 60th Vehicular Technology Conference, Los Angeles, CA, USA, 26–29 September 2004; pp. 1–4.
25. Effah, E.; Tiare, O. Survey: Faults, Fault Detection and Fault Tolerance Techniques in Wireless Sensor Networks. *Int. J. Comput. Sci. Inf. Secur.* **2018**, *16*, 1–14.
26. Ferentinos, K.; Tsiligiridis, T. Adaptive design optimization of wireless sensor networks using genetic algorithms. *Comput. Netw.* **2007**, *51*, 1031–1051.
27. World Bank. Access to electricity (% of population)—Sub-Saharan Africa. In *The World Bank and UN Data on SSA*; The World Bank: Washington, DC, USA, 2021; pp. 1–3.
28. Elleuchi, M.; Boujelben, M.; Saleh, M.S.B.; Obeid, A.M.; Abid, M. Tree based routing protocol in WSNs: A comparative performance study of the routing protocols DEEC and RPL. *Future Technol. Publ.* **2016**, *5*, 7–16.
29. Al-Fuqaha, A.; Guizani, M.; Mohammadi, M.; Aledhari, M.; Ayyash, M. Internet of Things: A Survey on Enabling Technologies, Protocols and Applications. *IEEE Commun. Surv. Tutor.* **2015**, *17*, 2347–2376. [CrossRef]
30. Loukatos, D.; Manolopoulos, I.; Arvaniti, E.-S.; Arvanitis, K.G.; Sigrimis, N.A. Experimental Testbed for Monitoring the Energy Requirements of LPWAN Equipped Sensor Nodes. *IFAC-PapersOnLine* **2018**, *51*, 309–313. [CrossRef]
31. Akyildiz, I.F.; Su, W.; Sankarasubramaniam, Y.; Cayirci, E. A survey on sensor networks. *IEEE Commun. Mag.* **2002**, *40*, 102–114. [CrossRef]
32. Jovanovic, M.D.; Djordjevic, G.L.; Nikolic, G.S.; Petrovic, B.D. Multichannel Media Access Control for Wireless Sensor Networks: A survey. In Proceedings of the 2011 10th International Conference on Telecommunication in Modern Satellite Cable and Broadcasting Services (TELSIKS), Nis, Serbia, 5–8 October 2011; pp. 741–744. [CrossRef]
33. Xu, L.; O'Hare, G.; Collier, R. A Smart and Balanced Energy-Efficient Multihop Clustering Algorithm (Smart-BEEM) for MIMO IoT Systems in Future Networks. *Sensors* **2017**, *17*, 1574.
34. Fei, Z.; Li, B.; Yang, S.; Xing, C.; Chen, H.; Hanzo, L. A Survey of Multi-Objective Optimization in Wireless Sensor Networks: Metrics, Algorithms, and Open Problems. *IEEE Commun. Surv. Tutor.* **2017**, *19*, 550–586.
35. McBratney, A.O. Future Directions of Precision Agriculture. *Precis. Agric.* **2005**, *6*, 7–23. [CrossRef]
36. Effah, E.; Thiare, O. Estimation of Optimal Number of Clusters: A New Approach to Minimizing Intra-Cluster Communication Cost in WSNs. *Int. J. Innov. Technol. Explor. Eng.* **2018**, *8*, 521–524.
37. Asim Zeb, A.K.M.; Islam, M.; Zareei, M.; Mamoon, I.A.; Mansoor, N.; Baharun, S.; Katayama, Y.; Komaki, S. Clustering Analysis in Wireless Sensor Networks: The Ambit of Performance Metrics and Schemes Taxonomy. *Int. J. Distrib. Sens. Netw.* **2016**, *12*, 4979142.
38. Younis, O.; Fahmy, S. HEED: A hybrid, energy-efficient, distributed clustering approach for ad hoc sensor networks. *IEEE Trans. Mob. Comput.* **2004**, *3*, 660–669.
39. Rajaram, M.L.; Kougianos, E.; Mohanty, S.P.; Choppali, U. Wireless Sensor Network Simulation Frameworks: A Tutorial Review: MATLAB/Simulink bests the rest. *IEEE Consum. Electron. Mag.* **2016**, *5*, 63–69. [CrossRef]

40. Gurpreet, K.; Sukhpreet, K. Enhanced M-Gear Protocol for Lifetime Enhancement in Wireless Clustering System. *Int. J. Comput. Appl.* **2016**, *147*, 30–34.
41. Yen, H. Optimization-based channel constrained data aggregation routing algorithms in multi-radio wireless sensor networks. *Sensors* **2009**, *9*, 4766–4788. [PubMed]
42. Le, T.T.T.; Moh, S. Link Scheduling Algorithm with Interference Prediction for Multiple Mobile WBANs. *Sensors* **2017**, *17*, 2231. [CrossRef]
43. Darwish, I.M.; Elqafas, S.M. Enhanced Algorithms for Fault Nodes Recovery in Wireless Sensors Network. *Int. J. Sens. Netw. Data Commun.* **2016**, *6*, 150.
44. Manisha, M.; Deepak, N. Fault Detection in Wireless Sensor Networks. *IPASJ Int. J. Comput. Sci.* **2015**, *3*, 6–10.
45. Banerjee, I.; Chanak, P.; Rahaman, H.; Samanta, T. Effective fault detection and routing scheme for wireless sensor networks. *Comput. Electr. Eng.* **2014**, *40*, 291–306. [CrossRef]
46. Sharma, P.; Kaur, I. A Comparative Study on Energy Efficient Routing Protocols in Wireless Sensor Networks. *Int. J. Comput. Sci. Issues* **2015**, *8*, 98–106.
47. Faniana, F.; Rafsanjanibc, M. Cluster-based routing protocols in wireless sensor networks: A survey based on methodology. *J. Netw. Comput. Appl.* **2019**, *142*, 111–142.
48. Iqbal, M.; Naeem, M.; Anpalagan, A.; Ahmed, A.; Azam, M. Wireless Sensor Network Optimization: Multi-Objective Paradigm. *Sensors* **2015**, *15*, 17572–17620. [PubMed]
49. Mamalis, B.; Gavalas, D.; Konstantopoulos, C.; Pantziou, G. Clustering in Wireless Sensor Networks. In *RFID and Sensor Networks*; CRC Press: Boca Raton, FL, USA, 2009; pp. 324–364.
50. Kalkha, H.; Satori, H.; Satori, K. Performance Evaluation of AODV and LEACH Routing Protocol. *Adv. Inf. Technol. Theory Appl.* **2016**, *1*, 113–118.
51. Dwivedi, A.K.; Kushwaha, S.; Vyas, O.P. Performance of Routing Protocols for Mobile Adhoc and Wireless Sensor Networks: A Comparative Study. *Int. J. Recent Trends Eng.* **2009**, *2*, 101–105.
52. Fjellin, J.E. Medium Access Control (MAC) in WSN. Unpublished Lecture Notes, 12 October 2018; pp. 1–27. Available online: https://www.uio.no/studier/emner/matnat/ifi/nedlagte-emner/INF5910CPS/h11/undervisningsmateriale/20111101_mac_in_wsn.pdf (accessed on 16 July 2023).
53. Ye, W.; Heidemann, J.; Estrin, D. Medium access control with coordinated adaptive sleeping for wireless sensor networks. *IEEE/ACM Trans. Netw.* **2004**, *12*, 493–506. [CrossRef]
54. Kabara, J.; Calle, M. MAC Protocols Used by Wireless Sensor Networks and a General Method of Performance Evaluation. *Int. J. Distrib. Sens. Netw.* **2012**, *8*, 834784. [CrossRef]
55. Buettner, M.; Yee, G.V.; Anderson, E.; Han, R. X-MAC: A Short Preamble MAC Protocol for Duty-Cycled Wireless Sensor Networks. 2006. pp. 307–320. Available online: http://portal.acm.org/citation.cfm?id=1182807.1182838 (accessed on 16 July 2023).
56. Kuntz, R.; Gallais, A.; Noel, T. Auto-adaptive MAC for energy efficient burst transmissions in wireless sensor networks. In Proceedings of the 2011 IEEE Wireless Communications and Networking Conference, Cancun, Mexico, 28–31 March 2011; pp. 233–238.
57. Polastre, J.; Hill, J.; Culler, D. Versatile low power media access for wireless sensor networks. In Proceedings of the Second International Conference on Embedded Networked Sensor Systems(SenSys'04), Baltimore, MD, USA, 3–5 November 2004; pp. 95–107.
58. Ergen, S.C.; Varaiya, P. PEDAMACS: Power efficient and delay aware medium access protocol for sensor networks. *IEEE Trans. Mob. Comput.* **2006**, *5*, 920–930. [CrossRef]
59. Tang, L.; Sun, Y.; Gurewitz, O.; Johnson, D.B. PWMAC: An energy-efficient predictive-wakeup MAC protocol for wireless sensor networks. In Proceedings of the 2011 IEEE INFOCOM, Shanghai, China, 10–15 April 2011; pp. 1305–1313.
60. Gautam, G.C.; Chand, N. A Novel Cluster Based Time Synchronization Technique for Wireless Sensor Networks. *Wirel. Sens. Netw.* **2017**, *9*, 145–165. [CrossRef]
61. Heinzelman, W.B.; Chandrakasan, A.P.; Balakrishnan, H. An application-specific protocol architecture for wireless microsensor networks. *IEEE Trans. Wirel. Commun.* **2002**, *1*, 660–670. [CrossRef]
62. Ben-Othman, J.; Mokdad, L.; Yahya, B. An energy efficient priority-based QoS MAC protocol for wireless sensor networks. In Proceedings of the 2011 IEEE International Conference on Communications (ICC), Kyoto, Japan, 5–9 June 2011; pp. 1–6.
63. Kumar, A. WiseMAC Protocol for Wireless Sensor Network-An Energy-Efficient Protocol. Master's Thesis, National Institute of Technology, Rourkela, India, 2014; pp. 1–63.
64. Karki, V.S.; Udupi, G.R.; Gadgil, A. Advanced WiseMAC Protocol for Wireless Sensor Network. *Int. Res. J. Eng. Technol.* **2015**, *2*, 771–778.
65. Pak, W. Ultra-low-power media access control protocol based on clock drift characteristics in wireless sensor networks. *Int. J. Distrib. Sens. Netw.* **2017**, *13*, 1550147717722155. [CrossRef]
66. Tang, L.; Sun, Y.; Gurewitz, O.; Johnson, D.B. EM-MAC: A dynamic multichannel energy-efficient MAC protocol for wireless sensor networks. In Proceedings of the Twelfth ACM International Symposium on Mobile Ad Hoc Networking and Computing, Paris, France, 17–19 May 2011; p. 23.

67. Lim, J.B.; Jang, B.; Sichitiu, M.L. MCAS-MAC: A Multichannel asynchronous scheduled MAC protocol for Wireless Sensor Networks. *Comput. Commun.* **2014**, *56*, 98–107.
68. Irandegani, M.; Bagherizadeh, M. Designing an asynchronous multi-channel media access control protocol based on service quality for wireless sensor networks. *Int. J. Adv. Comput. Res.* **2017**, *7*, 190–199. [CrossRef]
69. van Hoesel, L.F.W.; Havinga, P.J.M. A Lightweight Medium Access Protocol (LMAC) for Wireless Sensor Networks: Reducing Preamble Transmissions and Transceiver State Switches. In Proceedings of the 1st International Workshop on Networked Sensing Systems, Tokyo, Japan, 1–6 January 2004.
70. Incel, O.D. Multi-Channel Wireless Sensor Networks: Protocols, Design And Evaluation. Ph.D. Dissertation, University of Twente, Enschede, The Netherlands, 2009; pp. 1–162.
71. Zhang, Z.; Mehmood, A.; Shu, L.; Huo, Z.; Zhang, Y.; Mukherjee, M. A Survey on Fault Diagnosis in Wireless Sensor Networks. *IEEE Access* **2018**, *6*, 11349–11364. [CrossRef]
72. Parhami, B. *Fault-Tolerant Computing*; Lecture Notes; Electrical and Computer Engineering Department, University of California: Santa Barbara, CA, USA, 2018; pp. 1–2.
73. Raghunath, K.M.K.; Rengarajan, N. Investigation of Faults, Errors and Failures in Wireless Sensor Network: A Systematical Survey. *Int. J. Adv. Comput. Res.* **2013**, *3*, 2249–7277.
74. Jiang, P. A New Method for Node Fault Detection in Wireless Sensor Networks. *Sensors* **2009**, *9*, 1282–1294. [CrossRef]
75. Koushanfar, K.; Potkonjak, M.; Sangiovanni-Vincentell, A. Fault tolerance techniques for wireless ad hoc sensor networks. *Proc. IEEE Sens.* **2002**, *2*, 1491–1496.
76. Oyiza, O.S. Implementation of New Fault Tolerance Solution in Wireless Sensor Networks in A Multi-Channel Context. Master's Thesis, Department of Computer Science, African University of Science and Technology, Galadima, Nigeria, 2016; pp. 1–36.
77. Bhattacharya, R.; Chhanda, R. Wireless sensor networks—A study of fault detection and recovery based on OSI layers. *Int. J. Conceptions Comput. Inf. Technol.* **2013**, *1*, 7–14.
78. Yu, M.; Mokhtar, H.; Merabti, M. Self-Managed Fault Management in Wireless Sensor Networks. In Proceedings of the Second International Conference on Mobile Ubiquitous Computing, Systems, Services and Technologies (UBICOMM'08), Valencia, Spain, 29 September–4 October 2008; pp. 13–18.
79. Panda, M.; Khilar, P.M. Distributed Byzantine Fault detection technique in wireless sensor networks based on hypothesis testing. *Comput. Electr. Eng.* **2015**, *48*, 270–285. [CrossRef]
80. Paradis, L.; Han, Q. A Survey of Fault Management in Wireless Sensor Networks. *J. Netw. Syst. Manag.* **2007**, *15*, 171–190. [CrossRef]
81. Ding, M.; Chen, D.; Xing, K.; Cheng, X. Localized fault-tolerant event boundary detection in sensor networks. In Proceedings of the 24th Annual Joint Conference of the IEEE Computer and Communications Societies, Miami, FL, USA, 13–17 March 2005; Volume 2, pp. 902–913.
82. Lee, W.L.; Datta, A.; Cardell-Oliver, R. Network Management in Wireless Sensor Networks. In *Handbook of Mobile Ad Hoc and Pervasive Communication*; American Scientific Publishers: Valencia, CA, USA, 2006; pp. 1–201.
83. Zhang, Y.; Dragoni, N.; Wang, J. A framework and classification for fault detection approaches in Wireless Sensor Networks with an energy efficiency perspective. *Int. J. Distrib. Sens. Netw.* **2015**, *2*, 678029.
84. Asim, M.; Mokhtar, H.; Merabti, M. self-managing fault management mechanism for wireless sensor network. *Int. J. Wirel. Mob. Netw.* **2010**, *2*, 184–197. [CrossRef]
85. Heena, H.; Kapoor, S. Survey of Fault Detection Algorithm in WSN. *SSRG Int. J. Comput. Sci. Eng.* **2015**, *5*, 78–81.
86. Kaur, E.J.; Kaur, E.P. A Survey on Fault Detection and Recovery Techniques in Wireless Sensor Networks. *Int. J. Eng. Res. Gen. Sci.* **2015**, *3*, 638–642.
87. Zhang, Z.; Chong, E.K.P.; Pezeshki, A.; Moran, W.; Howard, S.D. Detection performance in balanced binary relay trees with node and link failures. *IEEE Trans. Signal Process.* **May 2013**, *61*, 2165–2177. [CrossRef]
88. Ho, J.; Tay, W.P.; Quek, T.Q.S.; Chong, E.K.P. Robust decentralized detection and social learning in tandem networks. *IEEE Trans. Signal Process.* **2015**, *63*, 5019–5032. [CrossRef]
89. Nardelli, P.H.J.; de Lima, C.H.M.; Alves, H.; Cardieri, P.; Latva-aho, M. Throughput analysis of cognitive wireless networks with Poisson distributed nodes based on location information. *Ad Hoc Netw.* **2015**, *33*, 1–18. [CrossRef]
90. Umebayashi, K.; Lehtomaki, J.J.; Yazawa, T.; Suzuki, Y. Efficient Decision fusion for cooperative spectrum sensing based on OR-rule. *IEEE Trans. Wireless Commun.* **2012**, *11*, 2585–2595. [CrossRef]
91. Luo, X.; Dong, M.; Huang, Y. On distributed fault-tolerant detection in wireless sensor networks. *IEEE Trans. Comput.* **2006**, *55*, 58–70. [CrossRef]
92. Kakamanshadi, G.; Gupta, S.; Singh, S. A survey on fault tolerance techniques in Wireless Sensor Networks. In Proceedings of the 2015 International Conference on Green Computing and Internet of Things (ICGCIoT), Greater Noida, India, 8–10 October 2015; pp. 168–173. [CrossRef]
93. Pedro, H.; Nardelli, J.; Ramezanipour, I.; Alves, H.; de Lima, H.M.C.; Latva-aho, M. Average Error Probability in Wireless Sensor Networks With Imperfect Sensing and Communication for Different Decision Rules. *arXiv* **2016**, arXiv:1508.02253v2.
94. Lau, B.C.; Ma, E.W.; Chow, T.W. Probabilistic fault detector for wireless sensor network. *Expert Syst. Appl.* **2014**, *41*, 3703–3711. [CrossRef]

95. Tang, P.; Chow, T.W. Wireless sensor-networks conditions monitoring and fault diagnosis using neighborhood hidden conditional random field. *IEEE Trans. Ind. Inform.* **2016**, *12*, 933–940. [CrossRef]
96. Dhal, R.; Torres, J.A.; Roy, S. Detecting link failures in complex network processes using remote monitoring. *Phys. Stat. Mech. Appl.* **2015**, *437*, 36–54. [CrossRef]
97. Titouna, C.; Ari, A.A.A.; Moumen, H. FDRA: Fault Detection and Recovery Algorithm for Wireless Sensor Networks. In Proceedings of the Mobile Web and Intelligent Information Systems, 15th International Conference, MobiWIS 2018, Barcelona, Spain, 6–8 August 2018; Springer: Cham, Switzerland, 2018; pp. 72–85.
98. Krishnamachari, B.; Iyengar, S. Distributed Bayesian algorithms for fault-tolerant event region detection in wireless sensor networks. *IEEE Trans. Comput.* **2004**, *53*, 1. [CrossRef]
99. Chen, Q.; Lam, K.-Y.; Fan, P. Comments on "Distributed Bayesian algorithms for fault-tolerant event region detection in wireless sensor networks". *IEEE Trans. Comput.* **2005**, *54*, 1182–1183. [CrossRef]
100. Ould-Ahmed-Vall, E.; Ferri, B.H.; Riley, G.F. Distributed Fault-Tolerance for Event Detection Using Heterogeneous Wireless Sensor Networks. *IEEE Trans. Mob. Comput.* **2012**, *11*, 1994–2007. [CrossRef]
101. Lee, M.; Choi, Y. Fault detection of wireless sensor networks. *Comput. Commun.* **2008**, *31*, 3469–3475.
102. Akbari, A.; Arash, A.D.; Khademzadeh, A.; Beikmahdavi, N. Fault Detection and Recovery in wireless Sensor Network Using Clustering. *Proc. Int. J. Wirel. Mob. Netw.* **2011**, *3*, 130–137.
103. Chen, J.; Kher, S.; Somani, A. Distributed Fault Detection of Wireless Sensor Networks. In Proceedings of the 2006 Workshop on Dependability Issues in Wireless ad Hoc Networks and Sensor Networks, Los Angeles, CA, USA, 26 September 2006; pp. 1–11.
104. Nandi, M.; Dewanji, A.; Roy, B.; Sarkar, S. Model Selection Approach for Distributed Fault Detection in Wireless Sensor Networks. *IEEE Trans. Comput.* **2014**, *55*, 1–12.
105. Guclua, S.O.; Ozcelebia, T.; Lukkiena, J. Distributed Fault Detection in Smart Spaces Based on Trust Management. *Procedia Comput. Sci.* **2016**, *83*, 66–73.
106. Ji, S.; Shen-fang, Y.; Ma, T.; Tan, C. Distributed Fault Detection for Wireless Sensor Based on Weighted Average. In Proceedings of the 2010 Second International Conference on Networks Security, Wireless Communications and Trusted Computing, Wuhan, China, 24–25 April 2010; pp. 57–60.
107. DePaola, A.; Gaglio, S.; Re, G.; Milazzo, F.; Ortolani, M. Adaptive distributed outlier detection for wsns. *IEEE Trans. Cybern.* **2015**, *45*, 888–899.
108. Li, W.; Bassi, F.; Dardari, D.; Kieffer, M.; Pasolini, G. Low-complexity distributed fault detection for wireless sensor networks. In Proceedings of the 2015 IEEE International Conference on Communications (ICC), London, UK, 8–12 June 2015; pp. 3469–3475.
109. Taleb, A.A.; Mathew, J.; Kocak, T.; Pradhan, D.K. A Novel Fault Diagnosis Technique in Wireless Sensor Networks. *Int. J. Adv. Netw. Serv.* **2009**, *2*, 230–240.
110. Myoupo, J.F.; Nana, B.P.; Tchendji, V.K. Fault-tolerant and energy-efficient routing protocols for a virtual three-dimensional wireless sensor network. *Comput. Electr. Eng.* **2018**, *72*, 949–964.
111. Titouna, C.; Gueroui, M.; Aliouat, M.; Ari, A.A.A.; Adouane, A. Distributed fault-tolerant algorithm for wireless sensor networks. *Int. J. Commun. Netw. Inf. Secur.* **2017**, *9*, 241–246.
112. Furquim, G.; Jalali, R.; Pessin, G.; Pazzi, R.W.; Ueyama, J. How to improve fault tolerance in disaster predictions: A case study about flash floods using IoT, ML and real data. *Sensors* **2018**, *18*, 907.
113. Titouna, C.; Aliouat, M.; Gueroui, M. FDS: Fault Detection Scheme for Wireless Sensor Networks. *Wirel. Pers. Commun.* **2016**, 86, 549–562. [CrossRef]
114. Tosic, T.; Thomos, N.; Frossard, P. Distributed sensor failure detection in sensor networks. *Signal Process.* **2017**, *93*, 399–410. [CrossRef]
115. Li, W.; Bassi, F.; Dardari, D.; Kieffer, M.; Pasolini, G. Defective Sensor Identification for WSNs Involving Generic Local Outlier Detection Tests. *IEEE Trans. Signal Inf. Process. Over Netw.* **2016**, *2*, 29–48. [CrossRef]
116. Viswanathan, R.; Varshney, P.K. Distributed detection with multiple sensors—Part I: Fundamentals. *Proc. IEEE* **1997**, *85*, 54–63. [CrossRef]
117. Bredin, J.; Demaine, E.; Hajiaghayi, M.; Rus, D. Deploying sensor networks with guaranteed capacity and fault tolerance. In Proceedings of the MobiHoc'05, Urbana-Champaign, IL, USA, 25–27 May 2005; pp. 309–319.
118. Smaragdakis, G.; Matta, I.; Bestavros, A. SEP: A Stable Election Protocol for Clustered Heterogeneous Wireless Sensor Networks. OpenBU, 2004. Available online: https://open.bu.edu/handle/2144/1548 (accessed on 16 July 2023).
119. Zhixiang, D.; Bensheng, Q. Three-layered routing protocol for WSN based on LEACH algorithm. In Proceedings of the 2007 IET Conference on Wireless, Mobile and Sensor Networks (CCWMSN07), Shanghai, China, 12–14 December 2007; pp. 72–75. [CrossRef]
120. Liu, T.; Li, F. Power-efficient clustering routing protocol based on applications in wireless sensor network. In Proceedings of the 2009 5th International Conference on Wireless Communications, Networking and Mobile Computing, Beijing, China, 24–26 September 2009.
121. Kang, S.; Nguyen, T. Distance based thresholds for cluster head selection in wireless sensor networks. *IEEE Commun. Lett.* **2012**, *16*, 1396–1399. [CrossRef]
122. Rajeev, K.; Rajdeep, K. Evaluating the Performance of DEEC variants. *Int. J. Comput. Appl.* **2014**, *97*, 9–16.

123. Sabet, M.; Naji, H. A decentralized energy-efficient hierarchical cluster-based routing algorithm for WSNs. *AEU Int. J. Electron. Commun.* **2015**, *69*, 790–799. [CrossRef]
124. Yi, D.; Yang, H. HEER—A delay-aware and energy-efficient routing protocol for WSNs. *Comput. Netw.* **2016**, *104*, 155–173. [CrossRef]
125. Cengiz, K.; Dag, T. Energy aware multi-hop routing protocol for WSNs. *IEEE Access* **2018**, *6*, 2622–2633. [CrossRef]
126. Sasikumar, P.; Khara, S. K-Means Clustering In Wireless Sensor Networks. In Proceedings of the 2012 Fourth International Conference on Computational Intelligence and Communication Networks, Mathura, India, 3–5 November 2012; pp. 140–144. [CrossRef]
127. Hassana, M.E.; Ziedanb, N.I. A Mobile BS and Multi-Hop LEACH-C Extension for WSNs. *Am. Sci. Res. J. Eng. Technol. Sci.* **2017**, *36*, 198–210.
128. Farooq, M.O.; Dogar, A.B.; Shah, G.A. MR-LEACH: Multi-hop routing with low energy adaptive clustering hierarchy. In Proceedings of the 2010 Fourth International Conference on Sensor Technologies and Applications, Venice, Italy, 18–25 July 2010; pp. 262–268.
129. Al-Sodairi, S.; Ounia, K. Reliable and energy-efficient multi-hop LEACH-based clustering protocol for WSNs. *Sustain. Comput. Inform. Syst.* **2018**, *20*, 1–13.
130. Amiri, A. Extending Network Lifetime of Wireless Sensor Networks. *Int. J. Comput. Netw. Commun.* **2015**, *7*, 1–17. [CrossRef]
131. Shanthi, G.; Sundarambal, M. Investigation of Multi Hop Sensor Node Data Aggregation in Building Management System. *Res. J. Biotech* **2017**, 324–330.
132. Akbar, M.; Javaid, N.; Imran, M.; Rao, A.; Younis, M.S.; Niaz, I.A. A multi-hop angular routing protocol for wireless sensor networks. *Int. J. Distrib. Sens. Netw.* **2016**, *12*, 1–7. [CrossRef]
133. Sert, S.A.; Alchihabi, A.; Yazici, A. A Two-Tier Distributed Fuzzy Logic Based Protocol for Efficient Data Aggregation in Multihop WSNs. *IEEE Trans. Fuzzy Syst.* **2018**, *26*, 3615–3629. [CrossRef]
134. Sert, S.A.; Yazici, A. Optimizing the performance of rule-based fuzzy routing algorithms in WSNs. In Proceedings of the 2019 IEEE International Conference on Fuzzy Systems (FUZZ-IEEE), New Orleans, LA, USA, 23–26 June 2019; pp. 1–6.
135. Mohrehkesh, S.; Weigle, M. Optimizing Communication Energy Consumption in Perpetual Wireless Nanosensor Networks. In Proceedings of the IEEE Globecom, Atlanta, GA, USA, 9–13 December 2013; pp. 545–550.
136. Basagni, S. Distributed Clustering for Ad Hoc Networks. 1999, pp. 310–315. Available online: https://ieeexplore.ieee.org/document/778957 (accessed on 16 July 2023).
137. Devi, G.Y.D. Clustering Algorithms In Wireless Sensor Networks—A Survey. *Int. J. Electr. Electron. Comput. Syst.* **2013**, *1*, 1–9.
138. Tandon, R.; Dey, B.; Nandi, S. Weight Based Clustering in Wireless Sensor Networks. In Proceedings of the 2013 National Conference on Communications (NCC), New Delhi, India, 1–3 February 2013; pp. 1–5.
139. Ducrocq, T.; Mitton, N.; Hauspie, M. Energy-based Clustering for Wireless Sensor Network Lifetime Optimization. In Proceedings of the WCNC—Wireless Communications and Networking Conference, Shanghai, China, 7–10 April 2013.
140. Wan, R.; Xiong, N.; Loc, N.T. An energy-efficient sleep scheduling mechanism with similarity measure for WSNs. *Hum. Cent. Comput. Inf. Sci.* **2018**, *8*, 1–6. [CrossRef]
141. Nanda, S.; Panda, G. Automatic clustering algorithm based on multi-objective Immunized PSO to classify actions of 3D human models. *Eng. Appl. Artif. Intell.* **2013**, *26*, 1429–1441. [CrossRef]

IoT

MDPI

Review

Global Models of Smart Cities and Potential IoT Applications: A Review

Ahmed Hassebo [1,*] and Mohamed Tealab [2]

1 School of Engineering, Wentworth Institute of Technology, Boston, MA 02115, USA
2 Faculty of Engineering, Al-Azhar University, Cairo 11651, Egypt; tealab2008@yahoo.com
* Correspondence: hasseboa@wit.edu

Abstract: As the world becomes increasingly urbanized, the development of smart cities and the deployment of IoT applications will play an essential role in addressing urban challenges and shaping sustainable and resilient urban environments. However, there are also challenges to overcome, including privacy and security concerns, and interoperability issues. Addressing these challenges requires collaboration between governments, industry stakeholders, and citizens to ensure the responsible and equitable implementation of IoT technologies in smart cities. The IoT offers a vast array of possibilities for smart city applications, enabling the integration of various devices, sensors, and networks to collect and analyze data in real time. These applications span across different sectors, including transportation, energy management, waste management, public safety, healthcare, and more. By leveraging IoT technologies, cities can optimize their infrastructure, enhance resource allocation, and improve the quality of life for their citizens. In this paper, eight smart city global models have been proposed to guide the development and implementation of IoT applications in smart cities. These models provide frameworks and standards for city planners and stakeholders to design and deploy IoT solutions effectively. We provide a detailed evaluation of these models based on nine smart city evaluation metrics. The challenges to implement smart cities have been mentioned, and recommendations have been stated to overcome these challenges.

Keywords: smart cities; IoT; ICT; urbanization; sensors; development; LTE; 5G

Citation: Hassebo, A.; Tealab, M. Global Models of Smart Cities and Potential IoT Applications: A Review. *IoT* **2023**, *4*, 366–411. https://doi.org/10.3390/iot4030017

Academic Editors: Antonio Cano-Ortega and Francisco Sánchez-Sutil

Received: 11 June 2023
Revised: 3 August 2023
Accepted: 4 August 2023
Published: 31 August 2023

1. Introduction

Emerging Internet of Things (IoT) applications and services, including smart health, smart grid, smart water, smart cities, and intelligent transportation systems (ITS), are set to transform and disrupt the way we live and work. IoT is currently enabling billions of smart devices and sensors which are communicated and remotely operated via the Internet. Currently, the number of IoT connected devices is greater than 15 billion devices. By 2030, it is expected that the IoT will incorporate more than 29 billion smart devices, which represents three times of that in 2020 [1]. According to the International Data Corporation (IDC), the IoT spending forecast will rapidly grow from $726 billion in 2019 to $1.1 trillion in 2023. According to IoT Analytics' global IoT enterprise spending dashboard, the IoT enterprise market size is forecasted to grow at a compound annual growth rate (CAGR) of 19.4% to $483 billion from 2022 until 2027 [2].

The IoT is denoted as machine-to-machine (M2M) communication; therefore, a staggering number of "things" will require ubiquitous connectivity [3]. These connected devices can range from household appliances and smart home devices to industrial equipment, vehicles, and even entire cities [4]. IoT devices are typically equipped with sensors that can collect data on their environment, such as temperature, humidity, and motion detection, as well as data on their own operation, such as power usage and performance [5]. This data is then transmitted to other devices or systems over the Internet, where it can be analyzed and used to inform decisions and actions.

There is an emerging consensus that cellular-based legacy fourth generation (4G) long-term evolution (LTE) and emerging 5G are the key technology candidates that can provide the required global IoT connectivity to such a staggering number of "things". The 5G cellular system will be distinct from previous generations. Basically, the fundamental merits will not be combinations of old and new radio access technologies (RAT); 5G will also enable new-use cases and mobile communication requirements beyond 4G cellular networks. It will be a combination of 4G cellular standards and technologies, as well as new disruptive technologies such as mmWave and spectrum sharing. 5G will be propelled by whole new services and demand [6]. By 2030, it is projected that there will be billions of interconnected things with data rates of several Gbps, allowing for a personalized user experience with minimal latency and great reliability [3].

Cellular-based machine-to-machine (M2M) communications are one of the key IoT enabling technologies with huge market potential for cellular service providers deploying LTE networks. Massive IoT (MIoT) refers to the tens of billions of M2M devices, objects, and machines that require ubiquitous connectivity [3]. According to the global standards body 3GPP, a massive scale means deploying at least 1 million devices per square kilometer [3]. IoT applications span a wide range of use cases ranging from mission-critical applications with strict latency and reliability requirements (e.g., driverless vehicles) to those that require support of massive number of connected M2M devices with relaxed latency and reliability requirements (e.g., smart meters) [3].

The prior cellular generations were essentially designed to meet the needs of human-type communications (HTC) (e.g., voice, video, and data). However, in order to accelerate industry digitalization, 5G networks are projected to enhance industrial communications, as well. As a result, new industry stakeholders will be able to use novel services and networking capabilities. 5G technology is projected to provide connection and communication needs in vertical industries (e.g., automotive, healthcare, manufacturing, entertainment, and so on) in a cost-effective manner [7].

According to the United Nations (UN) the world population has reached 8.0 billion in 2022, which is more than three times that in 1950. The world population is projected to reach 8.5 billion in 2030, and to further increase to 9.7 billion in 2050 and 10.7 billion in 2100 [8]. In 1950, approximately 70% of the world population were residing in rural areas. The turning point occurred in 2007; for the first time in history, the global urban population exceeded the global rural population. In 2018, 55% of world population (i.e., 4.2 billion) resided in urban areas. This percentage is expected to reach 60% in 2030. In 2050, more than two-thirds (68%) of global population are projected to reside in urban areas, which is the reverse of that in the mid-twentieth century [9].

Due to the projected increased demand for urbanization [9], it is expected that the demand for smart cities and their applications and services will increase accordingly. According to Allied Market Research, the global smart cities market size was valued at $648.4 B and is expected to reach $6.0 T in 2030, growing at a CAGR of 25.2% from 2021–2030 [10]. Consequently, the number of smart cities which contribute to enhancing the inhabitants' quality of life (QoL) will increase. According to a study produced by the smart city observatory, part of the IMD world competitive center (WCC), the number of smart cities has increased from 118 cities in 2021 to 141 cities in 2023 [11].

Adoption of IoT communication technologies such as low-power wide-area networks (LPWAN), for instance, LTE for machines (LTE-M), narrow band IoT (NB-IoT), cellular networks such as 4G/LTE and 5G are required for the realization of smart city concepts [3]. According to Global Suppliers Mobile Association (GSA), around 70 countries had 5G networks as of June 2022. 5G is partially deployed in approximately 15 countries [12]. 6G, the next generation mobile network, promises exponentially faster data speeds and lower latency than 5G [13]. As a result, the deployment of these connectivity technologies is a critical factor in the maturation of smart city projects. Such technologies also pose the groundwork for the IoT, a vast network of smart devices that collaborate to collect and analyze data and perform actions, making smart cities a reality [14].

The selection criteria of the model cities in the paper are based on the preliminary revision of studies, reports, and electronic websites that display the smart cities around the world. The interdependence on authentic studies from official administrations to obtain the information related to the smart cities' models were selected from Africa [15], Asia [16], and Europe [17], for instance, European Parliament studies, International Telecommunications Union (ITU) studies, and official websites which provide details of smart cities.

The selected smart cities must achieve the following standards:

- The recurrence of the city's appearance in more than one comprehensive study to analyze smart city projects.
- The availability of an official website or a special report for the city containing smart projects that have been implemented or are being implemented in the city.
- The same project within the smart city must have more than one source in the absence of an official website or report for the city.
- Cities should be as representative as possible of most geographical areas and different cultural and economic characteristics.
- Considering that the experiments under study include models for both directions of smart cities, existing cities that have already been developed to transform into smart cities, and new cities that already exist from first, second and third generation cities.

2. Smart City

Smart cities are urban areas that leverage technology and data to improve the quality of life for their citizens, increase efficiency and sustainability, and enhance economic development. These cities use interconnected technologies to collect and analyze data in real time, allowing them to make informed decisions and optimize resource use. Smart cities rely on a variety of technologies, such as sensors, IoT devices, artificial intelligence (AI), and machine learning, to collect data and automate processes as depicted in Figure 1. This data can then be used to improve city services, such as transportation, energy, waste management, and public safety [18].

Figure 1. Smart City Applications.

1. Smart Transportation: utilize IoT sensors to collect data on traffic flow, parking, and public transportation, optimizing routes and reducing congestion. Ultimately, this data could also be used to improve safety by detecting and alerting drivers of potential accidents [19].
2. Smart Energy: deploy IoT sensors to monitor and optimize energy usage in buildings and public spaces, reducing waste and carbon emissions [20,21].
3. Smart Waste Management: utilize IoT sensors to monitor waste levels in trash cans and dumpsters, optimizing garbage collection routes and reducing costs [22]. Predictive analytics: use machine learning algorithms to predict the amount of waste generated in different areas and schedule waste collection accordingly [23]. Recycling robots: deploy robots to sort and separate recyclable materials from general waste [24]. Smart bins: install smart bins that use sensors to detect when they are full and send alerts to waste collection teams [25]. Waste-to-energy systems: convert waste into energy through incineration, gasification, or anaerobic digestion [25].
4. Public Safety: use IoT sensors to monitor crime and traffic violations, as well as detecting natural disasters and emergencies, enabling faster response times and better disaster management [26].
5. Smart Water Management: use IoT sensors to monitor and optimize water usage in buildings and public spaces, reducing waste and conserving resources [27]. Smart water systems can be used by water utilities, businesses, and homeowners to monitor and control water usage. Overall, smart water systems offer a number of benefits that can help to conserve water, save money, and improve water management.
6. Smart Health: use of connected devices and sensors to monitor and manage various aspects of health. These devices can collect and transmit data to healthcare providers, caregivers, or the individuals themselves, allowing for better tracking and management of health conditions and improving overall health outcomes [28].
7. Smart Government: use of technology and data to improve the efficiency, effectiveness, and quality of government services and operations. It involves the integration of information and communication technologies (ICT) into government processes and services, with the aim of enhancing transparency, citizen engagement, and overall governance [29,30].
8. Smart Buildings: use of advanced technologies and systems to enhance their functionality, efficiency, and sustainability. These buildings are equipped with a wide range of sensors, control systems, and other IoT devices that enable them to collect and analyze data about their environment and occupants in real time [31].
9. Smart Manufacturing: utilize advanced technologies such as the IoT, big data analytics, AI, robotics, and automation to optimize the manufacturing process. It aims to create a more efficient and flexible manufacturing system that can adapt to changing market demands, reduce costs, and improve product quality [32].
10. Unmanned Aerial Vehicles (UAV): a type of aircraft that is operated remotely without a human pilot on board. UAVs can be either controlled by a human operator on the ground or can be programmed to operate autonomously. They are commonly used for military, commercial, scientific, and recreational purposes and have become increasingly popular in recent years due to advances in technology and lower costs [33].
11. Robotics: robotics and the Internet of Things (IoT) play a crucial role in shaping smart city applications by integrating physical devices and intelligent systems with the city's infrastructure [34]. These technologies enable the development of innovative solutions to improve efficiency, sustainability, and the overall quality of life for citizens. These are just a few examples of how robotics and IoT technologies are transforming urban environments into smarter, more sustainable, and efficient cities [34,35]. As technology continues to advance, we can expect even more innovative applications to emerge, improving the way we live and interact with our surroundings.

To implement smart cities, governments and businesses must work together to develop and deploy IoT technologies, as well as ensure the security, privacy, and interoperability of the systems [36]. Smart cities require investment in infrastructure and innovation, and they can create significant benefits for citizens, businesses, and the environment. Overall, the goal of smart cities is to create more livable, sustainable, and efficient urban environments for their citizens, while also reducing the environmental impact of cities and promoting economic growth [37].

2.1. Smart Transportation

As the number of vehicles increases, the transportation and logistics services represent a promising market for M2M communications. According to Statista, revenue in the passenger car market is expected to reach $1029B in 2023. The revenue in the passenger car market is expected to show an annual growth rate of +1.74%, resulting in a projected market volume of $2067 B in 2027 [38]. These emerging vehicles will be equipped with IoT sensors and actuators [3]. In addition, roads and transported goods are equipped with devices that enable the transportation and freight corporations to seamlessly track the vehicles and the goods by updating the status of delivery to the customers [39].

2.1.1. Logistic Services

The supply chain can work efficiently by enabling M2M sensors to track goods and vehicles in real time. M2M logistic services enable total surveillance on the status of raw materials, products, storage, transportation, and after-sell services by monitoring temperature, humidity, and light. If the status has a problem (i.e., an emergency event), the M2M sensor will transmit an alerting message to the M2M server to take the proper decision via the core network [40]. In addition, the M2M logistic services can also track the inventory in the warehouses by enabling the stakeholders to monitor the market dynamics and take the appropriate decision regarding either to launch sale or to refill. The aforementioned policies will help to reduce the occupied spaces in the warehouses, reducing the waiting times for the customers and consequently gaining the customers' satisfaction [41]. These services can help companies improve efficiency, reduce costs, and increase safety by providing real-time visibility and data-driven insights. The logistic transportation IoT services include:

1. Asset Tracking

Deploy IoT sensors to track the location and condition of cargo, containers, and other assets in real time. This can help companies optimize logistics and reduce the risk of theft or loss [42].

2. Condition Monitoring

Utilize sensors to monitor the condition of goods during transportation, including temperature, humidity, and other environmental factors. This can help companies ensure the quality and safety of their products [43].

3. Predictive Maintenance

IoT sensors can be used to monitor the condition of vehicles and equipment in real time and predict when maintenance is needed. This can help companies reduce downtime and extend the life of their assets [44,45]. Overall, logistic transportation IoT services can provide companies with a range of benefits, including improved efficiency, reduced costs, increased safety, and enhanced customer satisfaction.

2.1.2. Electric Vehicles

The intelligent transportation systems (ITS) depend on the M2M communications accompanied by the roads, which are equipped with IoT sensors and actuators which contribute to regulating the traffic flow, vehicle navigation and/or safety [46]. If the vehicle operator is sleeping, the sensors will alert the operator to avoid accidents. In addition, the M2M sensors equipped in the roads will guide the operators to the closest charging station, as it has all information of the vehicle including the battery charging level [47].

IoT sensors installed in the road and high definition (HD) cameras [14] equipped in the traffic lights are integrated together to regulate and control the vehicles' traffic flow in the street [48]. Furthermore, in case of an accident, the M2M sensors installed in the roads and vehicles will detect the location of the accident which enables the governmental agencies to act immediately. Electric vehicles (EVs) are a rapidly growing industry that is helping to reduce carbon emissions and improve air quality [49]. With the rise of the IoT, there are numerous IoT applications that can be used to enhance the capabilities of electric vehicles as illustrated in Figure 2.

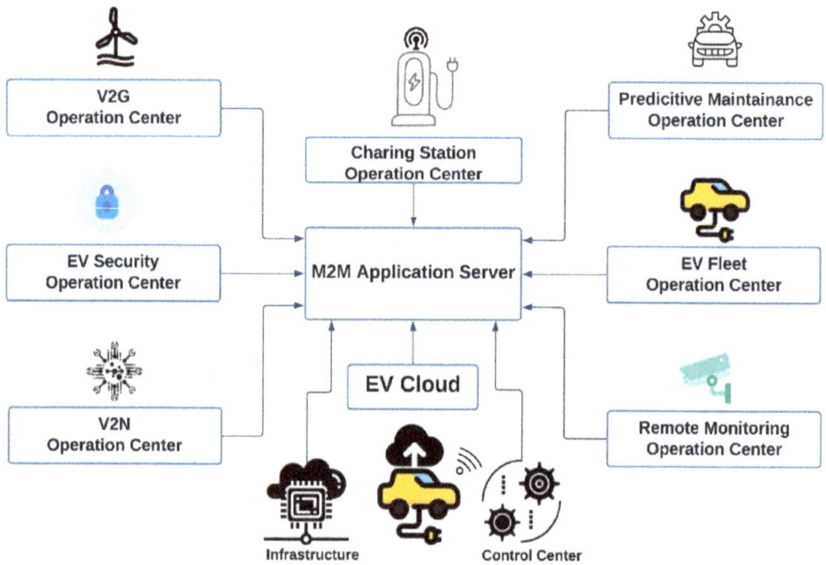

Figure 2. Electric Vehicle M2M Applications.

1. Remote monitoring:

   Monitor the health and status of electric vehicle components, such as batteries, motors, and charging systems using IoT sensors. This information can be transmitted to a central server for analysis, allowing vehicle owners to receive alerts if there are any issues with their vehicle [50].

2. Smart charging:

   Optimize the charging of electric vehicles, ensuring that they are charged at the most efficient times and using the most cost-effective energy sources. This can help to reduce the strain on the power grid and reduce the cost of charging for vehicle owners [51,52].

3. Vehicle-to-Grid (V2G) communication:

   Enable V2G communication, allowing electric vehicles to communicate with the power grid and provide energy back to the grid when it is needed. This can help to balance the load on the grid and reduce the need for additional power generation [53,54].

4. Predictive maintenance:

   Analyze data from electric vehicles to predict when maintenance is needed, allowing vehicle owners to schedule maintenance before issues arise. This can help to reduce downtime and extend the life of vehicle components [55].

5. Driver behavior monitoring:

   Deploy IoT sensors to monitor driver behavior, including acceleration, braking, and speed. This can help promote safer driving practices and optimize energy consumption. Ultimately, IoT applications can greatly enhance the capabilities of electric vehicles, improving their efficiency, reliability, and cost-effectiveness [56].

2.1.3. Smart Parking

Smart parking is the use of technology to improve the efficiency of parking management. This can include various tools and services, such as sensors, mobile apps, and real-time data analysis, to help drivers find parking spaces quickly and easily. Some of the benefits of smart parking services include reduced congestion, increased revenue for parking operators, and improved customer experience for drivers. For example, smart parking services are the use of sensors that detect whether a parking spot is occupied or not. This information can then be transmitted to a central database, which can be accessed by drivers through a mobile app. This allows drivers to find available parking spots in real time, reducing the amount of time spent searching for a spot and ultimately reducing congestion on the roads. These days, driving and parking a car in urban cities is challenging. For instance, the number of vehicles entering Manhattan (only the business district) in 2017 exceed 700 k per day [57,58]. Finding parking for these cars is becoming more difficult if the number of parking spots across New York City (NYC) is only between 3.4 M to 4.4 M. To overcome this problem, M2M sensors placed on roads will promptly guide the vehicle operators to the unoccupied spots on the street. This service will save energy (fuel) and money and will enable the government to monitor the occupancy level of available parking spots [59].

1. Smart vehicle counting

Within 2010 to 2050, the population of the urban cities will increase from 3.6 billion to 6.3 billion inhabitants with 80% increasing rate [60]. By 2035, the population of the US will exceed 370M [61], and the world population will approximately 8.8B inhabitants. On average, the number of vehicles running on the roads will reach 2B by 2035 [39,62]. According to a report by Allied Market Research, the global smart vehicle market is expected to reach $2.2 trillion by 2030, growing at a compound annual growth rate (CAGR) of 40.1% from 2021 to 2030 [63]. This absolutely will pose a serious challenge to manage the traffic control, intelligent transportation, and city management. In order to tackle this problem, employing the M2M sensors to collect data which provide accurate vehicle detection and measuring and controlling the traffic flow [64].

2. Passenger services

One of the utmost passenger services is the e-ticketing system [65]. Ticketing systems of traditional transportation systems are manual, and some of them may be semi-automatic and/or automatic systems used to collect fares. The near field communication (NFC)-based ticketing system will be utilized as the M2M node at the gates or exits in the airport terminals, or train stations scan the passenger identity using NFC-enabled user equipment (UE) [66]. The NFC-enabled UE is scanned at the M2M node; the M2M will send its code to the M2M server through the core network. Based on the tariff table, the traveled distance, and the class of the ticket (e.g., economy, business, and first class) the fare will be withdrawn from the passenger's bank account or credit card. Mobile ticketing raises customer satisfaction along with the ticketing system's effectiveness.

3. Fleet management

Nowadays, numerous cargo ships are voyaging through the open oceans. These containers' delivery services may be affected for many reasons, including piracy, physical damage, and delivery delay. To address this problem, M2M communications can be used in fleet management by developing a superior management system to deliver freight promptly between different regions [67]. Tracking of vehicles and cargos is enabled by installing M2M sensors that collect locations, delivery status, and climate data, which are used to reduce accidents and improve the fuel consumption efficiency, hence increasing the fleet management effectiveness [68].

2.2. Smart Energy

IoT services are a suite of technology-based solutions designed to optimize the production, distribution, and consumption of energy [21,69]. These services use sensors, data analytics, and machine learning to collect and analyze data on energy usage patterns and enable the automation and control of energy systems.

2.2.1. Smart Gri

A modern electricity grid that leverages IoT technologies to optimize energy distribution, consumption, and management. Smart grid solutions can help reduce energy consumption, increase efficiency, and minimize environmental impact [69,70].

2.2.2. Demand Response

Demand response is a change in the power consumption of an electric utility customer to better match the demand for power with the supply [71]. Demand response can be used to reduce peak demand, which can help to avoid blackouts and brownouts. IoT sensors can be deployed to manage energy demand by adjusting energy usage during peak demand times. This can help reduce the need for costly power generation and distribution infrastructure [3,69].

2.2.3. Distributed Energy Resources (DERs)

Managed distributed energy resources (DERs), such as solar panels and wind turbines, enabled energy providers to optimize energy distribution utilizing IoT sensors. [21,69]. DERs are becoming increasingly important as the electricity grid becomes more decentralized. As more and more people generate their own electricity, DERs will play a key role in ensuring the reliability and resilience of the grid [71].

2.2.4. Grid Monitoring

IoT sensors can be deployed to simultaneous monitor grid performance, enabling utilities to respond quickly to any problems or outages. This can help reduce downtime and improve reliability [69,72].

2.2.5. Power Quality Monitoring

The quality of electricity being distributed can be monitored, enabling utilities to identify and resolve any issues that may affect power quality using IoT sensors [72]. A combination of IoT devices, data analytics, and communication technologies are required to implement smart energy solutions. However, the potential benefits include reduced energy consumption, which improves efficiency, and environmental impact, which has a significant impact on both individuals and communities [3,69].

2.2.6. Smart Lighting

Smart lighting employs IoT sensors to adjust lighting levels based on occupancy and daylight levels. This can help reduce energy consumption and save costs while still providing adequate lighting. Another M2M application was implemented in smart homes, smart offices, smart cities, and streets [4,14]. Smart lighting systems impact energy savings in the cities around the world. Currently, 55% of the world population is residing in urban cities. As the growth accelerates rapidly, 68% of the population are predicted to live in urban cities by 2050 globally. By 2030, there are 43 megacities with more than 10 million residents predicted [73]. These statistics pose challenges to city management and smart buildings in terms of energy utilization efficiency. As smart cities tend to be green, the sources of greenhouse gas (GHG) emissions must be reduced [14]. Electricity is one of GHG emission sources which represents 27% of the total US GHG emission in 2018 [74]. The dominant electricity source of GHG emissions is the residential and commercial section, which represents 32% [74]. Reducing GHG emissions can be accomplished by reducing electricity usage and peak demand by increasing the utilization efficiency in residential and commercial buildings. Based on the report published by the World Council on City Data (WCCD) in 2017, GHG emissions were reduced by 63% in the city of Los Angeles due to switching to light emitting diode (LED) lighting [14]. By 2025, the global call of switching lighting systems to 100% smart LED lighting will contribute to the reduction of the lighting share of the global energy consumption from 15% to 8% [75,76].

2.2.7. Energy Storage

The use of energy storage solutions such as batteries and capacitors can be optimized [69]. This can help reduce energy waste and provide a more reliable source of energy in areas with unreliable energy supply. Implementing smart energy solutions requires a combination of IoT devices, data analytics, and communication technologies [3]. However, the potential benefits of smart energy, including reduced energy consumption, increased efficiency, and minimized environmental impact, can have a significant impact on both individuals and communities [77].

2.3. *Smart Waste Management*

IoT applications leverage the power of the Internet of Things (IoT) to enhance waste management systems, making them more efficient, cost-effective, and environmentally friendly [78]. These applications utilize various IoT technologies to monitor, collect, analyze, and manage waste in real time [79]. Overall, smart waste management IoT applications offer a holistic approach to handling waste, promoting sustainability, reducing operational costs, and contributing to a cleaner environment. As IoT technology continues to evolve, these applications are likely to become even more advanced and widespread.

2.3.1. Smart Waste Bins

IoT-enabled waste bins are equipped with sensors that can detect the fill level of the bin. These sensors can use ultrasonic, infrared, or weight-based technologies to measure the waste level. When the bin reaches a certain threshold, it sends an alert to waste collection teams, optimizing the collection process by reducing unnecessary trips and preventing overflow [80].

2.3.2. Route Optimization

IoT devices on waste collection vehicles enable real-time tracking of their location and status. These devices can use GPS technology to find the most efficient routes for waste collection, considering factors such as real-time traffic conditions and the fill levels of individual waste bins. Optimized routes reduce fuel consumption, emissions, and operational costs [81].

2.3.3. Environmental Monitoring

IoT sensors can be deployed in landfills to monitor environmental conditions such as methane emissions, temperature, and air quality. Real-time monitoring helps in early detection of potential issues and allows for timely mitigation actions, ensuring compliance with environmental regulations [82].

2.3.4. Recycling Management

IoT applications can facilitate the separation and sorting of recyclable materials. Smart recycling bins equipped with sensors can help users identify the correct recycling category for their waste items, promoting proper recycling practices [83].

2.3.5. Public Awareness and Education

Smart waste management IoT applications can also be used to raise public awareness about waste management and encourage responsible waste disposal behaviors. Interactive displays on waste bins or smartphone apps can provide information about recycling guidelines, waste reduction tips, and the environmental impact of different waste disposal methods [84].

2.3.6. Data Analytics

IoT-generated data from waste bins, collection vehicles, and landfill sensors can be analyzed to identify patterns and trends in waste generation, collection efficiency, and recycling rates. This data-driven approach allows municipalities and waste management companies to optimize their operations further and make informed decisions [85].

2.3.7. Remote Monitoring and Maintenance

IoT devices enable remote monitoring of waste management infrastructure, such as waste compactors and incinerators. This allows for proactive maintenance and reduces downtime, ensuring that waste management systems operate efficiently [86].

2.4. Public Safety

The goal of public safety is to create a safe and secure environment for all. This includes protecting people from crime, fire, and other hazards. Public safety also includes ensuring that people have access to emergency services when they need them. There are many different organizations that play a role in public safety. These include law enforcement agencies, fire departments, emergency medical services (EMS), and disaster response agencies [87]. These organizations work together to prevent crime, respond to emergencies, and protect the public from harm. Public safety IoT applications use connected devices and sensors to enhance public safety and security [88]. Some public safety IoT applications follow.

2.4.1. Traffic Management

Monitoring traffic conditions, detecting accidents or incidents, and optimizing traffic flow can improve response times for emergency services [88]. Traffic management applications are essential tools used to ensure the safety of road users and pedestrians, as well as to improve the efficiency of traffic flow. Here are some examples of traffic management and public safety applications:

1. Intelligent Traffic Systems (ITS)

ITS are systems that use advanced technologies such as sensors, cameras, and communication networks to monitor and manage traffic on roads [89]. These systems provide real-time information to drivers about traffic conditions, road closures, accidents, and other incidents, enabling them to make informed decisions about their routes and travel times.

2. Automated Traffic Enforcement (ATE)

Systems use cameras and sensors to automatically detect and enforce traffic violations such as speeding, red-light running, and illegal parking [90]. These systems are often used to improve public safety by reducing the number of accidents caused by irresponsible driving.

3. Emergency Vehicle Prevention (EVP)

EVP systems are used to provide emergency vehicles, such as ambulances and fire trucks, with high priority access to intersections. These systems use transmitters on emergency vehicles to communicate with traffic signals [76], allowing them to change to green lights and clear the way for the emergency vehicle.

4. Pedestrian Detection Systems (PDS)

PDS systems use sensors and cameras to detect pedestrians in crosswalks and alert drivers to their presence [91]. These systems are especially useful in areas with high pedestrian traffic, such as urban centers and school zones.

5. Variable Message Signs (VMS)

VMS are electronic signs that display real-time information about traffic conditions, road closures, and other relevant information to drivers. These signs are used to provide drivers with up-to-date information that can help them make immediate decisions about their routes and travel times [92]. In general, these traffic management and public safety applications are essential tools for ensuring the safety of road users and improving the efficiency of traffic flow.

2.4.2. Emergency Response

Emergency response applications automatically detect emergencies such as fires, gas leaks, or natural disasters and send alerts to emergency responders or the public. They are software tools designed to help emergency responders quickly and efficiently respond to emergencies, manage incidents, and communicate with each other during a crisis [93]. These applications can help emergency responders coordinate resources, share information, and make better decisions in real time. The emergency response public safety applications comprising:

1. Dispatch Systems

These systems allow dispatchers to receive emergency calls, quickly identify the location and nature of the emergency, and dispatch the appropriate resources to the scene [94]. These systems also provide real-time communication tools that enable responders to communicate with each other and share critical information.

2. Incident Management Systems

These systems help emergency responders manage incidents by providing a centralized platform for sharing information, tracking resources, and coordinating response efforts [95]. They can also provide in situ awareness tools that allow responders to monitor the status of the incident in real time.

3. Mapping and GIS Applications

These applications use geographic information system (GIS) technology to provide responders with detailed maps and location data that can help them navigate to the scene of an emergency, identify hazards and resources, and plan response strategies [96].

4. Emergency Mobile Apps

These apps provide emergency responders with onsite access to critical information and communication tools [97]. They can also provide real-time updates on incident status, resource availability, and other important information.

5. Social Media Monitoring Tools

These tools enable emergency responders to monitor social media platforms for information about emergencies, identify trends and patterns in user-generated content, and quickly respond to emerging threats or issues [98]. Eventually, emergency response public safety applications will help emergency responders work more efficiently and effectively, which can ultimately save lives and reduce the impact of disasters and emergencies.

2.4.3. Public Health Monitoring

Public health monitoring devices monitor air quality, detect pathogens, and track the spread of diseases, which can help prevent outbreaks and protect public health. These devices are essentially used to protect and improve the health and safety of individuals and communities [99]. The public health monitoring applications involve:

1. Disease Surveillance

Public health officials use disease surveillance systems to monitor the incidence and prevalence of infectious diseases, such as COVID-19, influenza, and tuberculosis [100]. These systems allow health officials to detect outbreaks early, track disease spread, and develop targeted interventions to prevent the further spread of disease.

2. Emergency Management

Emergency management systems are used to coordinate responses to natural disasters, terrorist attacks, and other emergencies [101]. These systems provide real-time information about the location and severity of incidents, as well as resources available to respond to them.

3. Environmental Monitoring

Environmental monitoring systems are employed to detect and respond to environmental hazards, such as air pollution, water contamination, and hazardous waste [102]. These systems allow officials to identify potential health risks and take measures to protect the public from harm.

4. Food Safety

Food safety systems are deployed to monitor food production, processing, and distribution to prevent foodborne illnesses. These systems track outbreaks of foodborne illness and identify the sources of contamination, allowing officials to take action to prevent future outbreaks [103].

5. Public Health Communication

Public health officials use communication systems to disseminate information about health risks, promote healthy behaviors, and encourage the public to take action to protect their health [104]. These systems use various channels, such as social media, traditional media, and public service announcements, to reach a broad audience and provide information that is timely and relevant. Ultimately, public health monitoring and public safety applications are essential tools that help protect and improve the health and safety of individuals and communities.

2.4.4. Infrastructure Monitoring

Infrastructure monitoring is an important aspect to ensure public safety. By monitoring infrastructure, such as bridges, roads, buildings, tunnels, pipelines, and other critical facilities, potential safety hazards can be detected early, and appropriate action can be taken to prevent accidents and minimize damage [105]. Every government must monitor and maintain a broad range of infrastructure, including bridges, tunnels, dams, parks, roads, cables, and pipes. IoT devices are utilized to efficiently monitor the public infrastructure equipped with sensors and RFID tags [106]; this contributes to a reduction of diurnal maintenance and inspection costs by utilizing the transportation traffic-rerouting strategies, for instance, management of parking facilities.

1. Bridge safety monitoring

Infrastructure monitoring systems can monitor bridges for structural damage, wear and tear, and other factors that can affect their safety [107]. This can help prevent accidents and ensure that bridges are safe for public use.

2. Building safety monitoring

Infrastructure monitoring systems can also be used to monitor buildings for safety hazards such as gas leaks, fires, or structural damage [108]. This can help prevent accidents and ensure that buildings are safe for occupancy.

3. Road safety monitoring

Infrastructure monitoring systems can monitor roads for potential hazards such as potholes, cracks, and other damage that can cause accidents. By detecting these hazards early, appropriate action can be taken to prevent accidents and ensure that roads are safe for public use [109].

4. Natural disaster monitoring

Infrastructure monitoring systems can also be used to monitor natural disasters such as earthquakes, hurricanes, and floods. By detecting these events early, appropriate action can be taken to minimize damage and ensure public safety. In general, infrastructure monitoring plays an important role in ensuring public safety [110]. By detecting potential safety hazards early and taking appropriate action, infrastructure monitoring systems can help prevent accidents, minimize damage, and ensure that critical infrastructure is safe for public use.

5. Asset tracking

Deploying IoT devices to track high-value assets such as vehicles, equipment, and cargo can help prevent theft and improve supply chain management [111]. Asset tracking has numerous public safety applications, comprising the following applications:

2.4.5. Tracking Emergency Response Vehicles

Asset tracking can be used to monitor the location of emergency response vehicles such as ambulances, police cars, and fire trucks [112]. This can help dispatchers to send the nearest available vehicle to an emergency situation and enhance response times.

1. Tracking Valuable Equipment

Monitoring valuable equipment, such as firearms, radios, and body cameras [113], can help ensure that these items are always accounted for and not lost or stolen. The best way to track valuable equipment will depend on the specific needs of the organization. If accuracy is critical, then GPS tracking is the best option. If affordability is important, then RFID tracking or barcode scanning may be better options.

2. Tracking Prisoners

Monitoring the location of prisoners who are on parole or probation can help ensure that they are complying with the terms of their release and not engaging in any criminal activity [114].

3. Tracking Stolen Vehicles

Monitoring the location of stolen vehicles can help law enforcement to quickly recover the stolen vehicle and apprehend the thief [115]. The best way to track a stolen vehicle will depend on the specific circumstances of the theft. If a vehicle has a built-in GPS tracker, it will be utilized to track the vehicle's location. If the vehicle does not have a built-in GPS tracker, a GPS tracker can be purchased and installed in the vehicle.

4. Tracking Evidence

Asset tracking can be used to monitor the location of evidence in criminal investigations. This can help ensure that the evidence is properly secured and not tampered with [116]. Ultimately, asset tracking can help public safety officials to better monitor and manage their resources, leading to improved response times, reduced losses, and better outcomes for the communities they serve. Ultimately, public safety IoT applications can help reduce crime, improve emergency response times, and enhance public safety and security.

2.5. Smart Water Management

Nowadays, the demand for water continues rapidly as the population grows significantly. Since the last century, global water consumption is more than twice the rate of world population increase. The water usage is predicted to escalate 50% from 2007 to 2025 in the emerging countries and 18% in the developed countries [117]. By 2025, two-thirds of the world's population may face water shortages. A variety of enterprises depend on water for manufacturing and management. Absolutely, a large percentage of the wasted water due to the aging and leakage of the pipelines, for instance, according to the world bank, the annual global value of water produced and lost by utilities is close to $14 billion [118]. To overcome this problem, smart cities must be capable of monitoring the water supply to ensure the delivery of adequate water supply to the residential and commercial buildings (i.e., water saving systems). The smart cities are equipped-M2M sensors which could be used to remotely control and report any leakage in the pipelines. M2M sensors measure the flow of the water inside the pipes regularly, if an emergency triggered, the sensor transmits an alarming message to the M2M server to take the proper decision if the water leakage is beyond a normal range.

2.6. Smart Health

Deploying technology, data analytics, and other advanced techniques to improve healthcare outcomes and make healthcare delivery more efficient, effective, and accessible. Smart health applications include wearable devices, remote monitoring tools, telemedicine platforms, electronic health records (EHRs), and artificial intelligence (AI) algorithms. Remote Patient Monitoring, Telemedicine, Medication Management, Remote Surgery [119], and Asset Tracking [120]. The smart health care applications may be categorized into four major types:

2.6.1. Tracking and Monitoring

Tracking is the function used to identify the moving patient (i.e., knowing the current position of the tracked person) [121]. For example, the tracking of the patients diagnosed with The Novel Corona Virus (COVID 19) pandemic and consequently those patients are subject to self-isolation [122]. For the government authorities to ensure that those patients are in self-isolating orders, they must be tracked using IoT sensors. In addition, to track the number of people interacting with the patient to contain the virus spreading. Regarding the monitoring, IoT smart health application allows the remote monitoring of the high-risk patients by attaching sensors [122]. These sensors send an alarm to the control center if the high-risk patient (who is vulnerable to fatal consequences due to critical conditions) has dangerous circumstances [123]. For example, if an elderly person falls or a diabetic person has hyperglycemia or hypoglycemia.

2.6.2. Authentication and Identification

Authentication and identification in smart health are needed in multiple forms. Accurate patient identification is crucial and critical to avoid the assignment of wrong medication in terms of (dosage, time, frequency, route, and procedure). Authentication is an important security issue, especially with devices attached to the human body, for instance, the pacemaker, which is a small device placed underneath the skin in the chest to regulate the heartbeats [124]. If a hacker has the ability to intervene in the operation of the pacemaker by increasing or decreasing the heart rate, it will result in a life-threatening issue. In addition, the real-time medical data record, privacy, and maintenance must be authenticated to avoid the leakage of the patient data and to ensure the patient's privacy protection [125].

2.6.3. Data Collection

Automatic data collection is required to reduce the processing and treatment time required to implement a medical treatment plan. If a medical device is attached to human body to measure the blood glucose for a diabetic patient, if the blood glucose in high, so sensor has to send the data automatically to the patient's physician (in the control center) to order the medication promptly. After that, physician sends the order back to the attached device to inject the enough insulin to consume the excessive glucose in the blood stream [126]. The previous situation is vital and represents instantaneous data collection and decision making. The energy consumption may be reported once at a time. The M2M applications may be delivered or reported in many ways:

1. Periodic reporting
2. On demand reporting
3. Scheduled reporting
4. Event-driven reporting

2.7. Smart Government

Smart government utilizes technology and data to enhance the efficiency and effectiveness of government services, improve citizen engagement, and promote transparency and accountability. This approach involves the integration of various technologies, such as big data, artificial intelligence, the IoT, and blockchain, to create a more responsive and proactive government that delivers better outcomes for citizens. Smart government initiatives can encompass a range of areas, including public safety, transportation, healthcare, education, and environmental sustainability [78]. For instance, governments can use big data and AI to analyze crime patterns and predict criminal activity, improve traffic flow and reduce congestion through the use of smart transportation systems, or use IoT sensors to monitor air and water quality in real time. Smart government can also promote citizen engagement and participation through the use of digital platforms and tools that enable citizens to access government services and information, provide feedback, and participate in decision-making processes. This can lead to a more transparent and accountable government that is more responsive to the needs of its citizens. Overall, smart government initiatives have the potential to improve the efficiency and effectiveness of government services, enhance citizen engagement and participation, and promote transparency and accountability in government [127].

2.8. Smart Building

We are living in an environment surrounded by many electric and electronic appliances, for instance, television sets, microwaves, refrigerators, dishwashing machines, air conditioners, etc. To remotely control and monitor these devices, IoT sensors and actuators will be installed to efficiently utilize the energy consumed by these appliances. Heating and cooling might be adjusted using the IoT sensor depending on the current meteorological conditions to sustain the desired temperature [128]. The lighting may be adjusted based on the number of people occupying the room utilizing the motion sensor. The street lighting may be adjusted by utilizing the light sensors to save energy consumption. The motion outside the property will be detected using motion sensors which can be utilized to detect any burglary activity [128]. The electric appliances could be automatically switched off in case of inactivity mode (i.e., idle mode) to reduce energy consumption especially in the prime time. At the prime time, IoT sensors will contribute to the money saving due to the high price of the electricity during this time. On the other hand, the price at other times will be much cheaper than the price the prime time [129]. On the consumer side, the customers may sell the electricity during the rush hour and buy it or consume it during the non-rush hour time.2.9. Smart Manufacturing

Smart manufacturing is the integration of advanced technologies, such as the IoT, AI, machine learning, robotics, and big data, into industrial processes to optimize efficiency, productivity, and flexibility [130]. The main goal of smart industry is to create a fully connected and automated production system that is more efficient, flexible, and customizable than traditional manufacturing processes [131]. By using real-time data analytics and advanced automation technologies, smart industry enables businesses to optimize their operations, reduce costs, and enhance product quality [132].

2.8.1. Predictive Maintenance

Predictive maintenance (PdM) is a maintenance strategy that uses data analysis to predict when equipment is likely to fail. This allows maintenance to be scheduled proactively before a failure occurs. PdM can be a valuable tool for smart manufacturing, as it can help to improve uptime, reduce costs, and increase reliability. Employing sensors and AI are used to predict when equipment will fail and contribute to preventing costly downtime [133].

2.8.2. Digital Twins

Creating virtual models of physical assets helps optimize performance and reduce maintenance costs [134]. Overall, smart industry represents a major shift in the way businesses approach industrial production and has the potential to transform entire industries.

2.8.3. Industrial Internet of Things (IIoT)

Connecting machines, sensors, and devices throughout the manufacturing process to collect and share data in real time enables better monitoring, predictive maintenance, and process optimization [135].

2.8.4. Big Data Analytics

Big data analytics is a powerful tool that can be used to improve a variety of processes in smart manufacturing. It is still important to have a good understanding of the manufacturing process and to use big data analytics in conjunction with other tools and techniques. Utilizing large volumes of data collected from various manufacturing processes helps gain insights, identify patterns, and make data-driven decisions [136].

2.9. *Unmanned Aerial Vehicle (UAV)*

UAV is a type of aircraft that is operated remotely without a human pilot on board. UAVs can be either controlled by a human operator on the ground or can be programmed to operate autonomously. They are commonly used for military, commercial, scientific, and recreational purposes and have become increasingly popular in recent years due to advances in technology and lower costs. UAVs equipped with IoT sensors can collect and transmit data in real time, enabling a variety of use cases across multiple industries [137].

2.9.1. Agriculture

UAVs equipped with sensors can be used to monitor crop health, collect data on soil moisture and nutrient levels, and identify areas in need of irrigation or fertilizer [138]. This can help farmers optimize their crop harvest, reduce costs, and minimize environmental impact. Eventually, UAVs can provide farmers with valuable data that can help them make more informed decisions about crop management, leading to higher yields, and increased sustainability. UAVs have a wide range of agricultural services including:

1. Crop Monitoring:

UAVs can be used to monitor crops for pests, diseases, and nutrient deficiencies. They can capture high-resolution images of the crops, which can be used to detect early signs of stress and other diseases based on AI algorithms [138]. In addition, these images may be utilized to identify the ideal time to harvest different crops.

2. Precision Agriculture:

UAVs equipped with specialized sensors can provide farmers with data on soil moisture, temperature, and other environmental factors. This data can be used to make more informed decisions about crop management, including planting and irrigation [139].

3. Crop Spraying:

UAVs can be used to spray crops with pesticides or fertilizers, which can reduce the need for manual spraying and minimize the risk of exposure to harmful chemicals [140]. Eventually, UAV-based crop spraying offers a number of potential benefits [141]. However, there are also some challenges that need to be addressed before they can become widely adopted [142].

4. Mapping:

UAVs can be used to create high-resolution maps of farms, including the layout of fields and buildings. This data can be used to plan future planting and construction projects [143]. They offer a number of advantages over traditional methods, such as ground-based surveying, and include accuracy, efficiency, and safety.

5. Irrigation Management:

UAVs can be used to monitor the effectiveness of irrigation systems. They can capture images of fields before and after irrigation, which can be used to determine the optimal amount of water to apply [144]. By using these systems, it will help to conserve water resources, which can be used in parallel with the smart water systems. Using such systems will contribute to high-quality crops.

2.9.2. Disaster Response

UAVs equipped with sensors can be used to assess damage and locate survivors in disaster zones, providing valuable information to first responders and enabling them to respond more effectively [145]. UAVs have become a valuable tool for disaster response and management due to their ability to quickly gather data, assess damage, deliver supplies to hard-to-reach areas, and deliver aid to those in need.

1. Search and Rescue:

UAVs equipped with thermal imaging cameras and other sensors can help search and rescue teams locate survivors in disaster zones, even in low-light or low-visibility conditions such as after a hurricane [146]. By determining the survivors' locations, it will give the exact information to the first responder teams who provide nutrition and medication resources.

2. Disaster Mapping and Assessment:

UAVs equipped with high-resolution cameras and LiDAR sensors can quickly and accurately map disaster areas, helping authorities assess the extent of damage and plan response efforts [147]. Additionally, UAVs can be used to create detailed maps of disaster zones, including infrastructure damage, flood levels, and road conditions. These maps can help authorities prioritize response efforts and allocate resources more efficiently [148].

3. Delivery of Aid and Supplies:

UAVs could be utilized to deliver emergency supplies to inaccessible areas, such as medical supplies, food, and water [148,149]. UAVs are used to deliver aid to people in need. They offer a number of advantages over traditional methods, such as land vehicles and helicopters.

4. Communication Support:

UAVs can be equipped with cellular repeaters and other communication devices to provide temporary connectivity in disaster zones where communication infrastructure has been damaged or destroyed [150].

2.9.3. Infrastructure Inspection

UAVs equipped with sensors can be used to inspect infrastructure such as bridges, roads, and power lines, identifying potential problems and enabling maintenance teams to respond quickly [151]. UAVs have a wide range of applications in infrastructure inspection [152] due to their ability to provide high-resolution imagery and data and ability to access hard-to-reach areas safely and efficiently [151]. Eventually, the use of UAVs in infrastructure inspection will allow for more effective and accurate maintenance and repair decisions.

1. Bridge inspections:

UAVs can fly underneath and alongside bridges to capture high-resolution images and video of their structural components, such as joints, beams, and cables. This helps identify any damage or wear and tear and helps prioritize repairs at certain and specific points of damage at bridges [153,154].

2. Power line inspections:

UAVs can fly along power lines to inspect them for damage or vegetation growth that could cause power outages. The UAVs can capture high-resolution images and video of the power lines and also use thermal imaging to detect hotspots and potential faults [155]. However, there are also some challenges associated with using UAVs to inspect power lines: government regulations, technology, reliability, and cost.

3. Wind turbine inspections:

UAVs can fly close to wind turbines to inspect their blades and towers for damage. This helps identify any wear and tear, erosion, or cracks, which could cause potential malfunctions, resulting in multiple problems in the distribution networks [156].

4. Roof inspections:

UAVs can inspect roofs for damage, such as missing shingles, cracks, or leaks. They can also capture images of the roof's overall condition and identify potential areas of weakness [157]. The problems resulting from roof damage are time sensitive and require quick decision making. For an effective decision, a big data-based AI algorithm may be utilized to provide these decisions.

2.9.4. Delivery Services

UAVs equipped with sensors can be used to deliver packages and goods in remote or hard-to-reach areas. This can provide faster and more efficient delivery services, especially in areas with limited transportation infrastructure [158]. Delivery services are increasingly turning to UAVs as methods of delivering goods quickly and efficiently. Overall, UAVs have the potential to revolutionize the delivery industry by providing a faster, more efficient, and cost-effective means of delivering goods. However, the technology is still relatively new and there are many regulatory and technical challenges that need to be addressed before UAVs can be widely used for delivery services [159]. Implementing UAV IoT applications requires a combination of technologies, such as sensors, communication systems, and data analytics. However, the potential benefits of UAV IoT applications, including improved efficiency, reduced costs, and increased safety, can have a significant impact on multiple industries.

1. Last-mile delivery:

UAVs can be used to deliver packages to customers' doorsteps [160] or other inaccessible locations in the last mile of the delivery process, reducing the time and cost of delivery [161]. UAVs offer a number of advantages over traditional methods, such as ground-based delivery and car-based delivery, including speed, accuracy, and efficiency.

2. Medical supply delivery:

UAVs can be used to transport medical supplies such as drugs, blood, and vaccines to hospitals or other medical facilities in emergency situations or in hard-to-reach areas [160,162]. The benefits of using UAVs to deliver medical supplies include saving lives, reducing costs, and increasing efficiency.

3. Retail delivery:

UAVs can be used to transport products from warehouses or distribution centers to retail stores, improving the efficiency of the supply chain [163]. The benefits of using UAVs for retail delivery may include increased customer satisfaction by delivering products faster, reducing delivery cost, and improving environmental impact [164].

4. Parcel delivery:

UAVs can be used to transport parcels of various sizes and weights, including small parcels for individual customers and larger parcels for businesses [161]. Overall, the use of UAVs for parcel delivery is still in its early stages. However, the potential benefits (e.g., speed, and accuracy) of this technology could make it a viable option for the future of parcel delivery.

2.10. Robotics

Robotics can affect our daily routine by improving our quality of life, conserving the budget, and minimizing expenditure. Nowadays, robots are intelligent machines and capable of interacting with other robots as well as human beings. Robotics has the potential to revolutionize various aspects of smart cities [165] by enhancing efficiency, safety, and sustainability. Robotic systems are increasingly being integrated with IoT technologies to enhance their capabilities and create new applications [166,167].

2.10.1. Automated Transportation

Autonomous vehicles and drones can be integrated into a city's transportation system, providing efficient and safe mobility solutions. Self-driving cars can reduce traffic congestion and emissions, while drones can be used for deliveries and surveillance [168].

2.10.2. Infrastructure Maintenance

Robots can be deployed to inspect and maintain critical infrastructure such as bridges, roads, and utility lines. They can quickly identify and repair issues, reducing downtime and improving infrastructure permanence [169].

2.10.3. Surveillance and Security

Robots equipped with cameras and sensors can monitor public spaces, enhancing security and surveillance in the city. They can patrol streets, parks, and other areas, detecting suspicious activities and helping law enforcement respond quickly to potential threats [170].

2.10.4. Environmental Monitoring

Robots equipped with sensors can monitor air and water quality, noise levels, and other environmental factors to provide real-time data for pollution control [171]. Environmental monitoring robotics is a rapidly growing field. As technology continues to develop, robots will become increasingly capable of collecting data and monitoring the environment in ways that were not possible before. This data can be used to improve our understanding of the environment and to develop better ways to protect it.

2.10.5. Agriculture and Urban Farming

In smart cities, rooftop gardens and vertical farms are becoming more popular. Robots can be used to automate planting, watering, and harvesting, ensuring sustainable and efficient food production. Robots can monitor crops and soil conditions. The sensors can measure factors such as temperature, humidity, soil moisture, and nutrient levels, which can be used to optimize crop growth and reduce water and fertilizer usage [172].

2.10.6. Healthcare Assistance

Robotics can assist in healthcare applications, such as delivering medications and supplies in hospitals or supporting elderly and disabled citizens with daily tasks in smart homes. Robotics IoT applications in healthcare include robotic assistants for surgery, patient monitoring, and medication management. These robots can be equipped with sensors to monitor vital signs and provide real-time feedback to healthcare professionals [173].

2.10.7. Disaster Response

In the event of natural disasters or emergencies, robots can be deployed for search and rescue operations, providing critical assistance to first responders, and minimizing human risk [174]. Despite the challenges, disaster response robotics is a promising field with the potential to save lives and improve the efficiency of disaster response. As technology continues to develop, robots are likely to play an increasingly important role in disaster response.

2.10.8. Education and Entertainment

Robots can be used in educational settings, offering interactive learning experiences. Additionally, they can be utilized in public spaces for entertainment purposes, like interactive art installations or robot-guided tours [175]. Educational and entertainment robots are becoming increasingly popular. They offer a fun and interactive way to learn about STEM concepts and to experience the latest in robotics technology.

2.10.9. Tourism

In smart cities, robots can act as tour guides, providing information about landmarks, historical sites, and tourist attractions [176]. Robotics is still a relatively new technology in the tourism industry, but it has the potential to revolutionize the way that tourists interact with destinations. Robots can provide a more personalized and engaging experience for tourists, and they can also help to reduce costs for businesses.

2.10.10. Smart Home Assistants

Domestic robots can help residents in their homes by performing household routines, managing appliances, and providing reminders and assistance to the elderly and disabled [177].The benefits of using smart home assistants and robotics in the home may comprise the following: convenience, security, and entertainment. However, there are also some potential drawbacks to using smart home assistants and robotics in the home: privacy, security, and cost.

3. Smart City Communication Systems

Due to the massive volume of sensors and their data, robust connectivity technology is a prerequisite for success coverage and reliability across the entire city is the key to launching any successful smart city [3,88]. Because cabling a massive number of sensors and smart IoT devices is cost prohibitive, wireless technology is the key and sole viable solution to the deployment of IoT networks across the city [88]. There is an emerging consensus that current fourth generation (4G) long term evolution (LTE) and current 5G are the key technological candidates that can provide the required global IoT connectivity to such a staggering number of "things" in a city [3,5,12]. Cellular-based machine-to-machine (M2M) communications is one of the key IoT-enabling technologies with huge market potential for cellular service providers deploying LTE networks [178].

4G and/or 5G cellular technologies can support a wide range of current and future smart city applications and services including video surveillance for public safety, intersection safety analytics (pedestrian safety), traffic management, traffic light controls, digital signage systems, EV charging, public Wi-Fi and much more [3,5,12,179]. Smart city IoT applications span a wide range-of-use cases ranging from mission-critical applications (e.g., traffic control, emergency response, video surveillance, and connected vehicles), which require ultra-reliability and ultra-low latency, to those that require support of a massive number of connected M2M devices with relaxed latency and reliability requirements (e.g., smart meters) [3,14].

There are numerous current competing solutions that can support network connectivity for such a wide range of IoT applications [3]. These solutions span a wide range-of-use cases ranging from relatively low cost and easily deployable solutions for basic services to the most expensive, high-performance systems suitable for the most demanding requirements [14].

Most massive scale deployment of IoT applications, however, requires low-cost devices that communicate infrequently, with a low data rate and low energy consumption so that they can deliver an extremely long ten-year battery life as well as good coverage [3]. This is where low-power wide area networks (LPWAN) technology is needed. LTE-M (LTE for machines) and narrowband IoT (NB-IoT) are the first cellular-based LPWAN technologies standard supporting massive IoT applications [180].

These simpler and lower cost cellular LPWA technologies support longer battery life (up to 10 years) and better coverage (for IoT devices underground and deep inside buildings) compared to traditional M2M-IoT cellular connectivity options. Though LP-WAN technologies have clear advantages over traditional IoT cellular connectivity options, they are only suited to meet very basic data requirements for IoT applications with limited/modest data needs and relaxed latency (in the order of few seconds) [3,181]. They are ideally suited for IoT applications, which require just extended coverage, but reliability, latency, and availability might be more or less important [181] as shown in Table 1.

Table 1. Summary of Communication technologies support in the Smart City.

Technology	Latency (s)	Frequency (Hz)	Coverage (m)	Data Rates (bps)	Use Cases	Power Usage
Bluetooth	100 m	2.4 G	10	25 M	Indoor e-health	Low
ZigBee	16 m	2.4 G	10	250 K	Smart Meter, indoor e-health	Low
WiFi	46 m	2.4 G	140	54 M	Smart cities, waste management	Medium
LORAWAN	1–16	125–500 K	<11 K	0.3–27 K	Healthcare, public safety	Low
NB-IoT	2–10	200 K	<25 K	26 K	Smart meter, smart city, smart home	Low
LTE-M	10–20 m	1.4–20 M	(1–10) K	200 K–1 M	Asset trackers, fleet tracking, alarms	Low
3G	100 m	850 M	(5–30) K	3 M	ITS, energy management, monitoring	High
LTE	5 m	700, 750,800,1900, 2500 M	(5–30) K	500 M–1 G	ITS, logistics, monitoring, mobile health, infotainment	High
5G	<1 m	24–68 G [mmWave]	(250–300) K	(3–20) G	Smart cities, healthcare, gaming, and entertainment	High

3.1. *Low Power Wide Area Networks (LPWANs)*

LPWAN technologies are a group of wireless technologies that are designed to provide long-range, low-power connectivity for IoT devices. LPWAN technologies are ideal for applications where battery life [3] is critical, such as smart metering, asset tracking, and environmental monitoring. Here are some of the most popular LPWAN technologies:

3.1.1. LORAWAN

LORAWAN (long range WAN) is a widely adopted LPWAN technology that enables long-range communication at low data rates [182]. It operates in unlicensed frequency bands, making it available for public use without the need for expensive licenses.

3.1.2. NB-IoT (Narrowband IoT)

NB-IoT is a cellular LPWAN technology standardized by the 3rd Generation Partnership Project (3GPP). It operates within existing cellular networks, providing deep coverage and compatibility with cellular infrastructure [3].

3.1.3. LTE-M (Long Term Evolution for Machines)

LTE-M is another 3GPP standardized LPWAN technology designed to operate within the LTE network infrastructure. It offers higher data rates compared to NB-IoT and is suitable for applications that require more bandwidth [3,183].

LPWANs are commonly used in various IoT applications, such as smart city solutions, industrial monitoring, agriculture, environmental monitoring, asset tracking, and more.

Their low power consumption allows IoT devices to operate on battery power for several years, reducing maintenance efforts and overall costs [3,184].

3.2. Fourth Generation (4G)

4G LTE stands for fourth generation long term evolution. It is the fourth generation of cellular network technology, succeeding 3G and preceding 5G [3,185]. 4G LTE networks offer significantly faster data speeds than 3G networks, with theoretical download speeds of up to 100 Mbps and upload speeds of up to 50 Mbps [186]. In practice, actual speeds will vary depending on a number of factors, including the carrier, the device, and the location.

4G LTE is the most widely deployed 4G technology in the world. It is supported by all major carriers in the United States, as well as many carriers in other countries. 4G LTE is also the technology used by most smartphones and tablets [3,187]. Here are some of the 4G LTE features:

3.2.1. Faster Data Speeds

4G LTE offers significantly faster data speeds than 3G networks. This means that you can download files, stream videos, and browse the web much faster [3,7].

3.2.2. Improved Reliability

4G LTE networks are more reliable than 3G networks. This means that you are less likely to experience dropped calls or slow data speeds [7,12].

3.2.3. Increased Capacity

4G LTE networks have a higher capacity than 3G networks. This means that they can support more devices and more data traffic [3,12,14].

3.2.4. Use Cases of 4G LTE

- Video streaming: 4G LTE is well-suited for video streaming, as it can deliver high-quality video without buffering. This makes it possible to watch movies and TV shows on mobile devices without any problems [3,148].
- VoLTE (Voice over LTE): VoLTE is a technology that allows users to make and receive calls over a 4G LTE network. This provides better voice quality and lower latency than traditional cellular networks [3].
- Mobile gaming: 4G LTE is ideal for mobile gaming, as it can provide the fast data speeds and low latency that are required for smooth gameplay [185].
- IoT (Internet of Things): 4G LTE is used to connect a wide variety of IoT devices, such as smart home devices, wearables, and industrial sensors. This allows these devices to communicate with each other and with the cloud [25].

3.3. Fifth Generation (5G)

5G is the fifth generation of cellular network technology. It is designed to offer significantly faster speeds, lower latency, and greater capacity than previous generations of cellular networks. 5G is still in its early stages of development, but it is expected to be widely deployed in the coming years [13,115,188]. As 5G networks become more widespread, we can expect to see a wide range of new and innovative applications and services that will change the way we live and work. Here are some of the benefits of 5G:

3.3.1. Faster Speeds

5G can offer download speeds up to 10 Gbps, which is 100 times faster than 4G. This will allow users to download large files in seconds, stream high-definition videos without buffering, and play online games with minimal lag [13].

3.3.2. Lower Latency

5G has a latency of just a few milliseconds, which is much lower than 4G [189]. This means that 5G is ideal for applications that require real-time communication, such as self-driving cars and remote surgery [3,190].

3.3.3. Greater Capacity

5G can support a much greater number of devices than 4G. This indicates that 5G networks will be able to handle the increasing number of connected devices that are being used today in multiple applications and services [3].

5G cellular system will support the following services enhanced mobile broadband (eMBB), massive machine type communications (mMTC), and ultra-reliable low-latency communications (URLLC). eMBB will enable the following applications: hotspot wide-area coverage with high capacity, enhanced connectivity, and high mobility. In the hotspot scenario (e.g., soccer game), serving a massive number of users requires low mobility and high traffic capacity. For the bus scenario, high mobility and a lower capacity than the hotspot are needed. mMTC is distinguished by a plethora of devices which transmit low volume data with less sensitivity to delay. URLLC has rigorous requirements for high throughput and low latency (e.g., remote medical surgery and driverless vehicles) [3,13], as shown in Figure 3.

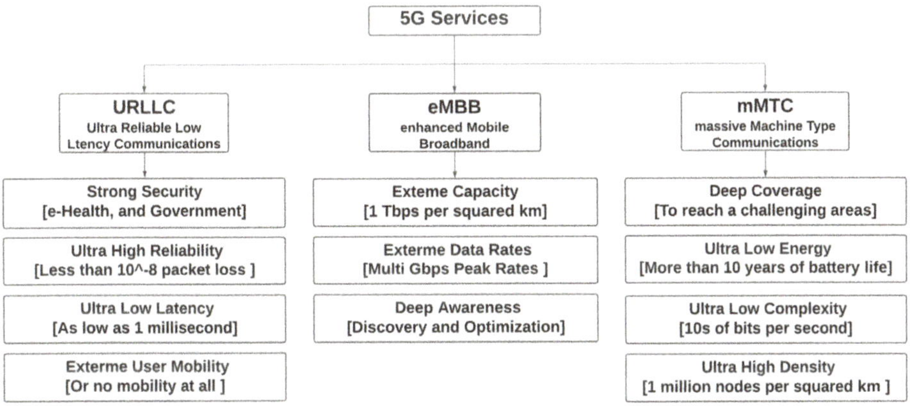

Figure 3. 5G Services.

URLLC services in 5G are fundamentally considered as IoT applications that must gather short packets (few bytes) from small sensors or robots with an uplink over the air latency of less than 1 ms. High reliability is described by the third generation partnership project (3GPP) for such IoT services as reaching a percentage of properly delivered packets within the application time limit constraints of 99.999 percent, depending on the application [188].

The phenomenal insurrection of the original internet of things (IoT) as a driver of machine-to-machine (M2M) communication. On the other hand, internet of everything (IoE) refers to a larger idea of connectivity in which network communication serves as the IoT's foundation. Wireless applications, such as fully autonomous cars, flying vehicles, drones, wireless brain computer interface (WBCI), and enhanced extended reality (XR) apps, will be part of the Internet of Everything. Augmented reality (AR), mixed reality (MR), and virtual reality (VR) are all examples of XR applications (VR) [188].

These new applications will have highly strict quality of service (QoS) requirements (for example, reliability and latency) and will blur the borders between 5G URLLC and eMBB services [13]. Despite the fact that 5G may be able to meet the QoS requirements of basic XR services or autonomous robotics, it will be unable to meet the QoS requirements of higher data rates (e.g., greater than 100 Gbps) for ultimate VR class of service (CoS) latency (e.g., 1 ms for wireless brain computer interface) and high reliability (near zero packet loss ratio (PLR) at low latency and extreme high reliability) [191].

There is no need to design a distinct sixth generation (6G) cellular technology to address the challenges of developing IoE applications. Traditionally, The architecture is tailored to the requirements of highly reliable, low latency, and high data rate services. 6G will be the outcome of classic communication technology developments (such as high data rates and massive antennas) combined with existing services and technological advancements such as new wireless devices (e.g., body implants, and XR equipment, etc.). The path to 6G must be able to overcome some of the 5G restrictions revealed in primary systems [188], as illustrated in Figure 4.

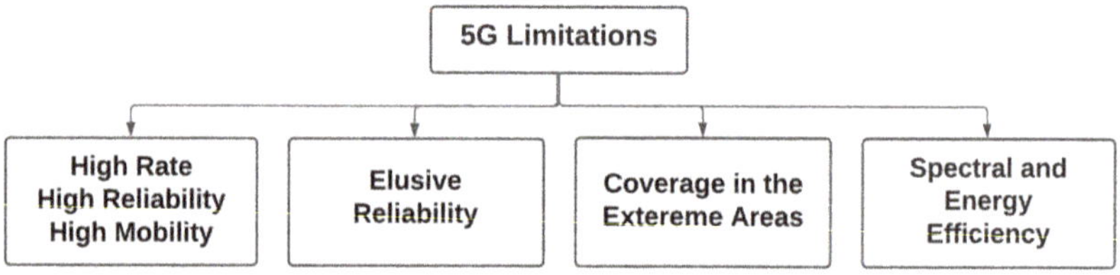

Figure 4. 5G Limitations.

3.4. Sixth Generation

The sixth generation (6G) of wireless communication refers to the next generation of mobile network technology that will follow the current 5G technology. 6G is expected to bring significant improvements in terms of data transfer rates, latency, reliability, and efficiency, as well as support for new applications and use cases. Some of the key features and capabilities that are expected to be part of 6G include the terahertz (THz) frequency band; 6G is expected to operate in the THz frequency band, which could enable much higher data transfer rates than the current 5G technology. Artificial Intelligence (AI) integration: 6G is expected to leverage AI to enhance network management, optimize network resources, and provide better user experiences [13]. Quantum communications: 6G could use quantum communication technologies to provide highly secure communication channels that are immune to eavesdropping. Ubiquitous connectivity: 6G is expected to enable seamless connectivity across a wide range of devices, including wearables, vehicles, drones, and smart home appliances. Holographic communication: 6G could enable holographic communication, allowing users to interact with virtual objects and environments in real-time. Enhanced mobile edge computing: 6G is expected to support the processing of data at the network edge, enabling low-latency, high-bandwidth applications such as virtual reality and augmented reality [13]. While 6G technology is still in the early stages of development, research is already underway to explore and develop some of these capabilities. It is expected to be commercially available sometime in the 2030s [192].

4. Examples of Smart Cities

There are many smart cities around the world [193] that are implementing advanced technologies to improve the Quality of lives of their citizens, increase sustainability, and enhance the efficiency of city. To develop smart cities, there are There are open-ended three phases defined as Smart City 1.0, Smart City 2.0, and Smart City 3.0 [194], and Smart City 4.0 [195] inspired by economics.

4.1. Smart City 1.0

Smart city 1.0 sets the foundation for the development of smarter and more sustainable cities. However, there is still much work to be done in terms of addressing the challenges facing urban areas, such as inequality, environmental degradation, and social exclusion. As such, the focus of smart city development has shifted towards more holistic and inclusive approaches in recent years. Smart city 1.0 refers to intelligent cities in the earliest phase of creation [196]. The utilization of modern technologies is initiated by ICT companies [197]. These companies implement various solutions regardless of whether they are essential for the cities or not.

4.2. Smart City 2.0

Smart city 2.0 refers to the second phase of smart city development, which is characterized by a more holistic and citizen-centric approach. This phase emphasizes the use of technology to enhance quality of life, promote sustainability, and foster social inclusion in urban areas. In this phase, the development of smart cities with a predominant role for public administration. The use of modern technologies is initiated by local authorities, and the introduction of new solutions which is aimed at improving the citizens' quality of life [198]. During Smart City 2.0, cities are focusing on creating more collaborative models of governance, which involve citizens and stakeholders in decision-making processes. This includes the use of digital platforms and social media to engage with citizens, as well as the establishment of innovation labs and spaces [199].

4.3. Smart City 3.0

Smart city 3.0 is the third phase of smart city development, which is characterized by a focus on innovation, resilience, and adaptability in the face of emerging challenges and opportunities. During Smart City 3.0, cities are leveraging emerging technologies, such as blockchain, the IoT, and autonomous systems, to create more resilient and adaptive urban environments [200]. This includes the use of smart infrastructure systems, such as self-healing power grids and automated water management systems, to enhance the reliability and resilience of critical urban services. Another key feature of Smart City 3.0 is the use of innovation ecosystems and digital innovation hubs to foster innovation and entrepreneurship in urban areas. This includes the establishment of co-working spaces, incubators, and accelerators to support startups and small businesses, as well as the integration of universities and research institutions into the urban innovation ecosystem [201].

4.4. Smart City 4.0

Smart city 4.0 is a theoretical concept that refers to the next phase of smart city development beyond smart city 3.0. While Smart City 4.0 is not yet fully defined, it is expected to build on the foundations of previous phases and further integrate emerging technologies, such as artificial intelligence, 5G networks, and edge computing, to create even more intelligent, responsive, and interconnected urban environments [202]. It is likely to focus on creating highly personalized and immersive experiences for citizens and visitors, through the use of virtual reality (VR) and augmented reality (AR) technologies, and the development of smart spaces that adapt to individual needs and preferences. It is also expected to involve greater collaboration between cities and private sector partners, as well as more decentralized and distributed models of governance [203].

4.5. Global Smart Cities

4.5.1. Singapore

Singapore is known for its advanced transportation system, which includes an extensive network of public buses, trains, and taxis. The city also uses smart sensors and data analytics to manage traffic flow and reduce congestion. Additionally, Singapore has implemented a number of smart solutions to improve energy efficiency and reduce its carbon footprint [204,205]. The following are the features of the smart city of Singapore:

1. Advanced technology infrastructure: the city has invested heavily in building an advanced technology infrastructure to support its smart city initiatives. It has a high-speed fiber optic network, a robust 4G LTE, and 5G cellular networks [204,206].
2. IoT sensors: Singapore has installed a large number of IoT sensors across the city to collect real-time data on various parameters such as traffic flow, air quality, and

energy consumption. This data is then analyzed to identify patterns and trends, which helps authorities to make informed decisions [179,204].
3. Intelligent transport system (ITS): Singapore has implemented an ITS that uses real-time data to optimize traffic flow and reduce congestion. It includes features such as electronic road pricing, which charges drivers for using congested roads during peak hours [63,204].
4. Smart buildings: many buildings in Singapore are equipped with smart systems that automate various functions such as lighting, temperature control, and security [204,207].
5. Smart nation initiative: the Singapore government has launched a smart nation initiative to leverage technology to improve the lives of its citizens. The initiative includes various projects such as the development of a national digital identity system, a cashless payment system, and a national sensor network [204].

4.5.2. Barcelona

Recognized for its use of smart technology to improve the quality of life for its citizens, Barcelona has implemented smart lighting systems that automatically adjust to the needs of pedestrians and cyclists. It has also implemented a smart waste management system that reduces waste and increases recycling rates [208,209].

1. Smart buildings: Barcelona has implemented smart building technologies that help reduce energy consumption and improve energy efficiency. For example, many buildings in the city use sensors to automatically adjust lighting and temperature based on occupancy levels [207,208].
2. Smart lighting: Barcelona has installed smart LED streetlights that can be controlled remotely and adjusted based on real-time data such as traffic flow and pedestrian activity, which reduces energy consumption and improves safety [206,208].
3. Citizen engagement: Barcelona has implemented various initiatives to engage its citizens and encourage them to participate in the city's decision-making process. For example, the city has developed a digital platform called "DECIDIM" that allows citizens to propose and vote on ideas for improving the city [208,210].
4. Open data: Barcelona has made a large amount of data publicly available, which enables researchers and developers to create innovative solutions to urban problems. The city has also established an open data portal that provides access to a wide range of datasets [208,211].

4.5.3. Amsterdam

Known for its focus on sustainability and livability, Amsterdam has implemented a number of smart solutions to reduce energy consumption, including smart buildings that use renewable energy sources and smart grids that optimize energy use [212]. Amsterdam is also known for its smart transportation systems, which include an extensive network of bike lanes and public transportation options [209,213].

1. Smart mobility: Amsterdam has developed an advanced mobility system that integrates various modes of transport such as bicycles, electric vehicles, and public transport. The city has also implemented a smart parking system that helps drivers find available parking spots using sensors and mobile apps [213,214].
2. Sustainable energy: Amsterdam has a strong focus on sustainable energy and has implemented various initiatives to reduce energy consumption and increase the use of renewable energy sources. For example, the city has developed a district heating system that uses waste heat from industrial processes to heat homes, buildings, and companies [213,215].
3. Circular economy: Amsterdam is committed to becoming a circular economy, which means reducing waste and reusing materials as much as possible. The city has implemented various initiatives to promote circular practices, such as a recycling

program for construction materials and a bike-sharing program that uses recycled bicycles [213,216].
4. Open data: Amsterdam has made a large amount of data publicly available, which enables researchers and developers to create innovative solutions to urban problems. The city has also established an open data portal that provides access to a wide range of datasets [211,213].

4.5.4. Copenhagen

Copenhagen is often considered a smart city due to its extensive implementation of technology and innovative solutions to enhance the quality of life, sustainability, and efficiency of the city [217].

1. Sustainable urban planning: Copenhagen has adopted a strong focus on sustainable urban planning. The city promotes compact development, mixed land use, and efficient transportation systems. It prioritizes cycling infrastructure, pedestrian-friendly streets, and public transportation, which contribute to reduced carbon emissions and improved mobility [209,218].
2. Renewable energy: Copenhagen aims to become carbon-neutral by 2025 and has made significant progress in utilizing renewable energy sources. The city has implemented wind turbines, district heating systems, and smart grid technologies to optimize energy production, distribution, and consumption [215,219].
3. Energy efficiency: Copenhagen has a robust smart grid infrastructure that allows for efficient management of energy resources. Smart grid technologies enable real-time monitoring, load balancing, and demand response, leading to better energy management and reduced wastage [220,221].
4. Integrated transport systems: Copenhagen has a well-integrated and multimodal transport system. It incorporates smart technologies for traffic management, intelligent traffic signals, and real-time public transport information, enabling smoother traffic flow and reducing congestion [214,222].
5. Data-driven decision making: Copenhagen utilizes data and digital technologies to make informed decisions and improve city services. The city collects and analyzes data on various aspects, including energy consumption, transportation patterns, and air quality, to identify areas for improvement and implement targeted solutions [211,222,223].
6. Citizen engagement: Copenhagen actively engages citizens in decision-making processes and encourages citizen participation through digital platforms. The city utilizes digital tools for public consultations, feedback collection, and collaborative problem-solving, fostering a sense of ownership and promoting a participatory approach [210,222].
7. Smart and connected infrastructure: Copenhagen leverages smart technologies to optimize the functioning of infrastructure. This includes smart street lighting, waste management systems, and sensor networks for monitoring environmental conditions, allowing for timely interventions and resource optimization [206,224].
8. Innovation ecosystem: Copenhagen has a thriving innovation ecosystem, with a focus on startups, research institutions, and industry collaborations. The city supports entrepreneurship, technology incubators, and innovation hubs, fostering the development and implementation of smart city solutions. Copenhagen's commitment to sustainability, use of technology for data-driven decision-making, and citizen-centric approach contribute to its reputation as a smart and livable city [224,225].

4.5.5. Tokyo

Tokyo is recognized for its advanced technology and efficient infrastructure. The city has implemented a number of smart solutions to manage traffic flow and improve transportation, including smart traffic lights and a high-speed train system. Tokyo is also known for its smart building technology, which includes energy-efficient systems that reduce waste and lower costs [226].

1. Advanced technology infrastructure: Tokyo has a highly advanced technology infrastructure that supports its smart city initiatives. The city has a strong focus on the development of 5G networks, IoT, and AI [206,226].
2. Energy efficiency: Tokyo has implemented various initiatives to improve energy efficiency and reduce carbon emissions. For example, the city has implemented a program to promote the installation of solar panels on buildings [221,226].
3. Disaster management: Tokyo is known for its advanced disaster management systems, which help the city respond quickly and effectively to natural disasters such as earthquakes and typhoons. The city has implemented various initiatives such as early warning systems and disaster drills [226,227].
4. Smart buildings: many buildings in Tokyo are equipped with smart technologies that help reduce energy consumption and improve energy efficiency. For example, many buildings use sensors to automatically adjust lighting and temperature based on occupancy levels [226,228].

4.5.6. Dubai

Renowned for its advanced infrastructure and use of cutting-edge technology, Dubai has implemented a number of smart solutions to improve transportation, including a smart traffic management system and a high-speed train system. Additionally, Dubai has implemented a number of smart solutions to improve the quality of life for its citizens, including smart lighting systems and a smart waste management system [229].

1. Advanced infrastructure: Dubai has invested heavily in advanced infrastructure to support its smart city initiatives. This includes the installation of high-speed fiber optic networks, the development of 5G networks, and the implementation of the Internet of Things (IoT) technologies [206,229].
2. Smart transportation: Dubai has implemented a comprehensive transportation system that integrates various modes of transport such as buses, trains, and taxis. The city has also developed a smart parking system that helps drivers find available parking spots using sensors and mobile apps [214,229].
3. Sustainable energy: Dubai has a strong focus on sustainable energy and has implemented various initiatives to reduce energy consumption and increase the use of renewable energy sources. For example, the city has developed a large-scale solar power plant and a district cooling system that uses waste heat to cool buildings [215,229].
4. Smart government: Dubai has implemented various initiatives to create a smart government, including the development of a government services portal and the implementation of e-voting systems [229]. Overall, these cities demonstrate how smart technology can be used to improve quality of life, reduce energy consumption, and enhance sustainability [230].

4.5.7. NEOM

NEOM is a $500 billion mega-city development project in Saudi Arabia. It is an ambitious initiative that aims to create a futuristic, sustainable city in the northwest part of the country, spanning over 26,500 square kilometers [231,232]. NEOM is envisioned as a hub for innovation, technology, and economic diversification, with a focus on various sectors such as energy, water, biotechnology, food, entertainment, and tourism [231,233]. Key features of NEOM include:

1. Sustainability: NEOM aims to be a model for sustainable development, with a focus on renewable energy sources, eco-friendly infrastructure, and efficient use of resources. The project seeks to minimize its environmental impact and promote sustainable practices [214,215,231].

2. Technology and innovation: NEOM plans to leverage cutting-edge technologies and innovations to create a smart city ecosystem. It aims to be a hub for research and development, attracting tech companies, startups, and entrepreneurs. The city intends to implement advanced technologies such as artificial intelligence, robotics, and automation [206,231].
3. Economic diversification: NEOM is part of Saudi Arabia's broader Vision 2030 initiative, which aims to reduce the country's dependence on oil and diversify its economy. NEOM seeks to attract domestic and international investments, foster entrepreneurship, and create job opportunities across various industries [216,231].
4. Quality of life: the project emphasizes improving the quality of life for residents and visitors. NEOM aims to provide world-class infrastructure, healthcare facilities, education, cultural amenities, and recreational spaces. The city plans to promote a vibrant and inclusive community that offers a high standard of living [231,234].
5. Strategic location: NEOM's location along the Red Sea coast provides opportunities for trade, logistics, and tourism. It aims to connect Asia, Europe, and Africa through its strategic position, enabling the development of a thriving economic zone [220]. It is important to note that NEOM is still in the development stage, and many aspects of the project are yet to be fully realized. As the project progresses, it will be essential to assess its implementation, sustainability efforts, economic impact, and the overall achievement of its goals.

4.5.8. New Administrative Capital

One example of a smart city in Egypt is the new administrative capital (NAC), which is currently under construction. The NAC is being built to alleviate the population and traffic congestion in Cairo and to serve as a model for future sustainable and smart cities in Egypt [235]. The NAC is planned to be a fully integrated smart city, using advanced technologies to enhance the quality of life for its residents [236].

1. Intelligent traffic management: the city will use sensors and cameras to monitor traffic flow and adjust traffic lights in real time to improve traffic flow and reduce congestion [206,214,235].
2. Renewable energy: the city is being designed to run on clean and renewable energy, with solar panels and wind turbines being installed throughout the city [215,235].
3. Smart buildings: the buildings in the NAC will be equipped with energy-efficient systems, smart lighting, and automated temperature control [228,235].
4. Smart waste management: the city will use sensors to monitor waste levels in bins and optimize collection schedules, reducing the amount of waste that goes to landfills [25,235].
5. Integrated public transportation: the NAC will have an integrated public transportation system, including buses, trams, and a metro line, with smart ticketing and real-time information for passengers [214,235].

5. Smart City Evaluation Metrics

5.1. Evaluation Metrics

Assessing smart city performance requires evaluating various aspects of the city's implementation of smart technologies and their impact on the quality of life, sustainability, and efficiency forecast in 2025 [237,238]. As depicted in Figure 5, the evaluation metrics may include:

1. Infrastructure

Evaluate the quality and coverage of the city's digital infrastructure, such as broadband connectivity, sensors, data centers, and communication networks. A robust infrastructure is essential for supporting smart city services [237,238].

As illustrated in Table 2, all of these cities have made significant investments in smart infrastructure. They have all implemented modern transportation systems, switched to renewable energy sources, and developed innovative waste management and water treatment systems. They have also made significant investments in healthcare, education, and security.

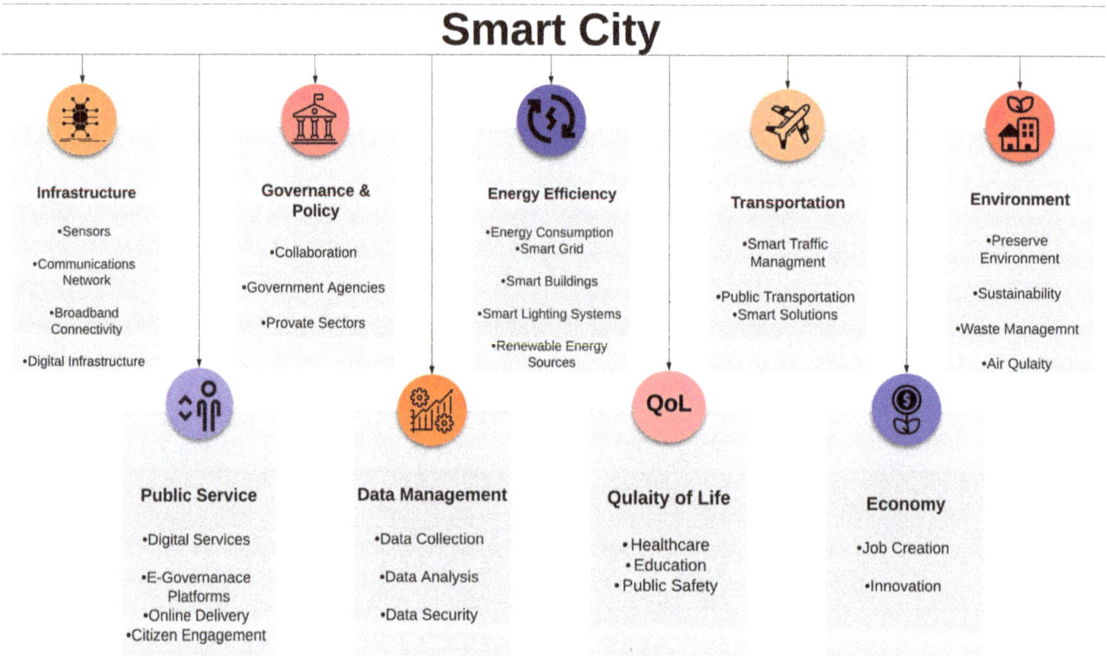

Figure 5. Smart City Evaluation Metrics.

Table 2. Infrastructure Metric [191].

City	Transportation	Energy	Water	Waste
Singapore	Highly developed	Highly efficient	Well-managed	Modern
Barcelona	Well-connected	Efficient	Well-managed	Modern
Amsterdam	Well-connected	Efficient	Well-managed	Sustainable
Copenhagen	Bike-friendly	Sustainable	Well-managed	Sustainable
Tokyo	Complex	Efficient	Well-managed	Modern
Dubai	Modern	Efficient	Well-managed	Modern
Neom	Innovative	Sustainable	Sustainable	Sustainable
NAC	Innovative	Sustainable	Sustainable	Sustainable

It is difficult to say which city has the best infrastructure, as they all have their own strengths and weaknesses. However, Singapore [239], Barcelona [193], Amsterdam [212], Copenhagen [217], and Tokyo are all leading the way in terms of smart city development. NEOM [231] and the NAC [236] are also promising new developments, but they are still under construction.

Overall, these eight cities are setting the standard for smart city infrastructure in 2025. They are all investing in technologies that will help them become more sustainable, efficient, and livable [240,241].

2. Governance and Policy

Examine the city's governance structure and policies related to smart city development. Assess the level of collaboration between different stakeholders, including government agencies, private sector partners, and the public. Search for evidence of effective planning, coordination, and regulation of smart programs [237,238].

As depicted in Table 3, there is no one-size-fits-all approach to smart city governance and policy. Each city has its own unique set of challenges and opportunities, and its government must tailor its approach accordingly [230].

Table 3. Government and Policy Metric [214].

City	Strategy	Key Areas	Data Sharing	Cybersecurity
Singapore	Yes	Sustainability, mobility, economy, QoL	Yes	Yes
Barcelona	Yes		Yes	Yes
Amsterdam	Yes	Sustainability, livability, economic growth	Yes	Yes
Copenhagen	Yes	Energy, water, waste, mobility, buildings	Yes	Yes
Tokyo	Yes	Transportation, energy, environment	Yes	Yes
Dubai	Yes	Mobility, energy, water, waste, environment	Yes	Yes
Neom	Yes	Mobility, energy, water, waste, environment, community	TBD	TBD
NAC	Yes		TBD	TBD

Another common theme is the focus on sustainability [213,214]. Many smart cities are using technology to reduce their environmental impact [225]. For example, they are using smart transportation systems [214] to reduce traffic congestion and pollution, and they are using smart water systems to conserve water [225].

Overall, the governance and policy frameworks of these smart cities are well-established and ambitious [230]. These cities are all working to develop and implement policies that will support their smart city visions. However, there is still some work to be done in terms of data sharing and cybersecurity [211].

3. Energy Efficiency

Assess the city's efforts to optimize energy consumption through smart grids, smart buildings, and energy management systems. Look for proposals that reduce energy waste, promote renewable energy sources, and improve overall energy efficiency [237,238].

The energy efficiency score is a measure of how efficiently a city uses energy by 2025. It is calculated by considering factors such as the city's energy consumption, its renewable energy production, and its energy efficiency policies [215].

As shown in Table 4, Singapore [239] has the highest energy efficiency score, followed by Barcelona and Amsterdam [212]. These cities are all leading the way in terms of sustainable urban development. They have implemented a number of policies and initiatives to reduce their energy consumption and increase their reliance on renewable energy [215].

Dubai and NEOM are also making significant progress in terms of energy efficiency. These cities are investing heavily in renewable energy projects and are developing new technologies to improve energy efficiency [215,231].

The new administrative city in Egypt is still in its early stages of development, but it has ambitious plans to become a sustainable city. The city is targeting an energy efficiency score of 70 by 2030 [236].

Overall, the energy efficiency of smart cities is improving. These cities are leading the way in terms of sustainable urban development and are setting an example for other cities around the world [242].

Table 4. Energy Efficiency Metric [199].

City	Target Energy Efficiency	Actual Energy Efficiency	Difference
Singapore	80%	85%	5
Barcelona	75%	78%	3
Amsterdam	70%	73%	3
Copenhagen	65%	68%	3
Tokyo	60%	63%	3
Dubai	55%	58%	3
Neom	50%	53%	3
NAC	45%	48%	3

4. Mobility and Transportation

Evaluate the city's transportation infrastructure, including smart traffic management systems, public transportation networks, and integration of emerging technologies like electric vehicles and autonomous vehicles, as illustrated in Table 5. Propose smart solutions to reduce traffic congestion, promote sustainable transportation alternatives, and improve mobility for residents [237,238].

Table 5. Mobility and Transportation Metric [198].

City	Key Smart Mobility Initiatives
Singapore	Public transportation, shared mobility, autonomous vehicles, smart parking
Barcelona	Bike sharing, autonomous vehicles, pedestrian-friendly streets
Amsterdam	Electric buses, underground train system, cycling culture
Copenhagen	Bike share program, cycle paths, electrifying public transportation
Tokyo	Autonomous buses, high-speed rail network, pedestrian-friendly streets
Dubai	Autonomous Vehicles, high-speed rail network, new airport
Neom	High-speed rail network, autonomous vehicles, new airport
NAC	Light rail system, electric buses, new airport

Overall, these cities are all making great strides in smart mobility [214]. They are investing in a variety of projects, such as autonomous vehicles, shared mobility, and sustainable transportation. These investments are helping to make these cities more livable and sustainable [225].

5. Environment and Sustainability

Assess the city's initiatives aimed at preserving the environment and promoting sustainability. Look for programs that focus on waste management, air quality monitoring, water conservation, green spaces, and the utilization of renewable energy sources. Consider the city's overall carbon footprint and its efforts to mitigate climate change [237,238].

As illustrated in Table 6, all of the cities have made significant progress in terms of environmental sustainability. However, there are some clear leaders, such as Singapore, Copenhagen, and Amsterdam. These cities have achieved high scores in all categories, and they are well-positioned to continue their progress in the years to come [78,218].

Table 6. Environment and Sustainability Metric [208].

City	Green Space	Renewable Energy	Public Transportation	Waste Management	Water Conservation	Overall Score
Singapore	52%	35%	90%	95%	98%	90
Barcelona	45%	70%	85%	90%	95%	87
Amsterdam	40%	80%	95%	95%	98%	89
Copenhagen	45%	90%	95%	98%	99%	93
Tokyo	35%	40%	90%	95%	97%	86
Dubai	25%	20%	80%	90%	95%	77
Neom	50%	50%	90%	95%	98%	88
NAC	55%	80%	95%	98%	99%	92

6. Public Services and Civic Engagement

Evaluate the availability and accessibility of digital services provided to residents, such as e-governance platforms, online service delivery, and citizen engagement tools as depicted in Table 7. Taking into consideration the city's efforts to involve the public in decision-making processes and the extent to which technology enhances civic participation and responsiveness [237,238].

Table 7. Public Service and Civic Engagement [195].

City	Public Services	Civic Engagement	Overall Score
Singapore	Excellent	Good	Excellent
Barcelona	Good	Excellent	Very Good
Amsterdam	Good	Good	Good
Copenhagen	Excellent	Good	Excellent
Tokyo	Very Good	Good	Very Good
Dubai	Good	Fair	Fair
Neom	Fair	Fair	Fair
NAC	Poor	Poor	Poor

Overall, these cities are all making significant progress in terms of public services and civic engagement [240]. They are using technology to improve the lives of citizens, and they are also encouraging citizens to get involved in local decision-making. These cities are setting a good example for other cities around the world, and they are showing how technology can be used to create more livable and sustainable cities.

7. Data Management and Privacy

As shown in Table 8, examine how the smart city collects, stores, and analyzes massive data generated by smart technologies. Assess the level of data security and privacy protection measures in place, as well as transparency in data usage and consent mechanisms [211,237,238].

Table 8. Data Management and Privacy [196].

City	Data Management	Privacy
Singapore	High	High
Barcelona	High	Medium
Amsterdam	Medium	Medium
Copenhagen	Medium	High
Tokyo	Medium	Low
Dubai	Low	Low
Neom	Low	Very Low
NAC	Low	Low

- *Data Management*

High: Cities with strong data management practices have clear policies and procedures for collecting, storing, and using data. They also have robust security measures in place to protect data from unauthorized access or disclosure.

Medium: Cities with medium-level data management practices have some policies and procedures in place, but they may not be as comprehensive or well-enforced as in high-level cities. They may also have some security measures in place, but they may not be as robust as in high-level cities.

Low: Cities with low-level data management practices have few or no policies and procedures in place for collecting, storing, and using data. They may also have very few security measures in place.

- *Privacy*

High: Cities with high levels of privacy protection have strong laws and regulations that protect the privacy of citizens. They also have transparent data collection and use practices, and they give citizens the ability to control their personal data.

Medium: Cities with medium levels of privacy protection have some laws and regulations in place, but they may not be as comprehensive or well-enforced as in high-level cities. They may also have some transparent data collection and use practices, but they may not give citizens as much control over their personal data.

Low: Cities with low levels of privacy protection have few or no laws and regulations in place to protect the privacy of citizens. They may also have opaque data collection and use practices, and they may not give citizens any control over their personal data.

8. Quality of Life (QoL)

Assess the overall impact of smart city initiatives on the quality of life for residents. Consider factors such as improved access to healthcare, education, safety, public amenities, and cultural opportunities [234] as it Table 9. Evaluate the level of inclusion and equity in the implementation of smart technologies to ensure that benefits are accessible to all sectors of the population [237,238]. The QoL score is based on a number of factors, including:

- Economy: the city's GDP per capita, unemployment rate, and job growth rate [225].
- Healthcare: the quality of the city's healthcare system, life expectancy, and infant mortality rate [123].
- Education: the quality of the city's schools, universities, and adult education programs [175].
- Environment: the city's air quality, water quality, and green space [230].
- Safety: the city's crime rate, traffic accident rate, and fire rate [212].
- Culture: the city's museums, art galleries, theaters, and other cultural attractions [175,218].
- Leisure: the city's parks, sports facilities, and other recreational opportunities [175,218].
- Transportation: the city's public transportation system, roads, and airports [214].

Table 9. Quality of Life (QoL) Metric [218].

City	QoL Score
Singapore	91.5
Barcelona	90.5
Amsterdam	90.0
Copenhagen	89.5
Tokyo	88.5
Dubai	80.0
Neom	78.5
NAC	77.5

9. Economic Development

Evaluate the impact of smart city initiatives on economic growth, job creation, and innovation. Look for evidence of attracting investment, supporting local businesses, fostering entrepreneurship, and creating a favorable environment for technology startups [237,238].

As depicted in Table 10, Singapore [239] has the highest GDP per capita of all the cities listed. NEOM has a 0% growth rate, while NAC has a negative growth rate. It is important to note that GDP growth rate is not the only factor that contributes to economic development. Other factors such as unemployment rate, inflation rate, and QoL also play a role [200].

Table 10. Economic Development Metric [200].

City	GDP Growth
Singapore	3.5%
Barcelona	2.5%
Amsterdam	2.0%
Copenhagen	1.5%
Tokyo	1.0%
Dubai	0.5%
Neom	0.0%
NAC	−1.0%

Overall, the economic development of the smart cities listed above is promising. These cities are all working to improve their infrastructure and to become more attractive to foreign investment. As a result, they have the potential to become major economic hubs in the years to come [200].

Consider whether the city has established appropriate metrics and indicators to measure the performance and effectiveness of smart city initiatives [237]. Assess the availability of data-driven feedback loops that allow continuous monitoring, evaluation, and iterative improvement of smart city projects [238]. It is important to note that assessing smart city performance is a complex task, and the evaluation criteria may vary depending on the specific goals, context, and priorities of each city [243].

5.2. Smart Cities' Evaluation

1. Singapore is a leading smart city in terms of its use of technology to improve the lives of its citizens. It has a well-developed smart transportation system, including a metro system, bus network, and public bike sharing program. The city also has a number of

smart buildings and homes that are equipped with sensors and other technology to monitor energy use and provide residents with information about their surroundings.
2. Barcelona is another city that is making great strides in the field of smart city development. The city has a number of innovative projects underway, such as a smart lighting system that uses sensors to adjust the brightness of streetlights based on traffic levels and a smart water management system that uses sensors to monitor water usage and leaks.
3. Amsterdam is a city that is known for its commitment to sustainability. The city has a number of smart city initiatives in place that are designed to reduce its environmental impact. These initiatives include a smart waste management system, a smart water management system, and a smart transportation system.
4. Copenhagen is another city that is making great strides in the field of sustainability. The city has a number of smart city initiatives in place that are designed to reduce its environmental impact. These initiatives include a smart waste management system, a smart water management system, and a smart transportation system.
5. Tokyo is a city that is known for its technological prowess. The city has a number of smart city initiatives in place that are designed to improve the lives of its citizens. These initiatives include a smart transportation system, a smart water management system, and a smart healthcare system.
6. Dubai is a city that is known for its ambition. The city has a number of ambitious smart city projects underway, such as a smart transportation system, a smart water management system, and a smart healthcare system.
7. NEOM. The city is designed to be a hub for innovation and technology. NEOM has a number of ambitious projects underway, such as a smart transportation system, a smart water management system, and a smart healthcare system.
8. NAC is a new smart city that is being built in Egypt. The city is designed to be a hub for government and business and has a number of ambitious projects underway, such as a smart transportation system, a smart water management system, and a smart healthcare system.

5.3. *Smart Cities' Implementation Challenges*

Smart cities are urban areas that use digital technologies and data to enhance the quality of life [218], efficiency, and sustainability [198,199] of their residents and stakeholders. However, there are a number of challenges to implementing smart city [244] solutions, including:

5.3.1. Funding

Smart city projects can be expensive, and funding can be a challenge [244], especially in developing countries.

5.3.2. Infrastructure

Smart city technologies require a good foundation of physical infrastructure, such as reliable power, water, and telecommunications networks [169]. In many cities, this infrastructure is outdated or lacking [244].

5.3.3. Data Privacy

The collection and use of data is a key part of smart city projects. However, there are concerns about data privacy and security [245].

5.3.4. Security

Smart city technologies are often connected to the internet, which makes them vulnerable to cyberattacks [245].

5.3.5. Lack of Coordination

Smart city projects often involve multiple stakeholders, such as government agencies, businesses, and citizens. It can be difficult to coordinate these stakeholders and ensure that they are working together towards a common goal [246].

5.3.6. Public Acceptance

Smart city projects need to be accepted by the public in order to be successful. If people do not trust or understand technology, they may be reluctant to participate [243].

Despite these challenges, there are a number of cities that are making progress in implementing smart city solutions. These cities are finding ways to overcome the challenges and gain the benefits of smart city technologies.

5.4. Recommendations

Here are some of the ways that cities are overcoming the challenges of implementing smart cities:

5.4.1. Finding New Sources of Funding

Cities are finding new ways to fund smart city projects, such as through public-private partnerships, impact investing, and crowdfunding.

5.4.2. Building New Infrastructure

Cities are investing in new infrastructure, such as smart grids, fiber optic networks, and sensor networks.

5.4.3. Protecting Data Privacy

Cities are developing new data privacy and security policies to protect the personal data of their citizens.

5.4.4. Securing Smart City Technologies

Cities are using security measures, such as encryption and firewalls, to protect smart city technologies from cyberattacks.

5.4.5. Building Consensus

Cities are working to build consensus among stakeholders and the public on the benefits of smart city technologies.

The challenges of implementing smart cities are significant, but they are not insurmountable. With careful planning and execution, cities can overcome these challenges and reap the benefits of smart city technologies.

6. Conclusions

In conclusion, this paper has provided a comprehensive review of global models of smart cities and explored the potential applications of the Internet of Things (IoT) in shaping the cities of the future. The analysis of various smart city initiatives from around the world has revealed common themes and key components that contribute to their success. These models encompass a holistic approach that integrates technology, governance, and citizen engagement to create sustainable, efficient, and livable urban environments. Through the examination of different IoT applications in smart cities, it has become evident that the deployment of connected devices and sensors has the potential to revolutionize urban systems and services. From transportation and energy management to healthcare and waste management, IoT technologies offer innovative solutions to address the complex challenges faced by cities. They enable real-time data collection, analysis, and decision-making, leading to improved efficiency, resource optimization, and enhanced quality of life for residents. However, while the potential benefits of IoT applications in smart cities are promising, several challenges and considerations need to be addressed. These include data privacy and security concerns, interoperability, and standardization issues, as well as social implications. Policymakers, urban planners, and technology providers must work collaboratively to address these challenges and ensure that the deployment of IoT in smart cities is done in a responsible and sustainable manner. Looking ahead, the evolution of smart cities will continue to be driven by technological innovations, evolving citizen needs, and the pursuit of sustainability. The integration of current technologies such as artificial intelligence, 5G, and the emerging 6G will further enhance the capabilities of smart cities and open new possibilities for innovation. Moreover, the ongoing collaboration and knowledge sharing among cities worldwide will accelerate the development and implementation of best practices, enabling cities to learn from each other and build upon successful models.

Author Contributions: Introduction—A.H. and M.T.; smart city—A.H.; communication systems—A.H.; smart city models—M.T.; conclusion—A.H. and M.T. All authors have read and agreed to the published version of the manuscript.

Funding: This research received no external funding.

Data Availability Statement: Not applicable.

Conflicts of Interest: The authors declare no conflict of interest.

References

1. Statista. Number of Internet of Things (IoT) Connected Devices Worldwide from 2019 to 2023, with Forecasts from 2022 to 2030. Available online: https://www.statista.com/statistics/1183457/iot-connected-devices-worldwide/ (accessed on 15 May 2023).
2. Global IoT Market Size. IoT Analytics. 2023. Available online: https://iot-analytics.com/iot-market-size/ (accessed on 15 May 2023).
3. Hassebo, A. *Commercial 4G LTE Cellular Networks for Supporting Emerging Mission-Critical IoT Applications*; ProQuest Dissertations Publishing: New York, NY, USA, 2018.
4. Tealab, M.; Hassebo, A.; Dabour, A.; AbdelAziz, M. Smart Cities Digital transformation and 5G—ICT Architecture. In Proceedings of the 2020 11th IEEE Annual Ubiquitous Computing, Electronics & Mobile Communication Conference (UEMCON), New York, NY, USA, 28–31 October 2020.
5. Hassebo, A.; Rezk, A.; Ali, M.A. A Hybrid Uplink Scheduling Approach for Supporting Mission-Critical Smart Grid Applications in Commercial 4G Cellular Networks. In Proceedings of the 2018 9th IEEE Annual Ubiquitous Computing, Electronics & Mobile Communication Conference (UEMCON), New York, NY, USA, 8–10 November 2018.
6. Stallings, W. *5g Wireless: A Comprehensive Introduction*; Addison-Wesley Professional: Boston, MA, USA, 2021.
7. Ali, N.A.; Taha, A.M.; Hassanein, H.S. *LTE, LTE-Advanced and Wi MAX Towards IMT-Advanced Networks*; Wiley: New York, NY, USA, 2012.
8. United Nations. *World Population Prospects 2022*; Department of Economic and Social Affairs, Population Division, The United Nations: New York, NY, USA, 2022.
9. United Nations. *World Urbanization Prospects 2018*; Department of Economic and Social Affairs, Population Division, The United Nations: New York, NY, USA, 2018.
10. Keshav, K.; Pramod, B.; Vineet, K. *Smart Cities Market Statistics: 2030*; Allied Market Research: Pune, India, 2022.
11. Lanvin, B. *Smart City Index 2023*; IMD World Competitiveness Center (WCC): Lausanne, Switzerland, 2023.
12. Hassebo, A.; Obaidat, M.; Ali, M.A. Commercial 4G LTE Cellular Networks for Supporting Emerging IoT Applications. In Proceedings of the 2018 Advances in Science and Engineering Technology International Conferences (ASET), Dubai, United Arab Emirates, 6 February 2018–5 April 2018.
13. Hassebo, A. The Road to 6G, Vision, Drivers, Trends, and Challenges. In Proceedings of the 2022 IEEE 12th Annual Computing and Communication Workshop and Conference (CCWC), Las Vegas, NV, USA, 26–29 January 2022.
14. Hassebo, A.; Ali, M.A. Robust Cellular Connectivity-Based Smart LED Street Lighting System: A Platform for Innovative Mission Critical Smart City IoT Applications. In Proceedings of the 2020 11th IEEE Annual Ubiquitous Computing, Electronics & Mobile Communication Conference (UEMCON), New York, NY, USA, 28–31 October 2020.
15. Achieng, M.; Ogundaini, O.; Makola, D.; Iyamu, T. The African Perspective of a Smart City: Conceptualisation of Context and Relevance. In Proceedings of the IST-Africa 2021 Conference Proceedings, South Africa, South Africa, 10–14 May 2021.
16. Joo, Y.M.; Tan, T.B. Smart Cities in Asia: An Introduction. In *Smart Cities in Asia: Governing Development in the Era of Hyper-Connectivity*; Edward Elgar Publishing: Cheltenham, UK, 2020.
17. Garrido-Marijuan, A.; Pargova, Y.; Wilson, C. *The Making of a Smart City: Best Practices Across Europe*; European Union: Brussels, Belgium, 2017.

18. Shahzad, F.; Rehman, S.u.; Javed, A.; Jalil, Z.; Bin-Zikria, Y. Future Smart Cities: Requirements, Emerging Technologies, Applications, Challenges, and Future Aspects. *Cities* **2021**, *129*, 103794.
19. Oladimeji, D.; Gupta, K.; Kose, N.A.; Gundogan, K.; Ge, L.; Liang, F. Smart Transportation: An Overview of Technologies and Applications. *Sensors* **2023**, *23*, 3880. [CrossRef] [PubMed]
20. Ceglia, F.; Marrasso, E.; Pallotta, G.; Roselli, C.; Sasso, M. The State of the Art of Smart Energy Communities: A Systematic Review of Strengths and Limits. *Energies* **2022**, *15*, 3462. [CrossRef]
21. Hassebo, A. xIoT-Based Converged 5G and ICT Infrastructure. In *Electric Grid Modernization*; Intech Open: London, UK, 2021.
22. Sosunova, I.; Porras, J. IoT-Enabled Smart Waste Management Systems for Smart Cities: A Systematic Review. *IEEE Access* **2022**, *10*, 73326–73363. [CrossRef]
23. Mihigo, I.N.; Zennaro, M.; Uwitonze, A.; Rwigema, J.; Rovai, M. On-Device IoT-Based Predictive Maintenance Analytics Model: Comparing TinyLSTM and TinyModel from Edge Impulse. *Sensors* **2022**, *22*, 14. [CrossRef]
24. Ikram, S.T.; Mohanraj, V.; Ramachandran, S.; Balakrishnan, A. An Intelligent Waste Management Application Using IoT and a Genetic Algorithm—Fuzzy Inference System. *Appl. Sci.* **2023**, *13*, 3943. [CrossRef]
25. Abdullah, N.; Al-wesabi, O.A.; Mohammed, B.A.; Al-Mekhlafi, Z.G.; Alazmi, M.; Alsaffar, M.; Baklizi, M.; Sumari, P. IoT-Based Waste Management System in Formal and Informal Public Areas in Mecca. *Int. J. Environ. Res. Public Health* **2022**, *19*, 20. [CrossRef]
26. Fraga-Lamas, P.; Fernández-Caramés, T.M.; Suárez-Albela, M.; Castedo, L.; González-López, M. A Review on Internet of Things for Defense and Public Safety. *Sensors* **2016**, *16*, 1644. [CrossRef]
27. AlGhamdi, R.; Sharma, S.K. IoT-Based Smart Water Management Systems for Residential Buildings in Saudi Arabia. *Processes* **2022**, *10*, 2022. [CrossRef]
28. Tian, S.; Yang, W.; Le Grange, J.M.; Wang, P.; Huang, W.; Ye, Z. Smart healthcare: Making medical care more intelligent. *Glob. Health J.* **2019**, *3*, 62–65. [CrossRef]
29. Anthopoulos, L.; Reddick, C.G. Smart City and Smart Government: Synonymous or Complementary? In Proceedings of the 25th International Conference Companion on World Wide Web, Geneva, Switzerland, 11–16 April 2016.
30. Nastjuk, I.; Trang, S.; Papageorgiou, E.I. Smart cities and smart governance models for future cities. *Electron. Mark.* **2022**, *32*, 1917–1924. [CrossRef]
31. Kim, D.; Yoon, Y.; Lee, J.; Mago, P.J.; Lee, K.; Cho, H. Design and Implementation of Smart Buildings: A Review of Current Research Trend. *Energies* **2022**, *15*, 4278. [CrossRef]
32. U.S. Department of Energy. *National Smart Manufacturing Strategic Plan*; U.S. Department of Energy: Washington, DC, USA, 2022.
33. Mohsan, S.A.H.; Khan, M.A.; Noor, F.; Ullah, I.; Alsharif, M.H. Towards the Unmanned Aerial Vehicles (UAVs): A Comprehensive Review. *Drones* **2022**, *6*, 147. [CrossRef]
34. Golubchikov, O.; Thornbush, M. Artificial Intelligence and Robotics in Smart City Strategies and Planned Smart Development. *Smart Cities* **2020**, *3*, 4. [CrossRef]
35. Artsen, A.M.; Burkett, L.S.; Duvvuri, U.; Bonidie, M. Surgeon satisfaction and outcomes of tele-proctoring for robotic gynecologic surgery. *J. Robot. Surg.* **2021**, *16*, 563–568. [CrossRef] [PubMed]
36. Syed, A.; Sierra-Sosa, D.; Kumar, A.; Elmaghraby, A. IoT in Smart Cities: A Survey of Technologies, Practices and Challenges. *Smart Cities* **2021**, *4*, 429–475. [CrossRef]
37. Gracias, J.; Parnell, G.; Specking, E.; Pohl, E.; Buchanan, R. Smart Cities-A Structured Literature Review. *Smart Cities* **2023**, *6*, 4. [CrossRef]
38. Dameri, R.P. *Smart City Implementation*; Springer International Publishing: Cham, Switzerland, 2017.
39. Statista. Passenger Cars—Worldwide. 2022. Available online: https://www.statista.com/outlook/mmo/passenger-cars/worldwide (accessed on 17 May 2023).
40. Masek, P.; Masek, J.; Frantik, P.; Fujdiak, R.; Ometov, A.; Hosek, J.; Andreev, S.; Mlynek, P.; Misurec, J. A Harmonized Per-spective on Transportation Management in Smart Cities: The Novel IoT-Driven Environment for Road Traffic Modeling. *Sensors* **2016**, *16*, 11. [CrossRef]
41. Zantalis, F.; Koulouras, G.; Karabetsos, S.; Kandris, D. A Review of Machine Learning and IoT in Smart Transportation. *Future Internet* **2019**, *11*, 94. [CrossRef]
42. Kadłubek, M.; Thalassinos, E.; Domagała, J.; Grabowska, S.; Saniuk, S. Intelligent Transportation System Applications and Logistics Resources for Logistics Customer Service in Road Freight Transport Enterprises. *Energies* **2022**, *15*, 4668. [CrossRef]
43. Dominguez, M.T. *Transportation Asset Management Plan*; NYS DoT: Albany, NY, USA, 2019.
44. Kostrzewski, M.; Melnik, R. Condition Monitoring of Rail Transport Systems: A Bibliometric Performance Analysis and Systematic Literature Review. *Sensors* **2021**, *21*, 4710. [CrossRef] [PubMed]
45. Ersöz, O.Ö.; Inal, A.F.; Aktepe, A.; Türker, A.K.; Ersöz, S. A Systematic Literature Review of the Predictive Maintenance from Transportation Systems Aspect. *Sustainability* **2022**, *14*, 21. [CrossRef]
46. Meldert, B.V.; Boeck, L.D. *Introducing Autonomous Vehicles in Logistics: A Review from a Broad Perspective*; Faculty of Business and Economics, KU Leuven: Leuven, Belgium, 2016.
47. IRENA. *Electric Vehicles Technology Brief*; International Renewable Energy Agency (IRENA): Masdar City, United Arab Emirates, 2017.

48. Muller, J.; Fregin, A.; Dietmayer, K. Multi-Camera System for Traffic Light Detection: About Camera Setup and Mapping of Detections. In Proceedings of the 2017 IEEE 20th International Conference on Intelligent Transportation Systems (ITSC), Yokohama, Japan, 16–19 October 2017.
49. IEA. *Global Electric Vehicle Outlook 2022*; International Energy Agency (IEA): Paris, France, 2022.
50. Ahmed, M.A.; El-Sharkawy, M.R.; Kim, Y.-C. Remote Monitoring of Electric Vehicle Charging Stations in Smart Campus Parking Lot. *J. Mod. Power Syst. Clean Energy* **2020**, *8*, 124–132. [CrossRef]
51. Brown, A.; Cappellucci, J.; Schayowitz, A.; White, E.; Heinrich, A.; Cost, E. *Electric Vehicle Charging Infrastructure Trends from the Alternative Fueling Station Locator: First Quarter 2022*; National Renewable Energy Laboratory (NREL): Golden, CO, USA, 2022.
52. US DoT. *Charging Forward*; US DoT: Washington, DC, USA, 2022.
53. Song, N.; Kwak, B. International Standard Trend of Vehicle to Grid (V2G) Communication Interface for Wireless Communication and RPT. In Proceedings of the 2019 IEEE Transportation Electrification Conference and Expo, Asia-Pacific (ITEC Asia-Pacific), Seogwipo, Republic of Korea, 8–10 May 2019.
54. Conti, M.; Donadel, D.; Poovendran, R.; Turrin, F. EVExchange: A Relay Attack on Electric Vehicle Charging System. In *European Symposium on Research in Computer Security*; Springer: Cham, Switzerland, 2022; pp. 488–508. [CrossRef]
55. Massaro, A.; Selicato, S.; Galiano, A. Predictive Maintenance of Bus Fleet by Intelligent Smart Electronic Board Implementing Artificial Intelligence. *IoT* **2019**, *1*, 12. [CrossRef]
56. Campos-Ferreira, A.E.; Lozoya-Santos, J.d.J.; Tudon-Martinez, J.C.; Mendoza, R.A.R.; Vargas-Martínez, A.; Morales-Menendez, R.; Lozano, D. Vehicle and Driver Monitoring System Using On-Board and Remote Sensors. *Sensors* **2023**, *23*, 814. [CrossRef] [PubMed]
57. DoT. *NYC Mobility Report*; NYC Department of Transportation (DoT): Washington DC, USA, 2019.
58. DoT. *Central Business District Tolling Program (CBDTP)*; NYS-DoT: NYC, NY, USA, 2022.
59. Kotb, A.O. *Smart Parking: Guidance, Monitoring and Reservations*; University of Liverpool: Liverpool, UK, 2016.
60. NYU Rudin Center for Transportation Policy. *Urban Mobility in the 21st Century*; NYU Rudin Center for Transportation Policy: New York, NY, USA, 2012.
61. Statista. Population Projections for the United States from 2015 to 2060. 2020. Available online: https://www.statista.com/statistics/183481/united-states-population-projection/ (accessed on 17 May 2023).
62. Green Car Reports. 2014. Available online: https://www.greencarreports.com/news/1093560_1-2-billion-vehicles-on-worlds-roads-now-2-billion-by-2035-report (accessed on 22 May 2023).
63. Autonomous Vehicle Market Statistics 2030. Allied Market Reserach, 2022. Available online: https://www.alliedmarketresearch.com/autonomous-vehicle-market (accessed on 23 May 2023).
64. Sacks, M. In Two Years, There Could Be 10 Million Self-Driving Cars on the Roads. Available online: https://stanfordmag.org/contents/in-two-years-there-could-be-10-million-self-driving-cars-on-the-roads (accessed on 27 May 2023).
65. Subramanya, K.; Kermanshachi, S.; Pamidimukkala, A. Evaluation of E-Ticketing Technology in Construction of Highway Projects: A Systematic Review of Adoption Levels, Benefits, Limitations and Strategies. *Front. Built Environ.* **2022**, *8*, 890024. [CrossRef]
66. Ivan, C.; Balag, R. An Initial Approach for a NFC M-Ticketing Urban Transport System. *J. Comput. Commun.* **2015**, *3*, 42–64. [CrossRef]
67. Hu, J.; Morais, H.; Sousa, T.; Lind, M. Electric vehicle fleet management in smart grids: A review of services, optimization and control aspects. *Renew. Sustain. Energy Rev.* **2016**, *56*, 1207–1226. [CrossRef]
68. Rojas, B.; Bolaños, C.; Salazar-Cabrera, R.; Ramírez-González, G.; Cruz, Á.P.D.L.; Molina, J.M.M. Fleet Management and Control System for Medium-Sized Cities Based in Intelligent Transportation Systems: From Review to Proposal in a City. *Electronics* **2020**, *9*, 9. [CrossRef]
69. Hassebo, A.; Mohamed, A.A.; Dorsinville, R.; Ali, M.A. 5G-based Converged Electric Power Grid and ICT Infrastructure. In Proceedings of the 2018 IEEE 5G World Forum (5GWF), Silicon Valley, CA, USA, 9–11 July 2018.
70. Stanelyte, D.; Radziukyniene, N.; Radziukynas, V. Overview of Demand-Response Services: A Review. *Energies* **2022**, *15*, 1659. [CrossRef]
71. Hargroves, K.; James, B.; Lane, J.; Newman, P. The Role of Distributed Energy Resources and Associated Business Models in the Decentralised Energy Transition: A Review. *Energies* **2023**, *16*, 4231. [CrossRef]
72. Mazhar, T.; Irfan, H.M.; Haq, I.; Ullah, I.; Ashraf, M.; Al Shloul, T.; Ghadi, Y.Y.; Imran; Elkamchouchi, D.H. Analysis of Challenges and Solutions of IoT in Smart Grids Using AI and Machine Learning Techniques: A Review. *Electronics* **2023**, *12*, 242. [CrossRef]
73. UN. Department of Economic and Social Affairs. 2018. Available online: https://www.un.org/development/desa/en/news/population/2018-revision-of-world-urbanization-prospects.html (accessed on 23 May 2023).
74. EPA—United States Environmental Protection Agency. Sources of Greenhouse Gas Emissions. Available online: https://www.epa.gov/ghgemissions/sources-greenhouse-gas-emissions (accessed on 23 May 2023).
75. Jansen, E.R. *GroningenPhilips Lighting*; Philips Lighting Corporate Communications: Eindhoven, The Netherlands, 2017.
76. Siddiqi, M.H.; Alruwaili, M.; Tarimer, İ.; Karadağ, B.C.; Alhwaiti, Y.; Khan, F. Development of a Smart Signalization for Emergency Vehicles. *Sensors* **2023**, *23*, 4703. [CrossRef] [PubMed]
77. Gellings, C.W. *The Smart Grid: Enabling Energy Efficiency and Demand Response*; CRC Press: Boca Raton, FL, USA, 2020.

78. Alonaizi, S.Y.; Manuel, P. IoT-Based Smart Government Enablers: An Exploration of Governments' Experiments. In Proceedings of the 2021 Fifth World Conference on Smart Trends in Systems Security and Sustainability (WorldS4), London, UK, 29–30 July 2021.
79. Cheema, S.M.; Hannan, A.; Pires, I.M. Smart Waste Management and Classification Systems Using Cutting Edge Approach. *Sustainability* **2022**, *14*, 10226. [CrossRef]
80. Guna, J.; Horvat, K.P.; Podjed, D. People-Centred Development of a Smart Waste Bin. *Sensors* **2022**, *22*, 1288. [CrossRef]
81. Lozano; Caridad, J.; Paz, J.F.D.; González, G.V.; Bajo, J. Smart Waste Collection System with Low Consumption LoRaWAN Nodes and Route Optimization. *Sensors* **2018**, *18*, 5. [CrossRef] [PubMed]
82. Lingaraju, A.; Niranjanamurthy, M.; Bose, P.; Acharya, B.; Gerogiannis, V.; Kanavos, A.; Manika, S. IoT-Based Waste Seg-regation with Location Tracking and Air Quality Monitoring for Smart Cities. *Smart Cities* **2023**, *6*, 3. [CrossRef]
83. Longo, E.; Sahin, F.A.; Redondi, A.E.C.; Bolzan, P.; Bianchini, M.; Maffei, S. A 5G-Enabled Smart Waste Management System for University Campus. *Sensors* **2021**, *21*, 8278. [CrossRef]
84. Debrah, J.; Vidal, D.; Dinis, M. Raising Awareness on Solid Waste Management through Formal Education for Sustaina-bility: A Developing Countries Evidence Review. *Recylcling* **2021**, *6*, 1.
85. Ahmed, S.; Mubarak, S.; Du, J.T.; Wibowo, S. Forecasting the Status of Municipal Waste in Smart Bins Using Deep Learning. *Int. J. Environ. Res. Public Health* **2022**, *19*, 16798. [CrossRef]
86. Veloz-Cherrez, D.; Lozada-Yanez, R.; Mayorga, J.R.P.; Panchi, J. Smart Waste Monitoring System as an Initiative to Develop a Digital Territory in Riobamba City. *Information* **2020**, *11*, 4. [CrossRef]
87. Ciaburro, G.; Iannace, G. Improving Smart Cities Safety Using Sound Events Detection Based on Deep Neural Network Algorithms. *Informatics* **2020**, *7*, 23. [CrossRef]
88. Alam, T. Cloud-Based IoT Applications and Their Roles in Smart Cities. *Smart Cities* **2021**, *4*, 1196–1219. [CrossRef]
89. Lilhore, U.K.; Imoize, A.L.; Li, C.-T.; Simaiya, S.; Pani, S.K.; Goyal, N.; Kumar, A.; Lee, C.-C. Design and Implementation of an ML and IoT Based Adaptive Traffic-Management System for Smart Cities. *Sensors* **2022**, *22*, 2908. [CrossRef] [PubMed]
90. Foremski, P.; Nowak, S.; Fröhlich, P.; Hernández-Ramos, J.L.; Baldini, G. Autopolicy: Automated Traffic Policing for Improved IoT Network Security. *Sensors* **2020**, *20*, 15. [CrossRef]
91. Toutouh, J.; Alba, E. A Low Cost IoT Cyber-Physical System for Vehicle and Pedestrian Tracking in a Smart Campus. *Sensors* **2022**, *22*, 6585. [CrossRef]
92. Lukasik, D.; Hale, D.; Ma, J.; Shibley, P.; Malone, T.; Chandler, A.; Cleary, C.; Matout, N.; Adebisi, A. *Enhancing Active Transportation and Demand Management (ATDM) with Advanced and Emerging Technologies and Data Sources*; US Department of Transportation (DoT): Washington, DC, USA, 2020.
93. Ouallane, A.A.; Bahnasse, A.; Bakali, A.; Talea, M. Overview of Road Traffic Management Solutions based on IoT and AI. *Procedia Comput. Sci.* **2022**, *198*, 518–523. [CrossRef]
94. Humayun, M.; Almufareh, M.F.; Jhanjhi, N.Z. Autonomous Traffic System for Emergency Vehicles. *Electronics* **2022**, *11*, 510. [CrossRef]
95. Mohamed, S.A.E.; AlShalfan, K.A. Intelligent Traffic Management System Based on the Internet of Vehicles (IoV). *J. Adv. Transp.* **2021**, *2021*, 1–23. [CrossRef]
96. Bazargani, J.S.; Sadeghi-Niaraki, A.; Choi, S.-M. A Survey of GIS and IoT Integration: Applications and Architecture. *Appl. Sci.* **2021**, *11*, 10365. [CrossRef]
97. Nellore, K.; Hancke, G.P. Traffic Management for Emergency Vehicle Priority Based on Visual Sensing. *Sensors* **2016**, *16*, 1892. [CrossRef] [PubMed]
98. NPSTC. *Public Safety Internet of Things (IoT) Use Case Report and Assessment Attributes*; NPSTC Technology and Broadband Committee—National Public Safety Telecommunications Council: Washington, DC, USA, 2019.
99. Franchi, F.; Marotta, A.; Rinaldi, C.; Graziosi, F.; Fratocchi, L.; Parisse, M. What Can 5G Do for Public Safety? Structural Health Monitoring and Earthquake Early Warning Scenarios. *Sensors* **2022**, *22*, 3020. [CrossRef] [PubMed]
100. Kostkova, P.; Saigí-Rubió, F.; Eguia, H.; Borbolla, D.; Verschuuren, M.; Hamilton, C.; Azzopardi-Muscat, N.; Novillo-Ortiz, D. Data and Digital Solutions to Support Surveillance Strategies in the Context of the COVID-19 Pandemic. *Front. Digit. Health* **2021**, *3*, 2021. [CrossRef] [PubMed]
101. Al-Kahtani, M.S.; Khan, F.; Taekeun, W. Application of Internet of Things and Sensors in Healthcare. *Sensors* **2022**, *22*, 5738. [CrossRef]
102. Rajagopal, S.; Lyngdoh, D.C.; Anathapadmanabhan, U.K.; Soangra, R.; Vasan, N. IoT-enabled Solutions for Environmental Monitoring in Hospitals. *Open Biomed. Eng. J.* **2023**, *17*, e187412072301020. [CrossRef]
103. Liu, Z.; Wang, S.; Zhang, Y.; Feng, Y.; Liu, J.; Zhu, H. Artificial Intelligence in Food Safety: A Decade Review and Bibliometric Analysis. *Foods* **2023**, *12*, 1242. [CrossRef]
104. Gunasekeran, D.V.; Tseng, R.M.W.W.; Tham, Y.; Wong, T.Y. Applications of digital health for public health responses to COVID-19: A systematic scoping review of artificial intelligence, telehealth and related technologies. *Nat. Partn. J.* **2021**, *4*, 40. [CrossRef]
105. García, L.; Garcia-Sanchez, A.; Asorey-Cacheda, R.; Garcia-Haro, J.; Zúñiga-Cañón, C. Smart Air Quality Monitoring IoT-Based Infrastructure for Industrial Environments. *Sensors* **2022**, *22*, 23. [CrossRef]
106. Wang, J.-M.; Yang, M.-T. Design a Smart Control Strategy to Implement an Intelligent Energy Safety and Management System. *Int. J. Distrib. Sens. Networks* **2014**, *10*, 312392. [CrossRef]

107. Michel, C.; Keller, S. Advancing Ground-Based Radar Processing for Bridge Infrastructure Monitoring. *Sensors* **2021**, *21*, 2172. [CrossRef]
108. Raoa, A.S.; Radanovicb, M.; Liu, Y.; Hu, S.; Fang, Y.; Khoshelham, K.; Palaniswamia, M.; Ngo, T. Real-time Monitoring of Construction Sites: Sensors, Methods, and Applications. *Autom. Constr.* **2021**, *136*, 4. [CrossRef]
109. Zhu, C.; Zhu, J.; Bu, T.; Gao, X. Monitoring and Identification of Road Construction Safety Factors via UAV. *Sensors* **2022**, *22*, 8797. [CrossRef] [PubMed]
110. Yu, M.; Yang, C.; Li, Y. Big Data in Natural Disaster Management: A Review. *Geosciences* **2018**, *8*, 165. [CrossRef]
111. Khalid, R.; Ejaz, W. Internet of Things-Based On-Demand Rental Asset Tracking and Monitoring System. In Proceedings of the 5th International Conference on Information and Computer Technologies (ICICT), New York, NY, USA, 4–6 March 2022.
112. Ortiz-Garcés, I.; Andrade, R.O.; Sanchez-Viteri, S.; Villegas-Ch, W. Prototype of an Emergency Response System Using IoT in a Fog Computing Environment. *Computers* **2023**, *12*, 81. [CrossRef]
113. Kumar, S.; Tiwari, P.; Zymbler, M. Internet of Things is a revolutionary approach for future technology enhancement: A review. *J. Big Data* **2019**, *6*, 1–21. [CrossRef]
114. Pan, T. Intelligent Monitoring System for Prison Perimeter Based on Cloud Intelligence Technology. *Wirel. Commun. Mob. Comput.* **2022**, *2022*, 2517446. [CrossRef]
115. 5G Communications for Automation in Vertical Domains. 5G Americas White Paper. 2018. Available online: https://www.5gamericas.org/wp-content/uploads/2019/07/5G_Americas_White_Paper_Communications_for_Automation_in_Vertical_Domains_November_2018.pdf (accessed on 25 May 2023).
116. Rajasekar, S.J.S. An Enhanced IoT Based Tracing and Tracking Model for COVID-19 Cases. *SN Comput. Sci.* **2021**, *2*, 42. [CrossRef]
117. Reuters. Water Use Rising Faster than World Population. 2011. Available online: https://www.reuters.com/article/us-population-water/water-use-rising-faster-than-world-population-idUSTRE79O3WO20111025 (accessed on 25 May 2023).
118. Forbes. With Annual Losses Estimated At $14 Billion, It's Time to Get Smarter About Water. 2013. Available online: https://www.forbes.com/sites/heatherclancy/2013/09/19/with-annual-losses-estimated-at-14-billion-its-time-to-get-smarter-about-water/#33d936c598a2 (accessed on 25 May 2023).
119. Mann, C.; Turner, A.; Salisbury, C. *The Impact of Remote Consultation on Personalized Care*; University of Bristol, Centre of Academic Primary Care: Bristol, UK, 2021.
120. Al-Rawashdeh, M.; Keikhosrokiani, P.; Belaton, B.; Alawida, M.; Zwiri, A. IoT Adoption and Application for Smart Healthcare: A Systematic Review. *Sensors* **2022**, *22*, 5377. [CrossRef]
121. Dash, S.; Shakyawar, S.K.; Sharma, M.; Kaushik, S. Big data in healthcare: Management, analysis and future prospects. *J. Big Data* **2019**, *6*, 45. [CrossRef]
122. Hassebo, A. A hybrid 5G/PON—Based E-Health Communication Infrastructure. In Proceedings of the 2021 IEEE 12th Annual Ubiquitous Computing, Electronics & Mobile Communication Conference (UEMCON), New York, NY, USA, 1–4 December 2021.
123. Griffin, A.C.; Chung, A.E. Health Tracking and Information Sharing in the Patient-Centered Era: A Health Information National Trends Survey (HINTS) Study. *AMIA Symp.* **2020**, *2*, 1041–1050.
124. Sodhro, A.H.; Sennersten, C.; Ahmad, A. Towards Cognitive Authentication for Smart Healthcare Applications. *Sensors* **2022**, *22*, 2101. [CrossRef] [PubMed]
125. Khan, M.A.; Din, I.U.; Majali, T.; Kim, B.-S. A Survey of Authentication in Internet of Things-Enabled Healthcare Systems. *Sensors* **2022**, *22*, 9089. [CrossRef] [PubMed]
126. Machado-Jaimes, L.-G.; Bustamante-Bello, M.R.; Argüelles-Cruz, A.-J.; Alfaro-Ponce, M. Development of an Intelligent System for the Monitoring and Diagnosis of the Well-Being. *Sensors* **2022**, *22*, 9719. [CrossRef] [PubMed]
127. Rejeb, A.; Rajeb, K.; Simske, S.; Treiblmaier, H.; Zailani, S. The big picture on the internet of things and the smart city: A review of what we know and what we need to know. *Internet Things* **2022**, *19*, 100565. [CrossRef]
128. Ahmed, M.A.; Chavez, S.A.; Eltamaly, A.M.; Garces, H.O.; Rojas, A.J.; Kim, Y.-C. Toward an Intelligent Campus: IoT Platform for Remote Monitoring and Control of Smart Buildings. *Sensors* **2022**, *22*, 9045. [CrossRef]
129. Floris, A.; Porcu, S.; Girau, R.; Atzori, L. An IoT-Based Smart Building Solution for Indoor Environment Management and Occupants Prediction. *Energies* **2021**, *14*, 10. [CrossRef]
130. Liu, Y.; Wang, L.; Makris, S.; Krüger, J. Smart robotics for manufacturing. *Robot. Comput. Integr. Manuf.* **2023**, *82*. [CrossRef]
131. Ryalat, M.; ElMoaqet, H.; AlFaouri, M. Design of a Smart Factory Based on Cyber-Physical Systems and Internet of Things towards Industry 4. *Appl. Sci.* **2023**, *13*, 2156. [CrossRef]
132. *Industry 4.0 and The Smart Factory*; InTech: Nord-Pas-de-Calais, France, 2022.
133. Pech, M.; Vrchota, J.; Bednář, J. Predictive Maintenance and Intelligent Sensors in Smart Factory: Review. *Sensors* **2021**, *21*, 1470. [CrossRef]
134. Mylonas, G.; Kalogeras, A.; Kalogeras, G.; Anagnostopoulos, C.; Alexakos, C.; Muñoz, L. Digital Twins from Smart Manufacturing to Smart Cities: A Survey. *IEEE Access* **2016**, *9*, 143222–143249. [CrossRef]
135. Li, K.; Zhang, Y.; Huang, Y.; Tian, Z.; Sang, Z. Framework and Capability of Industrial IoT Infrastructure for Smart Manufacturing. *Standards* **2023**, *3*, 1. [CrossRef]
136. Dani, S.; Rahman, A.; Jin, J.; Kulkarni, A. Cloud-Empowered Data-Centric Paradigm for Smart Manufacturing. *Machines* **2023**, *11*, 451. [CrossRef]

137. Sun, Y.; Fesenko, H.; Kharchenko, V.; Zhong, L.; Kliushnikov, I.; Illiashenko, O.; Morozova, O.; Sachenko, A. UAV and IoT-Based Systems for the Monitoring of Industrial Facilities Using Digital Twins: Methodology, Reliability Models, and Application. *Sensors* **2022**, *22*, 17. [CrossRef] [PubMed]
138. Maimaitijiang, M.; Sagan, V.; Sidike, P.; Daloye, A.M.; Erkbol, H.; Fritschi, F.B. Crop Monitoring Using Satellite/UAV Data Fusion and Machine Learning. *Remote Sens.* **2020**, *12*, 1357. [CrossRef]
139. Nduku, L.; Munghemezulu, C.; Mashaba-Munghemezulu, Z.; Kalumba, A.M.; Chirima, G.J.; Masiza, W.; De Villiers, C. Global Research Trends for Unmanned Aerial Vehicle Remote Sensing Application in Wheat Crop Monitoring. *Geomatics* **2023**, *3*, 115–136. [CrossRef]
140. Hanif, A.S.; Han, X.; Yu, S.-H. Independent Control Spraying System for UAV-Based Precise Variable Sprayer: A Review. *Drones* **2022**, *6*, 383. [CrossRef]
141. Rahman, M.F.F.; Fan, S.; Zhang, Y.; Chen, L. A Comparative Study on Application of Unmanned Aerial Vehicle Systems in Agriculture. *Agriculture* **2021**, *11*, 22. [CrossRef]
142. Dengeru, Y.; Ramasamy, K.; Allimuthu, S.; Balakrishnan, S.; Kumar, A.P.M.; Kannan, B.; Karuppasami, K.M. Study on Spray Deposition and Drift Characteristics of UAV Agricultural Sprayer for Application of Insecticide in Redgram Crop (*Cajanus cajan* L. *Mill* sp.). *Agronomy* **2022**, *12*, 3196. [CrossRef]
143. Christiansen, M.P.; Laursen, M.S.; Jørgensen, R.N.; Skovsen, S.; Gislum, R. Designing and Testing a UAV Mapping System for Agricultural Field Surveying. *Sensors* **2017**, *17*, 12. [CrossRef]
144. Samreen, T.; Ahmad, M.; Baig, M.T.; Kanwal, S.; Nazir, M.Z.; Muntaha, S.T. Remote Sensing in Precision Agriculture for Irrigation Management. *Environ. Sci. Proc.* **2023**, *23*, 31. [CrossRef]
145. Alawad, W.; Ben Halima, N.; Aziz, L. An Unmanned Aerial Vehicle (UAV) System for Disaster and Crisis Management in Smart Cities. *Electronics* **2023**, *12*, 1051. [CrossRef]
146. Alsamhi, S.H.; Shvetsov, A.V.; Kumar, S.; Shvetsova, S.V.; Alhartomi, M.A.; Hawbani, A.; Rajput, N.S.; Srivastava, S.; Saif, A.; Nyangaresi, V.O. UAV Computing-Assisted Search and Rescue Mission Framework for Disaster and Harsh Environment Mitigation. *Drones* **2022**, *6*, 154. [CrossRef]
147. Zwęgliński, T. The Use of Drones in Disaster Aerial Needs Reconnaissance and Damage Assessment—Three-Dimensional Modeling and Orthophoto Map Study. *Sustainability* **2020**, *12*, 6080. [CrossRef]
148. Porte, J.; Briones, A.; Maso, J.M.; Pares, C.; Zaballos, A.; Pijoan, J.L. Heterogeneous wireless IoT architecture for natural disaster monitorization. *EURASIP J. Wirel. Commun. Netw.* **2020**, *2020*, 184. [CrossRef]
149. Sharma, K.; Anand, D.; Sabharwal, M.; Tiwari, P.K.; Cheikhrouhou, O.; Frikha, T. A Disaster Management Framework Using Internet of Things-Based Interconnected Devices. *Math. Probl. Eng.* **2021**, *2021*, 9916440. [CrossRef]
150. Carreras-Coch, A.; Navarro, J.; Sans, C.; Zaballos, A. Communication Technologies in Emergency Situations. *Electronics* **2022**, *11*, 1155. [CrossRef]
151. Federici, F.; Martintoni, D.; Senni, V. A Zero-Trust Architecture for Remote Access in Industrial IoT Infrastructures. *Electronics* **2023**, *12*, 566. [CrossRef]
152. Halder, S.; Afsari, K. Robots in Inspection and Monitoring of Buildings and Infrastructure: A Systematic Review. *Appl. Sci.* **2023**, *13*, 2304. [CrossRef]
153. Galdelli, A.; D'imperio, M.; Marchello, G.; Mancini, A.; Scaccia, M.; Sasso, M.; Frontoni, E.; Cannella, F. A Novel Remote Visual Inspection System for Bridge Predictive Maintenance. *Remote Sens.* **2022**, *14*, 2248. [CrossRef]
154. Gabbar, H.A.; Chahid, A.; Isham, M.U.; Grover, S.; Singh, K.P.; Elgazzar, K.; Mousa, A.; Ouda, H. HAIS: Highways Automated-Inspection System. *Technologies* **2023**, *11*, 2. [CrossRef]
155. Zhang, Y.; Yuan, X.; Li, W.; Chen, S. Automatic Power Line Inspection Using UAV Images. *Remote Sens.* **2017**, *9*, 824. [CrossRef]
156. Dimitrova, M.; Aminzadeh, A.; Meiabadi, M.S.; Karganroudi, S.S.; Taheri, H.; Ibrahim, H. A Survey on Non-Destructive Smart Inspection of Wind Turbine Blades Based on Industry 4.0 Strategy. *Appl. Mech.* **2022**, *3*, 1299–1326. [CrossRef]
157. Pincheira, M.; Antonini, M.; Vecchio, M. Integrating the IoT and Blockchain Technology for the Next Generation of Mining Inspection Systems. *Sensors* **2022**, *22*, 899. [CrossRef] [PubMed]
158. Tokkozhina, U.; Mataloto, B.M.; Martins, A.L.; Ferreira, J.C. Decentralizing Online Food Delivery Services: A Blockchain and IoT Model for Smart Cities. *Mob. Networks Appl.* **2023**, *2023*, 1–11. [CrossRef]
159. Seth, A.; James, A.; Kuantama, E.; Mukhopadhyay, S.; Han, R. Drone High-Rise Aerial Delivery with Vertical Grid Screening. *Drones* **2023**, *7*, 300. [CrossRef]
160. Gambo, N.; Musonda, I. Impact of IoT on Achieving Smart Primary Healthcare Building Facilities in Gauteng, South Africa. *Int. J. Environ. Res. Public Health* **2022**, *19*, 11147. [CrossRef]
161. Yavari, A.; Bagha, H.; Korala, H.; Mirza, I.; Dia, H.; Scifleet, P.; Sargent, J.; Shafiei, M. ParcEMon: IoT Platform for Real-Time Parcel Level Last-Mile Delivery Greenhouse Gas Emissions Reporting and Management. *Sensors* **2022**, *22*, 7380. [CrossRef]
162. Bertino, E.; Choo, K.R.; Georgakopolous, D.; Nepal, S. Internet of Things (IoT): Smart and Secure Service Delivery. *ACM Trans. Internet Technol.* **2016**, *16*, 4. [CrossRef]
163. Khargharia, H.S.; Rehman, M.H.U.; Banerjee, A.; Montori, F.; Forkan, A.R.M.; Jayaraman, P.P. Towards Marketing 4.0: Vision and Survey on the Role of IoT and Data Science. *Societies* **2023**, *13*, 100. [CrossRef]
164. Sharma, V.; Gandhi, M.K. Internet of Things (IoT) on E-commerce Logistics: A Review. *J. Phys. Conf. Ser.* **2021**, *1964*, 062113. [CrossRef]

165. Irugalbandara, C.; Naseem, A.S.; Perera, S.; Kiruthikan, S.; Logeeshan, V. A Secure and Smart Home Automation System with Speech Recognition and Power Measurement Capabilities. *Sensors* **2023**, *23*, 5784. [CrossRef] [PubMed]
166. Dubey, A.K.; Kumar, A.; Kumar, S.R.; Gayathri, N.; Das, P. (Eds.) *AI and IoT-Based Intelligent Automation in Robotics*; Wiley: New York, NY, USA, 2021.
167. Derlukiewicz, D. Application of a Design and Construction Method Based on a Study of User Needs in the Prevention of Accidents Involving Operators of Demolition Robots. *Appl. Sci.* **2019**, *9*, 1500. [CrossRef]
168. Marin-Plaza, P.; Yagüe, D.; Royo, F.; de Miguel, M.; Moreno, F.M.; Ruiz-De-La-Cuadra, A.; Viadero-Monasterio, F.; Garcia, J.; Roman, J.L.S.; Armingol, J.M. Project ARES: Driverless Transportation System. Challenges and Approaches in an Unstructured Road. *Electronics* **2021**, *10*, 1753. [CrossRef]
169. Flögel, D.; Bhatt, N.P.; Hashemi, E. Infrastructure-Aided Localization and State Estimation for Autonomous Mobile Robots. *Robotics* **2022**, *11*, 82. [CrossRef]
170. Marchang, J.; Di Nuovo, A. Assistive Multimodal Robotic System (AMRSys): Security and Privacy Issues, Challenges, and Possible Solutions. *Appl. Sci.* **2022**, *12*, 4. [CrossRef]
171. Li, Y.; Fu, C.; Yang, H.; Li, H.; Zhang, R.; Zhang, Y.; Wang, Z. Design of a Closed Piggery Environmental Monitoring and Control System Based on a Track Inspection Robot. *Agriculture* **2023**, *13*, 1501. [CrossRef]
172. Oliveira, L.F.P.; Moreira, A.P.; Silva, M.F. Advances in Agriculture Robotics: A State-of-the-Art Review and Challenges Ahead. *Robotics* **2021**, *10*, 52. [CrossRef]
173. Kyrarini, M.; Lygerakis, F.; Rajavenkatanarayanan, A.; Sevastopoulos, C.; Nambiappan, H.R.; Chaitanya, K.K.; Babu, A.R.; Mathew, J.; Makedon, F. A Survey of Robots in Healthcare. *Technologies* **2021**, *9*, 8. [CrossRef]
174. Kiangala, K.S.; Wang, Z. An Experimental Safety Response Mechanism for an Autonomous Moving Robot in a Smart Manufacturing Environment Using Q-Learning Algorithm and Speech Recognition. *Sensors* **2022**, *22*, 941. [CrossRef]
175. López-Belmonte, J.; Segura-Robles, A.; Moreno-Guerrero, A.-J.; Parra-González, M.-E. Robotics in Education: A Scientific Mapping of the Literature in Web of Science. *Electronics* **2021**, *10*, 291. [CrossRef]
176. Ye, H.; Sun, S.; Law, R. A Review of Robotic Applications in Hospitality and Tourism Research. *Sustainability* **2022**, *14*, 10827. [CrossRef]
177. Rocha, A.P.; Ketsmur, M.; Almeida, N.; Teixeira, A. An Accessible Smart Home Based on Integrated Multimodal Interaction. *Sensors* **2021**, *21*, 5464. [CrossRef]
178. Gohar, A.; Nencioni, G. The Role of 5G Technologies in a Smart City: The Case for Intelligent Transportation System. *Sustainability* **2021**, *13*, 5188. [CrossRef]
179. Bellini, P.; Nesi, P.; Pantaleo, G. IoT-Enabled Smart Cities: A Review of Concepts, Frameworks and Key Technologies. *Appl. Sci.* **2022**, *12*, 3. [CrossRef]
180. García-Martín, J.P.; Torralba, A. Model of a Device-Level Combined Wireless Network Based on NB-IoT and IEEE 802.15.4 Standards for Low-Power Applications in a Diverse IoT Framework. *Sensors* **2021**, *21*, 11. [CrossRef]
181. Ogbodo, E.U.; Abu-Mahfouz, A.M.; Kurien, A.M. Enabling LPWANs for Coexistence and Diverse IoT Applications in Smart Cities Using Lightweight Heterogenous Multihomed Network Model. *J. Sens. Actuator Netw.* **2022**, *11*, 87. [CrossRef]
182. Basford, P.J.; Bulot, F.M.J.; Apetroaie-Cristea, M.; Cox, S.J.; Ossont, S.J. LoRaWAN for Smart City IoT Deployments: A Long-Term Evaluation. *Sensors* **2020**, *20*, 648. [CrossRef]
183. Fu, H.; Wang, X.; Zhang, X.; Saleem, A.; Zheng, G. Analysis of LTE-M Adjacent Channel Interference in Rail Transit. *Sensors* **2022**, *22*, 3876. [CrossRef]
184. Pathak, G.; Gutierrez, J.; Ghobakhlou, A.; Rehman, S.U. LPWAN Key Exchange: A Centralized Lightweight Approach. *Sensors* **2022**, *22*, 5065. [CrossRef]
185. Dahlman, E.; Parkvall, S.; Sköld, J. *4G: LTE/LTE-Advanced for Mobile Broadband*; Academic Press: Cambridge, MA, USA, 2014. [CrossRef]
186. Holma, H.; Toskala, A. (Eds.) *LTE for UMTS: Evolution to LTE-Advanced*; John Wiley & Sons, Ltd.: New York, NY, USA, 2011.
187. Rodriguez, J. *Fundamentals of 5G Mobile Networks*; John Wiley & Sons, Ltd.: New York, NY, USA, 2015.
188. Lin, X.; Lee, N. *5G and Beyond Fundamentals and Standards*; Springer Nature: Cham, Switzerland, 2021.
189. Dwivedi, J.; Mahgoub, I. Robotic Surgery—A Review on Recent advances in Surgical Robotic Systems. In Proceedings of the 2012 Florida Conference on Recent Advances in Robotics, Boca Raton, FL, USA, 10–11 May 2012.
190. Bramhe, S.; Pathak, S.S. Robotic Surgery: A Narrative Review. *Cureus* **2022**, *14*, 29179. [CrossRef]
191. Boccardi, F.; Heath, R.W.; Lozano, A.; Marzetta, T.L.; Popovski, P. Five disruptive technology directions for 5G. *IEEE Commun. Mag.* **2014**, *52*, 74–80. [CrossRef]
192. Alsharif, M.H.; Kelechi, A.H.; Albreem, M.A.; Chaudhry, S.A.; Zia, M.S.; Kim, S. Sixth Generation (6G) Wireless Networks: Vision, Research Activities, Challenges and Potential Solutions. *Symmetry* **2020**, *12*, 4. [CrossRef]
193. Miao, J.T.; Phelps, N. *The Evolution of Smart City: Case studies of Barcelona, Spain and Helsinki, Finland*; Melbourne School of Design: Melbourne, VIC, Australia, 2019.
194. Szarek-Iwaniuk, P.; Senetra, A. Access to ICT in Poland and the Co-Creation of Urban Space in the Process of Modern Social Participation in a Smart City—A Case Study. *Sustainability* **2020**, *12*, 2136. [CrossRef]
195. Yun, Y.; Lee, M. Smart City 4.0 from the Perspective of Open Innovation. *J. Open Innov. Technol. Mark. Complex.* **2019**, *5*, 4. [CrossRef]

196. EPRS. *Social Approach to the Transition to Smart Cities*; EPRS—European Parliamentary Research Service: Brussels, Belgium, 2023.
197. Buttazzoni, A.; Veenhof, M.; Minaker, L. Smart City and High-Tech Urban Interventions Targeting Human Health: An Equity-Focused Systematic Review. *Int. J. Environ. Res. Public Health* **2020**, *17*, 2325. [CrossRef] [PubMed]
198. Micozzi, N.; Yigitcanlar, T. Understanding Smart City Policy: Insights from the Strategy Documents of 52 Local Governments. *Sustainability* **2022**, *14*, 16. [CrossRef]
199. Golubchikov, O. *People-Smart Sustainable Cities*; The United Nations (UN): Geneva, Switzerland, 2020.
200. Hodson, E.; Vainio, T.; Sayún, M.N.; Tomitsch, M.; Jones, A.; Jalonen, M.; Börütecene, A.; Hasan, M.T.; Paraschivoiu, I.; Wolff, A.; et al. Evaluating Social Impact of Smart City Technologies and Services: Methods, Challenges, Future Directions. *Multimodal Technol. Interact.* **2023**, *7*, 3. [CrossRef]
201. Tobias, K. *Smart Cities: A snapshot of Australia in 2019*; KPMG: Sidney, NSW, Australia, 2019.
202. Makieła, Z.J.; Stuss, M.M.; Mucha-Kuś, K.; Kinelski, G.; Budziński, M.; Michałek, J. Smart City 4.0: Sustainable Urban Development in the Metropolis GZM. *Sustainability* **2022**, *14*, 6. [CrossRef]
203. Nugraha, Y. Building a Smart City 4.0 Ecosystem Platform: An Overview and Case Study. In Proceedings of the 2020 International Conference on ICT for Smart Society (ICISS), Bandung, Indonesia, 19–20 November 2020.
204. Rocque, M. Smart Cities World City Profiles—The Republic of Singapore. SmartCitiesWorld, 2017. Available online: https://www.smartcitiesworld.net/city-profile/smart-cities-reports/smartcitiesworld-profile{-}{-}singapore-full-report-1767 (accessed on 6 June 2023).
205. Das, D.; Lim, N.D.; Aravind, P. Developing a Smart and Sustainable Campus in Singapore. *Sustainability* **2022**, *14*, 14472. [CrossRef]
206. Kasznar, A.P.P.; Hammad, A.W.A.; Najjar, M.; Qualharini, E.L.; Figueiredo, K.; Soares, C.A.P.; Haddad, A.N. Multiple Dimensions of Smart Cities' Infrastructure: A Review. *Buildings* **2021**, *11*, 73. [CrossRef]
207. Xie, X.; Ramakrishna, S.; Manganelli, M. Smart Building Technologies in Response to COVID-19. *Energies* **2022**, *15*, 5488. [CrossRef]
208. Glasco, J. Smart Cities World City Profiles—Barcelona. SmartCitiesWorld, 2019. Available online: https://www.smartcitiesworld.net/city-profile/smart-cities-reports/smartcitiesworld-city-profile{-}{-}barcelona-4800 (accessed on 7 June 2023).
209. Kim, N.; Yang, S. Conceptually Related Smart Cities Services from the Perspectives of Governance and Sociotechnical Systems in Europe. *Systems* **2023**, *11*, 166. [CrossRef]
210. Berigüete, F.E.; Cantalapiedra, I.R.; Palumbo, M.; Masseck, T. Collective Intelligence to Co-Create the Cities of the Future: Proposal of an Evaluation Tool for Citizen Initiatives. *Sustainability* **2023**, *15*, 7956. [CrossRef]
211. Garcia, E.; Peyman, M.; Serrat, C.; Xhafa, F. Join Operation for Semantic Data Enrichment of Asynchronous Time Series Data. *Axioms* **2023**, *12*, 349. [CrossRef]
212. Somayya, M.; Ramaswamy, R. Amsterdam Smart City (ASC): Fishing village to sustainable city. In *The Sustainable City XI*; WIT Press: New Forest National Park, UK, 2016.
213. Mancebo, F. Smart city strategies: Time to involve people. Comparing Amsterdam, Barcelona and Paris. *J. Urban. Int. Res. Placemaking Urban Sustain.* **2020**, *13*, 133–152. [CrossRef]
214. Mitieka, D.; Luke, R.; Twinomurinzi, H.; Mageto, J. Smart Mobility in Urban Areas: A Bibliometric Review and Research Agenda. *Sustainability* **2023**, *15*, 6754. [CrossRef]
215. Łukasiewicz, K.; Pietrzak, P.; Kraciuk, J.; Kacperska, E.; Cieciora, M. Sustainable Energy Development—A Systematic Literature Review. *Energies* **2022**, *15*, 21. [CrossRef]
216. Fassio, F.; Chirilli, C. The Circular Economy and the Food System: A Review of Principal Measuring Tools. *Sustainability* **2023**, *15*, 10179. [CrossRef]
217. The SENSEable City Lab. *Copenhagen Sensible City Guide*; The Massachusetts Institute of Technology: Cambridge, MA, USA, 2011.
218. Bjørner, T. The advantages of and barriers to being smart in a smart city: The perceptions of project managers within a smart city cluster project in Greater Copenhagen. *Cities* **2021**, *114*, 103187. [CrossRef]
219. Katz, B.; Noring, L. *The Copenhagen City and Port: A Model for Regenerating Cities, Centennial Scholar Initiative at Brookings*; The Brookings Institution: Washington, DC, USA, 2017.
220. Secretariat for Management and Communication. *The Capital of Sustainable Development: The City of Copenhagen's action plan for the Sustainable Development Goals*; Department of Finance: Copenhagen, Denmark.
221. Cortese, T.T.P.; de Almeida, J.F.S.; Batista, G.Q.; Storopoli, J.E.; Liu, A.; Yigitcanlar, T. Understanding Sustainable Energy in the Context of Smart Cities: A PRISMA Review. *Energies* **2022**, *15*, 2382. [CrossRef]
222. Debnath, A.; Isaksen, S.A.; Mynster, A.S.; Nielsen, J.U. *White Paper Smart Cities and Communities Infrastructure*; FORCE Technology: Copenhagen, Denmark.
223. Taherdoost, H.; Madanchian, M. Multi-Criteria Decision Making (MCDM) Methods and Concepts. *Encyclopedia* **2023**, *3*, 77–87. [CrossRef]
224. Borruso, G.; Balletto, G. The Image of the Smart City: New Challenges. *Urban Sci.* **2022**, *6*, 5. [CrossRef]
225. Deeksha; Shukla, A. Ecosystem Services: A Systematic Literature Review and Future Dimension in Freshwater Ecosystems. *Appl. Sci.* **2022**, *12*, 17.
226. KPMG. *Smart Cities Awareness Survey Conducted Covering Five Major Cities in Japan*; KPMG: Tokyo, Japan, 2020.

227. Elvas, L.B.; Mataloto, B.M.; Martins, A.L.; Ferreira, J.C. Disaster Management in Smart Cities. *Smart Cities* **2021**, *4*, 819–839. [CrossRef]
228. Aliero, M.; Asif, M.; Ghani, I.; Pasha, M.; Jeong, S. Systematic Review Analysis on Smart Building: Challenges and Opportunities. *Sustainability* **2022**, *14*, 5. [CrossRef]
229. Khan, M.S.; Woo, M.; Nam, K.; Chathoth, P.K. Smart City and Smart Tourism: A Case of Dubai. *Sustainability* **2017**, *9*, 2279. [CrossRef]
230. Vujković, P.; Ravšelj, D.; Umek, L.; Aristovnik, A. Bibliometric Analysis of Smart Public Governance Research: Smart City and Smart Gov-ernment in Comparative Perspective. *Soc. Sci.* **2022**, *11*, 7. [CrossRef]
231. Farag, A.A. The Story of NEOM City: Opportunities and Challenges. In *New Cities and Community Extensions in Egypt and the Middle East*; Springer: Cham, Switzerland, 2019.
232. Saud, M.M.A. *Sustainable Land Management for NEOM Region*; Springer: Cham, Switzerland, 2020.
233. Minawi, M.; IPSOS. *Saudi Arabia's Mega-City: The NEOM Project*; IPSOS: Paris, France, 2019.
234. Chang, S.; Smith, M.K. Residents' Quality of Life in Smart Cities: A Systematic Literature Review. *Land* **2023**, *12*, 876. [CrossRef]
235. Im, J.-H.; Jang, S. Self-Sufficiency of New Administrative Capitals (NACs) Based on Types and Commuting Characteristics of Citizens: Case Study of Sejong. *Sustainability* **2022**, *14*, 13193. [CrossRef]
236. Ali, M.A. Smart city policy in developing countries: Case study of the new administrative capital in Egypt. *J. Public Aff.* **2021**, *22*, e2774. [CrossRef]
237. Toh, C.K. Smart city indexes, criteria, indicators and rankings: An in-depth investigation and analysis. *IET Smart Cities* **2022**, *4*, 211–228. [CrossRef]
238. Fang, Y.; Shan, Z. How to Promote a Smart City Effectively? An Evaluation Model and Efficiency Analysis of Smart Cities in China. *Sustainability* **2022**, *14*, 6512. [CrossRef]
239. Valeria, F. *Singapore: Smart City to Smart Nation*; Center for Digital Society: Yogyakarta, Indonesia, 2020.
240. Allam, Z. The Emergence of Anti-Privacy and Control at the Nexus between the Concepts of Safe City and Smart City. *Smart Cities* **2019**, *2*, 96–105. [CrossRef]
241. Van Geest, M.; Tekinerdogan, B.; Catal, C. Smart Warehouses: Rationale, Challenges and Solution Directions. *Appl. Sci.* **2021**, *12*, 219. [CrossRef]
242. Budka, K.C.; Deshpande, J.G.; Thottan, M. *Communication Networks for Smart Grids*; Springer: Cham, Switzerland, 2014. [CrossRef]
243. Hou, J.; Arpan, L.; Wu, Y.; Feiock, R.; Ozguven, E.; Arghandeh, R. The Road toward Smart Cities: A Study of Citizens' Acceptance of Mobile Applications for City Services. *Energies* **2020**, *13*, 2496. [CrossRef]
244. Paes, V.d.C.; Pessoa, C.H.M.; Pagliusi, R.P.; Barbosa, C.E.; Argôlo, M.; de Lima, Y.O.; Salazar, H.; Lyra, A.; de Souza, J.M. Analyzing the Challenges for Future Smart and Sustainable Cities. *Sustainability* **2023**, *15*, 7996. [CrossRef]
245. Fabrègue, B.F.G.; Bogoni, A. Privacy and Security Concerns in the Smart City. *Smart Cities* **2023**, *6*, 586–613. [CrossRef]
246. Ma, M.; Krawczyk, D.; Wodarski, K. Residents' Perceptions of Challenges Related to Implementation of Smart City Solutions by Local Government. *Sustainability* **2023**, *15*, 11.

sensors MDPI

Article

Deep Federated Adaptation: An Adaptative Residential Load Forecasting Approach with Federated Learning

Yuan Shi and Xianze Xu *

Electronic Information School, Wuhan University, Wuhan 430072, China; shiyuan@whu.edu.cn
* Correspondence: xuxianze@whu.edu.cn

Abstract: Residential-level short-term load forecasting (STLF) is significant for power system operation. Data-driven forecasting models, especially machine-learning-based models, are sensitive to the amount of data. However, privacy and security concerns raised by supervision departments and users limit the data for sharing. Meanwhile, the limited data from the newly built houses are not sufficient to support building a powerful model. Another problem is that the data from different houses are in a non-identical and independent distribution (non-IID), which makes the general model fail in predicting accurate load for the specific house. Even though we can build a model corresponding to each house, it costs a large computation time. We first propose a federated transfer learning approach applied in STLF, deep federated adaptation (DFA), to deal with the aforementioned problems. This approach adopts the federated learning architecture to train a global model without undermining privacy, and then the model leverage multiple kernel variant of maximum mean discrepancies (MK-MMD) to fine-tune the global model, which makes the model adapted to the specific house's prediction task. Experimental results on the real residential datasets show that DFA has the best forecasting performance compared with other baseline models and the federated architecture of DFA has a remarkable superiority in computation time. The framework of DFA is extended with alternative transfer learning methods and all of them achieve good performances on STLF.

Keywords: electric load forecasting; transfer learning; federated learning; domain adaptation

Citation: Shi, Y.; Xu, X. Deep Federated Adaptation: An Adaptative Residential Load Forecasting Approach with Federated Learning. *Sensors* **2022**, *22*, 3264. https://doi.org/10.3390/s22093264

Academic Editor: Hossam A. Gabbar

Received: 21 March 2022
Accepted: 21 April 2022
Published: 24 April 2022

1. Introduction

The International Energy Agency has identified energy efficiency in buildings as one of the five methods to secure long-term decarbonization of the energy sector [1]. In addition to environmental benefits, the improvement of the building energy efficiency also presents vast economic benefits. Buildings with efficient energy systems and management strategies have much lower operating costs [2]. The activities of humans in residences occupy a large portion of energy consumption and CO_2 emission [3]. Residential load forecasting can assist sectors in balancing the generation and consumption of electricity, which improves energy efficiency through the management and conservation of energy.

Several uncertain factors, such as historical load records, weather conditions, population mobilities, social factors and emergencies, influence electricity usage. Due to the high volatility and uncertainties involved, short-term load forecasting for a single residential unit may be more challenging than for an industrial building [4]. Machine-learning-based methods, driven by data, are applied to mitigate these challenges more and more frequently. However, the scope of machine-learning-based applications will be hindered due to the privacy and security concerns raised by more and more supervision departments and users. Even in some countries, many users refuse the installation of smart meters because users are reluctant to disclose their private data. In addition, newly built houses cannot provide sufficient data to build effective models. In summary, the data exist in the form of isolated islands, which makes it difficult to merge the data from different users to train a robust model. Hence, one of the problems in this paper we focused on is data availability and privacy.

A number of researches have achieved good results on STLF, such as support vector regression (SVR) [5], the artificial neural network [6] and boosted tree [7]. Additionally, some hybrid methods that combine artificial intelligence methods with traditional methods are proposed to achieve better forecasting performance, such as hybridizing extended Kalman Filter and ELM [8]. Fan et al. [9] proposed a SVR model hybridized with differential empirical mode decomposition (DEMD) method and auto regression (AR) for electric load forecasting. Transformer is a novel time series prediction model based on the encoder–decoder structure. Originating from this structure, many methods have yielded good results in the field of energy forecasting, such as STA-AED [10] and informer [11]. However, these approaches do not consider user privacy and modeling with limited data.

A lot of privacy-preserving solutions relying on data aggregation and obfuscation have been proposed to ensure privacy [12]. However, these solutions are not suitable for residential short-term energy forecasting since they often introduce extra procedures to obfuscate and reconstruct the data [13]. In addition, as the solutions based on machine learning are computationally intensive in the step of model training, most works consider only centralized training approaches. Clients' data should be collected onto a central server where the model is trained, which leads to a heavy burden on communication. Especially when the model needs to be constantly updated with new data, as the data from millions of distributed meters are required. Under this circumstance, federated learning has been proposed to overcome these challenges. Federated Learning is a distributed machine learning approach where a shared global model is trained, under the coordination of a central entity, by a federation of participating devices [14]. The peculiarity of the approach is that each device trains a local model with the data never leaving each local machine. Only the parameters of models are sent to the central computing server for updating the shared global model. Hence, the federated architecture can protect privacy effectively. Federated learning has been demonstrated to be effective in the area of load forecasting, federated learning with clustered aggregation is proposed in [15], and has good performance for individual load forecasting. Federated learning applied in heating load demand prediction of buildings also has a high capability of producing produce acceptable forecasts while preserving data privacy and eliminating the dependence of the model on the training data [16]. Furthermore, federated learning has been applied in several application successfully, such as human–computer interaction [17], natural language processing [18], healthcare classification [19], transportation [20,21], and so on, where privacy and scalability are essential.

Another critical problem for residential load forecasting is that the general model is not adapted to each house since the datasets are non-IID, which the federated architecture and conventional machine learning algorithms do not well handle with [22]. The problem is particularly acute in the case of newly built houses. Even though the dataset bias and unbalance are inevitable [23], many researchers classify users according to different attributes to deal with this challenge, but it does not fit well with a federated learning architecture [24]. This situation is particularly suitable for applying transfer learning. Transfer learning aims at establishing knowledge transfer to bridge different domains of substantial distribution discrepancies. In other words, data from different houses have domain discrepancies which is a major obstacle in adapting the predictive model across users. STLF models based on transfer learning are discussed in [4,25,26].

A representative transfer learning method is domain adaptation, which can leverage the data in the information-rich source domain to enhance the performance of the model in the data-limited target domain. As a well-known algorithm applied for domain adaptation, deep neural network [27] is capable of discovering factors of variations underlying the houses' historical data, and group features hierarchically in accordance with their relatedness to invariant factors, and it has been studied extensively. A lot of research has shown that deep neural networks can learn more transferable features for domain adaptation [28].

It is shown that deep features must eventually transition from general to specific in the network, with the transferability of features decreasing significantly at higher levels as domain discrepancies increase. In other words, the common features between different users are captured in lower layers, and the features of the specific user hide in higher layers which depend greatly on the target datasets and are not safely transferable to another user.

In this article, we address the aforementioned challenges within a novel user adaptative load forecasting approach. The approach is the combination of federated learning and transfer learning. The architecture of federated learning in this approach aims at building a CNN-LSTM based general model, which does not compromise privacy and works well with the limited data. Then, MK-MMD, a distance to measure domain discrepancies, is used to calculate the domain discrepancies between houses, then optimize the general network which can reduce the domain discrepancies effectively and reduce the forecasting error. The contributions of this paper are summarized as follows:

1. We propose a novel federated transfer approach DFA for residential STLF, which adopts a federated architecture to address the problems of data availability and privacy, and leverages transfer learning to deal with the non-IID datasets for improving forecasting performance;
2. DFA is investigated for STLF of residential houses and has shown remarkable advantages in forecasting performance over other baseline models. Especially, the federated architecture is superior to the centralized architecture in computation time;
3. The framework of DFA is extended with alternative transfer learning methods and all of them achieve good performances on STLF.

2. Technical Background

2.1. Federated Learning Concepts

Due to security and privacy concerns, data exist in the form of isolated islands, making it difficult for data-driven models to leverage big data. One possible approach is federated learning, which can train a machine learning model in a distributed way.

Let matrix $\mathcal{D}_i$ denote the data held by the partner i, each row of the matrix represents one sample, and each column is a feature. Since the feature and sample spaces of the data parties may not be identical, federated learning can be classified into three classes: horizontally federated learning, vertically federated learning and federated transfer learning.

Horizontal federated learning is applicable in the conditions in which different partners have the same or overlapped feature spaces but different spaces in samples. It is similar to the case of dividing data horizontally in a tabular view, hence horizontal federated learning is also known as sample-partitioned federated learning. Horizontal federated learning can be summarized as Formula (1):

$$\mathcal{X}_i = \mathcal{X}_j, \mathcal{Y}_i = \mathcal{Y}_j, \mathcal{I}_i \neq \mathcal{I}_j, \forall \mathcal{D}_i, \mathcal{D}_j, i \neq j \quad (1)$$

let $\mathcal{X}$, $\mathcal{Y}$, $\mathcal{I}$ denote the feature space, the label space and the sample ID space.

Different from horizontal federated learning, partners in the vertically federated learning share the same spaces in samples, but different ones in feature spaces. We can summarize vertically federated learning as shown in Formula (2):

$$\mathcal{X}_i \neq \mathcal{X}_j, \mathcal{Y}_i \neq \mathcal{Y}_j, \mathcal{I}_i = \mathcal{I}_j, \forall \mathcal{D}_i, \mathcal{D}_j, i \neq j \quad (2)$$

Federated transfer learning is applied in the conditions in which datasets differ not only in sample spaces but also in feature spaces. For example, a common representation or model is learned from different feature spaces and later used to make predictions for samples with only one-side features. Federated transfer learning is summarized as shown in Formula (3):

$$\mathcal{X}_i \neq \mathcal{X}_j, \mathcal{Y}_i \neq \mathcal{Y}_j, \mathcal{I}_i \neq \mathcal{I}_j, \forall \mathcal{D}_i, \mathcal{D}_j, i \neq j \quad (3)$$

In this paper, the federated learning framework is a horizontal federated learning architecture since the data collected by devices is in the same feature space. It uses a master–slave architecture, as shown in Figure 1. In this system, N participant devices collaborate to train a machine learning model with the help of the master server.

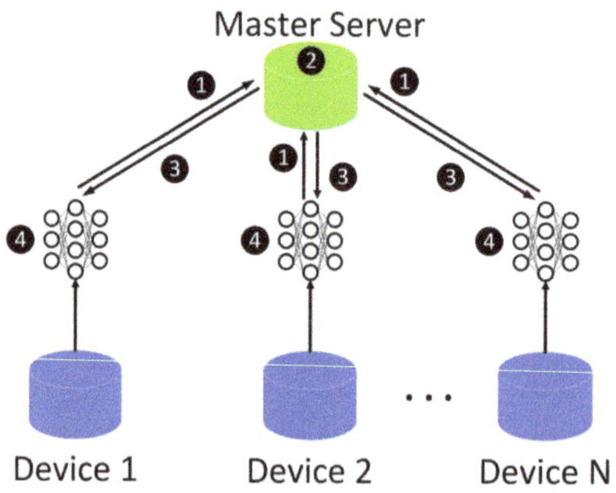

Figure 1. A horizontal federated learning architecture.

In step 1, each participant computes the model gradient locally and masks the gradient information using cryptographic techniques such as homomorphic encryption, and sends the results to the master server. In step 2, the master server performs a secure aggregation operation. In step 3, the server distributes the aggregated results to each participant. In step 4, each participant decrypts the received gradients and updates their respective model parameters using the decrypted gradients. The above steps continue iteratively until the loss function converges or the maximum number of iterations is reached. We can see that the data of the participants are not moved during the training process, so the federated learning can protect user privacy that distributed machine learning models trained on Hadoop do not have. In the training process, an arbitrary number of devices can concur to model training without the need of transferring collected data to a centralized location. The federated model can tackle the increasing data without consideration of communication bandwidth since only local gradients need to be sent.

2.2. Transfer Learning Concepts and MK-MMD

Firstly, it is hard to collect sufficient data from domains of interest, referred to as target domains. Meanwhile, a large number of data may be available for some related domains called source domains. Secondly, machine learning algorithms work well based on a fundamental assumption: the training and future data must be in the same feature space and follow the same distribution. However, this assumption is not held in real-world applications. For these reasons, transfer learning is introduced to address these problems. Transfer learning can leverage similarities between data, tasks, or models to conduct knowledge transfer from the source domain to the target domain. These similarities are considered a representation of the distance between domains. Then the key issue is to introduce the standard distribution distance metric and minimize the distance.

MK-MMD is a type of distance metrics. This distance is computed with respect to a particular representation $\phi(\cdot)$, a feature map function. This function can map the original data into a reproducing kernel Hilbert space (RKHS) endowed with a characteristic kernel k. The RKHS may be infinite dimensions that can transform non-separable data to linearly separable. The distance between the source domain with probability p and the target domain with probability q is defined as $d_k(p,q)$. The data distribution $p = q$ iff $d_k^2(p,q) = 0$. Then, the squared expression of MK-MMD distance [29] is denoted as Formula (4):

$$d_k^2(p,q) \triangleq \|E_p[\phi(x_S)] - E_q[\phi(x_T)]\|_{\mathcal{H}_k}^2 \quad (4)$$

where $\mathcal{H}_k$ denotes the RKHS endowed with a characteristic kernel k.

Kernel technique, as Formula (5) shows, can be used to compute Formula (4), which can convert the computation of the inner product of the feature map $\phi(\cdot)$ to computing the the kernel function $k(\cdot)$ instead.

$$k(x_S, x_T) = \langle \phi(x_S), \phi(x_T) \rangle \quad (5)$$

As mean embedding matching is sensitive to the kernel choices, MK-MMD uses multi-kernel k to provide better learning capability and alleviate the burden of designing specific kernels to handle diverse multivariate data. It provides more flexibility to capture different kernels and leads to a principled method for optimal kernel selection.

Multi-kernel $\mathcal{K}$ is defined as the convex combination of kernels $\{k_u\}$ as in Formula (6) [28]:

$$\mathcal{K} \triangleq \left\{ k = \sum_{u=1}^{m} \mathcal{B}_u k_u : \sum_{u=1}^{m} \mathcal{B}_u = 1, \mathcal{B}_u \geq 0, \forall u \right\} \quad (6)$$

where the constraints on coefficients $\{\mathcal{B}_u\}$ make the derived multi-kernel k characteristic.

3. The Proposed Method

3.1. The Overview of the Proposed Approach

The overview of the proposed approach is shown in Figure 2. Without loss of generality, there are 3 households which need to be predicted, the number can be extended without too much work. Each household has a device for computing models and communicating with the master server. The approach mainly consists of 6 procedures as follows:

Step 1: The master server constructs the initial global model with public datasets.

Step 2: The master server distributes the global model to all users.

Step 3: The master server selects a fraction of users, then the selected devices train models with their local data.

Step 4: The selected devices upload models to the master server.

Step 5: The master server updates the global model by aggregating the uploaded models. Repeat Step 2 to Step 5 until the global model convergence.

Step6: Each device fine-tunes the convergent global model using user adaptation with local data.

3.2. Federated Learning Process

Deep neural networks are selected in the federated learning process since neural networks update models based on gradient descent. The federated learning process can get a pre-trained model for the latter user adaptation. Firstly, in the training process, the model is initialized on the master server with public datasets. The initial global model is denoted as f_G, then the learning objective function is defined as shown in Formula (7):

$$\arg\min_{\Theta_G} \mathcal{L} = \sum_{i=1}^{n} \ell(y_i, f_G(x_i)) \quad (7)$$

where $\ell(\cdot)$ denotes the loss for the neural network, the loss used in this paper is mean squared error (MSE) loss since load forecasting problem is a regression problem. $\{x_i, y_i\}_{i=1}^{n}$ are samples from datasets, and Θ are the parameters learned.

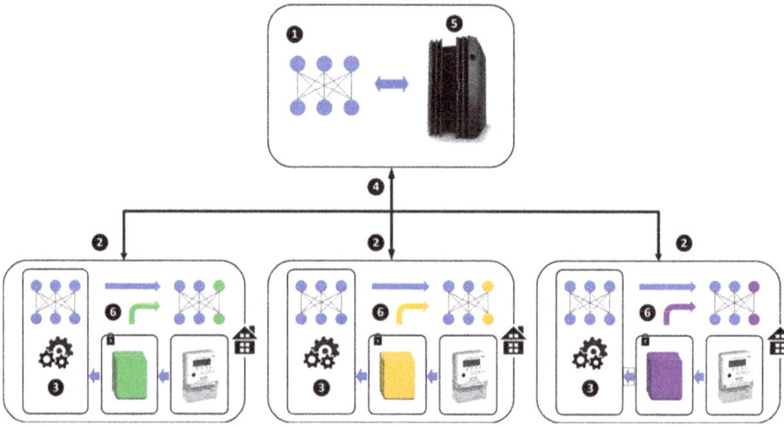

Figure 2. Overview of the deep federated adaptation. The top box is the master server while the 3 bottom boxes denotes 3 houses. Each house contains one computing device connected to the master server for processing the data. The data collected by the smart meter is locked and cannot be transmitted to the master server.

After the initial global model is trained, the master server will distribute the model to all remote devices. Then, a subset of remote devices are chosen for training user model f_u with local data. Let $\{x_i^u, y_i^u\}_{i=1}^{n^u}$ denote samples from datasets u. Technically, the learning objective function for each user is denoted as Formula (8):

$$\arg\min_{\Theta_u} \mathcal{L} = \sum_{i=1}^{n^u} \ell(y_i^u, f_u(x_i^u)) \tag{8}$$

Then, all the user models are uploaded to the master server for averaging based on the algorithm FedAVG [30], and the formulation of averaging is as Formula (9):

$$f_G{}'(w) = \frac{1}{K} \sum_{k=1}^{n} f_{u_k}(w) \tag{9}$$

where w are parameters of the network and K is the number of devices in the chosen subset. Then, let $f_G = f'_G$ on the master server, after adequate rounds of iterations, the updated server model f_G has better performance on generalization ability. When devices of newly built houses are connected to the federated system, the master server can distribute the global model to help new devices take part in the next iteration, hence, federated learning can deal with cold start problems and is extensible. It is worth noting that the network is trained by the federated learning using data from different houses, which expands the training data and makes the model more robust, and has better generalization ability.

3.3. User Adaptation with Multiple Kernel Variant of Maximum Mean Discrepancies

Federated learning solves problems of data availability and privacy. However, another important problem is personalization. Even if the cloud model can be directly used, it still performs poorly on a particular house. The weights of this network have been pre-trained by the federated learning process, then the user adaptation process will fine-tune the pre-trained network. Since the network does not need to update all weights from scratch for new tasks, it costs less in computation and time, which is especially suitable for edge devices.

Figure 3 shows the architecture of the proposed network. This is a classic hybrid model of convolutional neural network (CNN) and Bi-directional Long Short-Term Memory (BiLSTM), referred to as CNN-LSTM, more details can be found in [31]. This network is a two-stream architecture, thus the source data and the target data can be fed into the network simultaneously. Two streams of data go from the CNN layers to the Bi-LSTM layers and finally through the fully connected (FC) layers to compute the forward loss. We consider that sections of CNN can extract low-level features about a series of load values and BiLSTM aims at capturing sequential relationships.

To minimize domain discrepancy, domain loss is also introduced to optimize the network. MK-MMD is used to measure domain loss in which the source data is aligned with the target data for computing. Multi-kernel k is used to adapt to different feature domains and hidden representations of higher layers are embedded in a RKHS where the mean embeddings of distributions in different user data can be explicitly matched. The loss of MK-MMD is defined as shown in Formula (10) [23]:

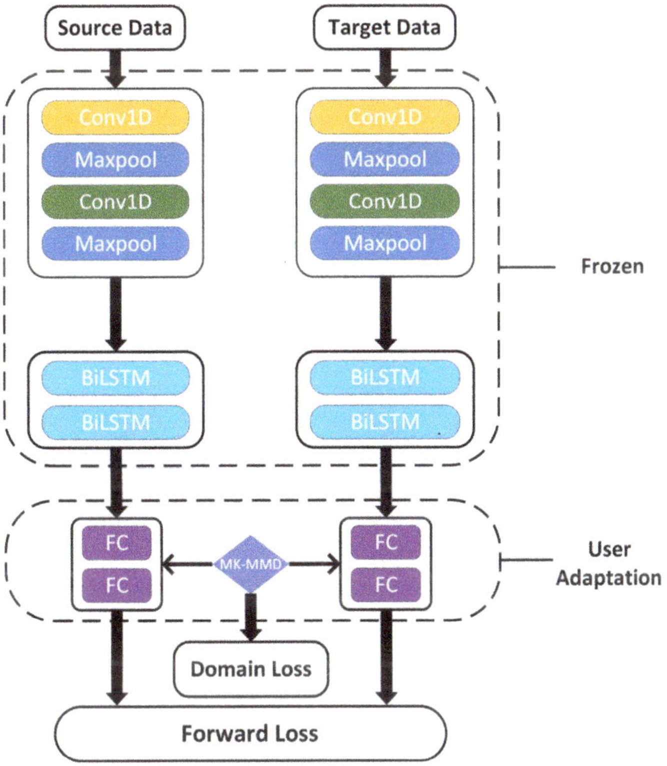

Figure 3. The architecture of proposed network, from top to bottom, consists of CNN layers, BiLSTM layers and fully connected layers.

$$\mathcal{L}_{MK-MMD}(X_S, X_T) = \left\| \frac{1}{|X_S|} \sum_{x_s \in X_S} \phi(x_s) - \frac{1}{|X_T|} \sum_{x_t \in X_T} \phi(x_t) \right\|_{\mathcal{H}}^2 \tag{10}$$

where $x_s \in X_S$ denote source data points from source datasets and $x_t \in X_T$ denote target points from datasets of houses need to be adapted. Gaussian kernels are selected as the kernel function k in this paper since they can map features to infinite dimensions. We use a combination of Gaussian kernels by varying bandwidth γ with a multiplicative step size of $2^{1/2}$. The Gaussian kernel function with the bandwidth γ is defined, as shown in Formula (11):

$$k(x_s, x_t) = e^{-\frac{\|x_s - x_t\|^2}{\gamma}} \tag{11}$$

let η denote the trade-off parameter, then the total loss function of the network during user adaptation is computed by Formula (12):

$$\arg\min_{\Theta_u} \mathcal{L}_u = \sum_{i=1}^{n} \ell(y_i, f_u(x_i)) + \sum_{i=1}^{n^u} \ell(y_i^u, f_u(x_i^u)) + \eta \mathcal{L}_{MK-MMD}(X_s, X_T) \tag{12}$$

Since learned features transition from general to specific along the network with increasing domain discrepancies, the lower CNN and BiLSTM layers catch general features that can transfer from different houses. Hence, the parameters in the first dashed box in Figure 3 are frozen during user adaptation, whereas the weights of FC layers are updated by the total loss, as shown in Formula (12).

3.4. Learning Process and Summary

The learning procedures of DFA are summarized in Algorithm 1. Furthermore, we can consider the algorithm as a general process applied in STLF and separate the procedures into two sections. Section 1 of step 1 to step 8 is a federated learning process, while Section 2 of step 9 is for transfer learning. Other federated learning methods (e.g., vertically federated learning) can replace the horizontal federated learning method in Section 1 to deal with heterogeneous features from diverse organizations. Meanwhile, other effective transfer learning methods can also be embedded in Section 2 for better personalization. The neural network used in this framework can also be replaced according to the computing power of real-world devices or the features of datasets.

Algorithm 1: DFA: Deep Federated Adaptation

Require: Data from different houses $\{\mathcal{D}_1, \mathcal{D}_2, \cdots, \mathcal{D}_N\}$
Ensure: Adaptative forecasting models f_u
1: Build an initial global model f_G with public datasets using (7)
2: **repeat**
3:   Distribute f_G to all computing devices
4:   Use local data to train f_u based on f_G with (8)
5:   Devices upload f_u to the model server
6:   Average models with (9)
7:   Update the global model $f_G = f_G{}'$
8: **until** $|\Theta_G - \Theta_G{}^*| < \varepsilon$
9: Use (12) to optimize the convergent global model f_G, get adaptative models f_u

4. Experiments

4.1. Datasets Description and Pre-Processing

The experimental datasets are electricity consumption readings for a sample of 5567 London Households that took part in the UK Power Networks led Low Carbon London project between November 2011 and February 2014. Readings were taken at half-hourly intervals. The customers in the trial were recruited as a balanced sample representative of the Greater London population. The dataset contains electricity consumption, in kWh (per half hour), unique household identifier, date and time [32]. As an example, a period of records from 4 households are shown in Figure 4. It can be seen that the records are in different patterns which means a general model is not suitable for forecasting electricity consumption for a particular house. Meteorological variables recorded in London collected from Dark Sky API [33] are introduced to enrich our datasets. We merge electricity consumption datasets and meteorological datasets in terms of timestamps to generate a new feature table for each household.

Some discrete features (e.g., 'weekday', 'icon') should be encoded to embedding features. Then, feature normalization is implemented for all features with min–max normalization, as shown in Formula (13):

$$\hat{x}_i^j = \frac{x_i^j - x_{i,min}}{x_{i,max} - x_{i,\min}} \quad (13)$$

where x_i^j denotes the value for feature i at the time step j, $x_{i,min}$ and $x_{i,max}$ denote the maximal and minimal value for feature i, respectively. $\hat{x}_i^j$ is the value for x_i^j after normalization.

We consider the feature table as time-series data according to the timestamps, each row of the table denotes a record sampled at half-hourly intervals. We implement a sliding window with a look back at 24 records to forecast the next record. Hence, the proposed network can give a half-an-hour-later load value prediction, one training sample consists of features of 24 records and the value of the next electricity consumption. The input dimension is $|\mathcal{X}| \times L$, where $|\mathcal{X}|$ is the number of expected features in the merged table $\mathcal{X}$, L denotes the width of the sliding window.

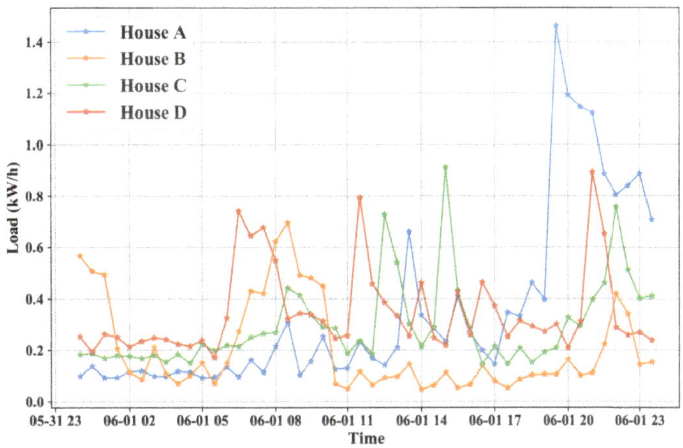

Figure 4. Load data of four houses for one day from the used datasets.

4.2. Implementation Information

The proposed network is composed of two convolutional layers, two pooling layers, two BiLSTM layers and two FC layers. The network adopts a convolution size of 1×17 and a kernel size of 3 for pooling layers. The proposed network is trained with the MSE loss and adopts stochastic gradient descent (SGD) with an initial learning rate of 0.01 and 0.9 momentum for optimization. Batch size is set to 32. The training process is early stopped within 10 epochs and the rate of dropout is set to 0.1 to prevent overfitting.

In the following experiments, cross validation and grid search are used to select the hyperparameters and the hyperparameters with the lowest average forecasting MAPE will be used. During the training process, we use 70% of the data for training while the rest 30% is for evaluation. All experiments are repeated five times to ensure reliability, implemented in Pytorch, and conducted on a single NVIDIA GeForce RTX 2080 GPU.

A single machine is used to simulate the federated learning process and we can set the number of user nodes N_{nodes} according to the experimental requirements. Table 1 shows some symbol definitions of the experiments. Since a single machine is used to simulate the federated process, the training process is serial. However, this has no effect on comparing model forecasting accuracy and computation time between the federated architecture and the centralized architecture. Centralized learning means data are gathered from all devices to train a single model on the central server, which does not secure the privacy of users.

Table 1. Symbol definitions of the experiments.

Name	Description
N_{nodes}	Number of computing nodes
N_{round}	Number of iterations
S_{data}	Size of the trained data
S_{model}	Size of the model

To evaluate the forecasting performance of DFA, four baseline models are used for comparison purposes. The following are simple introductions for these models.

1. LSTM network : the model is an artificial recurrent neural network (RNN) architecture with feedback connections used in the field of deep learning.
2. Double seasonal Holt–Winters (DSHW): DSHW is a kind of exponential smoothing method which can accommodate two seasonal patterns besides parts of trend and level.
3. Transformer: it is a deep learning model that adopts the mechanism of self-attention, differentially weighting the significance of each part of the input data.
4. Encoder–Decoder: the model encodes the input as embedding features which are decoded by the decoder, adopting a sequence-to-sequence architecture.

4.3. Model Evaluation Indexes

The mean absolute percentage error (MAPE) is used to evaluate forecasting accuracy. The evaluation equations are defined as shown in Formula (14):

$$\text{MAPE} = [\sum_{i=1}^{N} (\hat{y}_i - y_i)/y_i]/N \times 100\% \tag{14}$$

where $\hat{y}_i$ is the forecast load consumption value, y_i is the actual load consumption value and N is the total number of sampling points for evaluation.

To evaluate whether a particular model m has skill with respect to a baseline model r the MAE ratio, we use skill score, as shown in Formula (15):

$$s = 1 - \frac{\text{MAE}_m}{\text{MAE}_r} \tag{15}$$

where MAE is the mean absolute error. MAE is calculated as shown in Formula (16):

$$\text{MAE} = [\sum_{i=1}^{N} |\hat{y}_i - y_i|]/N \tag{16}$$

4.4. Experimental Forecasting Performance

The proposed DFA and four baseline models are evaluated on 10 randomly chosen target houses. For each target house, the load records from June 2012 to June 2013 are used as training data, and 720 load records in September 2013 for prediction to calculate MAPE values. DFA makes use of all datasets of ten houses in the federated process and leverages the datasets from the target house to operate user adaptation. Baseline models are trained with the data from the target house. Table 2 shows the MAPE values of DFA and baseline models for 10 houses. Figure 5 shows the MAPE values for direct observation.

Table 2. MAPE values of DFA and baseline models for 10 houses.

DataSets	DFA	Transformer	DSHW	Encoder–Decoder	LSTM
House 1	4.92%	8.86%	13.76%	13.59%	14.74%
House 2	4.56%	8.43%	18.94%	11.27%	15.47%
House 3	4.77%	9.02%	16.72%	13.34%	16.43%
House 4	5.23%	8.56%	15.36%	14.67%	16.86%
House 5	5.13%	8.31%	16.88%	15.67%	15.92%
House 6	4.88%	8.73%	18.23%	12.39%	15.13%
House 7	4.89%	8.64%	19.89%	13.17%	14.62%
House 8	4.78%	8.52%	21.88%	13.40%	14.33%
House 9	4.47%	9.16%	17.94%	15.17%	15.31%
House 10	4.67%	8.68%	15.46%	15.55%	15.26%
Average	4.83%	8.69%	17.51%	13.82%	15.41%

From Table 2, we can see that the proposed DFA consistently outperforms the baseline models for ten houses. On average, it shows 38.28%, 69.83%, 58.81% and 63.65% relative improvements over Transformer, DSHW, Encoder–Decoder and LSTM, respectively, based on skill scores. The performances of LSTM and Encoder–Decoder are similar to each other and worse than Transformer since the number of parameters is less compared to Transformer. Performances of DSHW fluctuate widely and are inferior to the other models based on deep learning. We believe this is due to the differences in the cyclical characteristics in different spans which are influenced by many uncertain factors in residential loads. In summary, DFA has the best performance. We conclude that one of the reasons for the remarkable superiorities is DFA uses all datasets from ten houses to learn a model in the federated architecture. Additionally, we calculate the curve of MAPE values by varying the number of houses as shown in Figure 6. It can be seen that MAPE values of DFA gradually decrease with the number of houses increasing whereas other models do not vary much. This means that the model will be more robust when more devices are connected to the systems in the reality. More discussions about the superiorities in forecasting performance can be found in Section 4.6.

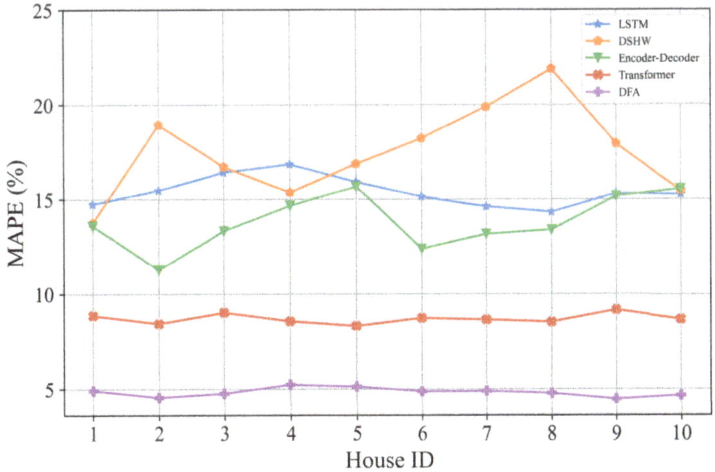

Figure 5. MAPE values of DFA and four baseline models for 10 houses.

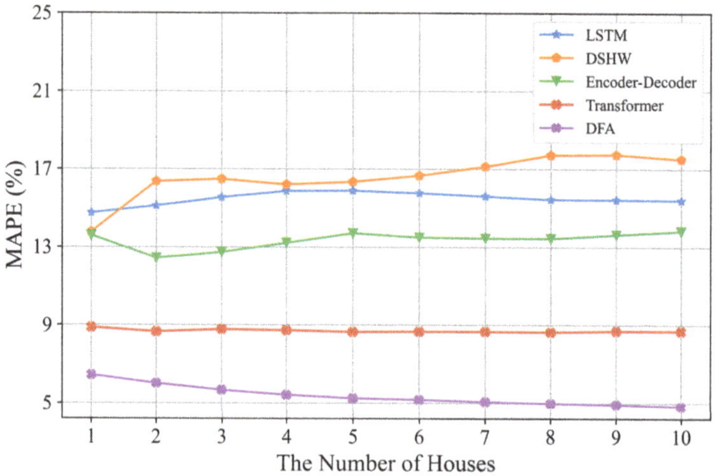

Figure 6. MAPE values of four baseline models and DFA with different numbers of houses connected to the federated system for 10 houses.

To evaluate the persistence of DFA, we conduct day-ahead and week-ahead forecasting tasks of DFA and four baseline models on one house, the results are shown in Table 3. It can be observed that although the forecasting performance of DFA decreases as the period goes from one day to one week, DFA outperforms all baseline models no matter how long is the forecasting period. We attribute this decline to the fact that DFA uses a sliding window for training and forecasting: the value forecasted by DFA will be added to the end of the sliding window for the next forecasting. Forecasting errors are cumulative as the period grows.

Table 3. Forcasting MAPE values of DFA and baseline models for day-ahead and week-ahead.

Model	Day 3	Day 11	Day 15–Day 21
DFA	6.43%	6.17 %	8.83 %
DSHW	16.85%	17.23%	17.61%
Transformer	8.77%	9.22 %	10.24%
Encoder–Decoder	13.64%	14.56 %	16.79%
LSTM	18.23%	17.64 %	21.54%

4.5. Performance of Federated and Centralized Architecture

Table 4 shows the forecasting performance and computation time comparison of the federated and centralized architecture. CNN-LSTM, as shown in Figure 3, is chosen as the test model. The different number of records in Table 3 means how many records for each node are used to train the model. For federated learning, the training time can only be estimated, the training time can be estimated as shown in Formula (17):

$$T_{training} = \bar{T}_{round} \cdot N_{round} \tag{17}$$

where $T_{training}$ denotes the training time, $\bar{T}_{round}$ indicates average computation time for all devices involved in each round.

From Table 3, it can be seen that the forecasting performance of the federated architecture is superior to the centralized while making predictions for STLF in the conditions of the different numbers of local records with $N_{nodes} = 10$. When N_{nodes} and records increase, the federated architecture can make use of more data to train the model, the forecasting performance will improve.

The federated architecture outperforms the centralized architecture on computation time comprehensively, with great advantage. As the federated architecture leverages devices involved in each round for training at the same time. It can be seen that the computation time fluctuates only slightly when N_{nodes} increases. This is because in each round, each device only processes the local data simultaneously and does not care about data from other devices in the system. Meanwhile, it also can be observed that the computation time rises at a lower rate than the centralized method as the number of records increases because for the centralized architecture, the incremental data of each device should be collected for training, the increment of data is decided by N_{nodes}.

Table 4. MAPE values and computation time of the federated and centralized architecture.

Approach	Forecasting Performance			Computation Time (s)		
Number of Local Records	5000	8000	15,000	5000	8000	15,000
Federated ($N_{nodes} = 1$)	13.82%	13.42%	13.38%	5.26	5.58	5.73
Federated ($N_{nodes} = 4$)	12.83%	12.33%	11.89%	5.38	5.73	5.84
Federated ($N_{nodes} = 7$)	11.57%	11.35%	10.87%	5.23	5.46	5.91
Federated ($N_{nodes} = 10$)	10.24%	10.13%	9.83%	5.47	5.62	5.88
Centralized ($N_{nodes} = 10$)	12.13%	12.24%	12.07%	8.32	11.74	25.63

Figure 7 shows the correlation between accuracy and the number of rounds for these two architectures. Federated learning shows a higher rate rise at the beginning of the iterations while the centralized accuracy rises slowly because multiple devices compute simultaneously in one round of iterations of federation learning. As can be seen from the trend of the curves, the federated architecture uses fewer iterations to achieve a satisfactory accuracy and achieve the state of convergence, which is also reflected in the shorter computation time in Table 3.

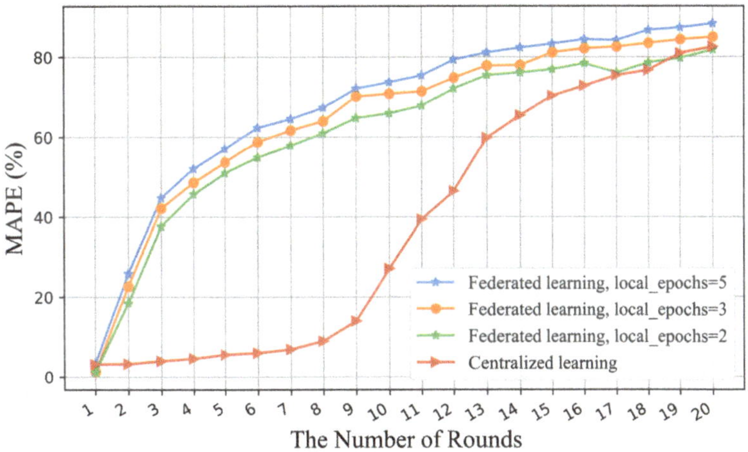

Figure 7. Correlation between accuracy and number of rounds for the federated and centralized architecture.

Now, we analyze the communication overhead of these two architectures. For the federated architecture, the calculation of communication overhead is defined in Formula (18):

$$Trans_{Fed} = 2N_{round} \cdot N_{nodes} \cdot S_{model} \tag{18}$$

meanwhile, the communication overhead of the centralized architecture is defined as shown in Formula (19):

$$Trans_{Cen} = N_{nodes} \cdot S_{data} \tag{19}$$

From the aforementioned formulas, the communication complexity of the federated architecture is $\mathcal{O}(N_{round} \cdot N_{nodes} \cdot S_{model})$ and the centralized architecture is $\mathcal{O}(N_{nodes} \cdot S_{data})$.

Since N_{nodes} is presented in both equations, it can be reduced. Therefore, the complexity is $\mathcal{O}(N_{round} \cdot S_{model})$ and $\mathcal{O}(S_{data})$ respectively. When the S_{data} is much larger than $N_{nodes} \cdot S_{data}$, the federated architecture has less communication burden than the centralized, which is common in practical applications. We can easily infer that the computation time will increase in the reality since the incremental communication overhead. In a summary, DFA is scalable with increasing data and has lower computation time and communication bandwidth requirements.

4.6. Ablation and Extensibility Experiments

To validate the superiority of DFA, we conduct the ablation and extensibility experiments based on datasets of five houses, other experiment settings are consistent with Section 4.4.

We use Fed to denote the DFA without MK-MMD optimization which is a CNN-LSTM network trained by the federated architecture. NoFed denotes the CNN-LSTM model trained by the centralized architecture with data only from the target house. We can see from Figure 8 that Fed achieves a better performance than NoFed on each target house. This indicates that each target house benefits from the federated architecture which makes it possible to leverage datasets from other houses, ensuring privacy simultaneously. It also can be seen that DFA has remarkable improvements in performance compared with Fed. We conclude that the transfer learning method can successfully conduct knowledge transfer from the federated model to the target houses to improve forecasting performance.

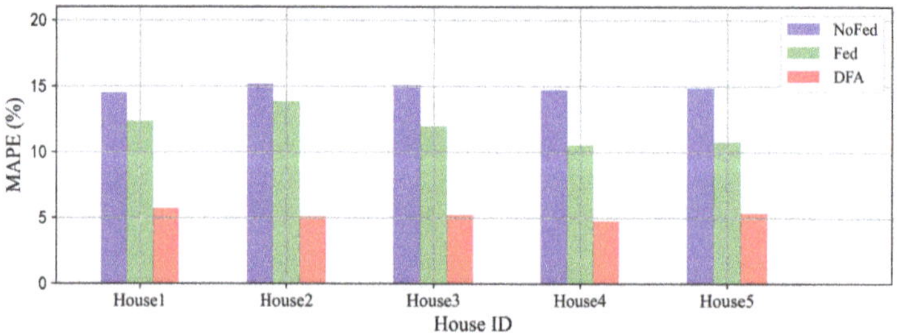

Figure 8. Ablation experiments of the federated architecture and MK-MMD optimization on 5 houses.

Furthermore, we extend DFA to different versions in which the part of MK-MMD is modified by the alternative transfer learning methods. Maximum mean discrepancy (MMD) is the single kernel version of MK-MMD. CORAL [34] is one of transfer learning methods that use the covariance matrices of the source and target features to compute the domain loss. It can be seen from Figure 9 that DFA can achieve satisfying performances on forecasting with different transfer learning methods. The results indicate that DFA is extensible with other transfer learning algorithms according to the real applications.

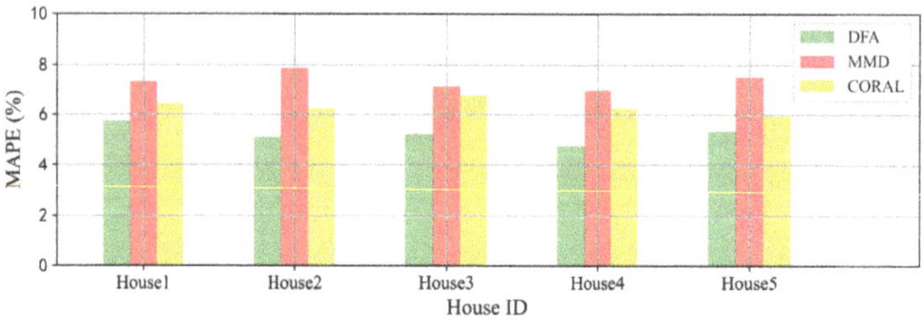

Figure 9. Extensibility experiments with alternative transfer learning methods on 5 houses.

5. Conclusions

In this paper, we propose a federated transfer learning approach for residential STLF. This approach addresses data availability and privacy by using a federated architecture. We implement a transfer learning method, multiple kernel variant of maximum mean discrepancies, to adapt to the non-IID data among different houses. The experimental results show that DFA shows a huge improvement in forecasting performance over other models. We also evaluate the federated architecture DFA used; it shows that the architecture is superior to the centralized architecture in computation time and has a small burden on communication. In the future, it would be promised for subsequent studies to adopt the state-of-the-art federated and transfer learning algorithms to achieve better forecasting performance with the framework of DFA.

Author Contributions: Conceptualization, Y.S. and X.X.; methodology, Y.S.; software, Y.S.; validation, Y.S.; formal analysis, Y.S.; investigation, Y.S.; resources, Y.S.; data curation, Y.S.; writing—original draft preparation,Y.S.; writing—review and editing, X.X.; visualization, Y.S.; supervision, X.X.; project administration, Y.S.; funding acquisition, X.X. All authors have read and agreed to the published version of the manuscript.

Funding: This research was funded by the National Natural Science Foundation of China, grant number 51975422.

Conflicts of Interest: The authors declare no conflict of interest.

References

1. Deb, C.; Zhang, F.; Yang, J.; Lee, S.E.; Shah, K.W. A review on time series forecasting techniques for building energy consumption. *Renew. Sustain. Energy Rev.* **2017**, *74*, 902–924. [CrossRef]
2. Amasyali, K.; El-Gohary, N.M. A review of data-driven building energy consumption prediction studies. *Renew. Sustain. Energy Rev.* **2018**, *81*, 1192–1205. [CrossRef]
3. Zhao, H.-x.; Magoulès, F. A review on the prediction of building energy consumption. *Renew. Sustain. Energy Rev.* **2012**, *16*, 3586–3592. [CrossRef]
4. Wu, D.; Wang, B.; Precup, D.; Boulet, B. Multiple Kernel Learning-Based Transfer Regression for Electric Load Forecasting. *IEEE Trans. Smart Grid* **2020**, *11*, 1183–1192. doi: [CrossRef]
5. Liu, D.; Chen, Q.; Mori, K. Time series forecasting method of building energy consumption using support vector regression. In Proceedings of the 2015 IEEE International Conference on Information and Automation, Lijiang, China, 8–10 August 2015, pp. 1628–1632.
6. Pavićević, M.; Popović, T. Forecasting Day-Ahead Electricity Metrics with Artificial Neural Networks. *Sensors* **2022**, *22*, 1051. doi: [CrossRef]
7. Ahmad, T.; Chen, H. Short and medium-term forecasting of cooling and heating load demand in building environment with data-mining based approaches. *Energy Build.* **2018**, *166*, 460–476. [CrossRef]
8. Liu, N.; Tang, Q.; Zhang, J.; Fan, W.; Liu, J. A hybrid forecasting model with parameter optimization for short-term load forecasting of micro-grids. *Appl. Energy* **2014**, *129*, 336–345. [CrossRef]
9. Fan, G.F.; Peng, L.L.; Hong, W.C.; Sun, F. Electric load forecasting by the SVR model with differential empirical mode decomposition and auto regression. *Neurocomputing* **2016**, *173*, 958–970. [CrossRef]
10. Zhou, H.; Zhang, S.; Peng, J.; Zhang, S.; Li, J.; Xiong, H.; Zhang, W. Informer: Beyond efficient transformer for long sequence time-series forecasting. In Proceedings of the AAAI, virtually, 2–9 February 2021.
11. Meng, Z.; Xu, X. A hybrid short-term load forecasting framework with an attention-based encoder–decoder network based on seasonal and trend adjustment. *Energies* **2019**, *12*, 4612. [CrossRef]
12. Savi, M.; Olivadese, F. Short-term energy consumption forecasting at the edge: A federated learning approach. *IEEE Access* **2021**, *9*, 95949–95969. [CrossRef]
13. Khwaja, A.S.; Anpalagan, A.; Naeem, M.; Venkatesh, B. Smart Meter Data Obfuscation Using Correlated Noise. *IEEE Internet Things J.* **2020**, *7*, 7250–7264. [CrossRef]
14. Yang, Q.; Liu, Y.; Chen, T.; Tong, Y. Federated machine learning: Concept and applications. *ACM Trans. Intell. Syst. Technol. (TIST)* **2019**, *10*, 1–19. [CrossRef]
15. Gholizadeh, N.; Musilek, P. Federated learning with hyperparameter-based clustering for electrical load forecasting. *Internet Things* **2022**, *17*, 100470 . doi: [CrossRef]
16. Moradzadeh, A.; Moayyed, H.; Mohammadi-Ivatloo, B.; Aguiar, A.P.; Anvari-Moghaddam, A. A Secure Federated Deep Learning-Based Approach for Heating Load Demand Forecasting in Building Environment. *IEEE Access* **2022**, *10*, 5037–5050. doi: [CrossRef]

17. Yang, T.; Andrew, G.; Eichner, H.; Sun, H.; Li, W.; Kong, N.; Ramage, D.; Beaufays, F. Applied federated learning: Improving google keyboard query suggestions. *arXiv* **2018**, arXiv:1812.02903.
18. Wu, X.; Liang, Z.; Wang, J. Fedmed: A federated learning framework for language modeling. *Sensors* **2020**, *20*, 4048. [CrossRef] [PubMed]
19. Chen, Y.; Qin, X.; Wang, J.; Yu, C.; Gao, W. Fedhealth: A federated transfer learning framework for wearable healthcare. *IEEE Intell. Syst.* **2020**, *35*, 83–93. [CrossRef]
20. Pokhrel, S.R.; Choi, J. Federated learning with blockchain for autonomous vehicles: Analysis and design challenges. *IEEE Trans. Commun.* **2020**, *68*, 4734–4746. [CrossRef]
21. Ye, D.; Yu, R.; Pan, M.; Han, Z. Federated learning in vehicular edge computing: A selective model aggregation approach. *IEEE Access* **2020**, *8*, 23920–23935. [CrossRef]
22. Abreha, H.G.; Hayajneh, M.; Serhani, M.A. Federated Learning in Edge Computing: A Systematic Survey. *Sensors* **2022**, *22*, 450 . doi: [CrossRef]
23. Tzeng, E.; Hoffman, J.; Zhang, N.; Saenko, K.; Darrell, T. Deep domain confusion: Maximizing for domain invariance. *arXiv* **2014**, arXiv:1412.3474.
24. Wang, H.; Lei, Z.; Zhang, X.; Zhou, B.; Peng, J. A review of deep learning for renewable energy forecasting. *Energy Convers. Manag.* **2019**, *198*, 111799. [CrossRef]
25. Xu, X.; Meng, Z. A hybrid transfer learning model for short-term electric load forecasting. *Electr. Eng.* **2020**, *102*, 1371–1381. doi: [CrossRef]
26. Cai, L.; Gu, J.; Jin, Z. Two-Layer Transfer-Learning-Based Architecture for Short-Term Load Forecasting. *IEEE Trans. Ind. Inform.* **2020**, *16*, 1722–1732. doi: [CrossRef]
27. Alhmoud, L.; Abu Khurma, R.; Al-Zoubi, A.M.; Aljarah, I. A Real-Time Electrical Load Forecasting in Jordan Using an Enhanced Evolutionary Feedforward Neural Network. *Sensors* **2021**, *21*. doi: [CrossRef] [PubMed]
28. Long, M.; Cao, Y.; Wang, J.; Jordan, M. Learning transferable features with deep adaptation networks. In Proceedings of the 32nd International Conference on Machine Learning, Lille, France, 7–9 July 2015; pp. 97–105.
29. Gretton, A.; Borgwardt, K.M.; Rasch, M.J.; Schölkopf, B.; Smola, A. A kernel two-sample test. *J. Mach. Learn. Res.* **2012**, *13*, 723–773.
30. McMahan, B.; Moore, E.; Ramage, D.; Hampson, S.; y Arcas, B.A. Communication-efficient learning of deep networks from decentralized data. In Proceedings of the 20th International Conference on Artificial Intelligence and Statistics, Fort Lauderdale, FL, USA, 20–22 April 2017; pp. 1273–1282.
31. Agga, A.; Abbou, A.; Labbadi, M.; Houm, Y.E. Short-term self consumption PV plant power production forecasts based on hybrid CNN-LSTM, ConvLSTM models. *Renew. Energy* **2021**, *177*, 101–112. doi: [CrossRef]
32. Smart Meter Energy Consumption Data in London Households. Available online: https://www.kaggle.com/jeanmidev/smart-meters-in-london (accessed on 10 January 2022).
33. DarkSky Service. Available online: ttps://darksky.net (accessed on 10 January 2022).
34. Sun, B.; Saenko, K. Deep coral: Correlation alignment for deep domain adaptation. In *European Conference on Computer Vision*; Springer: Berlin/Heidelberg, Germany, 2016, pp. 443–450.

energies

Article

Research on Energy Saving and Environmental Protection Management Evaluation of Listed Companies in Energy Industry Based on Portfolio Weight Cloud Model

Shanshan Li [1,2,†], Yujie Wang [3,*,†], Yuannan Zheng [1], Jichao Geng [2] and Junqi Zhu [2]

1 Joint National-Local Engineering Research Centre for Safe and Precise Coal Mining, Anhui University of Science and Technology, Huainan 232001, China; 2020029@aust.edu.cn (S.L.); yuannanzheng@163.com (Y.Z.)
2 School of Economics and Management, Anhui University of Science and Technology, Huainan 232001, China; gjcjsj@aust.edu.cn (J.G.); 2010005@aust.edu.cn (J.Z.)
3 School of Economics and Management, Taiyuan University of Technology, Taiyuan 030024, China
* Correspondence: wangyujiesx@163.com; Tel.: +86-13776589633
† These authors contributed equally to this work.

Abstract: Under the background of the "carbon peaking and carbon neutrality" strategy, energy saving and environmental protection (ESEP) management has become one of the most important projects of enterprises. In order to evaluate the ESEP management level of listed companies in the energy industry comprehensively, this study puts forward the evaluation framework of "governance framework-implementation process-governance effectiveness" for ESEP management level. Based on the comprehensive collection and collating of related information reports (e.g., sustainable development reports) of listed energy companies from 2009 to 2018, the ESEP information was extracted, and the portfolio weight cloud model was used to evaluate the ESEP management status of listed energy companies in China. It is of great theoretical innovation and practical significance to promote the evolution of the economy from "green development" to "dark green development". The results show that: (1) the number of SHEE information released by listed companies in the energy industry shows a steady increasing trend, but the release rate is low, and there are differentiation characteristics in different industries. (2) The ESEP management level of most listed companies in the energy industry is still at the low level, and only 17.19% (S = 65) of the sample companies are at the level of "IV level-acceptable" and "V level-claimable". (3) In terms of governance framework-implementation process-governance effectiveness, B1-governance framework (E_x = 3.4451) and B2-implementation process (E_x = 2.9480) are relatively high, but B3-governance effectiveness (E_x = 2.0852) and B4-public welfare (E_x = 2.0556) are relatively low. The expectation of most ESEP evaluation indexes fluctuates between "III level-transition level" and "II Level-improvement level". Finally, some suggestions are put forward to improve ESEP management levels.

Keywords: listed companies in the energy industry; ESEP management evaluation; analytic hierarchy process; entropy weight; cloud model

Citation: Li, S.; Wang, Y.; Zheng, Y.; Geng, J.; Zhu, J. Research on Energy Saving and Environmental Protection Management Evaluation of Listed Companies in Energy Industry Based on Portfolio Weight Cloud Model. *Energies* **2022**, *15*, 4311. https://doi.org/10.3390/en15124311

Academic Editors: Antonio Cano-Ortega, Francisco Sánchez-Sutil and Aurora Gil-de-Castro

Received: 11 May 2022
Accepted: 4 June 2022
Published: 13 June 2022

1. Introduction

Resources, environment and population are the three major problems that human society is facing, especially the environmental problem, which is posing a serious threat to human survival and development [1–3]. Since economic reform and opening up, China has made historic achievements in development, but also accumulated a large number of ecological and environmental problems; environmental pollution is on the rise, and the discharge of major pollutants is still serious, which has become a weakness in all-round well-off society [4]. In the new historical situation and background, the Chinese government is also positively changing its style of ruling, practicing green concept and actively carrying out the practice of building energy conservation and emissions reduction.

The Chinese government has introduced the "12th five-year plan for energy conservation and emissions reduction ", "13th Five-Year plan for energy conservation and emission reduction comprehensive work plan" and "the evaluation index system of ecological civilization construction" among other laws and regulations and clearly points to "vigorously developing the circulation economy", "the implementation of energy conservation and emissions reduction project", "strengthen the main pollutant emission reduction", etc., and putting forward "strive to achieve carbon dioxide emissions peak before 2030, per unit of GDP carbon dioxide emissions lower than in 2005 by more than 65%, strive to become carbon neutral before 2060" and other "carbon peaking and carbon neutrality" targets and specific indicators. In the face of energy conservation and environmental protection related indicators and enterprise sustainable development strategy demand, many enterprises, especially the energy industry (the main waste water, waste gas, solid waste emissions units) have implemented a series of energy conservation and emissions reduction environmental protection measures (hereinafter referred to as energy saving and environmental protection, ESEP) management measures [5,6]. However, developing these projects has became the burden of the enterprise to a certain extent, which leads to less attention being paid, limited implementation, limited investment and other phenomena. In this context, it has become an important topic to fully understand the implementation of ESEP management in energy industry enterprises, to mobilize enterprises to carry out ESEP management actively, and improve the weak links of enterprises' ESEP management.

Some institutions, organizations and scholars have actively explored the issues of energy saving and environmental protection from different perspectives. Current research mainly focuses on the influencing factors on energy saving and emission reduction [7,8], policies [9,10], efficiency [11,12], and environmental performance evaluation [13,14]. In terms of energy saving and emission reduction efficiency evaluation and environmental performance evaluation, scholars have mainly constructed an evaluation index system from the perspective of product life cycle, sustainable development, input-output and pressure-response framework. For example, Wu and Chen (2014), on the basis of analyzing the content of the whole-process environmental management, established an index system for the performance evaluation of the whole-process of the environmental management of the enterprise, which involves various activities and links between the whole process of the enterprise, including green procurement, ecological design, cleaner production, green transportation, green sales, green use and the construction of green corporate culture [15]. Xue et al. (2022) established a comprehensive evaluation framework based on life cycle assessment and the protection supply curve to evaluate the benefits of energy saving and emission reduction [16]. Wei et al. (2018) constructed an urban environmental performance evaluation indicator system from the four aspects of environmental health, ecological protection, environmental governance and sustainable utilization of resources and energy based on the "driving-pressure-state-impact-response (DPSIR)" model [17]. Li et al. (2019) focus on green behaviors of enterprises and constructed an evaluation index system of green governance from four aspects: green governance framework, green governance mechanism, green governance efficiency and green governance responsibility [18]. The strategies of energy enterprises are very important to their existence and development [19]. Although these studies have carried out a comprehensive evaluation on all aspects of ESEP, they focus more on evaluation research from the perspective of performance, and the measurement of management performance related to ESEP still heavily relies on lagging indicators such as energy consumption, pollutant emission and resource recycling. There is still a lack of systematic and comprehensive evaluation of the ESEP management status of energy industry enterprises from the perspective of management.

In terms of the measurement and evaluation methods of regional energy conservation and environmental protection, most studies adopt qualitative or semi-qualitative methods such as the expert scoring method, questionnaire survey method, analytic hierarchy process and life cycle assessment, etc. [19,20]. The Cloud model is a new evaluation method especially studying compound uncertainty proposed by Li et al. [21]. Compared with traditional assessment methods, cloud model evaluation methods can better describe the randomness and fuzziness of evaluation objects or variables (e.g., judge whether a variable is closer to 2 or to 3 when its primary experimental value is 2.5), and realize the mapping and conversion between qualitative and quantitative uncertainty, which has been widely applied to sustainability assessment, risk assessment and many other fields [22,23]. Based on this, this study constructs the integration of an assessment framework including "governance framework, implementation process, governance effectiveness", and uses the combination weighting method of the cloud model to evaluate the ESEP management ability of listed companies in the energy industry, in order to clarify the present situation of ESEP disclosure, the ESEP management situation and ESEP weak links, investigating ESEP benchmark enterprises and key indicators in various industries, and then putting forward countermeasures and suggestions for improving and strengthening ESEP relevant work.

The innovations of this study are as follows: (1) Focusing on listed companies in the energy industry, the ESEP management evaluation system based on the evaluation framework of "governance framework, implementation process, governance effectiveness" is constructed, which enriches the research on ESEP management evaluation; (2) Combine with the information disclosure measurement method, establish the qualitative index rating basis, and collect the evaluation index data information based on the ESEP information disclosed by listed companies, further enriching the relevant research on ESEP management evaluation; (3) The cloud model theory is applied to ESEP management evaluation, and a management evaluation model based on combination weight-cloud evaluation is constructed, which can provide guidance for ESEP management evaluation research.

2. Methods

2.1. Construction of Evaluation Index System

Although some studies have carried out a comprehensive evaluation of enterprise's ESEP management, these studies focus more on evaluation research from the perspective of performance, and the measurement of management performance related to ESEP still heavily relies on lagging indicators such as energy consumption, pollutant emission and resource recycling. There is still a lack of systematic and comprehensive evaluation of ESEP management status of energy industry enterprises from the perspective of management. By reading a large number of relevant laws and regulations and relevant literature, combined with the actual situation of the energy industry and following the principles of scientific, systematic, comparable and operable index design, this study constructs an ESEP management evaluation index system for listed companies in the energy industry. The system is divided into three layers: (1) The target layer is ESEP comprehensive evaluation of listed companies in the energy industry; (2) The criterion layer is divided into four categories: governance framework, implementation process, governance effectiveness, and others; (3) The index layer is composed of 20 first-level indicators reflecting "governance framework, implementation process, governance effectiveness, public welfare, etc.", and calculation and evaluation instructions are provided under each indicator (see Table 1). These indicators can reflect the performance of enterprises in energy saving and environmental management in a comprehensive and systematic way, and the indicators are described below.

Table 1. Energy saving and environmental protection (ESEP) management evaluation index system of listed companies in the energy industry.

Target Layer	Criterion Layer B	Index Layer C	Index Introduction
Comprehensive evaluation on energy saving and environmental protection	B1-governance framework	C1~ESEP institutional system	Degree of completeness of relevant management system, department and committee (1–5)
		C2~ESEP management system	Degree of completeness and systematicness of relevant management system certification and implementation (1–5)
		C3~ESEP management culture	Degree of emphasis on ESEP and richness of education activities
		C4~ESEP clauses and policies	Degree of completeness of environmental provisions for customers or suppliers (1–5)
	B2-implementation process	C5~clean production management	Degree of completeness of green raw material procurement and cleaner production audit (1–5)
		C6~pollution reduction management	Diversity of management measures for emission reduction of three wastes and perfection of implementation (1–5)
		C7~recycling management	Diversity and perfection of resource recycling management measures (1–5)
		C8~energy efficiency improvement management	Diversity and implementation of energy efficiency management measures (1–5)
		C9~tackling climate change management	Diversity and implementation of GHG emission management measures (1–5)
		C10~environmental protection management	Diversity and perfection of environmental protection management measures (1–5)
		C11~green office management	Diversity and perfection of green office management measures (1–5)
	B3-governance effectiveness	C12~environmental pollution events	The number of pollution incidents
		C13~discharge of three wastes	Discharge of COD, SO2, NOX and solid waste per ten thousand yuan of output value
		C14~energy consumption situation,	Comprehensive energy consumption per ten thousand yuan of output value (ton of standard coal/Ten thousand yuan)
		C15~resource recycling	Water resource/ waste resource recycling utilization rate
		C16~other greenhouse emissions	CO2, CH4, N2O and other greenhouse gas emissions per ten thousand yuan output value
		C17~ecological environment construction	Added green area or animal and plant protection per ten thousand yuan of output value
		C18~ESEP influence	Relevant awards/honors/patents/paper grades (1–5)
		C19~ESEP special investment index.	Energy saving per ten thousand yuan output value/environmental protection special fund input
	B4-public welfare and others	C20~ESEP public welfare activities	Degree of participation in environmental public welfare activities (1–5)

(1) Governance framework: A reasonable governance framework can determine the vision, culture, strategy and system of a company's ESEP from the top design level, which is the basis and key to improving a company's ESEP level and sustainable development. Tian et al. (2015) believe that forward-looking environmental strategy can effectively promote enterprise green innovation, enhance enterprise green image and improve enterprise environmental performance [24]. Liao et al. (2015) propose that establishing a social responsibility committee, an environmental protection committee and other organizations to coordinate stakeholder relations can improve corporate social responsibility performance [25]. Baboukardos (2018) emphasizes the importance of environmental clauses and points out that companies with recognized environmental clauses would help investors clarify the future economic benefits and costs related to the company's environmental performance by sending signals of strong future financial performance or improving the reliability of environmental performance information [26]. Therefore, this study believes that the ESEP management system should cover the dimension of governance framework, and sets up indicators such as C1-SEP institutional system, C2-ESEP management system, C3-ESEP management culture, and C4-ESEP clauses and policies to evaluate the governance framework.

(2) Implementation process: Greening production and operation activities of enterprises is an important link to improving ESEP management level and sustainable development ability. For example, Wu and Chen (2014) believe that effective prevention and control measures should be adopted to carry out environmental management across the whole process of procurement, design, production, transportation, sales and use [15]. Du (2013) believes that source management (clean production) and process control (improving resource efficiency) are the key points in the construction of a "environment-friendly and resource-conserving society" [27]. Therefore, this study suggests this dimension of the ESEP management system should cover the implementation process, and has set up C5-clean production management, C6-pollution reduction management, C7-recycling management, C8-energy efficiency improvement management, C9-tackling climate change management, C10-environmental protection management, C11-green office management and other indicators to evaluate the implementation situation.

(3) Governance efficiency: the ESEP governance efficiency index mainly reflects the situation of enterprises in energy conservation, "three wastes" emission reduction, resource recycling and waste reuse, which can intuitively measure the performance of enterprises from environmental aspects. Some scholars also introduced these indicators in their studies to measure the environmental performance of enterprises. For example, Qin et al. (2004) synthesize the emission indexes of important pollution factors such as SO2, NOX and COD into a comprehensive index to express the environmental performance of enterprises [28]. Hao et al. (2014) use CO2 emissions as a proxy variable to study the environmental impact of industrial enterprises [29]. Wang et al. (2018) select R&D investment per unit energy consumption to measure the level of green innovation of enterprises [30]. Therefore, this study believes that it is necessary to incorporate the dimension of governance effectiveness into the ESEP evaluation system. Specifically, it includes C12-environmental pollution events, C13-discharge of three wastes, C14-energy consumption situation, C15-resource recycling, C16-other greenhouse emissions, C17-ecological environment construction, C18-ESEP influence, and C19-ESEP special investment index.

(4) Others: The setting of other dimensions is mainly to measure the participation of enterprises in environmental public welfare activities. Wang et al. (2015) point out that enterprises' active participation in environmental protection and public welfare can convey signals of enterprises' green governance status to investors on the one hand, and objectively reflect the implementation status of enterprises' environmental management on the other hand. Therefore, in this study, some factors of ESG related evaluation are used for reference, and ESEP public welfare and other dimensions are incorporated into the ESEP evaluation system, so as to comprehensively measure the performance of enterprises in external environmental public welfare and other aspects.

2.2. Combination Weight-Cloud Evaluation Comprehensive Evaluation Model

2.2.1. Combination Weight Model

The analytic hierarchy process (AHP) is a method of subjective empowerment, and its basic idea is to use the systematic idea of decomposition followed by synthesis to organize and synthesize people's subjective judgments, realize the organic combination of qualitative and quantitative analysis, and complete quantitative decision-making [31,32]. The general steps of the research using this method are: (1) establishing the hierarchical structure model; (2) constructing the judgment matrix; (3) calculating the index weights; (4) testing the consistency of the judgment matrix. In the specific operation, due to the problems of large calculation workload and tedious testing process, this study uses Yahhp software for subjective weight measurement. The entropy weight (EW) method is a method of objective assignment of weights, the core of which is to use the amount of data information of each indicator to determine the weight; when the evaluation data value of an evaluation indicator differs greatly, its entropy value is smaller, indicating that the evaluator has a greater difference in the sensitivity degree of the indicator, that is, the indicator can provide more reference information for the evaluation of the merits, and has greater significance within the evaluation system [33,34]. The general steps when using this method for research are: (1) standardization of data; (2) calculation of the entropy value of each indicator; (3) calculation of the weight vector of each indicator. This study used AHP-EW for combined weighting to obtain more accurate and objective weights. The specific formula can be found in the related literature [35].

2.2.2. Cloud Evaluation Model

The cloud model is a kind of evaluation method based on probability statistics and fuzzy set theory, and its evaluation results can be expressed by cloud digital features (E_x, E_n, H_e), which is schematically shown in Figure 1. When cloud model evaluation method is used, it can be realized by the cloud generator (CG), and four types of each cloud generator algorithm are shown in Figure 2. Specific algorithms can be found in the related literature.

2.2.3. Comprehensive Evaluation Model

This study uses a combination of combined weights and cloud model to evaluate the energy saving and environmental protection management status of the company, and the specific steps are as follows. When using this method for evaluation, the general steps are: (1) establish the weight factor set $W = \{\omega_1, \omega_2, \ldots, \omega_n\}$ of indicators; (2) determine the indicator set and the evaluation language domain $V = \{V_1, V_2, \ldots, V_m\}$, in this study, the evaluation language is divided into five levels: vigilance-level, improvement-level, transition-level, acceptable-level, and declarable-level; (3) determine the cloud parameter matrix (E_x, E_n, H_e) for each level of each indicator; (4) calculate the affiliation degree of each sample and each indicator; and (5) determine the evaluation level. The specific formula for each step is referred to in the related literature [36].

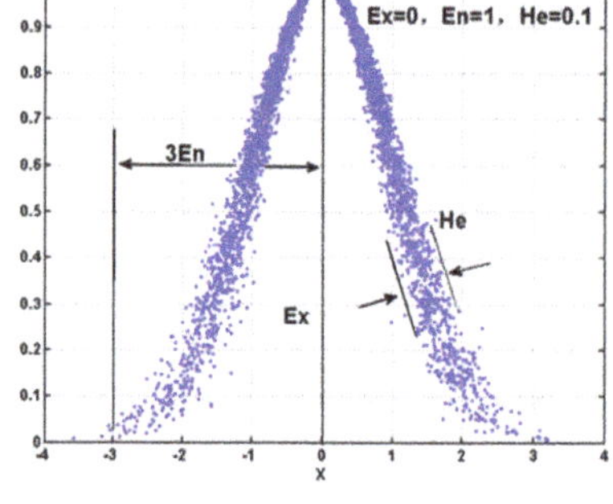

Figure 1. Normal cloud and digital features $(E_x = 0,\ E_n = 1,\ H_e = 0.1)$.

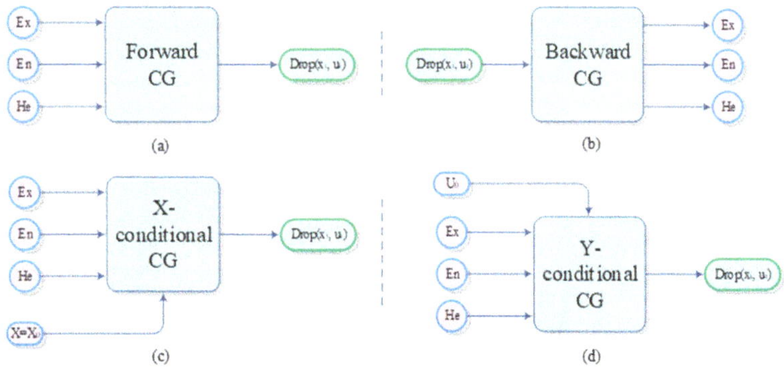

Figure 2. Schematic diagram of cloud generator. (**a**) Forward CG. (**b**) Backward CG. (**c**) X-conditional CG. (**d**) Y-conditional CG.

2.3. Data Collection and Samples

According to the Guidelines on Industry Classification of Listed Companies issued by CSRC, listed companies in the energy industry from 2006–2017 were selected for this study (industry codes 06, 07, 25, 44, 45, 46). The sample was also carefully screened (e.g., shaving off ST and *ST companies; shaving off companies listed after 2006, etc.), and after sample screening, 59 companies with 378 sample observations finally remained. It is worth noting that there are still 78 companies that did not release any ESEP-related reports during 2006–2018 and did not participate in this evaluation study.

The original data of this study can be divided into quantitative index data and qualitative index data. Quantitative indicators such as COD per ten thousand yuan output value, SO2, NOX, solid waste emissions, comprehensive energy consumption of ten thousand yuan output value (ton of standard coal/ten thousand yuan), etc. can be obtained or calculated through the social responsibility report, CSMAR database, enterprise official website and other channels. Quantitative indicators are difficult to be quantified by themselves, and they need to be quantified in combination with expert scoring and information disclosure measurement methods. Referring to relevant literature [37,38], this study uses 1–5 score points for quantification (see quantification standard of indicators in Table 2).

Table 2. Institutional indicators—Quantitative scoring standard.

Score	Specific Standard
5	The relevant institutions of ESEP are well established, such as systematic ESEP management system, specialized ESEP management department, ESEP management committee, and detailed text charts, data and information explanation
4	The relevant institutions of ESEP are relatively complete, such as ESEP management system, ESEP management department and ESEP management Committee.
3	The relevant institutions of ESEP are generally complete, with ESEP management system and departments, but no management committee.
2	The relevant institutions system of ESEP are not perfect, with only ESEP management system, no management department and management committee.
1	The relevant institutions system of ESEP is extremely imperfect, and there is no explanation on the construction of the institutional system of ESEP.

3. Results and Discussion

3.1. Analysis of Current Situation of Energy Saving and Environmental Protection Information

3.1.1. Quantitative Distribution of ESEP Information

Sorting out the quantity, quality and content of ESEP information released by listed companies in the energy industry is helpful for us to grasp its development status and trends as a whole. In order to investigate the quantity of ESEP information release, this study provides statistics on the ESEP information release of sample companies from 2006 to 2017, and the year-by-year change trend is shown in Figure 3.

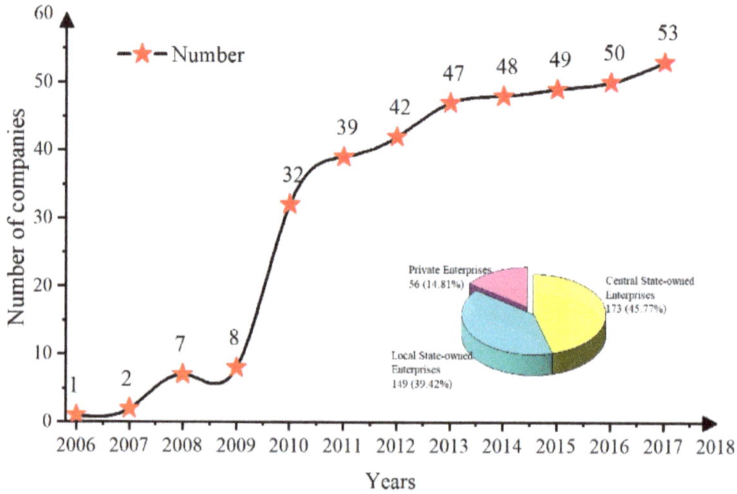

Figure 3. Quantity distribution of energy saving and environmental protection information disclosure.

As can be seen from Figure 4, the amount of ESEP information released in the energy industry showed a steady increasing trend during 2006–2017, but the release rate was still low. From 2006 to 2017, only 42.44% (N = 59) of enterprises in the energy industry released ESEP-related information reports (378 social responsibility reports/sustainability reports/employee responsibility reports), and 56.12% (N = 78) of enterprises did not release any ESEP-related information reports during this period. This shows that regular release of ESEP information has gradually become the consensus of listed companies in the energy industry, but there is still a big gap between the development of the national strategy of "beautiful China" and "healthy China". After further concluding ESEP related information of the company, it can be found that the central state-owned enterprises ESEP information release quantity (45.41%) was significantly better than that of local state-owned enterprises and private enterprises, reflecting that the central state-owned enterprise society responsibility consciousness is stronger and ESEP management level is higher, but the local state-owned enterprises and private enterprises release quantity remains to be further improved.

3.1.2. Industry Distribution of ESEP Information

This study further provides statistics on the industry of the company releasing ESEP information. It can be seen from Table 3 that different industrys' nature leads to great difference in the release rate of ESEP information. The oil and gas extraction industry has the highest release rate (80.00% in the last three years), while the gas production and supply industry has the lowest release rate (25.57%).

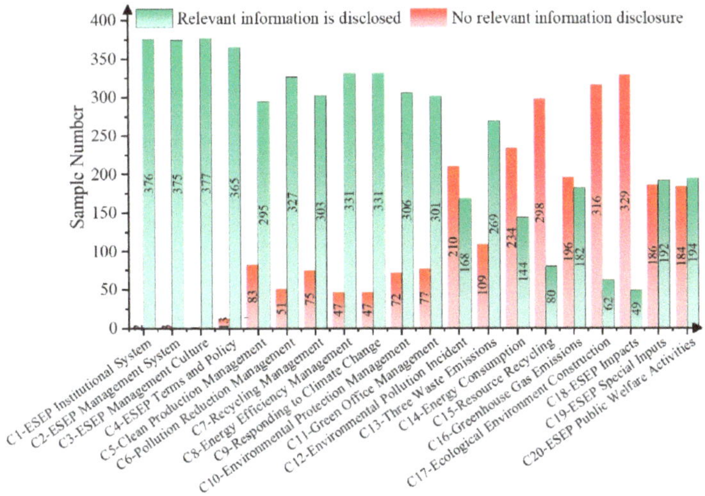

Figure 4. Distribution of energy saving and environmental protection information disclosure.

Table 3. Occupational safety and health related information industry distribution (in last three years).

The Name of the Industry	Release Quantity			Release Proportion		
	2015	2016	2017	2015	2016	2017
Coal mining and washing industry	13	14	13	48.15%	51.85%	48.15%
Oil and gas extraction	4	4	4	80.00%	80.00%	80.00%
Power and heat production and supply	23	22	23	32.86%	32.86%	32.86%
Gas production and supply	4	5	7	16.67%	20.83%	39.17%
Water production and supply industries	5	5	6	33.33%	33.33%	40.00%

3.1.3. Content Distribution of ESEP Information

In order to investigate the distribution of ESEP information content of sample companies, this study sorts out the distribution of ESEP information content based on the ESEP management evaluation system designed above (see Figure 4). As can be seen from Figure 3, on the whole, ESEP information content in the energy industry is relatively comprehensive, and the disclosure level of indicators that are represented by C1-management system, C2-management culture, C3-management system, and C4-clauses and policies reaches more than 90%. However, from the perspective of the disclosure quantity of each index, there are still problems such as the lack of standardization, systematization and comparability of ESEP information content. For example, the disclosure level of quantitative information of C18-ESEP influence, C17-ecological environment construction, C15-resource recycling and other indicators is low, and the disclosure is not scientific enough.

3.2. Evaluation Analysis of Cloud Model of Each Company

According to the ESEP evaluation framework constructed above, this study adopts the comprehensive evaluation cloud model to conduct equivalent evaluation of each sample company. The brief evaluation steps are as follows:

(1) AHP-EW method is selected to determine the factor subset of each index weight. (1) Firstly, on the basis of fully combing and referring to the ideas and methods of AHP, the subjective weight is obtained according to the operation steps of AHP; (2) Secondly, on the basis of obtaining relevant index data, the objective weight is obtained according to the operation steps of the entropy weight method (Formulas (1)–(4)). After getting the

subjective weight by AHP and the objective weight by entropy weight method, calculating according to Formula (5), the comprehensive weight of SHEE management evaluation of mineral resource-based listed companies can be obtained.

(2) According to the data value range of each index, determine the evaluation grade theory domain. By referring to relevant literature, this paper divides each indicator into five grades, which are used to evaluate the level of a company in a certain index: I-alert level, II-improvement level, III-transition level, IV-acceptable level, V-claimable level. Specific index levels are divided as follows: Taking index X1 (degree of perfection of mechanism system) as an example, the level I interval is [1, 1.5], the level II interval is [1.5, 2.5], the level III interval is [2.5, 3.5], the level IV interval is [3.5, 4.5], and the level V interval is [4.5, 5]. In the same way, all index grades can be obtained according to the Formula (10).

(3) According to Formula (6), the evaluation level corresponding to each indicator is represented by the corresponding cloud parameters (E_x, E_n, H_e). Taking indicator X1 (degree of perfection of mechanism system) as an example, the parameters of level I interval cloud model are $(E_x, E_n, H_e) = (1, 0.17, 0.05)$. The parameters of the II level interval cloud model are $(E_x, E_n, H_e) = (2, 0.17, 0.05)$. The parameters of level III interval cloud model are $(E_x, E_n, H_e) = (3, 0.17, 0.05)$. The parameters of IV level interval cloud model were $(E_x, E_n, H_e) = (4, 0.17, 0.05)$. The parameters of the V level interval cloud model were $(E_x, E_n, H_e) = (5, 0.17, 0.05)$; Similarly, according to Formula (10), cloud parameter matrices of all indicators at all levels can be obtained.

(4) Taking the screened indicator data and acquired cloud digital characteristic values as parameters, and the X-conditional cloud generator in the model is used to input the algorithm program into Matlab2014 software for calculation, so as to obtain the membership degree of an experiment. In order to improve the accuracy and credibility of the data, the number of experiments was set as K = 2000, and the final membership degree could be obtained according to Formula (7). Due to space limitations, the membership calculation results of SINOPEC in 2017 are taken as an example (see Table 4).

Table 4. Membership degree of each index of SINOPEC ESEP management in 2017.

Comments	I Level	II Level	III Level	IV Level	V Level	Conclusion
C1~ESEP institutional system	0.0000	0.0001	0.2984	0.3233	0.3781	III level
C2~ESEP management system	0.0000	0.0000	0.0021	0.0012	0.9967	V level
C3~ESEP management culture	0.0000	0.0000	0.0002	0.0002	0.9997	V level
C4~ESEP clauses and policies	0.0000	0.0000	0.0501	0.0373	0.9126	V level
C5~clean production management	0.0000	0.0011	0.5001	0.4988	0.0000	III level
C6~pollution reduction management	0.0000	0.0003	0.2686	0.3138	0.4173	V level
C7~recycling management	0.0000	0.0011	0.4983	0.5006	0.0000	IV level
C8~energy efficiency improvement management	0.0000	0.0001	0.3175	0.3439	0.3384	III level
C9~tackling climate change management	0.0000	0.0000	0.2646	0.2760	0.4594	V level
C10~environmental protection management	0.0000	0.0002	0.2770	0.2875	0.4353	V level
C11~green office management	0.0000	0.0002	0.3903	0.2849	0.3247	V level
C12~environmental pollution events	0.0000	0.0000	0.0000	0.0000	1.0000	V level
C13~discharge of three wastes	0.0000	0.0011	0.4991	0.4998	0.0000	IV level
C14~energy consumption situation,	0.0000	0.0000	0.0409	0.0371	0.9219	V level
C15~resource recycling	0.0025	0.9940	0.0015	0.0020	0.0000	V level
C16~other greenhouse emissions	0.0000	0.0000	0.0000	0.0000	1.0000	V level
C17~ecological environment construction	0.9977	0.0023	0.0000	0.0000	0.0000	I level
C18~ESEP influence	0.1714	0.8268	0.0008	0.0010	0.0000	II level
C19~ESEP special investment index.	0.0000	0.0000	0.0000	0.0000	1.0000	V level
C20~ESEP public welfare activities	0.0000	0.0000	0.0000	0.0000	1.0000	V level

Comprehensive evaluation results vector are obtained by computing Formula (8): {0.0000, 0.0000, 0.4582, 0.4657, 0.0761}, based on the principles of maximum membership degree, corresponding to the maximum membership degree of evaluation grade as a result of comprehensive evaluation, that is, the comprehensive evaluation results for IV SINOPEC in 2017 indicate that its ESEP management level is at an acceptable level.

Similarly, the evaluation cloud level of all sample companies can be obtained, and the company level can be visualized after quantitative processing, as shown in Figure 5.

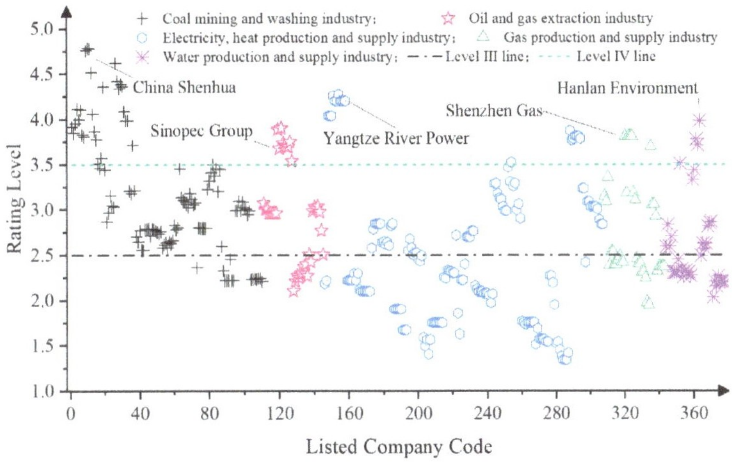

Figure 5. Comprehensive membership evaluation results of each company.

Figure 6 shows that the ESEP management level of most listed companies in the energy industry is between level II and III, indicating that the ESEP management level of most companies is between "transition level" and "improvement level". Further statistics on the number of samples at all levels showed that 1.32% (S = 5) of the samples belonged to class V, indicating that their ESEP management level reached the "claimable level"; 15.87% (S = 60) of the samples belonged to level IV, indicating that the ESEP management level reached the "acceptable level"; 56.611% (S = 214) of the samples belonged to level III, indicating that their ESEP management level reached the "transition level" level; 24.07% (S = 91) of the samples belonged to level II, indicating that their ESEP management level was at the "improvement level"; 2.11% (S = 8) of the samples belong to level I, indicating that their ESEP management level is at the "alert level". Further research shows that different industries have different ESEP management levels. The ESEP management levels from high to low are the coal mining and washing industry, oil and natural gas extraction industry, gas production and supply industry, water production and supply industry, power and heat production and supply industry. Among them, the coal mining and washing industry, oil and gas industry, electricity, heat production and supply industry, gas production and supply industry, water production and supply industry of 2018 ESEP management benchmarking enterprise respectively for China Shenhua (V), SINOPEC (IV), China Yangtze Power (IV), Shenzhen Gas (IV), Grandblue Environment (IV), etc. Some studies have found that the internationalization of the board of directors would enhance the tendency of listed companies' green business behavior [39], and the incentives of championships would also have a positive impact on the CEOs of listed companies to take environmental responsibility [40]. In the future, it can try to improve the level of energy saving and environmental protection practices of listed companies by guiding the internationalization of their boards of directors and actively carrying out ESEP activities in bidding competitions.

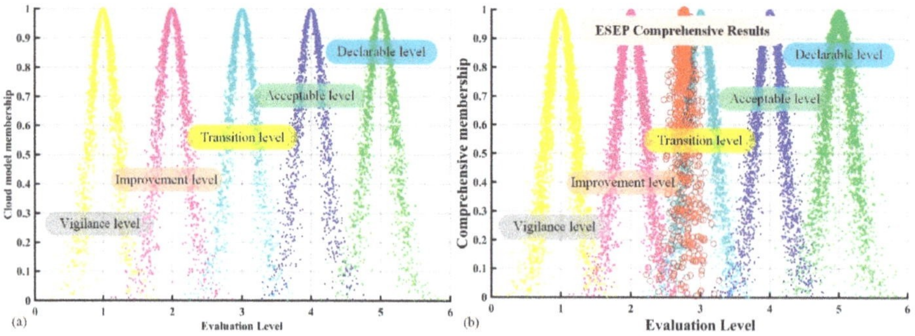

Figure 6. Evaluation cloud map of target layer and criterion layer. (**a**) Evaluation grade cloud scale; (**b**) ESEP comprehensive evaluation cloud chart.

3.3. Evaluation and Analysis of Each Indicator Cloud Model

Based on the screening index data, this study uses cloud generator in the cloud model, inputs the algorithm program operations into Matlab2014 software, sets all samples of each target cloud characteristic parameters (see Table 4), and sets cloud characteristic parameters of the criterion layer and target layer in turn by fuzzy arithmetic according to the Formula (8) (see Table 5).

Table 5. Characteristic values of cloud model.

Criterion Layer B	Index Layer C	Index Layer Cloud Model Parameter			Criterion Layer Cloud Model Parameter
		E_x	E_n	H_e	(E_x,E_n,H_e)
B1-governance framework	C1~ESEP institutional system	3.2078	0.0024	0.1440	3.4451, 0.0018, 0.1543
	C2~ESEP management system	3.6852	0.0009	0.1643	
	C3~ESEP management culture	3.5466	0.0018	0.1594	
	C4~ESEP clauses and policies	3.3735	0.0019	0.1509	
B2-implementation process	C5~clean production management	2.5137	0.0004	0.1022	2.9480, 0.0030, 0.1330
	C6~pollution reduction management	3.0846	0.0069	0.1504	
	C7~recycling management	2.9709	0.0032	0.1346	
	C8~energy efficiency improvement management	3.1635	0.0005	0.1363	
	C9~tackling climate change management	3.1196	0.0017	0.1376	
B3-governance effectiveness	C10~environmental protection management	2.9772	0.0001	0.1247	2.0852, 0.0019, 0.0776
	C11~green office management	2.9815	0.0066	0.1448	
	C12~environmental pollution events	2.5344	0.0082	0.0620	
	C13~discharge of three wastes	2.3127	0.0010	0.0854	
	C14~energy consumption situation,	1.9618	0.0032	0.0870	
	C15~resource recycling	1.6587	0.0022	0.0682	
	C16~other greenhouse emissions	1.9140	0.0030	0.0841	
	C17~ecological environment construction	1.4868	0.0016	0.0567	
	C18~ESEP influence	1.0831	0.0003	0.0214	
	C19~ESEP special investment index.	2.4651	0.0049	0.1161	
B4-others	C20~ESEP public welfare activities	2.0556	0.0035	0.0926	2.0556, 0.0035, 0.0926

After calculating, the cloud characteristic parameters of ESEP management are (2.7598, 0.0019, 0.1199). Based on the cloud characteristic parameters obtained above, combine with the cloud evaluation scale (Formula (6)), and use the forward cloud generator in the model to input the algorithm program into Matlab2014 software for calculation, so as to get the evaluation cloud map of target layer and criterion layer (see Figure 6).

As can be seen from Figure 6, the expected value of the comprehensive cloud of energy saving and environmental protection evaluation of listed companies in the energy industry $E_x = 2.7598$ falls between the "improvement level" and the "transition level", and it is more inclined to the "transition level". It can be seen that the energy conservation and environmental protection management of the energy industry is at the level between the "improvement level" and the "transition level". In addition, the entropy E_n of the evaluation result cloud is much smaller than that of the evaluation cloud, so it can be concluded that the evaluation result has a small range and good stability, reflecting that there is little difference between listed companies in energy conservation and environmental protection management, which may be caused by the fact that most companies are weak in energy conservation and environmental protection management. The result shows that H_e is relatively large, reflecting that cloud thickness is larger than the evaluation cloud, indicating that the energy conservation and environmental protection management of each company needs to be improved.

Similarly, cloud model graphs of B1-ESEP governance framework, B2-ESEP management implementation process, B3-ESEP governance efficiency, B4-ESEP public welfare and other criteria can be obtained, as shown in Figure 7.

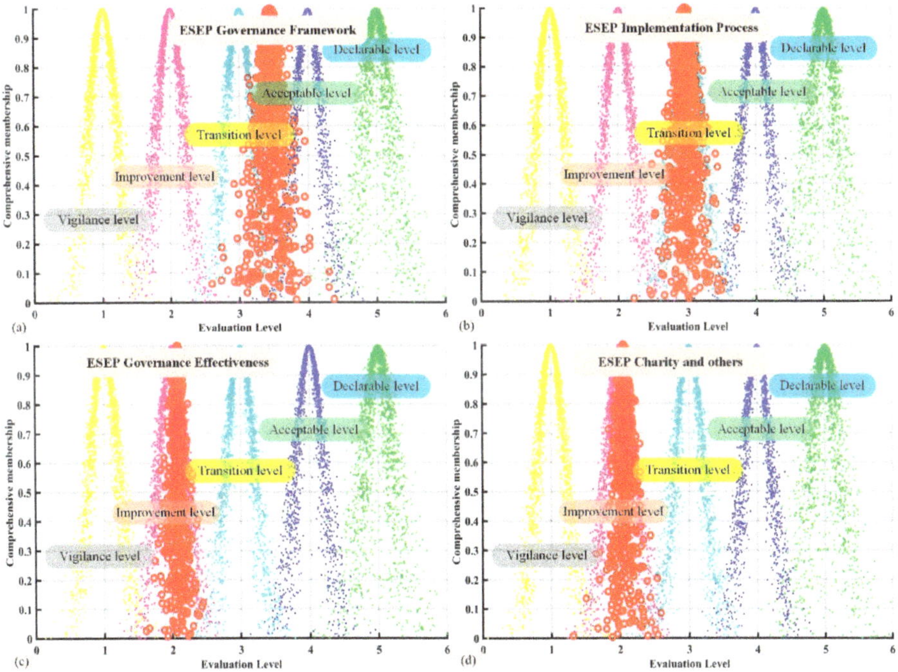

Figure 7. Evaluation cloud chart. (**a**) ESEP governance framework; (**b**) ESEP implementation process; (**c**) ESEP governance effectiveness; (**d**) ESEP charity and others.

This study further visualized E_x and its standard deviation in cloud model parameters for each indicator. It can be seen from Figure 8 that the cloud expectation value of most indicators fluctuated up and down the dividing line of level II~III, among which C2- management culture had the highest expectation value. This is followed by C3-management system, C4-clauses and policies, C1-institutional system, C8-energy efficiency management, and C9-tackling climate change management, indicating that most listed companies perform better in these aspects. It is worth noting that the C18-ESEP influence, C17-ecological environment construction, C16-greenhouse gas emissions, C14-energy consumption, and C13-waste emissions are weak. This indicates that the C18-ESEP influence, C17-ecological environment construction, C16-greenhouse gas emissions, C14-energy consumption, and C13-waste emissions are the key to further improving energy conservation and environmental protection management.

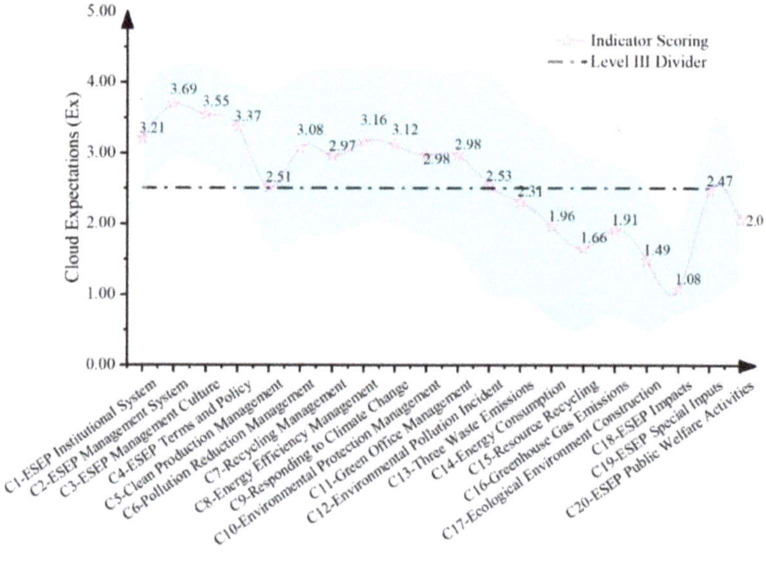

Figure 8. Expected value of each index cloud model.

This study further analyzes the original data of companies of all levels to clarify the focus of improvement of companies at all levels. The specific results are shown in Table 6.

3.4. Limitations

In the construction of the ESEP index system and quantitative research, this study strives to be scientific and rigorous, but there are still some deficiencies due to the limitations of many factors, and the specific limitations are as follows.

(1) The evaluation framework system and its indicators need to be further supplemented and modified. Due to the restriction of data availability, the index system itself cannot fully guarantee that it covers all the evaluation indicators reflecting ESEP management level, especially the evaluation of ESEP management performance. With the deepening of people's understanding of ESEP management, related evaluation indicators would be further expanded.

(2) The method of data acquisition needs to be further expanded. The ESEP management evaluation information in this paper mainly comes from the social responsibility report, sustainable development report, CSMAR database and company website issued by listed companies, which may lead to incomplete ESEP management information.

(3) The rationality of the evaluation results needs to be further verified. As some companies have adopted non-disclosure or selective disclosure in ESEP management, the evaluation results of this study may not fully represent the ESEP management level of these companies, and more comprehensive information can be collected by further combining questionnaire survey and other methods in subsequent research.

Table 6. Analysis of representative companies at each level.

Cloud Level	Representative Enterprise	Major Features etc.
V level	China Shenhua (2017)	(1) Perfect ESEP management system; Systematic energy conservation and environmental protection department; Environmental Protection Council; Attach great importance to environmental protection; Abundant energy conservation and environmental protection education activities; Implement ISO14001 environmental management; Systematic ESEP management system; Normative ESEP provisions; (2) Green procurement of raw materials; Environmentally friendly production; Clean production audit specification; Effective implementation of waste water, waste gas and solid waste reduction management, effective recycling of water resources, effective comprehensive utilization of solid waste; Diversification of energy efficiency measures, perfect implementation, diversification of climate change measures, effective management of greenhouse gas emissions, diversification of measures to reduce ecological environment damage, very effective restoration and governance of ecological environment, diversification of measures related to green office; (3) A large number of ESEP-related awards/honors/patents/papers with great influence; Ten thousand yuan output value environmental protection/high energy saving investment; (4) Participation in ESEP public welfare projects is general;
IV level	China Yangtze Power (2017)	(1) The institutional system, management culture, management system, terms and policies are relatively perfect; (2) Procurement of raw materials, product production, clean production, emission reduction of waste, water resources and solid waste recycling are all environmentally friendly, while energy efficiency improvement, tackling climate change and ecological environment recovery need to be further improved; (3) ESEP influence and ESEP investment are relatively weak; (4) Participation in ESEP public welfare projects needs to be improved;
III level	Datang International Power Generation (2017)	(1) The institutional system, management culture, management system, terms and policies are relatively perfect; (2) Procurement of raw materials, product production, clean production, emission reduction of waste, water resources and solid waste recycling are all environmentally friendly, while energy efficiency improvement, tackling climate change and ecological environment recovery need to be further improved; (3) ESEP influence and ESEP special investment are relatively weak; (4) Participation in ESEP public welfare projects needs to be improved;
II level	Guozhong Water (2017)	(1) Poor institutional system, management culture, management system and other aspects, and generally perfect terms and policies; (2) The procurement of raw materials, product production, clean production, emission reduction of three wastes, recycling of water resources and solid wastes are poor, and the implementation of energy efficiency improvement, tackling climate change and ecological environment restoration measures is mediocre; (3) There is no explanation of ESEP's influence and ESSP's input; (4) Poor participation in ESEP public welfare projects.
I level	Fuling Electric Power (2017)	C1-C20 are less disclosed, only indicating strict compliance with laws and regulations, implementation of some energy conservation and emission reduction measures, etc.

4. Conclusions and Suggestions

4.1. Conclusions

(1) The analysis results of the status quo of ESEP information indicate that the amount of ESEP information released shows a steady increasing trend, but the release rate is still low. Only 42.44% (N = 59) of energy enterprises released ESEP-related information reports (S = 378) from 2006 to 2017. The different nature of the industry leads to a great difference in the release rate of ESEP information, among which the release rate of the gas production and supply industry is the lowest (25.57% on average in recent 3 years). ESEP information content still has huge deficiencies in comparability, systematization and standardization. ESEP information content covers a wide range of areas, but quantitative information disclosure is less common.

(2) The results of cloud level analysis of all companies indicate that the energy conservation and environmental protection management level of most listed companies in the energy industry belongs to "III level-transition level" and "II level-improvement level", and only 17.19% of the sample enterprises are in the "IV level-acceptable level" and "V level-claim level". Further research shows that different industries have differences in ESEP management levels. The ESEP management levels from high to low are the coal mining and washing industry, oil and natural gas extraction industry, gas production and supply industry, water production and supply industry, power and heat production and supply industry. Among them, the coal mining and washing industry, oil and gas industry, electricity, heat production and supply industry, gas production and supply industry, water production and supply industry ESEP management benchmarking enterprise respectively for China Shenhua (V), SINOPEC (IV), China Yangtze Power (IV), Shenzhen Gas (IV), Grandblue Environment (IV), etc.

(3) Analysis results of cloud level of each indicator indicate that the expectation of most energy conservation and environmental protection management indexes fluctuates from level II to Level IV. C2-ESEP management culture, C3-ESEP management system, C4-ESEP clauses and policies, C1-ESEP institutional system, C8-energy efficiency management, C9-tackling climate change and other aspects perform well (reaching the "transition level" or above). In terms of C18-ESEP influence, C17-ecological environment construction, C16-greenhouse gas emissions, C14-energy consumption situation, and C13-discharge of three wastes, the performance is relatively weak (below the "transition level"). Further research shows that C17-ecological environment construction, C16- ecological environment construction, C15-greenhouse gas emissions, C14- energy consumption situation, and C13-discharge of three wastes are the key to further improve ESEP management level of level III to level IV enterprises. C2-ESEP management culture, C3-ESEP management system, C4-ESEP clauses and policies, C1-ESEP institutional system, C8-energy efficiency management and C9-tackling climate change are the key points in the construction of I~II level enterprises.

4.2. Suggestions

Based on the research conclusions, this study proposes the following improvement strategies for ESEP management.

(1) Strengthen the standards and supervision of ESEP information disclosure. At present, there is no systematic and authoritative framework and standard for enterprise's ESEP management disclosure, which leads to poor comparability, consistency and comprehensiveness of ESEP information disclosed by listed companies. As can be seen from the above results, there are some problems in ESEP management, such as low release rate of ESEP information and less quantitative disclosure of released content. In view of this, the government should establish and improve the relevant legal system to further regulate ESEP information disclosure. For example, enterprises can further improve ESEP management by setting minimum disclosure standards, standardizing disclosure formats, introducing authentication evaluation, including information disclosure in enterprise assessment, and imposing sanctions for false information.

(2) Actively carrying out ESEP management evaluation is an important measure to improve China's ESEP management level, but at present, no institution or scholar has conducted a systematic and comprehensive evaluation of ESEP management. Therefore, it is suggested that relevant departments establish a systematic, comprehensive, scientific, standardized, forward-looking and effective ESEP management evaluation system, actively carry out ESEP management evaluation work (such as establishing an ESEP management statistics system, etc.) and regularly release the evaluation results, so as to track and analyze the overall and sub-industry ESEP management status and change trend. It is expected to provide basic support for in-depth implementation of "energy conservation and emission reduction" and continuous improvement of the sustainable development capacity of enterprises.

(3) Give play to the exemplary role of benchmarking enterprises. As benchmarking enterprise of ESEP management coal mining and washing industry, oil and gas industry, electricity, heat production and supply industry, gas production and supply industry, water production and supply industry, China Shenhua, SINOPEC, China Yangtze Power, Shenzhen Gas, Grandblue Environment to enterprise are directional leaders in ESEP management reform and development. Relevant organizations should carry out ESEP management model selection activities, actively promote the ESEP management experience of model enterprises, promote these enterprises to maintain and improve ESEP management model image, and then influence and drive enterprises to improve ESEP management levels.

(4) Guide enterprises to continuously improve key links. Governance efficiency index is the core content of ESEP management, as well as the link that is most weak and most needs to improve. ESEP information disclosure of listed companies currently, including ESEP management influence, ESEP special investment, occupational disease incidence and other aspects, is respectively weak, and these weak links should be direction of further efforts for listed companies to improve their ESEP management level in the future. In view of this, it is feasible to increase the ESEP management impact by increasing the quality and quantity of awards/honors/papers/patents and to guide enterprises to increase ESEP special investment through green credit, green securities and other economic policies.

Author Contributions: S.L.: Conceptualization, Data curation, Writing-Original draft preparation. Y.W.: Methodology, Software. Y.Z.: Visualization, Investigation, Supervision, Software, Validation. J.G. and J.Z.: Writing-Reviewing and Editing. All authors have read and agreed to the published version of the manuscript.

Funding: This work was supported by the Key Research Project of Natural Science in Colleges and Universities of Anhui Department of Education in 2020 [grant number KJ2020A0302], and the Open Research Grant of Joint National-Local Engineering Research Centre for Safe and Precise Coal Mining [grant number EC2021009].

Institutional Review Board Statement: Not applicable.

Informed Consent Statement: Not applicable.

Data Availability Statement: Not applicable.

Conflicts of Interest: The authors declare no conflict of interest.

References

1. Fawcett, J. Thoughts about environment. *Nurs. Sci. Q.* **2022**, *35*, 267–269. [CrossRef] [PubMed]
2. Pal, R.; Banerjee, P.; Thakkar, P.; Hussain, A.M.T. Green firm, brown environment. *Manch. Sch.* **2022**, *90*, 107–121. [CrossRef]
3. Abbass, K.; Qasim, M.Z.; Song, H.M.; Murshed, M.; Mahmood, H.; Younis, I. A review of the global climate change impacts, adaptation, and sustainable mitigation measures. *Environ. Sci. Pollut. Res.* **2022**, *early access*. [CrossRef] [PubMed]
4. Li, Y. Level assessment of ecological environment of china and sustainable development strategies. *Nat. Environ. Pollut. Technol.* **2021**, *20*, 685–693. [CrossRef]
5. Lin, B.Q.; Du, Z.L. Promoting energy conservation in China's metallurgy industry. *Energy Policy* **2017**, *104*, 285–294. [CrossRef]
6. Shi, J. Research on enterprise performance management under the background of energy conservation and emission reduction. *Fresenius Environ. Bull.* **2022**, *31*, 4622–4629.
7. Yang, H.; Li, X.; Ma, L. Using system dynamics to analyse key factors influencing China's energy-related CO_2 emissions and emission reduction scenarios. *J. Clean. Prod.* **2021**, *320*, 128811. [CrossRef]
8. Px, A.; Fan, Y.; Zm, C. Influencing factors of the decoupling relationship between CO_2 emission and economic development in China's power industry. *Energy* **2020**, *209*, 118341. [CrossRef]
9. Lei, M.; Guo, J.; Chai, J. China's regional CO_2 emissions: Characteristics, inter-regional transfer and emission reduction policies. *Energy Policy* **2011**, *39*, 6136–6144. [CrossRef]
10. Yang, S.; Wei, J.; Cheng, P. Spillover of different regulatory policies for waste sorting: Potential influence on energy-saving policy acceptability. *Waste Manag.* **2021**, *125*, 112–121. [CrossRef]
11. Wu, J.; Lv, L.; Sun, J. A comprehensive analysis of China's regional energy saving and emission reduction efficiency: From production and treatment perspectives. *Energy Policy* **2015**, *84*, 166–176. [CrossRef]
12. Daioglou, V.; Mikropoulos, E.; Gernaat, D.; van Vuuren, D.P. Efficiency improvement and technology choice for energy and emission reductions of the residential sector. *Energy* **2022**, *243*, 0360544. [CrossRef]

13. Halkos, G.; Argyropoulou, G. Using environmental indicators in performance evaluation of sustainable development health goals. *Ecol. Econ.* **2022**, *192*, 107263. [CrossRef]
14. Kruger, M.; Muslubas, S.; Cam, E.; Lehmann, D.; Polenz, S.; Dreissigacker, V.; Klasing, F.; Knodler, P. Technical development and economic evaluation of the integration of thermal energy storage in steam power plants. *Energies* **2022**, *15*, 3388. [CrossRef]
15. Wu, L.H.; Chen, Y. Evaluation upon Enterprises' Environmental Management Performance from the Whole-course Perspective. *China Popul. Resour. Environ.* **2014**, *24*, 46–50.
16. Xue, R.; Wang, S.; Gao, G. Evaluation of symbiotic technology-based energy conservation and emission reduction benefits in iron and steel industry: Case study of Henan, China. *J. Clean. Prod.* **2022**, *338*, 130616. [CrossRef]
17. Wei, W.; Shang, Y.N.; Jiang, Y.J.; Wang, Q.; Wang, J.C.; Wen, B. Research on environmental performance evaluation in Chengdu. *China Popul. Resour. Environ.* **2018**, *28*, 80–85.
18. Li, W.A.; Zhang, Y.W.; Zheng, M.N.; Li, X.L.; Cui, G.Y.; Li, H. Research on Green Governance of Chinese Listed Companies and Its Evaluation. *Manag. World* **2019**, *5*, 126–133.
19. Borowski, P.F. Management of energy enterprises in zero-emission conditions: Bamboo as an innovative biomass for the production of green energy by power plants. *Energies* **2022**, *15*, 1928. [CrossRef]
20. He, K.; Zhu, N. Strategic emerging industry layout based on analytic hierarchy process and fuzzy comprehensive evaluation: A case study of Sichuan province. *PLoS ONE* **2022**, *17*, e0264578. [CrossRef]
21. Li, D.Y.; Meng, H.J.; Shi, X.M. Membership Clouds and Membership cloud Generators. *J. Comput. Res. Dev.* **1995**, *16*, 15–22.
22. Zhou, J.; Zhu, Y.Q.; Chai, X.D.; Tang, W.Q. Approach for analyzing consensus based on cloud model and evidence theory. *Syst. Eng.—Theory Pract.* **2012**, *32*, 2756–2763. [CrossRef]
23. He, J.P.; Gao, Q.; Shi, Y.Q. A multi-hierarchical comprehensive evaluation method of dam safety based on cloud model. *Syst. Eng.—Theory Pract.* **2016**, *36*, 2977–2983. [CrossRef]
24. Tian, H.; Pan, C.L. The Study of the Impact of Proactive Environmental Strategy on Corporate Green Image. *Chin. J. Manag.* **2015**, *12*, 1064–1071.
25. Liao, L.; Luo, L.; Tang, Q. Gender diversity, board independence, environmental committee and greenhouse gas disclosure. *Br. Account. Rev.* **2015**, *47*, 409–424. [CrossRef]
26. Baboukardos, D. The valuation relevance of environmental performance revisited: The moderating role of environmental provisions. *Br. Account. Rev.* **2018**, *50*, 32–47. [CrossRef]
27. Du, W.C. Whole Process Treatment for Industrial COD. *China Soft Sci.* **2013**, *7*, 77–85. [CrossRef]
28. Qin, Y.; Wu, C.Y.; Zhai, L.N. The Theoretical Study of the Relationship between the Environmental and Economic Performance of Firm and Model Construct. *Syst. Eng.-Theory Pract.* **2004**, *8*, 111–117. [CrossRef]
29. Hao, Z.Z.; Li, J.; Han, H.B. Measurement of and Empirical Study on Environmental Performance of China's Industry Sectors. *Syst. Eng.* **2014**, *32*, 1–11.
30. Wang, F.Z.; Jiang, T.; Guo, X.C. Government quality, environmental regulation and green technological innovation of enterprises. *Sci. Res. Manag.* **2018**, *39*, 26–33. [CrossRef]
31. Chen, X.J. Application of Analytical Hierarchical Process to Optimize the MSW Classification schemes in Pudong, China. *China Popul. Resour. Environ.* **2015**, *25*, 368–371.
32. Freeman, J.; Tao, C. Green supplier selection using an AHP-Entropy-TOPSIS framework. *Supply Chain. Manag.* **2015**, *20*, 327–340. [CrossRef]
33. Wang, Q.; Yuan, X.; Zhang, J.; Gao, Y.; Hong, J.; Zuo, J.; Liu, W. Assessment of the Sustainable Development Capacity with the Entropy Weight Coefficient Method. *Sustainability* **2015**, *7*, 13542–13563. [CrossRef]
34. Zhang, T.; Li, M.R.; Xu, Y.M. The Construction and Empirical Study of Rural Revitalization Evaluation Index System. *Manag. World* **2018**, *34*, 99–105. [CrossRef]
35. Wang, Y.; Chen, H.; Long, R.; Jiang, S.; Liu, B. Has the sustainable development planning policy promoted the green transformation in China's Resource-based cities? *Resour. Conserv. Recycl.* **2022**, *180*, 106181. [CrossRef]
36. Wang, Y.; Chen, H.; Long, R.; Liu, B.; Jiang, S.; Yang, X.; Yang, M. Evaluating green development level of mineral resource-listed companies: Based on a "dark green" assessment framework. *Resour. Policy* **2021**, *71*, 102012. [CrossRef]
37. Zeng, S.; Xu, X.; Dong, Z.; Tam, V.W. Towards corporate environmental information disclosure: An empirical study in China. *J. Clean. Prod.* **2010**, *18*, 1142–1148. [CrossRef]
38. He, L.M.; Hou, T. Determinants of Environmental Performance Information Disclosure in Chinese Listed Companies: Empirical Evidence Based on Social Responsibility Reports. *China Popul. Resour. Environ.* **2010**, *20*, 99–104. [CrossRef]
39. Usman, M.; Javed, M.; Yin, J. Board internationalization and green innovation. *Econ. Lett.* **2020**, *197*, 109625. [CrossRef]
40. Ullah, S.; Khan, F.U.; Cismaș, L.M.; Usman, M.; Miculescu, A. Does tournament incentives matter for CEOs to be environmentally responsible? Evidence from Chinese listed companies. *Int. J. Environ. Res. Public Health* **2022**, *19*, 470. [CrossRef]

sensors

MDPI

Article

Federated Deep Reinforcement Learning-Based Task Offloading and Resource Allocation for Smart Cities in a Mobile Edge Network

Xing Chen and Guizhong Liu *

School of Information and Communications Engineering, Xi'an Jiaotong University, Xi'an 710049, China; xing_chen@stu.xjtu.edu.cn
* Correspondence: liugz@xjtu.edu.cn

Abstract: Mobile edge computing (MEC) has become an indispensable part of the era of the intelligent manufacturing industry 4.0. In the smart city, computation-intensive tasks can be offloaded to the MEC server or the central cloud server for execution. However, the privacy disclosure issue may arise when the raw data is migrated to other MEC servers or the central cloud server. Since federated learning has the characteristics of protecting the privacy and improving training performance, it is introduced to solve the issue. In this article, we formulate the joint optimization problem of task offloading and resource allocation to minimize the energy consumption of all Internet of Things (IoT) devices subject to delay threshold and limited resources. A two-timescale federated deep reinforcement learning algorithm based on Deep Deterministic Policy Gradient (DDPG) framework (FL-DDPG) is proposed. Simulation results show that the proposed algorithm can greatly reduce the energy consumption of all IoT devices.

Keywords: smart city; mobile edge computing; task offloading; resource allocation; DDPG; federated learning

Citation: Chen, X.; Liu, G. Federated Deep Reinforcement Learning-Based Task Offloading and Resource Allocation for Smart Cities in a Mobile Edge Network. *Sensors* **2022**, 22, 4738. https://doi.org/10.3390/s22134738

Academic Editor: Antonio Cano-Ortega

Received: 26 May 2022
Accepted: 17 June 2022
Published: 23 June 2022

1. Introduction

With the full development of the fifth-generation mobile communication research, the information and intelligence of cities have been greatly developed. More and more smart facilities are deployed in every corner of the city, which enhance the quality of life for citizens. In the era of IoT, smart cities power and monitor a variety of intelligent IoT devices. Accompanied by intelligent devices, IoT applications are designed, such as smart parking, smart traffic, and smart security. These applications can generate some computation-intensive tasks such as camera tracking and object recognition. In the traditional central cloud network, these tasks will be offloaded to the central cloud server for execution. However, the central cloud network faces some challenges, as follows. (1) More users are served, which is easy to cause network congestion. (2) Since the central cloud server is far away from users, the data transmission process consumes a lot of time. As a main evolution technology in the 5G, MEC provides a good direction to solve these challenges [1,2]. MEC server is deployed at the edge of the core network, which is closer to users. The computation-intensive tasks can be offloaded to the MEC server for reducing the delay, network congestion, and energy consumption of IoT devices [3,4].

Based on the above description, how to make a reasonable offloading decision and resource allocation scheme subject to limited resources has become a key problem. The joint optimization of task offloading and resource allocation is a mixed-integer nonlinear programming problem [5,6]. Currently, scholars have some research works on task offloading and resource allocation. The joint problem is solved by splitting it into several sub-problems, relaxation variables, and deep reinforcement learning based on Deep Q Network (DQN) framework [7–9]. However, the first two algorithms simplify the original problem and do not directly solve the joint optimization problem of task offloading and resource allocation. With the development of deep neural network, deep reinforcement learning has a good effect on solving environmental decision-making problems. However, the algorithm based on DQN framework is difficult to deal with the problem of fine-grained space or continuous space. Therefore, the deep reinforcement learning algorithm based on the DDPG framework is adopted in this article. The DDPG algorithm has a good effect in dealing with spatial continuous decision-making problems.

In order to obtain a better Quality of Experience (QoE), some research works adopted the cooperation method between MEC servers, or the unified scheduling method on the central cloud [10,11]. However, these collaborative and centralized processing algorithms do not consider the privacy and security problems in the process of data migration and processing. Therefore, many users are reluctant to upload their private raw data to other MEC servers or the central cloud server. To tackle the problem, the federated learning technology is proposed by Google [12]. It is a distributed machine learning framework, which consists of one central server and a set of clients [13–15]. The main idea of federated learning is to enable the data on clients to train their respective network models. Then, the parameters of clients are aggregated to update the network model on the server side. A better training model is obtained by the iteration between distribution and aggregation without sharing the raw data. Therefore, the federated learning is introduced into the joint optimization problem of this article to obtain a better optimization performance.

In this article, we focus on the joint optimization problem of task offloading and resource allocation based on privacy protection in smart city. The optimization objective is to minimize the energy consumption of all IoT devices within the delay threshold. Since the joint optimization is a mixed-integer nonlinear programming problem, it is difficult to solve it by the traditional programming algorithms. Therefore, based on the above description, a two-timescale federated deep reinforcement learning algorithm based on DDPG framework is proposed to solve the problem. The small timescale is to optimize the offload decision and the resource allocation scheme in each MEC server by training DDPG network. The large timescale is to aggregate the parameters of MEC servers in order to obtain a better training performance. The contributions of this paper can be summarized as follows:

1. We investigate the joint optimization problem of task offloading and resource allocation subject to the delay threshold and the limited resources. In the existing literature, the joint optimization problem is generally decomposed into multiple sub-problems. Therefore, a deep reinforcement learning algorithm based on DDPG framework is proposed to solve the joint problem. The DDPG is the combination of DQN and Actor-Critic (AC), which can solve the decision-making problem of continuous action space.
2. The federated learning is introduced into the deep reinforcement learning to enhance the training performance while protecting privacy and security. In terms of privacy and security, the federated learning only needs to upload the network parameters without the raw data. In terms of training performance, the federated learning is a distributed machine learning algorithm, which can obtain a better convergence.
3. Extensive numerical experiments demonstrate that our proposed algorithm has better convergence than the centralized algorithm, and obtains better performance gain than other comparison algorithms.

The rest of this article is organized as follows: Section 2 presents the system model, including task model, communication model and computation model. Section 3 presents the optimization problem and solution. Section 4 provides the simulation results and evaluates the performance of the proposed algorithm. Section 5 concludes this article.

2. Related Work

The concept of MEC was put forward many years ago. In 2013, the world's first mobile edge computing platform was established by IBM and Nokia Siemens Network [16]. In 2014, the European Telecommunications Standards Institute (ETSI) proclaimed industry specifications for MEC, which was supported by IBM, Huawei, Intel, etc. Currently, most of the MEC research works focus on how to fully utilize the powerful computing and storage capacity of the MEC server to reduce delay and energy consumption of IoT devices [17]. Some popular contents are cached on the MEC server to reduce the delay and network backhaul load. Aung et al. [18] proposed a social-aware vehicular edge computing architecture that solves the content delivery problem by using some of the vehicles in the network as edge servers that can store and stream popular content to close-by end-users. The computation-intensive applications can be offloaded to the MEC server for execution [19]. Apostolopoulos et al. [20] proposed a joint problem of latency and energy minimization considering the data offloading characteristics of the end nodes. In this article, we only focus on the computing resource allocation of the MEC server.

The task offloading problem in the communication system will inevitably involve task scheduling, the allocation of computing and transmission resources [21,22]. Therefore, the problem can be easily regarded as a joint optimization problem of task offloading and resource allocation, which is a mixed-integer nonlinear programming problem. There are generally three types of algorithms to solve the problem. The first type of algorithm is to split the joint optimization problem into multiple sub-problems [7,23]. Zhao [24] formulated the joint optimization problem task offloading and resource allocation and decomposes it into three sub-problems named as offloading ratio selection, transmission power optimization, and sub-carrier and computing resource allocation. The joint optimization problem was decomposed into two-level sub-problems and solved by the iterative algorithm [25]. This type of algorithm is not a joint optimization algorithm for the original problem, and the efficiency of iterative optimization for several sub-problems is not high. The second type of algorithm is to relax the variables in the optimization problem [8]. Masoufdi [26] investigated the power minimization problem for the mobile devices by data offloading in a multi-cell multi-user Orthogonal Frequency Division Multiple Access (OFDMA) network. To solve the problem, it was converted to the convex form using variable changing, Difference of Convex (DC) approximation, adding a penalty factor, and relaxing the binary constraints. The lower bound and upper bound of the joint optimization problem were considered and the semi-definite relaxation and rounding methods were exploited to obtain the offloading decision [27]. The mixed integer nonlinear programming problem is transformed into a nonlinear programming problem by variable relaxation. Then, it is solved by iterative algorithm or genetic algorithm. Undoubtedly, the type of algorithm has a lower efficiency. The third type of algorithm is to use the deep reinforcement learning algorithm to solve the optimization problem. Li et al. [9] investigated the resource allocation scheme for vehicle-to-everything communications, and proposed the optimization problem of resource blocks allocation and vehicle transmission power allocation. A reinforcement learning based on DQN framework was designed to solve this problem. Suh et al. [28] proposed a DQN algorithm based network slicing technique to calculate the resource allocation policy, maximizing the long-term throughput while satisfying the Quality of Service (QoS) requirements in the beyond 5G systems. Since it is difficult for DQN algorithm to deal with the problem of fine-grained space or continuous space, a deep reinforcement learning algorithm based on DDPG framework is proposed to solve the joint optimization problem in this article.

To improve resource utilization and algorithm performance, some research works adopted the cooperation methods, such as Cloud-MEC, MEC-MEC, Cloud-MEC-Device. Naouri et al. [29] proposed a three-layer task offloading framework, which consisted of the device layer, cloudlet layer and cloud layer. A cloud-MEC collaborative computation offloading scheme was proposed in vehicular networks [24]. Chen et al. [30] studied an energy-efficient task offloading and resource allocation scheme for Augmented Reality (AR) in a multi-MEC collaborative system. Monia et al. [31] investigated the joint task assignment and power control problems for Device-to-Device (D2D) offloading communications with energy harvesting. A layered optimization method is proposed to solve this problem by decoupling the energy efficiency maximization problem into power allocation and offloading assignment. However, these collaborative and centralized processing algorithms do not consider the privacy and security problems in the process of data migration and processing. As a result, many users are reluctant to upload their private raw data to other MEC servers or the central cloud server. To solve this problem, federated learning is introduced in this article, which not only protects privacy but also improves the performance of the model.

3. System Model

In this article, a system model for the smart city in a mobile edge network is established, which consists of three layers: IoT device, MEC server and Central Cloud, as shown in Figure 1. The central cloud is an auxiliary role, which helps the MEC server obtain a better decision-making mechanism by aggregating the neural network parameters of each edge server. The MEC server has a powerful computing capacity, which can quickly process the tasks offloaded by IoT devices. The IoT devices can generate some tasks with strict computing requirements. Since the IoT devices have limited computing resources and limited energy, the computing tasks need to be offloaded to the MEC server for processing. In consideration of security and privacy issues, IoT devices can only offload their tasks to the trusted MEC server, not to the central cloud server. We denote the central cloud, the set of MEC servers and the set of IoT devices (the set of applications) by Γ, $k \in \{1, 2, \dots, K\}$ and $n \in \{1, 2, \dots, N\}$, respectively. We believe that IoT devices are special devices, and each IoT device corresponds to an application. We assume that each IoT device only requests one task at the same time and the network state is constant during task processing. The specific workflow of the system is as follows. First, IoT devices generate the tasks and send the relevant information to the MEC server through the base stations at the same time. Second, a decision on offloading and resource allocation is made according to the collected task information and network status. Finally, these tasks are executed according to the offloading decision and resource allocation schemes.

3.1. Task Model

In the smart city scenario, there are a large number of different types of applications (such as smart security, smart traffic, smart parking, smart lamp and so on). These applications have lower real-time requirements than AR applications. Therefore, we set the delay threshold of these applications to the same, which is denoted by T. To describe the parametric context of each application task, we define a tuple representation as $\phi_n = (\omega_n, \varphi_n)$. Specifically, ω_n and φ_n denote the data size (bit) and the computing workload (CPU cycles) of the task generated by IoT device n, respectively. The relationship between ω_n and φ_n is expressed as $\varphi_n = \eta_n \cdot \omega_n$, where η_n denotes the computing workload per bit. In this article, the offloading decision is denoted by $\alpha_n \in \{0, 1\}$. If $\alpha_n = 0$, the application task requested by IoT device n will not be offloaded to the edge server and will be processed on the IoT device n. If $\alpha_n = 1$, the application task requested by IoT device n will be offloaded to the MEC server. The import notations used in the rest of this article are summarized in Table 1.

Table 1. Parameter descriptions.

Notation	Definition
Γ	Central cloud
k	Index of MEC server
n	Index of IoT device
ω_n	Data size of IoT device n
φ_n	Total computing workload of IoT device n
η_n	Computing workload of IoT device n per bit
α_n	Offloading decision of IoT device n
B_n	Number of sub-bandwidth allocated to IoT device n
B	System bandwidth
D	Number of sub-bands
h_n	Uplink channel gain between the base station and IoT device n
p_n	Transmission power of IoT device n
f_n	Computing resources allocated by the MEC server to IoT device n
T	Delay threshold of all IoT device

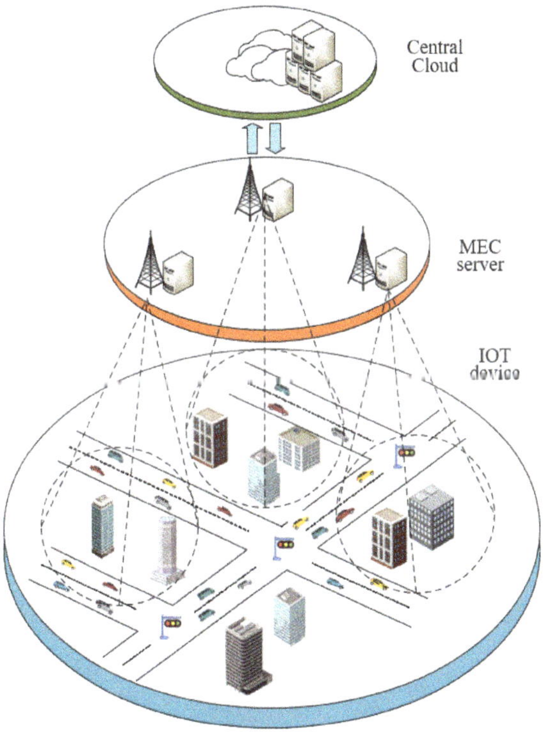

Figure 1. System Model.

3.2. Communication Model

In this article, we consider the system with the OFDMA as the multiple access technology, in which the system bandwidth B is divided into D equal orthogonal sub-bands.

In view of the OFDMA mechanism, interference is ignored due to the exclusive subcarrier allocation [25,32–34]. Therefore, we do not consider interference from other IoT devices in this article. A sub-band can only be allocated to one IoT device, but an IoT device can be allocated multiple sub-bands. Since the amount of data that needs to be returned to the IoT device after processing is very small, the time consumption in process of downlink transmission is not considered. Let B_n denotes the number of sub-bandwidths allocated to IoT device n. p_n denotes the transmission power of IoT device n. h_n denotes the uplink channel gain between the base station and IoT device n corresponding to a white Gaussian noise channel, which incorporates distance based path loss model and independent Rayleigh fading. Then, the uplink transmission rate r_n^{up} can be calculated by

$$r_n^{up} = B_n \frac{B}{D} log_2(1 + \frac{p_n h_n}{\delta^2}) \tag{1}$$

where δ^2 denotes the noise power. Therefore, the transmission time t_n^{up} and the energy consumption e_n^{up} of uplink transmission can be calculated by

$$t_n^{up} = \frac{\omega_n}{r_n^{up}} \tag{2}$$

$$e_n^{up} = t_n^{up} \cdot p_n \tag{3}$$

3.3. *Computation Model*

In this article, the task generated by IoT device can be offload to the MEC server in order to reduce the energy consumption of the IoT device when the network is in good state. If the network state is bad, the task can only be executed on the IoT device. Next, two situations are described in detail, respectively.

3.3.1. Processing at MEC Server

Let f_n denotes the computing resources allocated by the MEC server to the task generated by IoT device n. Then, the execution time t_n^{MEC} can be calculated by

$$t_n^{MEC} = \frac{\varphi_n}{f_n} \tag{4}$$

3.3.2. Processing at IoT Device

According to the optimization objective, if the task is processed on the IoT device, the energy consumption is the smallest when the delay is equal to the delay threshold. Therefore, the processing time t_n^{IoT} and the energy consumption e_n^{IoT} can be calculated by

$$t_n^{IoT} = T \tag{5}$$

$$e_n^{IoT} = \kappa \cdot (\frac{\varphi_n}{T})^2 \cdot \varphi_n \tag{6}$$

where κ is the energy coefficient, which depends on the chip architecture [35–37]. In this article, according to the work in [38], we set $\kappa = 10^{-25}$.

4. Two-Timescale Joint Optimization of Task Offloading and Resource Allocation

In this section, the joint optimization of task offloading and resource allocation is formulated, and it is considered as Markov Decision Process (MDP). A deep reinforcement learning algorithm based on DDPG framework is proposed to solve this problem. In order to protect user privacy and improve the training performance of the deep neural network, Federated learning is introduced into the training model. A two-timescale federated reinforcement learning algorithm is proposed. The small timescale is to train the scheme of task offloading and resource allocation on each MEC server. The large timescale is to aggregate the trained model parameters on the central cloud server. The two-timescale are executed alternately to obtain better training performance. In this article, since the central cloud server and MEC servers are connected by the wired network, the time consumption caused by parameters upload is not considered. The detail of problem formulation and solution are described as follows.

4.1. Problem Formulation

According to the above computation and communication models, the total time consumption and the energy consumption can be calculated by

$$t_n = \alpha_n \cdot (t_n^{up} + t_n^{MEC}) + (1 - \alpha_n) \cdot t_n^{IoT} \quad (7)$$

$$e_n = \alpha_n \cdot e_n^{up} + (1 - \alpha_n) \cdot e_n^{IoT} \quad (8)$$

The mathematical model with the objective of minimizing the energy consumption of all IoT devices subject to the latency requirement and the limited resources, is as follows:

$$\begin{aligned} &\min_{B_n, f_n, \alpha_n} \sum_{n=1}^{N} e_n \\ &s.t. \\ &(c1)\ \ t_n \leq T \\ &(c2)\ \ \sum_{n=1}^{N} B_n \leq B \\ &(c3)\ \ \sum_{n=1}^{N} f_n \leq F^{MEC} \\ &(c4)\ \ \alpha_n \in \{0, 1\} \end{aligned} \quad (9)$$

where F^{MEC} denotes the total computing resources of the MEC server. For the constraints, constraint $(c1)$ indicates that the execution time of the IoT device n cannot exceed the delay threshold to ensure the QoE. We believe that as long as the processing time of the IoT task is within the delay threshold, a satisfactory user experience can be obtained. For example, in the community access control system, if the delay threshold of the face recognition system is 0.1 s, the user experience can be satisfied as long as the face recognition is completed within 0.1 s. Since users have the same QoE for completing face recognition within 0.1 s and 0.01 s, there is no need to pursue a lower processing time, which is meaningless in real scenes. Constraint $(c2)$ indicates that the number of allocated sub-bandwidth cannot exceed the total bandwidth of base station. Constraint $(c3)$ indicates that the computing resources allocated to all IoT devices by the MEC server cannot exceed the total computing resources of the MEC server. Constraint $(c4)$ indicates that the task of IoT device is either processed on the MEC server or the IoT device n. If $\alpha_n = 0$, the task of IoT device will be processed on the IoT device. If $\alpha_n = 1$, the task of IoT device will be offloaded to the MEC server.

4.2. Small Timescale Policy Based on Deep Reinforcement Learning

In this subsection, the joint optimization problem is modeled as MDP, and a deep reinforcement learning based on DDPG framework is proposed to solve it. The common model of reinforcement learning is the standard MDP. Therefore, several elements of MDP are introduced in detail below.

4.2.1. State Space

State is the description of the environment, which will change after an action is generated by the agent. In this article, the MEC server is modeled as an agent to optimize the energy consumption of all IoT devices. Let $s_t = (s_t^1, s_t^2, \ldots, s_t^U)$ denotes the state of MDP at time t. The state includes four parts: (1) the task size, the computing workload, the channel state of all IoT devices; (2) the computing resources of the MEC server; (3) the bandwidth of the base station; (4) the resource allocation scheme at the current time. The value range of all data in the state is $[0, 1]$.

4.2.2. Action Space

Action is the description of agent behavior, which is the result of the optimization scheme. Let $a_t = (a_t^1, a_t^2, \ldots, a_t^L)$ denotes the action of MDP at time t, which includes the change of computing and communication resources. The action space corresponds to Part 4 of the state space one by one. The value range of all data in the action is $[-1, 1]$.

4.2.3. Reward

After the agent takes an action, reward is the feedback of environment to agent. Let r_t denotes the reward of MDP at time t. The objective of this article is to minimize the energy consumption of all IoT devices subject to the system resources and delay threshold. Therefore, the reward is set to two progressive steps. The first step is to ensure the system resources constraints, as follows:

$$\begin{aligned} r = {} & \chi_1 \cdot \sum_{u=1}^{U} \left((s^u - 1) \cdot \varepsilon(s^u - 1) - s^u \cdot \varepsilon(-s^u) \right) \\ & + \chi_2 \cdot \left(\sum_{n=1}^{N} B_n - B \right) \cdot \varepsilon\left(\sum_{n=1}^{N} B_n - B \right) \\ & + \chi_3 \cdot \left(\sum_{n=1}^{N} f_n - F^{MEC} \right) \cdot \varepsilon\left(\sum_{n=1}^{N} f_n - F^{MEC} \right) + b_1 \end{aligned} \tag{10}$$

The second step is to minimize the energy consumption of all IoT devices, as follows:

$$r = \chi_4 \cdot exp\left(-\sum_{n=1}^{N} e_n / N \right) \tag{11}$$

where $\chi_1, \chi_2, \chi_3, \chi_4, b_1$ are constants. The purpose is to make rewards develop in a good direction. Specifically, the reward setting algorithm is illustrated in Algorithm 1.

Algorithm 1: Reward calculation algorithm

```
Input: new state s_{t+1} ← environment(s_t, a_t)
Output: r_t
Initialize the reward r_t = 0
if c2, c3 of (9) then
    for each IoT n = 1, 2, ..., N do
        Calculate e_n (α_n = 0) according to Equation (8)
        Calulate t_n (α_n = 1) according to Equation (7)
        if t_n (α_n = 1) ≤ T then
            Calculate e_n (α_n = 1) according to Equation (8)
            if e_n (α_n = 0) > e_n (α_n = 1) then
                e_n = e_n(α_n = 1)
            else
                e_n = e_n(α_n = 0)
        else
            e_n = e_n(α_n = 0)
    Calculate the reward for energy consumption of all IoT devices according to
      Equation (11)
else
    Calculate the reward of resources constraint according to Equation (10)
```

4.2.4. DDPG-Based Solution

The DDPG algorithm is a paradigm of the reinforcement learning method, which is the combination of AC and DQN. The specific network structure is shown in Figure 2. The training process of the network is carried out according to the numbers in the Figure 2. The input of the actor network is the state, and the output is the determined action value.

The input of the critic network is the state and the action, the output is the Q value. The actor network consists of the evaluation network μ with parameter θ^{μ} and the target network μ' with parameter $\theta^{\mu'}$. The critic network consists of the evaluation network Q with parameter θ^{Q} and the target network Q' with parameter $\theta^{Q'}$. Since the experience replay method is adopted, the data (s_t, a_t, s_t', r_t) are stored in the replay buffer according to the format of (s, a, s', r). The parameters of critic network are updated by minimizing the loss,

$$Loss = \frac{1}{X}\sum_{j=1}^{X}\left(y_j - Q(s_j, a_j|\theta^{Q})\right)^2 \tag{12}$$

$$y_j = r_j + \gamma \cdot Q'(s_j', \mu'(s_j'|\theta^{\mu'})|\theta^{Q'}) \tag{13}$$

where X denotes the size of mini batch data, and γ denotes the discount factor. The actor network is updated according to the feedback of the critic network as follows:

$$\nabla_{\theta} J \approx \frac{1}{X}\sum_{j=1}^{X}\left(\nabla_a Q(s_j, a_j|\theta^{Q})|_{a_j=\mu(s_j)} \cdot \nabla_{\theta^{\mu}}\mu(s_j|\theta^{\mu})\right) \tag{14}$$

DDPG framework has the characteristics of centralized training and decentralized execution. After the training is completed, the state is input into the actor network to obtain the offloading decision and resource allocation scheme.

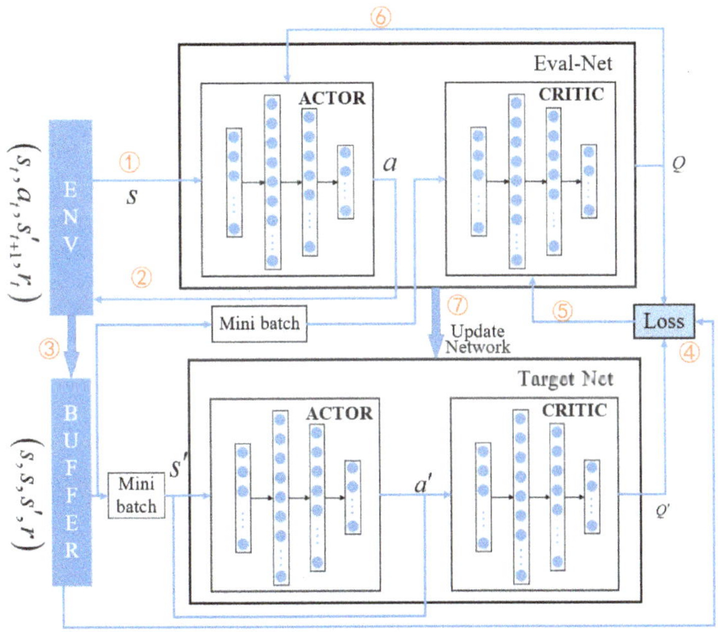

Figure 2. Convergence property of different algorithm.

4.2.5. Computational Complexity Analysis

Floating Point Operations (FLOPs) can be used to measure the computational complexity of the algorithm or model. The proposed algorithm is a reinforcement learning algorithm based on DDPG framework. The DDPG framework consists of an actor network and a critic network. In this article, the actor network is composed of three full connection layers, and the critic network is composed of four full connection layers. The FLOPs of a full connection layer is $2 \times I \times Q$, where I denotes the number of input neurons and

Q denotes the number of output neurons. Therefore, the FLOPs of the actor network is $\sum_{m=1}^{3} 2 \times I_m \times Q_m$, and the FLOPs of the critic network is $\sum_{m=1}^{4} 2 \times I_m \times Q_m$. Since DDPG has the characteristics of centralized training and decentralized execution, whether the proposed framework can be implemented in a real time manner depends on the execution time of the actor network. For example, for a single core computer (2 GHz), its computing capacity is about 2 billion FLOPs per second, which is more than enough to be used to process the computation of the actor network according to network settings in this article. The specific network parameters are set in Section 5.1. Therefore, the proposed framework can be implemented in a real time manner.

4.3. Large Timescale Policy Based on Federated Learning

In this subsection, for the purpose of protecting privacy and security, users are reluctant to send their data the central cloud server. However, in the process of neural network training, more data will generally bring better training performance. For the above two reasons, Federated Learning algorithm is introduced into reinforcement learning. Federated learning is essentially a distributed machine learning technology. Its goal is to realize joint modeling and improve the performance of Artificial Intelligence (AI) model on the basis of ensuring data privacy, security and legal compliance. Since different blocks of smart city have the characteristic of the same application types and different users, the horizontal federated learning is adopted in this article.

In horizontal federated learning, it can be regarded as a distributed training model based on samples, which distributes all data to different machines. Each machine downloads the model from the central server to train the model with local data, and then the training parameters are returned to the central server for aggregation. In this process, each machine is the same and complete model, which can work independently. The aggregation mode of network parameters is given by

$$\Theta = \frac{1}{\sum_{k \in K} D_k} \sum_{i=k}^{K} D_k \Theta_k \tag{15}$$

where D_k denotes the number of training samples on the k-th MEC server, Θ_k and Θ denote the parameter sets of the k-th MEC server and the central cloud, respectively. Specifically, the two-timescale training process is summarized in Algorithm 2.

Algorithm 2: Training process

```
Randomly initialize critic network Q and actor network μ
Initialize target network Q' and μ'
Initialize replay buffer φ
for episode t = 0,1,... do
    for MEC k = 0,1,...,K do
        Initialize state s
        Select action a_t = μ(s) + N, N is an exploration noise
        Execute action a_t and observe reward r_t and observe new state s_{t+1}
        Store transition (s_t, a_t, r_t, s_{t+1}) in Φ
        Sample a random mini batch of X transition (s^X, a^X, s'^X, r^X) from Φ
        Update critic network according to (12) and (13)
        Update actor network using the sampled policy gradient (14)
        Update the target networks:
        θ' = τθ + (1 − τ)θ'
    if episode t == t_aggregation then
        Upload the parameters to the central cloud server according to (15)
        Download the parameters from the central cloud server to each MEC server
```

5. Performance Evaluation

5.1. Parameter Setting

In this section, we evaluate the performance of our proposed algorithm for smart city. The experimental platform adopts DELL PowerEdge (DELL-R940XA, 4*GOLD-5117, RTX2080Ti). The simulation software is Pycharm (Professional Edition). The corresponding environment configuration is Python3.7.6, CUDA 11.4, Pytorch 1.5.0. The actor network is composed of three full connection layers (40×500, 500×128, 128×20), and the critic network is composed of four full connection layers (60×1024, 1024×512, 512×300, 300×1). Its activation function is RELU, and the output layer of actor network is tanh function to constraint the output value. Specifically, the simulation parameters of the system are presented in Table 2. The compared algorithms are as follows.

- Random Offload:the offloading scheme of each IoT device is determined randomly. If the task of IoT device is offloaded to the MEC server, the computing and communication resources are allocated according to the proportion of data size and computing workload, respectively. If the task is executed on the IoT device, the computing resource is allocated according to the delay threshold.
- Greedy: the task of the IoT device with good channel status is offloaded to the MEC server sequentially. Each IoT device occupies the least resources to ensure that more tasks can be offloaded to the MEC server subject to the delay threshold.
- DQN: DQN is a combination of Q-learning and deep neural network, which is used to deal with the discrete state and action problem. To solve the problem in this article, the continuous state space and action space need to be discretized [9,28].
- DDPG: DDPG is the basic algorithm of this article. It is a continuous reinforcement learning algorithm, which is composed of DQN and AC.
- FL-DDPG: FL-DDPG is an algorithm proposed in this article. Federated learning is introduced into reinforcement learning to solve the problem of resource allocation and offloading decision. Since FL-DDPG has the distributed characteristic, it can improve training performance while ensuring privacy and security.

5.2. Convergence Analysis

In this subsection, the convergence performance of FL-DDPG and DDPG is shown in Figure 3. In this article, the Adam optimizer is adopted to train the FL-DDPG and DDPG networks. In the training process of FL-DDPG, it needs about 240,000 training episodes (3 h) to achieve a better convergence performance. From Figure 3, we observe that the convergence performance of FL-DDPG is better than that of DDPG. Since FL-DDPG aggregates the parameters of three MEC servers, it is easy to jump out of the local optimal solution. DQN algorithm discretizes the resources, and decides where each resource block should be allocated. Therefore, DQN algorithm has no resource allocation constraints (the allocated resources will never exceed the total resources), and directly pursues the minimization of energy consumption. The value range of the reward of DQN algorithm is $0 < r < 1$. Since DQN algorithm is a coarse-grained resource allocation scheme, the convergence performance of DDPG is better than that of DQN. Figure 4 shows the training performance of different aggregation intervals in FL-DDPG algorithm. From Figure 4, it is observed that the training performance is the best when the aggregation interval is 30,000. When the aggregation interval is less or greater than 30,000, the training performance is not good. The reason for this is that there is not enough time to explore new environment when the aggregation interval is smaller. When the aggregation interval is larger, over-fitting is caused by too long exploration time. Therefore, the aggregation interval of 30,000 is adopted to train the network parameters in this article.

Table 2. Parameter descriptions.

Parameter	Value
Number of IoT devices	30
Number of base stations	3
Number of MEC servers	3
Uplink/Downlink system Bandwidth	10 MHz
Transmission powers of user terminal	1 W
Noise power	−100 dB
Size of task	[5, 90] Kb
Computing workload density	[200, 700] CPU cycles/bit
Path loss model	$PL = 127 + 30log(dis)$
Computing resources of local device	[2, 2.5] GHz
Computing resources of MEC server	15 GHz
Delay threshold of IoT task	100 ms
episode	240,000
Mini batch	100
Buffer size	20,000
Critic network learning rate	0.001
Actor network learning rate	0.0001
Optimizer	Adam

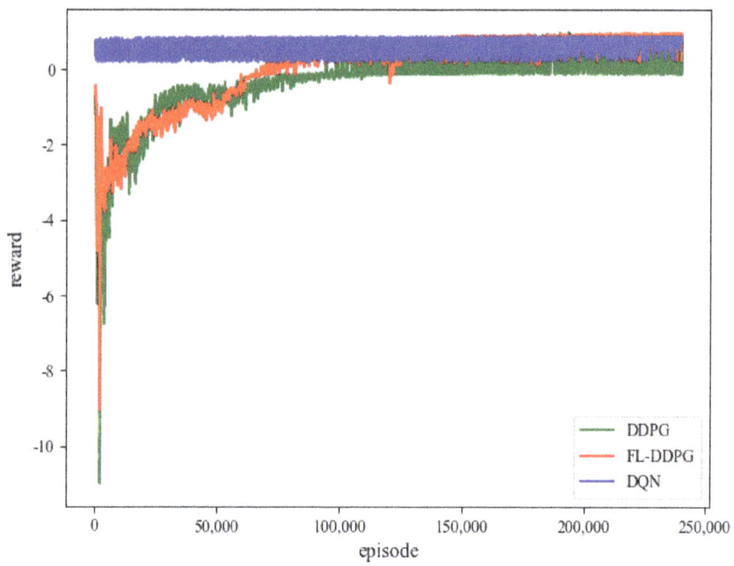

Figure 3. Convergence property of different algorithm.

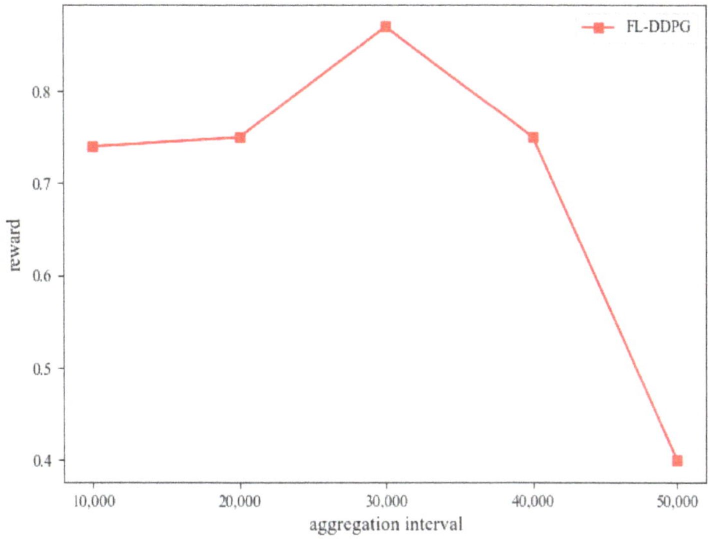

Figure 4. Performance evaluation on aggregation interval.

5.3. Performance Comparison

In this subsection, the performance evaluation of different algorithms is shown in Figures 5–7. Figure 5 shows the reward of different algorithms in terms of the system bandwidth. As the system bandwidth increases, more and more IoT tasks can be offloaded to the MEC server. Therefore, the energy consumption of IoT devices is reduced and the reward is increased in Figure 5. We can observe that the reward of FL-DDPG is higher than that of other algorithms. The reason can be obtained by analyzing each algorithm in detail, which is as follows. The DDPG algorithm only adopts one network model to train the decision-making scheme, which is easy to fall into local solution. Compared with DDPG and FL-DDPG, the DQN algorithm discretizes resources, which is a coarse-grained resource allocation scheme. Since DDPG algorithm is a fine-grained resource allocation scheme, the performance of DDPG algorithm is better than DQN algorithm. The GREEDY algorithm offloads the tasks generated by IoT devices with good network status to the MEC server. The algorithm only optimizes the communication resources, does not jointly optimize the communication resources and computing resources. The RANDOM algorithm is to randomly offload the tasks generated by IoT devices to the MEC server for execution. Further, we can observe that there is a little performance difference between DDPG and FL-DDPG algorithms when the system bandwidth is 5, 9, 10, 11 and 12. The reward of FL-DDPG is 1.3%, 1.1% and 1% higher than that of DDPG when the system bandwidth is 5, 9 and 10, respectively. The reason is that when the system bandwidth is very small, most of the tasks generated by IoT devices cannot be offloaded to the MEC server and can only be processed on the IoT devices. Since processing tasks on IoT devices do not involve the allocation of MEC computing resources and communication resources, the decision-making environment is simplified. Moreover, the energy consumption caused by a large number of IoT devices processing will drown out the energy consumption of transmission caused by offloading. When the system bandwidth is larger, most of the tasks generated by IoT devices can be offloaded to the MEC server. The reason for this is the same as above. Therefore, when the resources are in extreme situations, the exploration environment of reinforcement learning becomes relatively simple, resulting in a little performance difference between DDPG and FL-DDPG. In actual equipment deployment, these two extreme situations are generally not selected in terms of the cost and the quality of service. There is a large performance difference between DDPG and FL-DDPG algorithms when the system bandwidth is 6 and 7. The reward of FL-DDPG is 12% and 10% higher than that of DDPG when the system bandwidth is 6 and 7, respectively. When the system bandwidth is moderate, the decision-making environment becomes complex. The more complex the decision-making environment is, the greater the probability of DDPG algorithm falling into the local optimal solution is. Since the FL-DDPG algorithm aggregates the training parameters of three network models, it is easy to jump out of the local optimal solution.

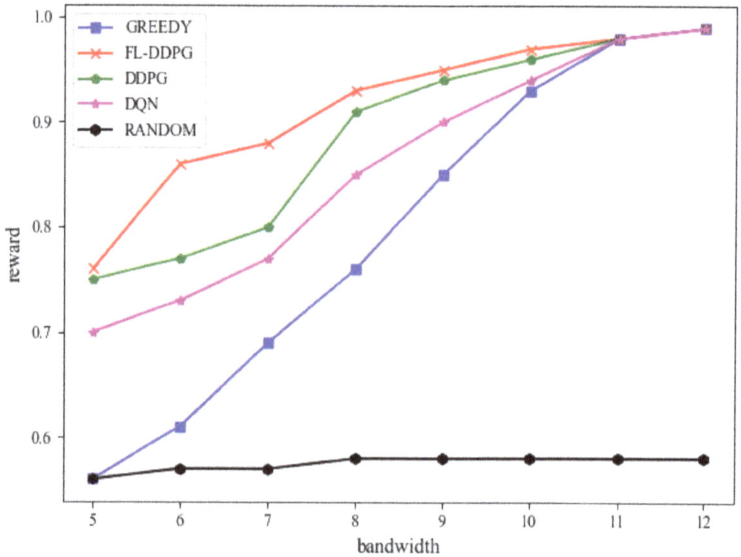

Figure 5. Performance evaluation on reward.

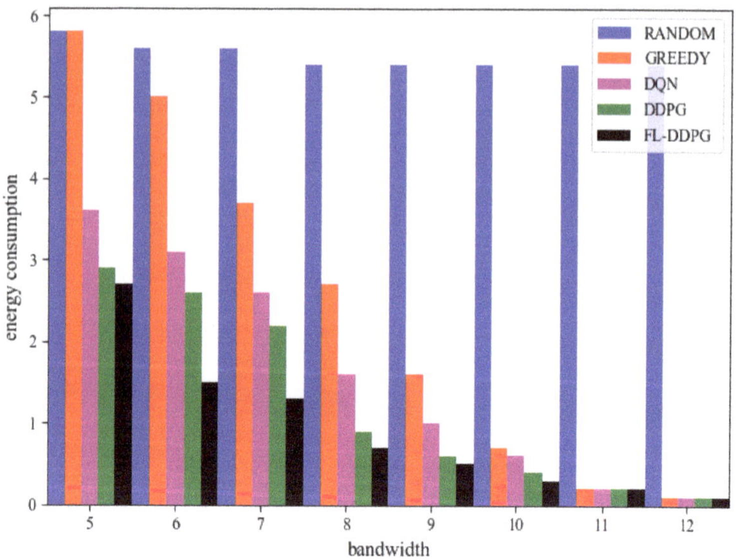

Figure 6. Performance evaluation on energy consumption.

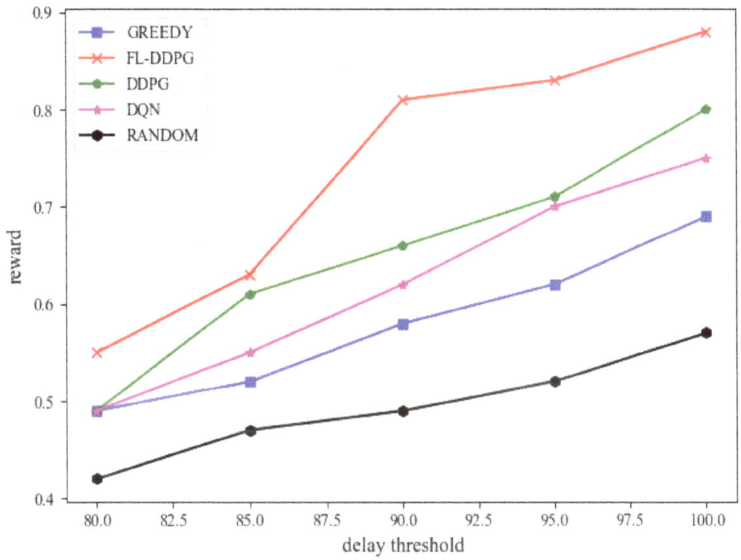

Figure 7. Performance evaluation on reward when the delay threshold is different.

Figure 6 shows the mean energy consumption of different algorithms in terms of the system bandwidth. From Figure 6, it is observed that the mean energy consumption of FL-DDPG algorithm is less than other algorithms. In the setting of reward, there is a negative exponential relationship between energy consumption and reward. Therefore, Figures 6 and 5 are one-to-one correspondence.

Figure 7 shows the reward of different algorithms in terms of the delay threshold. In this article, since these tasks generated by IoT devices are not very strict on the response time, the delay threshold is set to the same. From Figure 7, it is observed that the reward increases when the delay threshold increases. The reason is that when the delay threshold increases, more tasks can be offloaded to the MEC server and completed within the delay threshold. Therefore, the energy consumption of IoT devices is reduced and the reward is increased. Figure 8 shows the delay of different algorithms in the same environment configuration. The delay of five algorithms is less than the delay threshold (0.1 s).

5.4. *Analysis of Offload Location*

Figures 9 and 10 show the offloading location of FL-DDPG when the system bandwidth is 5 MHz and 10 MHz. In this experiment, the X-axis denotes the number of episodes, the Y-axis denotes the IoT device index, and the Z-axis denotes the offloading location. The value range of the offloading location is 0, 1. Value 0 indicates that the task is processed on the IoT device, value 1 indicates that the task is offloaded to the MEC server. From Figures 9 and 10, it is observed that the number of red points is less when the system bandwidth is 10 MHz. Figures 9 and 10 indicate that more tasks are offloaded to the MEC server when the system bandwidth increases. From Figure 9, we can observe that all tasks of IoT device 6 are not offloaded to the MEC server when the system bandwidth is 5 MHz. The reason is that the task of IoT device 6 has the characteristics of large amounts of data and low computing workload. If the task of IoT devices 6 is offloaded to the MEC server, it will consume a lot of bandwidth and a small amount of the MEC computing resources. Obviously, in the case of limited resources, it is unreasonable to offload the task to the MEC server. Therefore, all tasks of IoT device 6 are processed on the IoT device.

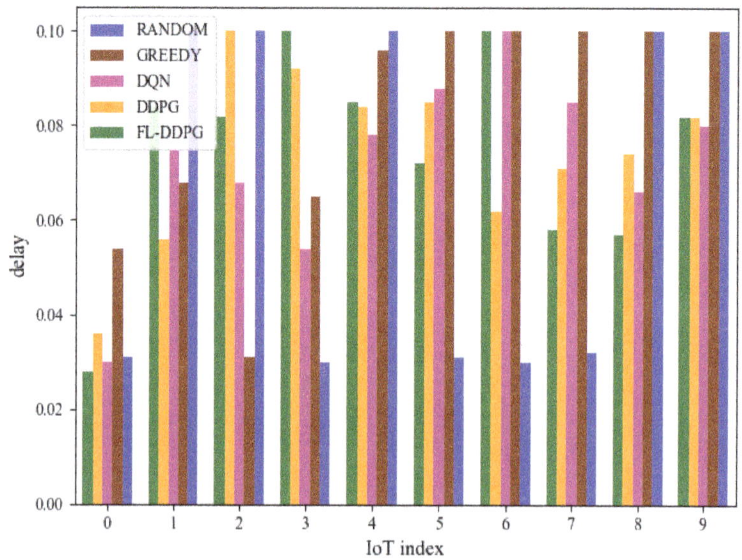

Figure 8. Delay of different algorithms.

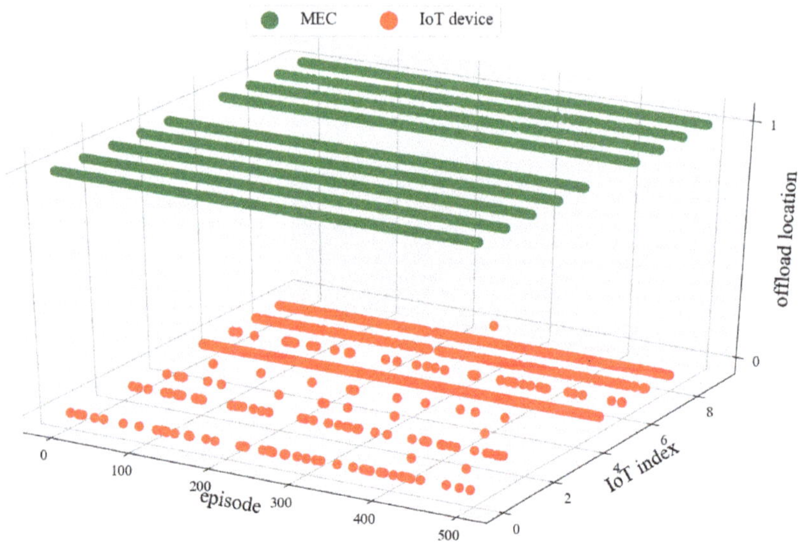

Figure 9. System bandwidth $B = 5$ MHz.

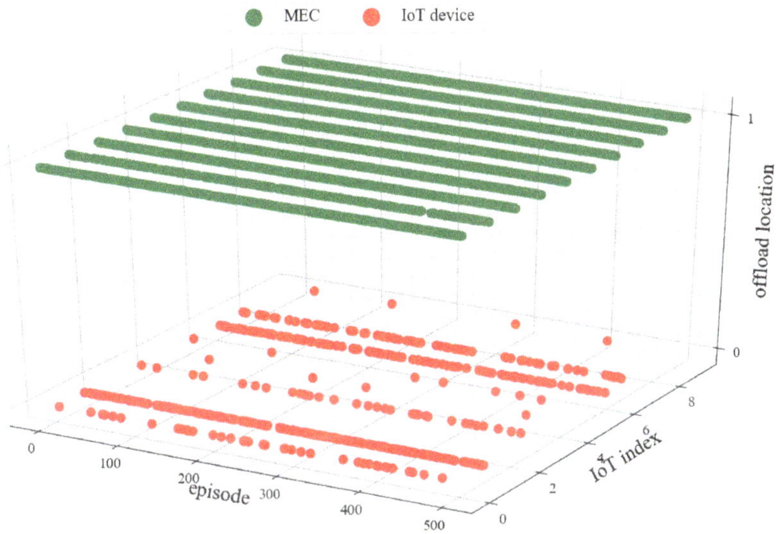

Figure 10. System bandwidth $B = 10$ MHz.

6. Conclusions

In this article, a joint optimization problem of task offloading and resource allocation based on privacy protection for smart city is formulated to minimize the energy consumption of all IoT devices. First, the deep reinforcement learning algorithm based on DDPG framework is proposed to solve the mixed-integer nonlinear programming problem. Then, in order to protect user privacy and improve training performance, the federated learning is introduced into the DDPG framework. To this end, the two-timescale FL-DDPG algorithm is proposed to optimize the above problem. Specifically, the small timescale is to train the DDPG network and the large timescale is to aggregate the parameters of DDPG network. In this way, the privacy of users is not only protected, but also the performance of the algorithm is improved. We provide numerical simulation results in terms of the convergence property, reward, and energy consumption, which shows that our proposed algorithm has better performance.

Author Contributions: X.C. performed the literature reading, algorithm coding and experiment. G.L. provided guidance, strcturing and proofreading of the article. All authors have read and agreed to the published version of the manuscript.

Funding: This work was supported by Shaanxi Key R&D Program (NO.2018ZDCXL-GY-04-03-02).

Institutional Review Board Statement: Not applicable.

Informed Consent Statement: Not applicable.

Data Availability Statement: Data available on request from the authors.

Conflicts of Interest: The authors declare no conflict of interest.

Abbreviations

The following abbreviations are used in this manuscript:

MEC	Mobile Edge Computing
IoT	Internet of Things
DDPG	Deep Deterministic Policy Gradient
FL-DDPG	Federated learning-Deep Deterministic Policy Gradient

OFDMA	Orthogonal Frequency Division Multiple Access
QoE	Quality of Experience
QoS	Quality of Service
DQN	Deep Q Network
AC	Actor-Critic
MDP	Markov Decision Process
FLOPs	Floating Point Operations
AI	Artificial Intelligence

References

1. Hu, Y.C.; Patel, M.; Sabella, D.; Sprecher, N.; Young, V. Mobile edge computing—A key technology towards 5G. *ETSI White Pap.* **2015**, *11*, 1–16.
2. ETSI; MECISG. *Mobile Edge Computing-Introductory Technical White Paper*; ETSI: Sophia Antipolis, France, 2014.
3. Ding, Y.; Liu, C.B.; Zhou, X.; Liu, Z.; Tang, Z. A code-oriented partitioning computation offloading strategy for multiple users and multiple mobile edge computing servers. *IEEE Trans. Ind. Inform.* **2019**, *16*, 4800–4810. [CrossRef]
4. Dab, B.; Aitsaadi, N.; Langar, R. Joint optimization of offloading and resource allocation scheme for mobile edge computing. In Proceedings of the 2019 IEEE Wireless Communications and Networking Conference (WCNC), Marrakech, Morocco, 15–18 April 2019; pp. 1–7.
5. Zhang, Z.Y.; Zhou, H.; Li, D.W. Joint Optimization of Multi-user Computing Offloading and Service Caching in Mobile Edge Computing. In Proceedings of the 2021 IEEE/ACM 29th International Symposium on Quality of Service (IWQOS), Virtual Conference, 10–12 June 2022; pp. 1–2.
6. Jin, Z.L.; Zhang, C.B.; Jin, Y.F.; Zhang, L.J.; Su, J. A Resource Allocation Scheme for Joint Optimizing Energy-Consumption and Delay in Collaborative Edge Computing-Based Industrial IoT. *IEEE Trans. Ind. Inform.* **2022**, *18*, 6236–6243. [CrossRef]
7. Zhang, Q.; Gui, L.; Hou, F.; Chen, J.C.; Zhu, S.C.; Tian, F. Dynamic task offloading and resource allocation for mobile-edge computing in dense cloud RAN. *IEEE Internet Things J.* **2020**, *7*, 3282–3299. [CrossRef]
8. Liang, Z.; Liu, Y.; Lok, T.M.; Huang, K.B. Multi-cell mobile edge computing: Joint service migration and resource allocation. *IEEE Trans. Wirel. Commun.* **2021**, *20*, 5898–5912. [CrossRef]
9. Li, J.H.; Zhao, J.H.; Sun, X.K. Deep Reinforcement Learning Based Wireless Resource Allocation for V2X Communications. In Proceedings of the 2021 13th International Conference on Wireless Communications and Signal Processing (WCSP), Changsha, China, 20–22 October 2021; pp. 1–5.
10. Chen, L.; Wu, J.G.; Zhang, J.; Dai, H.N.; Long, X.; Yao, M.Y. Dependency-aware computation offloading for mobile edge computing with edge-cloud cooperation. *IEEE Trans. Cloud Comput.* **2020**, *accepted*. [CrossRef]
11. Wu, H.M.; Wolter, K.; Jiao, P.F.; Deng, Y.J.; Zhao, Y.B.; Xu, M.X. EEDTO: An energy-efficient dynamic task offloading algorithm for blockchain-enabled IoT-edge-cloud orchestrated computing. *IEEE Internet Things J.* **2020**, *8*, 2163–2176. [CrossRef]
12. McMahan, B.; Moore, E.; Ramage, D.; Hampson, S.; Arcas, B.A. Communication-Efficient Learning of Deep Networks from Decentralized Data. In Proceedings of the 20th International Conference on Artificial Intelligence and Statistics, Fort Lauderdale, FL, USA, 20–22 April 2017; pp. 1273–1282.
13. Kulkarni, V.; Kulkarni, M.; Pant, A. Survey of personalization techniques for federated learning. In Proceedings of the 2020 Fourth World Conference on Smart Trends in Systems, Security and Sustainability (WorldS4), London, UK, 1 October 2020; pp. 794–797.
14. Messaoud, S.; Bradai, A.; Ahmed, O.B.; Quang, P.T.A.; Atri, M.; Hossain, M.S. Deep federated Q-learning-based network slicing for industrial IoT. *IEEE Trans. Ind. Inform.* **2020**, *17*, 5572–5582. [CrossRef]
15. Zhan, Y.F.; Zhang, J.; Hong, Z.C.; Wu, L.J.; Li, P.; Guo, S. A survey of incentive mechanism design for federated learning. *IEEE Trans. Emerging Top. Comput.* **2022**, *10*, 1035–1044. [CrossRef]
16. IBM. IBM and Nokia Siemens Networks Announce World's First Mobile Edge Computing Platform. 2013. Available online: http://www-03.ibm.com/press/us/en/pressrelease/40490.wss (accessed on 4 March 2013).
17. Dhelim, S.; Ning, H.S.; Farha, F.; Chen, L.M.; Atzori, L.; Daneshmand, M. IoT-enabled social relationships meet artificial social intelligence. *IEEE Internet Things J.* **2021**, *8*, 17817–17828. [CrossRef]
18. Aung, N.; Dhelim, S.; Chen, L.M.; Zhang, W.Y.; Lakas, A.; Ning, H.S. VeSoNet: Traffic-Aware Content Caching for Vehicular Social Networks based on Path Planning and Deep Reinforcement Learning. *arXiv* **2021**, arXiv:2111.05567.
19. Li, Y.Q.; Wang, X.; Gan, X.Y.; Jin, H.M.; Fu, L.Y.; Wang, X.B. Learning-aided computation offloading for trusted collaborative mobile edge computing. *IEEE Trans. Mob. Comput.* **2019**, *19*, 2833–2849. [CrossRef]
20. Apostolopoulos, P.A.; Tsiropoulou, E.E.; Papavassiliou, S. Cognitive data offloading in mobile edge computing for internet of things. *IEEE Access* **2020**, *8*, 55736–55749. [CrossRef]
21. Abd, E.M.; Abualigah, L.; Ibrahim, R.A.; Attiya, I. IoT Workflow Scheduling Using Intelligent Arithmetic Optimization Algorithm in Fog Computing. *Comput. Intell. Neurosci.* **2021**, *2021*, 9114113.
22. Attiya, I.; Abd, E.M.; Abualigah, L.; Nguyen, T.N.; Abd, E.; Ahmed, A. An improved hybrid swarm intelligence for scheduling iot application tasks in the cloud. *IEEE Trans. Ind. Inform.* **2022**, *18*, 6264–6272. [CrossRef]
23. Ning, Z.L.; Wang, X.J.; Rodrigues, J.J.; Xia, F. Joint computation offloading, power allocation, and channel assignment for 5G-enabled traffic management systems. *IEEE Trans. Ind. Inform.* **2019**, *15*, 3058–3067. [CrossRef]

24. Zhao, M.X.; Yu, J.J.; Li, W.T.; Liu, D.; Yao, S.W.; Feng, W.; She, C.Y.; Quek, T.Q.S. Energy-aware task offloading and resource allocation for time-sensitive services in mobile edge computing systems. *IEEE Trans. Veh. Technol.* **2021**, *70*, 10925–10940. [CrossRef]
25. Tan, L.; Kuang, Z.F.; Zhao, L.; Liu, A.F. Energy-Efficient Joint Task Offloading and Resource Allocation in OFDMA-based Collaborative Edge Computing. *IEEE Trans. Wirel. Commun.* **2022**, *21*, 1960–1972. [CrossRef]
26. Masoudi, M.; Cavdar, C. Device vs edge computing for mobile services: Delay-aware decision making to minimize power consumption. *IEEE Trans. Mob. Comput.* **2020**, *20*, 3324–3337. [CrossRef]
27. Ding, C.F.; Wang, J.B.; Zhang, H.; Lin, M.; Wang, J.Z. Joint MU-MIMO Precoding and Resource Allocation for Mobile-Edge Computing. *IEEE Trans. Wirel. Commun.* **2020**, *20*, 1639–1654. [CrossRef]
28. Suh, K.J.; Kim, S.W.; Ahn, Y.J.; Kim, S.Y.; Ju, H.Y.; Shim, B.H. Deep Reinforcement Learning-based Network Slicing for Beyond 5G. *IEEE Access* **2022**, *10*, 7384–7395. [CrossRef]
29. Naouri, A.; Wu, H.X.; Nouri, N.A.; Dhelim, S.; Ning, H.S. A Novel Framework for Mobile-Edge Computing by Optimizing Task Offloading. *IEEE Internet Things J.* **2021**, *8*, 13065–13076. [CrossRef]
30. Chen, X.; Liu, G.Z. Energy-Efficient Task Offloading and Resource Allocation via Deep Reinforcement Learning for Augmented Reality in Mobile Edge Networks. *IEEE Internet Things J.* **2021**, *8*, 10843–10856. [CrossRef]
31. Hamdi, M.; Hamed, A.B.; Yuan, D.; Zaied, M. Energy-Efficient Joint Task Assignment and Power Control in Energy Harvesting D2D Offloading Communications. *IEEE Internet Things J.* **2021**, *9*, 6018–6031. [CrossRef]
32. You, C.S.; Huang, K.B.; Chae, H.J.; Kim, B.-H. Energy-efficient resource allocation for mobile-edge computation offloading. *IEEE Trans. Wirel. Commun.* **2016**, *16*, 1397–1411. [CrossRef]
33. Saleem, U.; Liu, Y.; Jangsher, S.; Tao, X.M.; Li, Y. Latency minimization for D2D-enabled partial computation offloading in mobile edge computing. *IEEE Trans. Veh. Technol.* **2020**, *69*, 4472–4486. [CrossRef]
34. Saleem, U.; Liu, Y.; Jangsher, S.; Li, Y. Performance guaranteed partial offloading for mobile edge computing. In Proceedings of the 2018 IEEE Global Communications Conference (GLOBECOM), Abu Dhabi, United Arab Emirates, 9–13 December 2018; pp. 1–6.
35. Zhao, F.J.; Chen, Y.; Zhang, Y.C.; Liu, Z.Y.; Chen, X. Dynamic Offloading and Resource Scheduling for Mobile-Edge Computing With Energy Harvesting Devices. *IEEE Trans. Netw. Serv. Manag.* **2021**, *18*, 2154–2165. [CrossRef]
36. Li, S.; Sun, W.B.; Sun, Y.; Huo, Y. Energy-Efficient Task Offloading Using Dynamic Voltage Scaling in Mobile Edge Computing. *IEEE Trans. Netw. Sci. Eng.* **2020**, *8*, 588–598. [CrossRef]
37. Jiang, H.B.; Dai, X.X.; Xiao, Z.; Iyengar, A.K. Joint Task Offloading and Resource Allocation for Energy-Constrained Mobile Edge Computing. *IEEE Trans. Mob. Comput.* **2022**. [CrossRef]
38. Hao, Y.X.; Chen, M.; Hu, L.; Hossain, M.S.; Ghoneim, A. Energy efficient task caching and offloading for mobile edge computing. *IEEE Access* **2018**, *6*, 11365–11373. [CrossRef]

energies

MDPI

Article

Energy-Saving Routing Protocols for Smart Cities

Douglas de Farias Medeiros †, Cleonilson Protasio de Souza †, Fabricio Braga Soares de Carvalho *,† and Waslon Terllizzie Araújo Lopes †

Post-Graduation Program in Electrical Engineering, Department of Electrical Engineering, Federal University of Paraíba (UFPB), João Pessoa 5045, PB, Brazil
* Correspondence: fabricio@cear.ufpb.br
† These authors contributed equally to this work.

Abstract: In recent decades, expansion in urban areas has faced issues such as management of public waste, noise, mobility, and air quality, among others. In this scenario, Internet of Things (IoT) and Wireless Sensor Network (WSN) scenarios are being considered for Smart Cities solutions based on the deployment of wireless remote sensor nodes to monitor large urban areas. However, as the number of nodes increases, the amount of data to be routed increases significantly as well, meaning that the choice of the data routing process has great importance in terms of the energy consumption and lifetime of the network. In this work, we describe and evaluate the energy consumption of routing protocols for WSN-based Smart Cities applications in LoRa-based mesh networks, then propose a novel energy-saving radio power adjustment (RPA) routing protocol. The Cupcarbon network simulator was used to evaluate the performance of different routing protocols in terms of their data package delivery rate, average end-to-end delay, average jitter, throughput, and load consumption of battery charge. Additionally, a novel tool for determining the range of nodes based on the Egli propagation model was designed and integrated into Cupcarbon. The routing protocols used in this work are Ad Hoc On-Demand Distance Vector (AODV), Dynamic Source Routing (DSR), and Distance Vector Routing (DVR). Our simulation results show that AODV presents the best overall performance, DSR achieves the best results for power consumption, and DVR is the best protocol in terms of latency. Finally, the proposed RPA routing protocol presents power savings of 11.32% compared to the original DSR protocol.

Keywords: Smart Cities; Wireless Sensor Networks; routing protocols; Cupcarbon simulator; LoRa

Citation: Medeiros, D.d.F.; Souza, C.P.d.; Carvalho, F.B.S.d.; Lopes, W.T.A. Energy-Saving Routing Protocols for Smart Cities. *Energies* **2022**, *15*, 7382. https://doi.org/10.3390/en15197382

Academic Editors: Antonio Cano-Ortega, Francisco Sánchez-Sutil and Aurora Gil-de-Castro

Received: 29 July 2022
Accepted: 19 September 2022
Published: 8 October 2022

1. Introduction

Urban growth has been boosted in recent decades due both to economic factors and to political, social, and health trends. However, this growth brings with it problems in such different areas as waste management, mobility, scarcity of natural resources, noise and air pollution, and more.Air pollution, for example, is an environmental problem that can cause several diseases in humans and damage to both the environment and animals, with transportation and industry being the main sources of pollutants in the atmosphere [1].

Smart Cities are emerging as an alternative that can enable applications to deal with several problems associated with urban centers. Smart Cities are urban scenarios using Information, Technology, and Communications (ITC) to improve infrastructure and the quality of citizens' lives. Strongly linked with the concepts of the Internet of Things (IoT) and Wireless Sensor Networks (WSN), Smart Cities provide means to carry out acquisition, transmission, and processing of data to make more effective tools available for facing the challenges of the urban environment [2,3].

A WSN is comprised of several wireless sensor nodes distributed in a large area to perform control and monitoring tasks and to share sensor data with each other in order to solve specific problems [4].

To deploy a WSN, several stages are needed, including the development of sensors, which depends on the application, number of sensors, network topology, communication technology, and routing protocol. It is worth mentioning that the routing protocol is critical in any WSN design and that its performance needs to be evaluated for different scenarios and applications.

Computer simulations are an important design tool for evaluating routing protocols in high-density WSNs, thereby reducing costs and saving time during implementation. Simulations can support the choice of a particular routing protocol and help in evaluating new protocols, mainly in scenarios subjected to unfavorable conditions, such those in which device failure is highly probable.

Another important stage in WSN implementation is the wireless communication technology used by the network. Many of the main wireless communication technologies adopted in IoT and WSN applications are based on Low-Power Wide-Area Networks (LPWAN), 3G/4G/5G cellular networks, or ZigBee. LPWANs have gained importance compared to the others thanks to relevant characteristics such as low power consumption and transmission over long distances. Among LPWAN technologies, LoRa (Long Range) is being widely used worldwide, as it can achieve ranges up to 15 km in urban areas with a very low power consumption [5,6].

In this context, the main objective of this paper is to implement and evaluate the performance of routing protocols for the establishment of LoRa-based WSN applications. To this end, we chose the Cupcarbon network simulator, which was developed specifically for Smart Cities and IoT scenario, to evaluate the performance of different routing protocols in terms of data package delivery rate, average jitter, average end-to-end delay, throughput, and load consumption of battery power. In addition, we propose a novel tool for determining node ranges using the Egli propagation model inside the Cupcarbon simulator. In this paper, we consider the widely used WSN routing protocols Ad Hoc On-Demand Distance Vector (AODV), Dynamic Source Routing (DSR), and Distance Vector Routing (DVR). Additionally, a novel routing protocol based on radio power adjustment (RPA) is proposed as means of energy saving.

The rest of this paper is organized as follows. In Section 2, details about LoRa technology and routing protocols for WSN are presented. The Cupcarbon simulator and the Egli propagation loss model are highlighted within the methodology considered in this work. Simulation results are presented and evaluated in Section 3, then and the conclusions and next steps in the research derived from this paper are discussed in Section 4.

2. Materials and Methods

2.1. Related Works

In [7], the authors introduced an IoT application using a WSN distributed over a large geographic area in which the sensor nodes use LoRa technology. Their communication performance analysis was based on varying parameters related to the LoRa physical layer, such as the bandwidth (BW) and scattering factor (SF).

Performance comparisons of different routing protocols have been presented in [8–12] using simulation tools and metrics such as the packet delivery rate, average latency, average jitter, and throughput. Simulations were carried out under various scenarios, such as different node densities and variations in terms of mobility.

Routing protocols and LoRa networks have previously been detailed in [13–15]. In [13], a routing system protocol based on the AODV protocol was proposed for use in meshed LoRa networks. In [14], the development of a hybrid network based on a LoRa mesh topology and the LoRaWAN protocol was introduced. An emergency communication system using a mesh LoRa network and implementing a modified version of the AODV protocol was presented in [15] along with an evaluation of the system feasibility according to the package delivery rate.

2.1.1. LPWAN

Currently, the most widely adopted wireless communication technologies for IoT and WSN applications are Low-Power Wide-Area Network (LPWAN), 3G/4G/5G cellular networks, and ZigBee. LPWANs have gained prominence compared to the alternatives, as they feature low power consumption and transmission over very long distances. The main LPWAN wireless communication technologies are LoRa (Long Range), Sigfox, NB-IoT (Narrow-Band IoT), and Wi-SUN (Wireless Smart Ubiquitous Network).

2.1.2. LoRa

LoRa is a wireless communication technology patented by Semtech Corporation that can be applied on devices with battery restrictions, aiming at longer lifetimes and large transmission ranges [6]. The range of a LoRa-based network in an urban area is up to 15 km, and in rural areas it can be up to 30 km in normal conditions. Lora operates in the ISM (Industrial, Scientific, Medical) bands.

LoRa modulation is based on the Chirp Spread Spectrum (CSS), which is characterized by the spectral spread of the signal to be transmitted in a given frequency range (f_{low}, f_{high}), generating a signal called Compressed High Intensity Radar Pulse (Chirp) [6]. An unmodulated Chirp signal has constant amplitude, and its frequency varies inside the bandwidth (BW = $f_{high} - f_{low}$) by a given period of time (TS = Symbol time).

The parameters used in LoRa modulation are the Bandwidth (BW), Spreading Factor (SF), and Code Rate (CR). Each LoRa symbol spans an entire BW and can encode bits of data defined by the SF, which can be from six to twelve. A LoRa symbol is an up-chirp (from f_{low} to f_{high}), meaning that when a frequency related to the data being transmitted is reached, it is shifted to f_{low} while maintaining the same frequency slope, causing a discontinuity point.The discontinuity point position is responsible for the encoding of the transmitted data [6].

2.1.3. Sigfox

Sigfox, developed by the French company Sigfox, was the first LPWAN technology proposed by the IoT industry [5]. Sigfox physical layer modulation is based on an Ultra Narrow Band (UNB) modulation. However, there is limited documentation of its operation due to commercial protection, which becomes a relevant issue in academic studies on the network and the reproduction/simulation of results. A Sigfox network operates similarly to a cellular operator for the IoT industry [16], that is, there are service costs for subscribers to use the network. The coverage or range of Sigfox networks in urban areas is between 3 km and 10 km, and in rural areas it is between 30 km and 50 km. Sigfox operates in the ISM band (868 or 915 MHz), its communication rate is around 100 bps, and it supports up to 1,000,000 nodes per gateway [5,16].

2.1.4. NB-IoT

NB-IoT is a standard developed by the 3GPP (Third Generation Partnership Project), which is an international telecommunications standardization body. The operation of NB-IoT is performed by telecom operators and is an extension of the 4G cellular network infrastructure [17] (4G LTE service providers such as Verizon and AT&T in the United States). The data transfer rate can reach 234.7 kbps [17], and it supports up to 50,000 devices per cell [16]. An important feature is that the battery life of an NB-IoT radio can be as long as ten years [16].

2.1.5. Wi-SUN

Wi-SUN (Wireless Smart Ubiquitous Networks) technology is maintained by the Wi-SUN Alliance and consists of wireless communication networks that are based on the IEEE 802.15.4g standard and are designed to be reliable and have low power consumption. Wi-SUN allows the establishment of networks that integrate smart devices from different manufacturers and is able to implement different topologies, including star, mesh, or hybrid,making the coverage area wider [18]. Wi-SUN adopts a Gaussian FSK (GFSK) modulation scheme, operates in the ISM bands, has low latency when compared to other LPWANs technologies, and has a transmission rate of around 300 kbps [18,19].

2.2. Routing Protocols for WSN

With a higher the amount of sensor nodes, the amount of data exchanged over the WSN increases. This emphasizes the importance of an efficient data routing process when considering the mesh topology.

In short, a data routing process consists of verifying and evaluating available paths from a source node to a destination node, then determining the best path for forwarding data throughout the network based on a given criterion [20]. Based on this process, the data routing protocol specifies the technique by which routing tables are formed and maintained in order to aid in the forwarding of data [21].

In general, routing protocols fall into four categories:n Centralized vs. Distributed, Static vs. Adaptive, Flat vs. Hierarchical, and Proactive vs. Reactive vs. Hybrid [21].

In this work, we highlight proactive, reactive, and hybrid protocols, which differ in the way they operate according to the routing strategy [11]:

- Proactive: Routing tables are shared with neighboring nodes during network startup and at fixed times, meaning that all nodes know the paths to any destination nodes even before this information is needed [22]. Examples of proactive routing protocols include Distance Vector Routing (DVR), Destination-Sequenced Distance Vector Routing (DSDV), Optimized Link State Routing Protocol (OLSR), and Wireless Routing Protocol (WRP);
- Reactive: Routes to destination are established only when needed, that is, when there are data packets to be sent. Therefore, only active routes to destination nodes that are in use are stored. These protocols do not share data at network startup, and have periodic routing table sharing mechanisms [10]. Examples include Ad Hoc On-Demand Distance Vector (AODV), Dynamic Source Routing (DSR), and Temporally Ordered Routing Algorithm (TORA);
- Hybrid: These protocols combine resources from proactive and reactive protocols, enabling convenient use of the advantages of both; examples include Zone Routing Protocol (ZRP) and Fisheye State Routing (FSR).

When considering routing protocols for WSN applications, energy efficiency is an important characteristic. In [23], the authors analysed wireless network energy models based on five reactive and proactive routing protocols for WSNs, including AODV and DVR protocols. A WSN energy model was proposed in [24] using AODV and DVR routing protocols and considering the energy consumption at each node of the network. A genetic algorithm-based routing protocol for sensor networks was presented in [25]; the authors compared their proposed method with different routing protocols, including DVR and AODV.

2.3. Cupcarbon Simulator

Cupcarbon is an open-source Java-based network simulator with a focus on Smart Cities, WSN, and IoT [26–28]. It allows network designers to debug and validate network applications in a 2D/3D graphical environment [29]. Cupcarbon is composed of four main blocks:

1. The radio channel block: this comprises two Radio-Frequency (RF) propagation models, the first based on a visibility tree and the second one devoted to tracing 3D beans using a Monte Carlo algorithm [26].
2. 2D/3D environment: a graphical environment for implementing 2D/3D maps that can be displayed on Open Street Maps (OSM) or Google Maps [30].
3. Interference block: this can be divided into two layers, the physical layer and interference models layer. It can simulate realistic baseband models for wireless com-

munication technologies, even in the physical layer, including Wi-Fi, Zigbee, and LoRa [26].

4. Implementation block: the user-customization block of Cupcarbon is developed in a modular wa with the aim of simplifying the replacement and customization of specific part of the simulator.

In Cupcarbon, network devices are programmed in a script language called Senscript, proposed in [29], which allows the generation of code for the Arduino platform. An important feature of Cupcarbon is that the energy consumption of a sensor node can be analysed according to both the classic consumption model and the Heinzelman model.

2.4. Egli's Propagation Loss Model

Network simulation tools use computational models for their operations, from component and device models to environmental condition and mobility behaviour models, as well as for signal propagation. Propagation loss models are mathematical models used to estimate the attenuation between RF transmitters and RF receivers in order to obtain the received signal power according to specific conditions (frequency, antenna gain, etc.) [31].

The Egli model is a widely used propagation loss model derived from experimental results using actual measurements of television broadcast systems [32]; it is suitable for cellular communications where there are a number of both fixed and mobile devices. Furthermore, it is applicable in scenarios where transmission occurs across uneven terrain without the presence of vegetation in the communication link [33]. The Egli model can be adopted for frequency ranges between 40 MHz and 1 GHz and for distances up to 60 km [34], and takes into consideration the line of sight between the devices, The Path Loss is denoted as P_L, and is provided by the formula

$$P_L = G_t G_r \left(\frac{h_t h_r}{d^2} \right)^2 \left(\frac{40}{f_c} \right)^2 \tag{1}$$

where G_t and G_r are the gains of the transmitter and receiver antennas, respectively, h_t and h_r are the respective heights of the transmitter and receiver antennas, d is the distance between them, and f_c is the carrier frequency in MHz.

2.5. Range Calculation Tool

Cupcarbon provides default values for range depending on the wireless communication technology, for instance, 100 m, 400 m, and 5 km for Zigbee, WiFi, and LoRa, respectively.

This work proposes a slight modification to the Cupcarbon visual interface that allows for the computation of transmission range using the Egli model with the LoRa modulation parameters. The proposed modification, which is based on Java and integrated into the source code of Cupcarbon, allows the user to graphically select a node, choose LoRa, then enter the desired LoRa radio, LoRa parameters (SF, BW, frequency, etc.), and deployment parameters (G_t, G_r, h_t, h_r, radio power level, and receiver sensibility). Finally, after configuration, the user can apply it to all selected nodes in the simulation map.

It is important to highlight that when the user selects the LoRa radio module (SX1276, SX1277, SX1278, or SX1279), the radio parameters are automatically changed according to the datasheet [35].

2.6. Methodology

This work adopts as its main methodology the development and running of experimental simulations based on the DVR, AODV, and DSR routing protocols using the Cupcarbon simulator. Initially, these protocols were simulated while disregarding eventual errors that could occur in the network nodes or mobility situations; thus, several tests were carried out for each protocol with the purpose of applying the routing steps and their operation in a mobility scenario with both fixed and mobile sensors. Then, our new proposed radio power adjustment (RPA) routing protocol for energy saving was compared with different alternatives.

2.6.1. Simulation Scenario

In order to compare performance between the protocols in a near-real scenario, we considered a scenario based on a real city in which a source sensor node was represented as a bus traveling through the streets and avenues of an urban environment and performing air quality measurements while periodically sending the measured data to a central node (gateway). Thus, a mesh network topology was considered consisting of 25 nodes distributed among the streets and avenues of the city of João Pessoa, Brazil, as illustrated in Figure 1.

Figure 1. Network topology considered in the simulations.

2.6.2. Simulation Settings

Each node illustrated in Figure 1 was defined with the wireless communication module LoRa SX1276 in the frequency range of 915 MHz. The node range was calculated considering antennas with a gain of 3 dBi, antenna height of 2.5 m, and transmission power of 20 dBm. In addition, the LoRa parameters are shown in Table 1.

Considering that the DVR protocol is proactive, a period of 5 min for sharing the routing tables was defined, with the first sharing being initiated at the start of the simulation. In the case of AODV protocol simulations, a period of 90 s was defined for the route maintenance mechanism (sending 'Hello' messages). Finally, a period of 30 s was used to transmit general messages and data packets from the sensors.

Table 1. General parameters used in the simulations.

Total simulation time	3600 s (1 h)
Pause time	90 s
Radius of each node	1.56 km
Radio propagation model	Egli
Number of nodes	25
Wireless communication technology	LoRa
Transmission frequency	915 MHz
Spreading Factor (SF)	7
Code Rate (CR)	4/5
Bandwidth (BW)	125 kHz

2.6.3. Evaluation Metrics

To evaluate the performance of the protocols implemented in Cupcarbon, several quantitative metrics calculated by the gateway during the reception of data packets were considered.

- Packet Delivery Rate (PDR): this metric is the ratio between the number of packets sent by a source node and the number of packets received by the destination, as provided by [9]:

$$PDR = \frac{\sum N_{RX_i}}{\sum N_{TX_i}} \times 100\%, \tag{2}$$

  where N_{RX_i} represents the number of data packets successfully received by a destination node i and N_{TX_i} is the number of packets sent to this device.
- Throughput (THR): the number of data packets successfully transmitted to their final destination on a given communication channel per unit of time [11]; THR can be calculated as follows:

$$THR = \frac{N_{RX_i}}{Total\ simulation\ time}. \tag{3}$$

- End-to-End Delay (E2ED): the average time required for a number of data packets to be successfully transmitted over the network from a source node to a specific destination [9]; it can be defined mathematically by

$$E2ED = \frac{\sum T_{RX_i} - T_{TX_i}}{N_{RX_i}}, \tag{4}$$

  where N_{RX_i} is the number of data packets received by node i, T_{RX_i} represents the time the data packet was received, and T_{TX_i} is the time when the package was sent.
- Average Jitter (JIT): this represents the variation of the average end-to-end delay in the delivery of data packets in a network, which may cause a situation of non-regularity in the reception of data packets; the formula is provided by [36,37]:

$$Jitter = \frac{\sum E2ED_{i_k} - E2ED_{i_{k-1}}}{N_{RX_i}}, \tag{5}$$

  where $E2ED_{i_k}$ is the average delay of a given instant k, $E2ED_{i_{k-1}}$ is the average delay of the previous instant, and N_{RX_i} is the number of data packets received by node i.

2.7. Classic Energy Consumption Model

This is the standard energy consumption model adopted by the Cupcarbon simulator. To calculate how much battery energy a node consumes during data packet transmission, the following expression [29] can be used:

$$E_{TX} = \left(\frac{n}{8}\right) \times E_{TXb} \times P, \tag{6}$$

where n is the number of bits transmitted or received, E_{TXb} is the energy consumption required to transmit one byte, and P is the power of the transmission expressed as a percentage.

Similarly, the energy consumption during data reception can be obtained as follows:

$$E_{RX} = \left(\frac{n}{8}\right) \times E_{RXb}, \tag{7}$$

where n is the number of bits transmitted or received and E_{RXb} is the energy consumption required to receive one byte.

2.8. Proposed Radio Power Adjustment (RPA) Routing Protocol for Energy Savings

The proposed radio power adjustment (RPA) routing protocol is based on the DSR protocol, as DSR shows better overall performance, including the lowest energy consumption, as compared to DVR and AODV.

The first step of the proposed RPA routing protocol is to provide a way for Cupcarbon to compute the Radio Signal Strength Indicator (RSSI) from the distance between two nodes according to the log-distance model. This model was chosen because it is independent of the transmission frequency and antenna gain. The log-distance path loss model can be obtained by [38]:

$$\overline{PL}(d) = \overline{PL}(d_0) + 10n\log\left(\frac{d}{d_0}\right), \tag{8}$$

where n corresponds to the path loss coefficient (the value of 3.3 was adopted in this work), d is the distance between the transmitter and the receiver, and d_0 is the reference distance.

The choice of the value of n was carried out by taking into account the experimental results obtained in [39]. In these experiments, two devices were used to exchange data using LoRa modulation in an urban environment while evaluating the packet delivery rate and RSSI as a function of the transmission distance. Then, the RSSI curves versus the distance were obtained and compared to the experimental data and the data from the log-distance model, with the value of $n = 3.3$ representing the best fit between the curves.

The value of $\overline{PL}(d_0)$ represents the path loss for a direct line of sight with respect to the reference distance, and can be obtained by [38]:

$$\overline{PL}(d_0) = 20log\left(\frac{4\pi d_0}{\lambda}\right). \tag{9}$$

In Equation (9), d_0 is the reference distance and the parameter λ corresponds to the carrier wavelength.

Equations (8) and (9) were programmed in CupCarbon. When any data are received, the distance between the transmitting and receiving nodes which are exchanging data is obtained through the computed RSSI function.

The second step of the proposed RPA routing protocol is the definition of an equation that relates the computed RSSI value to the minimum P_{TX} necessary to maintain the communication link. To this end, simulations were performed in Cupcarbon varying the distance between two nodes.

In this way, a node RX was kept fixed and a node TX was initially moved to a distance equal to the maximum range of 1.7 km; P_{TX}, that is, the transmission power of node TX, was set to the maximum value (20 dBm). Thus, the distance between the nodes was reduced in 50-meter intervals until the minimum distance of 1 m was reached, with the RSSI value calculated for each distance at these intervals. At the same time as the distance was being reduced, at each new displacement point P_{TX} was reduced in 5% intervals until a minimum value at which the communication link was not lost was reached.

The results of this process were then imported into Octave software and an interpolation process was performed applying the *polyfit* function. To assess which polynomial obtained a better approximation of the original data, the Root Mean Square Error (RMSE) was used as a criterion, resulting in the following mathematical expression being obtained:

$$\begin{aligned} P_{TX}(RSSI) = & -1.161169468381691 \times 10^{-7} \cdot (RSSI)^5 \\ & -4.870470506506291 \times 10^{-5} \cdot (RSSI)^4 \\ & -0.008114743288230 \cdot (RSSI)^3 \\ & -0.660852476932444 \cdot (RSSI)^2 \\ & -25.953469491107786 \cdot RSSI \\ & -3.829454402717133 \times 10^2 \end{aligned} \tag{10}$$

Finally, this expression was inserted into the DSR protocol code to determine the appropriate P_{TX} for each RSSI value. To use the proposed RPA routing protocol, it is necessary to add a new column to the routing table that contains the value of P_{TX} for each destination, this value being initialized with 100% and automatically changed as the simulation is run. A safety margin of 2% was considered as well, and was added to the value calculated by Equation (10) to provide more reliability during data packet transmission.

Figure 2 shows the process used to compute the minimum level of P_{TX} needed to establish reliable communications. The first step is to identify the address of the node that sent the data and calculate the RSSI. This procedure is illustrated in Figure 2a, and the corresponding pseudocode is detailed in Algorithm 1 . This RSSI value is then used as an input parameter in Equation (10) to obtain the value of P_{TX}, which is added to the safety margin value and saved in the column of the routing table that stores the level information power relative to that destination node.

Algorithm 1: Identification of the P_{TX} level of a destination node.

```
for each data packet received do
    - Check the destination node;
    - Calculate RSSI;
    - Get the appropriate level for transmit power using the fit equation;
    - Adds the safety margin to the calculated power level;
    - Changes the transmit power value for the destination in routing table;
end
```

When it is needed to carry out a transmission, the source node consults the power level value in its routing table, updates it, and sends the data packet. This procedure is illustrated in Figure 2b, and the corresponding pseudocode is detailed in Algorithm 2.

Algorithm 2: Power adjustment before data transmission.

```
while True do
    if Need to send data then
        - Search the power level on the routing table;
        - Change transmission power;
        - Send the data packet;
    else
        - Wait;
    end
end
```

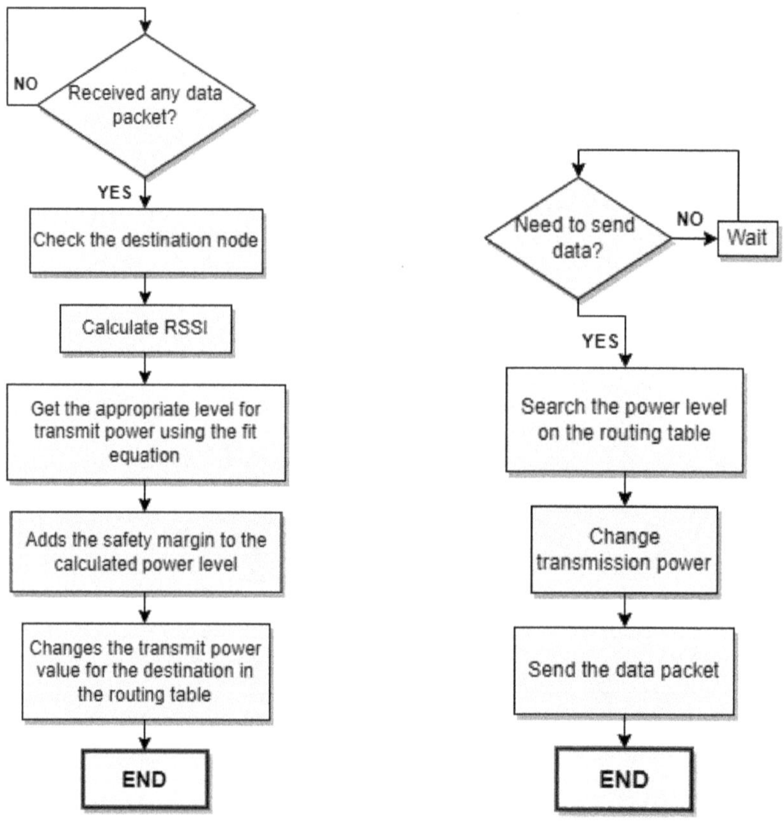

Figure 2. P_{TX} Adjustment Flowcharts: (**a**) identification of the P_{TX} level of a destination node and (**b**) power adjustment before data transmission.

3. Results

This section describes and discusses the main results obtained in this work. The simulation results for both metrics and total energy consumption refer to the average values obtained from five simulation trials.

3.1. Determining the Safety Margin of the proposed RPA Routing Protocol

With the aim of providing more reliability during the computation of P_{TX}, a safety margin in the range from 0 to 5% was added to the proposed RPA algorithm. Thus, the total energy consumption was evaluated as a function of the variation of the safety margin. For this purpose, three simulation experiments were performed for a scenario composed by 25 nodes, one source sensor node, 23 router nodes, and one gateway.

In this experiment, the source node moves at a fixed speed of 18 km/h and the others remain static. The source sensor node sends data to the gateway at every 30 s. At the end, the value of energy savings is calculated from the average consumption value comparing the simulations of the DSR protocol with and without the RPA algorithm.

As shown in Figure 3, when the adjustment margin is greater than 2.5% the energy savings are less than 10%. Therefore, to obtain energy savings of at least 10%, we decided to keep the safety margin fixed at 2%.

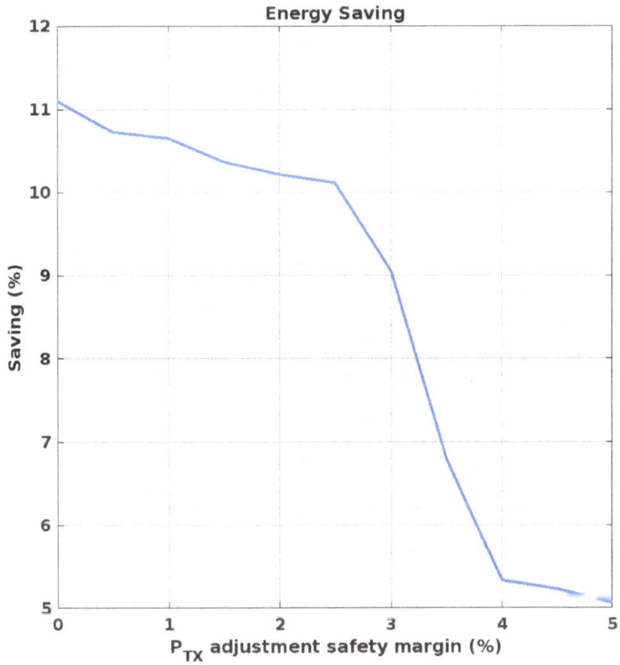

Figure 3. Power adjustment safety margin versus energy savings.

3.2. Simulation Results and Discussion

As mentioned before, the simulation results refer to the average values obtained from five simulation trials. Figure 4 shows the PDR metric for DVR, AODV, DSR, and modified DSR (that is, DSR with the proposed RPA routing protocol).

From these results, it can be seen that the reactive protocols present better overall performance in sending data; the AODV protocol has the highest rate of data packet delivery, followed by the DSR and modified DSR protocols. The result for AODV corresponds to a route verification mechanism that works periodically and during the movement of the source node, anticipating the discovery of new routes in case of unavailability.

For speeds below 20 km/h, all protocols maintained high data delivery rates. However, above this speed value the performance decay of the DVR protocol was quite high, with more than 50% of transmitted packets being lost when the source node moved at speeds greater than 60 km/h. The other protocols were able to maintain delivery rates above 80%

for the same speed range. Consequently, the impact of the source node speed variation on the proactive DVR protocol was much greater. This characteristic was expected, as new routes could be discovered only within the fixed routing interval of 5 min.

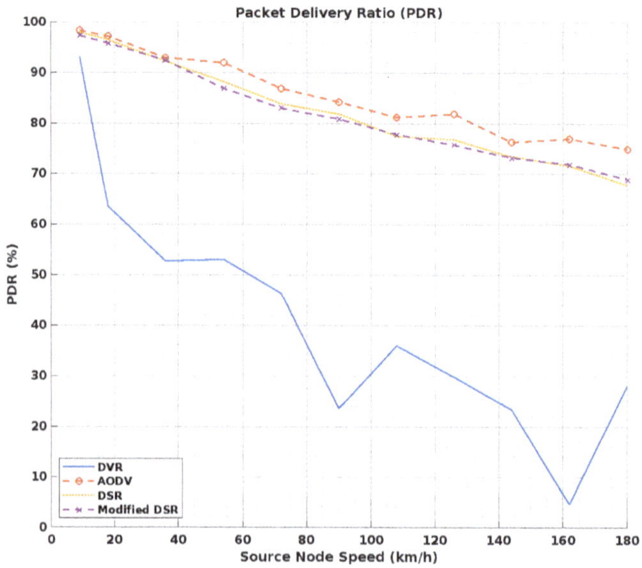

Figure 4. Packet delivery rate (PDR) as a function of the source node's speed for the DVR, AODV, DSR, and Modified DSR protocols.

In Figure 5, the performance results of the routing protocols in relation to the average end-to-end delay are shown. From these results, it can be observed that the DVR protocol had the lowest latency, followed by the AODV protocol. The DSR and modified DSR protocols had higher average latency compared to the others.

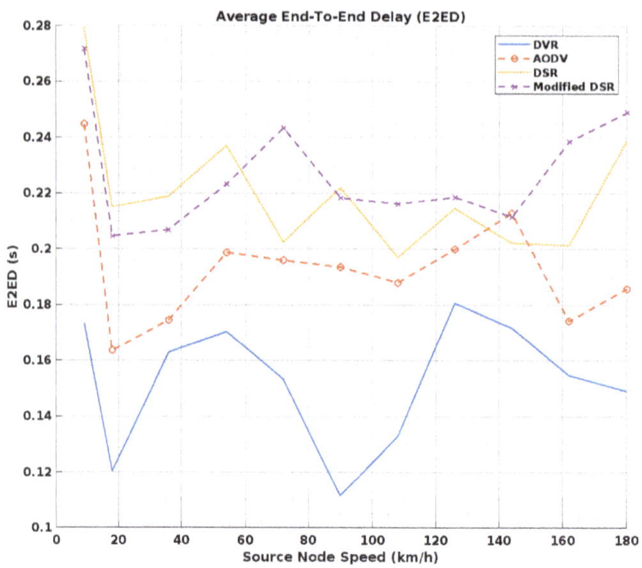

Figure 5. End-to-end delay (E2ED) as a function of the source node's speed for the DVR, AODV, DSR, and Modified DSR protocols.

It is important to mention that the performance of the DVR protocol in terms of latency shows expected behavior. As a proactive protocol, it is independent of unavailable routes, as the discovery of new paths to destinations takes place during fixed routing periods and is accessible in the routing tables prior to each transmission demand.

The average performance of the protocols according to the jitter metric is shown in Figure 6. Based on the curves in Figure 6, it can be seen that the DSR protocol was the only one that does not have a variation for latency higher than 4 ms. For speeds below 100 km/h, the DVR protocol presented similar behavior. However, at higher speeds, it is the protocol with the highest jitter.

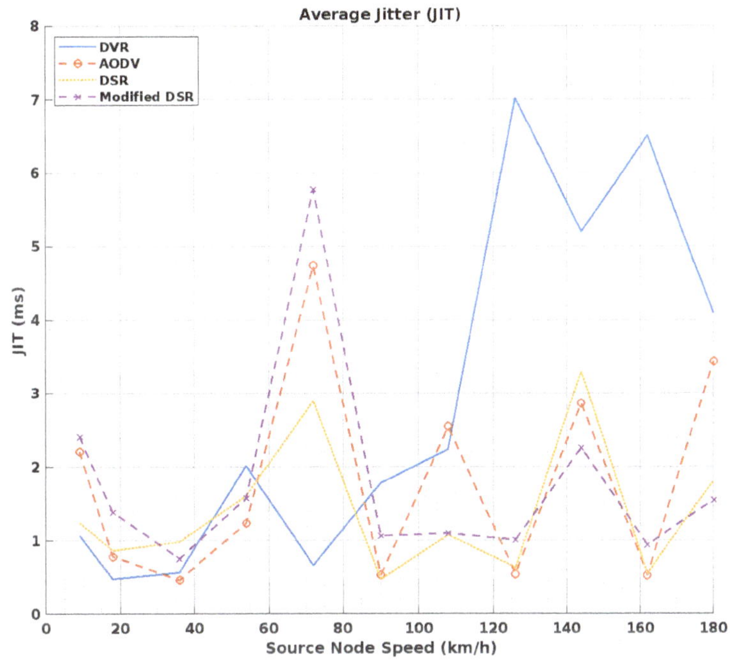

Figure 6. Average Jitter (JIT) as a function of the source node's speed for the DVR, AODV, DSR, and Modified DSR protocols.

Regarding the throughput performance, shown in Figure 7, a high correlation was observed with the packet delivery rate. The performance of the AODV protocol in terms of the packet delivery rate is reflected in the transfer rate metric, indicating that this protocol is capable of transmitting more data per unit of time than the others. Both the DSR protocol and the modified DSR had slightly lower performance than AODV.

As the DVR protocol obtains data regarding routes at a fixed interval of 5 min, this protocol has a higher packet loss rate when subjected to high speeds, causing a considerable reduction in its transfer rate, as can be seen in Figure 7.

Figure 8 shows the results of the average total battery charge consumption according to the routing protocol. From these results, the impact of the large number of re-transmissions on the total consumption of the DVR protocol can be observed, leading it to consume around 460 J for the whole simulation. Although the AODV protocol presents the best performance in the delivery of data packets, it consumed the second-highest energy, at almost 300 J, while the modified DSR protocol (DSR with RPA routing protocol) consumed the least energy over the total simulation time.

As this work is focused on applications for smart cities, our results take into account the existence of moving nodes (e.g., deployed in a vehicle) and node speeds from 0 to 180 km/h are considered. As expected, this scenario constitutes a very challenging environment for routing protocols. Consequently, all results in Figures 4–7 use the source node's speed as the independent variable. The simulation results demonstrate that the speed influences the adopted metric (Packet Delivery Rate (PDR), Throughput (THR), End-to-End Delay (E2ED), Average Jitter (JIT), and Energy Consumption). Therefore, these results prove the energy savings provided by our proposed DSR with RPA routing protocol has an average total power consumption that is 11.32% lower compared to the same protocol without the proposed RPA.

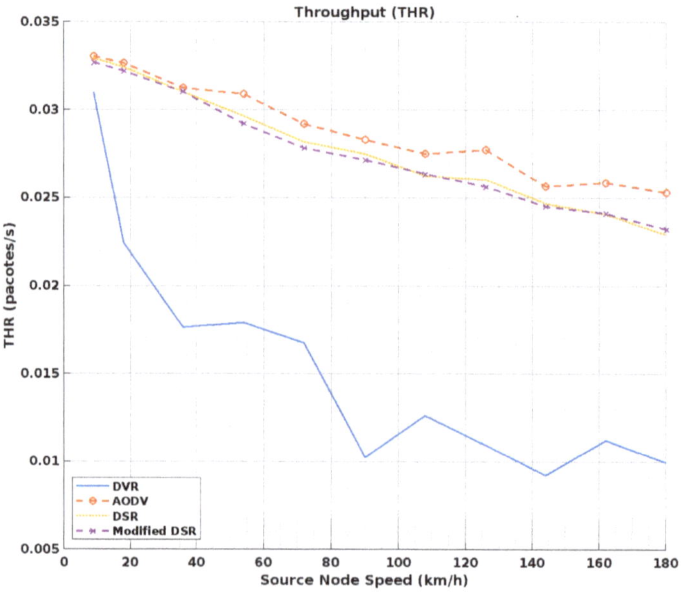

Figure 7. Throughput (THR) as a function of the source node's speed for the DVR, AODV, DSR, and Modified DSR protocols.

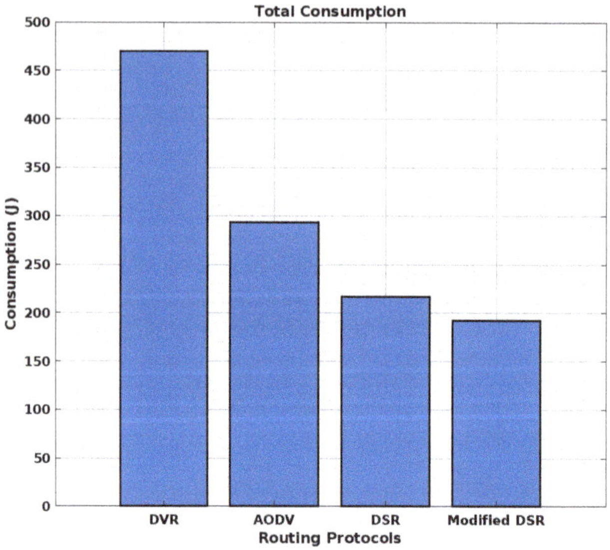

Figure 8. Total average consumption per protocol.

Table 2 summarizes the achievements of this work compared with other routing protocols considering a mobile WSN with 24 sensor nodes and one concentrator node at two speed values (18 km/h and 90 km/h).

Table 2. Comparative analysis for DVR, AODV, DSR, and modified DSR protocols.

Metric	Our Method		DVR		AODV		DSR	
	18 km/h	90 km/h	18 km/h	90 km/h	18 km/h	90 km/h	18 km/h	90 km/h
PDR (%)	95.78	80.84	63.47↓	23.62↓	97.14↑	84.20↑	96.47↑	81.84↓
E2ED (ms)	204.7	218.5	120.3↓	111.6↓	163.5↓	193.5↓	215.1↓	222.04↓
JIT (ms)	1.390	1.066	0.479↓	0.479↓	0.782↓	0.528↓	0.864↓	0.474↓
THR (package/min)	1.931	1.629	1.345↓	0.615↓	1.959↓	1.698↓	1.945↑	1.650↑
Average consumption by node (J)	8.00	-	19.58 145%↑	-	12.22 53%↑	-	9.02 13%↑	-
Total consumption (J)	192.17	-	470.08 145%↑	-	293.29 53%↑	-	216.70 13%↑	-

4. Conclusions and Future Works

In this work, we analyzed routing protocols in two distinct categories, namely, proactive and reactive protocols. The Cupcarbon network simulator was used to evaluate important metrics such as data package delivery rate, average end-to-end delay, average jitter, throughput, and load consumption of battery charge. Thus, the Ad Hoc On-Demand Distance Vector (AODV), Dynamic Source Routing (DSR), and Distance Vector Routing (DVR) routing protocols were implemented in the Cupcarbon simulator. In addition, a tool for calculating the range between devices according to the Egli propagation model was developed and integrated into the graphical interface of Cupcarbon.

The results showed that the DSR protocol was the most suitable option among those implemented for use in conjunction with the P_{TX} adjustment algorithm proposed in this work, providing energy savings of 11.32% compared to the original DSR. On the other hand, the AODV protocol had better overall performance and had the second-highest power consumption. While the DVR protocol consumed the most energy, it had the best performance in terms of latency; however, it led to high packet loss.

For this implementation, a mixed network topology was defined using the DSR protocol together with the LoRaWAN protocol. A cloud application was developed to monitor data reception, confirming the correct functioning of the network.

Future Works

The simulation results obtained here can be used as the basis for future experimental implementations, which we intend to carry out in our continuation of this work. It is important to note that Equation (10) cannot be generalized to every situation, and is specific to the simulation scenario used in this work. However, the proposed procedure can be replicated to obtain new equations for other simulation scenarios. In addition, while the algorithm proposed in this work does not take into consideration all of the factors that may influence the transmission of data in real scenarios, it is now possible to perform more realistic transmission modeling in Cupcarbon, as updates have been released in recent months that make it possible to use real devices in conjunction with simulations developed in the software.

In this context, we propose further experimental implementation of the proposed power adjustment algorithm; for example, it could be used with ESP32 devices for the purpose of evaluating its operation in a real scenario and making further improvements.

Author Contributions: Conceptualization, D.d.F.M., C.P.d.S., F.B.S.d.C. and W.T.A.L.; methodology, D.d.F.M., C.P.d.S., F.B.S.d.C. and W.T.A.L.; software, D.d.F.M.; validation, D.d.F.M., C.P.d.S., F.B.S.d.C. and W.T.A.L.; writing—original draft preparation, D.d.F.M., C.P.d.S., F.B.S.d.C. and W.T.A.L.; writing—review and editing, D.d.F.M., C.P.d.S., F.B.S.d.C. and W.T.A.L. All authors have read and agreed to the published version of the manuscript.

Funding: This work was supported by the Federal University of Paraiba (UFPB), by the PRONEX project funded by the Paraíba State Research Foundation (FAPESQ-PB/Brazil), by the Brazilian National Council for Scientific and Technological Development (CNPq-Brazil) under Grant Number 009/2019, by the CNPq Research Productivity fellowship (Proc. 309371/2019-8 PQ), and by the Institutional Scientific Initiation Scholarship Program (PIBIC/CNPq/UFPB). This work was funded by Public Call N. 03/2020 Research Productivity PROPESQ/PRPG/UFPB under Grant Number PVK13136-2020. Additional funding for this work was provided by Public Call N. 09/2021 *Demanda Universal* FAPESQ-PB/Brazil. Part of this work has been funded by the National Institute of Science and Technology of Micro and Nanoelectronics Systems (Proc. 573738/2008-4 INCT NAMITEC), by the Tutorial Education Program (PET/Electric) of the Brazilian Ministry of Education, and by the University of Washington Tacoma Endowment Professorship in Engineering Systems. This study was financed in part by the Coordenação de Aperfeiçoamento de Pessoal de Nível Superior Brasil (CAPES), Finance Code 001.

Institutional Review Board Statement: Not applicable.

Informed Consent Statement: Not applicable.

Data Availability Statement: The code developed for the preparation of this manuscript is available at https://github.com/douglasfmed/Cupcarbon_Energies_2022, accessed on 23 September 2022.

Acknowledgments: The authors would like to thank the Federal University of Paraíba (UFPB) for the support to this work.

Conflicts of Interest: The authors declare no conflict of interest.

Abbreviations

The following abbreviations are used in this manuscript:

AODV	Ad Hoc On-Demand Distance Vector
bps	Bits per second
BW	Bandwidth
Chirp	Compressed High-Intensity Radar Pulse
CR	Code Rate
CSS	Chirp Spread Spectrum
DRSSI	Differential Received Signal Strength Indication
DSR	Dynamic Source Routing
DVR	Distance Vector Routing
E2ED	End-to-End Delay
E_{RX}	Receiving Energy Consumption
E_{TX}	Sending Energy Consumption
f_c	Carrier Frequency
FSR	Fisheye State Routing
GFSK	Gaussian Frequency Shift Keying
G_r	Gain of the Receiver Antenna
G_t	Gain of the Transmitter Antenna
H_r	Height of the Receiver Antenna
H_t	Height of the Transmitter Antenna
IDE	Integrated Development Environment
IoT	Internet of Things
ISM	Industrial, Scientific, Medical
ITC	Information, Technology, and Communications
JIT	Jitter

LoRa	Long Range
LoRaWAN	Long-Range Wide-Area Network
LPWAN	Low-Power Wide-Area Network
LTE	Long-Term Evolution
MAC	Media Access Control
NB-IoT	Narrow-Band IoT
OLSR	Optimized Link State Routing Protocol
OSI	Open Systems Interconnection
OSM	Open Street Maps
RF	Radio Frequency
RSSI	Radio Signal Strength Indicator
PDR	Packet Delivery Rate
P_{TX}	Transmission Power
$\overline{PL}$	Path Loss
RMSE	Root Mean Square Error
SF	Scattering Factor
THR	Throughput
TORA	Temporally-Ordered Routing Algorithm
TTN	The Things Network
UNB	Ultra Narrow Band
Wi-Fi	Wireless Fidelity
Wi-SUN	Wireless Smart Ubiquitous Networks
WRP	Wireless Routing Protocol
WSN	Wireless Sensor Network
ZRP	Zone Routing Protocol
3G	Third Generation
3GPP	Third Generation Partnership Project
4G	Fourth Generation
5G	Fifth Generation
λ	Carrier Wavelength

References

1. Villarim, M.R.; Medeiros, D.D.F.; Souza, C.P.D.; Medeiros, L.C.d.S.; Pontieri, M.H.; dos Santos, N.A.; Baiocchi, O. Calibration of a Low Cost Sensor for PM2.5 using a Reference PM Monitoring Station. In Proceedings of the 6th International Conference on Sensors Engineering and Electronics Instrumentation Advances (SEIA' 2020), Porto, Portugal, 23–25 September 2020.
2. Zanella, A.; Bui, N.; Castellani, A.; Vangelista, L.; Zorzi, M. Internet of Things for Smart Cities. *IEEE Internet Things J.* **2014**, *1*, 22–32. [CrossRef]
3. Usman, N.; Alfandi, O.; Usman, S.; Khattak, A.M.; Awais, M.; Hayat, B.; Sajid, A. An Energy Efficient Routing Approach for IoT Enabled Underwater WSNs in Smart Cities. *Sensors* **2020**, *20*, 4116. [CrossRef] [PubMed]
4. Bhanu, K.N.; Jasmine, H.J.; Mahadevaswamy, H.S. Machine learning Implementation in IoT based Intelligent System for Agriculture. In Proceedings of the 2020 International Conference for Emerging Technology (INCET), Belgaum, India, 5–7 June 2020; pp. 1–5.
5. Centenaro, M.; Vangelista, L.; Zanella, A.; Zorzi, M. Long-range communications in unlicensed bands: The rising stars in the IoT and smart city scenarios. *IEEE Wirel. Commun.* **2016**, *23*, 60–67. [CrossRef]
6. Augustin, A.; Yi, J.; Clausen, T.; Townsley, W.M. A study of Lora: Long range & low power networks for the internet of things. *Sensors* **2016**, *16*, 1466. [CrossRef]
7. Lee, H.C.; Ke, K.H. Monitoring of Large-Area IoT Sensors Using a LoRa Wireless Mesh Network System: Design and Evaluation. *IEEE Trans. Instrum. Meas.* **2018**, *67*, 2177–2187. [CrossRef]
8. Medeiros, D.d.F.; Villarim, M.R.; de Carvalho, F.B.S.; de Souza, C.P. Implementation and Analysis of Routing Protocols for LoRa Wireless Mesh Networks. In Proceedings of the 2020 11th IEEE Annual Information Technology, Electronics and Mobile Communication Conference (IEMCON), Vancouver, BC, Canada, 4–7 November 2020; pp. 0020–0025. [CrossRef]
9. Al-Dhief, F.T.; Sabri, N.; Salim, M.S.; Fouad, S.; Aljunid, S.A. MANET Routing Protocols Evaluation: AODV, DSR and DSDV Perspective. *MATEC Web Conf.* **2018**, *150*, 06024. [CrossRef]
10. Matre, V.; Karandikar, R. Multipath routing protocol for mobile adhoc networks. In Proceedings of the 2016 Symposium on Colossal Data Analysis and Networking, CDAN 2016, Indore, India, 18–19 March 2016; pp. 1–5. [CrossRef]
11. Rahman, J.; Hasan, M.A.M.; Islam, M.K.B. Comparative analysis the performance of AODV, DSDV and DSR routing protocols in wireless sensor network. In Proceedings of the 2012 7th International Conference on Electrical and Computer Engineering, ICECE 2012, Dhaka, Bangladesh, 20–22 December 2012; pp. 283–286. [CrossRef]

12. Roshan, R.; Mishra, S.; Meher, C.P. A comparison between the AODV and DSDV routing protocols for mobile Ad-Hoc Network Using NS2. In Proceedings of the 2019 6th International Conference on Computing for Sustainable Global Development, INDIACom 2019, New Delhi, India, 13–15 March 2019; pp. 286–289.
13. Lundell, D.; Hedberg, A.; Nyberg, C.; Fitzgerald, E. A Routing Protocol for LoRA Mesh Networks. In Proceedings of the 19th IEEE International Symposium on a World of Wireless, Mobile and Multimedia Networks, WoWMoM 2018, Chania, Greece, 12–15 June 2018. [CrossRef]
14. Almeida, N.C.; Rolle, R.P.; Godoy, E.P.; Ferrari, P.; Sisinni, E. Proposal of a Hybrid LoRa Mesh / LoRaWAN Network. In Proceedings of the 2020 IEEE International Workshop on Metrology for Industry 4.0 and IoT, MetroInd 4.0 and IoT 2020, Roma, Italy, 3–5 June 2020; pp. 702–707. [CrossRef]
15. Macaraeg, K.C.V.G.; Hilario, C.A.G.; Ambatali, C.D.C. LoRa-based Mesh Network for Off-grid Emergency Communications. In Proceedings of the 2020 IEEE Global Humanitarian Technology Conference (GHTC) 2020, Seattle, WA, USA, 29 October–1 November 2020; pp. 1–4. [CrossRef]
16. Raza, U.; Kulkarni, P.; Sooriyabandara, M. Low Power Wide Area Networks: An Overview. *IEEE Commun. Surv. Tutorials* **2017**, *19*, 855–873. [CrossRef]
17. Tadayoni, R.; Henten, A.; Falch, M. Internet of Things—The battle of standards. In Proceedings of the 2017 Internet of Things Business Models, Users, and Networks, Copenhagen, Denmark, 23–24 November 2017; pp. 1–7. [CrossRef]
18. Mochizuki, K.; Obata, K.; Mizutani, K.; Harada, H. Development and field experiment of wide area Wi-SUN system based on IEEE 802.15.4g. In Proceedings of the 2016 IEEE 3rd World Forum on Internet of Things, WF-IoT 2016, Reston, VA, USA, 12–14 December 2016; pp. 76–81. [CrossRef]
19. Wi-SUN Alliance. 2022. Available online: https://wi-sun.org (accessed on 23 September 2022).
20. Gawiy, M.; Al-quh, A.; Lathaa, A.A.; Amran, Z.; Al-Hubaishi, M. Performance Analysis of Destination Sequenced Distance Vector Routing Protocol in MANET. *Int. J. Ad Hoc Veh. Sens. Netw.* **2014**, *6*, 10–23. .
21. Srivastava, A.; Mishra, A.; Upadhyay, B.; Yadav, A.K. Survey and overview of Mobile Ad-Hoc Network routing protocols. In Proceedings of the 2014 International Conference on Advances in Engineering and Technology Research, ICAETR 2014, Unnao, India, 1–2 August 2014; pp. 5–10. [CrossRef]
22. Parvathi, P. Comparative Analysis of CBRP, AODV, DSDV Routing Protocols in Mobile Ad-Hoc Networks. In Proceedings of the 2012 International Conference on Computing, Communication and Applications, Dindigul, India, 22–24 February 2012. [CrossRef]
23. Del-Valle-Soto, C.; Velázquez, R.; Valdivia, L.J.; Giannoccaro, N.I.; Visconti, P. An Energy Model Using Sleeping Algorithms for Wireless Sensor Networks under Proactive and Reactive Protocols: A Performance Evaluation. *Energies* **2020**, *13*, 3024. [CrossRef]
24. Del-Valle-Soto, C.; Mex-Perera, C.; Nolazco-Flores, J.A.; Velázquez, R.; Rossa-Sierra, A. Wireless Sensor Network Energy Model and Its Use in the Optimization of Routing Protocols. *Energies* **2020**, *13*, 728. [CrossRef]
25. Patel, Jatinkumar El-Ocla, H. Energy Efficient Routing Protocol in Sensor Networks Using Genetic Algorithm. *Sensors* **2021**, *21*, 7060. [CrossRef] [PubMed]
26. Bounceur, A.; Clavier, L.; Combeau, P.; Marc, O.; Vauzelle, R.; Masserann, A.; Soler, J.; Euler, R.; Alwajeeh, T.; Devendra, V.; et al. CupCarbon: A new platform for the design, simulation and 2D/3D visualization of radio propagation and interferences in IoT networks. In Proceedings of the CCNC 2018—2018 15th IEEE Annual Consumer Communications and Networking Conference, Las Vegas, NV, USA, 12–15 January 2018; pp. 1–4. [CrossRef]
27. Bounceur, A.; Mehdi, K.; Lounis, M. A CupCarbon Tool for Simulating Destructive Insect Movements. In Proceedings of the 1st IEEE International Conference on Information and Communication Technologies for Disaster Management (ICT-DM'14), Algiers, Algeria, 24–25 March 2014; pp. 24–25.
28. Mehdi, K.; Lounis, M.; Bounceur, A.; Kechadi, T. CupCarbon: A multi-agent and discrete event Wireless Sensor Network design and simulation tool. In Proceedings of the SIMUTools 2014—7th International Conference on Simulation Tools and Techniques, Lisbon, Portugal, 17–19 March 2014; pp. 126–131. [CrossRef]
29. Virtualys; IEMN; IRCICA; Xlim; Lab-STICC. *CupCarbon User Guide—Version U-One 5.1*; Manchester Metropolitan University: Manchester, UK, 2021.
30. *CupCarbon—A Smart City & IoT WSN Simulator*; Manchester Metropolitan University: Manchester, UK, 2020.
31. Onipe, J.A.; Alenoghena, C.O.; Salawu, N.; Paulson, E.N. Optimal Propagation Models for Path-loss Prediction in a Mountainous Environment at 2100 MHz. In Proceedings of the 2020 International Conference in Mathematics, Computer Engineering and Computer Science, ICMCECS 2020, Ayobo, Nigeria, 18–21 March 2020; Volume 231. [CrossRef]
32. da Silva, J.C. Influência da Vegetação no Desvanecimento e na Perda de Percurso de Enlaces de Radiocomunicação UHF na Faixa de 700 MHz. Master's Dissertation, Pontifical Catholic University of Rio de Janeiro (PUC-Rio), Rio de Janeiro, Brazil, 2014.
33. Mardeni, R.; Pey, L.Y. Path loss model development for urban outdoor coverage of Code Division Multiple Access (CDMA) system in Malaysia. In Proceedings of the 2010 International Conference on Microwave and Millimeter Wave Technology, Chengdu, China, 8–11 May 2010; Volume 6, pp. 441–444. [CrossRef]
34. Maloku, H.; Limani Fazliu, Z.; Sela, E.; Ibrani, M. Path loss model fitting for TV bands based on experimental measurements for urban environments in Kosovo. In Proceedings of the 2019 42nd International Convention on Information and Communication Technology, Electronics and Microelectronics, MIPRO 2019, Opatija, Croatia, 20–24 May 2019; pp. 480–485. [CrossRef]

35. Semtech Corporation. 2022 Available online: https://www.semtech.com/products/wireless-rf/lora-connect/sx1276 (accessed on 23 September 2022).
36. Ema, R.R.; Anik, A.; Nahar, N.; Rahman, M.A.; Eti, K.P.; Islam, T. Simulation Based Performance Analysis of Proactive, Reactive and Hybrid Routing Protocols in Wireless Sensor Network. In Proceedings of the 2020 11th International Conference on Computing, Communication and Networking Technologies, ICCCNT 2020, Kharagpur, India, 1–3 July 2020. [CrossRef]
37. Sampoornam, K.P.; Raaga, D.G. Performance Analysis of Bellman Ford, AODV, DSR, ZRP and DYMO Routing Protocol in MANET using EXATA. In Proceedings of the 2019 International Conference on Advances in Computing and Communication Engineering, ICACCE 2019, Sathyamangalam, Tamil Nadu, India, 4–6 April 2019. [CrossRef]
38. Rappaport, T. *Wireless Communications: Principles and Practice*, 2nd ed.; Prentice Hall PTR: Hoboken, NJ, USA, 2001.
39. Villarim, M.R.; De Luna, J.V.H.; De Farias Medeiros, D.; Pereira, R.I.S.; De Souza, C.P. LoRa performance assessment in dense urban and forest areas for environmental monitoring. In Proceedings of the INSCIT 2019—4th International Symposium on Instrumentation Systems, Circuits and Transducers, São Paulo, Brazil, 26–30 August 2019. [CrossRef]

smart cities

MDPI

Article

Efficient Communication Model for a Smart Parking System with Multiple Data Consumers

T. Anusha * and M. Pushpalatha

Department of Computing Technologies, SRM Institute of Science and Technology, Chennai 603203, India
* Correspondence: aa5293@srmist.edu.in

Abstract: A smart parking system (SPS) is an integral part of smart cities where Internet of Things (IoT) technology provides many innovative urban digital solutions. It offers hassle-free parking convenience to the city dwellers, metering facilities, and a revenue source for businesses, and it also protects the environment by cutting down drive-around emissions. The real-time availability information of parking slots and the duration of occupancy are valuable data utilized by multiple sectors such as parking management, charging electric vehicles (EV), car servicing, urban infrastructure planning, traffic regulation, etc. IPv6 wireless mesh networks are a good choice to implement a fail-safe, low-power and Internet protocol (IP)-based secure communication infrastructure for connecting heterogeneous IoT devices. In a smart parking lot, there could be a variety of local IoT devices that consume the occupancy data generated from the parking sensors. For instance, there could be a central parking management system, ticketing booths, display boards showing a count of free slots and color-coded lights indicating visual clues for vacancy. Apart from this, there are remote user applications that access occupancy data from browsers and mobile phones over the Internet. Both the types of data consumers need not collect their inputs from the cloud, as it is beneficial to offer local data within the network. Hence, an SPS with multiple data consumers needs an efficient communication model that provides reliable data transfers among producers and consumers while minimizing the overall energy consumption and data transit time. This paper explores different SPS communication models by varying the number of occupancy data collators, their positions, hybrid power cycles and data aggregation strategies. In addition, it proposes a concise data format for effective data dissemination. Based on the simulation studies, a multi-collator model along with a data superimposition technique is found to be the best for realizing an efficient smart parking system.

Keywords: smart parking system; data consumers; communication model; IPv6 Mesh; RPL; smart city; IoT

Citation: Anusha, T.; Pushpalatha, M. Efficient Communication Model for a Smart Parking System with Multiple Data Consumers. *Smart Cities* **2022**, *5*, 1536–1554. https://doi.org/10.3390/smartcities5040078

Academic Editors: Antonio Cano-Ortega and Francisco Sánchez-Sutil

Received: 5 October 2022
Accepted: 28 October 2022
Published: 2 November 2022

1. Introduction

Social, technological and economic factors contributed to the emergence of smart parking systems (SPS), and the recent advancements in electric and autonomous vehicles present a strong business case for intelligent parking services [1]. Currently, they are an urban requirement where users can search, navigate, reserve and pay for a free parking slot on a real-time basis. Countries across the world are turning to smart parking solutions for reducing traffic, minimize effort spent on parking, combat illegal parking, cutting down emissions as well as a business model to generate revenue. The global smart parking market is projected to grow at a compounded annual growth rate of 19.8% and is on its way to becoming a 16.3 billion dollar market in another six years [2]. As SPS matures, it fuels expansion of allied sectors such as sensor technology, Internet of Things (IoT) devices, communication access technologies, Machine to Machine (M2M) standards, smart city infrastructure projects and security solutions. As the roll out of 5G infrastructure facilitates real-time data availability with ultra low latency, IoT is excepted to realize its full potential [3].

All the literature on SPS concentrates on parking sensors that are data producers, access technologies and various software solutions. The data consumers, who access the generated parking data, are by default expected to be connected to cloud for their input. However, a smart city is analogous to the presence of heterogeneous IoT devices that consume data during M2M interactions [4]. Multiple data consumers are inevitable in an SPS, as various IoT devices are present in the parking lot. A workstation in the control room or a display device needs only local data. Whereas, an end user's mobile application may need more sophisticated data from a central cloud as it accesses the data over Internet. Hence, receiving data from the cloud may not be the best approach for on-site consumers because it takes up additional time in sending and receiving data through the Internet. A robust communication model is essential for establishing a quick, reliable communication between the data producers and consumers in a smart parking system.

Figure 1 shows the possible layout of a standalone parking lot equipped with different types of IoT devices.

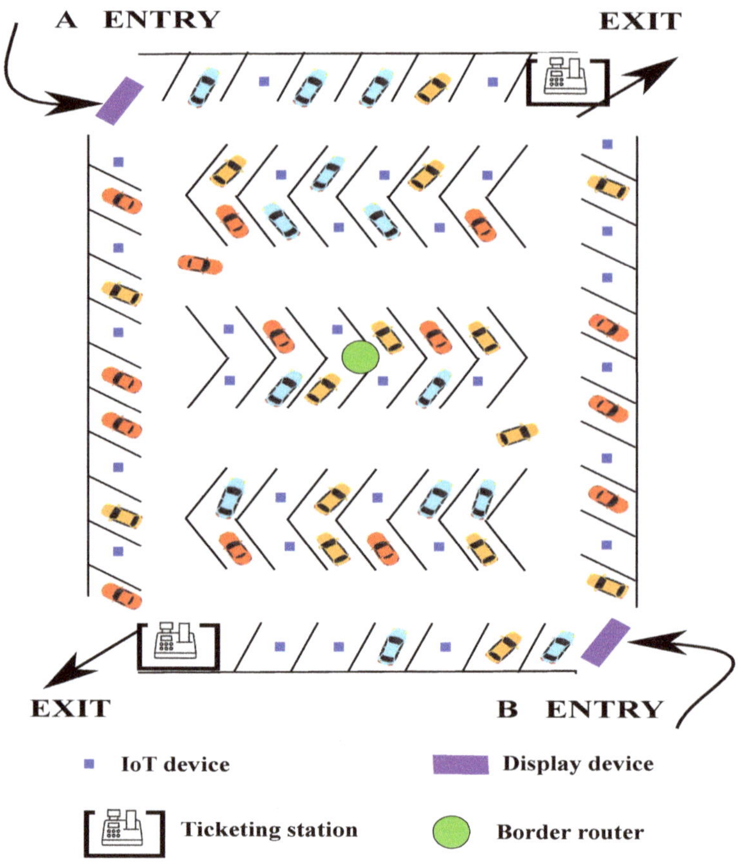

Figure 1. IoT infrastructure in a standalone parking lot.

IoT devices are installed at each parking space for gathering accurate data on availability, location and the duration of parking. These IoT devices are battery operated, simple to install and can last up to five or six years of operation without maintenance. The border router heads the mesh network and offers a global prefix for each device to equip them with a global IPv6 address. Parking availability data are locally consumed by the ticketing stations, which are present at the exit points and by the display screens, positioned at the entry points. As these devices depend only on the data collected from a standalone parking lot, they are labeled as on-site data consumers. Consumers that require global collated data from multiple parking lots are off-site data consumers, who access the same over the Internet or cloud.

A survey by J. J. Barriga et al. found that an SPS is predominantly implemented using Zigbee networks in 60% of the studies followed by 25% with IEEE 802.15.4 [5]. However, data collection networks are not the best candidates for M2M communication between IoT devices. RPL [6] is the IPv6 routing protocol for low-power lossy networks, whose directed cyclic graph (DAG) formation is best suited for networks incorporating local data consumers. In RPL's storage mode, heads of subtrees store the routes to nodes that are underneath them and hence provide fail-safe communication paths between thr various nodes in the network. This paper evaluates different types of communication models for an SPS with multiple local data consumers, using the RPL routing protocol, in an IEEE 802.15.4 mesh network.

The next section briefly identifies various data consumers that are present in a smart parking system and categories them as either on-site or over-the-Internet type. Section 3 summarizes the related research works, and Section 4 elaborates on the different aspects of efficiency for a communication model. The further section evaluates the models, discusses their relevance and converges on an efficient multi-collator communication model.

2. Data Consumers in an SPS

A smart parking system is a complete digital platform that manages city-wide parking resources in real-time and provides multiple services to end users [7,8]. Starting from searching for a nearby available parking slot, booking a parking slot in advance, navigating to an available parking lot, charging electric vehicles at the booked slot, and predicting the availability for a specific time until payment for parking or charging is possible with a SPS. An smart parking lot utilizes various sensors for the accurate identification of empty slots, parking boundaries and the automated counting of the number of entries and exits. Apart from these sensors, it may have other IoT devices such as overhead LEDs as indicators for vacancy, LCD displays showing layout/availability statistics, buzzers or alarms for indicating wrong parking, automated gates that open after payment verification or license plate identification. These IoT devices need data from the sensors for their intended operations and are data consumers in a smart parking system.

A cloud-based SPS collates availability data from the sensors and sends them to the cloud for storage and processing. The application(s) on cloud servers provide relevant decisions and inputs to the data consumers. In such a system, it is required that the data consumers are connected directly to the cloud. This may not be an economical solution, as a direct Internet connection is required for all the consumers. For instance, an overhead LED light, showing the occupancy of a specific parking lot, just needs input from the respective parking sensor. Such a local scope does not need a cloud SPS. A mesh with any-to-any traffic support is most suitable. Multi-hop mesh networks are a cost-effective way of connecting heterogeneous IoT devices and providing Internet connectivity to all nodes in a network. There is no gateway device involved in an IPv6 mesh as there are no protocol translations, and all the communication is IP based. In addition, local data consumers can be given instructions within the network by the border router itself in a scaled-down centralized approach, reducing the round-trip time taken by the data. Data consumers can be classified as on-site or over-the-Internet, depending on the scope of the data consumed by them.

2.1. On-Site Data Consumers

On-site data consumers are IoT devices that work with the data generated from devices that are in its proximity. Figure 2a–d depict some of the IoT devices that are employed in individual smart parking lots. Smart LED bulbs are used to provide visual clues for drivers in a closed parking system that has less day light visibility. These smart LEDs can be connected to other IoT devices through WiFi or BlueTooth technology. Similarly, smart alarms devices are also available and could be integrated to the central parking management system. The availability of systems on chips (SoC) supporting multiple radios in a single chip equips an IoT device to switch between different types of communication channels for device-to-device interaction. For instance, Qualcomm QCA4020 SoC provides intelligent multi-radio connectivity with WiFi, BlueTooth and 802.15.4 support [9].

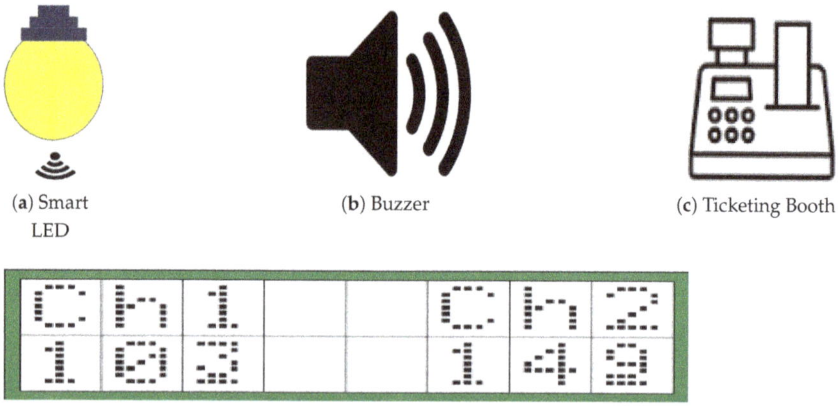

(**a**) Smart LED (**b**) Buzzer (**c**) Ticketing Booth

(**d**) LCD screen

Figure 2. Examples of on-site data consumers. (**a**) Smart bulb. (**b**) Buzzer for alarm. (**c**) Ticketing booth. (**d**) Display screen.

The ticketing booths could be a simple hand-held device with ticket or receipt-printing capability. They would need occupancy duration and timings for calculating relevant parking charges. They could also be a complex system, complete with an automated toll gate to allow passage for vehicles after verfication. Figure 3 shows a simple LCD screen display showing the aisle numbers and the respective numbers of lots that are vacant in them. Such a display screen, placed at the entrances of different levels, help users in a multi-level parking system. It could also be a complex system complete with a map in a very large parking lot.

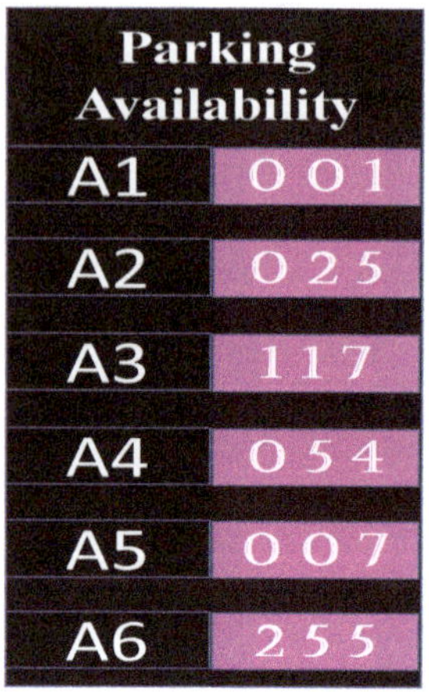

Figure 3. A display screen placed in a smart parking lot, showing parking availability.

2.2. Over-the-Internet Data Consumers

Off-site data consumers are remote devices that access the parking lot occupancy data combined with other systems such as maps or payments. Figure 4a,b show mobile applications for booking parking lots or viewing parking lots available in a particular area. In contrast, Figure 5 presents a web page that provides a passive view of the availability information from the parking sensors. These are good examples for remote data consumers that need Internet access to receive data from an SPS. In an IPv6 mesh, the BR advertises a global prefix, and hence, the nodes become accessible over the Internet. The IoT devices are capable of hosting a web page, and hence, it can supply the occupancy data to a web browser upon an HTTP request. The example for such a web page is as shown in Figure 5, where a laptop is connected to the BR through a Serial Line Internet Protocol (SLIP). SLIP allows IP datagrams to be encapsulated and exchanged over serial ports. The web page can be accessed through the IPv6 address of the data collator.

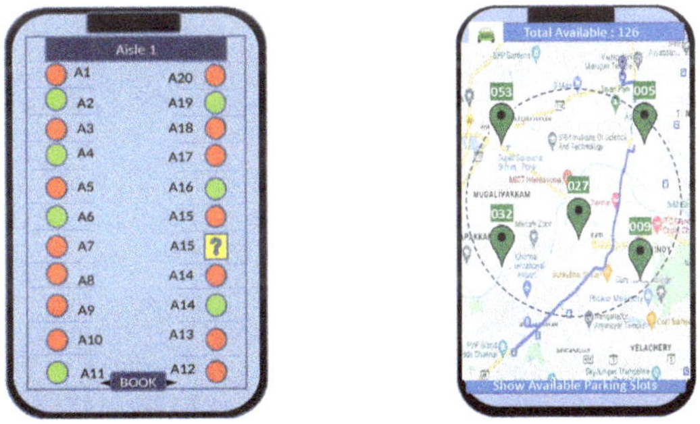

(**a**) Occupancy used for booking

(**b**) Occupancy used with map

Figure 4. Mobile application with multiple services. (**a**) Booking individual parking slot. (**b**) Searching for available parking slots in a map.

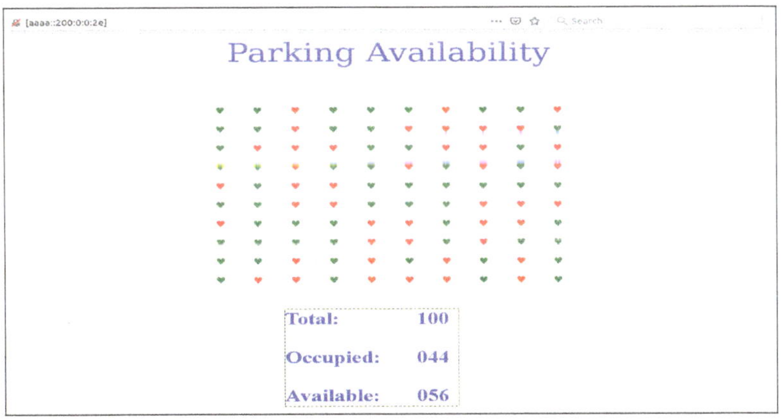

Figure 5. Over-the-Internet data consumer, a web browser showing parking occupancy.

Such direct BR to Internet connections work for a limited number of devices. However, if the number of simultaneous users increases, it is preferential to host the application in the cloud, as it provides elasticity in meeting user demand. In addition, web and mobile applications in the cloud have the privilege of collating data from multiple parking lots to show area-wide or city-wide car park availability. They can integrate the parking system data with other systems to offer seamless services.

3. Related Work in Literature

SPS systems are generally assisted or non-assisted where an assisted SPS allocates parking lots intelligently after considering various parameters such as the slot availability, user preference, closeness and traffic pattern of the route. This category strives to move forward toward autonomous vehicles and the navigation. A non-assisted SPS is a partially manual system where occupancy data are provided to end-users, and the actions are left to their decisions. Lin et al. classify SPS from another perspective based on the methods employed for information collection, the deployment technique of the system and their service dissemination model [8]. Paidi et al. show the gaps in deploying existing sensors and technologies for an open parking lot and the ways to design a robust multi-agent open parking lot [10]. Other works categorized parking systems based on the services offered by them such as parking reservation, guidance and crowdsourcing [11]. A survey by Fahim et al. identifies 12 different types of SPS systems depending on the technology used for sensing (Vision-based/GPS), communication (BlueTooth/WSN) or the learning models (ML/Fuzzy) employed [12]. In all these categories of SPS, a layered architecture is defined with the sensing layer at the bottom, the application layer at the top and a communication layer in between these two layers [13,14]. Al-Turjman et al. add one more layer for middleware to collate data from the sensors deployed at the parking lot [15].

The communication model of the available smart parking system perceives the sensing layer as a single unit where the sensors transmit the occupancy data to a central controller [16–18]. In rare cases, a parking guidance system considers a communication model with multiple wireless sensor systems for covering a single parking lot, and the unification of the data happens in the cloud-based application [19]. Another parking management system considers hierarchical occupancy data collation where sensor nodes communicate to group nodes and group nodes report to a central control node [20]. The parking systems do not measure the network performance of the sensing layer irrespective of its communication model being a flat network or clusters inside a single network or multiple networks. This paper studies different communication models for the sensing layer and proposes finding an efficient model for data collation and dissemination.

Access technologies inter-connect various subsystems, and the performance of the whole system is directly related to the performance of the communication layer. IoT systems have many options for access technologies depending on the required range of the wireless communication, the network topology, data-sharing models, open technologies versus proprietary and the availability of standardization. The most important support for IoT's access technology came in the form of a tiny IPv6 stack for the low power-constrained devices designed by the standard body Internet Engineering Task Force (IETF) [21]. 6LoWPAN and RPL are standard messaging protocols that could increase the utilization of open-source components in building a dependable information and communication technology for a sustainable smart city [22].

RPL is a mesh-routing protocol that supports IPv6 addresses for IoT devices and is used extensively in smart utility networks and smart grids. There are a huge number of studies to measure the performance of RPL for data-gathering applications based on a variety of parameters. Studies conclude that the combination of utilizing ETX as the link metric and a radio duty cycling mechanism to synchronously turn on and off radios empowers RPL in terms of lowest energy consumption [23,24]. Similarly, the network topology is found to have an impact on RPL network's energy consumption, and a circular topology is found to more effective than a grid or tree [25]. A study finds that compressed sensing and data aggregation in RPL reduces the data latency as well as cuts down packet loss [26]. In contrast, Pham et al. shows the need for a scheduling mechanism for delivering the aggregated data packet to reduce the latency and proposes a novel relative collision graph algorithm-based scheduler [27].

Lim, in a survey paper, categorizes multiple sinks as a viable method to reduce congestion and improve RPL's performance in an IoT network [28]. Many research works propose to increase network performance by defining more than one instance of RPL under a BR. Multi-sink approaches are proposed to handle high traffic volumes, offer safety against BR failure, combat congestion and balance traffic load across various forwarders [29–32]. A sink is a node that collates data; however, these works refer to the RPL root node as the sink.

The coordination between multiple sinks is proposed through a virtual root or through cooperative mechanisms between the different sinks [33–35]. Junior et al. argued that dynamism in invoking multiple instances is better than static multiple instances in handling different data traffic for multiple IoT applications [36]. Depending on the type of application, the node switches between stored instances to experience a reduction in control messages and power consumption. Hassani et al. show that combined metrics offer superior performance when compared to a single metric in a multi-sink scenario [37]. All such works incur modifications to RPL control messages, introduce new layers and increase the complexity. Furthermore, these works do not focus on exchanging data between multiple sinks, as they focus on a particular case of different sinks collating different type of data from the IoT network. Moreover, there is no need for the sink node to be the destination of data and any node in the network can act as a data server.

Tran et al. measure RPL's performance under different topologies such as linear, circular, random and grid. They conclude that the topology does not impact power consumption but influences latency [38]. The number of hops needed to reach the destination has an impact on the performance, as congestion is prevalent around the sink node. Hilmani et al. use a WSN for gathering occupancy data in the central gateway/sink node and apply a self-organizing algorithm for cluster formation [39]. In the clustering approach, there is no explicit insight on the exchange of data between clusters or the latency involved. Although there are a plethora of works in the literature to improve the performance of RPL [40], the simple effects of the position of root node or the usage of multiple servers to collate data are not studied.

4. Efficient Communication Model for an SPS with Multiple Data Consumers

IoT applications implemented with low-power personal area networks have a variety of requirements such as low power consumption, low latency, less traffic overload and high reliability [41]. In order to satisfy these requirements, an efficient communication model must:

- Provide reliable data collection in a large mesh network;
- Minimize the power consumption of the battery-operated IoT devices;
- Be quicker in collating and furnishing the data to consumer devices;
- Have a data format that compresses the volume of data.

A parking lot application that generates parking availability data has to forge effective communication paths between IoT devices that generate occupancy data, devices that collate the occupancy data and devices that consume occupancy data. The performance of a multi-hop network depends on the number of hops between the source and the destination. When data are transmitted through a minimal number of nodes, the latency and power consumption are optimal. On this basis, five different communication models are evaluated for implementing an SPS with multiple data consumers:

1. Border router with a single data collator at the perimeter of the parking lot;
2. Border router with a single data collator at the center of the parking lot;
3. Border router with multiple data collators distributed across the parking lot;
4. Border router with four data collators at the center of the parking lot;
5. Border router with four data collators at the center and each forwarder in the mesh aggregates occupancy data.

A border router (BR) facilitates connections between the mesh devices and Internet backbone. A BR aggregates routes to all mesh nodes and utilizes the same to connect them with hosts from other IP-based networks [6]. The wireless connection between all these entities forms the communication ecosystem of the SPS. In order to realize the goals of an efficient communication model, several aspects such as the position of border router and data collators, radio duty cycling, and data formats are considered in this work. Figure 6a–d represent a class of communication model where IoT devices simply forwards data toward the data collator. However, the position of the data collators vary among them. Except for the third model in Figure 6c, the BR and data collators are neighbors. The third model has split the entire network into four quadrants and has one data collator at the center of each quadrant. This places the data collator nested among the data producers. In contrast, Figure 6e shows a model where forwarders accept data from their children nodes, aggregate and then send out a single data packet with the consolidated occupancy data.

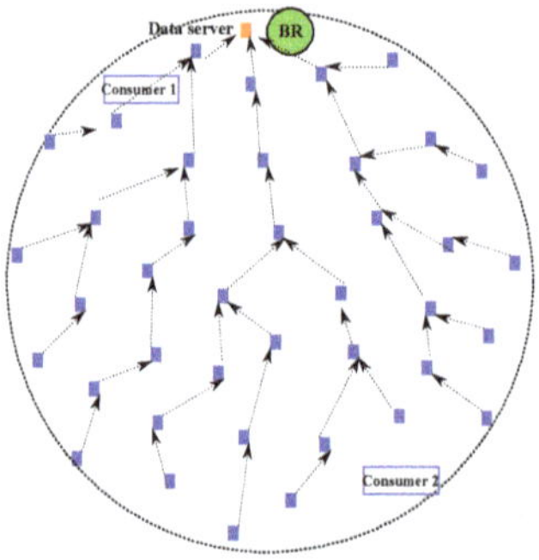

(**a**) BR at Top

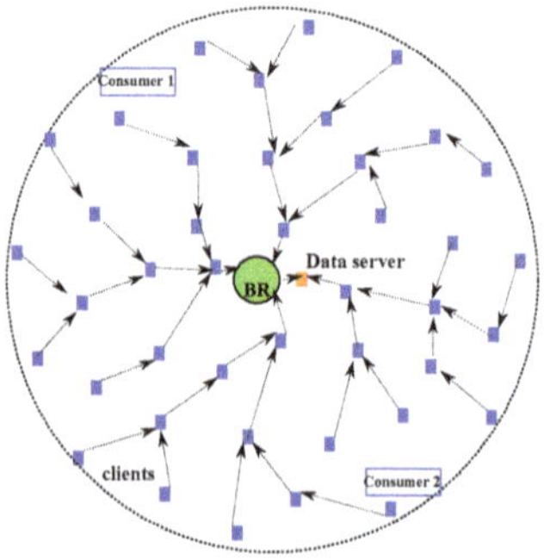

(**b**) BR at Middle

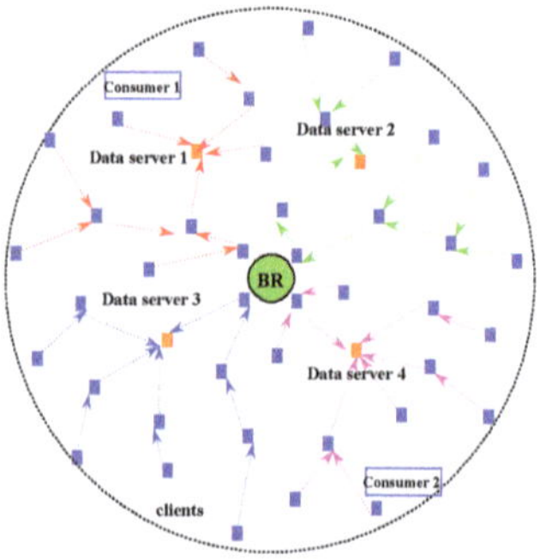

(**c**) Multiple Distributed Data Collators

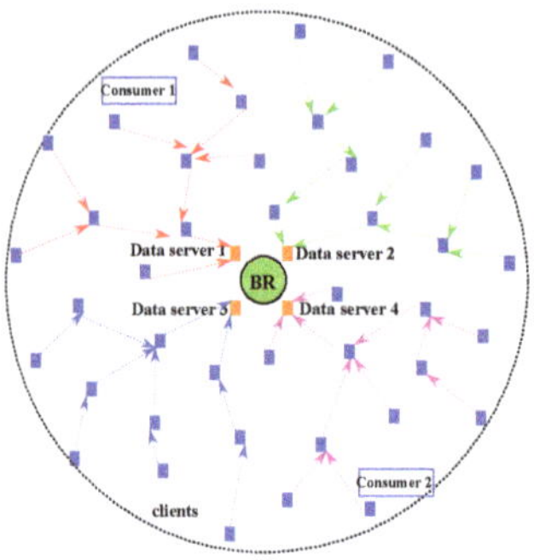

(**d**) Multiple Data Collators near BR

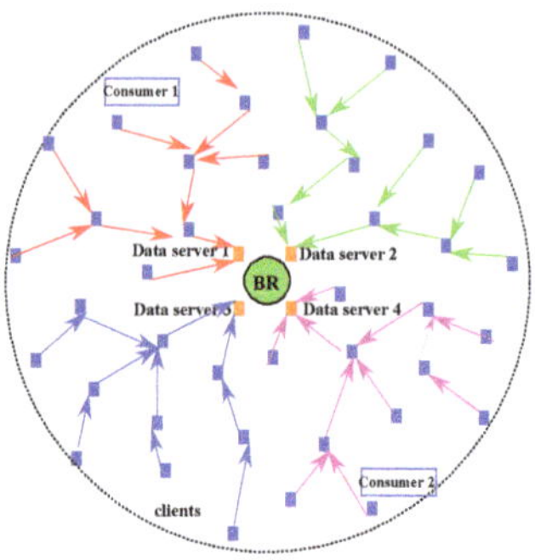

(**e**) Aggregate data at each hop

Figure 6. Flow of occupancy data (**a**) BR with one data collator at top. (**b**) BR with one data collator in middle (**c**) BR with four distributed data collators (**d**) BR with four data collators near BR. (**e**) Routers aggregate occupancy data at each hop.

The past research work on RPL's performance provides several pieces of vital information, including ETX for best link assessment, the significance of radio power consumption, duty cycling for reducing energy consumption, the relation between topology and performance and load balancing with multiple sinks. As multiple sinks mean more border routers, it involves high control overhead in maintaining more than one instance of RPL. Instead, this paper explores several alternate aspects such as multiple data collators, their positions, relative positions with data consumers, duty cycles of IoT devices, and data exchange format for arriving at a simple and efficient communication model.

4.1. Positioning BR for a Balanced Mesh Formation

BR is the root node in the mesh that initiates the mesh formation. RPL protocol forms a directed acyclic graph (DAG) that is destined to the BR by sending a DAG information object (DIO). DIO is an advertisement, and nodes hearing it join the DAG. It then furthers the transmission of DIO using trickle timers, and nodes join as in a ripple. Hence, nodes at the far end of the network perimeter takes more time to join the DAG. To have a uniform distribution of DIO along the perimeter of the network, it is necessary for the BR to be at the center of the network. This ensures that all nodes along the entire perimeter of the network have the smallest possible hops to reach the BR. With the number of hops directly proportional to the energy utilization and latency, positioning the BR at the center is the best approach. In addition, congestion around the BR node is quite low for a network having BR at the center when compared to a BR at the top (as in Figure 6a).

4.2. Positioning Data Collators for Reliable and Faster Data Collection

The BR itself can act as a data collecting point as RPL has a reliable DAG path to the root node. This introduces the funneling effect where forwarders close to the BR experience huge traffic. To reduce this effect, many researchers propose using multiple root nodes and collating the data outside the RPL network [42,43]. However, this deprives the on-site data consumers from directly accessing the occupancy data within the network and adds a dependency to the Internet connection besides increasing the delay in acquiring the data. The multi-sink approaches are complex with additional systems and modifications to the RPL control messages. As an alternate, multiple data servers are proposed in this work. It is essential that the data servers are stationed close to a BR so that a data server can reach another through the BR. This reduces the funneling effect and requires no complex improvisations to the RPL protocol. As the data servers are en route to BR, all the data-producing nodes already have an optimal path to reach the data collator. To illustrate this point, the third model has multiple data collators away from the BR.

4.3. Hybrid Power Cycling for Mesh Devices

Thread is an emerging routing protocol that is extensively used in smart home appliances [44]. The thread's communication model has mains-powered thread routers and duty-cycled sleepy end devices. Such a hybrid power cycling works for smart home mesh networks. However, the mains power is not suitable for mesh forwarders employed in a smart parking lot because a huge number of forwarders are required to cover the entire parking lot. However, the data collators are smaller in numbers and can be mains powered to maximize the packet reception rate of the data servers and also cater to high data volumes. They mains-powered devices do not switch off their radios. All other sensor nodes can have radio duty cycling to reduce energy consumption and switch off their radios most of the time except during packet transmissions. The BR, too, has its radio on so as to have a seamless connection to Internet. The positioning of BR makes it easier to extend the same to the data servers that are nearby. This hybrid power solution allows for lower energy consumption for the battery-powered nodes and ensures reliable occupancy data collection for the data servers.

4.4. Concise Data Format for Data Exchange

The occupancy data can be expressed in binary as they are two-state data, which are either occupied or not occupied. Hence, a single byte can represent eight parking slots and an 80-byte IP payload can effectively contain occupancy data for 640 parking slots. The occupancy data can be multiplexed at the data servers and is used for exchanging collated data between servers. The same is sent to data consumers. The concise string is then broken down back to occupancy data in the consumer node. The BR uses the compressed occupancy string in the HTTP data exchanged with the web browser. This short data exchange format reduces the load time of the web page.

The algorithm presented in Algorithm 1 takes the occupancy data array and converts the same to a single consolidated string. The index of the array is mapped to the position of the parking lot and is subsequently filled with either zero or one. The data collators fill the respective slots in the data array and convert the data to a concise string. This string is eight times compressed and can hold over 600 occupancy data in a single IP payload of IEEE 802.15.4 mesh.

Algorithm 1 Convert occupancy data to a concise string.

```
Input is ODA, Occupancy data array
Output is occupancy_str, Concised occupancy data as string
for i = 0 to no_of_parking_slots − 1 do
  for j = 0 to 7 do
    shifted_bit = ODA[i] << j
    combined_byte = combined_byte | shifted_bit
  end for
  byte_str = int_to_char(combined_byte)
  concatenate byte_str with occupancy_str
end for
```

The algorithm presented in Algorithm 2 takes the consolidated string and converts it back to occupancy data.

Algorithm 2 Convert the concise string back to occupancy data.

```
Input is occupancy_str, Concised occupancy data as array of characters
Output is ODA, Occupancy data array
bit_mask[] is {128, 64, 32, 16, 8, 4, 2, 1}
for i = 0 to no_of_parking_slots − 1 do
  int_value = char_to_int(occupancy_str[i])
  for j = 0 to 7 do
    occupancy_bit = bit_mask[j] & int_value
    if occupancy_bit > 0 then
      occupancy_bit = 1
    else
      occupancy_bit = 0
    end if
    ODA[i] = occupancy_bit
  end for
end for
```

All these decisions are expected to play a role in establishing efficient communication between all the concerned entities of the SPS system.

5. Evaluations

All the five different communication models referenced in the previous section are evaluated against each other for their efficiency in terms of data loss, packet latency, control overhead, energy consumption, time needed to obtain occupancy data for all the parking slots and the time taken for the occupancy data to reach the consumers. To this effect, an experimental study is carried out in a simulated IPv6 mesh network with 100 nodes and one BR. The Cooja simulator is a widely accepted simulator for conducting experimental studies of IEEE 802.15.4 based IoT networks [45].

5.1. Simulation Setup

In the experiments, the BR is the root node of the IPv6 mesh and creates a DODAG with one RPL instance. All the nodes are forwarders that are capable of forwarding data packets in the upward direction toward the root node (BR). The data collator is placed at a one-hop position from the BR so that it lies in the upward path en route to the BR for each node. The BR is connected through SLIP protocol to a laptop. The Firefox browser is used from the laptop to connect to the data collator to access the occupancy data over the Internet. The five networks to be examined are labeled as Top1, Mid1, Dist4, Mid4 and MidAgg as per their communication model. The model named Top1 denotes a network with a single BR and one data collator placed at the top of all the nodes. Mid1 refers to the network with a single BR and one data collator at the center of the network. The third model, Dist4, has the BR at the center, and its four data collators are distributed within the network and are away from the BR. In contrast, Mid4 refers to four data collators that are adjacent to the BR, at the middle of the network. The final MidAgg model denotes a network with four data collators in the center where each node aggregates the occupancy data. The final model is expected to consume less energy, as it reduces the total number of occupancy data packets transmitted in the network.

The grid network is considered for simulation, as the results are comparable across multiple studies. The channel check rate for a node is kept at 8HZ so as to reduce the power consumption of the nodes. A radio duty cycling ensures that the nodes remain in sleep mode for as long as possible. The data collators do not participate in radio duty cycling to ensure high reliability. The simulation parameters are summarized in Table 1.

Table 1. Configuration parameters for the simulation study.

Network Parameter	Value
Node placement	10 × 10 uniform grid
Radio medium	UDGM
Distance between nodes	30 m
TX Range/INT Range	50 m/100 m
IoT devices having parking sensors	96
IoT devices as on-site data consumers	2, top first node and bottom last node
Number of BR	1
BR position	Top or Center as per the models
Number of data collators	1 or 4 depending on the model
Data collators position	Top/Center of quadrants/Center
Mode of operation	Storing mode
Run Time	3600 s
Occupancy sensing interval	60 s
Radio duty cycling for parking sensors	ContikiMAC
RDC for data collators and BR	None
Channel check rate	8 HZ

After an initial delay of 120 s, each parking sensor node generates a data packet with occupancy data every 60 s. The data are either 0 or 1, depending on whether the respective slot is vacant or occupied. The data packet is addressed to the data server whose address is sought through service discovery. In case of multiple data collators, the address of the first discovered server is considered, since it would be the most nearest server. The data collators exchange data once every 60 s between them and also send the collated data to on-site consumers.

5.2. Simulation Results

The nodes record the number of occupancy data packets dispatched, the time of packet transmission, the number of packets received, the arrival time of the data packet, the number of different control messages transmitted for setting the mesh and the duration for its radio being active. The packet delivery ratio (PDR) is measured as the percentage of the number of occupancy data packets received at the collator(s) to that of the number of packets sent. A high PDR indicates reliable communication between the nodes and the data collator(s). The graph in Figure 7a shows the PDR of all the four communication models. As expected, it is 98.2% for the Top1 model, which has one BR and one data collector positioned at the top of all of the nodes. The model shows some initial packet loss for nodes with longer paths. The longer a data path, the more time it takes to stabilize. Dist4 also exhibits some packet loss, as nodes take relatively longer routes to data collators. The data packets have to travel upwards along the DAG to a common ancestor and then downwards toward the data collator. As the network becomes bigger, both Top1 and Dist4 would experience a further increase in the data path length. The PDR for all the other three models are almost the same and report negligible packet loss.

Figure 7b presents the average number of control packets transmitted by a node. RPL uses three different control packets, DAG information solicitation message (DIS) and DAG information object message (DIO) for forming upward routes, and (Destination advertisement object (DAO) for forming downward routes [6]. Nodes are required to send control packets in order to create and maintain the fail-safe routing paths. Less control overhead reflects the efficiency of the multi-hop mesh creation process and conserves energy in a network. The Mid4 model keeps the control overhead lowest among the models, which is closely followed by the Mid1 model.

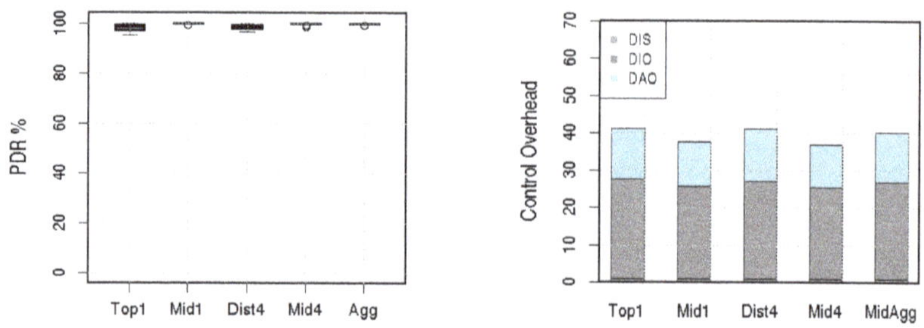

(**a**) Packet delivery ratio (**b**) Control packets for a node

Figure 7. Metrics for the communication models. (**a**) Data reliability in the network (**b**) Control overhead in the network.

The occupancy data packet latency is an average measure of the time duration for each data packet to reach the destination from its corresponding origin. Graph Figure 8a displays a 164.7 ms latency for MidAgg model and a comparable 471 ms and 420 ms for Mid1 and Mid4, respectively. The nodes in Dist4 model experience a latency of 823.2 ms even when there are multiple data collators. The higher latency reflects the longer routes along the DAG to the data collators. The packet latency is lowest in MidAgg because the occupancy data packets are not sent to the data collator but are sent to the immediate one-hop forwarder parent. Hence, the lowest average packet latency corresponds to one level of data aggregation. It must be noted that the occupancy data would take longer to reach the data collator as it has to cross multiple aggregation on its way.

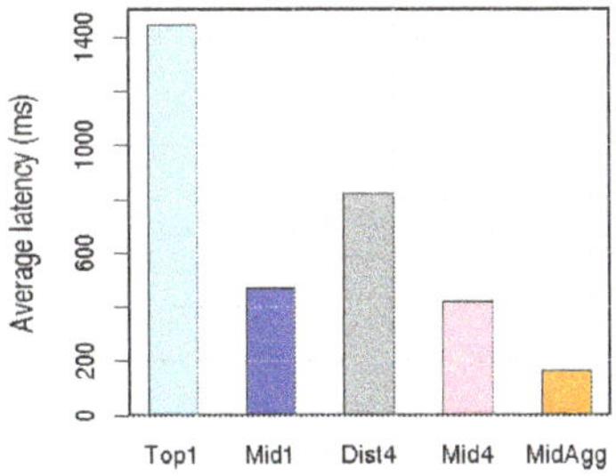

(a) Occpancy data latency

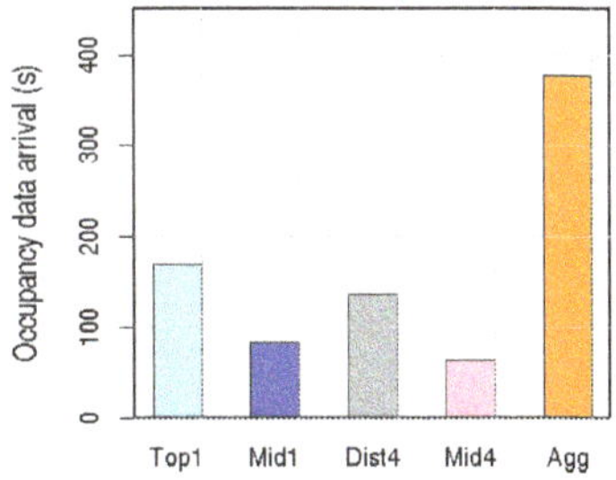

(b) Data arrival

Figure 8. Metrics for the communication models. (**a**) Latency for occupancy data packets. (**b**) Arrival of occupancy data from all nodes.

The next graph in Figure 8b shows the total time taken for the occupancy data from all the nodes in the network to reach the data collator. This is an important metric, as it shows the efficiency of the data collation in an SPS. The MidAgg model's under performance is because of the delay introduced by the aggregation at each hop. Both Mid1 amd Mid4 network's performances are lowest in the range of 83 s and 63 s, respectively.

The other metrics measured are the average packet latency for data packets between the data collator and the data consumer. Here, all the models have a similar delay under 1 s for one data consumer, but Top1 shows an elevated delay for one consumer, as shown in Figure 9a. The data consumers are placed at opposite sites of the network to simulate the presence of display screens at two far ends of a parking lot. So, when the data collator is at the top, it doubles the number of hops to reach a consumer at the far end. Dist4 exhibits a faster reach to consumers, since the distribution of data collators puts them closer to the consumer. This shows that the BR in middle is an efficient strategy to reach multiple consumers at the same time. Figure 9b visualizes the percentage of run time for which the radio was kept active. The first box shows the transmitter being active, and the second box shows the receiver active time. The transmitter is kept below 1% for models except the Top1 and Dist4, and receivers are kept active for less than 2%. Keeping the radios idle for longer helps conserve energy in the wireless network.

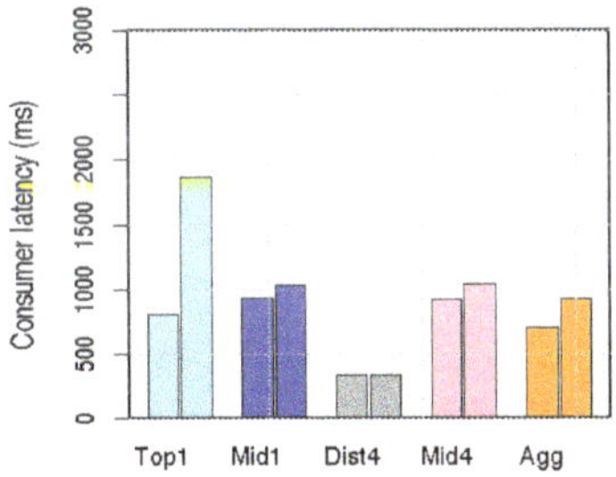

(a) Latency for local consumers

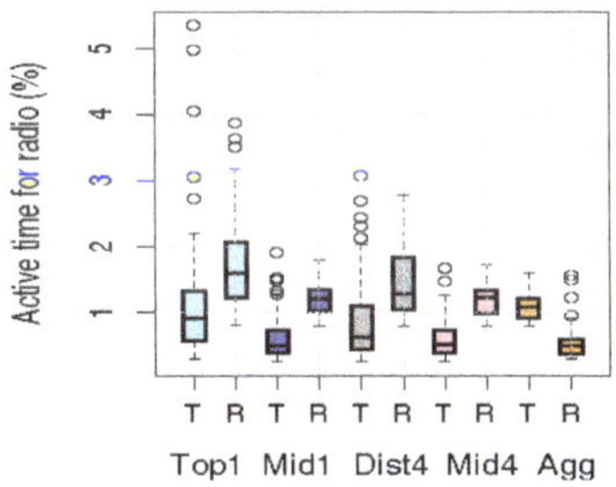

(b) Active time for radio

Figure 9. Metrics for the communication models. (**a**) Time to reach local consumers. (**b**) Percentage of time when radio was active.

In order to understand the energy utilization of nodes over time, the simulation is run for one hour with the energest metric report once every 5 min. The graph in Figure 10a showcases the average energy utilization of a node in different models. The initial spike is attributed to the network formation. There is a clear ranking in the energy consumption with Top1 being the highest with more packet transmission due to their longer distance to the BR. In the Dist4 model, energy utilization for a node is 22.7% higher than a node in the Mid1 or Mid4 models. Figure 10b displays the total charge consumed by a node in one hour. This also confirms the earlier findings and denotes that the Mid1 and Mid4 models outperform others in terms of efficient data collation and dissemination.

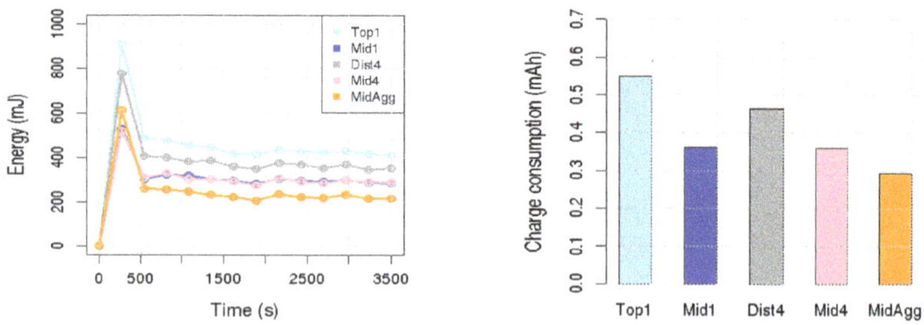

(**a**) Energy utilization along time (**b**) Battery charge consumption

Figure 10. Metrics for the communication models. (**a**) Average energy utilization of a single node. (**b**) Battery charge consumption for one hour.

6. Results Discussion

When comparing the performances of the five different communication models, it is evident that the Mid1 and Mid4 models are showing good packet delivery ratio along with low overhead. Although the MidAgg model exhibits a low packet latency along with the lowest power consumption, the time taken for the occupancy data to reach the data collator is six times over the time taken for the Mid4 model or five times over the time taken for the Mid1 model. It requires a specific scheduling algorithm for packet transmission that can reduce the delay introduced by data aggregation at each hop. The Top1 model apparently demonstrates a lower data reliability and a high packet latency as the data packets need to traverse a higher number of hops than the other models. Between these Mid1 and Mid4 models, the Mid1's packet delivery ratio has an edge over Mid4. However, latency wise, the Mid4 model holds an edge. To understand the advantage of these two models, the experiments are repeated in a larger 15 × 15 grid network.

Assessing the Scalability of Mid1 and Mid4 Models

The graph in Figure 11a presents the PDR for Mid1 and Mid4 models in a 10 × 10 grid network against the 15 × 15 grid network.

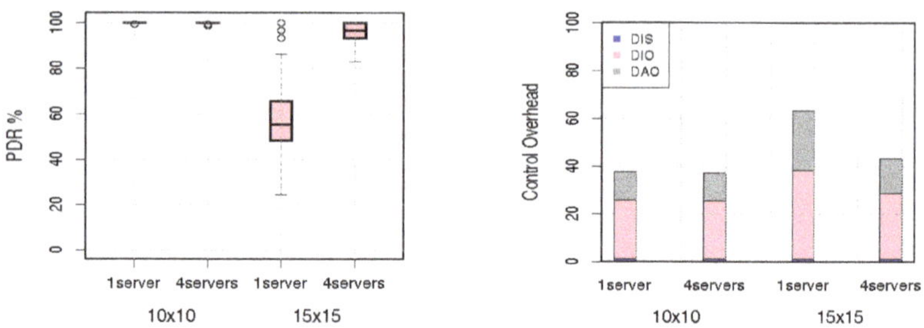

(**a**) Packet delivery ratio in a large network (**b**) Control packets for a node in a large network

Figure 11. Performance in a 10 × 10 grid vs. 15 × 15 grid. (**a**) Data reliability in the network. (**b**) Control overhead in the network.

In a larger network, the differences between the two models are evident. The Mid4 model outperforms and is 38.1% more efficient than the Mid1 model toward reliable data collation. Figure 11b shows a small rise in control packets for the Mid4 model in a scaled-up network. However, Mid1 suffers from a 40.9% increase in control overhead when compared to the same model in a smaller network.

The data latency metrics for the two models are presented in Figure 12. The average time taken by the occupancy data packet latency to reach the data collator is very high, clocking over 12 s for the Mid1 model. Mid4 takes about 2 s for reaching the data collators and shows a clear superior performance. As the number of nodes in the network increases, the congestion causes a severe funneling effect around the root node. Hence, the performance of the Mid1 model is very low in a large network.

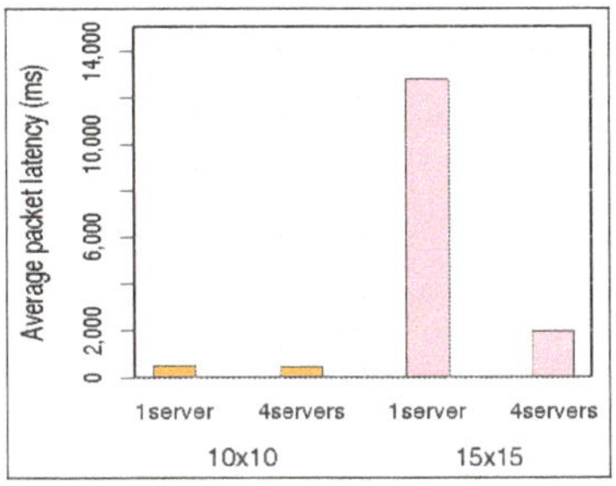

Figure 12. Occpancy data latency in a large network.

A similar trend is shown in Figure 13a for the time taken to reach the data consumers, and Mid4 outperforms the Mid1 model. The energy consumption is also lower for the Mid4 model, and the same is illustrated in Figure 13b. It can be concluded that a multi-data collator model with the BR at the center of the network fits the efficient communication model requirement for an SPS.

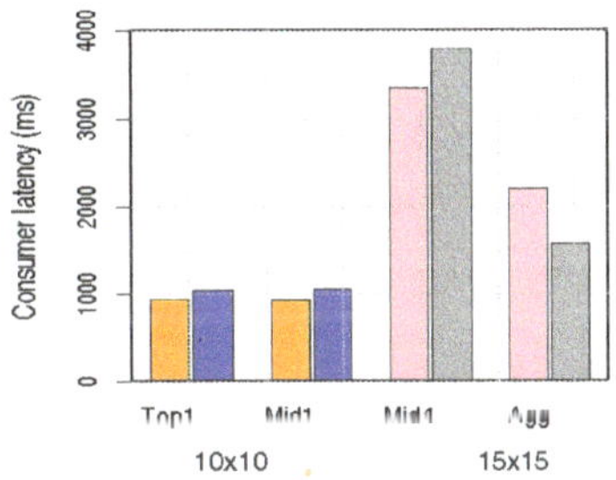

(**a**) Latency for local consumers in a large network

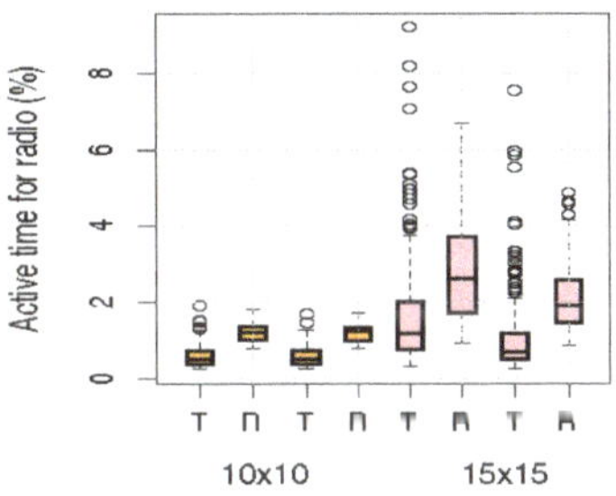

(**b**) Active time for radio in a large network

Figure 13. Performance in a 10×10 grid vs. 15×15 grid (**a**) Time to reach local consumers. (**b**) Percentage of time when radio was active.

7. Conclusions

The communication technology is a vital component of an SPS system, and it is necessary to have an effective communication model that provides reliable and faster occupancy data collation and dissemination between different entities. This paper explored various aspects such as the position of the BR in a mesh network, the presence of a single data collator against multiple data collators, their relative positions with respect to BR, consumers and the effects of hybrid radio duty cycling for mesh devices. It also proposed a concise data format that accommodates a large number of occupancy data (up to 640 parking slots) in a single data packet. This reduces the number of data packets exchanged between the data collators and data consumers. Lowering the radio activity directly improves the energy efficiency of the system. Along with that, the concise data format presents a short http message and improves the web page load time. Five different communication models are evaluated for their efficiency in providing low latency and energy efficient communication. The best two models were further subjected to a scalability test in a larger 15×15 grid network. A multiple data collator model where the data collators are adjacent to the BR and are positioned at the center of the network is identified as the best model for providing efficient communication between data producers and consumers. Having multiple data collators adjacent to BR reduces congestion around the BR in a large network and improves their reliability. Their position at a center point reduces the hop distance between the nodes and reduces latency. Congestion avoidance and shorter communication paths present an energy-efficient system. Thus, the strategic positioning of multiple data collators reduces data transit time, offers a higher data reliability and lowers the power consumption of the mesh devices.

Author Contributions: Conceptualization, T.A.; methodology, T.A. and M.P.; software, T.A.; validation, T.A. and M.P.; formal analysis, M.P. and T.A.; investigation, T.A.; resources, M.P.; data curation, T.A.; writing—original draft preparation, T.A.; writing—review and editing, M.P. and T.A.; visualization, T.A.; supervision, M.P.; project administration, M.P.; funding acquisition, Not Applicable. All authors have read and agreed to the published version of the manuscript.

Funding: This research received no external funding.

Data Availability Statement: The data are available with the corresponding author and can be provided on request.

Conflicts of Interest: The authors declare no conflict of interest.

References

1. Misra, P.; Vasan, A.; Krishnan, B.; Raghavan, V.; Sivasubramaniam, A. The Future of Smart Parking Systems with Parking 4.0: Creating Smarter Mobility Networks, White Paper, Tata Consultancy Solutions. *GetMobile Mob. Comput. Commun.* **2019**, *23*, 10–15. [CrossRef]
2. Global Smart Parking System Market Size & Share. Available online: https://www.prnewswire.com/news-releases/at-cagr-of-19-80-global-smart-parking-system-market-size--share--2022---2028--to-hit-usd-16346-57-million-mark-by-2028-industry-trends-value-analysis--forecast-report-by-zion-market-research-301543993.html (accessed on 10 August 2022).
3. Palattella, M.R.; Dohler, M.; Grieco, A.; Rizzo, G.; Torsner, J.; Engel, T.; Ladid, L. Internet of Things in the 5G Era: Enablers, Architecture, and Business Models. *IEEE J. Sel. Areas Commun.* **2016**, *34*, 510–527. [CrossRef]
4. Theoleyre, F.; Watteyne, T.; Bianchi, G.; Tuna, G.; Gungor, V.C.; Pang, A. Networking and communications for smart cities special issue editorial. *Comput. Commun.* **2015**, *58*, 1–3. [CrossRef]
5. Barriga, J.J.; Sulca, J.; León, J.L.; Ulloa, A.; Portero, D.; Andrade, R.; Yoo, S.G. Smart Parking: A Literature Review from the Technological Perspective. *Appl. Sci.* **2019**, *9*, 4569. [CrossRef]
6. Winter, T.; Thubert, P.; Brandt, A.; Hui, J.; Kelsey, R.; Levis, P.; Pister, K.; Struik, R.; Vasseur, J.P.; Alexander, R. *RFC 6550—RPL: IPv6 Routing Protocol for Low-Power and Lossy Networks*; Internet Engineering Task Force: Fremont, CA, USA, 2012. Available online: https://www.rfc-editor.org/rfc/pdfrfc/rfc6550.txt.pdf (accessed on 3 November 2016).
7. What Is a Smart Parking System? Functionalities and Benefits. Available online: https://tomorrow.city/a/smart-parking (accessed on 2 October 2022).
8. Lin, T.; Rivano, H.; Mouël, F.L. A survey of smart parking solutions. *IEEE Trans. Intell. Transp. Syst.* **2017**, *18*, 3229–3253. [CrossRef]
9. Qualcomm QCA4040—Multi-Mode Intelligent Connectivity Solution. Available online: https://www.qualcomm.com/products/technology/wi-fi/qca4020 (accessed on 2 September 2022).
10. Paidi, V.; Fleyeh, H.; Håkansson, J.; Nyberg, R.G. Smart parking sensors, technologies and applications for open parking lots: A review. *IET Intell. Transp. Syst.* **2018**, *12*, 735–741. [CrossRef]
11. Ogás, M.G.D.; Fabregat, R.; Aciar, S. Survey of Smart Parking Systems. *Appl. Sci.* **2020**, *10*, 3872. app10113872. [CrossRef]
12. Fahim, A.; Hasan, M.; Chowdhury, M.A. Smart parking systems: Comprehensive review based on various aspects. *Heliyon* **2021**, *7*, e07050. [CrossRef]
13. Ji, Z.; Ganchev, I.; O'Droma, M.; Zhao, L.; Zhang, X. A Cloud-Based Car Parking Middleware for IoT-Based Smart Cities: Design and Implementation. *Sensors* **2014**, *14*, 22372–22393. [CrossRef]
14. Biyik, C.; Allam, Z.; Pieri, G.; Moroni, D.; O'Fraifer, M.; O'Connell, E.; Olariu, S.; Khalid, M. Smart Parking Systems: Reviewing the Literature, Architecture and Ways Forward. *Smart Cities* **2021**, *4*, 623–642. [CrossRef]
15. Al-Turjman, F.; Malekloo, A. Smart parking in IoT enabled cities: A survey. *Sustain. Cities Soc.* **2019**, *49*, 101608. [CrossRef]
16. Yuan, C.; Fei, L.; Jianxin, C.; Wei, J. A smart parking system using WiFi and wireless sensor network. In Proceedings of the 2016 IEEE International Conference on Consumer Electronics-Taiwan (ICCE-TW), Nantou, Taiwan, 27–29 May 2016; pp. 1–2. [CrossRef]

17. Sahfutri, A.; Husni, N.L.; Nawawi, M.; Lutfi, I.; Silvia, A.; Prihatini, E. Smart Parking Using Wireless Sensor Network System. In Proceedings of the 2018 International Conference on Electrical Engineering and Computer Science (ICECOS), Pangkal, Indonesia, 2–4 October 2018; pp. 117–122. [CrossRef]
18. Islam, M.R.; Azam, S.; Shanmugam, B.; Karim, A.; El-Den, J.; DeBoer, F.; Jonkman, M.; Yadav, A. Smart Parking Management System to Reduce Congestion in Urban Area. In Proceedings of the 2020 2nd International Conference on Electrical, Control and Instrumentation Engineering (ICECIE), Kuala Lumpur, Malaysia, 28–28 November 2020; pp. 1–6. [CrossRef]
19. Cai, W.; Zhang, D.; Pan, Y. Implementation of smart Parking Guidance System based on parking lots sensors networks. In Proceedings of the 2015 IEEE 16th International Conference on Communication Technology (ICCT), Hangzhou, China, 18–20 October 2015; pp. 419–424. [CrossRef]
20. Chandra, H.; Michael; Hadisaputra, K.R.; Santoso, H.; Anggadjaja, E. Smart Parking Management System: An integration of RFID, ALPR, and WSN. In Proceedings of the 2017 IEEE 3rd International Conference on Engineering Technologies and Social Sciences (ICETSS), Bangkok, Thailand, 7–8 August 2017; pp. 1–6. [CrossRef]
21. Sobral, J.V.V.; Rodrigues, J.J.P.C.; Rabêlo, R.A.L.; Al-Muhtadi, J.; Korotaev, V. Routing Protocols for Low Power and Lossy Networks in Internet of Things Applications. *Sensors* **2019**, *19*, 2144. [CrossRef] [PubMed]
22. Tcholtchev, N.; Schieferdecker, I. Sustainable and Reliable Information and Communication Technology for Resilient Smart Cities. *Smart Cities* **2021**, *4*, 156–176. [CrossRef]
23. Barnawi, A.Y.; Mohsen, G.A.; Shahra, E.Q. Performance Analysis of RPL Protocol for Data Gathering Applications in Wireless Sensor Networks. In Proceedings of the 10th International Conference on Ambient Systems, Networks and Technologies (ANT), Leuven, Belgium, 29 April–2 May 2019. [CrossRef]
24. Hakeem, S.A.A.; Hady, A.A.; Kim, H.W. RPL Routing Protocol Performance in Smart Grid Applications Based Wireless Sensors: Experimental and Simulated Analysis. *Electronics* **2019**, *8*, 186. [CrossRef]
25. Sales, F.O.; Marante, Y.; Vieira, A.B.; Silva, E.F. Energy Consumption Evaluation of a Routing Protocol for Low-Power and Lossy Networks in Mesh Scenarios for Precision Agriculture. *Sensors* **2020**, *20*, 3814. [CrossRef]
26. Sennan, S.; Balasubramaniyam, S.; Luhach, A.K.; Ramasubbareddy, S.; Chilamkurti, N.; Nam, Y. Energy and Delay Aware Data Aggregation in Routing Protocol for Internet of Things. *Sensors* **2019**, *19*, 5486. [CrossRef]
27. Pham, V.; Nguyen, T.N.; Liu, B.; Thai, M.T.; Dumba, B.; Lin, T. Minimizing Latency for Data Aggregation in Wireless Sensor Networks: An Algorithm Approach. *ACM Trans. Sens. Netw.* **2022**, *18*, 1–21. [CrossRef]
28. Lim, C. A Survey on Congestion Control for RPL-Based Wireless Sensor Networks. *Sensors* **2019**, *19*, 2567. [CrossRef]
29. Ge, W.; Zheng, L.; Luo, P.; Liu, Z. Implementation of multiple border routers for 6LoWPAN with ContikiOS. In Proceedings of the 2015 International Conference on Information and Communications Technologies (ICT 2015), Xi'an, China, 24–26 April 2015; pp. 1–6. [CrossRef]
30. Carels, D.; Derdaele, N.; Poorter, E.D.; Vandenberghe, W.; Moerman, I.; Demeester, P. Support of multiple sinks via a virtual root for the RPL routing protocol. *EURASIP J. Wirel. Commun. Netw.* **2014**, *2014*, 91. [CrossRef]
31. Alreshoodi, M. Towards Effective Multisink Support in IPv6-based IoT Networks. *Int. J. Adv. Trends Comput. Sci. Eng.* **2020**, *9*, 2. [CrossRef]
32. Onwuegbuzie, I.U.; Razak, S.A.; Al-dhaqm, A. Multi-Sink Load-Balancing Mechanism for Wireless Sensor Networks. In Proceedings of the IEEE International Conference on Computing (ICOCO), Kuala Lumpur, Malaysia, 17–19 November 2021.
33. Zaatouri, I.; Alyaoui, N.; Guiloufi, A.B.; Kachouri, A. Performance Evaluation of RPL Objective Functions for multi-Sink. In Proceedings of the 18th International Conference on Sciences and Techniques of Automatic Control & Computer Engineering—STA'2017, Monastir, Tunisia, 21–23 December 2017.
34. Khelifi, N.; Nataf, E.; Oteafy, S.; Youssef, H. Rescue-Sink: Dynamic sink augmentation for RPL in the Internet of Things. *Trans. Emerg. Telecommun. Technol.* **2018**, *29*, e3278. [CrossRef]
35. Farooq, M.O.; Sreenan, C.J.; Brown, K.N.; Kunz, T. RPL-based routing protocols for multi-sink wireless sensor networks. In Proceedings of the 2015 IEEE 11th International Conference on Wireless and Mobile Computing, Networking and Communications (WiMob), Abu Dhabi, United Arab Emirates, 19–21 October 2015; pp. 452–459. [CrossRef]
36. Junior, S.; Riker, A.; Silvestre, B.; Moreira, W.; Oliveira-Jr, A.; Borges, V. DYNASTI—Dynamic Multiple RPL Instances for Multiple IoT Applications in Smart City. *Sensors* **2020**, *20*, 3130. [CrossRef] [PubMed]
37. Hassani, A.E.; Sahel, A.; Badri, A. Amalgamation of Novel Objective Function and Multi-sink Solution for a Reliable RPL in High Traffic Monitoring Applications. In *Digital Technologies and Applications, Lecture Notes in Networks and Systems*; Springer: Cham, Switzerland, 2021; ISBN 978-3-030-73881-5.
38. Tran, H.; Vo, M.T.; Mai, L. A Comparative Performance Study of RPL with Different Topologies and MAC Protocols. In Proceedings of the 2018 International Conference on Advanced Technologies for Communications (ATC), Ho Chi Minh City, Vietnam, 18–20 October 2018; pp. 242–247. [CrossRef]
39. Hilmani, A.; Maizate, A.; Hassouni, L. Designing and Managing a Smart Parking System Using Wireless Sensor Networks. *J. Sens. Actuator Netw.* **2018**, *7*, 24. [CrossRef]
40. Kim, H.-S.; Ko, J.; Culler, D.E.; Paek, J. Challenging the IPv6 Routing Protocol for Low-Power and Lossy Networks (RPL): A Survey. *IEEE Commun. Surv. Tutor.* **2017**, *19*, 2502–2525. [CrossRef]

41. Ghaleb, B.; Al-Dubai, A.Y.; Ekonomou, E.; Alsarhan, A.; Nasser, Y.; Mackenzie, L.M.; Boukerche, A. A Survey of Limitations and Enhancements of the IPv6 Routing Protocol for Low-Power and Lossy Networks: A Focus on Core Operations. *IEEE Commun. Surv. Tutor.* **2019**, *21*, 1607–1635. [CrossRef]
42. Foubert, B.; Montavont, J. Sharing is caring: A cooperation scheme for RPL network resilience and efficiency. In Proceedings of the IEEE Symposium on Computers and Communications (ISCC), Barcelona, Spain, 29 June–3 July 2019.
43. Khan, M.M.; Lodhi, M.A.; Rehman, A.; Hussain, F.B. A multi-sink coordination framework for low power and lossy networks. In Proceedings of the 2016 International Conference on Industrial Informatics and Computer Systems (CIICS), Sharjah, United Arab Emirates, 13–15 March 2016; pp. 1–5. [CrossRef]
44. Thread Mesh Networking Protocol, Thread Group, Inc. Thread 1.1.1 Specification, Feb. 2017. Available online: https://www.threadgroup.org/What-is-Thread/Overview (accessed on 20 October 2022).
45. The Cooja Network Simulator. Available online: https://github.com/contiki-ng/cooja (accessed on 15 January 2020).

electronics

MDPI

Article

A Platform for Analysing Huge Amounts of Data from Households, Photovoltaics, and Electrical Vehicles: From Data to Information

Antonio Cano-Ortega [1,*], Miguel A. García-Cumbreras [2], Francisco Sánchez-Sutil [1] and Jesús C. Hernández [1]

[1] Department of Electrical Engineering, Spain University of Jaen, 23071 Jaén, Spain
[2] Department of Computer Engineering, Spain University of Jaen, 23071 Jaén, Spain
* Correspondence: acano@ujaen.es; Tel.: +34-953-21-23-43

Abstract: Analytics is an essential procedure to acquire knowledge and support applications for determining electricity consumption in smart homes. Electricity variables measured by the smart meter (SM) produce a significant amount of data on consumers, making the data sets very sizable and the analytics complex. Data mining and emerging cloud computing technologies make collecting, processing, and analysing the so-called big data possible. The monitoring and visualization of information aid in personalizing applications that benefit both homeowners and researchers in analysing consumer profiles. This paper presents a smart meter for household (SMH) to obtain load profiles and a new platform that allows the innovative analysis of captured Internet of Things data from smart homes, photovoltaics, and electrical vehicles. We propose the use of cloud systems to enable data-based services and address the challenges of complexities and resource demands for online and offline data processing, storage, and classification analysis. The requirements and system design components are discussed.

Keywords: internet of things; data acquisition; cloud computing; big data analytics; load profile; smart meter

Citation: Cano-Ortega, A.; García-Cumbreras, M.A.; Sánchez-Sutil, F.; Hernández, J.C. A Platform for Analysing Huge Amounts of Data from Households, Photovoltaics, and Electrical Vehicles: From Data to Information. *Electronics* **2022**, *11*, 3991. https://doi.org/10.3390/electronics11233991

Academic Editor: Nikolay Hinov

Received: 17 October 2022
Accepted: 30 November 2022
Published: 1 December 2022

1. Introduction

Smart households are the future of new cities. The modernization of households involves the use of different Internet of Things (IoT) systems that allow monitoring and controlling the equipment installed in the households. These households generate a large amount of valuable data from the intelligent devices and appliances connected to an IoT system. The ability to use these data in real time makes it possible to analyse diverse information that has a significant impact on safety, the environment, and the economy of our society. Reports obtained from data in real time or stored over periods of time (days, weeks, months, and years) make it possible to study the behaviour of the household electricity demand.

Another consequence of this analysis is the adjustment of the term of power contracted with the supply company, which offers a significant reduction in the electricity bill. This leads to a more constant energy demand in the household. To achieve this feature, a priority system must be performed that in real time connects only those that do not exceed the contracted power limit, leaving on standby less priority equipment that would be connected when they are finished and have been assigned a higher priority.

To achieve these objectives, this study created a website with data from measurements obtained in different monitored households. In this sense, the main contributions provided in this paper are the following:

1. Develop a new prototype SMH for monitoring electrical variables, upload data to cloud using wireless network;
2. Design of a web platform that allows to analyse the different electrical magnitudes of the monitored houses through data that are measured by a smart meter (SM),

developed, and sent to the cloud. The design of the system permits for the massive processing of multiple sets of household data, which enables studying the information obtained by applying different algorithms;

3. The design components of the platform architecture were created. Specifically, the study of the data structure obtained from the households is developed, minimizing the communication overload with the cloud, and the design of the website where the data obtained appears for use by the research community;
4. The necessary technology to obtain data in real time as well as process and store data in the cloud for integrating into the control panels was developed; this shows user data in graphic form and allows downloading the data.

We designated a smart grid (SG) as a smart electricity distribution network. This is a two-way network capable of transmitting electricity in both directions, which allows households and different businesses to become small producers of electricity and not just consumers, as has been the case traditionally. Since SGs are combined with modern information technologies, they can provide data to both electricity distribution companies and consumers. One of the main tasks of SGs is the management and analysis of large amounts of generated information, known as big data.

The rest of the paper is organized as follows: Section 2 describes the related work; Section 3 shows the architecture of the system proposed; Sections 4 and 5 show the integration of the system and final screenshots, and finally our conclusions and future work are presented in Section 6.

2. Related Work

The study of information technologies (ICT) applied to Smart Cities and therefore, to smart grids and housing, is fundamental for the development of these paradigms.

IoT, cloud computing, and information analysis must use the optimal and highest speed ICT to achieve real-time data availability. In this respect Usman et al. [1] analysed the existing ICT adopted in SGs and their development over time. In this study, they analysed technologies such as Power Line Communication, Wireless Fidelity (Wi-Fi), Zigbee, Worldwide Interoperability for Microwave Access, Global System for Mobile General Packet Radio Service and DASH7.

2.1. Smart Grids and Meter Data

Numerous studies have integrated the use of SMs in SGs to monitor the behaviour of the different agents included within the network. Within this line of research, Munshi et al. [2] developed components based on big data for applications with SGs and the results obtained are transferred to a cloud computing platform. Kabalci [3] studied communication technologies and their security in data collection networks. Tanyali et al. [4] implemented a method for encrypting data taken by SMs since there is a risk of information theft when they are exposed on the network as well as finding out the user's habits. Khan et al. [5] dealt with the problem of integrating IoT sensors installed in households that have different communication protocols and technologies. To do this, they defined a standard in which all the sensors installed in the household were integrated (biometrics, security, electrical, etc.).

As for the cloud computing standards for SGs, Yigit et al. [6] defined the necessary architecture for cloud computing in SGs by using structures, methods, protocols, and algorithms. Al-Turjman et al. [7] reviewed the state-of-the-art in SGs, analysing the viability of using SGs to study and improve the quality and reliability of power. Al-Turjman et al. [8] studied the impact and efficiency of SMs, critical design factors, modifying and comparing parameters with real cases, and are categorized within this area. W. L. Rodrigues et al. [9] carried out a study on signal quality using a SM by means of cloud computing (fog computing). Recently, several studies have proposed systems and frameworks for the analysis of IoT data using various architectures related to cloud computing.

In this section, we discuss these studies, especially those that are representative of the state-of-the-art and similar to our work. It is important to note that a platform with cloud computing offers efficient resource processing of large IoT data in real time while providing information and processed data to the cloud for further processing and analysis. This integration design makes it easier for us to address cloud system latency issues that can have a significant impact on time-sensitive applications. The integration of SMs into IoT networks is another important part of the research related to SGs. Cano-Ortega et al. [10] developed equipment for power factor compensation using a TLBO optimization algorithm through a cloud data storage, control, and monitoring system. Cano-Ortega et al. [11] monitored the efficiency and the operating conditions of induction motors through an SM based on a LoRa LPWAN network. Sánchez-Sutil et al. [12] designed a measurement and control system for public lighting integrated in a LoRa LPWAN communication network. Asghari et al. [13] performed the current research techniques on IoT application approaches to analytically and statistically categorize this type of network. A. A. Mazhar Rathore, et al. [14] developed a combined system based on IoT for the development of Smart Cities using big data analysis. They used a complete system with several types of sensor deployment to make an SG. Naik et al. [15] designed an intelligent home management system based on IoT that uses sensors, actuators, smart phones, web services, and microcontrollers. This IoT platform and hardware are available through a mobile application. Pau et al. [16] made an intelligent metering infrastructure to automate and manage the distribution networks. The proposed architecture was based on a cloud solution, which allows communication with SMs and provides the necessary interfaces for the distribution of network services. Sánchez-Sutil et al. [17] developed and calibrated a low-cost SM to measure the electrical variables in homes supplied by photovoltaic solar energy. Moreover, in Ref. [18] the authors developed a smart plug to monitor and control electric load in a household with LoRaWAN network. Different web portals were analysed considering the electrical consumption measurements of households in different countries. As can be seen in Table 1, the measurements of the electrical variables have granularity varying between 1 s and 10 min; later, aggregations were made that can be used in different time horizons. Almost no time series below 1 s was used due to the large amount of data produced for each variable measured.

The websites shown in Table 1 store the recorded data and do not work in real time, the measurements of the electrical variables have a granularity that varies between 1 s and 10 min; the granularity times of the websites are less than 1 s [19–21], 1 s [19,20,22–27], between 1s and 1 min [19,22,28,29], and 10 min. Some allow downloading of the stored data for free and others for a fee. The websites do not display the data of all the monitored variables in real time. Only [22] can display data from the previous day, but this is paid. They do not allow comparisons between different households.

The SMs used in the websites are commercial devices where the measured data are sent every 1 s [20,22–25,27,28], every 1 min [29,30], and every 10 min. The websites do not indicate the costs associated with commercial SMs.

Table 1. Open access datasets of household power.

Web Site	Electrical Variables	Time Resolution	Number Houses	Country
PECANSTREET [22]	v, i, p, q	1 s–1 min	1115	USA
ACS-F1 and ACS-F2 [28]	v, i, p, q, f, PF	10 s (0.1 Hz)	225	Switzerland
AMPds [29]	v, i, f, pf, p, q, s, e	1 min	1	CANADA
BLUED [19]	v, i	8.33×10^{-5} s (12 kHz)	1	USA
DRED [20]	p	1 s (1 Hz)	1	Netherlands
ECO [23]	v, i, p, q, PF	1 s (1 Hz)	6	SUIZA
GREEND [24]	p	1 s (1 Hz)		Austria and Italy
ERC	p	10 min	255	UK
iAWE [25]	v, i, f, p, q, s, e, PF	1 s	1	India
REDD [26]	v, p	6.66×10^{-5} s	2	USA
REFIT [31]	p	8 s	20	UK
Smart [30]	v, f, p, s	1 min	400	USA
Tracebase [27]	p	1 s	15	Germany
UK-DALE [21]	v, i, p, s	6.25×10^{-5} s, 1 s, 6 s	3	UK

In the table, v is the voltage, i is the current, PF is the power factor, p is the active power, q is the reactive power, s is the apparent power, e is the energy and f is the frequency.

2.2. Meter Data Analytics

Data mining is the extraction of implicit information from other data. It can also be defined as the exploration and analysis, by automatic or semi-automatic means, of large amounts of data in order to discover meaningful patterns. Data mining techniques can be of two types: (i) descriptive, looking for interpretable patterns to describe data; (ii) predictive, using variables to predict future or unknown values of other variables.

The literature related to data mining, SGs, and SMs is varied. In this sense, Lui et al. [32], developed a big data system for data acquisition, processing, and analysis to create a database.

Other authors have studied big data applied to SGs [33] and performed a literature review on big data applied to electrical systems, defining the characteristics and future challenges. In addition, they analysed the characteristics of the SMs integrated into big data systems. Wilcox et al. [34], implemented a big data hardware/software system for the data analysis of household information stored in the cloud and with access to the data through a web portal. Yassine et al. [35] developed a big data analysis system based on an IoT network for measuring electricity consumption in households. Diamantoilakis et al. [36] applied big data-based methods for the real-time processing of data obtained by SMs. Tu et al. [37] proposed standards to be met by future big data systems applied to SGs.

In other investigations, the load profiles of households were analysed by applying big data-based techniques. Shyam et al. [38] studied data management techniques in the generation, transmission, distribution, and consumption of electrical energy. Saleh et al. [39] used measured data to obtain load predictions by applying filters for analysis. Guerrero et al. [40] developed a data mining algorithm to obtain an integrated database that reflects the consumption and load profiles of a household. Cano-Ortega et al. [41] developed a system for measuring electrical quantities to determine the load profiles of dwellings with a LoRa wireless network using an ABC optimization algorithm.

2.3. Big Data Architecture and Cloud Computing

Numerous studies have been carried out on SMs and big data. Lui et al. [32] applied a new development to ICT that allows reducing the data measured by SMs by utilizing analytical techniques. They developed a web portal and a scalable platform to process the measured data. Munshi et al. [2] implemented a platform of 6000 SMs with different data visualization and cloud computing scenarios. Yildiz et al. [42] performed methods for forecasting, clustering, classifying, and estimating the demand for electricity in households to optimize energy consumption. The paper by Funde et al. [43] was based on the unique combination of the symbolic aggregate approach (SAX), the discovery of temporal motifs, and the association mining rules to detect expected and unexpected patterns. The experimental data set obtained from the installed SMs supports the model developed in this research. Andreadou et al. [44] analysed parameters such as size, message transmission frequency, total transmission time, and buffer capacity and showed their effect on data obtained from medium voltage networks.

Meloni et al. [45] developed an architectural solution based on the Cloud-IoT for state estimation in SGs by combining cloud computing and the latest computer developments together with virtualization techniques for data processing. Razavi et al. [46] trained and developed genetic algorithms to predict the occupation status of households not only in the present but also in the future with a high degree of accuracy. Sial et al. [47] used heuristic techniques applied to data obtained with SMs to predict abnormal power consumption in campus residential buildings.

Araujo et al. [48] evaluated the performance of cloud storage systems. Yassine et al. [4] developed a platform for acquiring data from smart households using fog nodes and cloud computing to obtain the processing, analysis, and storage of the data measured. Forcan et al. [49] developed two communication models, cloud computing and fog nodes, to be used for estimating electricity losses in SGs and monitoring the voltage profile of a simulated IEEE system.

2.4. Big Data Applied to Smart Grids

Munshi et al. [2] presented innovative research for advancing smart grids through big data. They implemented a secure cloud-based platform. Tu et al. [37] conducted a state-of-the-art review of big data applied to smart grid integration. They reviewed big data applications for smart grids, focusing on the latest applications with the latest big data technologies. Kumar et al. [50] designed a circuit to help users take control of power consumption in their homes, improving energy savings through an intelligent method. The measured information from the monitored homes is stored in a big data server. Wang [51] proposed a localization oscillation scheme based on the theory and support of a vector machine, phase difference oscillation, and forced phase difference oscillation. Zang [52] improved the data analysis and data mining tool in energy control and improved the service quality of the electricity market through the computerization of power systems. Mostafa et al. [53] developed a framework for implementing big data analytics for smart grids and renewable energy, and implemented a five-step method to predict the stability of smart grids using five different machine learning methods.

2.5. Novel Contribution

Due to the existing limitations of the website that stores the data on the variables of electrical consumption homes, this research realizes a platform that allows to visualize the different electrical variables v, i, p, q, s and PF in real time with data upload every 0.5 s from the SMs installed in the different homes. In addition, each user has an app that allows to visualize the data in real time. Since the measured time series are below 1s, a large amount of data is produced for each of the measured variables, so they must be treated by NoSQL data management system and structured storage, which reduces the processing time. Included in the platform is the comparison between different households, which allows to analyse all the households that have been monitored since 2018.

The design of the measurement equipment and data upload to cloud are other advantages this research provides/achieved. The data uploads to the cloud every 5 s, and the data storage in both the cloud and measurement equipment have a data limit of one year. The SM was designed with a cost of EUR 46.28, considered as a low-cost equipment. Being an open-source equipment, it can be programmed according to the needs at any time and for each user.

After analysing the different existing platforms, this work adds the following innovations.

- Integration and storage of data from multiple sources. Traditional data usually deals with data from a single domain, it is essential to find a fusion method for the data set from multiple sources, which has different modalities, formats, and representations. In terms of big data storage, although some of the systems such as Hadoop (HDFS) seem to be feasible, it needs to be adapted and modified to fit the big data power grid;
- Real-time data processing technology for applications such as electricity consumption measurement with resolutions below 0.5 s demand estimation studies, residential occupancy, etc. Although the cloud system is able to provide a fast calculation service, the network congestion, the complicated algorithm, combined with the massive amount of data, results in latency;
- Big data visualization technology. Graph and chart visualization can present operators with granular and explicit changes in electrical variables. However, how to effectively find and represent correlations or trends among data from multiple sources is a major challenge;
- Data privacy and security. Data security is provided by 64-bit key encryption. Each user has their own account which makes them independent from others. On the other hand, a user with administrator role manages the system.

3. System Architecture

The SMH consists of a first data acquisition module that automatically collects data on electrical variables (voltage, current, power factor, active power, and reactive power) every half second. This information is dumped directly into a data storage system, allowing processing without performance problems, and can be performed by several devices at once. Subsequently, this large amount of data is processed asynchronously for analysis and selection. Interesting data for the different control panels (dashboards) are then loaded in a web portal so that any user can study, analyse, or even download them. Each of these system modules is detailed below. Figure 1 shows the system architecture.

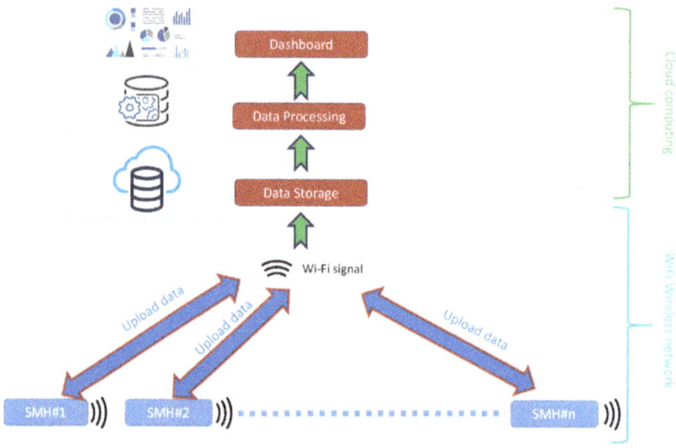

Figure 1. System architecture.

3.1. Data Acquisition

The flow chart in Figure 2 shows the data acquisition performed with the SMH by the Arduino Nano Rev3 (ANR3) microcontroller. The first task is the initialization of the system. Then the continuous measurement of the fundamental electrical variables (v, i) is performed using the analogue inputs A0 (v), A4, and A5 (i). Once the variables are processed, they are sent through the serial port to the Arduino Wemos D1 mini (WD1m) microcontroller and a backup copy is made to the SD memory card.

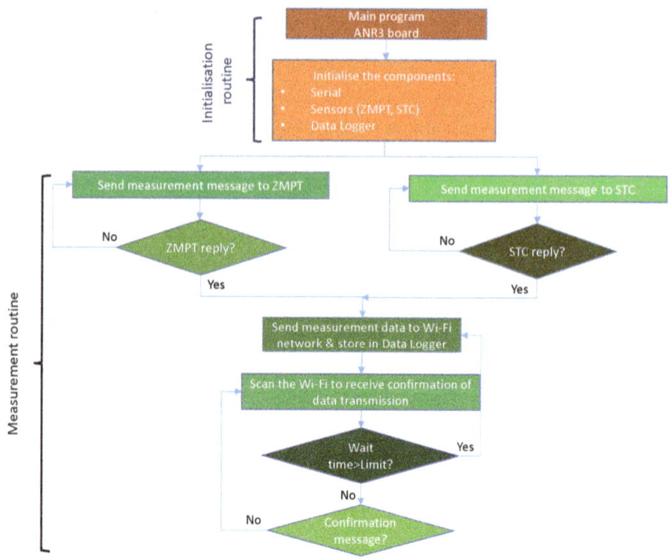

Figure 2. Flowchart for the measurement and computation of electric variable: ANR3 board.

Each of the tasks to be performed requires a processing time: (i) 10 cycle measurement of the input signals (200 ms); (ii) obtaining the fundamental and derived variables (30 ms); (iii) sending data to ANR3 (1 ms); (iv) storing data on the SD card (9 ms); and (v) waiting time until the next measurement (10 ms). The chosen sampling frequency is 1 kHz (1 ms). Since the measurement time is 200 ms, 200 samples are obtained in ten cycles of the measured signal. The timeline for the measurement process is shown in Figure 3. In the first part of the process, ANR3 obtains the fundamental and derived electrical variables. Then, they are sent to WD1m and finally, the data are saved on the card. In parallel, WD1m receives the data from ANR3 and uploads them to the cloud using the Wi-Fi connection.

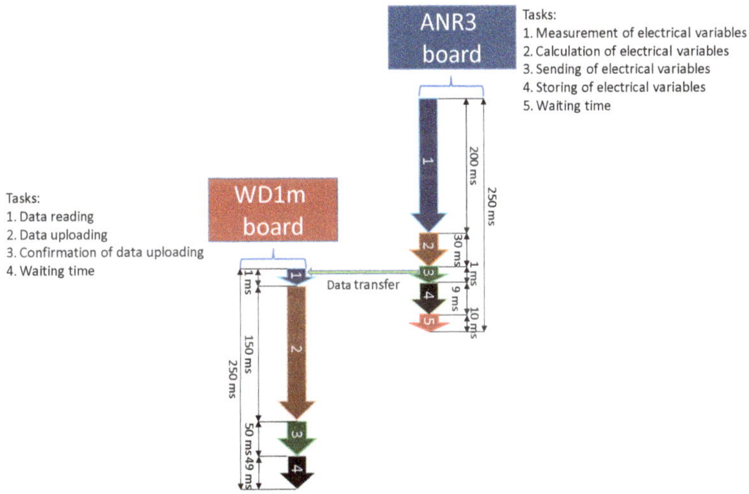

Figure 3. Process timeline for the SM.

Due to the data acquisition time (0.5 s), rise times of 0.25 s are required. The free version of Firebase offers storage times that meet the above requirements, with storage times of 0.1 s. Figure 4 shows the flow chart of data uploading to the cloud. The process is performed continuously by WD1m with the following steps: (i) system initialization; (ii) reading of the serial port; (iii) uploading to the cloud; (iv) and confirmation of the data upload.

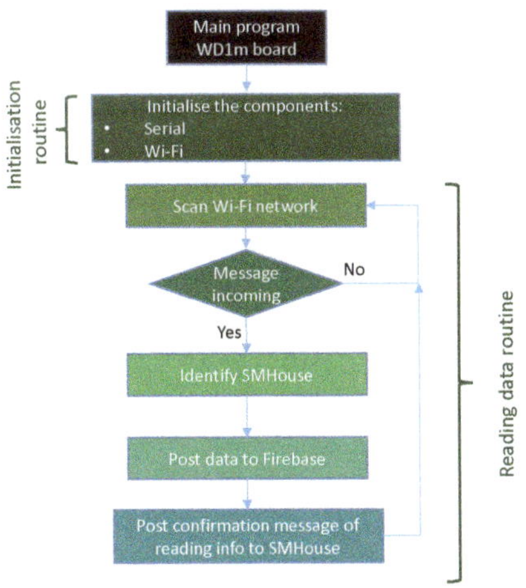

Figure 4. Flowchart for cloud data uploading: WD1m board.

The timeline for the process developed by WD1m is shown in Figure 3. The execution times are: (i) serial port data reading (1 ms); (ii) data upload to the cloud (150 ms); (iii) data upload confirmation (50 ms); (iv) and timeout (49 ms).

3.2. Data Processing

Huge amounts of data are generated every millisecond from thousands upon thousands of connected devices. This data, which constantly appears in the cloud, contains potentially great business value. For this reason, we need to perform effective data processing. In our system, once the data are obtained and stored in a "Not Only SQL" (NoSQL) database, it must be analysed and processed to obtain the maximum information possible. NoSQL is to talk about structures that allow us to store information in situations where relational databases generate certain problems. These issues are mainly due to the scalability and performance problems of relational databases, where thousands of concurrent users log on and millions of queries per day occur.

At this point, and when dealing with an information system that stores a huge amount of data, one of the first doubts is whether the data and information are all the same? Let us start by defining data as a symbolic representation of some situation or event, without any semantic sense, describing a concrete fact. However, we define information as a set of data that are properly processed so that they can provide a message that contributes to decision making when solving a problem or facing a situation in which any kind of decision making is required.

The main objective of this module is to move from that large amount of data obtained from the sensors of the data acquisition module to information that the end user can understand and process and that is oriented to decision making, among other things. The presented system automatically extracts and from time to time configures the data stored in the NoSQL system; it processes, filters, and loads them into the data system of the visualization platform described below.

3.3. Dashboard Design

The last module of the system, which serves the end user in visualizing aggregated information in various ways, is a set of dashboards with different purposes.

Dashboarding is a dynamic and purpose-based visual interface needed to display one-to-many database linkages so that the information can be presented for a single time period or dynamically monitored over time. This allows a user to quickly define focus areas of interest for their analysis.

Two dashboards were designed: a general dashboard and a device comparison dashboard. The purpose of the general dashboard is to analyse and display information about the load profiles in each household electrical vehicle (EV) consumption. This information can be used in studies related to consumption habits, load forecasts, and demand estimates. The purpose of the device comparison dashboard is to determine consumption patterns in different inhabited areas, analyse the different loads used, and offer studies for the development and planning of electrical networks.

4. System Integration

The technologies used in each module of the system and their integration are described below.

4.1. Automatic Data Collection

The SMH used is based on the Arduino ANR3 and WD1m boards with local and cloud storage. The data capture is performed with voltage sensors v and current i, which send data through the analogue inputs of ANR3. Subsequently, the active power variables p, reactive q, and power factor PF are calculated. The data are then uploaded by WD1m to the cloud. Figure 5 shows the SMH block diagram.

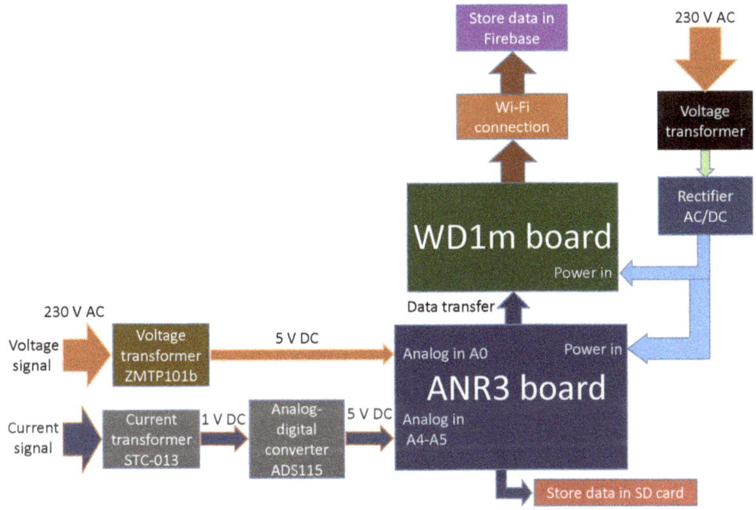

Figure 5. Hardware block diagram of the SMH.

The SMH is powered from the mains using a dual 12 V output transformer. The supply signal is rectified to 12 V DC, which is within the range of the supply voltages (range 7–12 V DC) supported by ANR3 and WD1m. The voltage sensor ZMTP101b is connected to the 230 V mains and has an output of 5 V DC, which is supported by the analogue input of ANR3. The current sensor used, STC-013, has a measuring range of 30 A, which translates to 1 V DC. Since the analogue input of Arduino is 5 V, the voltage is increased from 1 V to 5 V DC by means of the digital/analogue converter ADS1115, the voltage is increased from 1 V to 5 V DC.

Microcontroller: ANR3 is developed on the basis of the ATmega328P microcontroller as an open-source platform for prototype development. It has a clock speed of 16 MHz, which allows measurements in very short time intervals (0.25 s).

Wireless communication WD1m uses the ESP8266 platform for Wi-Fi communication access, which works with the 802.11 b/g/n wireless system standard. This board ensures upload times to the cloud of less than 0.15 s, being less than the expected 0.25 s.

Current sensor: The current sensor (STC-013) used in this research is of the non-invasive type. This means that it does not modify the monitored electrical installation. The 30 A version was chosen, which can be used in households up to 6600 W. For higher currents, there is a 100 A version with a maximum power of 23,000 W. To increase the 1 V DC output of the current sensor to the 5 V analogue input of ANR3, an ADS1115 digital/analogue converter is used.

Voltage sensor: The voltage sensor used is the ZMPT101b voltage transformer. It switches from 230 V AC from the mains to 5 V DC from the analogue input of ANR3, directly obtaining the reading without the need for any intermediate equipment.

Datalogger shield: Due to the loss of data through the Internet connection, an SD card mounted on a datalogger shield is used as a backup, ensuring storage without loss of data. In this case, an 8 GB SD card was used, which allows data storage for 3.76 years.

Storage: From time to time, configured in the Arduino device, the data are sent to the central data system. This massive data storage system is Firebase, a platform for the development of web and mobile applications created by Google in 2014.

Specifically, the Firebase Realtime Database is a NoSQL database hosted in the cloud that provides great performance for this type of connection. For the volume of data produced by these models, SQL-based data queries are not efficient.

The PCB board for the SMH was designed and built. The board allows the connection of all the components used, which are soldered and therefore a solid and resistant final system. The dimensions of the PCB board are 88 mm × 75 mm. Figure 6 shows the design of the PCB board front side and assembled with components of the SMH.

Figure 6. Printed circuit board (PCB) of the SMH with battery power supply: (a) front side: (b) back side and (c) assembled with real components.

In order to know whether the SMH is capable of mass production, it is necessary to make an economic assessment of the cost of the materials used. The budget also serves to validate the low-cost objective proposed earlier. The prices shown are from the manufacturers' official shops. As these are open-source components, compatible components can be found to further reduce the price. Table 2 shows the cost of the different components used.

Table 2. Cost of the components for a SMH with a battery power supply.

Description	Number	Unit Price (EUR)	Total (EUR)
Microcontroller Arduino Nano	1	20.00	20.00
ZMPT101b	1	10.20	10.20
Wemos D1 Mini	1	4.91	4.91
STC-013	1	5.40	5.40
Printed circuit board	1	0.40	0.40
Power supply unit	1	1.78	1.78
Box container	1	2.54	2.54
Auxiliary material and wiring	----	1.05	1.05
		Total cost	46.28

4.2. Characteristics of Big Data

- **Volume:** Big data systems are associated with large volumes of data. Thus, data are created by machines, networks, and social media. This means that the volume of data to be processed is high;
- **Variety:** It is related to data formats and the different sources that generate them. Thus, the data format can be structured, semi-structured and unstructured. Since data are generated in formats such as photos, videos, logs, sensor devices, etc., this implies facing challenges in terms of data storage, mining and analysis;
- **Velocity:** Refers to the pace of data generation and arrival. This section should include the processing and interpretation time of the data received to improve decision making;
- **Accuracy:** It is essential that the data be reliable and of high quality. To this end, the tools that manage data truthfulness within big data systems must eliminate noise and abnormal data, which are modified to reliable and trustworthy data.

4.3. Big Data Framework for Households

The framework presented comprises the data life cycle of the proposed network. It can be seen from the data generation phase to the analysis phase. Figure 7 shows the framework that serves as the basis for dealing with the big data of the proposed network.

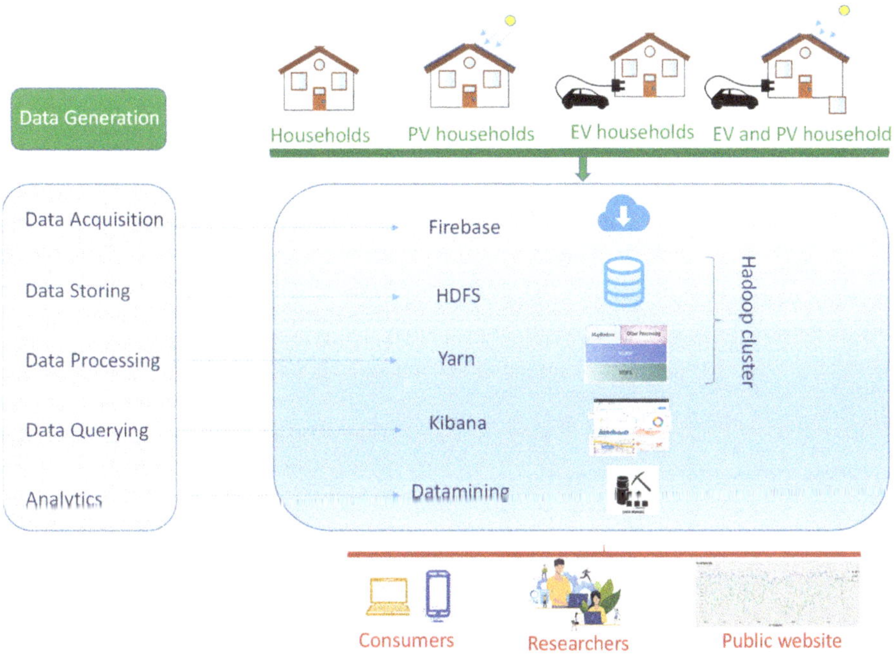

Figure 7. The framework to deal with electrical consumption in households with big data.

An integrated architecture based on big data and cloud computing is proposed. The system contains the following parts: the cloud environment, the big data tools and the database. Figure 1 shows the architecture of the proposed system.

4.3.1. Data Generation

The generated data stream comes from SMs installed in houses and PV installations, measured every 0.5 s. The location of the monitored data is diverse (house, houses with PV, house with PV and EV, and houses with EV). It is possible to add the information of electrical variables generated with sensors that record meteorological variables. In addition, of the application to houses, commercial buildings, factories, and other renewable sources, etc., can be included.

4.3.2. Data Acquisition

The data acquisition of the designed platform has in three tasks: data collection, data transmission and data pre-processing. The generated data were explained in the previous point that are collected by the developed SM.

The transmission of the collected data is performed through one or more master nodes that are part of the Hadoop cluster. The collected data are sent to the data storage system where it is processed in the following phases.

Data integration uses techniques that aim to combine data from different SMs in order to unify the information. The files can be transferred in different formats, such as csv files, json files, etc.

The information generated by the SMs contains the time stamp, the ID of each house, voltage, current, active power, apparent power, and PF. In the data pre-processing phase, erroneous information is modified or removed to improve data quality.

The data acquisition system must fulfil the collection function. In this sense, it must collect, aggregate, and send large volumes of data from the SMs to the Hadoop master node. The data are stored in files within an HDFS repository in the formats used.

4.3.3. Data Storage and Processing

HDFS performs the function of storing data for further processing. HDFS clusters consist of a NameNode with the responsibility of controlling the file system metadata. DataNode lists are used to store the actual data. Hadoop Yarn is the resource manager for data analysis. Yarn runs in conjunction with HDFS on the same node list. This allows processing of nodes with data that are part of the system.

4.3.4. Data Query

Once the data are stored in Firebase, a Python script is developed, which is executed from time to time in an automatic and configurable way, by extracting, processing, and filtering all the information. This information is necessary for the next module, the system analysis, and visualization of the information. Subsequently, the information is uploaded to HDFS and the data can be consulted from Kibana.

4.3.5. Data Analysis

The data of each home is sent to each of the users of the home, who can view it through the web and App. They can also be consulted anonymously by the researchers to analyse load profiles, estimate electricity demand, optimize electricity consumption, etc.

The main objectives of this phase are the improvement and stability of the system. The role of the consumers is fundamental in the system. The data visualization is performed with a dashboard, and the information can be accessed through the app and web portal.

4.4. Dashboard

A dashboard is a type of graphical user interface which often provides at-a-glance views of key performance indicators (KPIs) relevant to a particular objective or business process. In a dashboard, which summarizes the information to be shown to the user according to one or several objectives, we usually find three types of elements:

- **Timeline.** A timeline provides a visual representation of events that help you better understand history, a story, a process, or any other form of an event sequence arranged in chronological order and displayed along a line (usually drawn left to right or top to bottom);
- **KPIs.** A key performance indicator is a measurable value that demonstrates the effectiveness of a company in achieving its key business objectives. The proposed KPI model plays a key role in this work on monitoring residential electricity consumption.

The idea is to monitor the electrical variables of homes. Based on our experience in the implementation and creation of SM projects, we defined the most relevant characteristics to know the performance of this type of systems. The proposed KPI allows to measure the quality of the measurements made, the power and energy consumption of the house;

- **Data tables.** It is very common to find this element within a dashboard, which serves to analyse specific data and usually to download the set of data being displayed. The data table feature allows tabular representation of a database query to be displayed. It also allows you and consumers to manipulate the display of the data in the able while in the dashboard. For example, users can resize, sort, and change the order of columns.

Different methods were tested for information visualization, including systems developed from scratch, some commercial systems such as Microsoft Power Bi 3, a data analysis software solution with predefined graphs, etc. Finally, it was decided to use Elastic's ELK stack 4. Elastic Search is a NoSQL search engine built on top of Apache Lucene, a free and open-source information retrieval software library. Aligned with the Elastic Search project is the open-source visualization and analysis system, Kibana, which we utilized for custom stakeholder dashboarding. The dashboards create a visualization the stakeholder selected analysis and can be extended to recommend robust strategies to support decision making. All these tools were integrated to work together to create a complete SG system.

5. Results

This section shows the tests carried out with the SMH. For this purpose, data were collected from several households located in Jaén, Andalusia, Spain. The houses have different characteristics, such as EV electric vehicles, PV solar photovoltaic generation, etc. In each of them, the load profiles were monitored for comparison. The tests shown operation for two days in the year 2022. Figure 8 shows the device connected in the house.

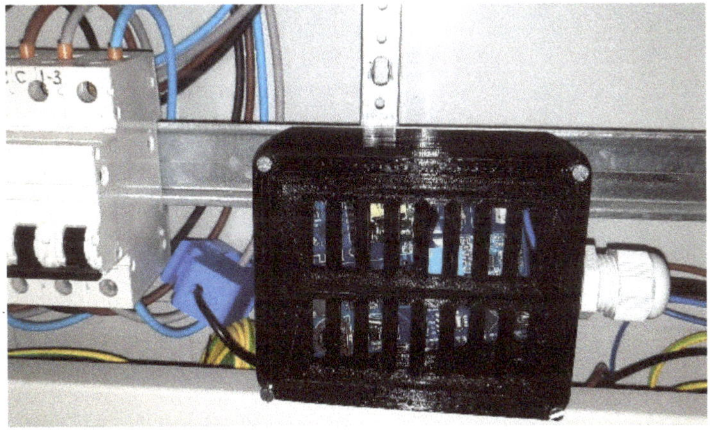

Figure 8. The device assembled with real components connected in the house.

The graphs show the average value of the electrical variable displayed (v, i, p, s, PF, and q). The aggregation is performed on the data taken by the meter every 0.5 s. The system automatically aggregates the data according to the time slice chosen for the display. This helps to understand the temporal evolution of the measurement in the shown time slice. As for the graphs on the left, the data are aggregated in an automated way according to the chosen time slice. The data representation corresponds to 6 and 7 October 2022, for a single-family house with four inhabitants.

The results of the entire system can be seen in the following screenshots. Figure 9 shows the first dashboard, specifically the general dashboard, where we can see a timeline with the values of reactive power q and voltage v.

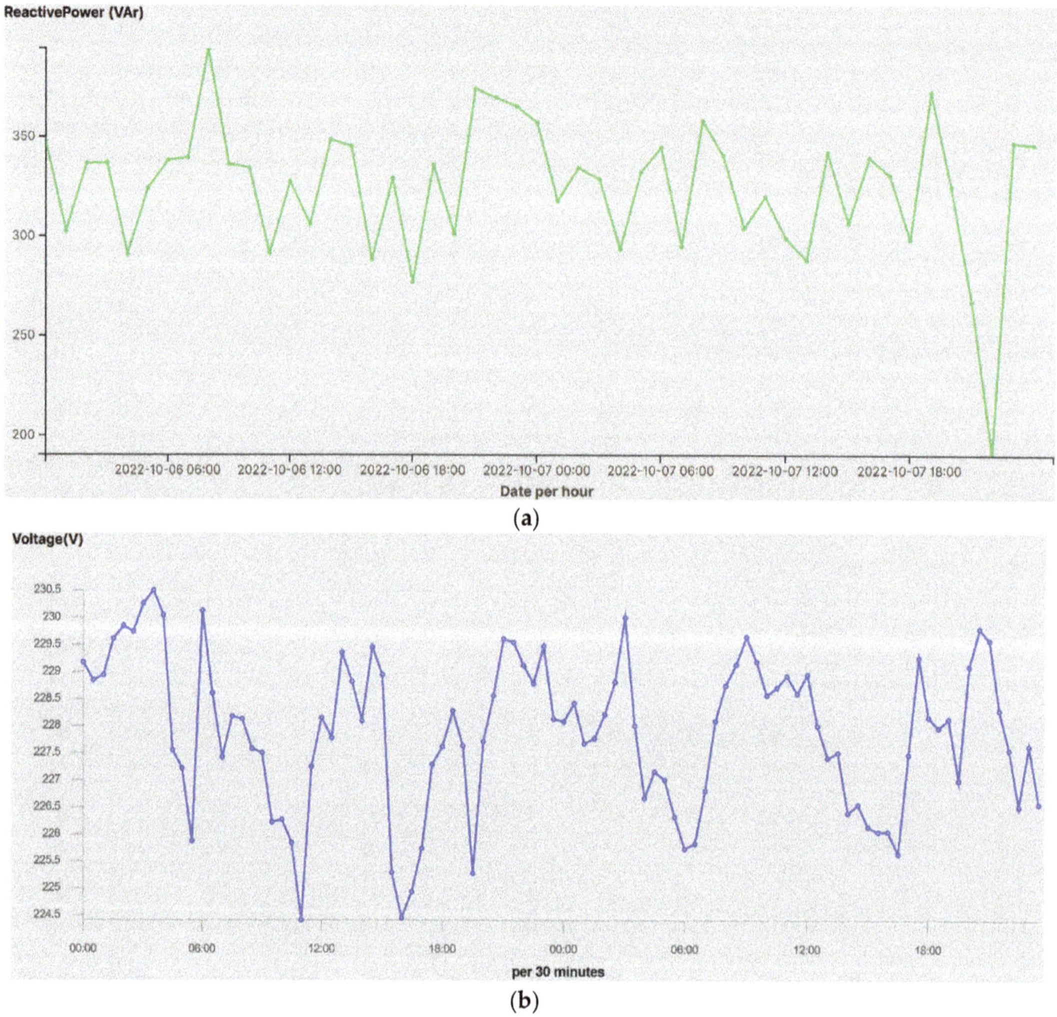

Figure 9. General dashboard (**a**) q and (**b**) v.

Figure 10 shows the second dashboard, specifically the general dashboard, where we can see a timeline with the values of apparent power p and current i. In Figure 10 it appears that i and p evolve with graphs that have similar shapes. This is because v varies in very narrow margins and remains almost constant within the time slice shown. It can be observed in Figure 9 that the voltage is very stable; this is due to the consistency provided by the electrical network. As a consequence, it has little influence on the variability of power. The user can observe that the maximum consumption peaks occur between 08 and 10, 14 and 16, and 20 and 22 h which correspond to the hours of maximum occupation in the household. These hours coincide with breakfast, lunch, dinner, and rest time for viewing TV. The behaviour is similar in the two days shown.

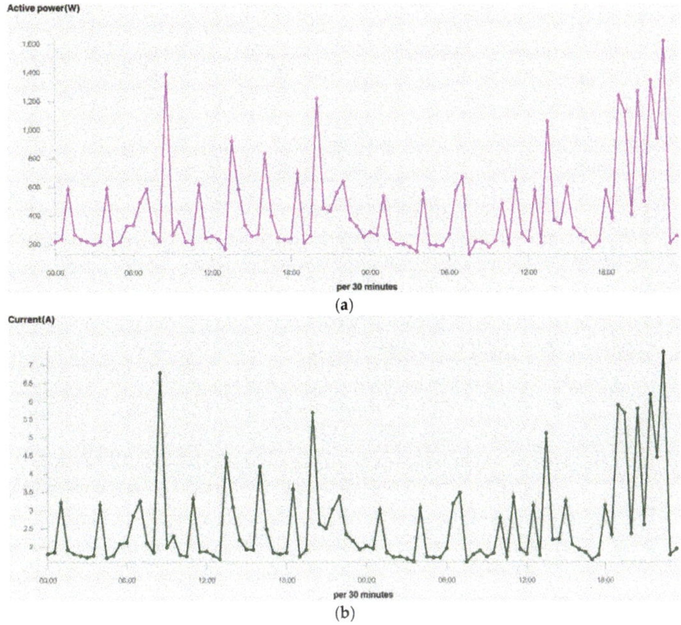

Figure 10. General dashboard (**a**) *p* and (**b**) *i*.

Figure 11 shows the power factor *PF* and apparent power *s* values. In all cases, the two graphs on the right show the average values that are being reached in the selected and configurable time period. The apparent power *s* is influenced by the variability of the electrical current consumed in the household due to the stability of the voltage as shows Figure 11.

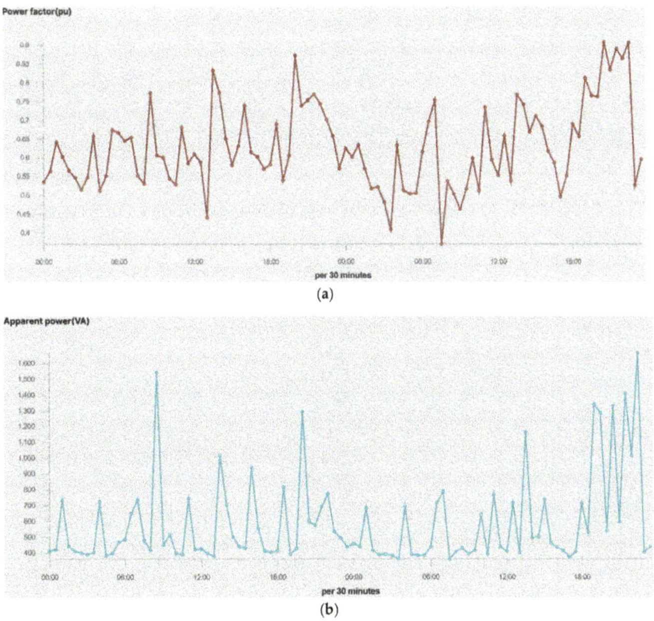

Figure 11. General dashboard (**a**) *PF* and (**b**) s.

Figure 11 shows the behaviour of the *PF* on the two days shown. The *PF* is quite high and has values close to the unit during the study period, which implies a high performance of the receivers connected in the house and low losses. This is due to the use of class A+++ electrical appliances and an aerothermal air conditioning system.

In addition to the graphs of the electrical variables, the values are used for data visualization and can be downloaded in CSV format. This feature is highly interesting for homeowners or researchers because it enables studying the energy behaviour of the household, photovoltaic (PV), and vehicle (EV). The CSV format was chosen because it is accepted by most software used for analysis, such as MATLAB, Excel, and most statistical suites.

By default, the dashboard loads the data from all the devices and shows these charts for the average and accumulated values of the devices. It was also designed so that filtering a component affects the rest of the displayed data. This allows filtering all the components at the same time, including the data table. This data table allows sorting and filtering each of the columns shown as well as downloading the data being displayed in CSV format.

Since everything is configurable, the platform allows different types of filtering for almost every one of the data fields. The most frequent filtering types are the following:

Filtering by date range: This filter can be applied in the platform in different ways. The most common is to use the date control that appears in the upper right corner, see Figure 12. Another option to filter by dates is to select, with the computer mouse, a specific area in the timeline charts. Automatically this range is selected and all the information and charts are updated.

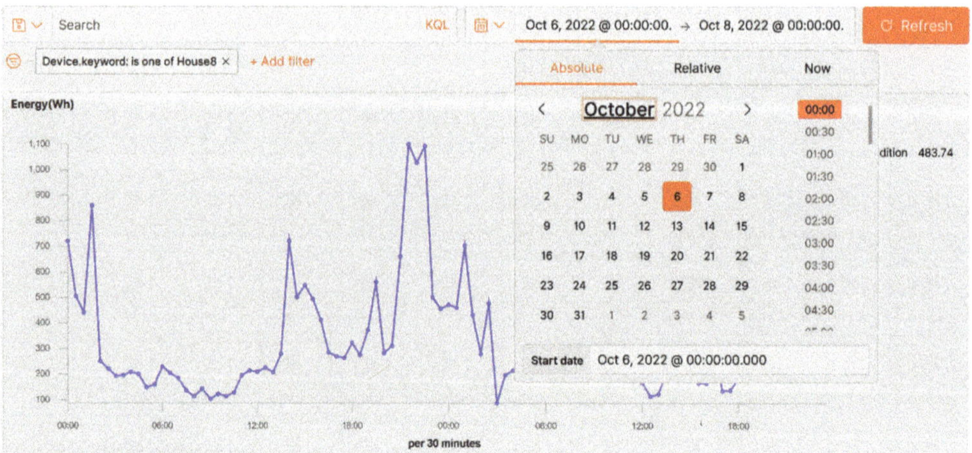

Figure 12. Filtering by date.

Filtering by device: We can use a filter control that the dashboard has in the upper left corner to filter by one or several devices. All the data and graphs of the dashboard are automatically updated. This filter control can be seen in the Figure 13.

The following designed and developed dashboard shows timeline-type graphs that visually compare all the devices loaded in the data with all the established measurements. From a visual point of view, it is interesting to see a great amount of information in very little space.

Figures 14 and 15 compare three households. The evolution of the three houses can be observed. One was analysed in previous figures and the other is a single-family household with two and three inhabitants in house#11 and house#1, respectively. As opposed to the house#13, in the second one the inhabitants are out of the house all day because of work, returning home after 9.30 pm when a higher consumption is observed. House#11 and house#1 are occupied almost all day with consumption peaks in the middle of the day, in the afternoon, and at dinner time. The use of these comparative graphs allows researchers to obtain analyses from which they can deduce conducts applicable to obtaining load profiles, demand estimates, energy study of the household, reduction in the electricity tariff, etc.

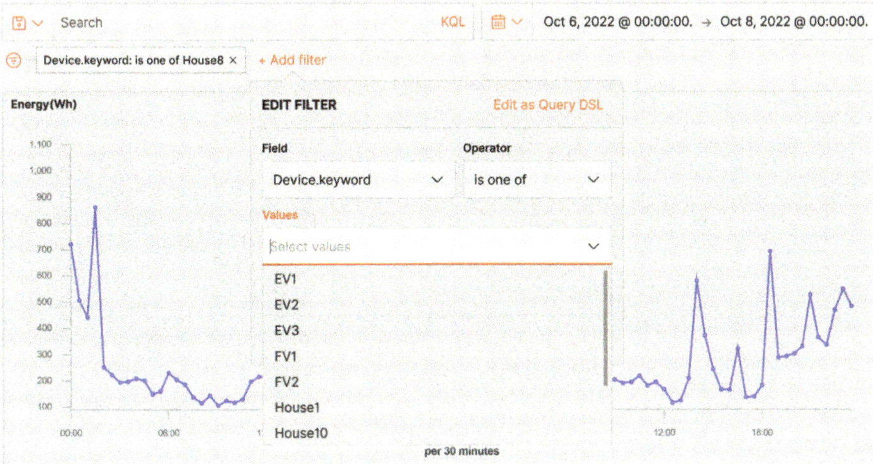

Figure 13. Filtering by device.

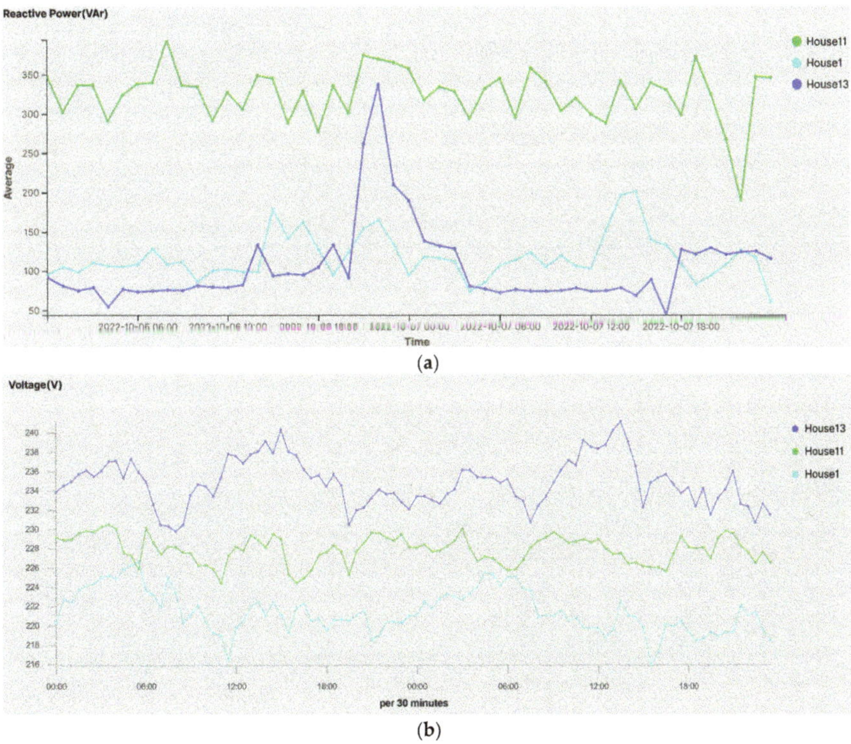

(a)

(b)

Figure 14. General dashboard comparison of (**a**) q and (**b**) v values.

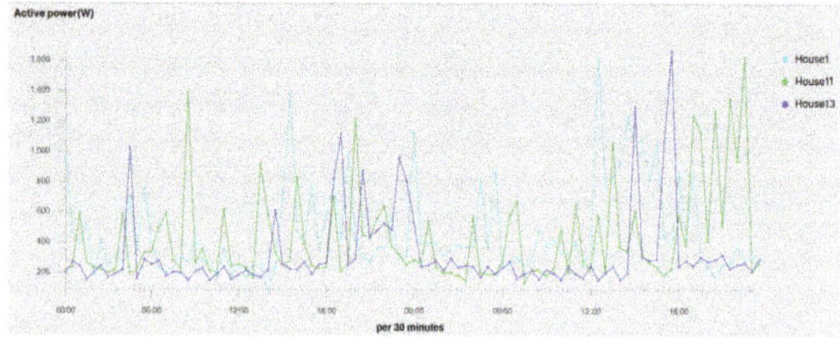

(**a**)

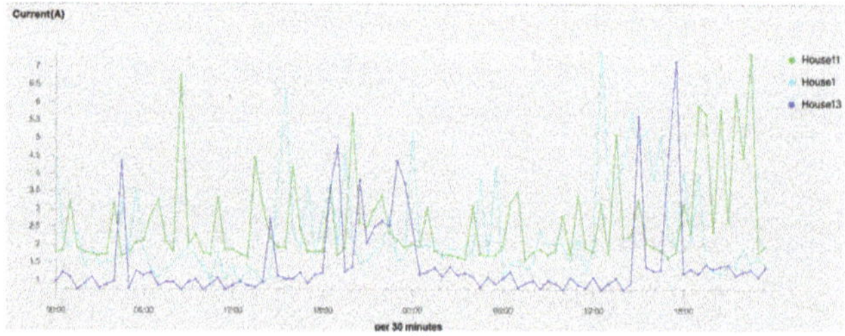

(**b**)

Figure 15. General dashboard comparison of (**a**) *p* and (**b**) *i* values.

Figure 16 shows the PF for the households, with EV and PV. House#13 and house#1 vary with respect to the PV, the work performed by the inverter for the grid connection, which maintains the PF at 1 for almost the entire generation time, is clearly noticeable. The EV, which is connected to the grid through a DC/AC converter, also maintains the PF at values above 0.95 during most of the charging process, making it very efficient. The different equipment installed in the households, especially the electrical appliances, make the first household present a much higher value of PF, above 0.9, making it very efficient. On the contrary, house#11 has equipment that reduces the average value of the PF to 0.65, thus presenting a less efficient behaviour.

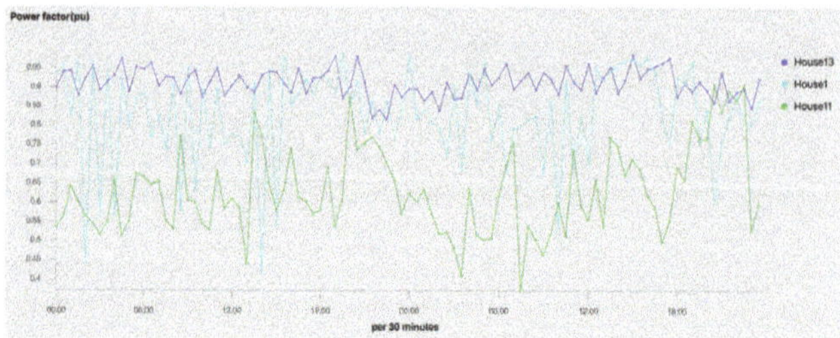

(**a**)

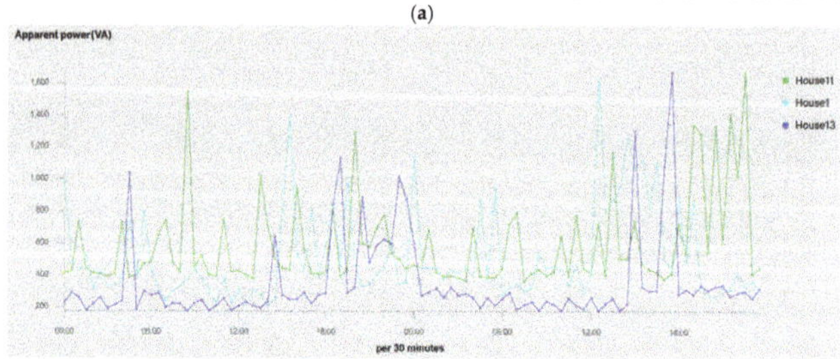

(**b**)

Figure 16. General dashboard comparison of (**a**) *PF* and (**b**) *s* values.

App for Non-Expert Users

For less experienced users who have difficulties in using the dashboards created, an app was created that allows real-time monitoring in a simpler and more interpretable way. Each user can only see their own home in their app, leaving the comparison and visualisation of other data to the dashboards. The App is personalised for each user and only the user can use it.

The app displays real-time data with a timestamp of voltage, current, and power. In addition, graphs of these three variables can be displayed, allowing the user to observe the evolution of the recorded values over time. Figure 17 shows the apps for houses 11 and 13.

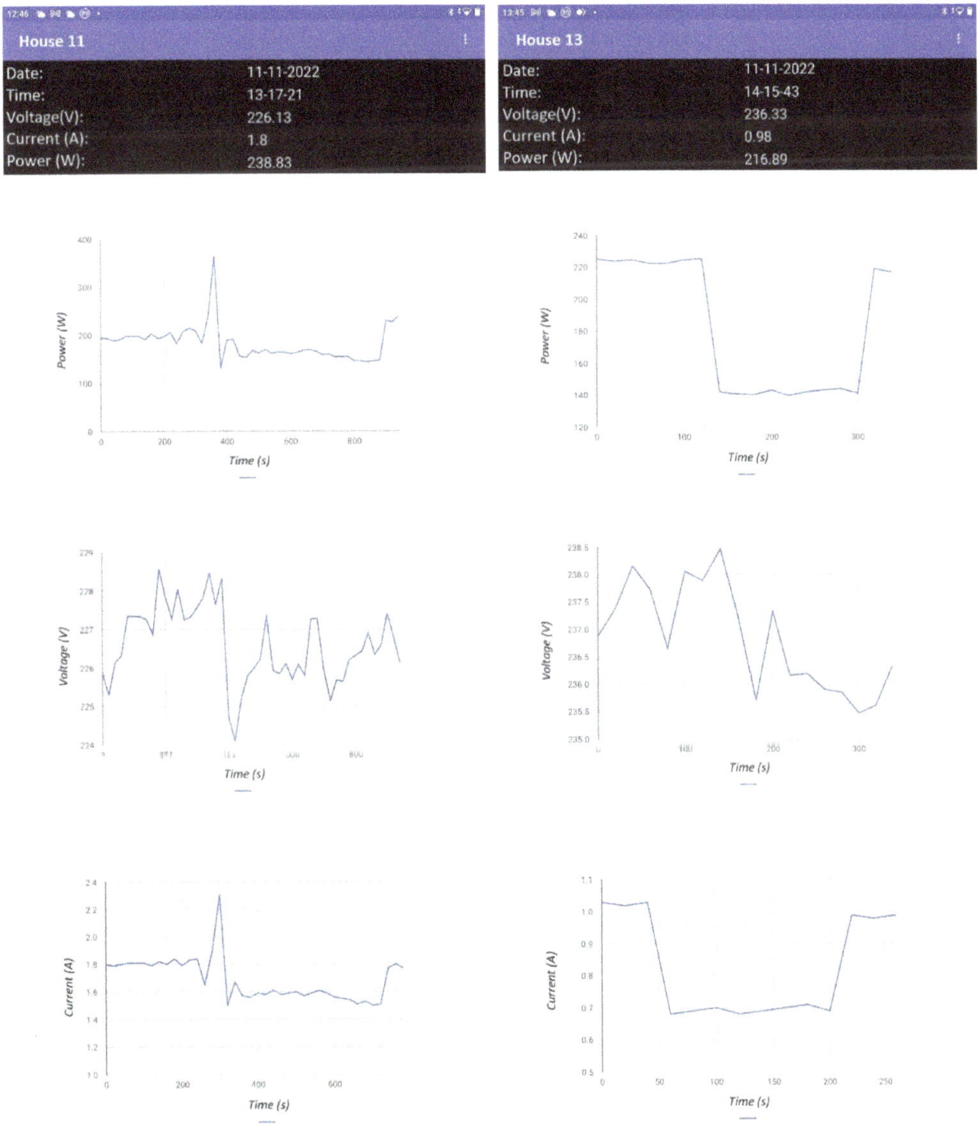

(**a**) (**b**)

Figure 17. App for non-expert users: (**a**) house#11 and (**b**) house#13.

6. Conclusions and Future Works

This research study successfully developed a prototype SMH for monitoring electrical variables and uploading data to cloud using a wireless connection. This SMH was based on the Arduino open-source electronic platform and input data were collected with a set of sensors. The data are uploaded to the cloud each 0.5 s.

In addition, this study presents a web platform based on a cloud system that allows an innovative analysis of data captured in the Internet of Things data from smart homes, photovoltaics, and electrical vehicles in real time. Each user can download the data from their SM and view them in the developed app. The data are also available for query and download for the scientific community.

After the analysis of several solutions, we choose a cloud storage service for the subsequent and automatic processing of the data and the visualization, going from data to information. To validate the platform and present significant results, a case study was presented using the data acquired. The results of the experiments clearly show the benefit and feasibility of the proposed platform. We provided a detailed requirement analysis and illustration of the platform components.

The authors refined the platform component and it was tested with datasets from different electrical devices, such as household appliances. This approach is crucial to validate the platform's applicability and robustness in dealing with all types of IoT data measurements. This research uses datasets with extensive temporal data and with different types of households, allowing for comparisons of load profiles, electrical variables, etc.

Future lines of research would include adding datasets from different electrical devices, such as household appliances. In addition, future research would include developing a demand–response algorithm to be included in the SM to work in conjunction with the proposed methodology. Another possible line is aggregated information derived from measurements with application of artificial intelligence algorithms, e.g., automatic prediction systems and development of tools to perform a multidimensional analysis of the data.

Author Contributions: Conceptualization, M.A.G.-C.; methodology, A.C.-O. and F.S.-S.; software, M.A.G.-C.; formal analysis, A.C.-O. and F.S.-S.; investigation, A.C.-O. and F.S.-S.; resources, J.C.H.; writing—original draft preparation, A.C.-O. and F.S.-S.; writing—review and editing, J.C.H. and M.A.G.-C.; visualization, M.A.G.-C.; supervision, J.C.H. and M.A.G.-C. All authors have read and agreed to the published version of the manuscript.

Funding: This research received no external funding.

Acknowledgments: The authors would like to thank the Department of Electrical Engineering of the University of Jaén for allowing the use of their laboratories and material in the development of this research.

Conflicts of Interest: The authors declare no conflict of interest.

References

1. Usman, A.; Shami, S. Evolution of communication technologies for smart grid applications. *Renew. Sustain. Energy Rev.* **2013**, *13*, 191–199.
2. Munshi, A.; Mohameda, Y.R. Big data framework for analytics in smart grids Electric Power Systems Research. *Electr. Power Syst. Res.* **2017**, *151*, 369–380.
3. Kabalci, Y. A survey on smart metering and smart grid communication. *Renew. Sustain. Energy Rev.* **2014**, *57*, 302–318.
4. Tonyali, S.; Akkaya, K.; Saputro, N.; Nojoumian, S.U.M. Privacy preserving protocols for secure and reliable data aggregation in iot-enabled smart metering systems. *Future Gener. Comput. Syst.* **2018**, *78*, 547–557.
5. Khan, M.; Jabbar, S.D.S.; Gohar, M.; Ghayvat, H.; Mukhopadhyay, S. Context-aware low power intelligent smarthome based on the internet of things. *Comput. Electr. Eng.* **2016**, *52*, 208–228.
6. Yigit, M.; Gungor, V.C.; Baktir, S. Cloud computing for smart grid applications. *Comput. Netw.* **2014**, *70*, 312–329.
7. Al-Turjman, F.; Altrjman, C.; Din, S.; Paul, A. Energy monitoring in iot-based ad hoc networks: An overview. *Comput. Electr. Eng.* **2019**, *76*, 133–142.
8. Al-Turjman, F.; Abujubbeh, M. Iot-enabled smart grid via sm: An overview. *Future Gener. Comput. Syst.* **2016**, *96*, 579–590.

9. Rodrigues, W.L.; Borges, F.A.; Veloso, A.S.; Rabelo, R.d.A.; Rodrigues, J. Low voltage smart meter for monitoring of power quality disturbances applied in smart grid. *Measurement* **2019**, *147*, 106890.
10. Cano-Ortega, A.; Sánchez-Sutil, F.; Hernández, J. Power factor compensation using teaching learning based optimization and monitoring system by cloud data logger. *Sensors* **2019**, *19*, 12172.
11. Cano-Ortega, A.; Sánchez-Sutil, F. Monitoring of the efficiency and conditions of induction motor operations by smart meter prototype based on a lora wireless network. *Electronics* **2019**, *8*, 1040.
12. Sánchez-Sutil, F.; Cano-Ortega, A. Smart public lighting control and measurement system using lora network. *Electronics* **2019**, *9*, 124.
13. Asghari, P.; Rahmani, A.M.; Javadi, H.S. Internet of things applications: A systematic review. *Comput. Netw.* **2019**, *148*, 241–261.
14. Rathore, M.M.; Ahmad, A.; Paul, A.; Rho, S. Urban planning and building smart cities based on the internet of things using big data analytics. *Comput. Netw.* **2016**, *101*, 63–80.
15. Naik, K.; Patel, S. An open source smart home management system based on IOT. *Wirel. Netw.* **2018**, 1–7. [CrossRef]
16. Pau, M.; Patti, E.; Barbierato, L.; Estebsari, A.; Pons, E.; Ponci, F.; Monti, A. A cloud-based smart metering infrastructure for distribution grid services and automation. *Sustain. Energy Grids Netw.* **2018**, *15*, 14–25.
17. Sánchez-Sutil, F.; Cano-Ortega, A.; Rus-Casas, J.H.C. Development and calibration of an open source, low-cost power smart meter prototype for pv household-prosumers. *Electronics* **2019**, *8*, 878.
18. Sutil, F.S.; Ortega, A.C. Smart plug for monitoring and controlling electrical devices with a wireless. *Expert Syst. Appl.* **2023**, *213*, 118976.
19. BLUED. [Online]. Available online: https://tokhub.github.io/dbecd/links/Blued.html (accessed on 11 November 2022).
20. DRED. [Online]. Available online: https://www.st.ewi.tudelft.nl/~{}akshay/dred/ (accessed on 11 November 2022).
21. UK-DALE. [Online]. Available online: https://ukerc.rl.ac.uk/DC/cgi-bin/edc_search.pl (accessed on 11 November 2022).
22. PECANSTREET. [Online]. Available online: https://www.pecanstreet.org (accessed on 11 November 2022).
23. ECO. ECO. [Online]. Available online: http://vs.inf.ethz.ch/res/show.html (accessed on 11 November 2022).
24. GREEND. [Online]. Available online: https://sourceforge.net/projects/greend (accessed on 11 November 2022).
25. iAWE. [Online]. Available online: https://www.iawe.org (accessed on 11 November 2022).
26. REDD. [Online]. Available online: https://www.reddprojectsdatabase.org (accessed on 11 November 2022).
27. TRACEBASE. [Online]. Available online: https://energy.duke.edu/content/tracebase-dataset (accessed on 11 November 2022).
28. ACS-F1 and ACS-F2. [Online]. Available online: https://icosys.ch/acs-f2 (accessed on 11 November 2022).
29. AMPds. [Online]. Available online: http://ampds.org (accessed on 11 November 2022).
30. Smart. [Online]. Available online: http://traces.cs.umass.edu/index.php/smart/smart (accessed on 11 November 2022).
31. REFIT. [Online]. Available online: https://pureportal.strath.ac.uk/en/datasets/refit-electrical-load-measurements (accessed on 11 November 2022).
32. Liu, X.; Nielsen, P. A hybrid ict-solution for smart meter data analytics. *Energy* **2016**, *115*, 1710–1722.
33. Wen, L.; Zhou, K.; Yang, L.; Li, S. Compression of smart meter big data: A survey. *Renew. Sustain. Energy Rev.* **2018**, *91*, 59–69.
34. Wilcox, T.; Jin, N.; Flach, P.; Thumim, J. A big data platform for smart meter data analytics. *Comput. Ind.* **2019**, *105*, 250–259.
35. Yassine, A.; Singh, S.; Hossain, M.S.; Muhammad, G. IoT big data analytics for smart homes with fog and cloud computing. *Future Gener. Comput. Syst.* **2019**, *91*, 563–573.
36. Diamantoulakis, P.; Kapinas, V.; Karagiannidis, G. Big data analytics for dynamic energy management in smart grids. *Big Data Res.* **2015**, *2*, 94–101.
37. Tu, C.; He, X.; Shuai, F.; Jiang, Z. Big data issues in smart grid—A review. *Renew. Sustain. Energy Rev.* **2017**, *79*, 1099–1107.
38. Shyam, R.; Ganesh, H.B.; Kumar, S.S.; Poornachandran, P.; Soman, K. Apache spark a big data analytics platform for smart grid. *Procedia Technol.* **2015**, *21*, 171–178.
39. Saleh, A.; Rabie, A.; Abo-Al-Ez, K. A data mining based load forecasting strategy for smart electrical grids. *Adv. Eng. Inform.* **2016**, *30*, 422–448.
40. Guerrero, J.; García, A.; Personal, E.; Luque, J.; León, C. Heterogeneous data source integration for smart grid ecosystems based on metadata mining. *Expert Syst. Appl.* **2017**, *79*, 254–268.
41. Cano-Ortega, A.; Sánchez-Sutil, F. Performance optimization lora network by artificial bee colony algorithm to determination of the load profiles in dwellings. *Energies* **2020**, *13*, 517.
42. Yildiz, B.; Bilbao, J.; Dore, J.; Sproul, A. Recent advances in the analysis of residential electricity consumption and applications of smart meter data. *Appl. Energy* **2017**, *208*, 402–427.
43. Funde, N.; Dhabu, M.; Paramasivam, A.; Deshpande, P. Motif-based association rule mining and clustering technique for determining energy usage patterns for smart meter data. *Sustain. Cities Soc.* **2019**, *46*, 101415.
44. Andreadou, N.; Kotsakis, E.; Masera, M. Smart meter traffic in a real lv distribution network. *Energies* **2018**, *11*, 1156.
45. Meloni, A.; Pegoraro, P.; Atzori, L.; Benigni, A.; Sulis, S. Cloud-based iot solution for state estimation in smart grids: Exploiting virtualization and edge-intelligence technologies. *Comput. Netw.* **2018**, *130*, 156–165.
46. Razavi, R.; Gharipour, A.; Fleury, M.; Akpan, I.J. Occupancy detection of residential buildings using smart meter data: A large-scale study. *Energy Build.* **2019**, *183*, 195–208.
47. Sial, A.; Singh, A.; Mahanti, A. Detecting anomalous energy consumption using contextual analysis of smart meter data. *Wirel. Netw.* **2019**, *27*, 4275–4292.
48. Araujo, V.; Mitra, K.; Saguna, S.; Åhlund, C. Performance evaluation of fiware: A cloud-based iot platform for smart cities. *J. Parallel Distrib. Comput.* **2019**, *132*, 250–261.

49. Forcan, M.; Maksimoví, M. Cloud-fog-based approach for smart grid monitoring. *Simul. Model. Pract. Theory* **2020**, *101*, 101988.
50. Kumar, A.; Alghamdi, S.A.; Mehbodniya, A.; haq, M.a.; Webber, J.L.; Shavkatovich, S.N. Smart power consumption management and alert system using IoT on big data. *Sustain. Energy Technol. Assess.* **2022**, *53*, 102555.
51. Wang, J. A novel oscillation identification method for grid-connected. In Proceedings of the 2021 International Conference on New Energy and Power Engineering (ICNEPE 2021), Sanya, China, 19–21 November 2021.
52. Zhao, X. Research on management informatization construction of electric. In Proceedings of the 2022 International Symposium on New Energy Technology Innovation and Low Carbon, Kunming, China, 21–23 January 2022.
53. Mostafa, N.; Ramadan, H.S.M.; Elfarouk, O. Renewable energy management in smart grids by using big data analytics. *Mach. Learn. Appl.* **2022**, *9*, 100363.

energies

MDPI

Article

Energy-Consumption Pattern-Detecting Technique for Household Appliances for Smart Home Platform

Matteo Caldera [1], Asad Hussain [2], Sabrina Romano [1,*] and Valerio Re [2]

1 Italian National Agency for New Technologies, Energy and Sustainable Economic Development (ENEA), Lungotevere Thaon di Revel, 76, 00196 Rome, Italy
2 Department of Engineering and Applied Sciences, University of Bergamo, Viale Marconi 5, 24044 Dalmine, Italy
* Correspondence: sabrina.romano@enea.it

Abstract: Rising electricity prices and the greater penetration of electricity consumption in end-uses have prompted efforts to set up data-driven methodologies to optimise energy consumption and foster user engagement in demand-side management strategies. The performance of energy-management systems is greatly affected by the consumer behaviors and the adopted energy-management methodology. Consequently, it is necessary to develop appliance-level, detailed energy-consumption information models to inform citizens to improve behaviors toward energy use. The goal of the Home Energy Management System (HEMS) is to foster an ecosystem that is energy-optimized and can manage Internet of things (IoT) equipment over its network. HEMS allows consumers to reduce energy costs by adapting their consumption to variable pricing over the day. With the use of descriptive data-mining techniques, we have developed a numerical model that gives consumers access to information on their domestic appliances with regard to the number and duration of operations, cycles disaggregation for appliances that have cyclic operation (e.g., washing machine, dishwasher), and energy consumption throughout various time periods basing on 15-min monitoring data. The model has been calibrated and validated on two datasets collected by ENEA by real-time monitoring of Italian dwellings and has been tested over several appliances showing effective analysis of the energy-consumption patterns. Therefore, it has been integrated in the DHOMUS IoT platform, developed by ENEA to monitor and analyse the energy consumption in dwellings in order to increase citizens' engagement and awareness of their energy consumption. The results indicate that the developed model is sufficiently accurate, and that it is possible to promote a more virtuous and sustainable use of energy by end users, as well as to reduce the energy demand as required by the current European Council Regulation (EU) 2022/1854.

Keywords: energy management; smart homes; smart appliances; patterns detecting algorithm

Citation: Caldera, M.; Hussain, A.; Romano, S.; Re, V. Energy-Consumption Pattern-Detecting Technique for Household Appliances for Smart Home Platform. *Energies* **2023**, *16*, 824. https://doi.org/10.3390/en16020824

Academic Editors: Antonio Cano-Ortega and Francisco Sánchez-Sutil

Received: 30 November 2022
Revised: 22 December 2022
Accepted: 23 December 2022
Published: 11 January 2023

1. Introduction

Domestic energy consumption is estimated to account for 30–40% of worldwide generation [1] and it is expected to increase as more appliances and electronic devices are utilized. Moreover, buildings contribute to one-third of worldwide final CO_2 emissions, thus urging a substantial reduction of both consumption and emissions in this sector [2]. Therefore, energy efficiency has become one of the most important challenges. According to statistics [3], customer engagement and awareness of better energy-consumption habits make a considerable difference in energy reduction [4]. Consumers should be given individual load-level consumption information rather than aggregated consumption information [5]. By raising customer knowledge on the usage habits of domestic appliances, appliance-level energy information can significantly contribute to reducing energy consumption [6]. Home energy-management systems (HEMS) can be an effective solution [7] by enabling a set of long-term smart energy-saving applications [8]. They provide visual feedback to consumers in the form of energy-use statistics, utility-driven automation and control, load forecasting, and optimum load-scheduling strategies [9]. With smart meters and sensors that can measure real-time power consumption, monitoring power consumption has become easier with the help of signal-processing methods [10]. The Internet of things (IoT) adds a new degree of machine-to-machine communication to the interactions between humans and apps. The advancement of new communication technologies and smart devices that function by parsing signals has increased the ability to manage different equipment. Smart home (SH) devices can store data, respond to users' prompts, provide feedbacks to users, and issue alarms. SH-IoT promises to create an energy-optimized environment by connecting devices, in addition to providing flexibility in home management and monitoring [11].

Unlike previous studies in which HEMS relies on raw data collected by energy meters from which models disaggregate desired appliances by relying on classification methods based on the prior information of appliance energy signatures, in this paper we propose a model based on 15-min aggregated raw data that relies on descriptive data-mining techniques. The model is quite simple and computationally less complex, yet effective. It can detect energy-consumption patterns, the appliance operation (i.e., on–off, standby), and the duration period in which the appliance is operative for all household appliances. Moreover, it provides consumers with the information on the total number of cycles, disaggregates cycles in close sequence, identifies cycles according to their duration (i.e., short, medium, and long cycles), and calculates energy consumption during customised time periods. The model has been calibrated and validated on two datasets collected by ENEA by real-time monitoring of Italian dwellings equipped with IoT devices. The former dataset contains measurements of 58 domestic appliances from 14 homes over a period of 2.2 years, and the latter contains measurements of 48 appliances from 10 homes over a period of 6 months. The key contributions of this work are as follows.

1. A pattern-detecting method for different appliances (cyclic and noncyclic) has been developed that is quite simple, computationally less complex, and that can accurately disaggregate energy-consumption cycles occurred in the close succession. Moreover, the algorithm developed herein could be applied to any device dataset.
2. The developed algorithms have been implemented in a code used to process the data acquired by a IoT platform developed by ENEA and currently operating in some dwellings, aimed at fostering user engagement and energy awareness by providing customised feedbacks to the users.
3. The DHOMUS platform uses open and interoperable protocols for data acquisition at the device and HEMS level and, therefore, can be easily replicated in broader contexts.

2. Literature Review

The diffusion of smart meters, low-cost sensors, and smart appliances has paved the way for novel energy management strategies including communication and interaction among users, devices, and the grid. Whereas greater resolution data is necessary for more precise energy analysis, raw consumption data by itself is not enough to provide details on the causes of energy demand and what can be done to reduce it.

Demand–response (DR) strategies can be used to boost the efficiency of smart grids. There are several DR initiatives in the literature aimed at lowering customers' bills and reducing grid loads. Authors in [11] devised a homomorphic encryption-based alternate direction technique of multipliers approach to address cost-aware appliance scheduling optimization in a distributed way and scheduled home appliances without compromising users' privacy. They show that the suggested secure appliance scheduling for a flexible and efficient energy-consumption method, termed SAFE, significantly decreases power costs while protecting users' privacy through an intensive simulation research by using real-world datasets. Appliances' operating mode identification, which uses the cycle-clustering technique proposed in [12] is a smart home energy-management system primary approach based on sensed power consumption values, which enables DR by allowing users to employ energy-saving appliance operation modes. Cycles from an appliance single-usage profile, are extracted and reshaped into characteristics in the form of clusters of cycles. By using K-nearest neighbours, these attributes are then used to determine the operating mode during each occurrence. Identification of operation modes is deemed fundamental for several possible smart DR applications inside HEMS. Authors in [13] presented the K-means clustering technique to classify different modes of operations of home appliances. To validate the proposed method, a case study based on a refrigerator was presented. By combining an automatic switch-off system with load balancing and a scheduling algorithm, authors in [14] proposed a smart-home energy-management system that reduces energy waste. The load-balancing technique operates within defined limitations to keep the household's total energy consumption within a certain limit. The least slack time (LST) method is used to schedule appliances, by considering the user's comfort. The simulation results demonstrated a considerable reduction in residential energy consumption aided by an automatic switch-off mechanism thanks to the suggested LST-based energy-management system (EMS). Support vector machines were used in [15] to identify load characteristics of appliances in various operating modes, and a smart home energy-conservation system was constructed to test the approach and validate the proposed EMS. The authors of [16] applied an artificial neural network to identify usage patterns of some common household electric appliances from daily profiles of energy records obtained with 15-min granularity, and the proposed method showed effectiveness when applied to several daily load diagrams captured by a few residential meters. The iterative disaggregation approach based on consumption patterns presented in [17] is a method that merges the fuzzy C-means clustering technique, which provides an initial working condition, with subsequence searching dynamic time warping, which recovers single-source energy usage based on typical power-consumption patterns. The results demonstrate that the suggested method properly disaggregates power usage and is suited for situations when many appliances are used at the same time. To decrease peak power and improve users' comfort, authors of [18] proposed an efficient home energy-management controller (EHEMC) based on the genetic harmony search algorithm. With real-time electricity pricing and critical peak pricing tariffs, they evaluated EHEMC for a single house and multiple dwellings. The study [19] provided a nonintrusive affinity propagation clustering algorithm based on factor graph model and the belief propagation theory, and the results revealed that the algorithm correctly recognizes the basic and combination classes of household appliances. This strategy laid the groundwork for power-management firms to efficiently and effectively allocate electricity. The study [20] proposed a novel method for assessing the energy consumption of various appliances, and it was utilized to develop a recommendation system that would advise renters on how to reduce their usage.

Clustering, association rule mining, and artificial neural networks were all used to perform data-mining tasks. Authors in [21] presented various unsupervised machine learning techniques to find energy-usage patterns in a smart home. Moreover, they provide solutions for smart metering systems that may help to: (1) raise energy awareness; (2) assist accurate use forecasts; and (3) give input for demand response systems in homes that provide users with timely energy saving advice. In [22], three data-analysis processes and a data-mining framework, i.e., data classification, cluster analysis and association rule mining, were proposed for efficient data analysis of buildings. A systematic approach for providing immediate feedback and recommendations to building occupants to assist them in taking appropriate action to minimize energy use was developed in [23]. For a specific building, the notion of a reference building (RB), i.e., an energy-efficient building, was established, and RB was created by using data-mining techniques that included clustering analysis and neural networks. The performance of a building was determined in comparison to a reference building, which provided feedbacks to suggest occupants the appropriate actions to enhance building energy performance (e.g., turning off lights or heating and air conditioning, etc.). In some other studies [24,25], the nonintrusive load-monitoring (NILM) method was designed to identify the activation of a target appliance by analysing the recorded active power transient response and estimating its consumption in real time.

Smart metering, on the other hand, has been promising in several applications, such as energy modelling [26–31], behaviour characterization [32–34], grid infrastructure technical evaluations [35–39], and end-user engagement [40–42]. Power and consumption data for household appliances used in a two-person family in Turkey was gathered with high resolution in one-second intervals in [43]. Moreover, authors of [44] reported a dataset of 15 households in Germany over a 3.5-year period, with a total of 50 appliances monitored at 1 Hz.

From the literature, it has been observed that most of work reported is related to overall load-level consumption information, cost-aware appliance scheduling, load balancing, and a scheduling algorithm based upon energy meter data. To the best of our knowledge, we have not found simple models aimed at extracting patterns such as the number and duration of operations, cycle disaggregation for appliances that have cyclic operation (e.g., washing machine, dishwasher), and energy consumption throughout various time periods, which can be easily integrated into IoT platforms dedicated to residential users.

3. DHOMUS Platform

Data Homes and Users (DHOMUS), is a platform developed by ENEA (https://dhomus.smartenergycommunity.enea.it, accessed on 1 December 2022) that is dedicated to residential users for smart home applications. The layout of the smart home is depicted in Figure 1 [45]. The main objective of DHOMUS is to make users aware of their energy data, to let them understand how much energy they consume and why, to support reduction of both consumption and costs, thereby contributing to decrease their impact on the environment, to increase energy awareness, and to transform residential users into active subjects that contribute to the stability of the grid. The DHOMUS platform currently collects and analyses real-time data from smart homes located in Rome. They consist of 24 homes equipped with a kit of commercial sensors (their characteristics are described in Table 1), which monitor the electrical consumption of the home meter and of selected appliances, as well as the presence of people and indoor comfort. The management of all these devices is wireless, based on Z-Wave protocol, and the gateway is connected to the Internet through the energy box, which collects, integrates, and sends sensors' data to the cloud platform. As the electrical data acquisition time varies from sensor to sensor, postprocessing is performed to calculate their 15-min averages, similar to the method used for smart meters.

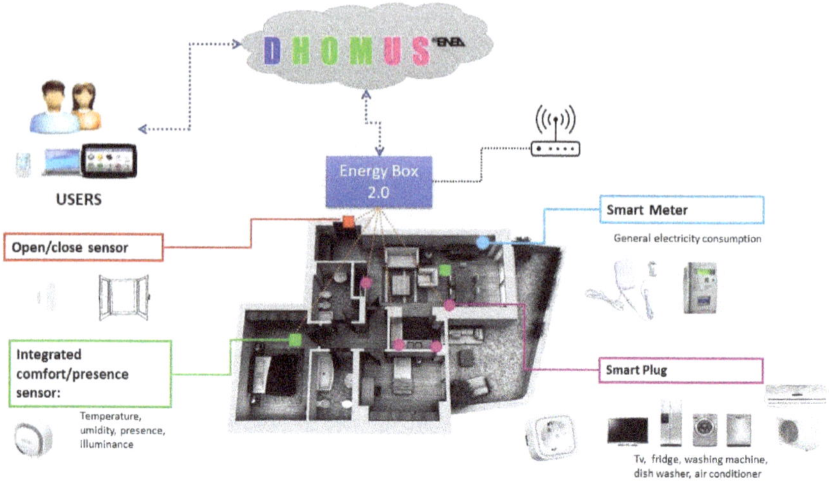

Figure 1. Smart home: representation of sensors.

Table 1. Sensors used in the experimentation.

Sensor	Manufacturer/Model	Measured Quantity
Home energy meter	Aeotec/ZW095-C	power, energy
Smart plug	Aeotec/switch 7	power, energy, current, voltage
Sensor and integrated or comfort-presence	Aeotec/Multisensory 6	temperature, illuminance, humidity, movement
Door/window contact	Aeotec/Door window sensor	opening/closing

3.1. Datasets

The model has been calibrated and validated based on two datasets of electrical consumption of homes and single appliances monitored by DHOMUS platform, in the following Dataset A and Dataset B. Dataset A includes energy measurements of 58 appliances in 14 homes over 2.2 years, collected with a 15-min granularity. The start and end times of the measurements are not the same for all dwellings, as reported in Table A1 in Appendix A. Figure 2 depicts the available data for the appliances included in Dataset A. For most of the appliances, the data are recorded over the period from February 2018 to March 2020. Fridges, dishwashers, washing machines, and TVs are the most commonly monitored appliances.

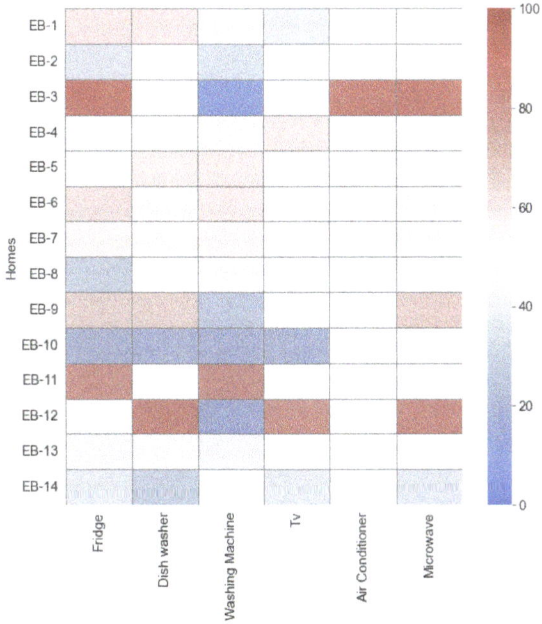

Figure 2. Percentage of available data for Dataset A.

Dataset B includes measurements of 48 appliances from 10 homes from June 2021 to November 2021 and data were collected with a 15-min granularity as reported in Table A2. Figure 3 shows the available data for the appliances in dataset B. Dishwashers and washing machines have the most data available.

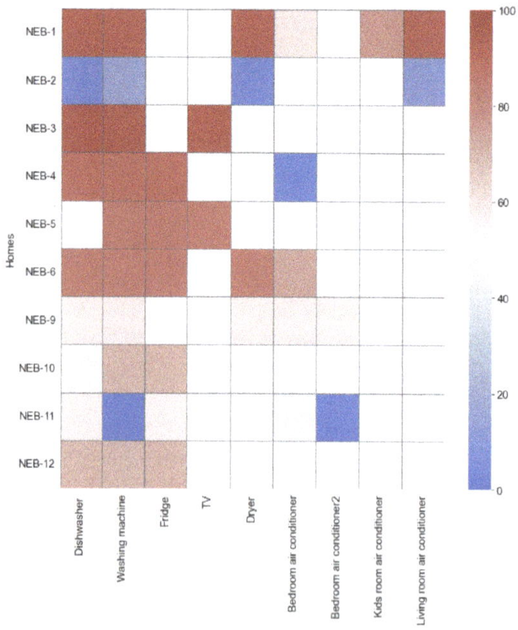

Figure 3. Percentage of available data for Dataset B.

4. Methodology

4.1. Appliance Data Analysis Model

A model has been implemented in the MATLAB® environment to analyse the data collected by the data acquisition system interacting with ENEA's DHOMUS platform. The model performs data mining and a statistical analysis of the energy use of the dwelling energy meter and of single appliances, distinguishing between those that operate with cycles (e.g., washing machine, dishwasher, tumble dryer) and those that operate continuously (e.g., air conditioners, TV, etc.). The model can detect the operation mode (standby, operating, off, monitoring sensor disconnected/not working) of all domestic appliances, as well as the start, end, and length of operation. Moreover, a statistical analysis of electricity consumption is also performed for each operation, determining the total consumption, the average energy consumption on the reference time step, which is equal to 15 min, and the maximum and minimum energy. Additionally, consumption is split into time slots set by the ARERA Authority [46], as well as customised time slots between 8 a.m. and 8 p.m. (with a minimum resolution of one hour). The model then carries out a monthly analysis of the operation and calculates the number of times the appliance is working, the duration of operation, and the total monthly energy, and then it compares these outputs with reference values, based on similar users or benchmarks derived from the technical literature, in order to make assessments on the actual use of the appliance. Additional outputs are calculated for domestic appliances that operate in a cyclic mode. The model is simple, fast, and it can distinguish nearby cycles in which one cycle runs immediately after another. It calculates the number of times the appliance has been used and the corresponding consumption; this information is used to provide households with feedback aimed at improving rational energy use. The algorithms are detailed as follows.

The flow chart in Figure 4 represents the structure of the numerical model. First, the algorithm imports monitoring data from .csv or .txt files exported from the data acquisition system. Data for various homes, different household appliances, and for an arbitrary time span could well be stored in the file. Input files have the following structure (taken from an example file), and the description of parameters is shown in Table 2:

```
home_id, sensor, date,sum_of_energy_of_power, delta_energy, last_value_switch
EnergyBox1, fridge plug,2019-01-01 00:30:00.0000,8.8,0,0
EnergyBox1, fridge plug,2019-01-01 00:45:00.0000,8.8,0,0
EnergyBox1, fridge plug,2019-01-01 01:00:00.0000,8.8,0,0
```

The quantities `sum_of_energy_of_power` and `delta_energy` provide similar information on the consumption of the appliance; therefore, the model uses the former for the energy calculations.

Module Smart Home Analyzer (SHAM) (i.e., module 1 in Figure 4) imports the monitoring data in a table format. Then, it sequentially calls the modules explained below. The call syntax is as follows:
`smart_home_analyser_v2(B_id, B_name, elettrodom, energy_data).`

The required inputs are `B_id`, the ID of the building which represents the energy box (e.g., EB-1 to EB-14 for dataset A, and NEB-1 to NEB-12 for dataset B), `B_name`, the acronym of the building used to plot the results and save the results, `elettrodom`, the name of the appliance (e.g., "dishwasher"), and `energy_data`, a table with the imported quarter-hour monitoring data. The code can recognize all possible appliances installed in the home.

Table 2. Input data for the model.

Parameters	Description
home_id	Name of the energy box
sensor	Name of the sensor associated with the appliance
date	Datetime
sum_of_energy_of_power	Energy provided in Wh on a quarter-hour basis
delta_energy	Energy counter, in Wh or kWh
last_value_switch	sensor parameter (not used)

The module detects the dwelling and the appliance provided as input, and if it cannot find the combination "house, appliance", then it provides an error message and ends as shown in Figure 4.

The function "smart home" (module 2 in Figure 4) is automatically called if the dwelling and the appliance are found in the dataset. This function sorts the imported energy data over time and removes any duplicate; moreover, it verifies the unit of measurement of the energy quantities and, if necessary, converts from kWh to Wh. The module then produces a table-formatted array containing the input quantities sorted through time, checked, and with required units.

The function "appliances" (module 3 in Figure 4) automatically receives the following inputs from the previous function: `B_name` is the acronym of the dwelling (e.g., "EB-1"), `B_appl` and `appl_name` are the abbreviation and label of the appliance, respectively, and they are used to plot and save the results, `cyc_on` is a boolean variable that indicates whether the appliance has a cyclic operation or not, and `t_cdz_i` is the table array produced by the "smart home" function.

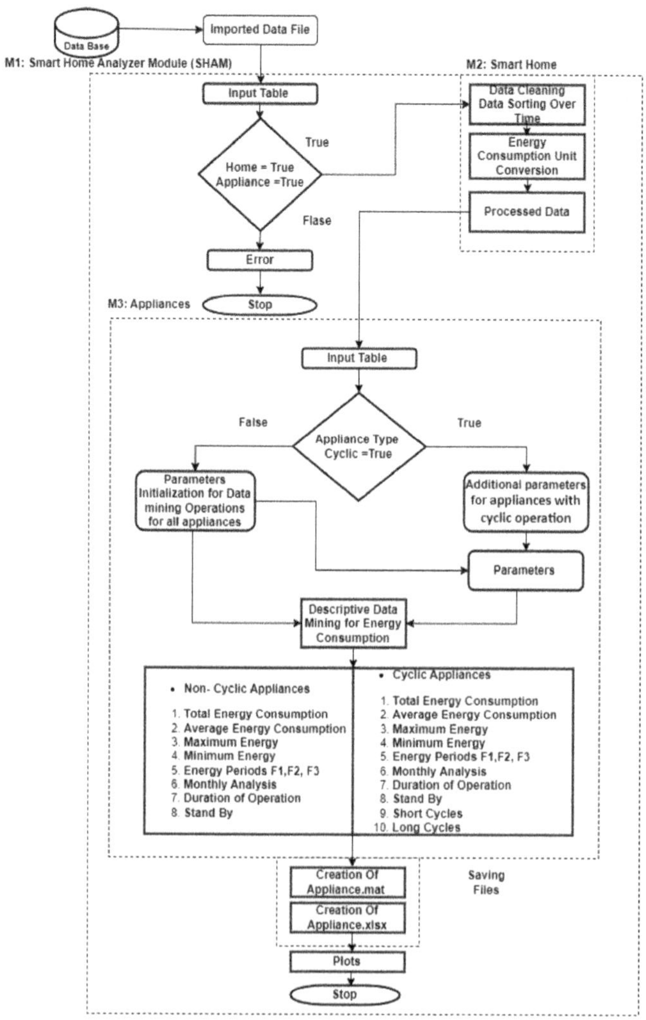

Figure 4. Flow chart of the numerical model.

Furthermore, the "appliance" module performs data analysis and data mining. First, it provides a set of operating features that are used to calculate the appliance standby energy and to setup customised daytime periods. Additional parameters are adopted for the study of individual cycles if the appliance has a cyclic operation:

- Minimum cycle duration, i.e., shorter operations are neglected;
- Minimum cycle energy consumption, i.e., operations with a smaller consumption are neglected;
- Parameters for the separation of nearby cycles;
- Threshold to skip the analysis for those months when insufficient data are available;
- Energy threshold to distinguish low consumption cycles;
- Time threshold for short cycles; and
- Time threshold for long cycles.

These parameters depend on the type of appliance, and therefore they require calibration.

The module creates a table with energy consumption and corresponding datetime, which starts on the first day of the first month and ends on the last day of the last month recorded in the imported dataset. This operation is useful to manage full months, especially to account for new sensors and for relying on a comparable data structure because sensors may be connected to different appliances in different moments.

4.2. Information Analysis

The model relies on consumption during standby mode in order to determine whether the appliance is operative or not. Standby energy (U) is calculated by the following equations,

$$\widehat{F}_n(t) = \frac{1}{n}\sum_{i=1}^{n} E_{i\leqslant t} \tag{1}$$

$$F = \nabla \widehat{F}_n(t) \tag{2}$$

$$U = max(F), \tag{3}$$

where $\widehat{F}_n(t)$represents the empirical cumulative distribution of energy consumption $E_{i\leqslant t}$ function over time, and F represents the gradient of $\widehat{F}_n(t)$. The algorithm assumes that switching from standby to normal operation corresponds to the maximum energy gradient; therefore, the standby consumption is calculated as the energy that determines the maximum gradient of the cumulative distribution of measured consumption.

In actual operation, the appliance may perform several cycles in close succession (e.g., two washes, one immediately after the other for the washing machine). The model can separate cycles in close succession based on their energy-consumption profile. The disaggregation of cycles is computed with the following equations,

$$Y = \nabla \theta_A \tag{4}$$

$$\Delta Y = ZCI(Y), \tag{5}$$

where θ_A is the energy consumption of aggregated cycles, Y is the gradient of consumption and ΔY is the position over time of the relative maxima of energy consumption in aggregated cycles. The $ZCI(Y)$ function interpolates from the points it identifies near the zero crossings to linearly approximate the actual zero crossings. Maxima characterized by energy consumption above the value that corresponds to the percentile set as a parameter of the model (e.g., 85% of the consumption of the aggregated cycle) identify separate cycles and are used by the algorithm to disaggregate nearby cycles. Typically, energy consumption is higher during the first stage of the cycle (i.e., water heating) for washing machines and dishwashers.

As anticipated, the model neglects operations that last less than or that consume less than the thresholds set as input parameters in order to exclude those situations that may not represent real operating cycles. The corresponding data (spurious data) are stored separately in the results.

Then the model allocates energy consumption in time slots. In particular, hours between 8 a.m. and 8 p.m. can be grouped in a customized number of time slots provided as input. Moreover, the algorithm accounts for the standard time slots defined by the Italian Energy Authority (ARERA) as follows [46]:

- F1: from Monday to Friday, from 8 a.m. to 7 p.m. (except holidays);
- F2: from Monday to Friday from 7 a.m. to 8 a.m. and from 7 p.m. to 11 p.m., Saturday from 7 a.m. to 11 p.m., (except holidays); and
- F3: from Monday to Saturday, from 12 a.m. to 7 a.m. and from 11 p.m. to 12 a.m., Sundays and holidays.

For each individual period and cycle of operation, the following quantities are calculated: total energy consumption, average 15-min energy consumption, maximum and minimum consumption, and maximum/minimum ratio as an indicator of the variability of the consumption during operation.

The code then performs a monthly analysis, and calculates the following quantities:

a. Number of records in the dataset;
b. Number of NULL records, i.e., with sensor disconnected or inactive;
c. Number of records with zero consumption;
d. Number of records when the appliance is in standby mode;
e. Number of spurious records, i.e., characterized by either duration or consumption below the input thresholds;
f. Number of records when the appliance is operative;
g. Number of records when the appliance is operative during F1 time slot;
h. Number of records when the appliance is operative during F2 time slot;
i. Number of records when the appliance is operative during F3 time slot; and
j. Number of records when the appliance is operative during the i-th custom time slot, for all daytime slots set as input.

Similar quantities are calculated with reference to the monthly energy consumption. For household appliances with cyclic operation, the following additional quantities are calculated by differentiating cycles according to energy consumption and duration:

- Cycles with consumption above the minimum input value;
- Cycles with duration less than the input threshold S_{d1} (short cycles);
- Cycles with duration between boundaries S_{d1} and S_{d2} (medium cycles); and
- Cycles with duration greater than the input threshold S_{d2} (long cycles).

Moreover, for single cycles and for aggregated cycles (i.e., cycles before the application of the algorithm that distinguish cycles in close sequence) the following monthly quantities are calculated:

- Percentage of records in the month;
- Number of cycles recorded;
- Number of cycles extrapolated for every month, in the presence of missing data (i.e., the sensor returns NULL) and within the limits of acceptability set as input parameter (e.g., months with only 10% of records are discarded, because extrapolation over the month would not be reliable);
- Number of cycles extrapolated on annual basis, that comply similar condition on missing data as for monthly extrapolation;
- Average cycle duration;
- Average cycle consumption;
- Maximum consumption; and
- Minimum consumption.

Finally, the code performs a series of energy balance checks and then results are printed and plotted in graphs, and the relevant variables are saved in .xlsx and .mat files.

5. Results and Discussion

In this section, the analysis on the operation of a washing machine is used as an example to describe the results and the logic of the model. Indeed, this electrical appliance has many features that can be representative of other appliances, i.e., it is cyclic, and it can be used to illustrate the main features of the model. Similar remarks apply to other appliances.

5.1. Calibration of Parameters

As the algorithm can analyse all domestic appliances, the control parameters need to be tuned before deploying the model to the analysis of real data. Hence, Dataset A has been used for calibration of these parameters, which are then are applied to Dataset B in order to validate the model. Table 3 shows the range of parameters and the calibrated values obtained for washing machines. In detail, parameter `st_by_prc` is the reference percentile of energy used to evaluate the usability of standby energy. Parameter `min_dur` represents the minimum length of useful operative cycles, and the calibrated value (30 min) has been obtained from all cyclic appliances. Parameter `min_Wh` represents the minimum energy of useful operative cycles (i.e., 100 Wh for washing machines). All the other parameters are equal for cyclic appliances.

Table 3. Parameter tuning for washing machines.

Parameter	Calibrated Value	Tuning Range
st_by_prc	98	[90–99]
min_dur	30 min	[15 30 45]
min_Wh	100 Wh	[50 100 150 . . . 500]
thr_en	800 Wh	[500 600 700 800]
thr_d1	45 min	[15 30 45]
thr_d2	120 min	[90 105 . . . 500]

5.2. Algorithm Results

The energy analysis starts with the calculation of the standby consumption of the appliance. Figure 5 depicts the cumulative distribution of the energy consumption for a washing machine from which the standby energy is equal to 0.15 Wh. The model uses this value to distinguish among periods of operation and inactivity for the specific appliance.

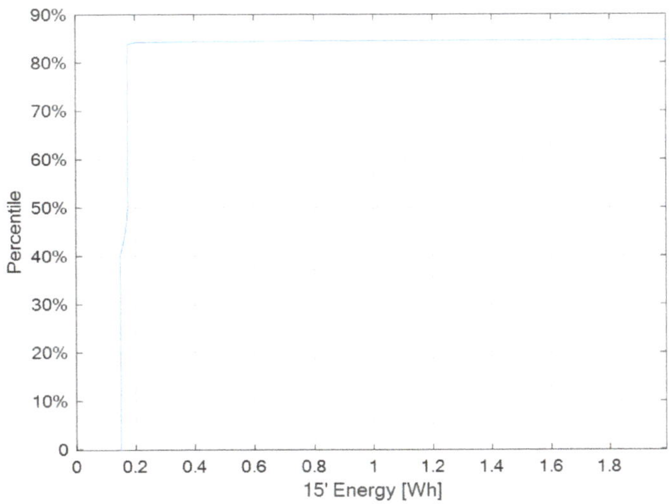

Figure 5. Cumulative distribution of the energy consumption.

The energy consumption profile in a typical cycle for washing machines is shown in Figure 6: an initial peak of consumption corresponding to water heating is followed by a longer phase (depending on the type of cycle set by the user as well as the type, model, and age of the appliance) with lower consumption. The algorithm calculates the gradient of energy demand and finds consumption peaks, which correspond to relative maxima (i.e., zero gradient), for each aggregate cycle (multiple nearby cycles). The consumption of the peaks that separate individual nearby cycles (i.e., the first phase of operation) exceeds a specific percentile of the cycles' consumption, the value of which is selected among the parameters of the model. Once the various cycles have been recognized according to this methodology, the model determines the start and the end time (which corresponds to the lowest consumption before the peak of consumption of the next cycle), and duration of the single cycles. Tests have demonstrated that the algorithm is reliable in distinguishing nearby cycles, as shown in Figure 7. A finer sample period could produce better accuracy, but for the purposes of this investigation, the quarter-hour interval is appropriate.

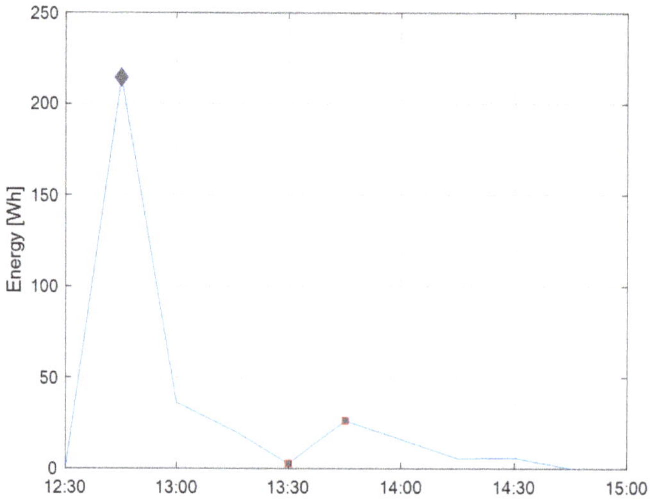

Figure 6. Automatic identification of the phases in the consumption profile of a single washing cycle of a washing machine.

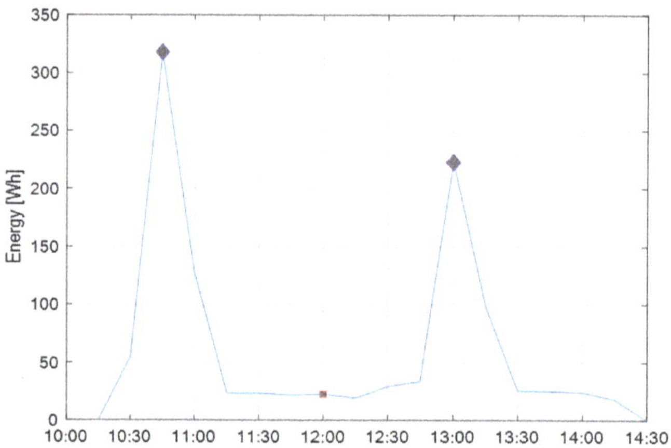

Figure 7. Automatic detection of two consecutive washing cycles.

Figure 8 illustrates a bubble chart with relevant information on cycle length and consumption of the same washing machine. The start hour during the day is displayed (abscissa) versus duration (ordinate), and the consumption is represented by the coloured scale of bubbles, whose size is proportional to the ratio between the highest and lowest quarter-hour consumption in the cycle. Larger bubble sizes in washing machines (and dishwashers) indicate washing cycles at a medium-high temperature.

Figure 9 depicts the subdivision of records (upper chart) and consumption (lower chart) based on the type of operation: appliance operative, standby mode, off, spurious data, and sensor not active (i.e., n/a label). Washing cycles correspond to a small fraction of the time but are associated with most of the consumption, unlike standby mode. Moreover, Figure 10 shows the subdivision of the records on a monthly basis, whereas Figure 11 illustrates the monthly (histogram on the left) and overall subdivision of consumption (pie chart on the right) for ARERA time slots. The upper graphs in Figure 12 show the monthly (histogram on the left) and overall (pie chart on the right) allocation of consumption according to the customized daytime slots, in the specific case three time slots (i.e., 8 a.m. to 12 p.m., 12 p.m. to 4 p.m., 4 p.m. to 8 p.m.), whereas the bottom graphs represent the number of operating hours (histogram on the left) and the overall operating hours (pie chart on the right).

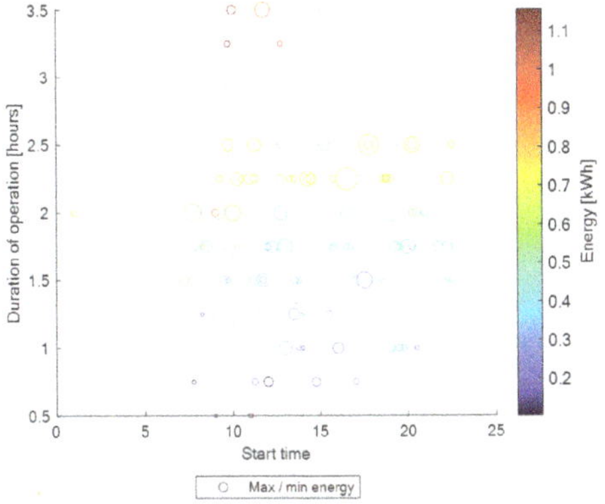

Figure 8. Duration, start time, and energy consumption of a washing machine's cycles.

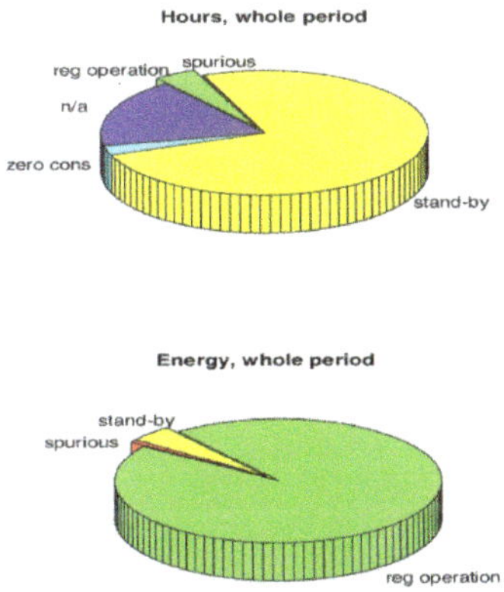

Figure 9. Subdivision of records (upper graph) and consumption (lower graph) based on the type of operation for a washing machine.

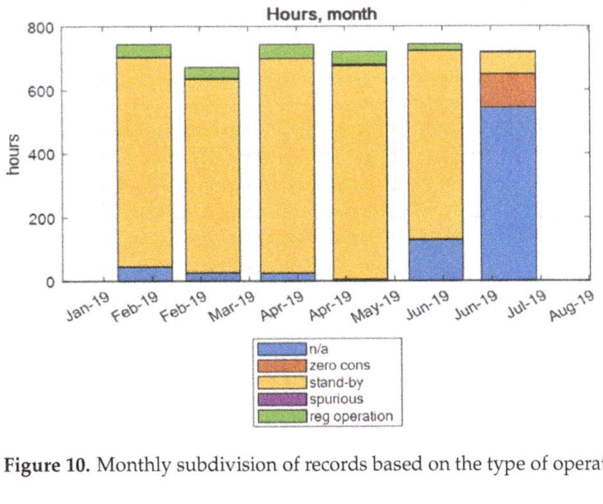

Figure 10. Monthly subdivision of records based on the type of operation for a washing machine.

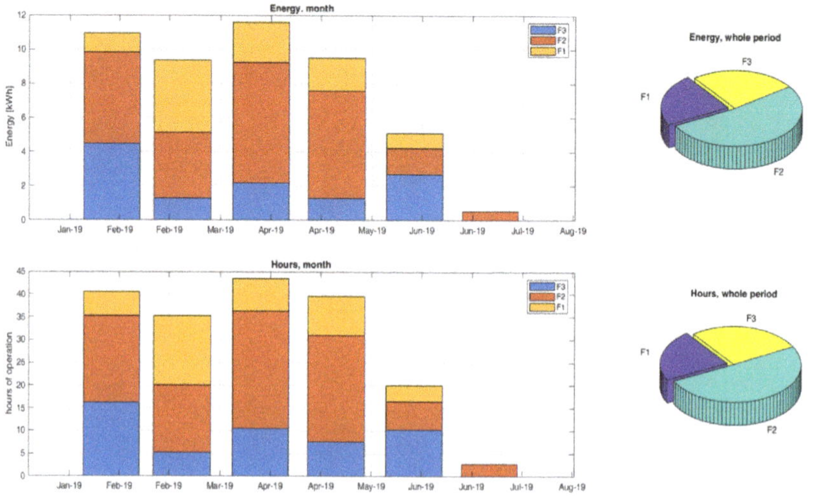

Figure 11. Monthly allocation of consumption in the three time slots defined by ARERA.

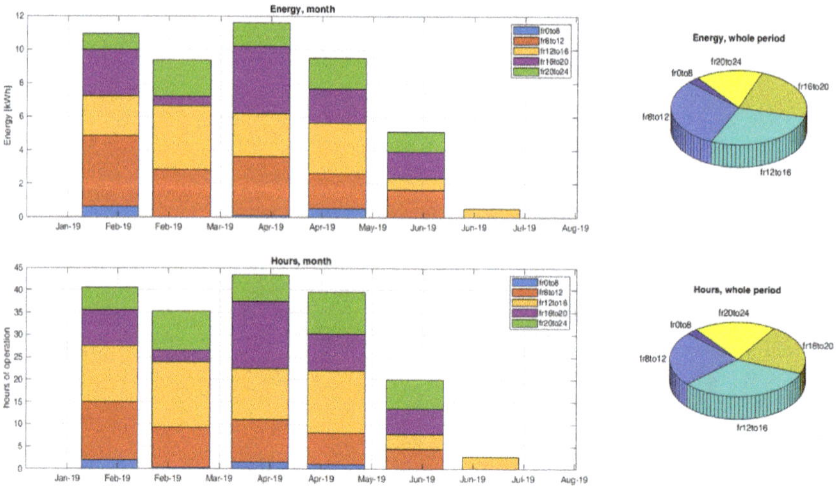

Figure 12. Monthly allocation of consumption according to the user-defined daytime slots.

5.3. *Comparative Analysis*

This section illustrates the comparison of appliances characterized by different energy classes. As described, the proposed algorithm is able to detect many features, and some of them have been selected in order to compare the performance of appliances: total number of cycles (recorded, extrapolated), percentage of consumption during peak hours, percentage of short cycles (recorded, extrapolated), percentage of long cycles (recorded, extrapolated), average energy consumption of short cycles and average energy consumption of long cycles.

Table 4 compares seven washing machines representing three energy classes (A, A++, and A+++) belonging to both datasets described in Section 3.1. Washing machines installed in homes EB-2 and NEB-4 are class A, those installed in homes EB-3, NEB-9, and NEB-10 are class A++, whereas homes EB-9 and NEB1 use washing machines of class A+++. Homes have a different number of residents, and they use the appliance in different manners. The model can determine the usage pattern of households in such different contexts. Furthermore, Table 4 presents the relevant quantities considered in the comparative analysis:

the total number of cycles (Tn_θ) that is subdivided into short and long cycles, the energy consumption during peak hours (PEC), and the percentage of short cycles (P_{θ_s}) and long cycles (P_{θ_L}). The washing machine installed in home EB-3 carried out the largest number of cycles (Tn_θ) in both datasets, as confirmed by the energy consumption of both short cycles (AEC_{θ_s}) and long cycles (AEC_{θ_L}). As regards the energy consumption during peak hours, the washing machine in NEB-9 operated at 96% in this time slot, whereas the household of EB-2 showed a wiser behavior since he run his washing machine only for 28% of the time during peak hours. The percentage of short cycles for the washing machines of NEB-4 and NEB-1 is approximately the same even though the number of short cycles is 23 for the former and 52 for the latter, and the corresponding energy consumption during the analyzed period is 2.263 kWh and 3.264 kWh, respectively. Therefore, the average energy consumption in a single short cycle calculated for these appliances is consistent with their energy labels, i.e., class A and A+++ respectively.

Table 4. Comparative analysis on energy consumption for washing machines.

Home	EC	NoP	Tn_θ		PEC	P_{θ_s}		AEC_{θ_s}	P_{θ_L}		AEC_{θ_L}
			R_θ	Ex_θ	Peak Hours	PR_{θ_s}	PEx_{θ_s}	(kWh)	PR_{θ_L}	PEx_{θ_L}	(kWh)
EB-2	A	2	99	114	28%	6%	6%	1.140	54%	53%	8.523
NEB-4	A	2	31	36	45%	65%	64%	2.263	0	0	0
EB-3	A++	4	340	391	72%	7%	6%	2.062	42%	41%	18.295
NEB-9	A++	3	31	26	96%	9%	88%	3.759	0	0	0
NEB-10	A++	2	80	74	44%	86%	85%	3.662	0	0	0
EB-9	A+++	3	82	21	65%	12%	19%	1.492	4%	0	2.521
NEB-1	A+++	4	76	82	60%	64%	65%	3.264	1%	1%	1.558

Similarly, the results obtained with seven dishwashers monitored in the two datasets are reported from Figures 13–15. Figure 13 depicts the total number of cycles (recorded and extrapolated) in both datasets. The model calculated the highest number of cycles for the dishwasher in EB-12, while the total number of cycles for dishwashers in EB-1, EB-9 and in NEB-1 are similar. Figure 14 illustrates the percentage of short and long cycles for the dishwashers. The highest number of long cycles occurred in EB-1 while dishwashers in NEB-1, NEB-4, NEB-9, and NEB-11 operated only short cycles. Figure 15 depicts the total energy consumption of long (orange bar) and short (green bar) cycles during the analyzed period. The dishwasher in EB-12 consumed more energy during long cycles. By comparing the usage of appliances with their energy labels, it can be seen that even though dishwasher in EB-9 run fewer cycles than the dishwasher in EB-12 (see Figure 13), they consumed a similar amount of energy (see Figure 15). This outcome is consistent with the energy class of those appliances, i.e., class D for the former and class A++ for the latter.

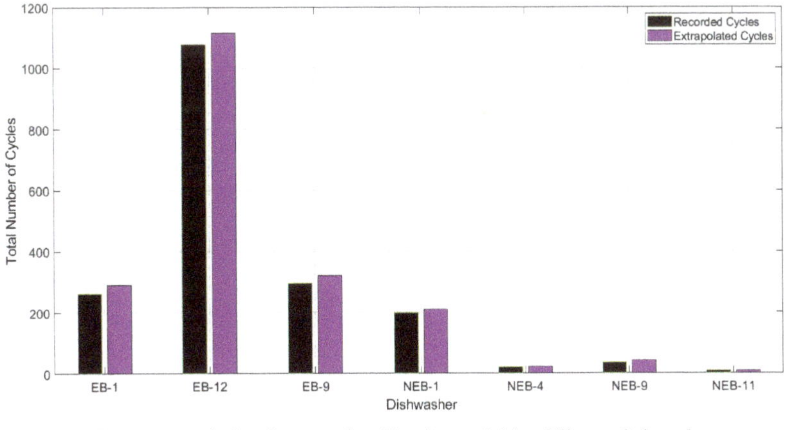

Figure 13. Energy recorded and extrapolated by the model for different dishwashers.

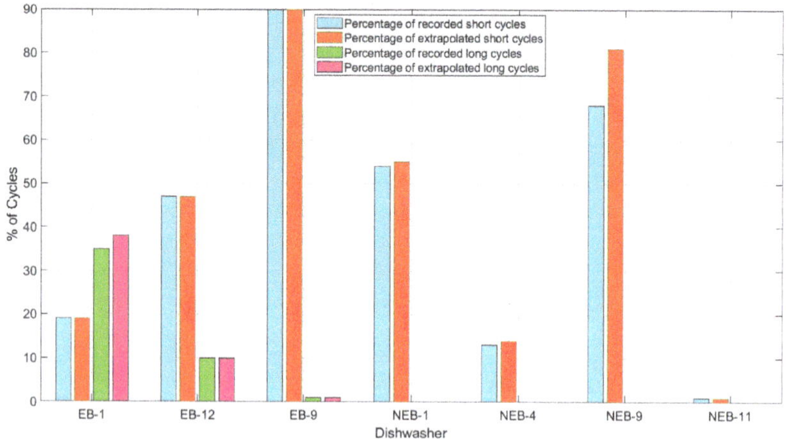

Figure 14. Percentage of short and long cycles for dishwashers.

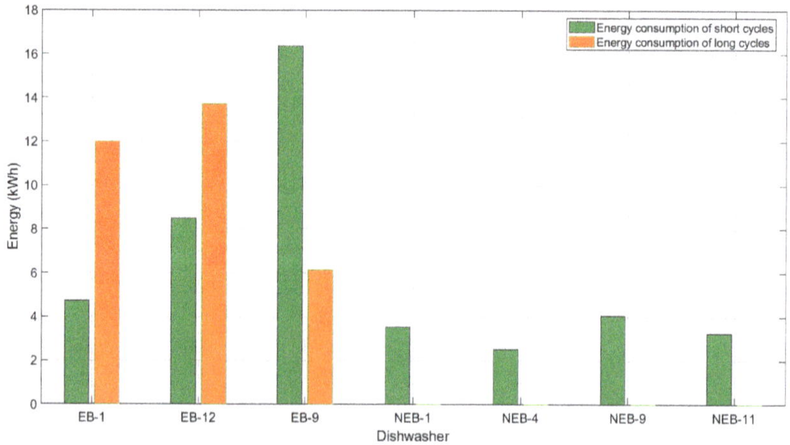

Figure 15. Energy consumption in short and long cycles for dishwashers.

In order to test the model on other datasets found in the literature, we applied it on the GreenD Energy database [47], containing detailed information on energy consumption obtained through a measurement campaign in households in Austria and Italy (December 2013 to November 2014). The results show that our implemented algorithms can extract all patterns for the available appliances in the dataset, including disaggregation of energy consumption cycles occurred in close succession, which is one of the distinctive outcomes of the model, as shown in Figure 16.

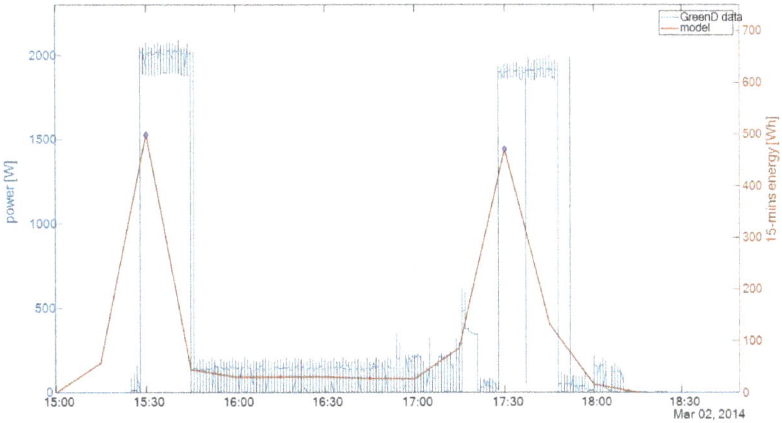

Figure 16. Comparison of energy consumption cycles occurred in close succession for the washing machine.

5.4. Feedbacks to the Consumer in DHOMUS Platform

The implemented model has been integrated in the DHOMUS platform in order to provide users with feedback on the electrical consumption and to compare their energy habits and patterns with other users participating in the experimentation, and with benchmarks as well. The platform provides a web interface for user feedback to encourage their virtuous and conscious behaviours. Moreover, a monthly report presented in Figure 17 contains summary of the calculations carried out by model and some suggestions to promote energy saving. In particular, the monthly report shows the detailed results obtained from the model and related to consumption during peak hours and to the shares of short cycles ("cicli brevi" in Italian) and of long cycles ("cicli lunghi" in Italian). Furthermore, tips are provided to the users, e.g., those reported in Figure 18 for the washing machine (infographics are in Italian since the DHOMUS platform is native for Italian users), which shows a comparison with similar users. Moreover, tips are provided to assess the impact of virtuous behaviours, e.g., the reduction of the number of cycles by using the appliance at full load in order to reduce water consumption as well. On the other hand, Figure 18 shows the comparison of the average consumption with simular customers: washing machine, refrigerator, and dishwasher are compared in a bar graph on the left side of the illustration. The orange bar reflects the average consumption of other users, while the blue bar represents the average consumption of the specific user.

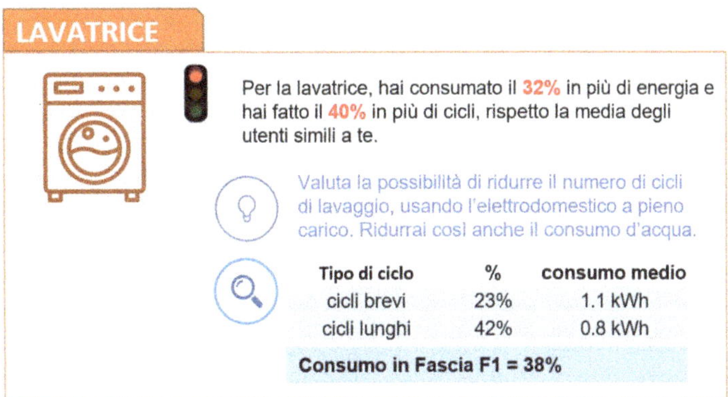

Figure 17. Feedback on the DHOMUS platform regarding the use of washing machine (in Italian).

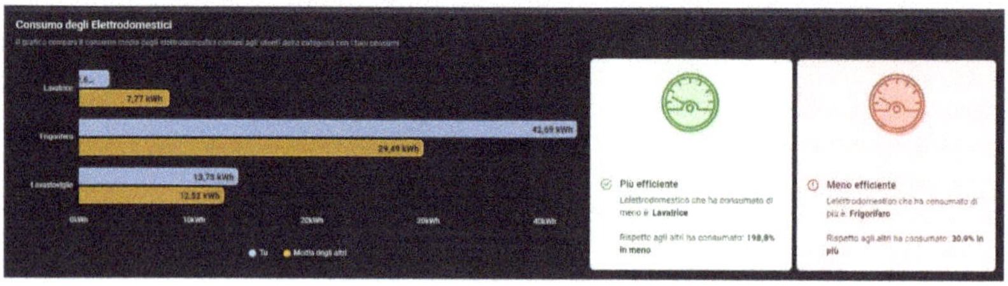

Figure 18. Section of the DHOMUS dashboard reporting the user's consumption vs. average consumption of other similar users (in Italian).

6. Conclusions

In this paper, a method for the detection of energy-consumption patterns from household appliances data for a smart home platform is described. The model is quite simple, computationally not intensive, but effective in detecting patterns, such as the number of times the appliance is on, off, in standby mode, start time, end time, etc. The method can handle both cyclic and noncyclic appliances. For cyclic appliances, by using descriptive data-mining techniques, the model provides quantitative information on the total number of cycles and the disaggregation of cycles in close succession. Moreover, energy consumption computation over customized time periods and disaggregation of aggregated cycles is a distinguishing feature of the proposed model. The model has been applied on a set of appliances for training and for parameter calibration. Validation has been done with a different dataset, demonstrating that the model can efficiently detect energy consumption and usage patterns. Results from two cyclic appliances (washing machine, dishwasher) are presented to show how the model manages energy-consumption patterns, disaggregates independent washing cycles, and performs monthly statistics. Based on these findings, the implemented model has been integrated into DHOMUS, an IoT platform developed by ENEA, to offer users feedback on their consumption patterns and enable them to effectively compare their energy habits with those of similar customers.

The proposed model can be used with sufficient accuracy for all kinds of household appliances. However, the current release of the model can only handle batch processing data with 15-min granularity. Furthermore, in the case of high-frequency data sets (e.g., with 1-s data sampling), a frequency conversion algorithm is required before applying our developed algorithm, because the current version of the model automatically imports dataset where data have been already aggregated every 15 min by the IoT devices. Moreover, the current version of the model cannot be used in a real-time processing environment; the latter feature has been selected among future developments of the numerical code.

Author Contributions: Conceptualization, M.C. and S.R.; methodology, M.C.; software, M.C. and A.H.; validation, M.C. and A.H.; formal analysis, M.C.; investigation, A.H.; resources, S.R. and V.R.; data curation, S.R.; writing—original draft preparation, A.H.; writing—review and editing, M.C. and V.R.; visualization, S.R.; supervision, M.C.; project administration, S.R.; funding acquisition, S.R. All authors have read and agreed to the published version of the manuscript.

Funding: This research and the APC was funded by the "Electrical System Research" Programme Agreement 2019–2021 between ENEA and the Italian Ministry of Ecological Transition, funding number I34I19005780001.

Data Availability Statement: The data presented in this study are available on request from the corresponding author. The data are not publicly available due to privacy restrictions.

Acknowledgments: We thank the anonymous reviewers for providing helpful feedback.

Conflicts of Interest: The authors declare no conflict of interest.

Abbreviations and Symbols

The following abbreviations and symbols are used in this manuscript:

DHOMUS	ENEA's Data Homes and Users Platform
DR	Demand response
EC	Energy class
HEMS	Home Energy Management System
IoT	Internet of Things
NoP	Number of people
SH	Smart home
NILM	Non-intrusive load monitoring
$Tn\theta$	Total Numbers of cycles
$R\theta$	Recorded cycles
$Ex\theta$	Extrapolated cycles
$PEC(F1)$	Percentage of energy consumption during peak hours
$PR\theta_s$	Percentage of recorded short cycles
$PEx\theta_s$	Percentage of extrapolated short cycles
S_{d1}	Lower cycle duration limit (short cycles)
S_{d2}	Higher cycle duration limit (long cycles)
$AEC\theta_s$	Energy consumption of short cycles
$PR\theta_L$	Percentage of recorded long cycles
$PEx\theta_L$	Percentage of extrapolated long cycles
$AEC\theta_L$	Energy consumption of long cycles
$E_{i\leqslant t}$	Energy consumption function over time
$\widehat{F}_n(t)$	Empirical cumulative density function
U	Standby energy
θ_A	Aggregated cycles
Y	Gradient over aggregated cycles

Appendix A

Table A1. Details of Dataset A.

Home	Appliance	Start Date	End Date	Time Span Days	Available Data	Data Holes
EB-1	Fridge	21/02/2018	28/03/2020	766	61.89%	July–August 2019
EB-1	Dish washer	20/02/2018	28/03/2020	767	61.37%	July–August 2019
EB-1	Washing machine	21/02/2018	28/03/2020	766	45.30%	July–November 2019
EB-1	Tv	21/02/2018	28/03/2020	766	37.77%	July–November 2019
EB-2	Fridge	22/02/2018	28/03/2020	765	34.73%	March 2019–March 2020
EB-2	Bimby	22/02/2018	28/03/2020	765	28.60%	March2019–March 2020
EB-2	Washing machine	22/02/2018	28/03/2020	765	33.76%	March2019–March 2020
EB-2	Lamp	22/02/2018	28/03/2020	765	28.57%	March 2019–March 2020
EB-2	Fan heater	22/02/2018	28/03/2020	765	15.89%	March 2019–March 2020
EB-3	Fridge	14/04/2018	28/03/2020	714	87.83%	Nov-18
EB-3	Air conditioner	14/04/2018	28/03/2020	714	85.16%	Nov-18
EB-3	Washing machine	01/03/2018	28/03/2020	758	10.26%	March 2018, November 2018
EB-3	Microwave	14/04/2018	28/03/2020	714	86.41%	Nov-18
EB-4	Vacuum cleaner	02/03/2018	28/03/2020	757	5.58%	February 2020–March 2020
EB-4	Kettle	02/03/2018	28/03/2020	757	44.95%	February 2020–March 2020
EB-4	Washing machine	02/03/2018	28/03/2020	757	52.46%	February 2020–March 2020
EB-4	Tv	02/03/2018	28/03/2020	757	58.91%	February 2020–March 2020
EB-5	Dish washer	02/03/2018	28/03/2020	757	58.27%	March 2018, 2019, January 2020–March 2020
EB-5	Washing machine	02/03/2018	28/03/2020	757	58.68%	March 2018,2019, January 2020–March 2020
EB-5	Coffee machine	02/03/2018	28/03/2020	757	58.96%	March 2018,2019, January 2020–March 2020

Table A1. *Cont.*

Home	Appliance	Start Date	End Date	Time Span Days	Available Data	Data Holes
EB-6	Fridge	09/05/2018	28/03/2020	689	61.57%	June 2019–August 2019, January 2020–March 2020
EB-6	Dish washer	09/05/2018	28/03/2020	689	52.80%	June 2019–August 2019, December 2019, January 2020–March 2020
EB-6	Washing machine	09/05/2018	28/03/2020	689	59.37%	June 2019–August 2019, January 2020–March 2020
EB-6	Microwave	05/03/2018	28/03/2020	754	48.17%	March 2018–April 2018, June 2019–August 2019, December 2019, January 2020–March 2020
EB-7	Fridge	06/03/2018	28/03/2020	753	55.41%	July 2019–March 2020
EB-7	Dish washer	06/03/2018	28/03/2020	753	47.23%	July 2019–March 2020
EB-7	Washing machine	06/03/2018	28/03/2020	753	54.79%	July 2019–March 2020
EB-7	Microwave	06/03/2018	28/03/2020	753	52.13%	July 2019–March 2020
EB-8	Vacuum cleaner	09/03/2018	28/03/2020	750	29.95%	March 2018–April 2018, August 2018–September 2018, February–March 2020
EB-8	Washing machine	06/03/2018	28/03/2020	753	47.61%	March 2018–April 2018, August 2018–September 2018, February–March 2020
EB-8	TV	09/03/2018	28/03/2020	750	29.37%	March 2018–April 2018, August 2018–September 2018, February–March 2020
EB-9	Fridge	05/04/2018	28/03/2020	723	65.16%	April 2018, January 2020–March 2020
EB-9	Dish washer	06/04/2018	28/03/2020	722	65.10%	April 2018, January 2020–March 2020
EB-9	Washing machine	05/04/2018	28/03/2020	723	26.81%	April 2018, January 2020–March 2020
EB-9	Microwave	19/10/2018	28/03/2020	526	66.39%	January 2020–March 2020
EB-9	Fan heater	05/04/2018	28/03/2020	723	64.51%	April 2018, January 2020–March 2020
EB-10	Fridge	04/04/2018	28/03/2020	724	20.77%	August 2018–September 2018, March 2019–May 2019, August 2019–March 2020
EB-10	Dish washer	05/04/2018	28/03/2020	723	20.67%	August 2018–September 2018, March 2019–May 2019, August 2019–March 2020
EB-10	Washing Machine	04/04/2018	28/03/2020	724	20.56%	August 2018–September 2018, March 2019–May 2019, August 2019-March 2020
EB-10	TV	05/04/2018	28/03/2020	723	20.75%	August 2018–September 2018, March 2019–May 2019, August 2019–March 2020
EB-11	Fridge	19/06/2018	08/06/2020	720	79.84%	April–June 2020
EB-11	Washing machine	20/06/2018	08/06/2020	719	79.46%	April–June 2020
EB-11	Lamp	19/06/2018	08/06/2020	720	75.31%	April–June 2020
EB-12	Oven	27/02/2018	31/01/2020	703	82.25%	No Holes
EB-12	Dryer	27/02/2018	31/01/2020	703	15.84%	No Holes
EB-12	Dishwasher	02/03/2018	31/01/2020	700	83.45%	No Holes
EB-12	Microwave	27/02/2018	31/01/2020	703	82.37%	No Holes
EB-12	Washing machine	02/03/2018	31/01/2020	700	17.72%	No Holes
EB-12	TV	27/02/2018	31/01/2020	703	79.11%	No Holes
EB-13	Fridge	20/06/2018	08/06/2020	719	44.85%	July 2019–June 2020
EB-13	Dish washer	13/06/2018	08/06/2020	726	45.74%	July 2019–June 2020
EB-13	Washing machine	13/06/2018	08/06/2020	726	43.07%	July 2019–June 2020
EB-13	Warmer	07/09/2018	08/06/2020	640	41.11%	July 2019–June 2020
EB-13	Oven	07/09/2018	08/06/2020	640	29%	April 2019–June 2018
EB-13	Fridge	12/06/2018	08/06/2020	727	38.06%	April 2019–June 2018
EB-13	Dishwasher	07/09/2018	08/06/2020	640	29.49%	April 2019–June 2018
EB-13	Microwave	12/06/2018	08/06/2020	727	36.31%	April 2019–June 2018
EB-13	TV	12/06/2018	08/06/2020	727	38.01%	April 2019–June 2018

Table A2. Details of Dataset B.

Home	Appliance	Start Date	End Date	Time Span Days	Available Data	Data Holes
NEB-1	Dryer	01/06/2021	30/11/2021	182	92.20%	No Holes
NEB-1	Kids' computer	01/06/2021	30/11/2021	182	40.80%	June–August 2021
NEB-1	Computer studio	01/06/2021	30/11/2021	182	40.81%	June–August 2021
NEB-1	Bedroom air conditioner	01/06/2021	30/11/2021	182	59.20%	October–November 2021
NEB-1	Kids' room air conditioner	01/06/2021	30/11/2021	182	77.08%	No Holes
NEB-1	Living room air conditioner	01/06/2021	30/11/2021	182	92.20%	No Holes
NEB-1	Whole house lighting	01/06/2021	30/11/2021	182	39.86%	June–August 2021
NEB-1	Dishwasher	01/06/2021	30/11/2021	182	92.22%	No Holes
NEB-1	Washing machine	01/06/2021	30/11/2021	182	92.23%	No Holes
NEB-1	Water heater and heat pump	01/06/2021	30/11/2021	182	92.26%	No Holes
NEB-1	Tv Bedroom	01/06/2021	30/11/2021	182	19.22%	June–August 2021, November 2021
NEB-1	Tv living room	01/06/2021	30/11/2021	182	40.81%	June–August 2021
NEB-2	Dryer	01/06/2021	30/11/2021	182	4.80%	June–August 2021, October–November 2021
NEB-2	Bedroom air conditioner	01/06/2021	30/11/2021	182	0%	Whole Period
NEB-2	Living room air conditioner	01/06/2021	30/11/2021	182	10.61%	June–September 2021
NEB-2	Dishwasher	01/06/2021	30/11/2021	182	4.82%	June–August 2021, October–November 2021
NEB-2	Washing machine	01/06/2021	30/11/2021	182	15.21%	June–August 2021, June–August 2021, November 2021
NEB-2	Water heater and heat pump	01/06/2021	30/11/2021	182	4.79%	June–August 2021, June–August 2021, November 2021
NEB-3	Dishwasher	01/06/2021	30/11/2021	182	92.90%	No Holes
NEB-3	Washing machine	01/06/2021	30/11/2021	182	92.94%	No Holes
NEB-3	Tv	01/06/2021	30/11/2021	182	92.87%	No Holes
NEB-4	Bedroom air conditioner	01/06/2021	30/11/2021	182	3.32%	June–July 2021, September–November 2019
NEB-4	Fridge	01/06/2021	30/11/2021	182	87.77%	No Holes
NEB-4	Dishwasher	01/06/2021	30/11/2021	182	87.70%	No Holes
NEB-4	Washing machine	01/06/2021	30/11/2021	182	87.21%	No Holes
NEB-5	Coffee machine	01/06/2021	30/11/2021	182	83.43%	No Holes
NEB-5	Fridge	01/06/2021	30/11/2021	182	84.11%	No Holes
NEB-5	Washing machine	01/06/2021	30/11/2021	182	84.11%	No Holes
NEB-5	TV	01/06/2021	30/11/2021	182	84.07%	No Holes
NEB-6	Dryer	01/07/2021	30/11/2021	152	83.75%	No Holes
NEB-6	Fridge	01/07/2021	30/11/2021	152	83.85%	No Holes
NEB-6	Washing machine	01/07/2021	30/11/2021	152	83.85%	No Holes
NEB-6	Dishwasher	01/07/2021	30/11/2021	152	83.84%	No Holes
NEB-6	Bedroom air conditioner	01/07/2021	30/11/2021	152	73.49%	No Holes
NEB-9	Washing machine	01/06/2021	30/11/2021	182	56.65%	November 2021
NEB-9	Dishwasher	01/06/2021	30/11/2021	182	56.57%	November 2021
NEB-9	Bedroom air conditioner	01/06/2021	30/11/2021	182	56.65%	November 2021
NEB-9	Bedroom air conditioner2	01/06/2021	30/11/2021	182	56.63%	November 2021
NEB-9	Dryer	01/06/2021	30/11/2021	182	56.60%	November 2021
NEB-10	Washing machine	01/06/2021	30/11/2021	182	69.66%	No Holes
NEB-10	Dishwasher	01/06/2021	30/11/2021	182	48.88%	No Holes
NEB-10	Fridge	01/06/2021	30/11/2021	182	69.63%	No Holes
NEB-11	Washing machine	01/06/2021	30/11/2021	182	2.70%	July–November 2021
NEB-11	Dishwasher	01/06/2021	30/11/2021	182	55.68%	No Holes
NEB-11	Fridge	01/06/2021	30/11/2021	182	55.65%	No Holes
NEB-11	Bedroom air conditioner	01/06/2021	30/11/2021	182	45.57%	October–November2021
NEB-11	Air conditioner hobby room	01/06/2021	30/11/2021	182	2.47%	July–November2021
NEB-12	Fridge	01/06/2021	30/11/2021	182	69.75%	No Holes
NEB-12	Washing machine	01/06/2021	30/11/2021	182	68.89%	No Holes
NEB-12	Dishwasher	01/06/2021	30/11/2021	182	69.72%	No Holes

References

1. Gan, P.; Li, Z. An econometric study on long-term energy outlook and the implications of renewable energy utilization in Malaysia. *Energy Policy* **2008**, *36*, 890–899. [CrossRef]
2. Herrero, S.; Nicholls, L.; Strengers, Y. Smart home technologies in everyday life: Do they address key energy challenges in households? *Curr. Opin. Environ. Sustain.* **2018**, *31*, 65–70. [CrossRef]
3. Jorde, D.; Jacobsen, H. Event detection for energy consumption monitoring. *IEEE Trans. Sustain. Comput.* **2020**, *6*, 703–709. [CrossRef]
4. Armel, K.; Gupta, A.; Shrimali, G.; Albert, A. Is disaggregation the holy grail of energy efficiency? The case of electricity. *Energy Policy* **2013**, *52*, 213–234. [CrossRef]
5. Lopez, J.; Pouresmaeil, E.; Canizares, C.; Bhattacharya, K.; Mosaddegh, A.; Solanki, B. Smart residential load simulator for energy management in smart grids. *IEEE Trans. Ind. Electron.* **2018**, *66*, 1443–1452. [CrossRef]
6. Li, P.; Wang, H.; Zhang, B. A distributed online pricing strategy for demand response programs. *IEEE Trans. Smart Grid* **2017**, *10*, 350–360. [CrossRef]
7. AlHammadi, A.; AlZaabi, A.; AlMarzooqi, B.; AlNeyadi, S.; AlHashmi, Z.; Shatnawi, M. Survey of IoT-based smart home approaches. In Proceedings of the 2019 Advances In Science And Engineering Technology International Conferences (ASET), Dubai, United Arab Emirates, 26 March–10 April 2019; pp. 1–6.
8. Barsim, K.; Streubel, R.; Yang, B. An approach for unsupervised non-intrusive load monitoring of residential appliances. In Proceedings of the 2nd International Workshop On Non-Intrusive Load Monitoring, Austin, TX, USA, 3 June 2014; pp. 1–5.
9. Liu, Y.; Qiu, B.; Fan, X.; Zhu, H.; Han, B. Review of smart home energy management systems. *Energy Procedia* **2016**, *104*, 504–508. [CrossRef]
10. Yang, J.; Zhao, J.; Wen, F.; Kong, W.; Dong, Z. Mining the big data of residential appliances in the smart grid environment. In Proceedings of the 2016 IEEE Power And Energy Society General Meeting (PESGM), Boston, MA, USA, 17–21 July 2016; pp. 1–5.
11. Errapotu, S.; Wang, J.; Gong, Y.; Cho, J.; Pan, M.; Han, Z. SAFE: Secure appliance scheduling for flexible and efficient energy consumption for smart home IoT. *IEEE Internet Things J.* 2018, *5*, 4380–4391. [CrossRef]
12. Jaradat, A.; Lutfiyya, H.; Haque, A. Appliance Operation Modes Identification Using Cycles Clustering. *arXiv* **2021**, arXiv:2101.10472.
13. Wang, Z.; Srinivasan, R. Classification of household appliance operation cycles: A case-study approach. *Energies* **2015**, *8*, 10522–10536. [CrossRef]
14. Silva, B.; Khan, M.; Han, K. Load balancing integrated least slack time-based appliance scheduling for smart home energy management. *Sensors* **2018**, *18*, 685. [CrossRef] [PubMed]
15. Chen, M.; Lin, C. Standby power management of a smart home appliance by using energy saving system with active loading feature identification. *IEEE Trans. Consum. Electron.* **2018**, *65*, 11–17. [CrossRef]
16. Prudenzi, A. A neuron nets based procedure for identifying domestic appliances pattern-of-use from energy recordings at meter panel. In Proceedings of the 2002 IEEE Power Engineering Society Winter Meeting. Conference Proceedings (Cat. No. 02CH37309), New York, NY, USA, 27–31 January 2002; Volume 2, pp. 941–946.
17. Wang, H.; Yang, W. An iterative load disaggregation approach based on appliance consumption pattern. *Appl. Sci.* **2018**, *8*, 542. [CrossRef]
18. Hussain, H.; Javaid, N.; Iqbal, S.; Hasan, Q.; Aurangzeb, K.; Alhussein, M. An efficient demand side management system with a new optimized home energy management controller in smart grid. *Energies* **2018**, *11*, 190. [CrossRef]
19. Du, S.; Li, M.; Han, S.; Shi, J.; Li, H. Multi-pattern data mining and recognition of primary electric appliances from single non-intrusive load monitoring data. *Energies* **2019**, *12*, 992. [CrossRef]
20. Ashouri, M.; Haghighat, F.; Fung, B.; Lazrak, A.; Yoshino, H. Development of building energy saving advisory: A data mining approach. *Energy Build.* **2018**, *172*, 139–151. [CrossRef]
21. Gajowniczek, K.; Ząbkowski, T. Data mining techniques for detecting household characteristics based on smart meter data. *Energies* **2015**, *8*, 7407–7427. [CrossRef]
22. Yu, Z.; Fung, B.; Haghighat, F. Extracting knowledge from building-related data—A data mining framework. *Build. Simul.* **2013**, *6*, 207–222. [CrossRef]
23. Ashouri, M.; Fung, B.; Haghighat, F.; Yoshino, H. Systematic approach to provide building occupants with feedback to reduce energy consumption. *Energy* **2020**, *194*, 116813. [CrossRef]
24. Athanasiadis, C.; Doukas, D.; Papadopoulos, T.; Chrysopoulos, A. A scalable real-time non-intrusive load monitoring system for the estimation of household appliance power consumption. *Energies* **2021**, *14*, 767. [CrossRef]
25. Athanasiadis, C.; Papadopoulos, T.; Doukas, D. Real-time non-intrusive load monitoring: A light-weight and scalable approach. *Energy Build.* **2021**, *253*, 111523. [CrossRef]
26. Ge, L.; Liao, W.; Wang, S.; Bak-Jensen, B.; Pillai, J. Modeling daily load profiles of distribution network for scenario generation using flow-based generative network. *IEEE Access* **2020**, *8*, 77587–77597. [CrossRef]
27. Fischer, D.; Härtl, A.; Wille-Haussmann, B. Model for electric load profiles with high time resolution for German households. *Energy Build.* **2015**, *92*, 170–179. [CrossRef]
28. Widen, J.; Lundh, M.; Vassileva, I.; Dahlquist, E.; Elleg, K.; Wäckelg, E. Constructing load profiles for household electricity and hot water from time-use data—Modelling approach and validation. *Energy Build.* **2009**, *41*, 753–768. [CrossRef]

29. Wijaya, T.; Vasirani, M.; Humeau, S.; Aberer, K. Cluster-based aggregate forecasting for residential electricity demand using smart meter data. In Proceedings of the 2015 IEEE International Conference On Big Data (Big Data), Santa Clara, CA, USA, 29 October–1 November 2015; pp. 879–887.
30. Khan, Z.; Jayaweera, D.; Alvarez-Alvarado, M. A novel approach for load profiling in smart power grids using smart meter data. *Electr. Power Syst. Res.* **2018**, *165*, 191–198. [CrossRef]
31. Sousa, J.; Neves, L.; Jorge, H. Assessing the relevance of load profiling information in electrical load forecasting based on neural network models. *Int. J. Electr. Power Energy Syst.* **2012**, *40*, 85–93. [CrossRef]
32. Hayn, M.; Bertsch, V.& Fichtner, W. Electricity load profiles in Europe: The importance of household segmentation. *Energy Res. Soc. Sci.* **2014**, *3*, 30–45. [CrossRef]
33. Yilmaz, S.; Chambers, J.; Patel, M. Comparison of clustering approaches for domestic electricity load profile characterisation-Implications for demand side management. *Energy* **2019**, *180*, 665–677. [CrossRef]
34. Craig, T.; Polhill, J.; Dent, I.; Galan-Diaz, C.; Heslop, S. The north east scotland energy monitoring project: Exploring relationships between household occupants and energy usage. *Energy Build.* **2014**, *75*, 493–503. [CrossRef]
35. Martinez-Pabon, M.; Eveleigh, T.; Tanju, B. Smart meter data analytics for optimal customer selection in demand response programs. *Energy Procedia* **2017**, *107*, 49–59. [CrossRef]
36. Zhou, K.; Fu, C.; Yang, S. Big data driven smart energy management: From big data to big insights. *Renew. Sustain. Energy Rev.* **2016**, *56*, 215–225. [CrossRef]
37. Zhou, K.; Yang, S. Understanding household energy consumption behavior: The contribution of energy big data analytics. *Renew. Sustain. Energy Rev.* **2016**, *56*, 810–819. [CrossRef]
38. Laicane, I.; Blumberga, D.; Blumberga, A.; Rosa, M. Reducing household electricity consumption through demand side management: The role of home appliance scheduling and peak load reduction. *Energy Procedia* **2015**, *72*, 222–229. [CrossRef]
39. Gyamfi, S.; Krumdieck, S.; Urmee, T. Residential peak electricity demand response—Highlights of some behavioural issues. *Renew. Sustain. Energy Rev.* **2013**, *25*, 71–77. [CrossRef]
40. Azarova, V.; Cohen, J.; Kollmann, A.; Reichl, J. Reducing household electricity consumption during evening peak demand times: Evidence from a field experiment. *Energy Policy* **2020**, *144*, 111657. [CrossRef]
41. Weber, S.; Puddu, S.; Pacheco, D. Move it! How an electric contest motivates households to shift their load profile. *Energy Econ.* **2017**, *68*, 255–270. [CrossRef]
42. Crosbie, T.; Broderick, J.; Short, M.; Charlesworth, R.; Dawood, M. Demand response technology readiness levels for energy management in blocks of buildings. *Buildings* **2018**, *8*, 13. [CrossRef]
43. Issi, F.; Kaplan, O. The determination of load profiles and power consumptions of home appliances. *Energies* **2018**, *11*, 607. [CrossRef]
44. Wenninger, M.; Maier, A.; Schmidt, J. DEDDIAG, a domestic electricity demand dataset of individual appliances in Germany. *Sci. Data* **2021**, *8*, 176. [CrossRef]
45. Clerici Maestosi,P. *Sustainable Urban Transitions*; Maggioli Publisher: Santarcangelo di Romagna, Italy, 2021.
46. Aggiornamento delle Fasce Orarie Con Decorrenza 1 Gennaio 2007. ARERA. Available online: https://www.arera.it/it/docs/06/181-06.html (accessed on 2 August 2022).
47. Monacchi, A.; Egarter, D.; Elmenreich, W.; D'Alessandro, S.; Tonello, A. GREEND: An energy consumption dataset of households in Italy and Austria. In Proceedings of the 2014 IEEE International Conference On Smart Grid Communications (SmartGridComm), Venice, Italy, 3–6 November 2014; pp. 511–516.

smart cities

MDPI

Article

A Cloud-Based Data Storage and Visualization Tool for Smart City IoT: Flood Warning as an Example Application

Victor Ariel Leal Sobral [1,2,*], Jacob Nelson [3], Loza Asmare [4], Abdullah Mahmood [4], Glen Mitchell [3], Kwadwo Tenkorang [1], Conor Todd [4], Bradford Campbell [1,2] and Jonathan L. Goodall [3]

1 Department of Electrical and Computer Engineering, University of Virginia, Charlottesville, VA 22904, USA; bradjc@virginia.edu (B.C.)
2 Department of Computer Science, University of Virginia, Charlottesville, VA 22904, USA
3 Department of Civil and Environmental Engineering, University of Virginia, Charlottesville, VA 22904, USA; jn8vc@virginia.edu (J.N.); goodall@virginia.edu (J.L.G.)
4 Department of Systems and Information Engineering, University of Virginia, Charlottesville, VA 22904, USA
* Correspondence: sobral@virginia.edu

Abstract: Collecting, storing, and providing access to Internet of Things (IoT) data are fundamental tasks to many smart city projects. However, developing and integrating IoT systems is still a significant barrier to entry. In this work, we share insights on the development of cloud data storage and visualization tools for IoT smart city applications using flood warning as an example application. The developed system incorporates scalable, autonomous, and inexpensive features that allow users to monitor real-time environmental conditions, and to create threshold-based alert notifications. Built in Amazon Web Services (AWS), the system leverages serverless technology for sensor data backup, a relational database for data management, and a graphical user interface (GUI) for data visualizations and alerts. A RESTful API allows for easy integration with web-based development environments, such as Jupyter notebooks, for advanced data analysis. The system can ingest data from LoRaWAN sensors deployed using The Things Network (TTN). A cost analysis can support users' planning and decision-making when deploying the system for different use cases. A proof-of-concept demonstration of the system was built with river and weather sensors deployed in a flood prone suburban watershed in the city of Charlottesville, Virginia.

Keywords: Internet of Things; smart cities; environmental monitoring; LoRaWAN; cloud computing; AWS; data management; cost analysis

Citation: Leal Sobral, V.A.; Nelson, J.; Asmare, L.; Mahmood, A.; Mitchell, G.; Tenkorang, K.; Todd, C.; Campbell, B.; Goodall, J.L. A Cloud-Based Data Storage and Visualization Tool for Smart City IoT: Flood Warning as an Example Application. *Smart Cities* **2023**, *6*, 1416–1434. https://doi.org/10.3390/smartcities6030068

Academic Editors: Francisco Sánchez-Sutil and Antonio Cano-Ortega

Received: 15 April 2023
Revised: 13 May 2023
Accepted: 16 May 2023
Published: 19 May 2023

1. Introduction

Recent advances in information and communication technologies (ICT) are enabling Internet of Things (IoT) smart city projects to collect and analyze vast amounts of data in an effort to support more environmentally and economically sustainable communities [1,2]. For instance, smart stormwater projects have shown successful IoT-based infrastructure-monitoring applications to address communities' operation and planning challenges [3–5]. As IoT devices become more pervasive, the collected data is expected to play an increasingly central role to inform communities' decisions and, therefore, it is critical to develop and maintain cyber infrastructure to collect, store, and visualize sensor data.

However, as a growing number of new ICT technologies become available, the task of developing and integrating hardware and software solutions for IoT smart city projects can demand extensive specialized knowledge in different ICT domains [6,7], which can be challenging for IoT system designers. To reduce IoT systems' design effort and to make IoT solutions more accessible, The Things Industry (TTI) [8] created and sponsored The Things Network (TTN) [9], a set of open-source tools to provide the basic software infrastructure to deploy IoT sensors based on LoRaWAN [10,11], a low-power and wide-area network (LPWAN) wireless communication protocol. This open-source project enables contributors around the globe to publicly share TTN compatible gateways that can connect LoRaWAN sensors to a network server known as The Things Stack, which is maintained by TTI. The use of TTN for smart city projects has been successfully demonstrated in the literature for different applications (e.g., [3,12]), while it also benefits communities by creating an open LoRaWAN communication infrastructure that can be leveraged by other IoT projects such as air quality monitoring [13].

Although deploying an IoT system is greatly simplified by using TTN tools, their goal is to provide only the network server infrastructure and leave the application server to be developed by users. For instance, long-term data storage, graphical user interfaces (e.g., plotting tools), and the capacity to send alarm notifications are functionalities not supported by TTN's network server. To achieve such functionalities, users need to develop their own application server or adopt third-party service providers such as Ubidots [14] and myDevices [15]. Another possible solution is to develop a custom server using a TTN open-source networking solution and modify it to include application layer functionalities; however, this solution implies an increased server workload and code maintenance requirements when compared to only developing and hosting application layer functions.

While third-party application servers might provide great value to many applications, users might still decide to develop their own application server solution to achieve more control over their data, to create customized application solutions, or to reduce recurring costs. However, developing an application server implies selecting, developing, and integrating software modules to achieve the application's goals, which can be challenging due to the large diversity of architecture options and software solutions currently available as commercial products and open-source modules. In this context, IoT application case studies can offer users a valuable insight into developing and integrating software systems to meet application goals. To help guide users on the path of creating integrated IoT smart city applications, we introduce our use case of a flood warning system for a suburban watershed in Virginia, USA. Our system uses a pressure sensor and two ultrasonic sensors to monitor water levels at three locations on the stream network, and a weather station to monitor precipitation rates. All our monitoring devices use LoRAWAN to communicate to TTN's network server. We developed and integrated a scalable set of cloud-based application tools to perform long-term data storage, data visualization, and automated alarm notification functionality. We discuss the implementation challenges and insights for our system, as well as a cost analysis using Amazon Web Services (AWS). To support users' planning and decision-making, we included a cost analysis section where we evaluate how costs currently evolve with time, number of sensors, and data storage requirements.

This paper's main contributions can be summarized as: (1) our work provides practical insights on the development of cloud-based tools for IoT applications, an emerging area that is frequently overlooked in empirical IoT research; (2) we propose a general cloud backend system architecture that can guide IoT developers to quickly prototype smart city applications by using our demonstrated tools such as serverless data ingestion for IoT historical backup data storage, on-demand MySQL database and Grafana servers, and a RESTful API for programmatical data access; and (3) we perform a cost analysis for the first few years of using AWS cloud services in an IoT application, highlighting the cost-effectiveness of our proposed solution, and providing to IoT developers a cost estimate of these cloud services under a varying number of sensors and data rate.

This work is organized as follows: in Section 2, we present an overview of related works on IoT for smart city projects that share similarities with our solution; in Section 3, we introduce our example application along with its objective and goals; in Section 4, we present the use case, the adopted overall system architecture, and the design requirements; in Section 5, we present our main results and discussion; finally, in Section 6, we present our final conclusions.

2. Related Work

To illustrate some of the possible IoT architecture solutions to smart city projects, we selected two works, the first one targeting Radon gas concentration monitoring [16] and the second one a smart stormwater system using LoRaWAN and TTN [3]. The Radon gas concentration monitoring work was selected to represent a typical IoT project, where the study made use of available components and tools to build its own remote sensing solution. The smart stormwater system work was selected as an example of a similar application goal using LoRaWAN and TTN, but one adopting alternative design components to our system.

2.1. Radon Gas Monitoring Application

To monitor the concentrations of Radon gas in indoor locations, the authors of [16] presented an IoT system that collects and transmits data to a remote server where values are stored. We summarize the IoT system architecture used for the Radon gas application in Figure 1.

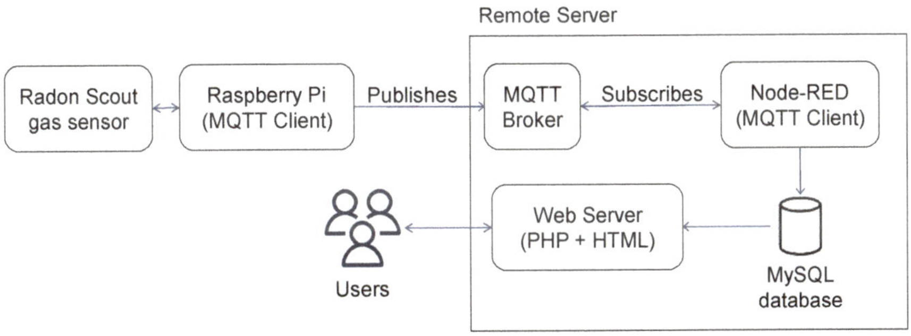

Figure 1. IoT system architecture diagram for the Radon gas monitoring application.

As sensing devices, the authors adopted a commercially available Radon Scout gas sensor connected to a Raspberry Pi 3 device, used as a controller, and connected to the internet. The Raspberry Pi was programmed to read and transmit sensor data to their remote server through a message queuing telemetry transport protocol (MQTT) [17] communication interface. The server receives sensor data through a MQTT broker that publishes received messages to a subscribed MQTT client managed by a Node-RED [18] application responsible for parsing and storing the data in a MySQL database [19]. Finally, a web server interface was created to read the database and display a table of stored sensor readings to users.

For our flood warning use case, we adopted a commercially available LoRaWAN gateway and sensors. While our LoRaWAN gateway required internet connectivity similarly to the Raspberry Pi controller used in this related work [16], the LoRaWAN sensors can be deployed hundreds of meters away from the gateway, which allowed us to reach our desired deployment locations. We used TTN as our network server to register and manage devices, reducing development time and enabling better scalability as new sensors only need to be registered to our TTN application. For this related work [16], authors needed to individually configure the MQTT clients in each one of their Raspberry Pi devices to publish sensor measurements to their server's MQTT broker, as well as individually manage any security key. Instead of using Node-RED to parse and ingest data as adopted in this related work [16], we used a python script to manage the data ingestion and parsing system that periodically receives data from our TTN application through a MQTT client. As our data storage solution, we also adopted a MySQL database, as similarly presented in [16], but we also decided to create a dedicated long-term cloud-based storage solution using AWS S3 [20] as a backup to the MySQL database. For this long-term data storage backup, we used AWS Lambda [21] service to create a serverless and independent data ingestion solution to periodically request data from TTN storage integration and store it in AWS S3.

Instead of displaying sensor data through a website server, we created a dashboard on a Grafana application [22] to plot relevant sensor information such as measurements and battery voltage level.

Although the authors of this related work were targeting an indoor Radon gas monitoring application, some of their system components could be adopted by other IoT applications such as in collecting and displaying data from LoRaWAN sensors connected to TTN. For instance, TTN offers a MQTT Broker service that can publish received LoRaWAN messages to subscribed clients, making it possible to re-use the server infrastructure described in [16] by updating the broker address, client credentials, parsing function, database configuration, and sensor measurement variables.

2.2. Smart Stormwater System Application

For the stormwater monitoring system introduced in [3], the authors deployed a set of sensors around the Illawarra-Shoalhaven region in Australia. Their sensors relied on either the LoRaWAN or 4G cellular network to communicate, depending on each sensor's required data rate. Sensing devices included water-level sensors, tipping bucket rain gauges sensors, pressure and humidity sensors, lagoon monitoring devices, and a culvert blockage monitoring system. To receive data collected by the LoRaWAN based sensors, the authors deployed a network of TTN gateways in the study region. This gateway infrastructure was also seen by the authors as an investment to support other future applications including education-related projects. We summarize the IoT system architecture used for the smart stormwater system application in Figure 2.

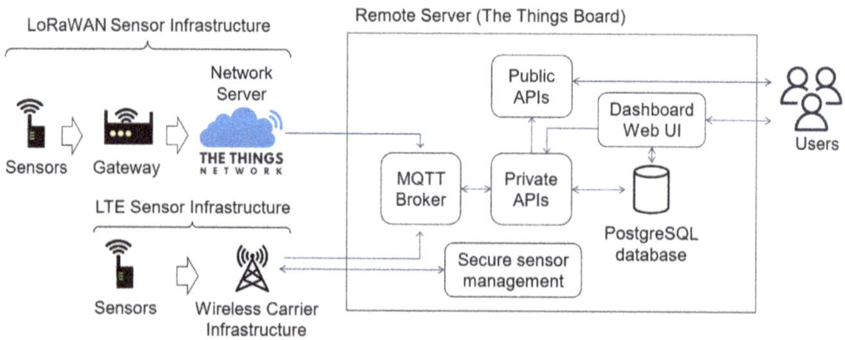

Figure 2. IoT system architecture diagram for the stormwater monitoring application.

To store and display the collected data, authors of [3] adopted the open-source solution provided by the ThingsBoard [23], using the MQTT protocol to receive LoRaWAN sensor data from TTN and store it in a PostgreSQL database. ThingsBoard also provides alerting and graphical interface tools to generate custom dashboards to display sensor data in real time and send automated alert messages. The authors did not specify whether the server solution was hosted in a computer owned by them or a cloud solution, however, ThingsBoard offers a platform as a service solution where they host their system in the cloud with pricing currently ranging from USD 10/month (Maker) to USD 749/month (Business).

Despite offering data storage, API access, and visualization tools, Thingsboard is a turnkey software solution that requires users to have an always-running server to ingest, store, and visualize data. On the other hand, our solution leverages cloud services to break down data ingestion from other on-demand uses, namely: (1) a serverless data ingestion and storage cloud application using AWS Lambda [21] and AWS S3 [20]; (2) a virtual machine instance with a MySQL database to provide responsive data access; and (3) a second virtual machine hosting a Grafana server [22] to provide data visualization. Our serverless data ingestion solution requires only a few lines of code to query sensor data from TTN, parse, and store data as a csv file in AWS S3, thus reducing the complexity to manage bugs and update the system when compared to full servers such as ThingsBoard [23]. Using dedicated virtual machines for a database and visualization allows for tailored resource allocation based on the application needs, the flexibility to provide only the needed service, and code isolation to facilitate upgrading, adding, or switching services (e.g., replacing MySQL with PostgreSQL).

3. Example Application Motivation and Objectives

With the increase in weather variability and flooding [24], it is vital that communities launch flood mitigation initiatives for the safety and quality of life of their residents. To create a sensing and alert system, we need to collect real-time sensor data from various locations around a city, and parse, store, provide responsive visualization, and transmit alert messages. For preemptive flood management strategies, we also need to collect data about existing infrastructure and land features to model stormwater flow and forecast future flood conditions.

This example application's main goal was to demonstrate cloud-based application solutions to support the monitoring and alerting of flooding events. The basic features of our application system include data collection, storage, visualization, and alert creation as well as a RESTful API to provide data access to data-driven environmental forecasting and physics-based stormwater flow simulation. Although this use case is focused on flood warning, we describe each component and lessons learned in a general way, so it can be easily translated to other smart city use cases.

4. Methodology

4.1. System Architecture Overview

Data flows from sensors to our cloud-based software solution as depicted in the system architecture diagram in Figure 3. We built our cloud-based system using Amazon Web Services (AWS) to take advantage of their latest resources and capabilities such as serverless functions (AWS Lambda [21]), data storage (Amazon S3 [20]), API gateway interfaces (Amazon API Gateway [25]), and computing instances (Amazon EC2 [26]).

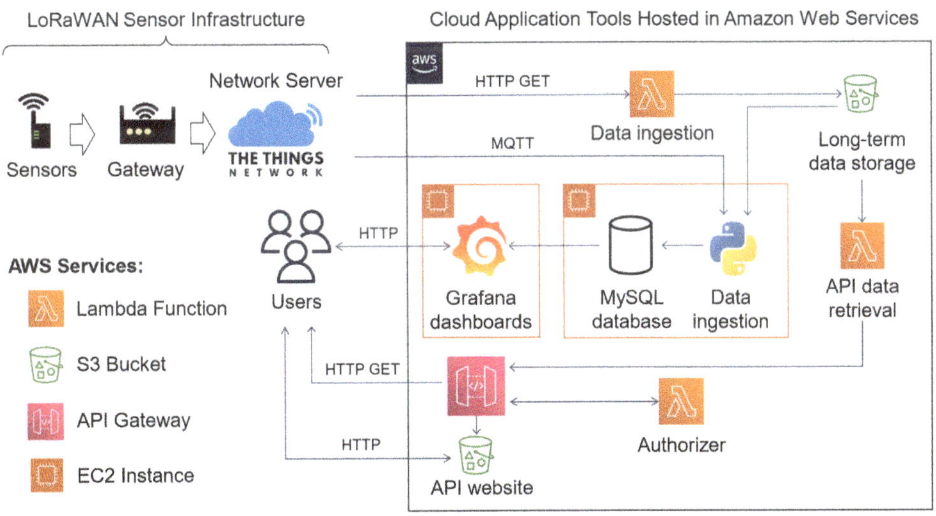

Figure 3. Our system architecture diagram using Amazon Web Services and The Things Network.

Using AWS Lambda [21], sensor measurements are queried from our TTN application, transformed, and uploaded as csv files to our long-term cloud data storage solution in an AWS S3 bucket [20]. We adopted a MySQL database server to provide responsive data access to our application. The MySQL database is hosted in an AWS EC2 instance [26] alongside a python script that ingests historical sensor data when the virtual machine starts up, and another script that connects to our application's TTN MQTT broker to receive and ingest real time sensor data. The data is then queried for visualization, monitoring, and alerts through a graphical user interface (GUI) tool, Grafana [22]. Both MySQL and Grafana EC2 instances are only started under demand if users need fast access to structured data or a monitoring dashboard, respectively. Sensor data can also be programmatically downloaded using our RESTful API, as, for instance, in scripts to perform data analysis tasks in Jupyter notebooks [27], or to perform modeling tasks with Storm Water Management Model (SWMM) software [28]. We also hosted a static website to document the API interface and offer users direct access to data download using Swagger UI [29]. While not explicitly shown, we assume simulation and modeling tasks will be performed by users in their own servers that could either be hosted by EC2 instances in AWS, by other cloud providers, or also hosted on their own computer machines.

4.2. Design Requirements

Our cloud-based system requires several different components to work in conjunction to meet application requirements. Firstly, the deployed IoT sensors must successfully relay messages to the TTN network server to deliver real time data. The data must then be received, processed, and stored in our MySQL database and the S3 long-term data storage for backup purposes. To make the system simpler to develop and manage, we adopted a single cloud service provider to develop our application's services and tools. In this system's case, it was hosted by provisioning services through Amazon Web Services (AWS). Next, for this system to be sustainable and meet different users' cost constraints, it must operate at minimum cost and have efficient resource consumption. The system must also be intuitive and straightforward to deploy, use, maintain, and modify.

4.3. System Components

4.3.1. Sensors, TTN, and Ingestion to Cloud Platform

As proof-of-concept, we deployed three water level monitoring sensors and one weather station in a flood prone watershed in Charlottesville, VA. All four devices were connected to The Things Network (TTN) through a LoRaWAN gateway installed in the same neighborhood region as the devices. We utilized commercial sensors from Decentlab [30] to focus efforts on data gathering, storage, and analysis systems rather than on the sensor's hardware and software. Another motivation behind this decision was to make our solution more general and easily translatable to other smart city projects based on sensor hardware compatible with The Things Network (TTN). We have left sensor deployment details out of the scope of this work, since our main goal is to advance the software backend infrastructure of IoT systems.

The sensors communicated using LoRaWAN [10,11] with TTN-compatible gateways that interfaced with TTN network server through an internet connection. The sensors were connected to TTN to enable cost effective interfacing and management, and to utilize the platform's available single-day storage via TTN's data storage integration service. To query data from TTN and upload it to the Amazon Web Service (AWS) stack, we wrote a python function to perform a HTTP GET request to retrieve data for a particular application. This data querying python function runs as an Amazon Lambda service that is periodically executed, set initially to run in one-hour intervals. To ingest real-time data to our MySQL database, we used MQTT clients connected to our TTN applications' MQTT brokers. TTN network server MQTT broker publishes new sensor data to our MQTT clients as soon as it is available in their server, providing our application with timely access to information.

4.3.2. Cloud Platform and Used Services

We decided to develop our application using AWS tools, but the same application architecture can be reproduced using equivalent services from other cloud providers. For regions impacted by flooding, high availability of the computing backend is imperative due to the need for quick analysis of the incoming weather and real time water level data. AWS offers high availability, which includes regional failovers in case a data center is taken offline. Deploying and redeploying resources on AWS can also be quickly automated using AWS CloudFormation [31], a tool used to provision specified resources (such as Lambda, EC2, RDS, etc.) through a provided script. The code written for the backend of the cloud-based system can be found at [32].

Amazon Lambda

AWS provides a serverless computing platform known as Amazon Lambda [21], which allows users to run their custom functions on demand. The underlying infrastructure of Lambda is maintained by AWS, which means the system developer must only worry about choosing the correct runtime environment to deploy their code. Using Lambda, the sensors are queried for uplink data at specified intervals. The uplink is then parsed, and the data is transformed to only include information pertinent to the application. The sensors' uplink data is uploaded to S3 for long-term storage and becomes available to be queried into the MySQL database when needed. After the Lambda function finishes uploading the transformed data, it automatically shuts off, allowing the user to pay only for the computing time and memory resources used rather than provisioning a continuously running machine (e.g., EC2). Lambda was chosen for our solution due to ease of scalability with future added devices, monitoring, high availability, and resource efficiency. For instance, if a new TTN application is added to the system, the existing Lambda function can be promptly updated to query sensor data. Should multiple applications need to report data in overlapping intervals, the same Lambda function can run in parallel of up to 1000 instances if needed.

The Lambda functions for this use case require modification from the default settings. We used the AWS SDK Pandas Lambda Layer [33] to query from TTN, parse data, and to store or read data from a S3 bucket. Python's Pandas module is used to quickly transform and manipulate data. The urllib3 module is used to send HTTP requests to The Things Network's storage integration and to retrieve sensor data. Other configurations for the Lambda function include setting the allocated memory to 192 MB (determined by AWS Compute Optimizer [34]), a timeout limit of 1 min, and being triggered to run once every hour. The lambda function triggering period can be adjusted based on application needs, where shorter periods translate to lower latency between data being available on TTN and stored in the S3 bucket but also resulting in higher costs for the lambda function computing service.

Another use we make of AWS Lambda is to return stored sensor data requested by our RESTful API and to manage user authentication. When receiving a query from Amazon Gateway API, a lambda function is initially executed to check an authorization token provided in the API request and authorize or deny the API request. If authorization is granted, a second lambda function reads, and parse data stored in the long-term data storage solution in the S3 bucket to return the required data to the API gateway. This lambda function to query data from S3 and return to the API gateway is configured to allocate 512 MB of data as a compromise between cost and performance to serve the API functionality and timeout limit of 1 min. The authorizer function uses the default settings of 128 MB memory allocation and 3 s timeout due to the simplicity of our currently adopted solution that only checks if the authorization key input matches a hardcoded string value.

Amazon S3 Data Storage

Amazon Simple Storage Solution (S3) is a cost-effective way to store data for extended periods. Data collected by sensors are uploaded in S3 for long-term storage as a read-only resource of the raw data feed. These readings can be used to repopulate the database in case of a database failure or migration and can be performed using the python library created for this system. AWS also maintains a python module (BOTO3 [35]) that allows users to download a copy of the readings from S3 to a local machine. All readings in S3 are currently stored as the AWS Standard tier for regular access for this application example.

Another use for the S3 storage is hosting static websites. We used a S3 bucket to store our RESTful API documentation using the Swagger UI interface [29]. Our website is based on the Swagger UI demonstration provided in their github page, and adapted to read an OpenAPI 3.0 description of our API service. The static website contains the API server address, a description of the required header, all accepted parameters, and the possibility to perform an API GET request trial with parameters provided by the user.

Amazon Elastic Cloud Compute

To host MySQL and Grafana, two Amazon Elastic Compute Cloud (Amazon EC2) instances were provisioned. Amazon EC2 allows for a continuous computing platform on the cloud, which allows access to the database and Grafana when needed. The developed system uses t3.micro instances with 10 GB storage, which fits the needs of this example application by minimizing costs while still maintaining a reliable performance for the relatively low number of sensors currently in the system. A more capable instance could be used to serve a larger number of users or for a use case requiring quicker response times. For this study, MySQL and Grafana were hosted on two separate EC2 instances for simpler management and increased flexibility, allowing, for example, the easy replacement of visualization software or on-demand use of MySQL database to allow fast data access to applications. It may also be worthwhile to adopt an AWS Relational Database Service (RDS) [36] instead of an EC2 instance running MySQL as the system database solution and then scale the RDS database based on the application's requirements for maintainability and access speed. This was considered, but not implemented in this study because RDS comes at a higher cost. However, RDS has the advantage that it provides built-in scalability as data volumes and users grows. The Section 5 includes a cost comparison between these alternatives for hosting the database and a discussion of pros and cons of each alternative.

4.3.3. Relational Database Design and Implementation

As our relational database, we selected MySQL as a simple solution with wide community support. We deployed MySQL on the cloud through Bitnami [37], which provides a pre-configured virtual machine image which is ready to be loaded to an Amazon EC2 instance. We created an entity relationship diagram (ERD) to normalize the sensor readings as shown in Figure 4. The ERD is centered around the Measurements entity, which stores the value of individual data points along with the time of data collection (Received_at). The Devices entity stores the device's unique identifier (Device_ID), the device's model (Device_model), the last received battery reading of the device (Last_battery), and the last activity timestamp (Last_activity). Similarly, the Locations entity contains data on the latitude, longitude, and altitude for each location that data is collected from, along with a unique identifier for each location. For each value in the values table, the Variables entity stores the data points' unique display name and the unit of the variable. The Measurements entity has a one-to-many relationship with the three other entities, meaning that each value data point can only have one device, variable, and location, while the remaining entities can have many values for each data point in their tables. This ERD was developed by advancing an approach from previous related research [38]. This design of the database allows for easy further advancement and change as additional devices and variables can be more easily incorporated.

4.3.4. Graphical User Interface

This system allows users to visualize and monitor data through Grafana, an open-source analytics platform for querying, visualizing, and alerting on data metrics. Grafana was selected as the software solution to visualize incoming data due to its dynamic dashboards, built-in alerting capabilities, and its specialization in time series data.

Grafana was deployed on the cloud through Bitnami [37], which provides a system image of a pre-configured Grafana stack on AWS. A connection was then made to the MySQL database in Grafana to access the data for visualization. Dashboards of each monitoring station were created to display relevant information for users. In Figure 5 we show an example of the dashboard for the water depth monitoring station. This dashboard includes a graph of the water depth over time, statistics on the water depth values for the set time range, a water level gauge of the current depth, a map of the sensor station location, and a gauge of the sensor battery level. The water depth graph and gauge allow users to view the current and past water levels in relation to a threshold of 0.4 m to signify flooding. Grafana's built-in alert system can send alert notifications if the incoming data triggers a set alert rule. As an example, the water depth dashboard has alert rules set to send a notification through the messaging application Slack [39] if the 0.4 m threshold is met, although sending alerts to other systems or via email is also possible.

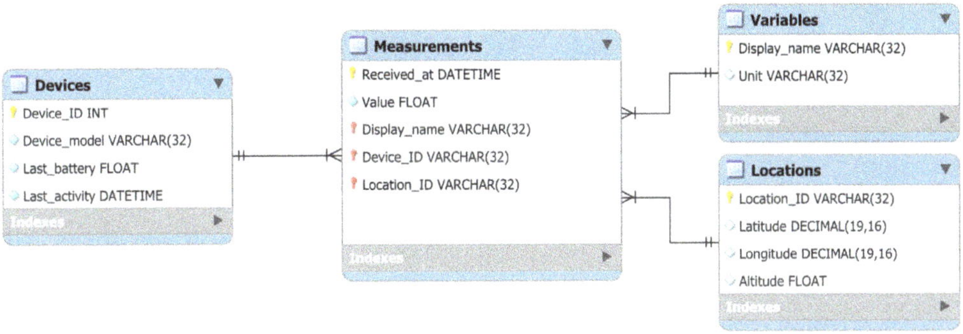

Figure 4. Entity relationship diagram for database design.

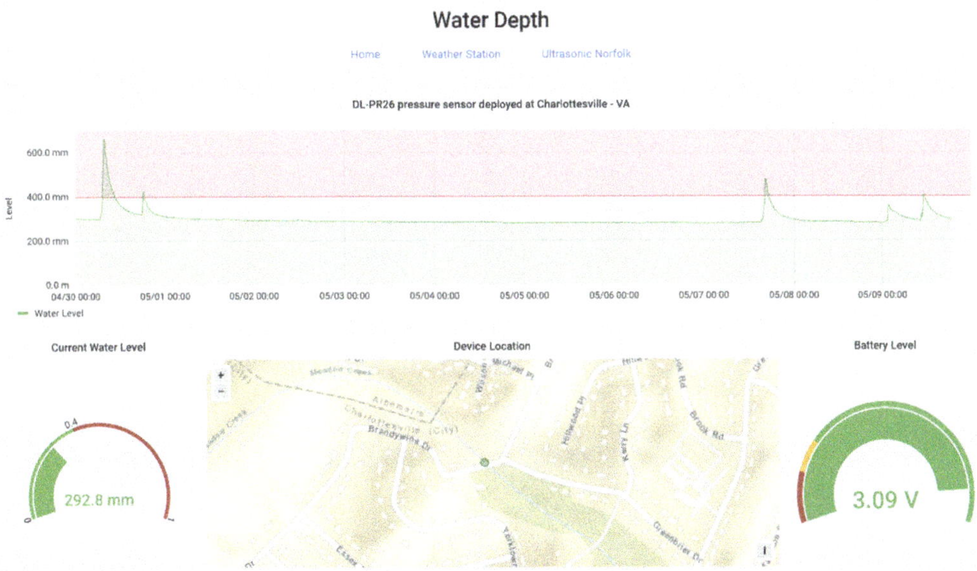

Figure 5. Grafana decision support dashboard of a water depth monitoring sensor.

4.3.5. RESTful API

Our RESTful API serves as a programmatic interface for users to quickly download data from sensors. We created the API using the Amazon API Gateway service [25] and lambda functions, both to manage API access and to read, parse, and return data from our long-term data storage solution in AWS S3. To document our RESTful API and provide easy access to sensor data, we created a static website using Swagger UI and it is currently hosted using AWS S3 buckets. We also enabled CORS in our API Gateway service, and we added a custom header with an authorization token for access control.

We described our API following the Open API 3.0 framework and stored it as a json file loaded to a specification variable in the javascript code for our documentation website. To download data using the API, the user will be required to input a valid authorization token to be granted access. Although we are currently using a simple custom lambda function to grant access, other more comprehensive user access management tools can be used in future versions, such as Amazon Cognito [40]. Other available parameters to customize the sensor data request are: "application", which selects which TTN application to download data from; "device_id" which selects devices using a unique identifier; and "last" or "start_date" and "end_date" which allow the selection of periods of time to download data. Using the "last" parameter, users can retrieve data collected by the sensors from the time of querying to the day specified. Using the "start_date" parameter, users can specify the beginning of the time range of the dataset to download. By default, if only one of "start_date" or "end_date" parameters are provided, data from the single specified day will be returned. Using the API, the user can request datasets for any of the available sensors. In Figure 6, we illustrate a typical use of the API to request data from a pressure sensor by using the Swagger graphical user interface. In Figure 7, we show a typical API call with parameters and the response.

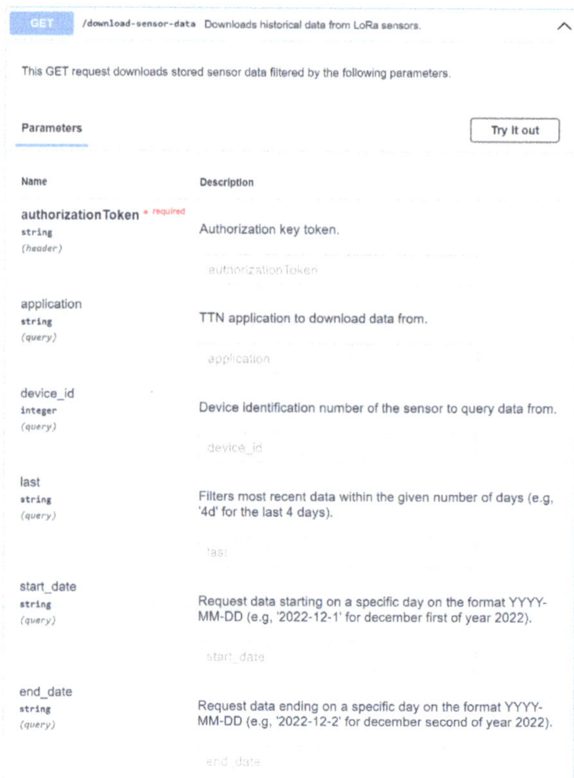

Figure 6. Example of parameters for the sensor data download API, with the asterisk representing the required authorization token field.

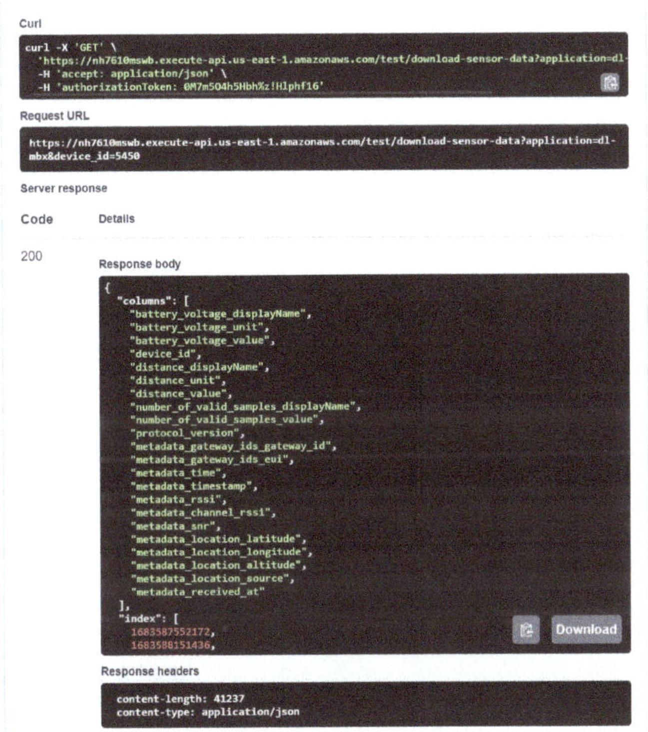

Figure 7. Response from API using example parameters.

5. Results and Discussion

5.1. Discussion of Alternative System Components and Potential System Enhancements

5.1.1. Cost Analysis of Cloud Services

The first two versions of the databases created for this application example were hosted in a MySQL database using Amazon Aurora [41] and then Amazon RDS [36]. For the application example needs, Aurora and RDS costs presented a constraint, which is the reason we chose two EC2 instances to host MySQL and Grafana that meet user requirements at a lower average cost. The current virtual machine cyber infrastructure costs between USD 24 and USD 210/year, depending on how long the EC2 instances will be required to be available. However, the database hosted this way may require maintenance such as updating software, or services to fix bugs, along with providing no regional failover. In the event that an AWS region experiences an outage, regional failover allows a copy of the database hosted in a separate region to quickly take over operations. Since in our use case we might not need Grafana and the MySQL database to be always available, the EC2 instances can be shut down and only started under demand, for example when users expect an incoming storm. Turning off the EC2 instances reduces the recurring costs to only the instance's storage units, which costs around USD 12/year for each instance using currently 10 GB of memory space or around USD 24/year for both EC2 instances. Should the application require seamless regional failover and high database performance, one alternative solution is the provision of two redundant instances running Amazon RDS for MySQL with multi-availability zone support. This configuration's estimated costs are USD 623.28/year, considering on-demand instance base costs and 10 GB of SSD storage. Memory storage calculations and their associated costs with S3 and the database configurations were based on the sensors used in the proof-of-concept system (Table 1).

Table 1. Adopted Sensors for the Proof-of-Concept IoT System.

Device Type	Model	Measured Variables	Readings/Month
Atmospheric	DL-ATM-41	18	4800
Pressure	DL-PR-26	3	4800
Ultrasonic (unit 1)	DL-MBX	3	4800
Ultrasonic (unit 2)	DL-MBX	3	4800

Our calculations for Tables 2–7 were carried out based on the current sensor device configuration of the system (Table 1), pricing rates at the development time (January 2023), and a projected 5-year use. The default measurement frequency for the system is 1 measurement every 10 min, averaging 4380 readings per month. To account for temporary measurement frequency increases during storm events, calculations instead used a figure of 4800 readings per month. One csv file is uploaded every hour to S3 for each registered TTN application, with each write request to S3 costing USD 0.000005. Sensor devices currently in use are one eleven parameter weather station (DL-ATM41), one pressure/liquid level and temperature sensor (DL-PR26), and two ultrasonic distance/level sensors (DL-MBX), with one TTN application for each sensor model type, resulting in a total of three TTN applications. The average payload size for these four sensors is 343 bytes after parsing and transforming, and the csv file header average size is 822 bytes. Since the weather station contains more measurements per reading than the other two sensor types, its sampling frequency has the most significant impact in the used data storage space. It is important to note that, when data is stored in the MySQL database, the weather station requires almost five times as much storage capacity as either of the other two sensor devices. Since the current system is based on these four sensor devices, AWS storage configurations may need to be readjusted based on the chosen sensors for the application's system.

Table 2. MySQL database storage (MB) requirements over time per device type (4800 readings/ month).

Device	1 Month	1 Year	5 Years
Atmospheric	7.13	85.58	427.92
Pressure	1.50	18.02	90.098
Ultrasonic	1.50	18.02	90.09
Current Config (CC)	11.63	139.64	698.19
Average (CC)	2.91	34.91	174.54

Table 3. S3 storage costs calculations for generic sensor devices in the first year (343 Bytes/sensor payload, 828 Bytes/csv header, 4800 sensor payloads/month, and 3 TTN applications).

Number of Devices		Q1	Q2	Q3	Q4	Total
1 [1]	Storage (GB)	0.006	0.012	0.018	0.025	-
	Cost (USD)	0.03	0.03	0.03	0.03	0.12
4	Storage (GB)	0.023	0.046	0.070	0.093	-
	Cost (USD)	0.03	0.03	0.04	0.04	0.14
50	Storage (GB)	0.23	0.47	0.70	0.93	-
	Cost (USD)	0.04	0.06	0.07	0.09	0.26
100	Storage (GB)	0.46	0.47	0.70	0.93	-
	Cost (USD)	0.05	0.08	0.11	0.14	0.38

[1] Only one TTN application was considered for this case.

Table 4. MySQL database storage (GB) requirements over time based on number of generic IoT devices (1 kB/reading and 4800 readings/month).

Number of Devices	1 Month	1 Year	5 Years
1	0.01	0.05	0.27
5	0.02	0.27	1.37
25	0.11	1.37	6.87
50	0.23	2.75	13.73
100	0.46	5.49	27.47

Table 5. Database cost on single EC2 instance (t3.micro) assuming a single instance, storage requirements for 5 years, and generic IoT devices (1 kB/reading and 4800 readings/month).

Number of Devices	Storage Requirement (GB)	Cost/Month (USD)	Cost/Year (USD)
1	5	8.09	97.10
4	5	8.09	97.10
25	10	8.59	103.10
50	15	9.09	109.10
100	30	10.59	127.10
Every 5 new devices	+2	+0.20	+2.40

Table 6. Database cost on 2 separate RDS EC2 instance (db.t3.micro) with multi-availability zone deployment and assuming generic IoT devices (1 kB/reading and 4800 readings/month).

Number of Devices	Storage Requirement (GB)	Cost/Month (USD)	Cost/Year (USD)
1	5	51.94	623.28
4	5	51.94	623.28
25	10	54.24	650.88
50	15	56.54	678.48
100	30	63.44	761.28
Every 5 devices	+2	+0.40	+2.40

Table 7. Database cost on Aurora (t3.small) [1] and assuming generic IoT devices (1 kB/reading and 4800 readings/month).

Number of Devices	Storage Requirement (GB)	Cost/Month (USD)	Cost/Year (USD)
1	5	61.39	736.68
4	5	61.39	736.68
25	10	62.39	748.68
50	15	64.44	773.28
100	30	67.44	809.28
Every 5 devices	+2	+0.30	+3.60

[1] For 50 devices and more, we estimated higher IOPS to handle the average measurement writing load. The cost also includes running 2 EC2 instances by default for regional failover.

The overall yearly system costs can also be lowered by configuring S3 and EC2 instance provisioning and by using built-in AWS cost optimization tools. For S3, if the backup data will not be frequently accessed, it is recommended to change the access tiers of the data. For this application example, the data is stored under Standard tier, which costs USD 0.023 per GB. In future iterations of the system, it is recommended to use Intelligent-Tiering: Standard-Infrequent Access (USD 0.0125 per GB), One Zone-Infrequent Access (USD 0.01 per GB), or even Glacier tiers (USD 0.004 per GB). For the Infrequently Accessed and Glacier tiers, there is a retrieval fee for every gigabyte retrieved. Infrequently Accessed will allow for millisecond latency to the user when requesting data, whereas with Glacier it can take minutes or hours. Deleting data from non-standard S3 tiers before their minimum storage durations will charge the user for the respective minimum storage durations. Infrequently Accessed and Glacier tiers also have a minimum capacity charge per object, so it is recommended to combine individual readings into larger datasets (i.e., monthly readings per sensor) to store as one file in these tiers. To reduce costs of EC2 instance provisioning (including for RDS and Aurora), AWS allows for reserving instances in 1 and 3 year increments instead of using on-demand instances, bringing costs down by up to 38%. The costs calculated in this paper use the current configuration of the system which uses on-demand EC2 instances and consider they will remain always on.

As shown in Table 2, our current configuration of four sensors reporting on average between six and seven samples per hour (4800 samples/month) results in a MySQL database of less than 140 MB of data at the end of the first year of operation. In Table 3, we show that storing this amount of data in AWS S3 service would cost USD 0.14 for the first year and even scaling to 100 sensors with the same average data rate would result in USD 0.38 storage costs. This indicates that many small to medium scale applications could benefit from this data storage service to backup sensor data at low costs.

In Table 4, we estimate the size of a MySQL database for the first five years, assuming generic sensor samples of 1 kB size being uploaded at the rate of 4800 samples/month as we adopted in our example application. The estimated MySQL database size is then used to inform the storage requirement of the virtual machines hosting the respective MySQL databases as shown in Table 5. Our system with 4 sensors would cost about USD 97.10/year with each one GB increase in storage space resulting in an additional cost of USD 1.20/year. This analysis shows that the uptime of EC2 servers has the greatest impact on the overall system cost and turning them off while they are not required can result in substantial savings. To reduce costs even further, MySQL server disk images can be saved in the S3 data storage service, eliminating EC2 server costs while they are shut down for long periods.

As a brief exploration of the alternative robust database services offered by AWS, we assume, in Table 6, two Amazon RDS EC2 instances with multi-availability zone deployment, and, in Table 7, the Amazon Aurora managed database on a more powerful EC2 instance. Both solutions result in total costs over USD 600/year, representing six times the cost of running a database in a single EC2 MySQL server. Therefore, we recommend using our proposed EC2 MySQL server solution when a failover system is not critical to the application due to the substantial cost savings.

In Figure 8, we estimate how the cost of S3 data storage varies with sampling rate, operation total duration, number of sensors, and sampling rate. For these calculations we used a simplified estimation model considering only a fee of USD 0.023 per GB stored, and USD 0.000005 fee of per write request. As in the tables previously introduced, we assume up to three TTN applications and one data request and ingestion operation per hour.

With the S3 storage costs curves depicted in Figure 8, IoT application developers can estimate how the number of sensors and data rate parameters influence the total S3 storage costs, as well as how these costs accumulate with time. For instance, in Figure 8d, we can verify that the cost of S3 data storage of an application with 50 sensors for the first ten years is comparable to an application with 200 sensors for the first five years.

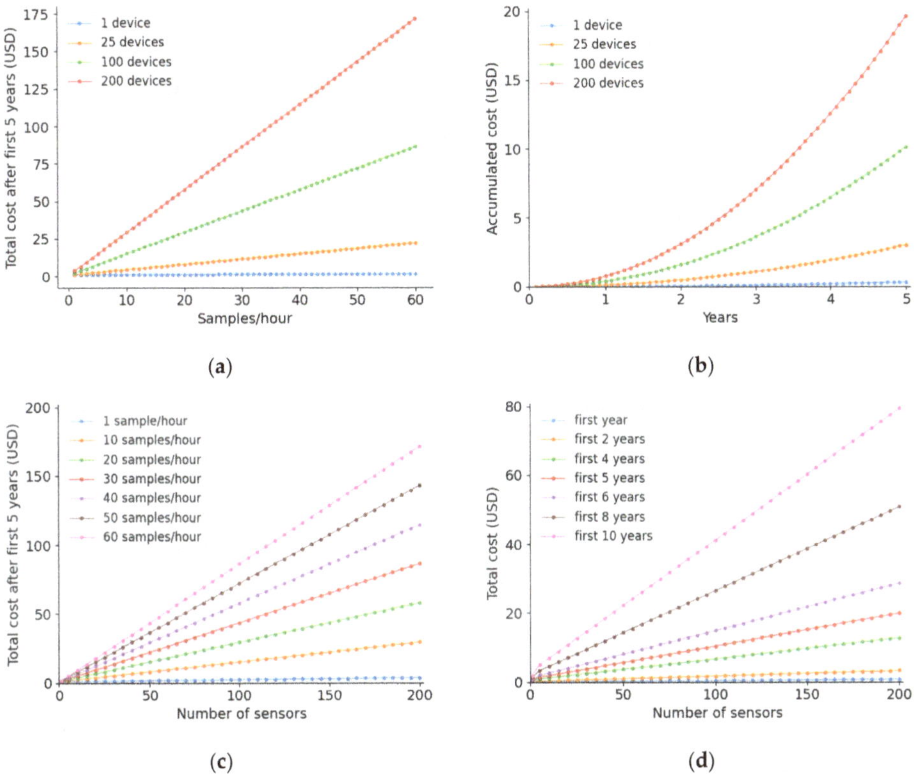

Figure 8. S3 storage costs with varying parameters. Plots (**a**,**c**) evaluate the total cost of S3 data storage at the end of 5 years. Plot (**b**,**d**) assume devices with sampling rate of 4800 samples per month.

5.1.2. Discussion about the RESTful API Limitations and Data Access

We identified some limitations when testing the sensor data download application programming interface (API). Through an endpoint provided by the Amazon API Gateway, a user request is passed to the Lambda function to retrieve datasets from the S3 storage, which needs to be parsed before being returned to the user. The first limitation of the solution adopted in this example application is the maximum 30 s timeout on API Gateway requests when large datasets are requested. Even after the retrieval code was optimized to run faster, there was a second limitation through Lambda, which is a payload limit size of 6 MB. For large datasets (e.g., 1 month of data from the weather sensor), the Lambda is not able to send to the user their requested dataset. Therefore, we recommend only using the RESTful API to download data for a few days at each GET request. An alternative and faster solution to download a large amount of data is using the AWS provided BOTO3 python library [35] and downloading the raw csv files directly. We recommend downloading the raw csv files when data is needed in order of a few months of sensor data. Another available alternative solution to perform more responsive data exchange in larger sizes is to query data directly from the MySQL database running in the EC2 instance. We recommend using the MySQL database when data in order of a few weeks is needed for applications such as dynamic websites requiring fast responses or time sensitive simulations.

5.1.3. Security Considerations

Cloud service providers such as AWS acknowledge that security is a major concern for users and provide management tools to support the creation of secure applications. For instance, when deploying a cloud-based system, it is recommended to create an AWS organization with trusted users to manage AWS identity and access management (IAM) roles and policies. Although, in hindsight, we agree that creating an AWS organization from the beginning would be best, our research team initially used separate AWS accounts to create and manage Lambda, S3, and EC2 instances based on who was working in each part of the system, resulting in a poor managing practice. Therefore, we recommend access privileges to AWS services to be tailored to the developers and systems administrators that oversee each subsystem.

We utilized a secure shell (SSH) with a key pair generated by AWS to access the EC2 instances, using SSH port forwarding to access the Grafana user interface. Although this approach limits the number of EC2 instance ports accessible through the web, it also results in a worse user experience due to the increased number of required steps to access the Grafana dashboards. For future versions of the system, we recommend creating a user access webpage using AWS Cognito service and reverse proxy to serve the Grafana application, without having the need to use SSH tunnels and still avoiding directly exposing ports of the EC2 instance to the web.

5.1.4. Alternatives for Graphical User Interface

Providing users with easily understandable information in a clear and efficient manner is paramount when working with large amounts of time series data. In this application example, three data visualization platforms were compared in order to find the best tool to effectively communicate information, namely, Grafana, AWS QuickSight [42], and AWS SageMaker [43]. QuickSight was initially determined as the platform that best met cost, visualization, analysis, and alerting capabilities requirements. However, after creating a QuickSight account and working with the platform, we found that it does not support embedding visualizations in websites without assigning each user with permissions to view. We then determined that QuickSight was not a suitable tool as it did not meet some of our envisioned uses for the application. After conducting more research on data visualization platforms, we decided that Grafana would be the best tool for this application due to its ability to easily share and embed visualizations. Grafana allows for the creation of snapshots of dashboards which can then be used to share interactive dashboards publicly through snapshot links. Additionally, Grafana is designed for time series data and allows for alerts to be sent out through many alert notifiers such as text message, email, and Slack.

5.1.5. Opportunities for Forecasting and Advanced Analytics

The long-term data gathered by this monitoring system can support the generation of accurate forecasting real time models in the areas of interest. Developing such models with longer observation periods would better assess seasonality effects and, therefore, could reduce the uncertainty arising from precipitation effects, creating more accurate forecasts. Users can feed sensor data from our RESTful API to simulate the generated models and provide real time forecasts on demand. Another potential study that could benefit the creation of forecasting models would be an evaluation of the optimized sampling intervals for each location, as the wide variation of water depth between collection intervals can hide patterns and result in less accurate statistical analysis.

5.2. Discussion of the System Performance

To analyze the system performance, we can break down the proposed system to a few main data paths, namely: (1) serverless data ingestion, receiving data from TTN, and saving to the S3 bucket; (2) MySQL server startup and historical data ingestion from the S3 bucket; (3) MySQL live data ingestion through MQTT; (4) Grafana data query from MySQL; and (5) RESTful API data query.

The serverless data ingestion operates independently from the other system components and its latency is dominated by the lambda function execution time, which takes up to 4.8 s. The MySQL server startup includes the EC2 instance boot up, queries to historical data from the S3 bucket, and most recent data from TTN storage integration, leading to a startup latency of up to 10 min in the current version of the system. This startup time can be improved, but we assume that the MySQL server can typically be turned on hours before an event of interest (in our example application, triggered by a storm forecast). For the live data ingestion, data is received from TTN through MQTT and a python script ingests data to the database within milliseconds. More in depth study is still required to analyze the impact of high sensor data rates, but EC2 instance computational power can be upgraded to avoid possible bottlenecks. For the Grafana query from the MySQL database, the co-location of EC2 servers in the same availability zone results in overall good performance. Again, more in depth study is required to analyze performance degradation when scaling the number of users logged to the Grafana server. Finally, as previously discussed in Section 5.1.2, the RESTful API has some significant limitations, and its use should be restricted to accessing small batches of data. RESTful API latency can be improved by optimizing the S3 querying lambda function and creating larger S3 objects aggregating a larger number of measurements.

5.3. Broader Impacts of This Study

In this work, we introduced a cloud-based data storage and visualization tool for smart city IoT projects that can be leveraged by researchers in academia and industry to quickly prototype applications, allowing them to promptly evaluate the impact of their solutions in the real world. The low cost and maintenance requirements of cloud solutions can enable a higher range of experimentation and collaboration between smart city projects, combining IoT data accessibility with computational resources for modeling and simulation. Furthermore, lowering the barrier-to-entry of cloud systems can foster the development of new smart city solutions, supporting more environmentally and economically sustainable communities.

6. Conclusions

While data collected by IoT smart city applications are a central asset in supporting management and planning decisions for many communities, designing and deploying IoT solutions is still challenging due to system integration complexity, reliability limitations, and cost. We presented a cloud data storage and visualization system for smart cities, leveraging reliable existing technology to integrate a complete IoT monitoring solution hosted in AWS and costing under USD 26/year for long-term data storage and USD 0.0204/hour of use for MySQL database and Grafana servers. By using this cloud-based solution together with TTN infrastructure and commercial LoRaWAN sensors, users can collect, store, and visualize datasets to address their needs and integrate their own services. We demonstrated the use of the system for a flood warning system application example with river and weather LoRaWAN sensors. The cloud-based system design uses serverless data ingestion to provide a simple and cost-effective data storage solution that is independent of other services such as data visualization. An on-demand database and visualization servers offer flexibility to adapt to application needs while saving costs and simplifying maintenance operations. Furthermore, we explored the different AWS tiers and their respective reliability/cost tradeoff so users can make informed decisions when tailoring our system to their own application. As opposed to focusing mainly on the example application, as commonly seen in the literature, we highlight common tasks that are required by an IoT project and share our insights in leveraging modern cloud services to simplify IoT backend system design and optimize costs.

As a future research avenue, we intend to explore the use of new serverless cloud backend architectures in smart city IoT applications and investigate practical tradeoffs to server solutions. We intend to analyze in particular the on-demand allocation of computational resources as we scale the number of sensors, total sensor data rate, and number of clients connecting to user interfaces in cloud IoT backend systems. We also intend to explore the integration of modeling and simulation tools with IoT data acquisition systems while efficiently allocating computational resources.

Author Contributions: Conceptualization, V.A.L.S., B.C. and J.L.G.; methodology, V.A.L.S., L.A., A.M., G.M., K.T. and C.T.; software, V.A.L.S., L.A., A.M., G.M., K.T. and C.T.; validation, V.A.L.S.; data curation, V.A.L.S.; writing—original draft preparation, L.A., A.M., G.M., K.T. and C.T.; writing—review and editing, V.A.L.S., J.N., B.C. and J.L.G.; supervision, V.A.L.S., B.C. and J.L.G.; funding acquisition, B.C. and J.L.G. All authors have read and agreed to the published version of the manuscript.

Funding: We acknowledge support from the US National Science Foundation through the award number 1735587.

Data Availability Statement: Code and instructions to setup the cloud infrastructure described in this paper are available at https://github.com/uva-hydroinformatics/iot-cloud-platform (accessed on 14 April 2023).

Acknowledgments: The authors would like to recognize Ruchir Shah for his valuable comments on and assistance with this work.

Conflicts of Interest: The authors declare no conflict of interest.

References

1. Shahat Osman, A.M.; Elragal, A. Smart Cities and Big Data Analytics: A Data-Driven Decision-Making Use Case. *Smart Cities* **2021**, *4*, 286–313. [CrossRef]
2. Tcholtchev, N.; Schieferdecker, I. Sustainable and Reliable Information and Communication Technology for Resilient Smart Cities. *Smart Cities* **2021**, *4*, 156–176. [CrossRef]
3. Barthelemy, J.; Amirghasemi, M.; Arshad, B.; Fay, C.; Forehead, H.; Hutchison, N.; Iqbal, U.; Li, Y.; Qian, Y.; Perez, P. Problem-Driven and Technology-Enabled Solutions for Safer Communities: The case of stormwater management in the Illawarra-Shoalhaven region (NSW, Australia). In *Handbook of Smart Cities*; Springer: Berlin, Germany, 2020; pp. 1–28. [CrossRef]
4. Powar, V.; Post, C.; Mikhailova, E.; Cook, C.; Mayyan, M.; Bapat, A.; Harmstad, C. Sensor Networks for Hydrometric Monitoring of Urban Watercourses. In Proceedings of the 2019 IEEE 16th International Conference on Smart Cities: Improving Quality of Life Using ICT IoT and AI (HONET-ICT), Charlotte, NC, USA, 6–9 October 2019; pp. 85–89.
5. Ebi, C.; Schaltegger, F.; Rüst, A.; Blumensaat, F. Synchronous LoRa Mesh Network to Monitor Processes in Underground Infrastructure. *IEEE Access* **2019**, *7*, 57663–57677. [CrossRef]
6. Syed, A.S.; Sierra-Sosa, D.; Kumar, A.; Elmaghraby, A. IoT in Smart Cities: A Survey of Technologies, Practices and Challenges. *Smart Cities* **2021**, *4*, 429–475. [CrossRef]
7. Iqbal, A.; Olariu, S. A Survey of Enabling Technologies for Smart Communities. *Smart Cities* **2021**, *4*, 54–77. [CrossRef]
8. The Things Industries. Available online: https://www.thethingsindustries.com/ (accessed on 11 August 2021).
9. The Things Network. Available online: https://www.thethingsnetwork.org/ (accessed on 11 August 2021).
10. Mekki, K.; Bajic, E.; Chaxel, F.; Meyer, F. A Comparative Study of LPWAN Technologies for Large-Scale IoT Deployment. *ICT Express* **2019**, *5*, 1–7. [CrossRef]
11. Shanmuga Sundaram, J.P.; Du, W.; Zhao, Z. A Survey on LoRa Networking: Research Problems, Current Solutions, and Open Issues. *IEEE Commun. Surv. Tutor.* **2020**, *22*, 371–388. [CrossRef]
12. Drenoyanis, A.; Raad, R.; Wady, I.; Krogh, C. Implementation of an IoT Based Radar Sensor Network for Wastewater Management. *Sensors* **2019**, *19*, 254. [CrossRef] [PubMed]
13. Basford, P.J.; Bulot, F.M.J.; Apetroaie-Cristea, M.; Cox, S.J.; Ossont, S.J. LoRaWAN for Smart City IoT Deployments: A Long Term Evaluation. *Sensors* **2020**, *20*, 648. [CrossRef] [PubMed]
14. IoT Platform | Internet of Things | Ubidots. Available online: https://ubidots.com/ (accessed on 14 September 2022).
15. MyDevices—Cayenne. Available online: https://developers.mydevices.com/cayenne/features/ (accessed on 14 September 2022).
16. Medina-Pérez, A.; Sánchez-Rodríguez, D.; Alonso-González, I. An Internet of Thing Architecture Based on Message Queuing Telemetry Transport Protocol and Node-RED: A Case Study for Monitoring Radon Gas. *Smart Cities* **2021**, *4*, 803–818. [CrossRef]
17. MQTT—The Standard for IoT Messaging. Available online: https://mqtt.org/ (accessed on 20 September 2022).
18. Node-RED. Available online: https://nodered.org/ (accessed on 20 September 2022).
19. MySQL. Available online: https://www.mysql.com/ (accessed on 20 September 2022).
20. Cloud Object Storage—Amazon S3—Amazon Web Services. Available online: https://aws.amazon.com/s3/ (accessed on 20 October 2022).
21. Serverless Computing—AWS Lambda—Amazon Web Services. Available online: https://aws.amazon.com/lambda/ (accessed on 14 September 2022).
22. Grafana: The Open Observability Platform. Available online: https://grafana.com/ (accessed on 14 September 2022).
23. ThingsBoard—Open-Source IoT Platform. Available online: https://thingsboard.io/ (accessed on 13 August 2021).
24. Hodgkins, G.A.; Whitfield, P.H.; Burn, D.H.; Hannaford, J.; Renard, B.; Stahl, K.; Fleig, A.K.; Madsen, H.; Mediero, L.; Korhonen, J.; et al. Climate-Driven Variability in the Occurrence of Major Floods across North America and Europe. *J. Hydrol.* **2017**, *552*, 704–717. [CrossRef]

25. Amazon API Gateway—API Management—Amazon Web Services. Available online: https://aws.amazon.com/api-gateway/ (accessed on 6 January 2023).
26. Secure and Resizable Cloud Compute—Amazon EC2—Amazon Web Services. Available online: https://aws.amazon.com/ec2/ (accessed on 20 October 2022).
27. Project Jupyter. Available online: https://jupyter.org (accessed on 14 September 2022).
28. United States Environmental Protection Agency. Storm Water Management Model (SWMM). Available online: https://www.epa.gov/water-research/storm-water-management-model-swmm (accessed on 14 September 2022).
29. REST API Documentation Tool | Swagger UI. Available online: https://swagger.io/tools/swagger-ui/ (accessed on 6 January 2023).
30. Decentlab. Available online: https://www.decentlab.com (accessed on 26 August 2021).
31. Provision Infrastructure as Code—AWS CloudFormation—AWS. Available online: https://aws.amazon.com/cloudformation/ (accessed on 23 January 2023).
32. Uva-Hydroinformatics/Iot-Cloud-Platform: Cloud IoT Platform. Available online: https://github.com/uva-hydroinformatics/iot-cloud-platform (accessed on 23 January 2023).
33. Quick Start—AWS SDK for Pandas 2.18.0 Documentation. Available online: https://aws-sdk-pandas.readthedocs.io/en/stable/ (accessed on 6 January 2023).
34. AWS Compute Optimizer. Available online: https://aws.amazon.com/compute-optimizer/ (accessed on 14 September 2022).
35. Boto3—The AWS SDK for Python 2022. Available online: https://github.com/boto/boto3 (accessed on 20 October 2022).
36. Fully Managed Relational Database—Amazon RDS—Amazon Web Services. Available online: https://aws.amazon.com/rds/ (accessed on 14 September 2022).
37. Bitnami. Available online: https://bitnami.com/ (accessed on 14 September 2022).
38. Carlson, K.; Chowdhury, A.; Kepley, A.; Somerville, E.; Warshaw, K.; Goodall, J. Smart Cities Solutions for More Flood Resilient Communities. In Proceedings of the 2019 Systems and Information Engineering Design Symposium (SIEDS), Charlottesville, VA, USA, 26 April 2019; pp. 1–6. [CrossRef]
39. Slack. Available online: https://slack.com/ (accessed on 14 September 2022).
40. Customer Identity and Access Management—Amazon Cognito—Amazon Web Services. Available online: https://aws.amazon.com/cognito/ (accessed on 9 January 2023).
41. Fully MySQL and PostgreSQL Compatible Managed Database Service | Amazon Aurora | AWS. Available online: https://aws.amazon.com/rds/aurora/ (accessed on 14 September 2022).
42. Amazon QuickSight—Business Intelligence Service—Amazon Web Services. Available online: https://aws.amazon.com/quicksight/ (accessed on 14 September 2022).
43. Machine Learning—Amazon Web Services. Available online: https://aws.amazon.com/sagemaker/ (accessed on 14 September 2022).

energies

MDPI

Article

Faults Feature Extraction Using Discrete Wavelet Transform and Artificial Neural Network for Induction Motor Availability Monitoring—Internet of Things Enabled Environment

Muhammad Zuhaib [1], Faraz Ahmed Shaikh [2], Wajiha Tanweer [1,2], Abdullah M. Alnajim [3,*], Saleh Alyahya [4], Sheroz Khan [4], Muhammad Usman [5], Muhammad Islam [4] and Mohammad Kamrul Hasan [6]

1 Department of Electrical Engineering, PNEC, National University of Science and Technology, Karachi 75350, Pakistan
2 Department of Electrical Engineering, Nazeer Hussain University (NHU), Karachi 75950, Pakistan
3 Department of Information Technology, College of Computer, Qassim University, Buraydah 51452, Saudi Arabia
4 Department of Electrical Engineering, College of Engineering & Information Technology, Onaizah Colleges, Unayzah 51911, Saudi Arabia
5 Department of Mechatronics and Control Engineering, Faisalabad Campus, University of Engineering and Technology (UET), Lahore 39161, Pakistan
6 Center for Cyber Security, Faculty of Information Science & Technology, Universiti Kebangsaan Malaysia, Bangi 43600, Selangor, Malaysia
* Correspondence: najim@qu.edu.sa

Citation: Zuhaib, M.; Shaikh, F.A.; Tanweer, W.; Alnajim, A.M.; Alyahya, S.; Khan, S.; Usman, M.; Islam, M.; Hasan, M.K. Faults Feature Extraction Using Discrete Wavelet Transform and Artificial Neural Network for Induction Motor Availability Monitoring—Internet of Things Enabled Environment. *Energies* **2022**, *15*, 7888. https://doi.org/10.3390/en15217888

Academic Editors: Antonio Cano-Ortega and Francisco Sánchez-Sutil

Received: 19 September 2022
Accepted: 19 October 2022
Published: 24 October 2022

Abstract: Motivation: This paper presents the high contact resistance (HCR) and rotor bar faults by an extraction method for an induction motor using Discrete Wavelet Transform (DWT) and Artificial Neural Network (ANN). The root mean square (RMS) and mean features are obtained using DWT, and ANN is used for classification using activation functions. Activation provides output by assigning the specific input with respect to the transfer function according to the nature and type of the activation function. **Method:** The faulty conditions are induced using MATLAB by adopting the motor current signature analysis (MCSA) method to achieve current signature signals of the healthy and faulty motors. **Results:** The DWT technique has been applied to obtain fault-specific features of the average continuously varying signal (RMS) and an average of the data points (mean) at levels 5, 7, 8, and 9, followed by ANN to classify the faults for condition monitoring. **Utility:** The utility of the results is to reduce unscheduled downtime in the industry, thus saving revenue and reducing production losses. This work will help provide support to ensure early indication of faults in induction motors under operating conditions, enabling in-service engineers to take timely preventive measures as part of the availability of resources in IoT-enabled systems. **Application:** Resource availability and cybersecurity are becoming vital in an environment that supports the Internet of Things (IoT) as the essential components of Industry 4.0 scenarios. The novelty of this research lies in the implementation of high contact resistance and rotor bar faults using DWT and ANN with different activation functions to achieve accuracy up to 98%.

Keywords: induction motors; MCSA; DWT; ANN; IoT security; resources availability

1. Introduction

Background: Availability in a secure IoT-enabled environment refers to the ease with which an organization's critical data, applications, and resources can be accessed by authorized users. Industry motors bear the features of being simple in design, robust in construction, reliable in operation, low in cost, and easy to maintain, and accordingly, are widely used in industry. However, induction motors are also prone to various faults [1–3], leading to their failure, which put the availability of resources at risk. If not detected in time, these faults will lead to irremediable mechanical failures. The objective of this research is, therefore, to develop a technique for detecting faults that appear on operational motors. The main faults detected and analyzed during induction-motor processing are related to stator faults, broken rotor bars, and air-gap eccentricity-related faults that have been investigated and reported extensively in the literature [4–7]. The motivation of this paper is derived from Internet resources and articles exploring faults using none-ANN techniques.

Motivation: Operational motors, as resources, generate data, and it is important that induction motors as data resources are easily accessible to authorized users bearing IoT-enabled features. In particular, high contact resistance (HCR) and rotor bar faults have been considered in this study in order to investigate their availability as data resources. Because of these potential faults, it is crucial to monitor motor health and maintenance issues. Thus, the presentation as a platform for safe operation from the IoT-enabled scenario perfective is important, and refers to the ease with which organizations' critical data, applications, and resources can be accessed by their respective users [8].

State-of-art works: Multiple machine condition-monitoring strategies have already been investigated in the literature [9,10], including partial discharge, vibration analysis, current signature investigation, magnetic flux, and acoustic noise monitoring. However, the Motor Current Signature Analysis (MCSA) is considered to be the most commonly reported technique applied to fault analysis [11,12]. The phase current signal contains current motor operation-dependent components, which are the by-product of unique rotating flux components. In the supply current, the occurrence of faults brings about changes that have specific harmonic contents which are specific to the faults that occur. The MCSA technique works on stator current measurements to detect the harmonics generated under faulty conditions, which, although unwanted, are used for fault analysis. Furthermore, the MCSA provides current spectra that possess potential information for the detection of electrical and mechanical faults. In this investigation, motor parameters, mainly current, have been acquired for the three phases of a motor in operation.

Novelty and contribution: The novelty of this work is the current spectrum that has been analyzed over a wide range of distinct current patterns that could serve as signature values for fault detection. The two faults under investigation in this work are: (1) the high resistance connection fault (21%), and (2) the broken rotor bar fault (7%), both categories are caused by manufacturing defects [13,14]. The fault analysis of the induction motor is done through high resolution spectral analysis of the low frequency signals associated with various fault categories [15]. Statistical features extraction followed by a mix of classification algorithms are used in [16]. Induction motors make up between 40 to 60% of any industrial site's electrical loads employed. The online sensor-based detection of induction motors faults through the sensing of stray flux components containing information of relevance for the diagnosis of faults in induction motors is shown in [17]. The unbalanced supply voltage and inter-turns short circuits are studied in [18] through thermal behavior of the motors. The analysis of supply current signals under faulty condition of specific number of rotor bars as time–frequency features is commonly used in recent work [19].

The remainder of the paper is organized as follows. Section 2 consists of the work related to the two faults. Section 3 describes induction motors. Discrete Wavelet Transform (DWT) for fault detection is explained Section 4. Section 5 explains the ANN Classifier. Results are given in Section 6. The paper's conclusions are presented in Section 7.

2. Related Work

From a review of the literature, it was concluded that many different techniques have been used to diagnose machine faults in industry, and various classification methods have been implemented to obtain satisfactory results. Table 1 has been adopted from the state-of-the-art paper, tabulated in [20], and presents the results obtained through artificial neural network (ANN) classification. A maximum of 95% accuracy was previously achieved, but details of the hidden layers and activation function were not mentioned. Conversely, in the present study, more than 96% accuracy has been attained at different Debauches levels, with detailed information reported on the hidden layers and activation functions.

Moreover, stator winding faults, bearing faults, broken rotor bar faults, and eccentricity are the primary research interests in previous studies [21], whereas in this current research, HCR and broken rotor bar faults are discussed.

Table 1. Accuracy tabulated from the literature review versus a classification approach.

Reference	Fault Considered in Induction Motor	Artificial Intelligence (AI) Method Adopted	Accuracy
[6]	Stator Winding Bearing	ANN	Over 95% accuracy
[8]	Broken Rotor Bar	ANN	
[8]	Broken Rotor Bar Stator Winding Eccentricity	Fuzzy logic	93% accuracy
[22,23]		Neural networks Fuzzy logic Neuro-fuzzy	-
[18]	Overloads Single phasing Unbalance Supply voltage	ANN	-
This present paper	High Contact Resistance (HCR) Fault Rotor Bar Fault	ANN	Over 96% accuracy

2.1. High Connection-Resistance (HCR) Fault

As the name implies, an HCR fault relates to poor and loose contacts with localized heat due to unbalance, and failures are associated with this fault, resulting in reduced efficiency and reliability. This fault can be located in circuit breakers, motor starter terminals, or other similar points in the power circuit where contacts and connections are involved. As the above-mentioned faults increase, the motor temperature also rises, and is associated with onset of insulation damage. According to standard electrical testing, an effective method of measuring HCR is to measure phase-to-phase resistance between the three phases.

In this research, a HCR fault has been designed in MATLAB software by connecting a 1 kΩ, 10 W resistor in series in one phase of the induction motor model, as shown in Figure 1. It may be noted that HCR faults can be modelled by resistance in the kΩ to MΩ range.

2.2. Rotor Bar Fault

When a rotor bar fault occurs in the motor and the current flowing through the broken bars stops, the broken bars stop sharing torque. This results in a burden on the healthy bars. In cases where there are many broken rods, the motor will not start. Broken rotor bar (BRB) fault occurs due to magnetic stresses, thermal stresses, contamination, or abrasion and wearing away of the rotor due to a lack of maintenance. Cracks from high temperatures are also a possible consequence. To investigate the rotor bar fault induced in this study, the MATLAB Simulink environment was designed, as shown in Figure 2.

Artificial intelligence (AI) procedures may be developed to diagnose faults in their early stages. The ANN model was preferred in this study, and discrete wavelet transform (DWT) was used to obtain features at different levels, for which the ANN classifier was trained accordingly. Finally, the test data were fed into the ANN classifier, which, along with the algorithmic details and steps, are illustrated in Figure 3.

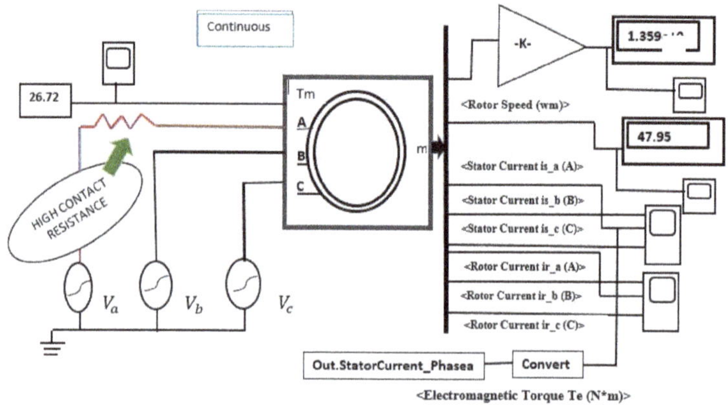

Figure 1. High contact resistance (HCR) fault in a MATLAB Simulink environment. A–C shows stator phases.

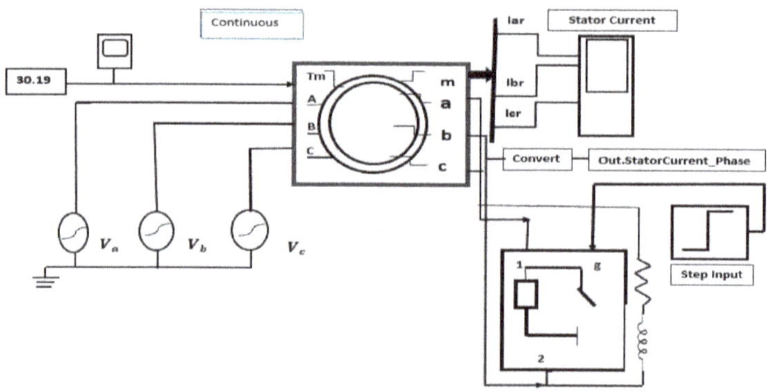

Figure 2. Rotor bar fault in MATLAB Simulink Environment A–C stator phases and a–c rotor phases.

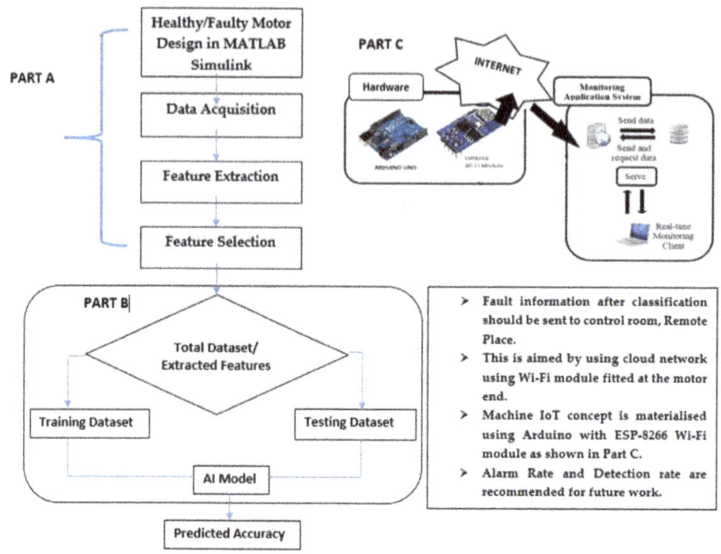

Figure 3. Proposed Algorithm Flowchart (Quality Improved also enlarged-300 dpi).

Figure 3 represents the proposed steps that were followed in this research. Part A represents the steps of signal analysis for feature extraction and selection. The features are acquired using wavelet db-4 transform technique, as shown in Figure 3. Part B shows how to extract relevant features to train the ANN model to predict the results, while Part C shows the feature of enabling IoT with a Wi-Fi module with the motor by adapting the technology [24] with less interference [25] and also with fast computational performance [26].

In an IoT-enabled environment, communication between the induction motors (resources) and central monitoring station is facilitated by two-way communication via a radio communication system that operates in the cloud, as shown in Figure 3. The features signifying faults become recognizable as indicators of the availability of inductors for use in industry. Alternatively, this makes it possible for employees with expert knowledge to gain easy access to motors in order to assess their availability as resources.

3. Induction Motor Modelling

Induction motor modelling MATLAB software was used to design the motor model in this research. The 5.4 HP (4 kW), 400 V, 50 Hz, 1430 RPM induction motor was designed in a MATLAB Simulink environment, as shown in Figure 4. HCR and rotor bar faults are illustrated in Figures 1 and 2, respectively. From the simulated motor design, the data were acquired from a healthy motor and a fault-induced motor.

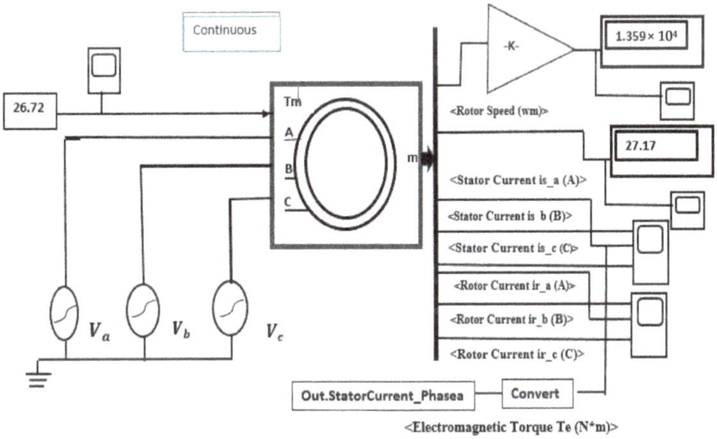

Figure 4. Healthy Motor SIMULINK model with A–C stator phases.

4. Discrete Wavelet Transform (DWT)

This paper focuses on extracting features of the mean and RMS for fault diagnosis using DWT. These features were used as input into the ANN model [27,28]. They were investigated in this study, as shown in Figure 5, then analyzed to obtain the most relevant information from the signals for further processing. The main contributions of this study are that we propose a method to convert time-series data, such as current signals, into grayscale images, using the EWT transformation coupled with the deep CNN model for fault diagnosis [29]. The use of ML algorithms and the DL to machine failure diagnosis is reported in [22]. Discrete wavelet transform is mostly used where analysis of the time–frequency domain signal is required. It is the most efficient tool for extracting transient-based nature signals. Discrete wave transform computes the series of wavelet coefficients, approximation for low frequency feature points, and detail coefficients for high frequency feature points [30]. In a very closely related report, an ensemble machine learning-based fault classification scheme using DWT to compare both raw and filtered separately through wavelet decomposition was reported [31]. In practical terms, DWT can be implemented by using a pair of low-pass and high-pass filters. Wavelet transform is formed from a wavelet of a prototype signal—called a 'mother' or 'single modeling' wavelet y (t)—through shifting operation and dilation. This relationship is shown in the following equation. For the case of Discrete Wavelet Transform, the mother wavelet is shifted and scaled by the powers of two, where j and k are scale and shift parameters, both representing integer values. Here, the mother wavelet is shifted and scaled by a power of two.

$$\psi(j,k)(t) = \frac{1}{\sqrt{2^j}}\psi\left(\frac{(t-k2^j)}{2^j}\right) \tag{1}$$

where j is scaling parameter, k represents shifting parameter, t is time and ψ (t) represents the function of mother wavelet. Mathematically, DWT can be computed by convolving the signal $x(t)$ with the dilated, reflected, and normalized version of the mother-wavelet $\psi(j,k)$. The Equation (3) is result of convolution of the samples with the mother-wavelet.

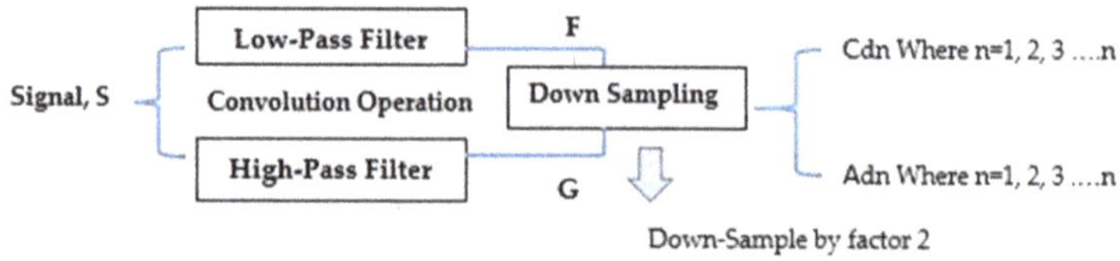

Figure 5. Decomposition procedure for the DWT: representation of the low-pass and high-pass filters, convolving with signal 'S'. (Figure Reduced in size-300 dpi).

Recalling the wavelet coefficient γ of a signal $x(t)$ is a projection of $x(t)$ onto a wavelet, and let $x(t)$ be a signal of length 2^N, in the case of child wavelet in the discrete family above.

$$\gamma_{j,k} = \int_{-\infty}^{\infty} x(t)\frac{1}{\sqrt{2^j}}\psi(\frac{(t-k2^j)}{2^j})dt \tag{2}$$

By fixing j at a scale, so that $\gamma_{j,k}$ is a function of k only, by the Equation (2), $\gamma_{j,k}$ can be viewed as a convolution of $x(t)$ with reflected, dilated and normalized version of mother-wavelet, $h(t) = \frac{1}{\sqrt{2^j}}\psi(\frac{(-t)}{2^j})$, sampled at the points 1, $2^j, 2^{2j} \ldots 2^N$. But this is precisely what the detail coefficients give at different levels of j for DWT as shown in the Figures 6–8. Therefore an appropriate choice for the signals, g[n] and h[n], the detail coefficients of the filter bank correspond exactly to a wavelet coefficient of discrete set of child wavelet for a given mother wavelet $\psi(j,k)(t)$.

$$dwt(j,k)(t) = y[n] = (x * \psi(j,k)(t))[n] = \frac{1}{\sqrt{2^j}}\int x(t)\psi\left(\frac{(t-k2^j)}{2^j}\right)dt \tag{3}$$

Filters decompose the signals into approximation and detail coefficients, and their computation is given as [17]

$$y_{low}[n] = (x * g)[n] = \sum_{k=-\alpha}^{\alpha} x[k]g[2n-k] \tag{4}$$

$$y_{high}[n] = (x * h)[n] = \sum_{k=-\alpha}^{\alpha} x[k]h[2n-k] \tag{5}$$

The above equations can be more precisley described by the given convolution approach (6), (7).

$$y_{low} = (x * g) \downarrow 2 \tag{6}$$

$$y_{high} = (x * h) \downarrow 2 \tag{7}$$

The DWT decomposition process of 2-level DWT is illustrated in Figure 5, Equation (2), presented above, helps compute DWT, and Equations (4) and (5) represent the low- and high-pass filters, respectively, showing the high-frequency and low-frequency components. The outputs give the detail coefficients (from the high-pass filter) and approximation coefficients (from the low-pass filter).

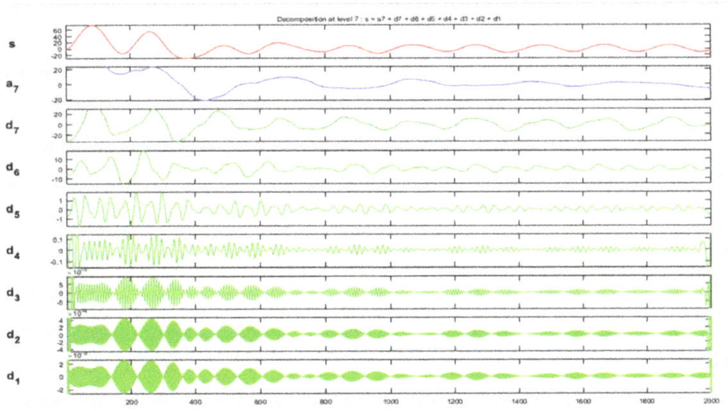

Figure 6. Representation of DWT Approximation Coefficients and Detail Coefficients of Healthy Phase-A Currents with 2000 samples.

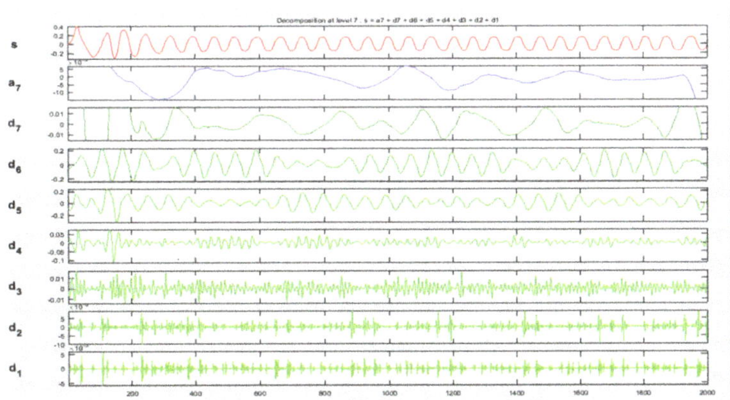

Figure 7. Representation of DWT approximation coefficients and detail coefficients of HCR fault in Phase-A currents with 2000 samples.

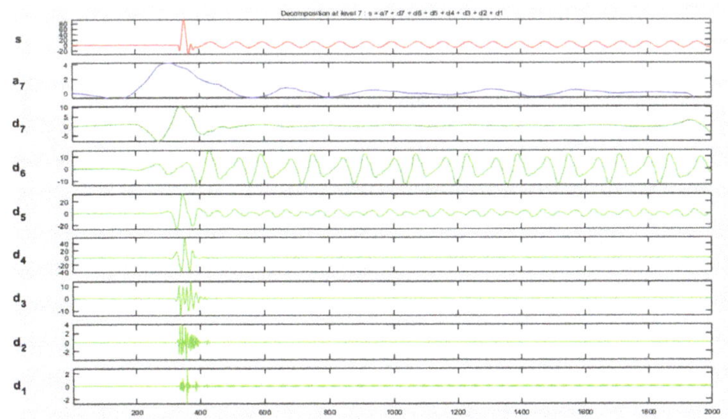

Figure 8. Representation of DWT approximation coefficients and detail coefficients of a faulty signal rotor bar fault in Phase-A currents with 2000 samples.

Designation of Frequency Band

Each Debauches wavelet db-4 contains multiple levels. As a 10 kHz frequency was used to sample the current signals in this research, a single sample of 10 kHz was distributed across 9 db-4 wavelet levels. The distribution of frequency band is tabulated in Table 1. All wavelet levels except for 7, 8 and 9 were discarded, because these levels possessed no valuable information. Moreover, the signals with line noise at 50 Hz were also discarded, as noise reduction clarified the target signals more effectively. Table 2 shows distribution frequency bands.

Table 2. Distribution of frequency bands.

Wavelet Level	Frequency Band (Hz)
9	5000–10,000
8	2500–5000
7	1250–2500
6	625–1250
5	312.5–625
4	156.25–312.5
3	78.125–156.25
2	39.0625–78.125
1	19.531–39.0625

In this research, DWT 1-dimension analysis was used for the healthy and faulty motor signals, and the following graphs were derived from the 2000 data samples gathered. Figure 6 represents the DWT approximation levels (a7 only, using level-7 alone) and detail coefficient levels (d1–d7) of a current healthy conditions signal in Phase A, using just seven levels of decomposition. It may be concluded that on starting, the induction motor current showed greater amplitude, due to higher torque, which gradually settled down into a steady state from 800 samples onward. Signals were decomposed into seven levels to obtain high- and low-frequency information about the current signal. The authors in [31] report successful implementation of the DWT technique, with accuracy more than 99%. In comparison, this research work also achieves more than 98% accuracy, possibly because of the dataset used. The authors in [23] also present successful fault diagnosis using DWT and achieve claimable accuracy. The authors in Refs. [32,33] show techniques other than the DWT, such as data augmentation-based, feature learning-based and classifier designing-based, using different datasets, calling them intelligent diagnosis models with deep networks to conclude how to reduce the consumption of computing time. However, these strategies are complex and need improvement for data generation ability and data dimensionality issues based on the different dataset conditions. Finally, the authors in [34] report on the identification of faults using coupled neuron-based approach, mainly focusing on enhancing the weak useful signature, and not applying a classifier-based approach to find accuracy using a relatively complex approach compared to the proposed approach, where the condition of fault is the main interest.

Figure 7 displays the DWT coefficient levels (a7 only) and detail coefficient levels (d1–d7) for the HCR fault in the Phase-A current. The fault was analyzed and the detail coefficients, d1, d2, d3, and d4, were found to contain less information and fewer features, as most of the noise was contained by d1, d2, d3, and d4. All detail coefficients, representing most of the high-frequency content and the approximate coefficients containing low-frequency information about the current signals, were obtained at level 7, with a clearly visible difference between the healthy and damaged motor signals.

Figure 8 displays the DWT coefficients of Phase-A currents with rotor bar fault, wherefrom, it may be concluded that initially, the motor was running normally. A rotor bar fault was subsequently introduced at 350 samples, due to which, the motor displayed the fault with spikes at corresponding coefficients. This indicated that the fault had been detected, and the DWT coefficients had received frequency information about the currents. The signal contamination was due to the rotor bar fault. Moreover, feature points were extracted using DWT. The signal and approximation coefficients at level 1 are very identical.

However, they become gradually different at higher and higher levels. A similar tendency can be visibly noted in the case of the bearing fault using DWT and ensemble machine learning algorithms [31].

Figure 9 represents the feature points for a healthy motor in Phase-A at level 9, and similar features were extracted for the remaining faults at different levels, using DWT. In this research, the mean and RMS features were investigated. In addition, Figures 6 and 8 show the detail and approximation coefficients for healthy motor and rotor bar fault signals, containing both low-frequency and high-frequency content information about the approximation and detail coefficients, respectively.

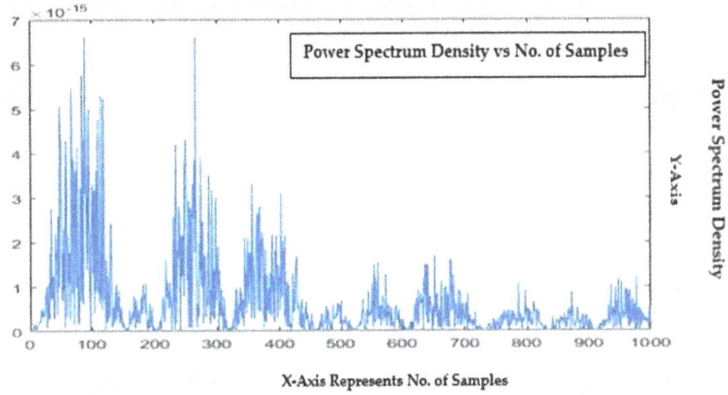

Figure 9. Healthy motor phase: a feature point representation with 1000 samples. (Figures Titled-300 dpi).

Figures 10 and 11 display the features obtained using DWT at level 9. These features are the composite form of different detail levels, as represented in Figure 6.

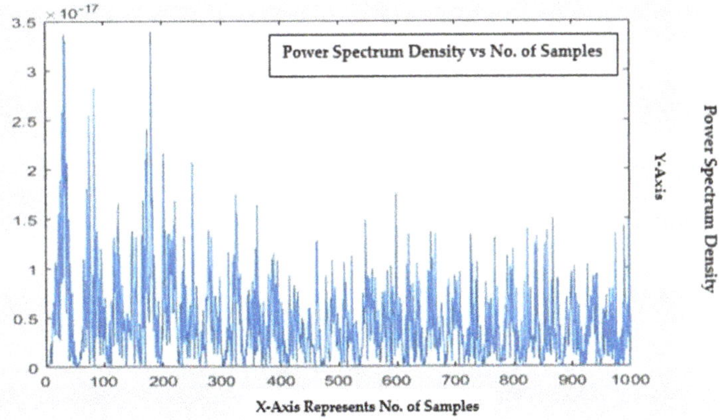

Figure 10. High contact resistance fault in Phase-A: Representation of feature points with 1000 samples. (Figures Titled-300 dpi).

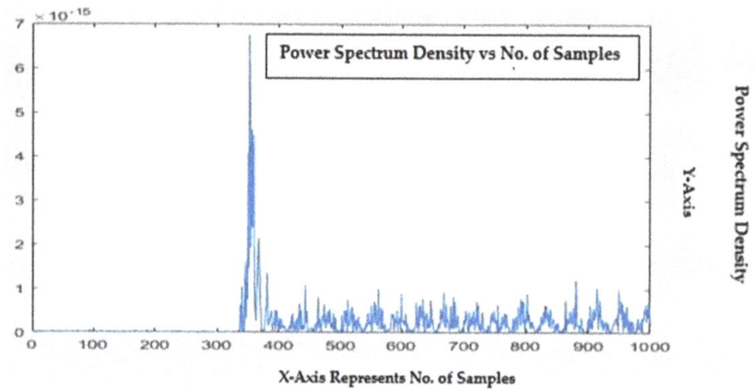

Figure 11. Rotor bar fault in Phase-A: Representation of feature points with 1000 samples (Figures Titled-300 dpi).

5. ANN Classifier

The concept of ANN is to classify deniable features that characterize target parameters in real-world applications. Artificial neural networks are brain-inspired neural network (NN) systems. The Neural Network (NN) works similar to neurons in our nervous system, and is capable of learning from past data; the ANN is able to learn from data and provide responses in the form of classifications. It is the most widely used form of AI [27], mostly preferred for fault detection and diagnosis. An ANN can be trained by causing the interconnected nodes (called neurons) to learn in a way that resembles that of the human brain by recognizing speech patterns and tendencies in images. The behavior is defined by the weights assigned to these interconnections, ensuring that the desired task is performed correctly. Artificial neurons networks are excellent tools for identifying complex problems. The multi-layer feed-forward NN with a back-propagation training algorithm is the classic NN proposed in this study. Neural Networks have already been successfully applied and proven suitable for fault detection [22,28–31], and pseudocode has been provide in Appendix A.

In this research, supervised learning was applied to train both healthy and faulty motors [22,27–31]. There are three main components of an ANN: the input layer, hidden layers, and output layers, as shown in Figure 12. In an ANN, layers possess nodes, and these nodes relate to the nodes in the next layer. Moreover, each connection has proportionated weights, considered as the effect of the nodes on the layers. The hidden layers contain information that is transformed onto higher order node layers. Each node is associated with a predefined activation function, which will determine whether the node will be activated on a node in a higher layer. The way in which it is activated will depend on the summarized values obtained by multiplying the values of the nodes with their associated weights. The neuron in an ANN is the computational unit that takes x1, x2, and x3 as inputs, as well as an intercept term. The output 'y' is obtained by

$$y = f\left(W^T x\right) = f\left(\sum_{i=1}^{3} W_i x_i + b\right) \tag{8}$$

where the activation function is represented by 'f', and often chosen to be the sigmoid function (as in this research), different activation functions are investigated by taking 21 hidden layers and 11 epochs. 'W' is the ANN model weight, while 'b' is the scalar value, known as the bias. Here, Equation (8) was used to compute output 'y' of the ANN. Performance indicator parameters for NN model are shown in Figure 13, below, where it can be seen that the overall cross entropy error of the trained NN is 0.59872 at epoch 10. The model was trained for an HCR fault at level 9, with RMS features. During the analysis of results for different activation functions and size of layers, it was observed that the designing of Neural Network could be accomplished with a lesser number of hidden layers without affecting the accuracy. In this work, at 21 hidden layers, good accuracies are obtained for all transfer functions while the rest of the parameters are kept the same.

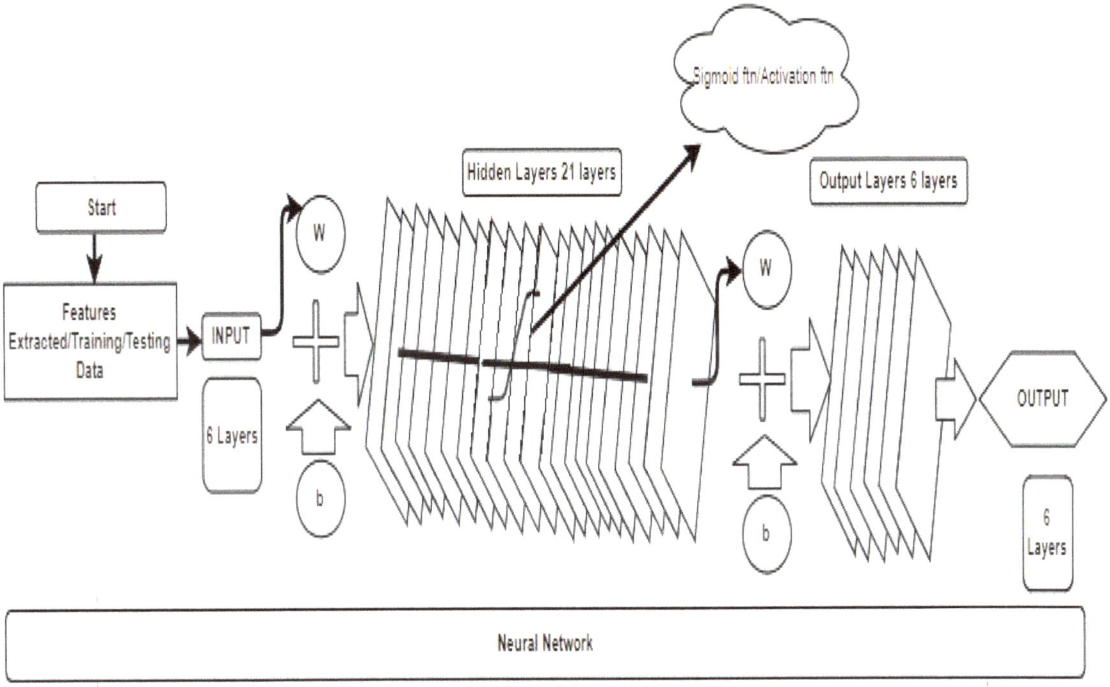

Figure 12. Neural network (NN) model.

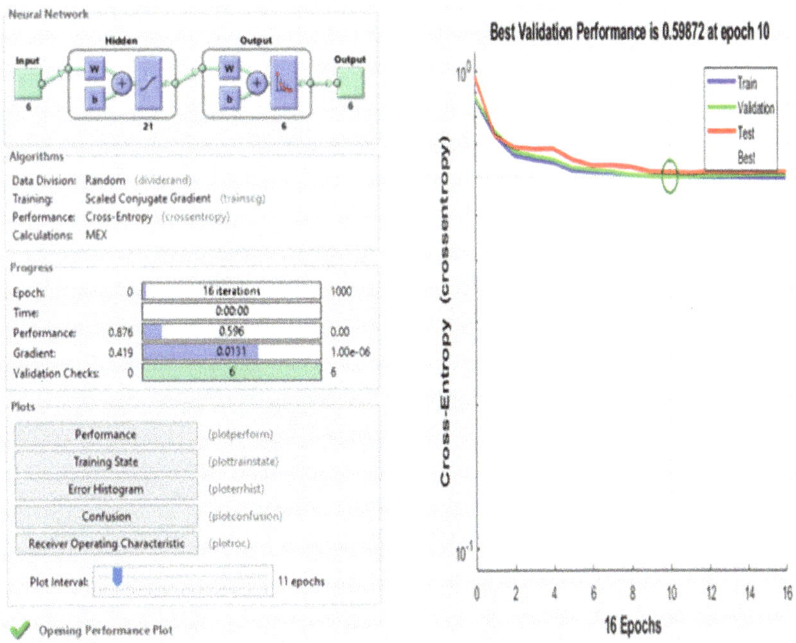

Figure 13. Neural network model performance parameters and cross-entropy error.

Figure 13 shows that validation and training have similar characteristic curves with efficient testing. As in NNs, the hyper-parameters were tuned to provide a high quality solution for the ANN model and the number of epochs consisted of numerical figures—indicating the number of times that the entire dataset was propagated forward and backward through an NN. The epochs were chosen to control the number of complete passes through the training dataset. These epochs were divided into batches for processing.

It may be pointed out that all the ROC's figures (Train, Test, and Validation, etc.) if accommodated in a few plots, will result in the cramping of many of our interests, considering we do not have any ZOOMING option in the tool used.

From the above figure, it is concluded that all faults are classified with 76% accuracy, All ROC curves point out that the grey diagonal line of ROC means that the samples lying on this line show that the proportion of correctly classified samples is the same as the proportion of incorrectly classified samples.

Class-1 (Healthy Phase A)
Class-2 (Healthy Phase-B)
Class-3 (Healthy Phase-C)
Class-4 (Faulty Phase-A) Rotor Bar/High-Contact Resistance
Class-5 (Faulty Phase-B) Rotor Bar/High-Contact Resistance
Class-6 (Faulty Phase-C) Rotor Bar High-Contact Resistance

In the ROC-training section, the six different classes have been classified. During the validation process, each class has transmit data according to its specific feature type. Different colours represent the different phases and types of classes. The third section describes the testing of ROC using a different scheme of classes, and the classification was categorized according to features. The tangent/diagonal line was received in ALL-ROC sections, which shows that if the line of different Phases exists above the ROC diagonal grey coloured line, it means the samples are correctly and properly classified with a true positive rate, else, if the lines of Phases exist below the diagonal lines, it means the samples are classified incorrectly, and do not belong to the same Phases. From Figure 14, it can be concluded that some of the samples of the Phases (features) are correctly classified and some samples are incorrectly classified, therefore, the accuracy obtained is 76.0%, as shown in the Confusion Matrix.

From Figure 15, it is clear that with an accuracy of 82.6%, compared to Figure 14, with 76.0%, we have obtained a more appropriate classification of the different faults and Phases with the same number of samples and activation function. Almost all Phases lines drawn are received above the grey diagonal line, and this line is supposed to represent the threshold line for the ROC Curves.

Figure 16 makes the things more clear by showing the grey diagonal line and almost all Phase lines with their samples very appropriately classified with a good True Positive rate by obtaining an accuracy level of about 92.2%, which, as compared to the results obtained in Figures 14 and 15, are better. Moreover, a single Phase line is below that diagonal line, meaning that only the Phase samples are incorrectly classified.

Similarly, Figure 17 represents less accurate accuracy results compared to Figure 16, and the ROC curves also represent that some Phases are classified appropriately, while others have incorrect classification. However, this time, the training/activation function is 'lm'.

Figure 18 clearly represents the ROC's result, showing very proper classification results as compared to Figures 17 and 19, with the same training/activation function and number of samples. All Phases results are above the grey diagonal line with good accuracy of 93.1%, which means the faults are classified more accurately, with a true positive rate. As far as accuracy is concerned, from the previous last two figures, we obtain better accuracy in comparison, and therefore, receive good results for the ROCs.

From the above figures, it is shown that, with 91.0% accuracy, most of the Phases are classified with good results, and in Figure 20, with am accuracy of 85.8%, some faults are incorrectly classified, with Phase lines below the diagonal line. In addition to this, the CGP training/activation function here is used to obtain results with same number of samples.

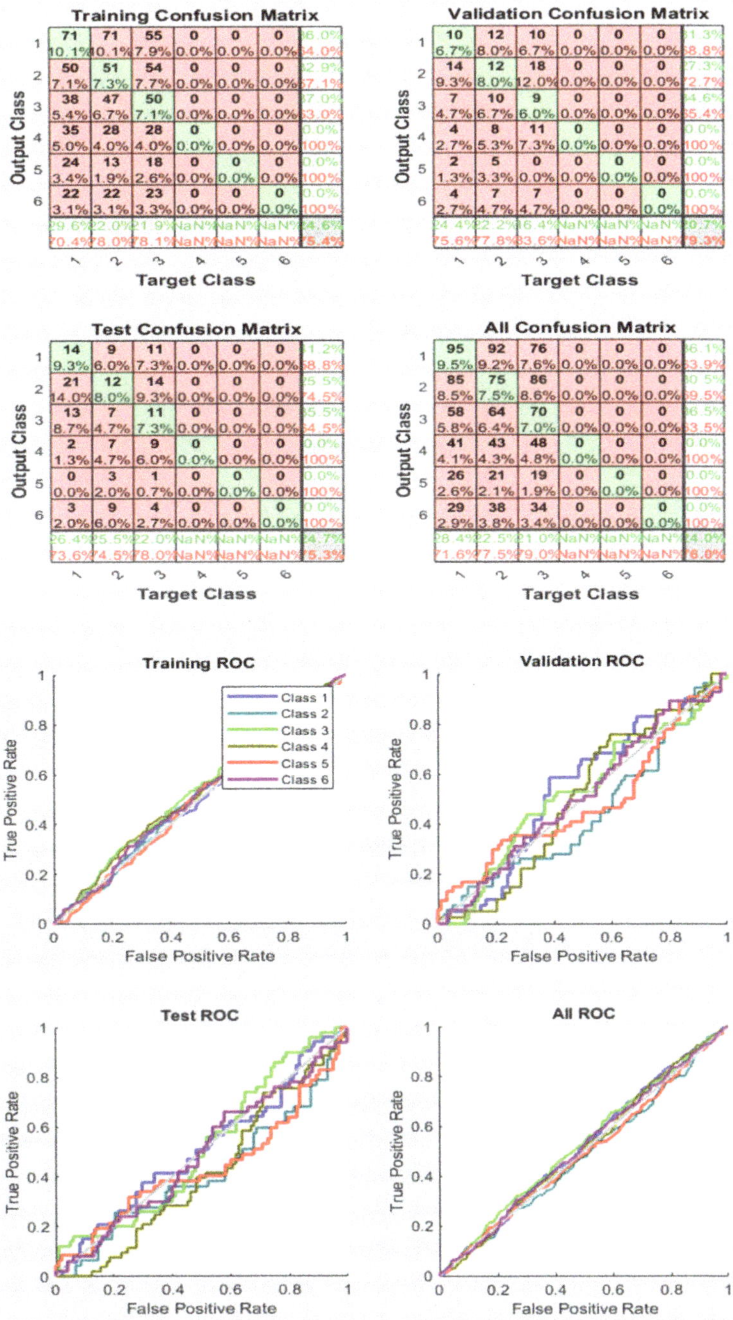

Figure 14. Neural Network Model ROC (Receiver Operating Characteristics) and Confusion Matrix for OSS Train Fcn/Activation Function at level 9 (RMS) Features for High Contact Resistance Fault with 76.0% Accuracy. Grey line shows diagonal of the ROC for indicating Class 1–6 lines are lying above or below or on it.

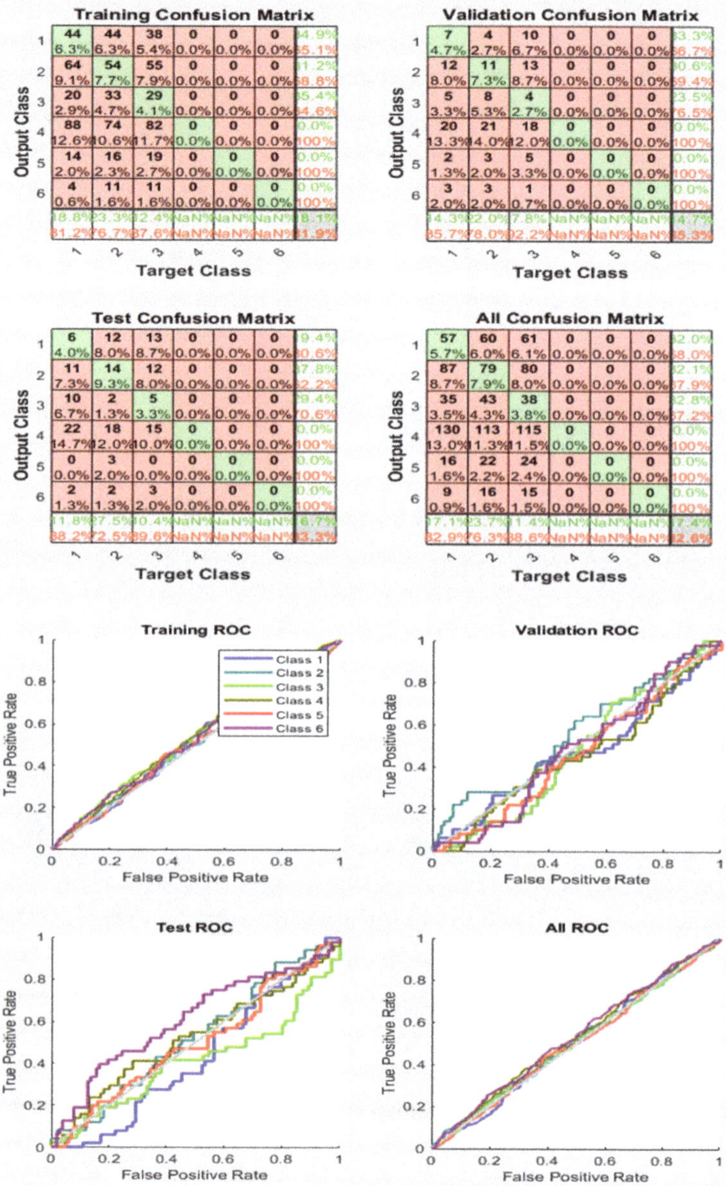

Figure 15. Neural Network Model ROC (Receiver Operating Characteristics) and Confusion Matrix for OSS TrainFcn/Activation Function at level 9 (RMS) Features for High Contact Resistance Fault with 82.6% Accuracy. Grey line shows diagonal of the ROC for indicating Class 1–6 lines are lying above or below or on it.

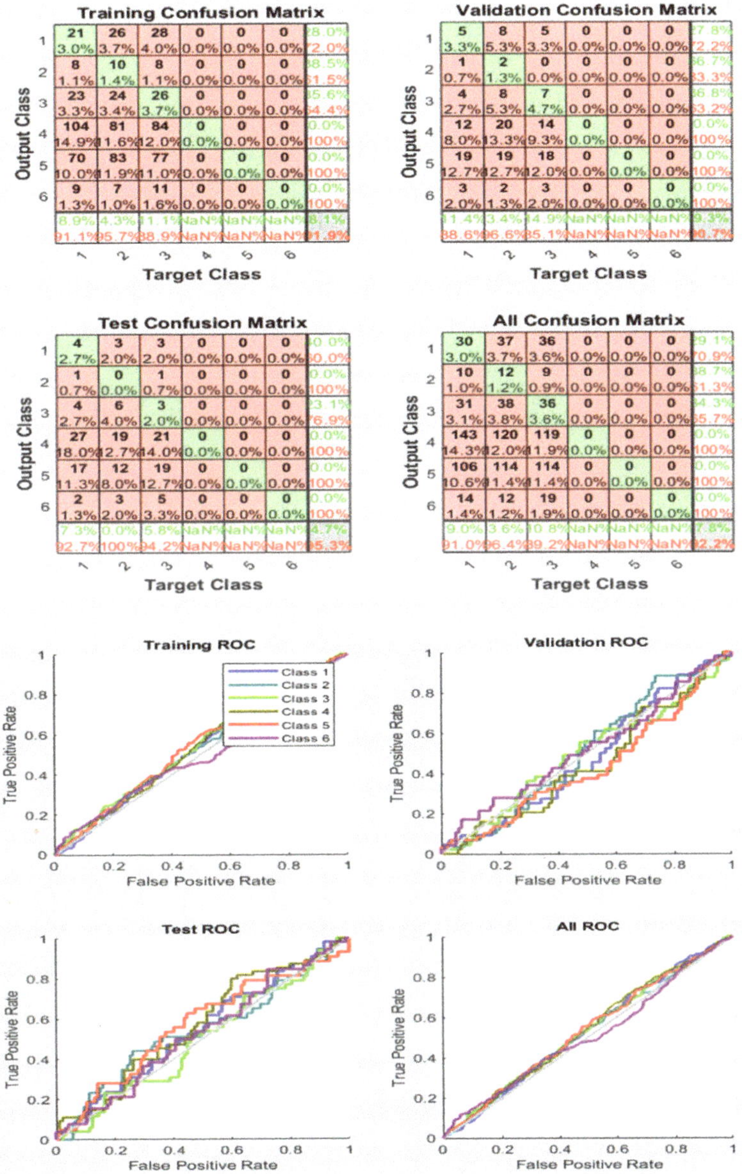

Figure 16. Neural Network Model ROC (Receiver Operating Characteristics) and Confusion Matrix for OSS TrainFcn/Activation Function at level 9 (RMS) Features for High Contact Resistance Fault with 92.2% Accuracy. Grey line shows diagonal of the ROC for indicating Class 1–6 lines are lying above or below or on it.

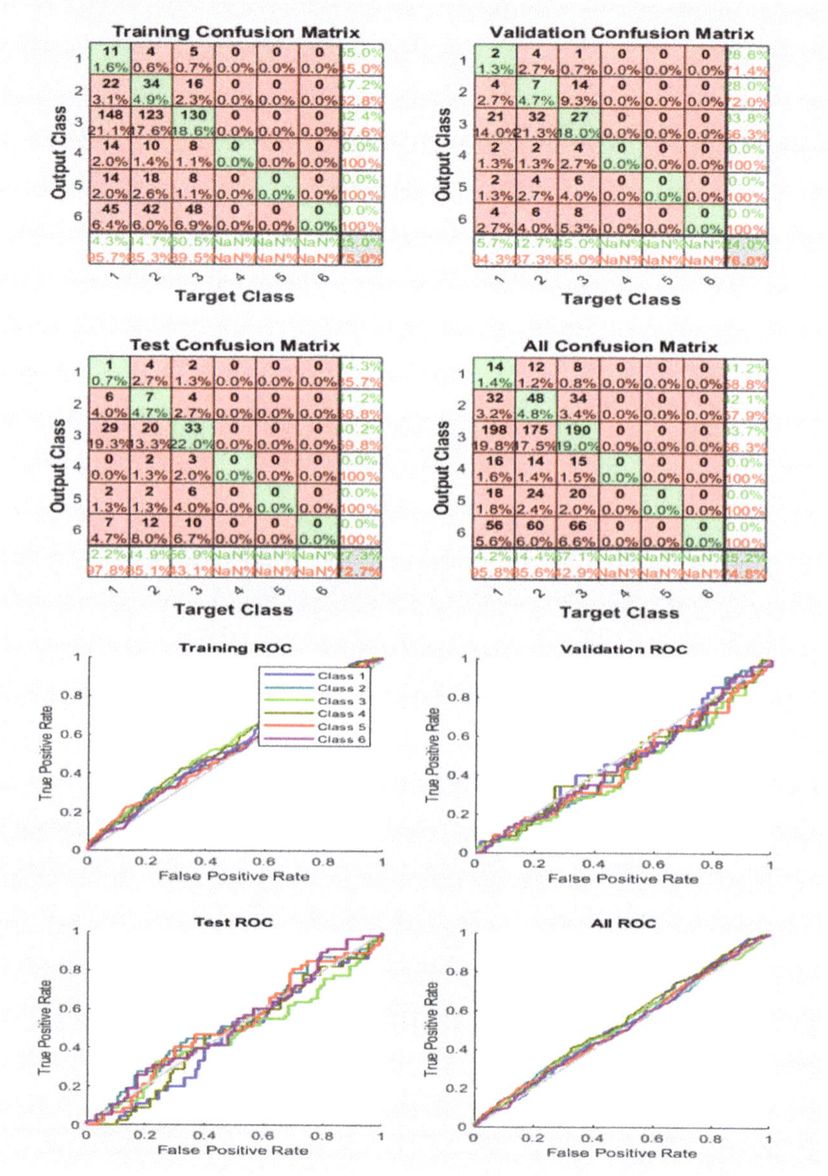

Figure 17. Neural Network Model ROC (Receiver Operating Characteristics) and Confusion Matrix for LM TrainFcn/Activation Function at level 9 (RMS) Features for High Contact Resistance Fault with 74.8% Accuracy. Grey line shows diagonal of the ROC for indicating Class 1–6 lines are lying above or below or on it.

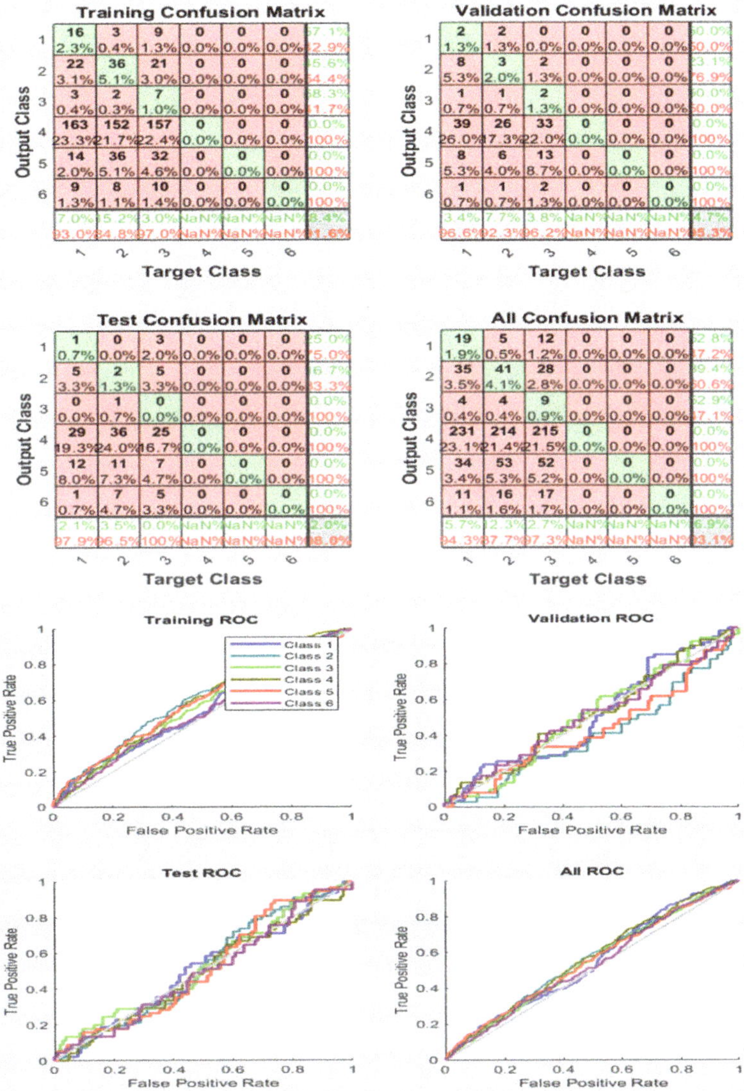

Figure 18. Neural Network Model ROC (Receiver Operating Characteristics) and Confusion Matrix for LM TrainFcn/Activation Function at level 9 (RMS) Features for High Contact Resistance Fault with 93.1% Accuracy. Grey line shows diagonal of the ROC for indicating Class 1–6 lines are lying above or below or on it.

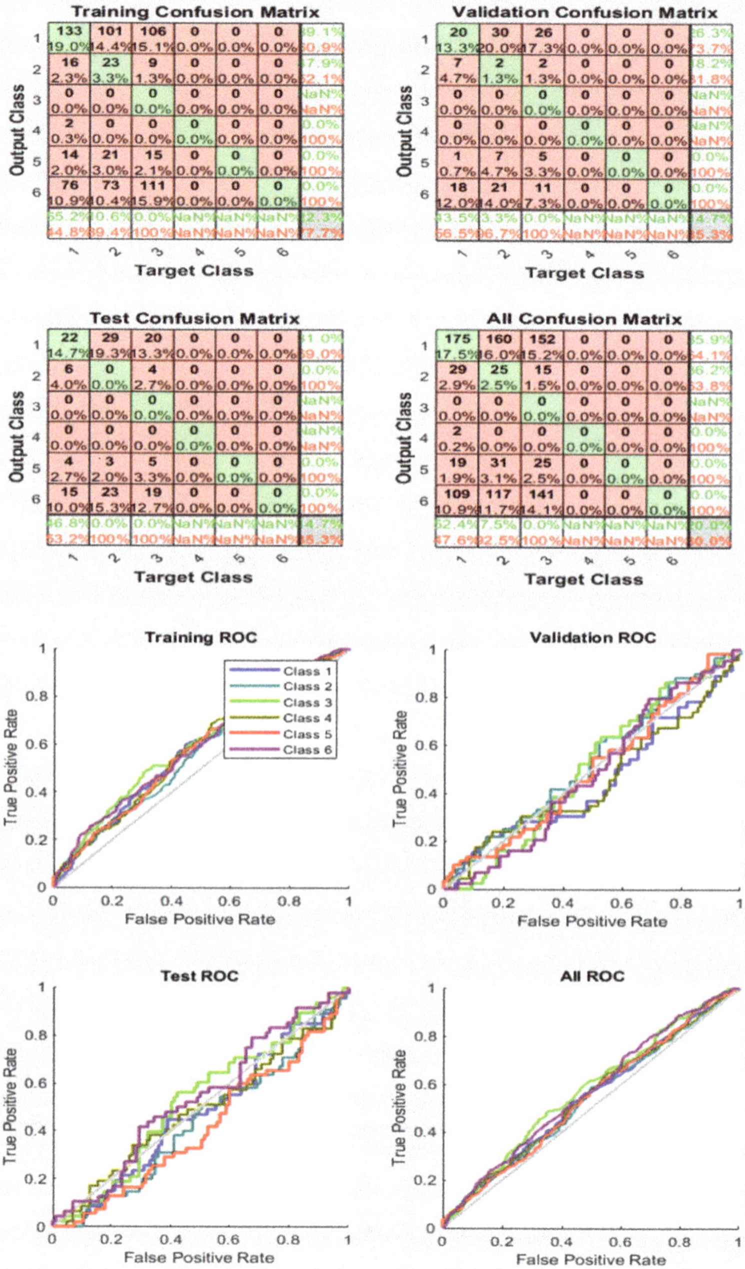

Figure 19. Neural Network Model ROC (Receiver Operating Characteristics) and Confusion Matrix for LM TrainFcn/Activation Function at level 9 (RMS) Features for High Contact Resistance Fault with 80.0% Accuracy. Grey line shows diagonal of the ROC for indicating Class 1–6 lines are lying above or below or on it.

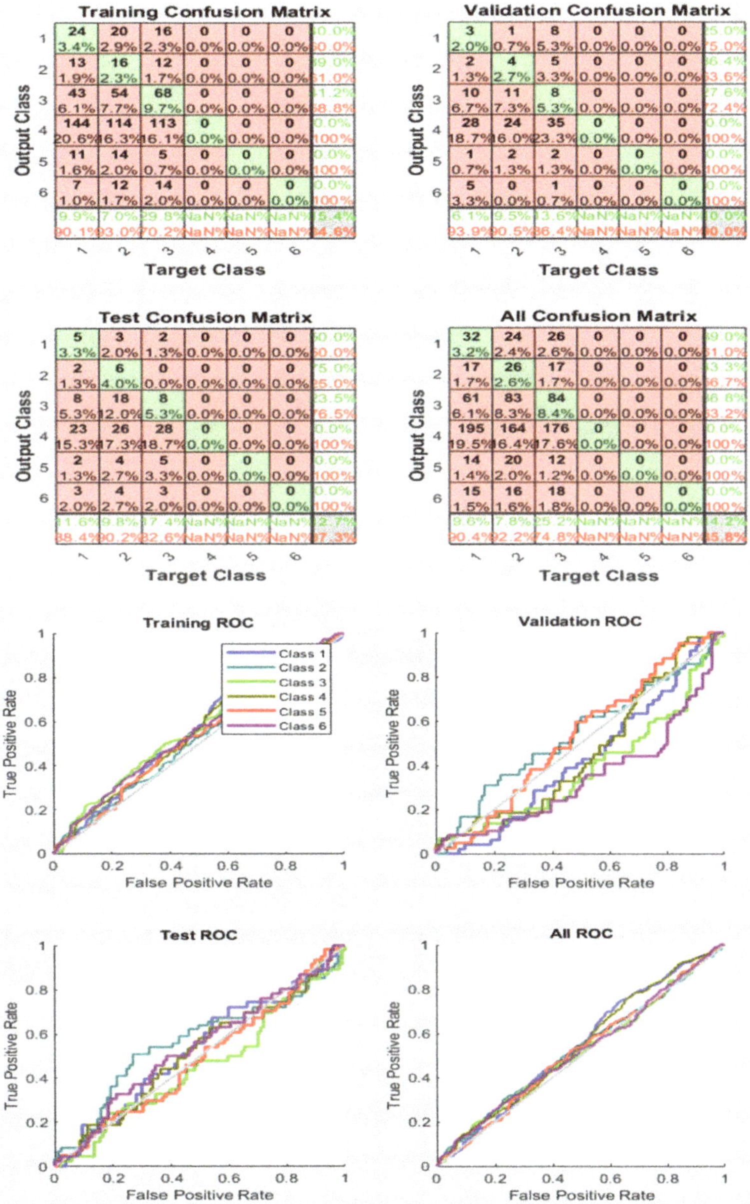

Figure 20. Neural Network Model ROC (Receiver Operating Characteristics) and Confusion Matrix for CGP TrainFcn/Activation Function at level 9 (RMS) Features for High Contact Resistance Fault with 85.8% Accuracy. Grey line shows diagonal of the ROC for indicating Class 1–6 lines are lying above or below or on it.

The rate value of a full load current of 5.5 HP (4 kW) three-phase motor is between 8–10 A under normal working conditions. Motors typically generate 3.6 times the normal rate current under short circuit conditions. In this paper, a short circuit current of approximately ($9.2 \times 3.6 = 33.12$ A) is obtained. The results are validated by iteration three times. In the open circuit, there is no current flow. In this case, there are no characteristic signs of motor current, as our algorithm takes a single-phase current as the input for the balanced three-phases. The high contact resistance accuracy results for RMS features are tabulated at different levels and with different accuracies, and the derived tabulated results are shown.

The ROC's curves and confusion matrix are as shown in Figures 14–21, accordingly. Table 3 shows 76.0%-TrainFcn = 'oss', 82.6%-TrainFcn = 'oss', 92.2%-TrainFcn = 'oss', 74.8%-TrainFcn = 'lm', 80.0%-TrainFcn = 'lm', 93.1%-TrainFcn = 'lm', 85.8%-TrainFcn = 'cgp' and 91.0%-TrainFcn = 'cgp' accuracies using different TrainFcn are obtained, respectively, for High Contact resistance faults, as presented in Table 3 for level 9 RMS features. Similarly, for the accuracies obtained from the rest of the levels, as shown in the table for RMS features of HCR Faults, or for any other type of fault at different levels for different features, we can get the ROCs and Confusion matrix results, as shown in the Figures 14 and 15, and so on.

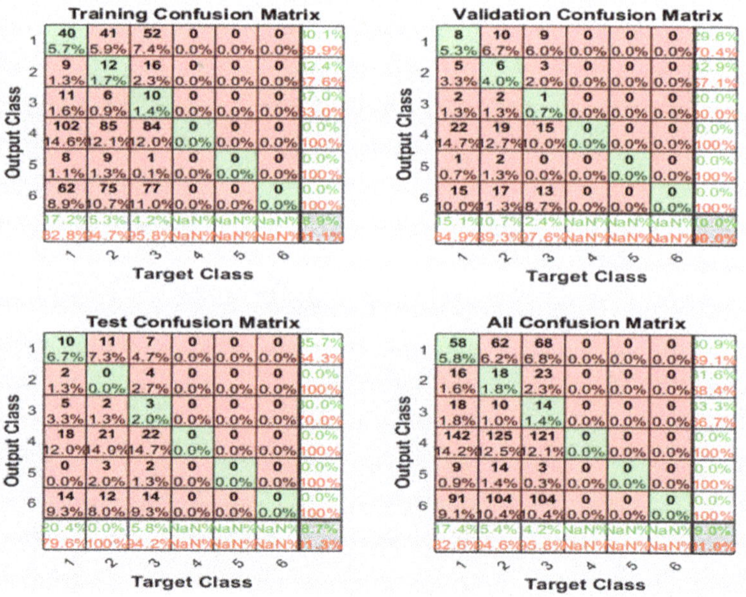

Figure 21. *Cont.*

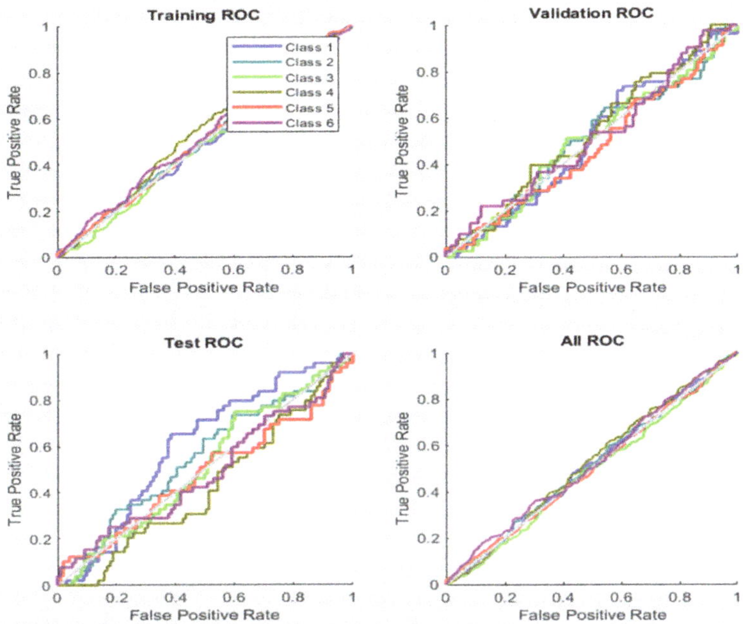

Figure 21. Neural Network Model ROC (Receiver Operating Characteristics) and Confusion Matrix for CGP TrainFcn/Activation Function at level 9 (RMS) Features for High Contact Resistance Fault with 91.0% Accuracy. Grey line shows diagonal of the ROC for indicating Class 1–6 lines are lying above or below or on it.

Table 3. High Contact Resistance Fault with RMS Feature Accuracies under different Activation Functions.

Fault	Activation Function	Characteristic Feature	Accuracy				Successive Rate
			Level 5	Level 7	Level 8	Level 9	
High Contact Resistance (HCR) Fault	Trainscg	RMS	80.1%	74.5%	81.2%	85.2%	(Short-Circuit Current) $I_{SC} \cong$ 33.12 A
	Trainbfg		83.4%	85.4%	87.5%	77.8%	
	Traincgp		88.8%	88.0%	88.4%	85.6%/85.8%/91.0% (By Retraining the System, Highest Accuracy Result is preferred.)	
	Traincgb		80.1%	79.7%	77.0%	79.1%	
	Traingda		84.1%	77.3%	88.1%	77.3%	
	Traingdx		77.1%	82.3%	82.2%	86.1%	
	Traingdm		80.9%	82.5%	87.2%	85.2%	
	Trainlm		96.4%	65.4%	83.6%	74.6%/74.8%/80.0%/93.1%	
	Trainoss		85.1%	81.0%	84.5%	83.3%/76.0%/82.6%/92.2%	

Figures 22–27 have ROCs with good positive rates and all Phases with good classification, which means that good accuracy results have good ROCs with Phase lines above the grey diagonal line.

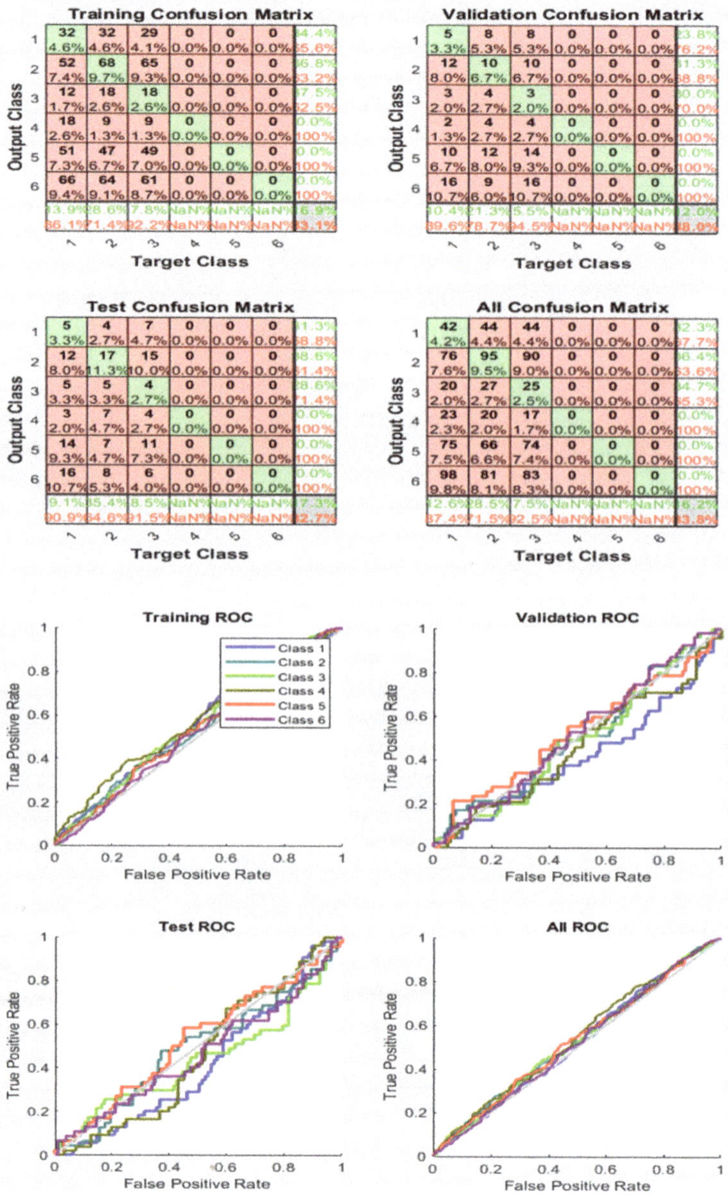

Figure 22. Neural Network Model ROC (Receiver Operating Characteristics) and Confusion Matrix for OSS TrainFcn/Activation Function at level 9 (RMS) Features for Rotor Bar Fault with 83.8% Accuracy. Grey line shows diagonal of the ROC for indicating Class 1–6 lines are lying above or below or on it.

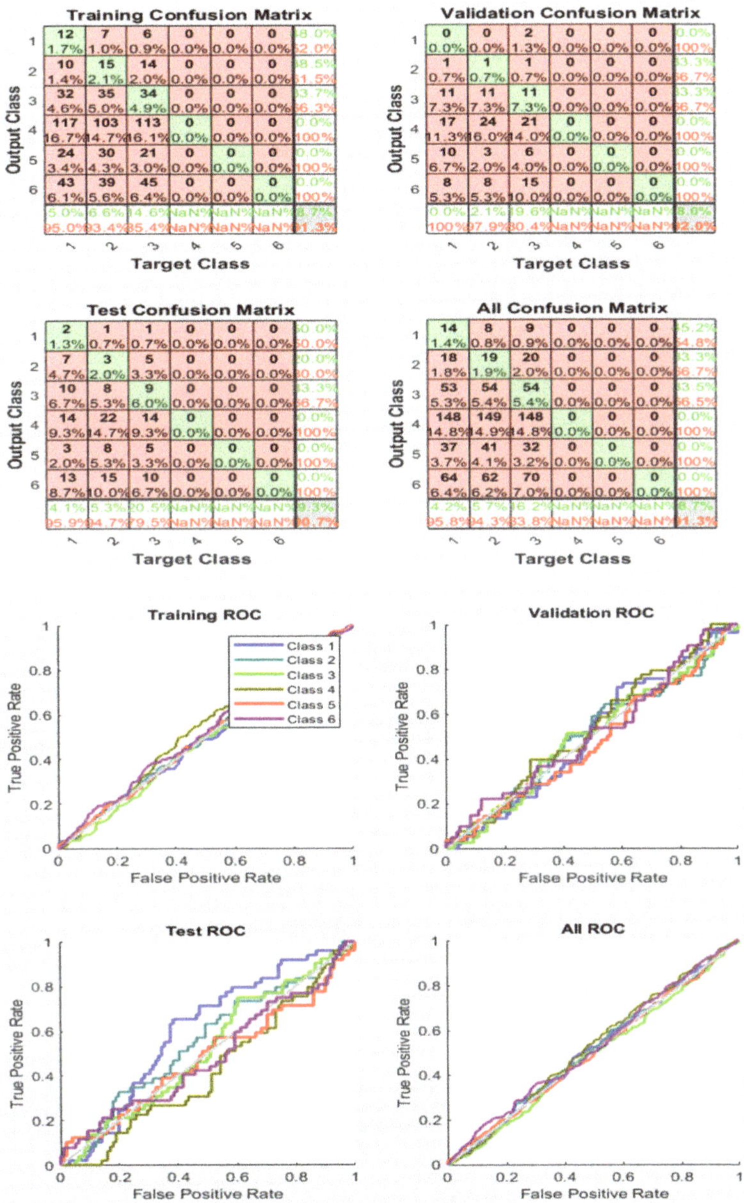

Figure 23. Neural Network Model ROC (Receiver Operating Characteristics) and Confusion Matrix for CGP TrainFcn/Activation Function at level 9 (RMS) Features for Rotor Bar Fault with 91.3% Accuracy. Grey line shows diagonal of the ROC for indicating Class 1–6 lines are lying above or below or on it.

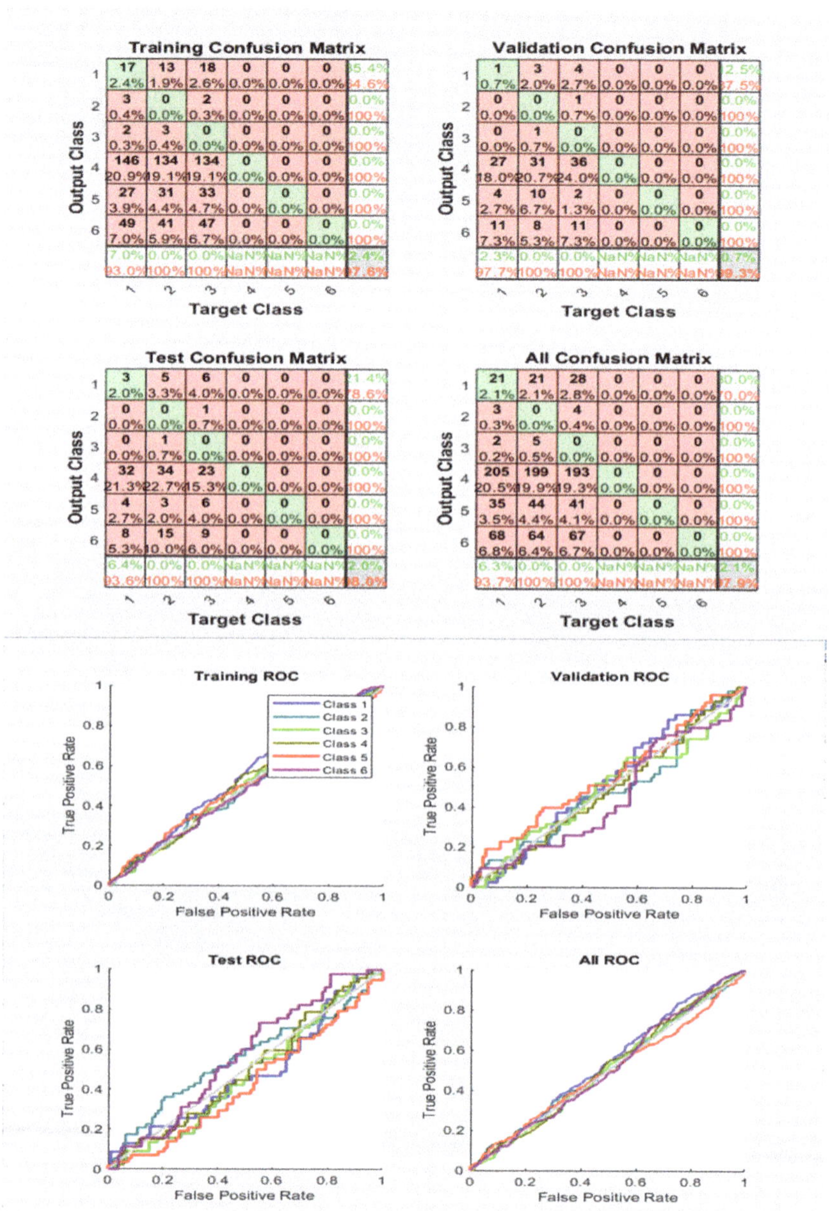

Figure 24. Neural Network Model ROC (Receiver Operating Characteristics) and Confusion Matrix for CGP TrainFcn/Activation Function at level 9 (RMS) Features for Rotor Bar Fault with 97.9% Accuracy. Grey line shows diagonal of the ROC for indicating Class 1–6 lines are lying above or below or on it.

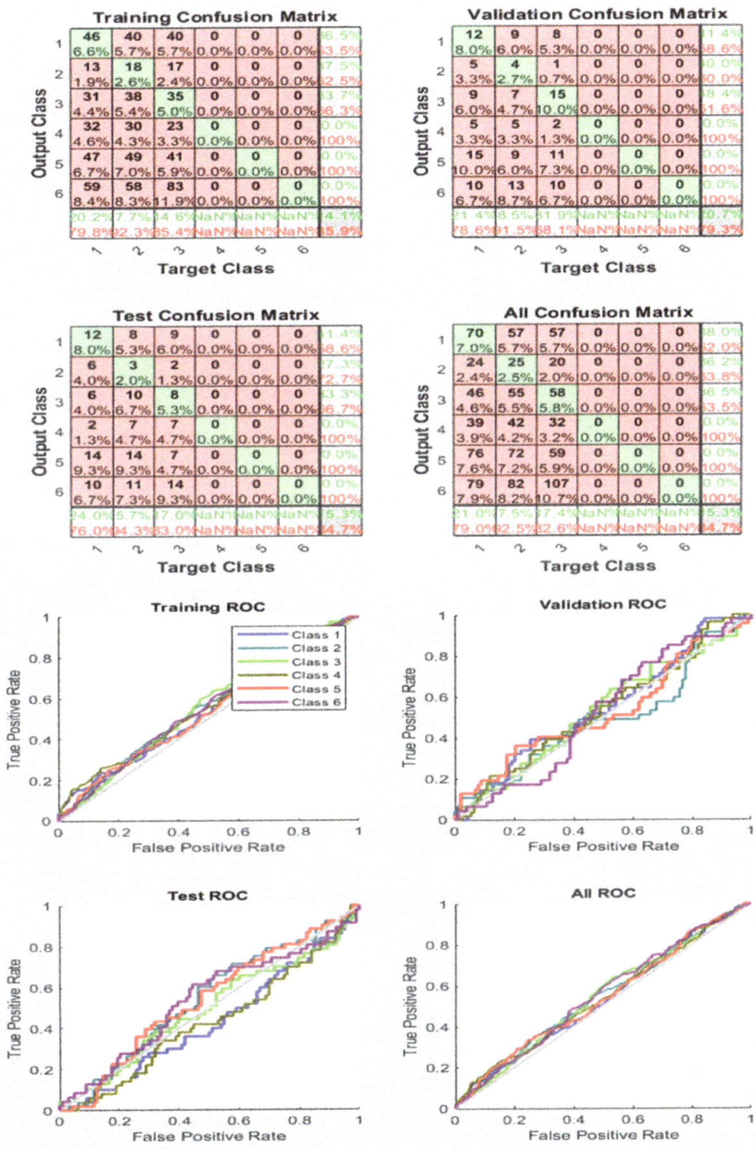

Figure 25. Neural Network Model ROC (Receiver Operating Characteristics) and Confusion Matrix for LM TrainFcn/Activation Function at level 9 (RMS) Features for Rotor Bar Fault with 84.7% Accuracy. Grey line shows diagonal of the ROC for indicating Class 1–6 lines are lying above or below or on it.

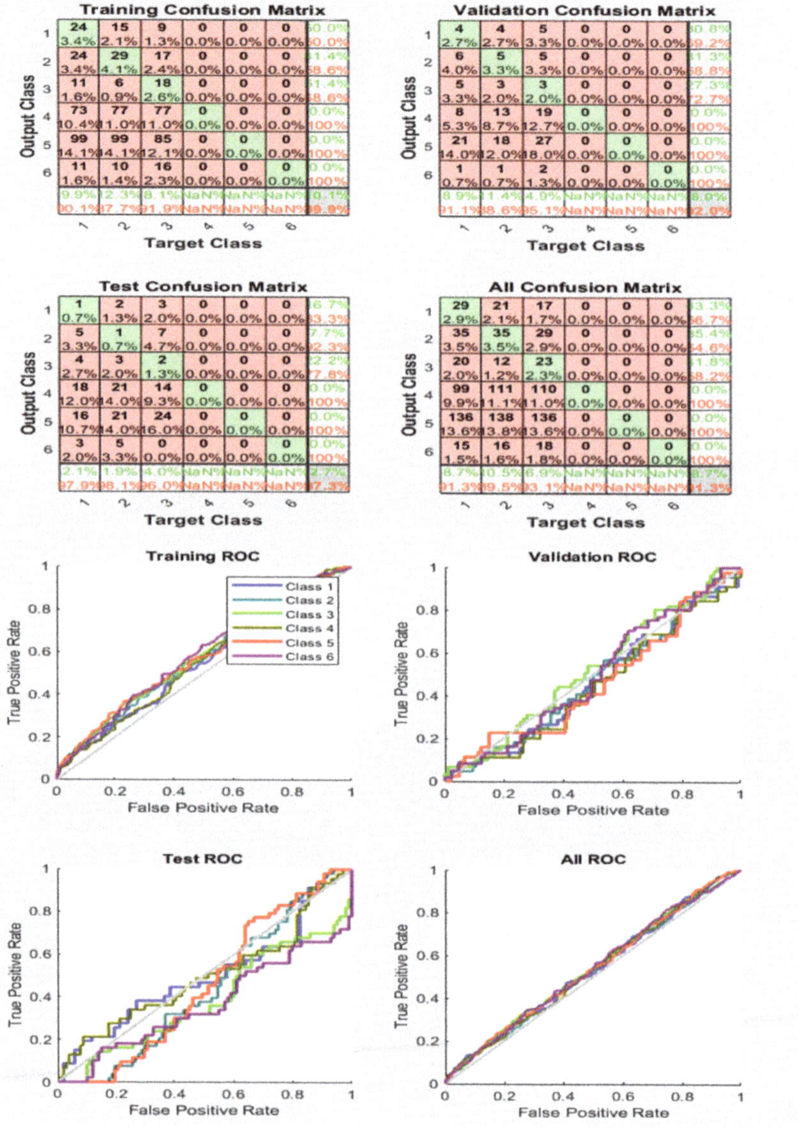

Figure 26. Neural Network Model ROC (Receiver Operating Characteristics) and Confusion Matrix for LM TrainFcn/Activation Function at level 9 (RMS) Features for Rotor Bar Fault with 91.3% Accuracy. Grey line shows diagonal of the ROC for indicating Class 1–6 lines are lying above or below or on it.

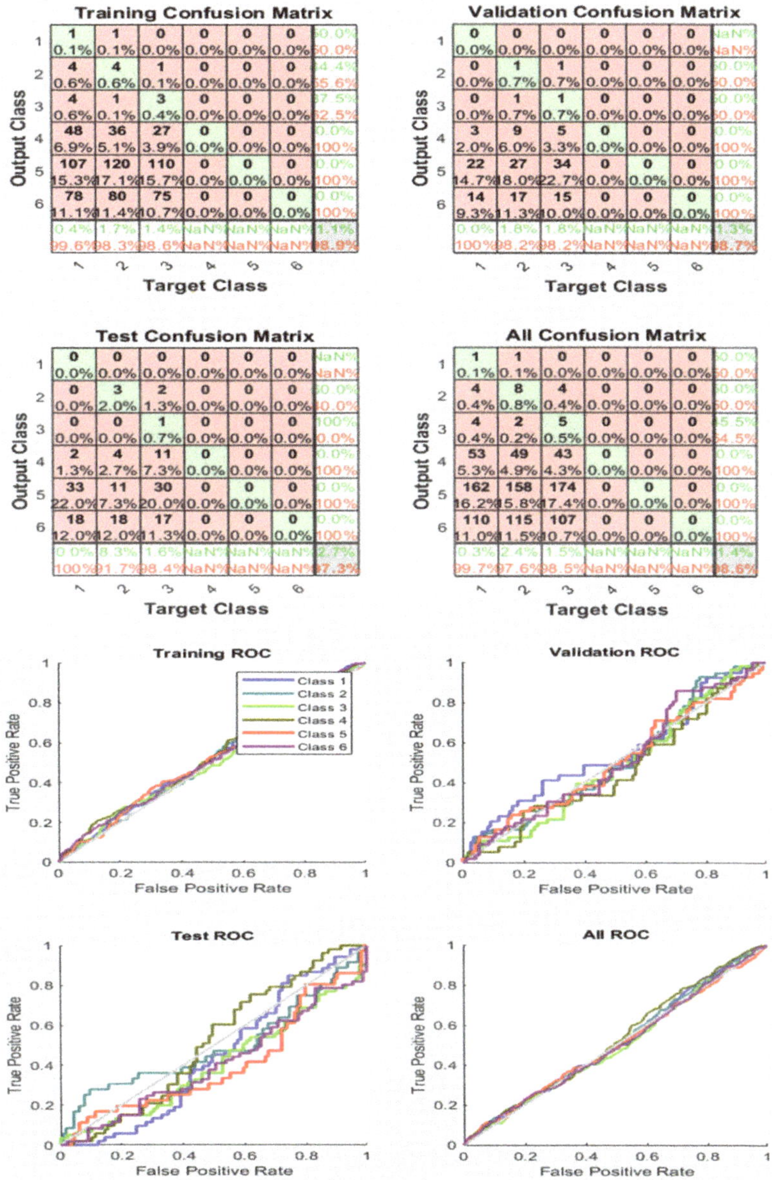

Figure 27. Neural Network Model ROC (Receiver Operating Characteristics) and Confusion Matrix for LM TrainFcn/Activation Function at level 9 (RMS) Features for Rotor Bar Fault with 98.6% Accuracy. Grey line shows diagonal of the ROC for indicating Class 1–6 lines are lying above or below or on it.

Table 4, presents the accuracy results for rotor bar fault with RMS features at different levels, and it is concluded that ROCs curves and confusion matrix of Figures 22–27 with accuracies and respective training functions are obtained for rotor bar fault, as shown in Table 4 for level 9 RMS features. By a similar approach, we can get the results of ROCs and confusion matrix for accuracy perspective for different faults at different levels for different features. Tables 5 and 6 show HCR fault and Rotor Bar faults respectively with Mean features accuracies under different actions functions.

Table 4. Rotor Bar Fault with RMS Feature Accuracies under different Activation functions.

Fault	Activation Function	Characteristic Feature	Accuracy			
			Level 5	Level 7	Level 8	Level 9
Rotor Bar Fault	Trainscg	RMS	89.5%	72.8%	82.6%	79.2%
	Trainbfg		83.4%	75.0%	85.8%	83.2%
	Traincgp		80.6%	81.5%	82.2%	94.1%/91.3%/97.9%
	Traincgb		87.5%	82.2%	91.9%	85.7%
	Traingda		90.5%	76.3%	80.1%	77.3%
	Traingdx		87.8%	86.6%	77.8%	87.6%
	Traingdm		89.3%	90.6%	87.3%	83.6%
	Trainlm		91.1%	92.0%	85.2%	93.3%/84.7%/91.3%/98.6%
	Trainoss		90.0%	87.3%	68.0%	94.1%/83.8%

Table 5. High Contact Resistance Fault with Mean Feature Accuracies under different Activation Functions.

Fault	Activation Function	Characteristic Feature	Accuracy			
			Level 5	Level 7	Level 8	Level 9
High Contact Resistance (HCR)	Trainscg	MEAN	77.3%	81.5%	84.9%	85.3%
	Trainbfg		86.3%. At 23 layers, 11 epochs: 89.2%	**88.9%. At 23 layers and 11 epochs: 96.3%.**	83.5%. At 23 layers, 11 epochs: 90.0%	82.0%
	Traincgp		87.2%. At 15 layers, 11 epochs: 94.7%	84.1% At 23 layers, 11 epochs: 79.0%	76.2%	81.2%. At 23 layers and 11 epochs: 88.3%
	Traincgb		-	80.2%	79.5%	82.7%
	Traingda		-	81.9%	85.8%	87.4%
	Traingdx		86.5%	79.9%	81.9%	85.3%
	Traingdm		87.2%	83.5%	80.9%	79.7%
	Trainlm		88.8%	82.5%. At 23 layers, 11 epochs: 85.7%.	**95%**	**94.2%**
	Trainoss		-	82.7%	-	82.4%

Table 6. Rotor Bar Fault with Mean Feature Accuracies under different Activation Functions.

Fault	Activation Function	Characteristic Feature	Accuracy			
			Level 5	Level 7	Level 8	Level 9
Rotor Bar Fault	Trainscg	MEAN	89.4%	85.2%	92.4%	90.2%
	Trainbfg		88.0%	83.8%	87.7%	90.2%
	Traincgp		85.6%	85.2%	82.8%	81.0%
	Traincgb		79.4%	87.5%	88.5%	83.7%
	Traingda		84.8%	87.7%	82.2%	87.2%
	Traingdx		75.5%	82.1%	80.7%	88.1%
	Traingdm		83.3%	83.0%	81.5%	85.7%
	Trainlm		85.7%	87.6%	81.3%	**97.1%**
	Trainoss		80.0%	**98.0%**	89.2%	90.4%

6. Results

The results show that different activation functions have been used to obtain greater accuracy by changing the optimizing parameters, as compared to the results obtained in [21], where it may be noted in Table 1 that 95% accuracy had been achieved using an ANN. Moreover, in this research, 96.3%, 97.1%, and 98.0% accuracy are achieved, respectively, using Trainbfg, Trainlm, and Trainoss activation functions for the HCR and rotor bar faults at different levels, specifically for the mean and RMS features. The mean features show the functioning or working features for the fault detection, as well as contain the overall important and useful information (features/data points) of the faults. Moreover, mean covers all the values of the dataset. From the Table 1 in [23], this research contributes mainly regarding two faults (high-contact resistance fault and rotor bar fault) with the implementation of an ANN classifier. Although the ANN classifier has been used with different type of faults, particularly for stator winding faults, as tabulated in Table 1 in [24] and discussed in the 'Related Work' section of this paper, the accuracies of fault detection and classification for the mentioned faults with different techniques are lower than the accuracies achieved in this research.

Discussion

Technically speaking, some other approaches are also recommended for detection and classification of faults for the features that are fed as an input to the system. Linear Discriminant Analysis (LDA) and Principle Component Analysis (PCA), supervised and unsupervised machine learning techniques for dimensionality reduction approaches of data, both look for linear combinations of the features which best explain the data. Furthermore, ANN as a classifier uses an iterative-based approach, and accordingly, by retraining the classifier we get higher accuracy. In that case, LDA is preferred for obtaining perfect accuracy without further iterations, and LDA is trained on the dataset one time to get good accuracy. Both LDA and PCA are recommended for future work. Moreover, the ANN approach is iterative and retrains the system several times to achieve the highest accuracy, and preferably, for good classification results. On the contrary, this is not a mechanism for the LDA and PCA classifiers.

No doubt, IoT enables the exchange of data and connection of technologies with other devices using cloud-based networks and systems over the Internet, however, the most recent advancements in technology have also supported the opportunity and facility of Mobile Edge Computing (MEC) approaches [24], which can manage the time delay and energy consumption of IoT devices more appropriately.

7. Conclusions

The results show that mean features had greater accuracy than RMS features. Furthermore, when a fault appeared, the variance in detail coefficients had increased beyond the coefficients of healthy motor signal details. Moreover, in this research, the ANN classifier was used; by retraining the classifier and optimizing the parameters, even 100% accuracy could be obtained. However, so far, an accuracy of 98.0% was achieved using mean rotor bar fault features. Table 7 shows entropy features for the rotor bar fault, and it is found that with entropy features, 0.0% accuracies are obtained. Here, an ANN classifier with 21 hidden layers and 11 epochs was trained. The results obtained are based on an iterative method, since repeatedly training the algorithm can yield more accurate results. Accordingly, the best result is considered and achieved by training the algorithm several times. As presented in the tables, different accuracies are obtained using the ANN classifier at various levels. In addition, 98.0% accuracy is obtained using mean features for the rotor bar fault at level 7, while 98.6% accuracy is obtained for rotor bar fault with RMS features at level 9. Furthermore, 96.4% accuracy was achieved for the HCR fault, and was classified using RMS features at level 5. Therefore, it may be concluded that the faults have been classified according to different features with good accuracy. Figure 13 presents the tested results for the HCR fault at level 9 with RMS features, using the 'npr' tool in MATLAB.

Meanwhile, the ANN was trained to obtain results and for training and testing purposes. The utility of this work lies in its application in monitoring induction motors to ensure their availability in an IoT environment. Availability, as the core component of cyber-security, is the ease with which an organization's critical data, applications, induction machines, and system resources can be accessed by authorized users in times of need.

Table 7. Rotor Bar Fault with Entropy Features Accuracies with different Activation Functions.

Fault	Activation Function	Characteristic Feature	Accuracy			
			Level 5	Level 7	Level 8	Level 9
Rotor Bar Fault	Trainscg	**Entropy**	0.0%	0.0%	0.0%	0.0%
	Trainbfg		0.0%	0.0%	0.0%	0.0%
	Traincgp		0.0%	0.0%	0.0%	0.0%
	Traincgb		0.0%	0.0%	0.0%	0.0%
	Traingda		0.0%	0.0%	0.0%	0.0%
	Traingdx		0.0%	0.0%	0.0%	0.0%
	Traingdm		0.0%	0.0%	0.0%	0.0%
	Trainlm		0.0%	0.0%	0.0%	0.0%
	Trainoss		0.0%	0.0%	0.0%	0.0%

Author Contributions: Conceptualization, M.Z., F.A.S. and S.K.; methodology, M.Z., F.A.S. and S.K.; software, M.Z., F.A.S., S.K. and M.I.; validation, M.Z., F.A.S., S.K., A.M.A., M.U., M.K.H. and W.T.; formal analysis, M.Z., W.T., F.A.S., S.K. and M.I.; investigation, M.Z., F.A.S. and A.M.A.; writing—original draft preparation, M.Z., F.A.S., W.T. and M.U.; writing—review and editing, M.Z., F.A.S., S.K., M.I. and A.M.A.; supervision, F.A.S., W.T., M.U. and S.K.; funding acquisition, A.M.A., M.I., S.K. and S.A. All authors have read and agreed to the published version of the manuscript.

Funding: This research received no external funding.

Institutional Review Board Statement: Not applicable.

Informed Consent Statement: Not applicable.

Data Availability Statement: Not applicable.

Acknowledgments: The researchers would like to thank the Deanship of Scientific Research Qassim University for funding the publication of this project.

Conflicts of Interest: The authors declare no conflict of interest.

Appendix A

Definition of Parameters and Legend:

```
Close all;
Clear all;
% This script assumes these variables are defined:
% FeaturesRMSHealthyHCRlevel9—input data.
% Output10006Inputs—target data.
x = FeaturesRMSHealthyRotorlevel9;
%(x is input data-trained in NN Classifier)
t = Output10006Inputs;
%(t represents output data)
% The names can be replaced as per demand or according to the dataset names at user-side
% Choose a Training Function
% For a list of all training functions type: help nntrain
% 'trainlm' is usually fastest.
% 'trainbr' takes longer but may be better for challenging problems.
% 'trainscg' uses less memory. Suitable in low memory situations.
```

```
trainFcn = 'trainlm';
% Scaled conjugate gradient backpropagation.
%(trainFcn(Activation-Funcation)can be replaced with, i-e 'oss','lm','cgp' and so on)
% Create a Pattern Recognition Network
hiddenLayerSize = 21;
net = patternnet(hiddenLayerSize, trainFcn);
% Setup Division of Data for Training, Validation, Testing
net.divideParam.trainRatio = 70/100;
net.divideParam.valRatio = 15/100;
net.divideParam.testRatio = 15/100;
% Train the Network
[net, tr] = train (net, x, t);
% Test the Network
y = net(x);
e = gsubtract(t, y);
performance = perform (net, t, y)
tind = vec2ind (t);
yind = vec2ind(y);
percentErrors = sum(tind ~= yind)/numel(tind);
% View the Network
view(net)
% Plots
% Uncomment these lines to enable various plots.
%figure, plotperform(tr)
%figure, plottrainstate(tr)
%figure, ploterrhist(e)
%figure, plotconfusion(t, y)
%figure, plotroc (t, y)
end
```

References

1. Jigyasu, R.; Sharma, A.; Mathew, L.; Chatterji, S. A Review of Condition Monitoring and Fault Diagnosis Methods for Induction Motor. In Proceedings of the 2018 Second International Conference on Intelligent Computing and Control Systems (ICICCS), Madurai, India, 14–15 June 2018; pp. 1713–1721. [CrossRef]
2. Reddy, B.K. Condition Monitoring and Life extension of Induction Motor. *TEST Eng. Manag.* **2020**, *82*, 8645–8651.
3. Kumar, P.; Hati, A.S. Review on machine learning algorithm based fault detection in induction motors. *Arch. Comput. Methods Eng.* **2021**, *28*, 1929–1940. [CrossRef]
4. Fontes, A.S.; Cardoso, C.A.; Oliveira, L.P. Comparison of techniques based on current signature analysis to fault detection and diagnosis in induction electrical motors. In Proceedings of the 2016 Electrical Engineering Conference (EECon), Colombo, Sri Lanka, 15 December 2016; p. 74.
5. Deeb, M.; Kotelenets, N.F. Fault diagnosis of 3-phase induction machine using harmonic content of stator current spectrum. In Proceedings of the 2020 International Youth Conference on Radio Electronics, Electrical and Power Engineering (REEPE), Moscow, Russia, 12–14 March 2020; pp. 1–6.
6. AlShorman, O.; Irfan, M.; Saad, N.; Zhen, D.; Haider, N.; Glowacz, A.; AlShorman, A. A review of artificial intelligence methods for condition monitoring and fault diagnosis of rolling element bearings for induction motor. *Shock Vib.* **2020**, *2020*, 8843759. [CrossRef]
7. Deeb, M.; Kotelenets, N.F.; Assaf, T.; Sultan, H.M. Three-Phase Induction Motor Short Circuits Fault Diagnosis using MCSA and NSC. In Proceedings of the 2021 3rd International Youth Conference on Radio Electronics, Electrical and Power Engineering (REEPE), Moscow, Russia, 11–13 March 2021; pp. 1–6.
8. Sridhar, S.; Rao, K.U.; Jade, S. Detection of broken rotor bar fault in induction motor at various load conditions using wavelet transforms. In Proceedings of the 2015 International Conference on Recent Developments in Control, Automation and Power Engineering (RDCAPE), Noida, India, 12–13 March 2015; pp. 77–82. [CrossRef]
9. Zolfaghari, S.; Noor, S.B.; Rezazadeh Mehrjou, M.; Marhaban, M.H.; Mariun, N. Broken rotor bar fault detection and classification using wavelet packet signature analysis based on fourier transform and multi-layer perceptron neural network. *Appl. Sci.* **2018**, *8*, 25. [CrossRef]
10. Maciejewski, N.A.; Treml, A.E.; Flauzino, R.A. A Systematic Review of Fault Detection and Diagnosis Methods for Induction Motors. In Proceedings of the 2020 FORTEI-International Conference on Electrical Engineering (FORTEI-ICEE), Bandung, Indonesia, 23–24 September 2020; pp. 86–90.

11. Sharma, A.; Mathew, L.; Chatterji, S. Analysis of Broken Rotor bar Fault Diagnosis for Induction Motor. In Proceedings of the IEEE International Conference on Innovations in Control, Communication and Information Systems (ICICCI), Greater Noida, India, 12–13 August 2017; pp. 492–497.
12. Purushottam, G.; Tiwari, R. Online diagnostics of mechanical and electrical faults in induction motor using multiclass support vector machine algorithms based on frequency domain vibration and current signals. *ASCE-ASME J. Risk Uncertain. Eng. Syst. Part B Mech. Eng.* **2019**, *5*, 031001.
13. Liang, X.; Ali, M.Z.; Zhang, H. Induction Motors Fault Diagnosis Using Finite Element Method: A Review. *IEEE Trans. Ind. Appl.* **2020**, *56*, 1205–1217. [CrossRef]
14. Terron-Santiago, C.; Martinez-Roman, J.; Puche-Panadero, R.; Sapena-Bano, A. A Review of Techniques Used for Induction Machine Fault Modelling. *Sensors* **2021**, *21*, 4855. [CrossRef]
15. Merizalde, Y.; Hernández-Callejo, L.; Duque-Perez, O. State of the art and trends in the monitoring, detection and diagnosis of failures in electric induction motors. *Energies* **2017**, *10*, 1056. [CrossRef]
16. Toma, R.N.; Prosvirin, A.E.; Kim, J.M. Bearing fault diagnosis of induction motors using a genetic algorithm and machine learning classifiers. *Sensors* **2020**, *20*, 1884. [CrossRef]
17. Zamudio-Ramírez, I.; Osornio-Ríos, R.A.; Antonino-Daviu, J.A.; Quijano-Lopez, A. Smart-sensor for the automatic detection of electromechanical faults in induction motors based on the transient stray flux analysis. *Sensors* **2020**, *20*, 1477. [CrossRef]
18. Adouni, A.; Marques Cardoso, A.J. Thermal analysis of low-power three-phase induction motors operating under voltage unbalance and inter-turn short circuit faults. *Machines* **2020**, *9*, 2. [CrossRef]
19. Martinez-Herrera, A.L.; Ferrucho-Alvarez, E.R.; Ledesma-Carrillo, L.M.; Mata-Chavez, R.I.; Lopez-Ramirez, M.; Cabal-Yepez, E. Multiple Fault Detection in Induction Motors through Homogeneity and Kurtosis Computation. *Energies* **2022**, *15*, 1541. [CrossRef]
20. Albattah, W.; Khel, M.H.K.; Habib, S.M.; Khan, S.; Kadir, K.A. Hajj Crowd Management Using CNN-Based Approach. *Comput. Mater. Contin.* **2020**, *66*, 2183–2197. [CrossRef]
21. Purushottam, G.; Tiwari, R. Signal based condition monitoring techniques for fault detection and diagnosis of induction motors: A state-of-the-art review. *Mech. Syst. Signal Process.* **2020**, *144*, 106908.
22. Esakimuthu Pandarakone, S.; Mizuno, Y.; Nakamura, H. A comparative study between machine learning algorithm and artificial intelligence neural network in detecting minor bearing fault of induction motors. *Energies* **2019**, *12*, 2105. [CrossRef]
23. Mohamed, M.A.; Hassan, M.A.M.; Albalawi, F.; Ghoneim, S.S.M.; Ali, Z.M.; Dardeer, M. Diagnostic Modeling for Induction Motor Faults via ANFIS Algorithm and DWT-Based Feature Extraction. *Appl. Sci.* **2021**, *11*, 9115. [CrossRef]
24. Cui, Y.Y.; Zhang, T. New Quantum-Genetic Based OLSR Protocol (QGOLSR) for Mobile Ad hoc Network. *Appl. Soft Comput.* **2019**, *80*, 285–296. [CrossRef]
25. Zhang, T.; Zhang, D.; Qiu, J.; Zhang, X.; Zhao, P.; Gong, C. A Kind of Novel Method of Power Allocation with Limited Cross-Tier Interference for CRN. *IEEE Access* **2019**, *7*, 82571–82583. [CrossRef]
26. Li, T.; Dong, Q.; Wang, Y.; Gong, X.; Yang, Y. Dual buffer rotation four-stage pipeline for CPU–GPU cooperative computing. *Soft Comput.* **2017**, *23*, 859–869. [CrossRef]
27. Abid, A.; Khan, M.T.; Iqbal, J. A review on fault detection and diagnosis techniques: Basics and beyond. *Artif. Intell. Rev.* **2021**, *54*, 3639–3664. [CrossRef]
28. Chen, S. Review on Supervised and Unsupervised Learning Techniques for Electrical Power Systems: Algorithms and Applications. *IEEJ Trans. Electr. Electron. Eng.* **2021**, *16*, 1487–1499. [CrossRef]
29. Hsueh, Y.M.; Ittangihal, V.R.; Wu, W.B.; Chang, H.C.; Kuo, C.C. Fault diagnosis system for induction motors by CNN using empirical wavelet transform. *Symmetry* **2019**, *11*, 1212. [CrossRef]
30. Valtierra-Rodriguez, M.; Rivera-Guillen, J.R.; Basurto-Hurtado, J.A.; De-Santiago-Perez, J.J.; Granados-Lieberman, D.; Amezquita-Sanchez, J.P. Convolutional neural network and motor current signature analysis during the transient state for detection of broken rotor bars in induction motors. *Sensors* **2020**, *20*, 3721. [CrossRef]
31. Nishat Toma, R.; Kim, J.M. Bearing fault classification of induction motors using discrete wavelet transform and ensemble machine learning algorithms. *Appl. Sci.* **2020**, *10*, 5251. [CrossRef]
32. Zhang, T.; Chen, J.; Li, F.; Zhang, K.; Lv, H.; He, S.; Xu, E. Intelligent fault diagnosis of machines with small & imbalanced data: A state-of-the-art review and possible extensions. *ISA Trans.* **2022**, *119*, 152–171.
33. Tang, S.; Yuan, S.; Zhu, Y. Convolutional neural network in intelligent fault diagnosis toward rotatory machinery. *IEEE Access* **2020**, *8*, 86510–86519. [CrossRef]
34. Qiao, Z.; Shu, X. Coupled neurons with multi-objective optimization benefit incipient fault identification of machinery. *Chaos Solitons Fractals* **2021**, *145*, 110813. [CrossRef]

MDPI
St. Alban-Anlage 66
4052 Basel
Switzerland
www.mdpi.com

MDPI Books Editorial Office
E-mail: books@mdpi.com
www.mdpi.com/books

www.ingramcontent.com/pod-product-compliance
Lightning Source LLC
Chambersburg PA
CBHW041934260926
51850CB00018B/152
9783036591520